Methods in Enzymology

Volume 71

LIPIDS

Part C

METHODS IN ENZYMOLOGY

EDITORS-IN-CHIEF

Sidney P. Colowick   Nathan O. Kaplan

Methods in Enzymology

Volume 71

Lipids

Part C

EDITED BY

John M. Lowenstein

DEPARTMENT OF BIOCHEMISTRY
BRANDEIS UNIVERSITY
WALTHAM, MASSACHUSETTS

1981

ACADEMIC PRESS

A Subsidiary of Harcourt Brace Jovanovich, Publishers

New York London Toronto Sydney San Francisco

ACADEMIC PRESS, INC.
111 Fifth Avenue, New York, New York 10003

United Kingdom Edition published by
ACADEMIC PRESS, INC. (LONDON) LTD.
24/28 Oval Road, London NW1 7DX

Library of Congress Cataloging in Publication Data
Main entry under title:

Lipids.

   (Methods in enzymology, v. 14, 35)
   Includes bibliographical references.
   1. Lipids. I. Lowenstein, John M., Date   ed.
II. Series: Methods in enzymology, v. 14 [etc.].
[DNLM: 1. Lipids. W1 ME9615K v. 14, etc.]
QP601.M49 vol. 14, etc.   574.19'25s   77-26907
ISBN 0-12-181971-X (v. 71)   [574.19'247]

PRINTED IN THE UNITED STATES OF AMERICA

81 82 83 84   9 8 7 6 5 4 3 2 1

Table of Contents

Section I. Fatty Acid Synthesis

v

    * Unknown to the authors and Editor articles 21 and 41 were switched. This should have been article 41 in Section II.

Section II. Fatty Acid Activation and Oxidation

    * Unknown to the authors and Editor articles 21 and 41 were switched. This should have
been article 21 in Section I.

Section III. Hydroxymethylglutaryl-CoA Enzymes

Section IV. Enzymes of Phospholipid Synthesis and Related Enzymes

Section V. Hydrolases

Section VI. Miscellaneous

Contributors to Volume 71

Article numbers are in parentheses following the names of contributors.
Affiliations listed are current.

V. P. AGRAWAL (50), *Department of Chemistry, Tribhuvan University, Kirtipur Kathmandu, Nepal*

FAZAL AHMAD (2), *Papanicolaou Cancer Research Institute, Miami, Florida 33101*

PATRICIA M. AHMAD (2), *Papanicolaou Cancer Research Institute, Miami, Florida 33101*

BERNARD AXELROD (53), *Department of Biochemistry, Purdue University, West Lafayette, Indiana 47907*

TERRY A. BAKER (55), *Lipid Metabolism Laboratory, William S. Middleton Memorial Veterans Hospital, Madison, Wisconsin 53705*

CLINTON E. BALLOU (29), *Department of Biochemistry, University of California, Berkeley, California 94720*

C. J. BEDORD (27), *Syntex Corporation, 3401 Hillview Avenue, Palo Alto, California 94304*

PER BELFRAGE (74, 75), *Department of Physiological Chemistry, University of Lund, S-220 07 Lund, Sweden*

WILLIAM R. BENSCH (56), *Lilly Research Laboratories, Indianapolis, Indiana 46206*

ROLF KRISTIAN BERGE (28), *Laboratory of Clinical Biochemistry, University of Bergen, 5016 Haukeland Sykehus, Norway*

BRADLEY BERGER (58), *University of Chicago Medical School, Chicago, Illinois 60612*

JOHN T. BERNERT, JR. (30), *United States Department of Health, Education and Welfare, Center for Disease Control, Atlanta, Georgia 30333*

L. L. BIEBER (42), *Department of Biochemistry, Michigan State University, East Lansing, Michigan 48824*

JUDITH F. BINSTOCK (49), *Department of Chemistry, Manhattan College, Riverdale, New York 10471*

HOWARD L. BROCKMAN (72), *The Hormel Institute, University of Minnesota, Austin, Minnesota 55912*

J. S. BUCKNER (13), *Metabolism and Radiation Research Laboratory, Fargo, North Dakota 58102*

THOMAS M. CHEESBROUGH (53), *Department of Biochemistry, Purdue University, West Lafayette, Indiana 47907*

PATRICK C. CHOY (68), *Department of Biochemistry, University of Manitoba, Winnipeg, Manitoba R3E OW3, Canada*

PHILIP COHEN (3), *Biochemistry Department, Dundee University, Dundee DD1 4HN, Scotland*

ROBERT S. CONWAY (51), *Department of Microbiology, Emory University, Atlanta, Georgia 30322*

R. C. COTTRELL (80), *The British Industrial Biological Research Association, Carshalton, Surrey SM5 4DS, England*

JOHN E. CRONAN, JR. (18, 21, 41), *Department of Microbiology, University of Illinois, Urbana, Illinois 61801*

WILLIAM C. DEAL, JR. (10), *Department of Biochemistry, Michigan State University, East Lansing, Michigan 48824*

RAYMOND A. DEEMS (81), *Department of Chemistry, University of California, San Diego, La Jolla, California 92093*

EDWARD A. DENNIS (81), *Department of Chemistry, University of California, San Diego, La Jolla, California 92093*

RAYMOND DILS (26), *Department of Physiology & Biochemistry, University of Reading, Whiteknights, Reading RG6 2AJ, England*

PETER F. DODDS (11), *Department of Chemistry, Georgetown University, Washington, D.C. 20057*

YUKIO DOI (87), *Burnsides Research Laboratory, Department of Food Sciences, University of Illinois, Urbana, Illinois 61801*

WILLIAM DOWHAN (65, 66), *Department of Biochemistry and Molecular Biology, University of Texas Medical School, Houston, Texas 77025*

ROGER F. DRONG (36), *Lipid Metabolism Laboratory, William S. Middleton Memorial Veterans Hospital, Madison, Wisconsin 53705, and Department of Physiological Chemistry, University of Wisconsin, Madison, Wisconsin 53706*

RICHARD E. DUGAN (55), *Lipid Metabolism Laboratory, William S. Middleton Memorial Veterans Hospital, Madison, Wisconsin 53705*

PAUL C. ENGEL (43), *Department of Biochemistry, University of Sheffield, Sheffield S10 2TN, England*

MARY LOU ERNST-FONBERG (8, 22), *Department of Biochemistry, College of Medicine, East Tennessee State University, Johnson City, Tennessee 37601*

R. RAY FALL (91, 92), *Department of Chemistry, University of Colorado, Boulder, Colorado 80309*

MIKAEL FARSTAD (28), *Laboratory of Clinical Biochemistry, University of Bergen, 5016 Haukeland Sykehus, Norway*

JIM C. FONG (47), *Department of Psychiatry, New York University Medical Center, New York, New York 10016*

RICHARD FRANSON (78), *Department of Biochemistry, Medical College of Virginia, Richmond, Virginia 23219*

GUDRUN FREDRIKSON (74), *Department of Physiological Chemistry, University of Lund, S-220 07 Lund, Sweden*

EUGENE P. FRENKEL (38), *Department of Internal Medicine, Evelyn L. Overton Hematology-Oncology Research Laboratory, University of Texas Health Science Center at Dallas, Southwestern Medical School, Dallas, Texas 75235, and Veterans Administration Medical Center, Dallas, Texas 75216*

JEANIE FRYE (78), *Life Sciences Division, Meloy Laboratories Inc., 6715 Electronic Drive, Springfield, Virginia 22151*

LINDA L. GALLO (77), *Department of Biochemistry, George Washington University, Washington, D.C. 20037*

R. E. GARCIA (88), *Department of Biochemistry, University of California, Riverside, California 92521*

SHIMON GATT (60), *Laboratory of Neurochemistry, Department of Biochemistry, The Hebrew University-Hadassah Medical School, Jerusalem, Israel*

J. G. GAVILANES (17), *Department of Biochemistry, Faculty of Sciences, Complutensis University, Madrid 3, Spain*

JAMES L. GAYLOR (32), *Department of Biochemistry, University of Missouri, Columbia, Missouri 65212*

DAVID M. GIBSON (57), *Department of Biochemistry, Indiana University School of Medicine, Indianapolis, Indiana 46223*

TAMAR GOLDFLAM (19), *Department of Pharmacology, Case Western Reserve University, Cleveland, Ohio 44106*

ALAN G. GOODRIDGE (19), *Department of Pharmacology, Case Western Reserve University, Cleveland, Ohio 44106*

HARRY GRIFFIN (78), *Agricultural Research Council, Poultry Research Centre, Edinburgh EH9 3JS, Scotland*

MICHAEL J. GRIFFITH (37), *Dental Research Center, The University of North Carolina, Chapel Hill, North Carolina 27514*

DENNIS W. GROGAN (18), *Department of Microbiology, University of Illinois, Urbana, Illinois 61801*

INGER GRUNNET (26), *Institute of Biochemistry, University of Odense, DK-5230 Odense M, Denmark*

PAUL S. GUY (3), *Biochemistry Department, Dundee University, Dundee DD1 4HN, Scotland*

CAROLE L. HALL (45, 46), *School of Chemistry, Georgia Institute of Technology, Atlanta, Georgia 30332*

D. GRAHAME HARDIE (3), *Biochemistry Department, Dundee University, Dundee DD1 4HN, Scotland*

TAKASHI HASHIMOTO (1), *Department of Biochemistry, Shinshu University Faculty of Medicine, Asahi, Matsumoto 390, Japan*

EDWARD HAWROT (67), *Department of Pharmacology, Yale University School of Medicine, New Haven, Connecticut 06510*

TAKASHI HIRABAYASHI (65), *Central Research Institute, Suntory Limited, Osaka, Japan*

KOHEI HOSAKA (39, 40, 61), *Department of Biochemistry, Gunma University School of Medicine, Showa-cho, Maebashi 371, Japan*

ANTHONY H. C. HUANG (93), *Department of Biology, University of South Carolina, Columbia, South Carolina 29208*

HIROH IKEZAWA (84), *Faculty of Pharmaceutical Sciences, Nagoya City University, 3-1 Tanabedori, Mizuhoku, Nagoya 467, Japan*

THOMAS S. INGEBRITSEN (57), *Biochemistry Department, Medical Sciences Institute, Dundee University, Dundee DD1 4HN, Scotland*

ROBERT A. JENIK (12), *Lipid Metabolism Laboratory, William S. Middleton Memorial Veterans Hospital, Madison, Wisconsin 53705, and the Department of Physiological Chemistry, University of Wisconsin, Madison, Wisconsin 53706*

DEAN P. JONES (32), *Department of Biochemistry, Emory University School of Medicine, Atlanta, Georgia 30322*

VASUDEV C. JOSHI (31), *Marrs McLean Department of Biochemistry, Baylor College of Medicine, Houston, Texas 77030*

TATSUYUKI KAMIRYO (5, 39), *Department of Medical Chemistry, Kyoto University Faculty of Medicine, Yoshida, Sakyo-ku, Kyoto 606, Japan*

J. N. KANFER (70, 86), *Department of Biochemistry, Faculty of Medicine, University of Manitoba, Winnipeg, Manitoba R3E OW3, Canada*

HIDEO KANOH (62), *Department of Biochemistry, Niigata University School of Medicine, Niigata 951, Japan*

SARVAGYA S. KATIYAR (36), *Department of Physiological Chemistry, University of Wisconsin, Madison, Wisconsin 53706*

AKIHIKO KAWAGUCHI (15, 16), *Institute of Applied Microbiology, The University of Tokyo, Bunkyo-ku, Tokyo 113, Japan*

JOHN C. KHOO (73), *Department of Medicine, Division of Metabolic Disease, University of California, San Diego, La Jolla, California 92093*

I. C. KIM (10), *Department of Biochemistry, Michigan State University, East Lansing, Michigan 48824*

YU SAM KIM (20), *Department of Biochemistry, College of Medicine, Yonsei University, Seoul, Korea*

RICHARD L. KITCHENS (38), *Department of Internal Medicine, Evelyn L. Overton Hematology-Oncology Research Laboratory, University of Texas Health Science Center at Dallas, Southwestern Medical School, Dallas, Texas 75235, and Veterans Administration Medical Center, Dallas, Texas 75216*

DON A. KLEINSEK (55), *Department of Physiological Chemistry, University of Wisconsin, Madison, Wisconsin 53706*

JENS KNUDSEN (26), *Institute of Biochemistry, University of Odense, DK-5230 Odense M, Denmark*

P. E. KOLATTUKUDY (13, 20, 27, 33, 50, 76), *Institute of Biological Chemistry and Biochemistry/Biophysics Program, Washington State University, Pullman, Washington 99164*

RUTH KRAMER (79), *51 Park Street, Brookline, Massachusetts 02146*

SOMA KUMAR (11), *Department of Chemistry, Georgetown University, Washington, D.C. 20057*

SIMO LAAKSO (53), *Department of Biochemistry, University of Turku, SF-20500 Turku 50, Finland*

J. D. LARSON (33), *Institute of Biological Chemistry, Washington State University, Pullman, Washington 99164*

TIMOTHY LARSON (66), *Fachbereich Biologie, University of Konstonz, Konstonz, Federal Republic of Germany*

EDWARD P. LAU (92), *Department of Chemistry, University of Colorado, Boulder, Colorado 80309*

CLIVE LITTLE (83), *Institute of Medical Biology, University of Tromsø, N-9001 Tromsø, Norway*

FRANK A. LORNITZO (36), *Lipid Metabolism Laboratory, William S. Middleton Memorial Veterans Hospital, Madison, Wisconsin 53705*

MARTIN G. LOW (85), *Department of Biophysics, Medical College of Virginia, Virginia Commonwealth University, Richmond, Virginia 23298*

FEODOR LYNEN* (90), *Max-Planck Institut für Biochemie, 8033 Martinsried, Federal Republic of Germany*

TOM MCKEON (23, 34), *Department of Vegetable Crops, University of California, Davis, California 95616*

I. B. MAITI (76), *Department of Microbiology, Sherbrooke University, Sherbrooke, Quebec J1H 5N4, Canada*

PAUL MANDEL (60), *Centre de Neurochimie du CNRS, Université Louis Pasteur, Strasbourg, France*

M. A. K. MARKWELL (42), *Molecular Biology Institute, University of California-Los Angeles, Los Angeles, California 90024*

HARUKO MEYER (7), *Department of Microbiology, College of Medicine, State University of New York, Upstate Medical Center, Syracuse, New York 13210*

FRANZ MEYER (7), *Department of Microbiology, College of Medicine, State University of New York, Upstate Medical Center, Syracuse, New York 13210*

YOSHINOBU MIKI (61), *Department of Medical Chemistry, Kyoto University Faculty of Medicine, Yoshida, Sakyo-ku, Kyoto 606, Japan*

CRAIG MILLER (78), *Department of Biochemistry, Bowman Gray School of Medicine, Winston-Salem, North Carolina 27103*

MASAYOSHI MISHINA (5, 39), *Department of Medical Chemistry, Kyoto University Faculty of Medicine, Yoshida, Sakyo-ku, Kyoto 606, Japan*

T. MIURA (70), *The National Institute for Environmental Studies, Yatabe-machi, Tsukuba-gun, Ibaraki Prefecture, Japan*

T. S. MOORE, JR. (71), *Department of Botany, University of Wyoming, Laramie, Wyoming 82071*

ROBERT A. MOREAU (93), *Department of Biochemistry and Biophysics, University of California, Davis, California 95616*

SIDNEY M. MORRIS, JR. (19), *Department of Pharmacology, Case Western Reserve University, Cleveland, Ohio 44106*

J. B. MUDD (88), *Department of Biochemistry, University of California, Riverside, California 92521*

A. M. MUNICIO (17), *Department of Biochemistry, Faculty of Sciences, Complutensis University, Madrid 3, Spain*

SHIGETADA NAKANISHI (1), *Department of Medical Chemistry, Kyoto University Faculty of Medicine, Yoshida, Sakyo-ku, Kyoto 606, Japan*

NENAD M. NESKOVIC (60), *Centre de Neurochimie du CNRS, Université Louis Pasteur, Strasbourg, France*

* Deceased.

GARY NEUDAHL (10), *Department of Biochemistry, Michigan State University, East Lansing, Michigan 48824*

NIELS C. NIELSEN (6), *USDA-SEA, Agronomy Department, Purdue University, West Lafayette, Indiana 47907*

JUN-ICHI NIKAWA (1), *Department of Biochemistry, Gunma University School of Medicine, Showa-cho, Maebashi 371, Japan*

NILS ÖSTEN NILSSON (74), *Department of Physiological Chemistry, University of Lund, S-220 07 Lund, Sweden*

SUKANYA NIMMANNIT (54), *Department of Biochemistry, Faculty of Pharmaceutical Sciences, Chulalongkorn University, Bangkok 5, Thailand*

TOSHIRO NISHIDA (87), *Burnsides Research Laboratory, Department of Food Science, University of Illinois, Urbana, Illinois 61801*

JONATHAN S. NISHIMURA (37), *Department of Biochemistry, The University of Texas Health Science Center at San Antonio, San Antonio, Texas 78284*

SHOSAKU NUMA (1, 5, 39, 40, 61, 64), *Department of Medical Chemistry, Kyoto University Faculty of Medicine, Yoshida, Sakyo-ku, Kyoto 606, Japan*

HIDEO OGIWARA (1), *Department of Biochemistry, Gunma University School of Medicine, Showa-cho, Maebashi 371, Japan*

KIMIYOSHI OHNO (62), *Department of Biochemistry, Sapporo Medical College, Chuo-Ku, West-17, South-1, Sapporo 060, Japan*

A. OHSAKA (82), *The Second Department of Bacteriology, National Institute of Health, 10-35 Kamiosaki-2-Chome, Shinagawa-ku, Tokyo 141, Japan*

SHIGENOBU OKUDA (15, 16), *Institute of Applied Microbiology, The University of Tokyo, Bunkyo-ku, Tokyo 113, Japan*

SATOSHI ŌMURA (15), *School of Pharmaceutical Science, Kitasato University, Minato-ku, Tokyo 108, Japan*

STEVEN D. PELECH (68), *Department of Biochemistry, University of British Columbia, Vancouver, British Columbia V6T 1W5, Canada*

DAVID O. PETERSON (14), *Department of Biochemistry and Biophysics, University of California, San Francisco, California 94143*

REGINA PIETRUSZKO (89), *Center of Alcohol Studies and Department of Biochemistry, Rutgers University, New Brunswick, New Jersey 08903*

MICHAEL R. POLLARD (35), *Biochemistry Department, Dundee University, Dundee DD1 4HN, Scotland*

JOHN W. PORTER (12, 36, 54, 55), *Lipid Metabolism Laboratory, William S. Middleton Memorial Veterans Hospital, Madison, Wisconsin 53705, and the Department of Physiological Chemistry, University of Wisconsin, Madison, Wisconsin 53706*

A. J. POULOSE (13, 20, 27), *Institute of Biological Chemistry, Washington State University, Pullman, Washington 99164*

M. RENUKA PRASAD (31), *Department of Biological Chemistry, University of Illinois Medical Center, Chicago, Illinois 60612*

R. E. PURDY (76), *Environmental Sciences, Chesapeake Bay Program, 2083 West Street, Annapolis, Maryland 21401*

NILOFER QURESHI (54), *Mycobacteriology Laboratory, William S. Middleton Memorial Veterans Hospital, Madison, Wisconsin 53705, and the Institute for Enzyme Research, University of Wisconsin, Madison, Wisconsin 53706*

R. HANUMANTHA RAO (78), *Department of Chemistry, Rust College, Holly Springs, Mississippi 38635*

SAMUEL M. RAPOPORT (52), *Institut für Physiologische und Biologische Chemie, Humboldt Universität, DDR-1040 Berlin, German Democratic Republic*

CHARLES O. ROCK (21, 41), *Department of Biochemistry, St. Jude Children's Research Hospital, Memphis, Tennessee 38101*

VICTOR W. RODWELL (56), *Department of Biochemistry, Purdue University, West Lafayette, Indiana 47907*

LINDA ROGERS (27, 33), *Institute of Biological Chemistry, Washington State University, Pullman, Washington 99164*

DANIEL A. K. RONCARI (9), *Institute of Medical Science and Department of Medicine, Toronto Western Hospital, University of Toronto, Toronto M5S 1A8, Canada*

TANKRED SCHEWE (52), *Institut für Physiologische und Biologische Chemie, Humboldt Universität, DDR-1040 Berlin, German Democratic Republic*

ULRICH SCHIELE (90), *Hormon Chemie München GmbH, 8 Munich 45, Federal Republic of Germany*

WOLFGANG J. SCHNEIDER (69), *Department of Molecular Genetics, University of Texas Health Science Center, Dallas, Texas 75235*

ANN W. SCHONGALLA (22), *College of Medicine, Cornell University, New York, New York 10021*

HORST SCHULZ (47, 48, 49), *Department of Chemistry, City College of the City University of New York, New York, New York 10031*

YOUSUKE SEYAMA (16), *Department of Biochemistry, Faculty of Medicine, The University of Tokyo, Bunkyo-ku, Tokyo 113, Japan*

PATRICIA SISSON (78), *Department of Biochemistry, Bowman Gray School of Medicine, Winston-Salem, North Carolina 27103*

STUART SMITH (24, 25), *Bruce Lyon Memorial Research Laboratory, Children's Hospital Medical Center of Northern California, Oakland, California 94609*

MARTIN D. SNIDER (63), *Center for Cancer Research and Department of Biology,* *Massachusetts Institute of Technology, Cambridge, Massachusetts 02139*

HOWARD SPRECHER (30), *Department of Physiological Chemistry, Ohio State University, Columbus, Ohio 43210*

K. SREEKRISHNA (31), *Department of Biochemistry, University of Kentucky College of Medicine, Lexington, Kentucky 40506*

HAROLD STAACK (48), *American Minitor Company, Indianapolis, Indiana 46268*

DANIEL STEINBERG (73), *Department of Medicine, Division of Metabolic Disease, University of California, San Diego, La Jolla, California 92093*

PETER STRÅLFORS (74), *Department of Physiological Chemistry, University of Lund, S-220 07 Lund, Sweden*

PAUL K. STUMPF (23, 34), *Department of Biochemistry and Biophysics, University of California, Davis, California 95616*

T. SUGAHARA (82), *The Second Department of Bacteriology, National Institute of Health, 10-35 Kamiosaki-2-Chome, Shinagawa-ku, Tokyo 141, Japan*

MANFRED SUMPER (4), *Institut für Biochemie, Genetik und Mikrobiologie, Lehrstuhl Biochemie, Universität Regensburg, 8400 Regensburg, Federal Republic of Germany*

RYO TAGUCHI (84), *Faculty of Pharmaceutical Sciences, Nagoya City University, 3-1 Tanabedori, Mizuhoku, Nagoya 467, Japan*

T. TAKAHASHI (82), *Department of Microbiology, Hoshi College of Pharmacy, Ebara-2-Chome, Shinagawa-ku, Tokyo 142, Japan*

T. TAKI (70, 86), *Department of Biochemistry, Shizuoka College of Pharmacy, 2-2-1, Oshika, Shizuoka-Shi 422, Japan*

TADASHI TANABE (1), *Department of Biochemistry, National Cardiovascular Center Research Institute, Fujishiro-dai, Suita 565, Japan*

TAKAO TANAKA (40), *Third Division, Department of Internal Medicine, Osaka Medical College, Daigaku-cho, Takatsuki 569, Japan*

FREDERICK R. TAYLOR (18), *Department of Microbiology, Oregon State University, Corvallis, Oregon 97331*

COLIN THORPE (44), *Department of Chemistry, University of Delaware, Newark, Delaware 19711*

HIROSHI TOMODA (15), *Institute of Applied Microbiology, The University of Tokyo, Bunkyo-ku, Tokyo 113, Japan*

HANS TORNQVIST (75), *Department of Clinical Chemistry, University of Lund, Malmö General Hospital, S-214 01 Malmö, Sweden*

DENNIS E. VANCE (68, 69), *Department of Biochemistry, University of British Columbia, Vancouver, British Columbia V6T, 1W5, Canada*

H. VAN DEN BOSCH (59), *Laboratory of Biochemistry, State University of Utrecht, 3508 TB Utrecht, The Netherlands*

G. P. H. VAN HEUSDEN (59), *Laboratory of Biochemistry, State University of Utrecht, 3508 TB Utrecht, The Netherlands*

G. M. VIANEN (59), *Laboratory of Biochemistry, State University of Utrecht, 3508 TB Utrecht, The Netherlands*

MOSELEY WAITE (78), *Department of Biochemistry, Bowman Gray School of Medicine, Winston-Salem, North Carolina 27103*

THERESA A. WALKER (22), *Department of Biology, Yale University, New Haven, Connecticut 06520*

ROBERT M. WATERSON (51), *Department of Microbiology, Emory University, Atlanta, Georgia 30322*

RAINER WIESNER (52), *Institut für Physiologische und Biologische Chemie, Humboldt Universität, DDR-1040 Berlin, German Democratic Republic*

JACK S. WOLPERT (8), *Department of Biochemistry, College of Medicine, East Tennessee State University, Johnson City, Tennessee 37601*

WILLIAM I. WOOD (14), *Laboratory of Molecular Biology, National Institute of Arthritis, Metabolism and Digestive Diseases, National Institutes of Health, Bethesda, Maryland 20205*

KENICHI K. YABUSAKI (29), *Hana Biologics Inc., 1400 53rd Street, Emeryville, California 94608*

TAMIO YAMAKAWA (16), *Department of Biochemistry, Faculty of Medicine, The University of Tokyo, Bunkyo-ku, Tokyo 113, Japan*

SATOSHI YAMASHITA (61, 64), *Department of Biochemistry, Gunma University School of Medicine, Showa-cho, Maebashi 371, Japan*

TAKASHI YONETANI (89), *Department of Biochemistry and Biophysics, University of Pennsylvania, Philadelphia, Pennsylvania 19174*

NANCY L. YOUNG (58), *Department of Medicine, Cornell University Medical College, New York, New York 10021*

PETER ZAHLER (79), *Institute of Biochemistry, University of Berne, 3012 Berne, Switzerland*

Preface

Let me haue men about me, that are fat
Sleek-headed men, and such as sleep a'nights
Julius Caesar Act I, Scene II*

Fat people are no longer the preferred stereotype of placidity and health, and sleep is a somewhat overestimated commodity. Fat books, on the other hand, proliferate. A single volume was planned originally to cover recent developments in the biochemistry of lipids. It would have been too obese for comfort, so it was decided to divide the book into two volumes. If only obesity could be controlled so easily in the real world!

Volume 71 deals with enzymes while Volume 72 covers methods useful to the lipid biochemist. Each section starts with a list of related articles previously published in the *Methods in Enzymology* series. In general, the enzymes included have been purified highly. A notable exception are enzymes of lipid metabolism from plants, a number of which are included even though they have only been purified partially. It is hoped that their inclusion will provide a stimulus to their isolation in homogeneous form.

The borderline between fatty acid and sterol metabolism has been set at β-hydroxy-β-methylglutaryl-CoA. The inclusion of some enzymes may strike the reader as odd, but there was in each case good reason. For example, 3-methylcrotonyl-CoA carboxylase is included because of its relation to acetyl-CoA carboxylase and its ability to carboxylate free biotin, a valuable property for understanding acetyl-CoA carboxylase.

I welcome suggestions for future volumes. Please do not hesitate to draw my attention to errors of omission or commission.

JOHN M. LOWENSTEIN

* H. M. Furness, Jr. (ed.). (1913). J. B. Lippincott Co., Philadelphia and London, p. 45.

METHODS IN ENZYMOLOGY

EDITED BY

Sidney P. Colowick and Nathan O. Kaplan

VANDERBILT UNIVERSITY
SCHOOL OF MEDICINE
NASHVILLE, TENNESSEE

DEPARTMENT OF CHEMISTRY
UNIVERSITY OF CALIFORNIA
AT SAN DIEGO
LA JOLLA, CALIFORNIA

I. Preparation and Assay of Enzymes
II. Preparation and Assay of Enzymes
III. Preparation and Assay of Substrates
IV. Special Techniques for the Enzymologist
V. Preparation and Assay of Enzymes
VI. Preparation and Assay of Enzymes (*Continued*)
Preparation and Assay of Substrates
Special Techniques
VII. Cumulative Subject Index

METHODS IN ENZYMOLOGY

EDITORS-IN-CHIEF

Sidney P. Colowick   Nathan O. Kaplan

VOLUME VIII. Complex Carbohydrates
Edited by ELIZABETH F. NEUFELD AND VICTOR GINSBURG

VOLUME IX. Carbohydrate Metabolism
Edited by WILLIS A. WOOD

VOLUME X. Oxidation and Phosphorylation
Edited by RONALD W. ESTABROOK AND MAYNARD E. PULLMAN

VOLUME XI. Enzyme Structure
Edited by C. H. W. HIRS

VOLUME XII. Nucleic Acids (Parts A and B)
Edited by LAWRENCE GROSSMAN AND KIVIE MOLDAVE

VOLUME XIII. Citric Acid Cycle
Edited by J. M. LOWENSTEIN

VOLUME XIV. Lipids
Edited by J. M. LOWENSTEIN

VOLUME XV. Steroids and Terpenoids
Edited by RAYMOND B. CLAYTON

VOLUME XVI. Fast Reactions
Edited by KENNETH KUSTIN

VOLUME XVII. Metabolism of Amino Acids and Amines (Parts A and B)
Edited by HERBERT TABOR AND CELIA WHITE TABOR

VOLUME XVIII. Vitamins and Coenzymes (Parts A, B, and C)
Edited by DONALD B. MCCORMICK AND LEMUEL D. WRIGHT

VOLUME XIX. Proteolytic Enzymes
Edited by GERTRUDE E. PERLMANN AND LASZLO LORAND

VOLUME XX. Nucleic Acids and Protein Synthesis (Part C)
Edited by KIVIE MOLDAVE AND LAWRENCE GROSSMAN

VOLUME XXI. Nucleic Acids (Part D)
Edited by LAWRENCE GROSSMAN AND KIVIE MOLDAVE

VOLUME XXII. Enzyme Purification and Related Techniques
Edited by WILLIAM B. JAKOBY

VOLUME XXIII. Photosynthesis (Part A)
Edited by ANTHONY SAN PIETRO

VOLUME XXIV. Photosynthesis and Nitrogen Fixation (Part B)
Edited by ANTHONY SAN PIETRO

VOLUME XXV. Enzyme Structure (Part B)
Edited by C. H. W. HIRS AND SERGE N. TIMASHEFF

VOLUME XXVI. Enzyme Structure (Part C)
Edited by C. H. W. HIRS AND SERGE N. TIMASHEFF

VOLUME XXVII. Enzyme Structure (Part D)
Edited by C. H. W. HIRS AND SERGE N. TIMASHEFF

VOLUME XXVIII. Complex Carbohydrates (Part B)
Edited by VICTOR GINSBURG

VOLUME XXIX. Nucleic Acids and Protein Synthesis (Part E)
Edited by LAWRENCE GROSSMAN AND KIVIE MOLDAVE

VOLUME XXX. Nucleic Acids and Protein Synthesis (Part F)
Edited by KIVIE MOLDAVE AND LAWRENCE GROSSMAN

VOLUME XXXI. Biomembranes (Part A)
Edited by SIDNEY FLEISCHER AND LESTER PACKER

VOLUME XXXII. Biomembranes (Part B)
Edited by SIDNEY FLEISCHER AND LESTER PACKER

Methods in Enzymology

Volume 71
LIPIDS
Part C

Section I

Fatty Acid Synthesis

[1] Acetyl-CoA Carboxylase from Rat Liver

EC 6.4.1.2 Acetyl-CoA : carbon-dioxide ligase (ADP-forming)

By Tadashi Tanabe, Shigetada Nakanishi, Takashi Hashimoto, Hideo Ogiwara, Jun-ichi Nikawa, and Shosaku Numa

$$ATP + HCO_3^- + \text{acetyl-CoA} \rightleftharpoons ADP + P_i + \text{malonyl-CoA}$$

Assay Methods

The principles underlying the various assays of acetyl-CoA carboxylase have been described in previous articles in this series.[1-4] Most conveniently, the enzyme activity is determined by the $^{14}CO_2$-fixation assay or by the spectrophotometric assay in combination with the pyruvate kinase and lactate dehydrogenase reactions. The $^{14}CO_2$-fixation assay can be used for enzyme preparations from all steps, whereas the spectrophotometric assay is applicable to preparations from the DEAE-cellulose chromatography step and subsequent steps.

$^{14}CO_2$-Fixation Method

Reagents

> Tris-HCl buffer, 0.5 M, pH 7.5
> Potassium citrate, 0.1 M
> $MgCl_2$, 0.1 M
> Reduced glutathione, 0.1 M, pH 7.5
> Bovine serum albumin, 3%
> ATP, 0.15 M
> Acetyl-CoA, 10 mM
> $KH^{14}CO_3$ (0.25 $\mu Ci/\mu mol$), 0.2 M
> HCl, 5 M
> Scintillator solution: 4 g of 2,5-diphenyloxazole and 0.1 g of 1,4-bis[2-(4-methyl-5-phenyloxazolyl)]benzene in 1 liter of toluene plus 0.5 liter of Triton X-100

Procedure. When the crude extract is assayed, it is passed through a Sephadex G-50 column to remove endogenous substrates. Because rat

---

[1] M. Matsuhashi, this series, Vol. 14 [1].
[2] S. Numa, this series, Vol. 14 [2].
[3] H. Inoue and J. M. Lowenstein, this series, Vol. 35 [1].
[4] A. L. Miller and H. R. Levy, this series, Vol. 35 [2].

liver acetyl-CoA carboxylase requires preincubation with citrate to attain its full activation,[5] the enzyme is first preincubated at 37° for 30 min in a mixture containing 50 mM Tris-HCl buffer, pH 7.5, 10 mM potassium citrate, 10 mM MgCl$_2$, 3.75 mM glutathione, and 0.75 mg of bovine serum albumin per milliliter. The reaction is then initiated by adding an aliquot of the preincubated enzyme (up to 0.2 mU) to an assay mixture (final volume, 0.8 ml) containing 50 mM Tris-HCl buffer, pH 7.5, 10 mM potassium citrate, 10 mM MgCl$_2$, 3.75 mM glutathione, 0.75 mg of bovine serum albumin per milliliter, 3.75 mM ATP, 0.125 mM acetyl-CoA, and 12.5 mM KH^{14}CO$_3$ (0.25 μCi/μmol). After incubation at 37° for 10 min, the reaction is terminated with 0.2 ml of 5 M HCl. The reaction mixture is allowed to stand in a vacuum desiccator for 30 min to remove the unreacted H^{14}CO$_3^-$ and is centrifuged at 1500 g for 10 min to eliminate the insoluble material. A 0.5-ml aliquot of the supernatant is taken to dryness at 60° in a counting vial in a vacuum desiccator. After addition of 0.5 ml of distilled water and 10 ml of the scintillator solution, the radioactivity is determined with the use of a liquid scintillation spectrometer. Under the assay conditions described, the reaction follows zero-order kinetics, and the initial rate of reaction is proportional to enzyme concentration.

Spectrophotometric Method

Reagents

> KHCO$_3$, 1 M
> Potassium phosphoenolpyruvate, 40 mM
> NADH, 5 mM, pH 8
> Pyruvate kinase (rabbit muscle; Boehringer), 10 mg/ml
> Lactate dehydrogenase (rabbit muscle; Boehringer), 5 mg/ml
> Other reagents, as for the ^{14}CO$_2$-fixation method

Procedure. The assay mixture contains 50 mM Tris-HCl buffer, pH 7.5, 10 mM potassium citrate, 10 mM MgCl$_2$, 3.75 mM glutathione, 0.75 mg of bovine serum albumin per milliliter, 3.75 mM ATP, 0.125 mM acetyl-CoA, 25 mM KHCO$_3$, 0.5 mM potassium phosphoenolpyruvate, 0.125 mM NADH, 15 μg of pyruvate kinase and 6 μg of lactate dehydrogenase per milliliter, and enzyme (up to 5 mU) in a final volume of 0.8 ml. A mixture (0.76 ml) containing all ingredients except ATP and KHCO$_3$ is preincubated at 37° for 10 min in a cuvette with 1 cm light path. The oxidation of NADH is followed at 37° with a recording spectrophotometer at 340 nm (or at 334 nm). After addition of ATP, the consumption of NADH is followed for 1 min, and the reaction is then started by addition of KHCO$_3$.

[5] S. Numa and E. Ringelmann, *Biochem. Z.* **343**, 258 (1965).

Initial velocities are obtained from the initial slopes of the recorder traces. Under the assay conditions described, the reaction follows zero-order kinetics for at least 3 min, and the initial rate of reaction is proportional to enzyme concentration.

Units. One unit of acetyl-CoA carboxylase activity is defined as that amount which catalyzes the formation of 1 μmol of malonyl-CoA or ADP per minute under the assay conditions described. Essentially identical activities are measured by the $^{14}CO_2$-fixation method and by the spectrophotometric method. Specific activity is expressed as units per milligram of protein. Protein is determined by the method of Lowry *et al.*[6] with crystalline bovine serum albumin as a standard or, for the purified enzyme, by absorbance at 280 nm. The relation between both values for the purified enzyme is $A_{280 \text{ nm}}^{1 \text{ cm}} \times 0.70$ = milligrams of protein per milliliter.

Purification Procedure[7,8]

All operations except Sepharose 2B chromatography are carried out at 0–4°. All phosphate buffers used are potassium phosphate buffer, pH 7.5, containing 5 mM 2-mercaptoethanol and 1 mM EDTA, unless otherwise specified. As the enzyme is susceptible to endogenous protease(s), rapid purification is necessary to avoid proteolytic modification (see later).

Preparation of Crude Extract. Thirty Wistar strain rats, each weighing 250–300 g, are fasted for 48 hr and subsequently refed a fat-free diet (56% glucose, 22% casein, 6% salts, 4% vitamins, and 12% cellulose) for 48 hr to induce the enzyme. The animals are killed by decapitation, and the livers (340 g) are quickly removed and dipped in ice-cold saline. The pooled livers are cut into small pieces and homogenized in two volumes of 0.25 M sucrose with a Teflon–glass homogenizer by two strokes under cooling in an ice-water bath. The homogenate is centrifuged at 13,000 g for 45 min. The fat layer at the top is discarded, and the supernatant is collected through cheesecloth. This is further centrifuged at 100,000 g for 1 hr, and the resulting supernatant is collected through cheesecloth (590 ml).

First Ammonium Sulfate Fractionation (0–30%). The crude extract is brought to 30% saturation by adding 176 g of solid ammonium sulfate per liter of the extract; the pH is maintained at 7.3–7.4 by addition of 5 M KOH. After further stirring for 30 min, the resulting precipitate is collected by centrifugation at 16,000 g for 20 min and dissolved in 10 mM phosphate buffer (final volume, 300 ml).

[6] O. H. Lowry, N. J. Rosebrough, A. L. Farr, and R. J. Randall, *J. Biol. Chem.* **193**, 265 (1951).
[7] S. Nakanishi and S. Numa, *Eur. J. Biochem.* **16**, 161 (1970).
[8] T. Hashimoto and S. Numa, *Eur. J. Biochem.* **18**, 319 (1971).

Calcium Phosphate Gel Fractionation and Second Ammonium Sulfate Fractionation (0–30%). The first ammonium sulfate fraction is stirred into 170 ml of a calcium phosphate gel suspension[9] (13.4 mg/ml; protein : gel = 1 : 1.8). After further stirring for 20 min, the gel is collected by centrifugation at 1000 g for 5 min and washed three times, each time with 120 ml of 33 mM phosphate buffer (0.7 volume of the calcium phosphate gel suspension added). The enzyme is then eluted twice, each time with 120 ml of 0.2 M phosphate buffer. The enzyme in the combined eluates is precipitated by adding 176 g of ammonium sulfate per liter of eluate; the pH is kept at 7.3–7.4 by addition of 5 M KOH. After further stirring for 30 min, the resulting precipitate is collected by centrifugation at 16,000 g for 20 min and dissolved in 10 mM phosphate buffer (final volume, 20 ml). The enzyme preparation of this step can be stored at $-15°$ for at least 3 months. In this case, the precipitate is dissolved in 0.1 M phosphate buffer containing 0.25 M sucrose, in which the enzyme is more stable.

DEAE-Cellulose Chromatography. The second ammonium sulfate fraction is dialyzed against 500 ml of 10 mM phosphate buffer for 1 hr. The dialyzed solution is diluted fivefold with 10 mM phosphate buffer and applied to a column (4 × 24 cm) of DEAE-cellulose (Serva) equilibrated with 10 mM phosphate buffer. The column is washed with 1 column volume of 10 mM phosphate buffer. Then, elution is accomplished with a linear concentration gradient established between 1.5 column volumes of 10 mM phosphate buffer and 1.5 column volumes of 0.5 M phosphate buffer. The enzyme appears in the eluate when the phosphate concentration reaches approximately 0.15 M. Fractions exhibiting specific activities higher than 0.6 unit/mg are combined, and the enzyme in the pooled eluate is precipitated by adding 0.67 volume of a saturated (at 0°) ammonium sulfate solution, pH 7.4. After centrifugation at 16,000 g for 20 min, the pellet is suspended in approximately 2 ml of 0.1 M phosphate buffer containing 0.25 M sucrose.

Sepharose 2B Chromatography. The enzyme suspension from the DEAE-cellulose chromatography step is dialyzed against 250 ml of 0.1 M phosphate buffer containing 10 mM potassium citrate and 0.25 M sucrose for 12 hr; the suspended enzyme is solubilized during the dialysis. After incubation at 37° for 30 min, the enzyme is chromatographed at 24° on a Sepharose 2B column (2 × 80 cm) equilibrated with 0.1 M phosphate buffer containing 10 mM potassium citrate and 0.25 M sucrose. Elution is carried out with the same solution. Fractions exhibiting the highest specific activities are combined, placed in a dialysis bag, and dialyzed at 4° for 5 hr against 0.1 M phosphate buffer containing 0.25 M sucrose and a sufficient amount of ammonium sulfate to give 50% saturation at equilib-

[9] L. A. Heppel, this series, Vol. 2 [91, footnote 4].

TABLE I
PURIFICATION OF ACETYL-CoA CARBOXYLASE FROM RAT LIVER[a]

Fraction	Volume (ml)	Protein (mg)	Total activity (units)	Specific activity (units/mg)	Yield (%)
Crude extract	590	13,000	122	0.0094	100
First $(NH_4)_2SO_4$	300	1,270	144	0.113	118
$Ca_3(PO_4)_2$ gel and second $(NH_4)_2SO_4$	20	350	61	0.174	50
DEAE-cellulose	138	60	38	0.633	31
Sepharose 2B	1.7	3.8	23	6.05	19

[a] From 340 g of liver.

rium. The enzyme precipitated is collected by centrifugation at 23,000 g for 20 min and dissolved in a minimal volume of 0.1 M phosphate buffer containing 0.25 M sucrose (final volume, 1.7 ml).

The results of the purification of acetyl-CoA carboxylase are summarized in Table I. The overall purification is 640-fold with a yield of 19%. The final enzyme preparations thus obtained are homogeneous as evidenced by sedimentation velocity experiments in an analytical ultracentrifuge and electrophoresis on an agar plate as well as by the fact that Ouchterlony double diffusion analysis and immunoelectrophoresis with the use of antibody raised against them yield a single precipitation line. The final enzyme preparations have a specific activity of 5–7 units/mg at 37°, which is considerably lower than that of the preparations obtained by our earlier procedure (7.55 units/mg at 25° corresponding to 15.1 units/mg at 37°),[7] but exhibit a greater dependency on citrate than the previous preparations. The previous purification procedure involves a dialysis step at 24° after the second ammonium sulfate fractionation step. During the dialysis, the total enzyme activity increases 1.6-fold. In view of the fact that incubation of the enzyme with the second ammonium sulfate fraction results in the cleavage of the native subunit into smaller polypeptides[10] and that limited treatment of the enzyme with trypsin elevates the enzyme activity[11] and reduces the dependency of the enzyme on citrate,[12] the higher specific activity of the previous preparations seems to be due to proteolytic modification of the enzyme.

Purification of [14C]*Biotin-Labeled Enzyme.*[10] Young male Wistar strain rats, each weighing about 60 g, are fed a biotin-deficient diet composed of 25% egg white, 62% sucrose, 4% corn oil, 5% salts, 1% vitamins exclud-

[10] T. Tanabe, K. Wada, T. Okazaki, and S. Numa, *Eur. J. Biochem.* 57, 15 (1975).
[11] T. Hashimoto and S. Numa, unpublished results.
[12] N. Iritani, S. Nakanishi, and S. Numa, *Life Sci.* 8, (pt2) 1157 (1969).

ing biotin, 0.1% choline chloride, and 3% cellulose. After 6–8 weeks, the animals, weighing 85–120 g, are fasted for 48 hr and subsequently refed a corn oil-free, biotin-deficient diet for 1 week. During the refeeding, the animals received daily intraperitoneal injections of 10 μg (1.87 μCi) of d-[carbonyl-^{14}C]biotin dissolved in 0.2 ml of saline. After this treatment, 6 rats, weighing 125–150 g, are sacrificed by decapitation. The livers are removed and used for the purification of [^{14}C]biotin-labeled acetyl-CoA carboxylase. The purification procedure is the same as described above, except that sucrose density gradient centrifugation is substituted for the final Sepharose 2B chromatography step. The concentrated DEAE-cellulose fraction is dialyzed against 50 mM Tris-HCl buffer, pH 7.5, containing 10 mM potassium citrate, 4% (w/v) sucrose, 5 mM 2-mercaptoethanol, and 1 mM EDTA at 4° for 6 hr and then incubated at 37° for 10 min. The enzyme preparation thus treated is applied onto three sucrose gradients (5 to 20%, w/v) containing 50 mM Tris-HCl buffer, pH 7.5, 10 mM potassium citrate, 5 mM 2-mercaptoethanol, and 1 mM EDTA; each gradient receives 3.7 mg of protein (0.27 ml). The tubes are centrifuged in a Beckman SW65L Ti rotor at 19° and 40,000 rpm for 100 min. Twelve 0.25-ml fractions are collected from each tube, and fractions exhibiting the highest specific activities are combined.

At the DEAE-cellulose chromatography step as well as at the final sucrose density gradient centrifugation step, the elution and sedimentation profile of the radioactivity coincides precisely with that of the enzyme activity. In addition, more than 95% of the radioactivity present in enzyme preparations from these steps is precipitated by antibody raised against pure unlabeled acetyl-CoA carboxylase. The radioactive material resulting from the digestion of the labeled enzyme with proteinase K is identified as biocytin. Thus, the radioactivity found in enzyme preparations purified through the DEAE-cellulose chromatography step is ascribed solely to the isotope in the biotinyl prosthetic group of acetyl-CoA carboxylase.

Properties

Storage of Enzyme. The second ammonium sulfate fraction as well as the final enzyme preparation, dissolved in 0.1 M potassium phosphate buffer, pH 7.5, containing 0.25 M sucrose, 5 mM 2-mercaptoethanol, and 1 mM EDTA, can be stored at −15° for at least 3 months, but loses activity on repeated freezing and thawing.

Biotin Content. [10] The colorimetric assay of Green[13] yields a value of 1.03 μg of biotin per milligram of protein (equivalent to 1 mol of biotin per

---

[13] N. M. Green, this series, Vol. 18A [74].

237,000 g of protein) in agreement with the value reported by Inoue and Lowenstein.[3,14] On the other hand, the microbiological assays with *Saccharomyces cerevisiae* and *Lactobacillus arabinosus* gave 93% and 65%, respectively, of the value found by the colorimetric method of Green. Bioautography of the enzyme hydrolyzate used for the microbiological assays showed that the sample contained, in addition to biotin and biotin sulfoxide, a "biocytin-like" compound, which was active in promoting the growth of *S. cerevisiae* but inactive as a biotin-vitamer for *L. arabinosus*. Therefore, the apparently lower values[7,15] found by the bioassay with *L. arabinosus* can be ascribed to the presence of the "biocytin-like" compound.

Subunit Structure and Proteolytic Modification.[10] Analysis by dodecyl sulfate–polyacrylamide gel electrophoresis has shown that the enzyme, unlike bacterial and plant biotin enzymes,[16,17] has only one kind of subunit, which has a molecular weight of 230,000 and contains one molecule of biotin. Because this subunit apparently carries the functions of biotin carboxylase, biotin carboxyl carrier protein, and carboxyl transferase as well as the regulatory function, the rat liver enzyme exhibits a highly integrated structure, representing a multifunctional polypeptide.[18] In some enzyme preparations, the native subunit was proteolytically cleaved into two polypeptides with molecular weights of 124,000 and 118,000, respectively, which were reported by Inoue and Lowenstein[14]; biotin was contained in the larger polypeptide, but not in the smaller polypeptide. The [^{14}C]biotin-labeled enzyme in crude rat liver extracts, when immunoprecipitated with specific antibody, invariably shows only the native subunit. The susceptibility of the enzyme to proteolysis by extracts of rat liver lysosomes is enhanced by palmitoyl-CoA, an allosteric inhibitor, whereas it is diminished by citrate, an allosteric activator.[19] Limited treatment of the enzyme with trypsin results in a partial loss of the sensitivity of the enzyme to both allosteric effectors.[12]

Molecular Forms. The enzyme sediments in a sucrose gradient with sedimentation coefficients ($s_{20,w}$) of 40–50 S and 20–25 S, respectively, in the presence and in the absence of citrate.[7,20] Centrifugation of the enzyme in an analytical ultracentrifuge under conditions of higher pH and higher salt concentration, which favor depolymerization of the enzyme, exhibits an $s_{20,w}$ of 13–16 S.[10,15] This molecular form represents a mixture of the

[14] H. Inoue and J. M. Lowenstein, *J. Biol. Chem.* **247**, 4825 (1972).
[15] S. Numa, T. Hashimoto, S. Nakanishi, and T. Okazaki, *Biochem. Soc. Symp.* **35**, 27 (1972).
[16] M. Obermeyer and F. Lynen, *Trends Biochem. Sci.* **1**, 169 (1976).
[17] H. G. Wood and R. E. Barden, *Annu. Rev. Biochem.* **46**, 385 (1977).
[18] K. Kirschner and H. Bisswanger, *Annu. Rev. Biochem.* **45**, 143 (1976).
[19] T. Tanabe, K. Wada, H. Ogiwara, and S. Numa, *FEBS Lett.* **82**, 85 (1977).
[20] S. Numa, E. Ringelmann, and F. Lynen, *Biochem. Z.* **343**, 243 (1965).

multifunctional monomeric subunit (molecular weight, 230,000) and its dimer. Thus, the 20–25 S form is assumed to arise from partial aggregation of the monomeric subunit. On the basis of electron microscopic[21] and light-scattering studies[22] of the chicken liver enzyme, the 40–60 S form is known to be a large filamentous polymer with a molecular weight of 4,000,000 to 11,000,000. The correlation between the sedimentation coefficient and catalytic activity of the enzyme indicates that the "large" molecular form (40–60 S) represents the active conformation, whereas the "small" molecular form (13–25 S) represents the inactive conformation.[23]

Steady-State Kinetics.[8,15,24] The enzyme reaction proceeds through an ordered Bi Bi Uni Uni Ping-Pong mechanism, the order of addition of substrates being ATP, HCO_3^-, and acetyl-CoA in the forward reaction, and malonyl-CoA, P_i, and ADP in the reverse reaction. The hybrid (or two-site) Ping-Pong character[25] of the enzyme reaction is also noted. The kinetic data corroborate the two-step mechanism of the enzyme reaction with intermediate formation of the carboxylated biotin enzyme.[2] The kinetic constants are as follows[8]: $K_m(ATP) = 15 \mu M$, $K_m(HCO_3^-) = 2.5$ mM, $K_m(\text{acetyl-CoA}) = 25 \mu M$, $K_m(\text{malonyl-CoA}) = 16 \mu M$, $K_m(P_i) = 7$ mM, $K_m(ADP) = 10 \mu M$. The enzyme carboxylates acetonyldethio-CoA, in which the sulfur is replaced by methylene[26]; the terminal methyl group is carboxylated as in the case of acetyl-CoA. The apparent K_m value for acetonyldethio-CoA is approximately five times as high as that for acetyl-CoA, while the V_{max} value for this analog is approximately one-seventh that for acetyl-CoA.

Activators and Inhibitors.[2,3,23,27] The enzyme exhibits a nearly absolute requirement for tricarboxylate activators such as citrate and isocitrate. Kinetic studies[8,24] on the activation mechanism indicate that, of the obligatory enzyme forms, only the carboxylated form of the enzyme (E-biotin ~ CO_2) is dependent on citrate. The apparent citrate-independence of the uncarboxylated form of the enzyme (E-biotin) during the catalysis can be accounted for by the fact that the dissociation constant of the E-biotin ~ CO_2 · citrate complex (3–6 mM) is three orders of magnitude larger than that of the E-biotin · citrate complex (3–14 μM). The main role of citrate in the catalysis is to keep the carboxylated form of the enzyme in active conformation by shifting the equilibrium between the active and inactive species of this enzyme form.

[21] A. K. Kleinschmidt, J. Moss, and M. D. Lane, *Science* **166**, 1276 (1969).
[22] G. Henniger and S. Numa, *Hoppe-Seyler's Z. Physiol. Chem.* **353**, 459 (1972).
[23] S. Numa, *Ergeb. Physiol.* **69**, 53 (1974).
[24] T. Hashimoto, H. Isano, N. Iritani, and S. Numa, *Eur. J. Biochem.* **24**, 128 (1971).
[25] D. B. Northrop, *J. Biol. Chem.* **244**, 5808 (1969).
[26] J. Nikawa, S. Numa, T. Shiba, C. J. Stewart, and T. Wieland, *FEBS Lett.* **91**, 144 (1978).
[27] S. Numa and S. Yamashita, *Curr. Top. Cell. Regul.* **8**, 197 (1974).

The enzyme is inhibited by long-chain acyl-CoA thioesters such as palmitoyl-CoA. The mechanism of the inhibition has been extensively studied by means of the kinetics of tight-binding inhibitors and [^{14}C]palmitoyl-CoA binding experiments.[28] One mole of palmitoyl-CoA completely inhibits 1 mol of the enzyme, the inhibition constant being as low as 5.5 nM; this value is about three orders of magnitude smaller than the critical micellar concentration of palmitoyl-CoA. The enzyme, when preincubated with palmitoyl-CoA in low molar ratios (inhibitor: enzyme $\leq$ 5), is associated with an equimolar amount of the inhibitor, assuming the "small" molecular form. The equimolar enzyme–inhibitor complex is formed even in the presence of phosphatidylcholine, which is known to bind palmitoyl-CoA. The activator citrate, which competes kinetically with palmitoyl-CoA,[20] not only prevents this equimolar association but also dissociates the equimolar enzyme–inhibitor complex in the presence of an acceptor for long-chain acyl-CoA, such as alkylated cyclodextrin or phosphatidylcholine. The enzyme thus freed from the inhibitor assumes the "large" molecular form and regains its full activity. In contrast, the enzyme, when preincubated with palmitoyl-CoA in high molar ratios (inhibitor: enzyme $\geq$ 20), binds a large molar excess of the inhibitor and is further dissociated into the monomeric subunit, as is the enzyme treated with sodium dodecyl sulfate. Phosphatidylcholine protects the enzyme from binding an excess of palmitoyl-CoA. When the enzyme associated with a large excess of palmitoyl-CoA is deprived of the inhibitor by treatment with citrate and alkylated cyclodextrin, the enzyme becomes nonspecifically aggregated, as in the case of the sodium dodecyl sulfate-treated enzyme, and does not regain activity. Thus, it is concluded that palmitoyl-CoA binds tightly and reversively to the enzyme in an equimolar ratio to inhibit it. The inhibition constants for various structural analogs of palmitoyl-CoA are listed in Table II.[29] The 3'-phosphate of the CoA moiety is essential for the inhibition of the enzyme by palmitoyl-CoA. Modification of the pantoic acid, adenine, or thioester moiety of palmitoyl-CoA does not appreciably influence its inhibitory effect. The CoA thioesters of saturated fatty acids with 16–20 carbon atoms inhibit acetyl-CoA carboxylase more effectively than those of saturated fatty acids of shorter or longer chain lengths. The CoA thioesters of unsaturated fatty acids are less inhibitory than those of saturated fatty acids of corresponding chain lengths. The rather strict structural requirement for the inhibitory effect of long-chain acyl-CoA indicates that the inhibitor binds to a specific site on the acetyl-CoA carboxylase molecule. This, together with the reversible formation of the equimolar enzyme-inhibitor

[28] H. Ogiwara, T. Tanabe, J. Nikawa, and S. Numa, *Eur. J. Biochem.* **89**, 33 (1978).
[29] J. Nikawa, T. Tanabe, H. Ogiwara, T. Shiba, and S. Numa, *FEBS Lett.* **102**, 223 (1979).

TABLE II
INHIBITION CONSTANTS OF ACETYL-CoA CARBOXYLASE FOR LONG-CHAIN ACYL-CoA AND ITS ANALOGS[a]

Inhibitor	K_i (nM)
Palmitoyl-CoA	6.5
Palmitoyl-CoA (L)	22
Palmitoyl-keto-CoA	21
Palmitoyl-1,N^6-etheno-CoA	15
Palmitoyl-inosino-CoA	14
Palmitoyl-dephospho-CoA	260
Palmitoyl-4′-phosphopantetheine	650
S-Cetyl-CoA	10
Lauroyl-CoA	—[b]
Myristoyl-CoA	680
Stearoyl-CoA	1.3
Arachidoyl-CoA	<1
Docosanoyl-CoA	40
Tetracosanoyl-CoA	150
Palmitoleoyl-CoA	130
Oleoyl-CoA	44
Linoleoyl-CoA	27
Linolenoyl-CoA	66
Arachidonoyl-CoA	48

[a] From Nikawa et al.[29]
[b] Essentially no inhibition was observed at concentrations up to 10 μM.

complex, supports the concept that long-chain acyl-CoA is a physiological regulator of the enzyme. The allosteric modulation of the catalytic efficiency of the enzyme appears to play an essential role in the short-term regulation of fatty acid synthesis.[23,27,30] It is also indicated that long-chain acyl-CoA acts not only as an allosteric inhibitor but also as a corepressor (or its eukaryotic equivalent) for the enzyme[31] (see below).

Kynurenate and xanthrenate[24] as well as hypolipidemic agents such as 2-methyl-2-[p-(1,2,3,4-tetrahydro-1-naphthyl)phenoxy]propionate and 2-(p-chlorophenoxy)-2-methylpropionate[32] inhibit the enzyme. The apparent inhibition constants for these compounds are four to five orders of magnitude higher than those for long-chain acyl-CoA.

Phosphorylation. The enzyme has been reported to contain 2.1 mol of phosphate per 215,000 g of protein.[3,14] Incorporation of ^{32}P into the rat

[30] K. Nishikori, N. Iritani, and S. Numa, *FEBS Lett.* **32**, 19 (1973).
[31] T. Kamiryo, Y. Nishikawa, M. Mishina, M. Terao, and S. Numa, *Proc. Natl. Acad. Sci. U.S.A.* **76**, 4390 (1979).
[32] M. E. Maragoudakis and H. Hankin, *J. Biol. Chem.* **246**, 348 (1971).

liver as well as adipose tissue enzyme has been shown both *in vitro*[33-36] and *in vivo*.[37] No agreement has been reached, however, with respect to the physiological role of this phosphorylation in the regulation of the catalytic efficiency of the enzyme.[34,35,37,38]

Synthesis and Degradation. The hepatic content of the enzyme varies in accord with the rate of fatty acid synthesis under a variety of dietary, hormonal, developmental, and genetic conditions.[23,27] Fasted or diabetic rats exhibit a subnormal enzyme content, whereas rats fasted and subsequently refed a fat-free diet show an enzyme content higher than the normal level. It has been demonstrated by combined immunochemical and isotopic techniques that the increase or decrease in the enzyme content in refed or diabetic rats is ascribed solely to a corresponding change in the rate of synthesis of the enzyme, whereas the decrease in the enzyme content in fasted rats is due to both diminished synthesis and accelerated degradation of the enzyme.[7,39] The elevated enzyme content in obese mice is due mainly to elevated synthesis of the enzyme and, in a minor degree, to diminished degradation of the enzyme.[40] ^{125}I-Labeled antibody binding experiments[41,42] and cell-free translation studies[43] have demonstrated that the hepatic content of acetyl-CoA carboxylase-synthesizing polysomes parallels the rate of hepatic synthesis *in vivo* of the enzyme under the different metabolic conditions, thus indicating that the different rates of synthesis of the enzyme is attributed to changes in the amount of translatable mRNA coding for the enzyme. Recently, mRNA from the hydrocarbon-utilizing yeast *Candida lipolytica,* which is used as a suitable eukaryotic model system, has been translated in a cell-free system to synthesize complete acetyl-CoA carboxylase.[44,45] The level of acetyl-CoA carboxylase mRNA activity varies in accord with the rate of synthesis of the enzyme.[44,45] Furthermore, the long-chain acyl-CoA to be utilized for

[33] K.-H. Lee and K.-H. Kim, *J. Biol. Chem.* **252,** 1748 (1977).
[34] R. W. Brownsey, W. A. Hughes, R. M. Denton, and R. J. Mayer, *Biochem. J.* **168,** 441 (1977).
[35] P. H. Pekala, M. J. Meredith, D. M. Tarlow, and M. D. Lane, *J. Biol. Chem.* **253,** 5267 (1978).
[36] L. A. Witters, E. M. Kowaloff, and J. Avruch, *J. Biol. Chem.* **254,** 245 (1979).
[37] K.-H. Lee and K.-H. Kim, *J. Biol. Chem.* **254,** 1450 (1979).
[38] C. A. Carlson and K.-H. Kim, *Arch. Biochem. Biophys.* **164,** 490 (1974).
[39] P. W. Majerus and E. Kilburn, *J. Biol. Chem.* **244,** 6254 (1969).
[40] S. Nakanishi and S. Numa, *Proc. Natl. Acad. Sci. U.S.A.* **68,** 2288 (1971).
[41] S. Nakanishi, T. Tanabe, S. Horikawa, and S. Numa, *Proc. Natl. Acad. Sci. U.S.A.* **73,** 2304 (1976).
[42] T. Tanabe, S. Horikawa, S. Nakanishi, and S. Numa, *FEBS Lett.* **66,** 70 (1976).
[43] S. Horikawa, S. Nakanishi, and S. Numa, *FEBS Lett.* **74,** 55 (1977).
[44] S. Horikawa, T. Kamiryo, S. Nakanishi, and S. Numa, *Eur. J. Biochem.* **104,** 191 (1980).
[45] M. Mishina, T. Kamiryo, and S. Numa, this volume [5].

lipid synthesis, but not that to be degraded via β-oxidation, is involved in the repression of acetyl-CoA carboxylase.[31] It seems probable that the long-chain acyl-CoA for lipid synthesis, or a compound metabolically related to it, acts as a corepressor or its eukaryotic equivalent to suppress the expression of the acetyl-CoA carboxylase gene. Such a regulatory mechanism is evidently of teleological significance in view of the homeostasis of lipid synthesis.

[2] Acetyl-CoA Carboxylase from Rat Mammary Gland (Including an Additional Step Yielding Fatty Acid Synthase)

EC 6.4.1.2 Acetyl-CoA : carbon-dioxide ligase (ADP-forming)

By Fazal Ahmad and Patricia M. Ahmad

$$\text{Acetyl-CoA} + \text{HCO}_3^- + \text{ATP} \rightleftharpoons \text{malonyl-CoA} + \text{ADP} + \text{P}_i$$

Assay of Acetyl-CoA Carboxylase

A number of methods, spectrophotometric and isotopic, for assaying acetyl-CoA carboxylase have been described.[1-4] Most spectrophotometric methods, though convenient, are useful only when the enzyme is available in a purified state. It is somewhat difficult to determine accurately the activity of acetyl-CoA carboxylase in crude tissue extracts. The main difficulty arises because of the presence of inhibitor(s) of acetyl-CoA carboxylase and other interfering substances in the unfractionated tissue homogenates.[3,5,6] This difficulty can be readily overcome by centrifuging crude extracts at high speeds (100,000 g for 1–2 hr) and then chromatographing the supernatant solution over Sephadex G-25 columns prior to assay. During purification from lactating rat mammary gland, the activity of acetyl-CoA carboxylase is determined by the isotopic method. In this method acid-labile [^{14}C]bicarbonate is converted to the carboxyl group of malonyl-CoA. The ^{14}CO$_2$ fixed into malonyl-CoA is acid stable and this property is used to determine the activity of the enzyme.

[1] M. Matsuhashi, this series, Vol. 14 [1].
[2] S. Numa, this series, Vol. 14 [2].
[3] H. Inoue and J. M. Lowenstein, this series, Vol. 35 [1].
[4] A. L. Miller and H. R. Levy, this series, Vol. 35 [2].
[5] C. Gregolin, E. Ryder, and M. D. Lane, *J. Biol. Chem.* **234**, 4227 (1968).
[6] J. Moss, M. Yamagishi, A. K. Kleinschmidt, and M. D. Lane, *Biochemistry* **11**, 3779 (1972).

Reagents

Reaction mix A
    Imidazole-HCl buffer, 1.0 M, pH 7.5
    Magnesium acetate, 0.4 M
    EDTA, 40 mM, pH 7.0
    EGTA, 2 mM, pH 7.0
    Potassium citrate, 1.0 M
    Bovine serum albumin (BSA), 10 mg/ml in H_2O
Reaction mix B
    Acetyl-CoA, 15 mM
    Sodium [^{14}C]bicarbonate, 0.5 M, 4.5 μCi/ml
    ATP, 144 mM, pH 6.8
Creatine phosphate, 0.35 M
Creatine phosphokinase, 66 units/ml, dissolved in 50 mM
    imidazole-HCl, pH 7.5, containing 10 mg/ml BSA

Two previously described procedures[4] with some modifications[7] have been employed to assay acetyl-CoA carboxylase activity. In one of these procedures an ATP-regenerating system is used whereas in the other it is not.

Reaction mix A is made fresh daily. The mix, sufficient for 30 assays, contains the following (in milliliter): imidazole-HCl buffer, 0.18; magnesium acetate, 0.18; EDTA, 0.03; EGTA, 0.15; citrate, 0.09; BSA, 0.15, and H_2O, 0.12. Each assay tube contains 0.03 ml of reaction mix A and 0.5–3.0 mU of acetyl-CoA carboxylase in a final volume of 0.085 ml. At this point the contents are preincubated at 37° for 15 min if citrate activation is necessary (in post BioGel samples no activation is required). Reaction mix B is made by mixing equal parts of the three ingredients (acetyl-CoA, sodium bicarbonate, and ATP) immediately prior to use. After addition of 0.015 ml of reaction mix B, and incubation at 37° for 5 min, the reaction is terminated by adding 0.025 ml of 4 N HCl. An aliquot (0.05 ml) of the sample is transferred to a glass scintillation minivial and taken to dryness in a forced-air oven at 85°. The residue containing [^{14}C]malonyl-CoA is dissolved in 0.5 ml H_2O. After addition of 5 ml of scintillation fluid,[8] the radioactivity is determined in a scintillation spectrometer. Whenever necessary, appropriate controls (minus ATP or minus acetyl-CoA) are included.

For assay of less pure samples, especially prior to chromatography over DE-52 columns, an ATP-regenerating system is included during the preincubation period. To 0.03 ml of mix A containing carboxylase, 0.015 ml of creatine phosphokinase is added and the volume is adjusted to 0.08

[7] F. Ahmad, P. M. Ahmad, L. Pieretti, and G. T. Watters, *J. Biol. Chem.* **253**, 1733 (1978).
[8] U. Fricke, *Anal. Biochem.* **63**, 555 (1975).

ml. After preincubation, 0.02 ml of modified reaction mix B (equal volumes of acetyl-CoA, sodium bicarbonate, ATP, and creatine phosphate) is added. All other steps are as described above.

Units. One unit of enzyme is defined as that amount which catalyzes the synthesis of 1 μmol of malonyl-CoA per minute at 37°. Specific activity is defined as units per milligram of protein.

Purification Procedure

In the purification scheme described here the most predominant protein copurifying with the carboxylase is fatty acid synthase. The constituent polypeptide(s) arising from the dissociation of rat mammary gland acetyl-CoA carboxylase and fatty acid synthase possess almost identical molecular weights as judged by sodium dodecyl sulfate (SDS)–polyacrylamide gel electrophoresis.[7] Therefore, acetyl-CoA carboxylase preparations of comparatively low specific activity and containing sizable amounts of fatty acid synthase may appear to be quite pure when examined by this technique. The successful purification of acetyl-CoA carboxylase from rat mammary gland depends upon the removal of fatty acid synthase. To achieve this, certain modifications to the existing purification scheme[4] have been made, which result in excellent recoveries of acetyl-CoA carboxylase and also allow concomitant purification of fatty acid synthase.

The results on the purification of acetyl-CoA carboxylase from lactating rat mammary gland are summarized in Table I.[7,9,10]

Buffers Employed. The buffers (pH 6.5) employed during the purification scheme contain 20% glycerol and are of the following compositions[4]: buffer I: 50 mM imidazole HCl, 1 mM EDTA, 7 mM β-mercaptoethanol; buffer II: 50 mM imidazole HCl, 20 mM potassium citrate, 0.5 mM dithiothreitol (DTT), 0.1 mM EDTA; buffer III: 5.0 mM imidazole-HCl, 7 mM β-mercaptoethanol, 0.1 mM EDTA, 50 mM MgSO$_4$.

Step 1. Crude Extract. Ten to fifteen female rats of Fischer strain, 15–18 days postpartum, each of which has nursed seven healthy pups, are sacrificed by cervical dislocation. Mammary glands are excised and rinsed thoroughly with ice cold 0.25 M sucrose until free of milk. After finely mincing with scissors, glands are suspended in two volumes of buffer I and homogenized in a VirTis homogenizer at 22,500 rpm (or in a Waring blender at top speed) and 4° for three periods of 45 sec each interspersed with 1 min rest periods. This homogenate is centrifuged for 10 min at 2000 g at

[9] O. H. Lowry, N. J. Rosebrough, A. L. Farr, and R. J. Randall, *J. Biol. Chem.* **193**, 265 (1951).
[10] O. Warburg and W. Christian, *Biochem. Z.* **310**, 384 (1941).

TABLE I

PURIFICATION OF ACETYL-CoA CARBOXYLASE FROM LACTATING RAT MAMMARY GLAND[a]

Step	Volume (ml)	Total activity (units)	Total protein (mg)	Specific activity
1. Crude extract	230	196	3,933[b]	0.040
2. First $(NH_4)_2SO_4$ (0–50%) precipitation	93	185	2,325[b]	0.079
3. 100,000 g supernatant after dialysis	91	372	1,885[b]	0.199
4. Eluate from DE-52	383	294	613[b]	0.48
5. $(NH_4)_2SO_4$ fractionation	3	202	218[b]	0.93
6. BioGel A-5m				
Pool A	53	77.6	5.03[c]	15.34
Pool B	51	33.0	6.35[c]	5.18
7. $(NH_4)_2SO_4$ precipitation of BioGel pools				
Pool A	0.8	40.4	2.37[b]	17.0
Pool B	0.8	35.6	2.92[b]	12.2

[a] The mammary gland tissue (~ 100 g wet weight) derived from eight lactating female rats served as a source for the purification scheme.

[b] Protein determined by the method of Lowry et al.[9]

[c] Protein determined by the ultraviolet-absorption method.[10]

4°. After removal of the fat layer with a spatula, the supernatant is filtered The 2000 g pellet is suspended in one volume of buffer I, homogenized, and treated as above. The combined supernatants are centrifuged at 23,500 g for 30 min at 4°.

Step 2. Precipitation with Ammonium Sulfate. Solid ammonium sulfate is added to the enzyme solution slowly with gentle stirring at 4° to give 50% saturation. The precipitate is allowed to settle for 1–2 hr at 4° and collected by centrifugation at 23,500 g at 4° for 30 min. The pellet contains both the acetyl-CoA carboxylase and fatty acid synthase.

Step 3. Dialysis and Centrifugation at 100,000 g. The pellet from step 2 is suspended in buffer I to give one-third the volume of the original crude extract. It is stirred gently on a magnetic stirrer at 4° for 15–30 min and the resulting suspension is dialyzed overnight against 4 liters of buffer III containing 0.02% sodium azide.

The dialyzed turbid protein solution is centrifuged at 100,000 g at 4° for 2 hr. The floating fat layer is removed, and the supernatant solution is filtered through cheesecloth. The solution is dialyzed against 4 liters of buffer III containing 0.02% sodium azide until the conductivity of the dialyzate is equal to that of buffer III.

It is generally observed that after the above-described dialysis and centrifugation at 100,000 g there is a marked increase in the total carboxylase activity over that present in the initial extract. However, if the crude extract is dialyzed and centrifuged at 100,000 g prior to assay, its specific activity is approximately 1.5–2 times higher than given in step 1. This type of phenomenon has been observed previously with the crude enzyme preparations derived from avian liver[5] and bovine adipose tissue.[6] This enhancement of activity during the early stages of purification is probably due to the removal of hydrophobic inhibitor(s) from the carboxylase.

Step 4. Negative Adsorption on DE-52. Prior to use, 400 g of DE-52 are cycled as recommended by the manufacturer. After cycling, the resin is suspended in 2 liters of buffer containing imidazole HCl, pH 6.5, 0.1 M; β-mercaptoethanol, 7 mM; EDTA, 0.1 mM; MgSO$_4$, 50 mM; it is allowed to equilibrate for 15 min. The resin is filtered through a Büchner funnel. This procedure is repeated until the pH of the filtrate is 6.5. The resin may be stored at 4° in buffer III containing 0.02% sodium azide. Before packing the column, the resin slurry is warmed to room temperature and degassed. To maintain acceptable flow rates, glass wool (~0.5 cm) may be inserted into the bottom of the column prior to adding resin slurry. The packed column is equilibrated with buffer III containing sodium azide. The column is used only after the conductivity and the pH of the effluent are identical to those of the eluent.

The dialyzed turbid enzyme solution (100–150 ml) is applied to a 5 × 25 cm column at room temperature. After the solution has entered the resin, the column is eluted with buffer III at a flow rate of 1 ml/min, and fractions of 5–7 ml each are collected. The protein solution must emerge as a clear solution for the successful resolution of acetyl-CoA carboxylase from fatty acid synthase in the succeeding steps. The protein concentration of the fractions is monitored at 280 nm. The elution is continued until the absorbance reaches ~0.5. The $A_{280}:A_{260}$ ratio of most fractions is greater than 1.3, but, on continued elution as the protein concentration decreases, generally there is a reversal in this ratio. This reversal is coincident with the low levels of enzymatic activity in the eluate. Only the leading and trailing fractions of protein need be assayed for enzymatic activity. Fractions of $A_{280} \geq 0.7$ are pooled and then precipitated by adding, with gentle stirring at 4°, an equal volume of a saturated and neutralized solution of ammonium sulfate. The precipitate is allowed to settle for 1–2 hr and then collected by centrifugation at 23,500 g for 30 min at 4°. The protein pellet contains acetyl-CoA carboxylase and fatty acid synthase.

Step 5. Separation of Fatty Acid synthase from Acetyl-CoA Carboxylase by Ammonium Sulfate Fractionation. The protein pellet from step 4 is suspended in 30 ml of buffer I containing 40% saturated ammonium sulfate and stirred at 4° for 15 min, allowed to stand for 15 min, and then centrifuged at 23,500 g for 15 min. The supernatant solution from this step is discarded. The precipitate is then extracted as before with buffer I containing the following saturation levels of ammonium sulfate: 33%, 30%, 28%, 25%, 10%, and 0%. The supernatants from these steps are monitored for carboxylase and synthase activities.

Most of the fatty acid synthase is extracted into buffer I containing 28% and 30% ammonium sulfate. It may be purified further as given below.

Approximately two-thirds of the acetyl-CoA carboxylase and very little synthase is obtained in the 0% extract. The remaining one-third of the carboxylase is present in the 25% and 28% extractions: however, these fractions contain higher levels of fatty acid synthase. They are combined and brought to 30% saturation by the addition of solid ammonium sulfate. The precipitate is dissolved in buffer I to a protein concentration of 15 mg/ml. Saturated and neutralized ammonium sulfate solution is added to achieve saturation levels (generally between 17 and 25%) that allow separation of the carboxylase from synthase.

Fractions (usually 10% and 0%) containing maximal acetyl-CoA carboxylase activity and low levels of fatty acid synthase (specific activity $\simeq$ 0.01–0.02) are brought to 30% saturation with respect to ammonium sulfate, allowed to stand overnight at 4°, and then centrifuged at 23,500 g for 15 min. The precipitates are dissolved in a minimal volume of buffer II and allowed to incubate at 4° for 24–48 hr prior to chromatography.

Step 6. Chromatography over BioGel A-5m Column. The enzyme in buffer II (containing 0.02% sodium azide) from the previous step is layered over a BioGel A-5m column (2.5 × 90 cm), which is eluted with the same buffer at room temperature. Fractions of ~2.5 ml each are collected at a flow rate of 9–10 ml/hr. Each fraction is monitored for enzymatic activity and protein concentration. Fractions containing the highest specific activity are pooled and precipitated by dialysis at 4° against buffer II containing 10% glycerol and 50% ammonium sulfate. The precipitated enzyme is separated by centrifugation at 23,500 g for 30 min and dissolved in a minimal volume of buffer II. Small aliquots are frozen in liquid N_2 and stored at −70°.

During BioGel chromatography in buffer II, the enzyme elutes in the breakthrough volume indicating that the polymeric form of the enzyme behaves like a protein of molecular weight $\geq 5 \times 10^6$. The asymmetric distribution of protein during chromatography also suggests that the polymeric form of the enzyme is not weight homogeneous but may consist of a heterogeneous population of filaments containing different numbers of protomers. This suggestion is consistent with the electron microscopic analysis of the enzyme, where filaments of different sizes are seen.[7]

Alternative Procedure for the Purification of Rat Mammary Gland Acetyl-CoA Carboxylase

The purification of rat mammary gland acetyl-CoA carboxylase may also be achieved by a simplified procedure. This method is similar to that employed by Hardie and Cohen[11] for the purification of rabbit mammary gland acetyl-CoA carboxylase. During purification by this method the enzyme carried through either step 3 or step 5 of the above-described scheme may be employed. The results on the purification of rat mammary gland acetyl-CoA carboxylase by this procedure are given in Table II.[12]

The enzyme solution from either of the steps is brought to 35% ammonium sulfate saturation by adding 0.64 volume of a 3.6 M neutralized solution of ammonium sulfate with stirring at 4°. The precipitate is allowed to stand for 1 hr at 4° and collected by centrifugation at 20,000 g for 10 min at 4°. The precipitate is suspended in buffer containing 100 mM potassium phosphate, 25 mM potassium citrate, 1 mM EDTA, 15 mM β-mercaptoethanol, pH 7.0, and dialyzed against this buffer for 9–12 hr at room temperature (25–27°). All subsequent steps are also carried out at room temperature. Any insoluble material is removed by centrifugation at 20,000 g for 2 min. To the supernatant solution, polyethylene glycol 6000 (50%, w/v) is added slowly to give a final concentration of 3% (w/v). After standing for at least 3 hr the precipitate is collected by centrifugation at 20,000 g for 10 min, suspended in the above buffer, and then incubated at 37° for 10 min. If insoluble material is present, it is removed by centrifugation at 20,000 g for 2 min. The enzyme solution is again brought to 3% (w/v) polyethylene glycol concentration, allowed to stand for 3 hr, centrifuged at 20,000 g for 10 min, redissolved in the above buffer, and after quick freezing in an acetone–Dry Ice bath is stored at −70° in small aliquots.

Purification of Rat Mammary Gland Fatty Acid Synthase[12]

The fatty acid synthase from lactating rat mammary gland copurifies with acetyl-CoA carboxylase through many steps of the purification scheme given in Table I. Most of the fatty acid synthase activity separates

---

[11] D. G. Hardie and P. Cohen, *FEBS Lett.* **91**, 1 (1978).
[12] P. M. Ahmad and F. Ahmad, unpublished results.

TABLE II

PURIFICATION OF ACETYL-CoA CARBOXYLASE FROM LACTATING RAT MAMMARY
GLAND BY PRECIPITATION WITH POLYETHYLENE GLYCOL

	Source of carboxylase			
	Fresh mammary gland extract[a]		Post DE-52 column chromatography fraction[b]	
Step	Units	Specific activity	Units	Specific activity
1. (NH₄)₂SO₄ precipitation (35% saturation)	16.0	0.23	17.6	2.8
2. First 3% (w/v) polyethylene glycol precipitate	16.1	2.0	16.4	9.5
3. Second 3%(w/v) polyethylene glycol precipitate	10.6	5.4	10.0	18.9

[a] Enzyme from step 3 of Table I.
[b] Enzyme from step 5 of Table I.

from acetyl-CoA carboxylase during step 5, which involves extractions of the precipitated enzyme with Buffer I containing 28% and 30% saturation of ammonium sulfate. Each extract is precipitated separately by adding saturated and neutralized solution of ammonium sulfate to 33% final concentration at 4°. The precipitated enzyme is collected by centrifugation at 20,000 g for 20 min at 4°. Each precipitate is dissolved separately in 250 mM potassium phosphate buffer (pH 6.5) containing 10 mM DTT, 0.1 mM EDTA, and 50% glycerol and assayed for protein concentration[9] and enzymatic activity. The specific activity at this stage generally equals and in some cases exceeds 1 μmol of NADPH oxidized per minute per milligram of protein when assayed at 37° by the previously described method.[13] In this respect the synthase purified as a by-product of acetyl-CoA carboxylase purification is comparable to the highly purified enzyme obtained by a somewhat lengthy procedure.[13] If required, the synthase may be purified further by chromatography on a BioGel A-1.5m column using buffer containing 0.25 M potassium phosphate, pH 6.5, 10 mM DTT, and 0.1 mM EDTA at room temperature. Fractions of highest specific activity ($\geq$1.3) are pooled, precipitated with ammonium sulfate to 35%

[13] S. Smith and S. Abraham, this series, Vol. 35 [8].

saturation at 4°, dissolved in phosphate buffer containing 50% glycerol, and stored at −70°.

Properties

Stability of Acetyl-CoA Carboxylase. Acetyl-CoA carboxylase preparations of varying degrees of purity can be quick frozen in liquid nitrogen or in an acetone–Dry Ice bath and stored at −70° for up to 4 years without noticeable loss of activity. The purified enzyme (specific activity ≥ 10) is stable for up to 1 week at room temperature when stored in buffer (pH 7.5) containing 50 mM imidazole HCl, 0.1 mM EDTA, 0.02% sodium azide, 20% glycerol, and 25 mM citrate. Unlike the polymeric form, the protomeric form of mammary gland acetyl-CoA carboxylase is considerably less stable both at 4° and at room temperature. This perhaps is due to the presence of trace amounts of proteolytic activity, which readily cleaves the protomer rather than the polymer.

Acetyl-CoA carboxylase purified by polyethylene glycol precipitation from mammary gland extract (step 3 of Table I) is unstable when stored at room temperature in a buffer (pH 7.0) containing 100 mM potassium phosphate, 25 mM sodium citrate, 1 mM EDTA, and 15 mM β-mercaptoethanol, losing 60% of its activity in 16 hr. In contrast, the enzyme purified by an identical procedure starting with the enzyme that has been carried through step 5 of Table I is stable when stored under identical conditions. The reasons why the former enzyme preparation is less stable are not yet clear.

Purity. Like other carboxylases of animal origin the enzyme from rat mammary gland displays a single hypersharp peak during sedimentation in the presence of citrate in the analytical ultracentrifuge. The sedimentation constant is ≥45 S. It also reveals a filamentous structure when negatively stained preparations of the enzyme are viewed in the electron microscope.[7] During glycerol density gradient centrifugation the protomeric form sediments as a symmetrical peak (~12–13 S). A small amount of dimer is also present.

When the purified enzyme in buffers containing citrate is dissociated in 1% SDS containing 8 M urea and 5 mM DTT (in a boiling H$_2$O bath for 5 min), electrophoresed in 4% polyacrylamide gels in the presence of 0.1% SDS and 6 M urea, predominantly a single protein band is observed, which has a molecular weight of approximately 240,000–260,000.

When the enzyme is converted to the protomeric form by dialysis against 50 mM Tris-HCl buffer, pH 7.5, containing 0.5 mM DTT and 0.15 M NaCl, dissociated and electrophoresed as described above, a variety of polypeptides with molecular weights less than 240,000–260,000 is observed. This indicates that the highly purified preparations of rat mammary gland carboxylase may be contaminated with proteolytic activity.

Fatty acid synthase from rat mammary gland is also composed of polypeptides of molecular weight approaching 240,000–260,000.[7] Since fatty acid synthase copurifies with acetyl-CoA carboxylase, it is essential that methods other than SDS–polyacrylamide gel electrophoresis be used to validate the purity of the carboxylase. Immunochemical studies show that acetyl-CoA carboxylase and fatty acid synthase do not share common antigenic determinants.[7] Hence, in this laboratory, besides specific activity and SDS gel electrophoresis, immunological methods are routinely employed to validate the purity of acetyl-CoA carboxylase. Preparations of acetyl-CoA carboxylase are chromatographed over Sepharose columns' to which anti-fatty acid synthase has been covalently attached to remove contaminating fatty acid synthase.[12]

Acetyl-CoA carboxylase purified from rat mammary gland extracts by polyethylene glycol precipitation has a specific activity of 5.4 (Table II); it appears to be as pure as preparations made by other methods which have a specific activity $\geq$ 15.0. It is not yet clear why the polyethylene glycol preparations have such low specific enzymatic activities. Attempts to activate these preparations have been unsuccessful thus far.[12]

Activators and Inhibitors. The crude and the pure enzyme is activated by citrate. As stated previously the crude, but not the purified, enzyme requires preincubation with citrate prior to assay. The purified enzyme shows almost complete dependence upon citrate for activity. Maximal activation occurs in the presence of 25–30 mM citrate with 30 mM magnesium acetate present. The enzyme, under certain conditions can be inactivated,[14] when incubated with Mg^{2+}-ATP prior to assay. It has been difficult to establish that the inhibition of enzyme activity is due to incorporation of the terminal phosphate of ATP into acetyl-CoA carboxylase molecules.

The enzyme is strongly inhibited by sulfhydryl reagents, such as p-chloromercuribenzoate. The presence of acetyl-CoA prior to addition of the inhibitor completely protects the enzyme from inactivation.[15] S-4-Bromo-2,3-dioxobutyl-CoA, an affinity labeling agent for rat mammary gland fatty acid synthase,[16] also inhibits acetyl-CoA carboxylase under certain conditions.[15] The citrate induced polymeric form of the carboxylase is not inhibited by this acyl-CoA analog.[15] Acetyl-CoA carboxylase is also inhibited when the protomer is incubated with biotin-binding antibodies prior to assay. Citrate prevents this type of inhibition.[17]

Biotin and Phosphate Content. The enzyme contains 1 μg of biotin per milligram of protein (three different preparations assayed in triplicate) as

[14] P. Ahmad, E. Greenfield, and F. Ahmad, unpublished results.
[15] J. E. Rodriguez and F. Ahmad, unpublished results.
[16] P. Clements, R. E. Barden, P. M. Ahmad, and F. Ahmad, *Biochem. Biophys. Res. Commun.* **86**, 278 (1979).
[17] P. Ahmad, R. Dickstein, and F. Ahmad, *Fed. Proc.* **37**, 1403 (1978).

determined by the colorimetric method.[18] This biotin content corresponds to 1 mol of biotin per 240,000 g of carboxylase protein. The biotin content of rat mammary gland carboxylase is equivalent to that reported for rat liver[3,19] and bovine adipose tissue[6] enzymes.

The phosphate content has been determined by a previously described method[20] using two different preparations, and each milligram of the carboxylase protein is estimated to contain 25.5 nmol of phosphate. Therefore, approximately 6 mol of phosphate are present per 240,000 g of protein. The phosphate content is based upon protein determined by the method of Bradford,[21] using ovalbumin as standard. The phosphate content of the rat liver enzyme was reported to be 2.1 mol of phosphate per 215,000 g of protein,[22] or 2.3 mol of phosphate per 240,000 g of protein.

NOTE ADDED IN PROOF. We have recently determined the biotin[18] and phosphate[20] content of acetyl-CoA carboxylase preparations of low and high specific activity of Table II. The biotin content of both the preparations corresponded to 1 mol of biotin per 240,000 g of carboxylase protein. The phosphate content of the low and high specific activity preparations was estimated at 15 and 13 mol per 240,000 g of carboxylase protein, respectively.

Acknowledgments

The published and unpublished work in the authors' laboratory is supported by the National Institutes of Health Public Health Service Grant No. CA-15196.

[18] N. M. Green, this series, Vol. 18A [74].
[19] T. Tanabe, K. Wada, T. Okazaki and S. Numa, *Eur. J. Biochem.* **57**, 15 (1975).
[20] H. H. Hess and J. E. Derr, *Anal. Biochem.* **63**, 607 (1975).
[21] Bradford, M. M., *Anal. Biochem.* **72**, 248 (1976).
[22] H. Inoue and J. M. Lowenstein, *J. Biol. Chem.* **247**, 4825 (1972).

[3] Acetyl-CoA Carboxylase and Fatty Acid Synthase from Lactating Rabbit and Rat Mammary Gland

EC 6.4.1.2 Acetyl-CoA: carbon-dioxide ligase (ADP-forming)

By D. GRAHAME HARDIE, PAUL S. GUY, and PHILIP COHEN

Acetyl-CoA carboxylase

$$\text{Acetyl-CoA} + CO_2 + \text{ATP} \rightarrow \text{malonyl-CoA} + \text{ADP} + P_i$$

Fatty acid synthase

$$\text{Acetyl-CoA} + 7 \text{ malonyl-CoA} + 14 \text{ NADPH} + 14 \text{ H}^+ \rightarrow$$
$$\text{palmitic acid} + 8 \text{ CoA} + 7 CO_2 + 6 H_2O + 14 \text{ NADP}^+$$

The purification of acetyl-CoA carboxylase and fatty acid synthase from a number of different sources has been described in previous volumes (XIV, XXXV).

In this chapter we describe a method that allows the simultaneous purification of both enzymes from lactating mammary gland.[1,2] The procedure for acetyl-CoA carboxylase is novel in that it avoids all time-consuming ion exchange or gel filtration steps and results in a much higher yield than other methods. The procedures were originally developed using lactating rabbit mammary gland, but they have now been applied successfully also to lactating rat mammary gland.

Enzyme Assays

Acetyl-CoA Carboxylase

Principle. The assay used is based on the method of Manning *et al.*[3] and is a coupled assay using fatty acid synthase. In the presence of NADPH and excess fatty acid synthase, two molecules of NADPH are converted to NADP for each molecule of malonyl-CoA produced by the acetyl-CoA carboxylase reaction. In some tissue extracts contaminating activities that reconvert NADP to NADPH interfere with the assay, but this is not a problem with extracts of lactating mammary gland. Acetyl-CoA carboxylase is completely inactive in the absence of the allosteric activator citrate. Bovine serum albumin stimulates the activity, probably by binding traces of long-chain acyl-CoA esters, which are extremely potent inhibitors of the enzyme.[4]

The concentration of Mg^{2+} in the assay is critical since the $Mg\text{-}ATP^{2-}$ complex is the substrate for the enzyme, and citrate also chelates Mg^{2+}. The Mg^{2+} must therefore be sufficient to allow for chelation by both ATP and citrate. However, free Mg^{2+} is a potent inhibitor of the enzyme, and the total Mg^{2+} is maintained at a concentration slightly lower than the sum of the concentrations of ATP and citrate.

Reagents

Tris-acetate buffer, pH 7.2, $I = 0.2$ (29.8 g of Tris base and 0.2 mol of acetic acid per liter) containing 50 mM Na_3 citrate and 40 mM Mg(acetate)$_2$ (0.25 ml)

[1] D. G. Hardie and P. Cohen, *FEBS Lett.* **91**, 1 (1978).
[2] D. G. Hardie and P. Cohen, *Eur. J. Biochem.* **92**, 25 (1979).
[3] R. Manning, R. Dils, and R. J. Mayer, *Biochem. J.* **153**, 463 (1976).
[4] H. Ogiwara, T. Tanabe, J. Nikawa, and S. Numa, *Eur. J. Biochem* **89**, 33 (1978).

Acetyl-CoA, 3.0 mM (0.05 ml)
NADPH, 2.4 mM (0.05 ml)
Bovine serum albumin, 10 mg/ml (0.05 ml)
NaHCO$_3$, 0.4 M (0.025 ml)
ATP, 80 mM, Mg(acetate)$_2$, 40 mM, pH 7.0 (0.025 ml)
H$_2$O (0.02 ml)

Procedure. The reagents listed above are incubated at 37° in a microcuvette. The rate of decrease of absorbance at 340 nm in the absence of added enzyme (which should be less than 0.01 per minute) is recorded. The reaction is initiated with 0.01 ml of purified fatty acid synthase (20 mg/ml) in buffer E and 0.02 ml of acetyl-CoA carboxylase (0.001–0.01 unit) diluted in buffer A (see below). The rate of decrease of absorbance at 340 nm is monitored, and the rate of decrease obtained in the absence of acetyl-CoA carboxylase is subtracted to give the corrected rate ($\Delta A_{340}/$ min).

Unit. One unit of acetyl-CoA carboxylase is that amount which produces 1 μmol of malonyl-CoA per minute at 37°. In the assay described above the concentration of acetyl-CoA carboxylase in units per milliliter is given by $\Delta A_{340}/$min $\times$ 2.0 $\times$ d, where d is the dilution of acetyl-CoA carboxylase before assay.

Fatty Acid Synthase

Principle. The malonyl-CoA-dependent oxidation of NADPH is followed by monitoring the reaction at 340 nm. As discussed above for acetyl-CoA carboxylase, interfering activities that reconvert NADP to NADPH are not significant in extracts of lactating mammary gland, although they may cause problems in other tissue extracts.

Reagents

Sodium phosphate buffer, 1.0 M, pH 6.6, containing 5.0 mM EDTA (0.1 ml)
Acetyl-CoA, 0.3 mM (0.05 ml)
Malonyl-CoA, 1.08 mM (0.025 ml)
β-Mercaptoethanol, 0.3 M (0.025 ml)
NADPH, 2.4 mM (0.05 ml)
Fatty acid synthase (0.05 ml)
Water (0.2 ml)

Procedure. All the components except malonyl-CoA are incubated in a microcuvette at 37°, and the rate of decrease of absorbance at 340 nm is recorded. The reaction is initiated with malonyl-CoA, and the rate of decrease in absorbance at 340 nm is again recorded. The rate obtained in the absence of malonyl-CoA (which should be less than 0.01 per minute) is then subtracted from the rate obtained in the presence of malonyl-CoA to give the corrected rate (ΔA_{340}/min).

Unit. One unit of fatty acid synthase is that amount which will catalyze the malonyl-CoA-dependent oxidation of 1 μmol of NADPH per minute at 37°. In the assay described above, the fatty acid synthase concentration in units per milliliter is given by ΔA_{340}/min $\times$ 1.6 $\times$ d, where d is the dilution of the enzyme before assay.

Purification of Enzymes

Buffers

The buffers and solutions used in the preparation of acetyl-CoA carboxylase contain proteinase inhibitors at the following concentrations: leupeptin (0.4 mg/liter), antipain (0.4 mg/liter), pepstatin (0.4 mg/liter), L-1-tosylamide-2-phenylethylchloromethyl ketone (0.1 mM), N-α-p-tosyl-L-lysine chloromethyl ketone (0.1 mM). Phenylmethylsulfonyl fluoride cannot be used, as it irreversibly inhibits fatty acid synthase by inactivating the thioesterase activity.[5]

All buffers and solutions also contain 1.0 mM EDTA and 0.1% v/v β-mercaptoethanol, except buffer E, in which dithiothreitol replaces mercaptoethanol.

   Buffer A: 100 mM sodium phosphate, 25 mM sodium citrate, pH 7.0
   Buffer B: 50 mM sodium phosphate, pH 7.0
   Buffer C: 125 mM sodium phosphate, pH 7.0
   Buffer D: 250 mM sodium phosphate, pH 7.0
   Buffer E: 250 mM sodium phosphate, pH 7.0, containing 10% w/v glycerol and 10 mM dithiothreitol

Preparation of Extract

Lactating New Zealand white rabbits (15–20 days postpartum) are killed by injection of phenobarbitone and exsanguinated by cutting the jugular vein. The animals are skinned, and the mammary tissue is removed using a blunt scalpel. The tissue is chopped into small pieces (<1.0 cm in diameter), rinsed in 0.25 M sucrose to wash away excess milk, and homogenized in 2.5 ml per gram of 0.25 M sucrose. The homogenization is carried out at 4° using 3 $\times$ 30 second bursts at top speed in a Waring blender. The homogenate is squeezed through cheesecloth and centrifuged at 4° for 60 min at 70,000 g. The supernatant is decanted through glass wool to trap the floating fat cake (step 1).

[5] S. Kumar, *J. Biol. Chem.* **250,** 5150 (1975).

Purification of Acetyl-CoA Carboxylase

Added are 0.64 volumes of 3.6 M ammonium sulfate (475 g/liter): after stirring for 15 min the suspension is centrifuged at 4°. The rest of the procedure is carried out at room temperature (18–20°). The supernatant is discarded, and the precipitate is redissolved in buffer A and dialyzed against the same buffer for 4 hr (step 2). Then 0.064 volume of 50% (w/w) polyethylene glycol 6000 is added, giving a final concentration of 3% (w/w). After standing for 2–3 hr, the suspension is centrifuged at 10,000 g for 10 min. The supernatant is decanted and can be used for the purification of fatty acid synthase (see later). The precipitate containing acetyl-CoA carboxylase is resuspended in buffer A, and a large amount of insoluble material that does not redissolve is removed by centrifugation. The precipitate is resuspended in buffer A and recentrifuged as before. The pellet is discarded, the two supernatants are pooled (step 3), and made 3% w/w in polyethylene glycol as before. After standing for at least 2 hr, the suspension is centrifuged. The precipitate is redissolved in the minimum volume of buffer A, and the activity is checked. The solution is then adjusted to 6–9 U/ml and centrifuged for 10 min at 15,000 g to remove turbidity. The supernatant, which represents pure acetyl-CoA carboxylase (step 4), is collected.

Purification of Fatty Acid Synthase

The preparation of this enzyme is carried out at room temperature (18–20°). The first step in the purification is chromatography on DEAE-cellulose. If acetyl-CoA carboxylase is not required, the 70,000 g supernatant (step 1) is applied directly to DEAE-cellulose. If, however,

TABLE I

PURIFICATION OF ACETYL-CoA CARBOXYLASE FROM 300 G OF MAMMARY TISSUE (TWO RABBITS)[a]

Step	Activity (U)	Protein (mg)	Specific activity (units/mg)	Purification (fold)	Yield (%)
1. Supernatant, 70,000 g	168	12,400	0.014	1	100
2. (NH₄)₂SO₄ precipitate, 35%	176	4,800	0.037	2.6	105
3. First 3% polyethylene glycol precipitate	132	90	1.5	105	79
4. Second 3% polyethylene glycol precipitate	85	20	4.2	300	51

[a] Data from Hardie and Cohen.[1]

TABLE II

PURIFICATION OF FATTY ACID SYNTHASE FROM 120 G OF MAMMARY TISSUE
(ONE RABBIT)[a]

Step	Activity (U)	Protein (mg)	Specific activity (U/mg)	Purification (fold)	Yield (%)
1. Supernatant, 70,000 g	428	3227	0.13	1	100
2. DEAE-cellulose, pH 7.0	370	938	0.39	3	86
3. $(NH_4)_2SO_4$ precipitate, 30%	201	174	1.15	9	47
4. Sepharose 4B	177	101	1.75	13	41
5. $(NH_4)_2SO_4$ precipitate, 30%	183	105	1.74	13	42

[a] Data from Hardie and Cohen.[2]

acetyl-CoA carboxylase is also being purified, the first 3% w/w polyethylene glycol supernatant (see above) diluted threefold with 1.0 mM EDTA, 0.1% v/v β-mercaptoethanol, pH 7.0, is used as the source of fatty acid synthase.

The solution containing fatty acid synthase is loaded on a column (15 × 7 cm) of DEAE-cellulose (Whatman DE-52) equilibrated in buffer B. The column is washed with buffer B until the absorbance at 280 nm of the effluent falls to zero; fatty acid synthase is then eluted with buffer C. Fractions with $A_{280} > 0.2$ are pooled, and 0.5 volume of 3.6 M ammonium sulfate (475 g/liter) is added. After standing for 15 min, the suspension is centrifuged for 10 min at 10,000 g. The pellet is resuspended in buffer D to give a volume of <5.0 ml and dialyzed for 2 hr against buffer D. The solution is clarified by centrifugation for 10 min at 10,000 g and loaded on a column of Sepharose 4B (120 × 3 cm) equilibrated in buffer D. Fatty acid synthase elutes at a V_e/V_o value of 2.0 and is the major ultraviolet absorbing peak. Fractions with an absorbance at 280 nm > 0.8 are pooled, and 0.5 volume of 3.6 M ammonium sulfate (475 g/liter) is added. After standing for 15 min, the suspension is centrifuged for 10 min at 15,000 g, and the supernatant is discarded. The pellet is redissolved in buffer E at a final concentration of 20 mg/ml. The solution is dialyzed overnight against buffer E and stored in aliquots at $-20°$.

Comments

Summaries of typical purification procedures for acetyl-CoA carboxylase and fatty acid synthase are shown in Tables I and II. Typically 5–10 mg of acetyl-CoA carboxylase and 70–100 mg of fatty acid synthase are obtained from one rabbit. The preparation itself is highly reproducible, although the specific activity and amount of each enzyme in the 70,000 g supernatant (step 1) is somewhat variable from rabbit to rabbit: 0.005–0.015 unit/mg for acetyl-CoA carboxylase; and 0.05–0.15 unit/mg for fatty acid synthase.

The activity of acetyl-CoA carboxylase is regulated by phosphorylation and dephosphorylation.[1,6] In the 70,000 g supernatant (step 1), the enzyme is in a highly phosphorylated state of low activity. During the preparation there is a partial dephosphorylation of the enzyme, which is accompanied by an activation of twofold or more and occurs mainly during the dialysis of the resuspended ammonium sulfate pellet. This activation is not apparent in the data of Table I, because the aliquots removed for assay at each stage of the preparation were incubated at 37° for 30 min to allow complete activation by endogenous protein phosphatases to take place. If this incubation is omitted and the enzyme is assayed immediately, overestimates for the yield and purification are obtained. The activation can be prevented by the inclusion of 50 mM sodium fluoride (a protein phosphatase inhibitor) in all the buffers[6] or by carrying out the entire purification procedure at 4°. This results in the purification of acetyl-CoA carboxylase in a highly phosphorylated form (see Table III) of low specific activity (1.2 units/mg).

Isolation of Acetyl-CoA Carboxylase and Fatty Acid Synthase from Rat Mammary Gland

These purification procedures can be applied also to lactating rat mammary gland using animals at 14 days postpartum. The only modification for acetyl-CoA carboxylase is the use of 3.5% rather than 3% (w/w) polyethylene glycol 6000. Typically 50 mg of acetyl-CoA carboxylase and 100 mg of fatty acid synthase can be obtained from 10 rats (200 g of tissue) corresponding to yields of 70% and 46%, respectively. Acetyl-CoA carboxylase is purified 120-fold with a final specific activity of 6–7 units/mg; fatty acid synthase is purified 16-fold with a final specific activity of 1.6 units/mg. The amount of acetyl-CoA carboxylase in rat mammary gland is therefore 4- to 5-fold higher in terms of enzyme protein, and 8-fold higher in terms of activity, than the rabbit.

Properties

Storage and Stability. Acetyl-CoA carboxylase can be stored in buffer A and fatty acid synthase in buffers D or E for several days at room temperature in the presence of 0.02% sodium azide and the proteinase

[6] D. G. Hardie, and P. Cohen, *FEBS Lett.* **103,** 333 (1979).

TABLE III

PHYSICOCHEMICAL PROPERTIES OF ACETYL-COA CARBOXYLASE AND FATTY ACID
SYNTHASE FROM LACTATING RABBIT MAMMARY GLAND[a]

Property	Acetyl-CoA carboxylase	Fatty acid synthase
$A_{280\,nm}^{1\%}$	14.5	10.0
Partial specific volume	0.734	0.735
Sedimentation coefficient, $s_{20,w}$(S)	50.5	13.3
Subunit molecular weight	252,000	252,000
Molecular weight of native enzyme	Several million	515,000
Subunit structure	Linear polymer	Dimer
Alkali-labile phosphate (mol/mol)	4.8 ± 0.3^b	0.2 ± 0.1
	6.2 ± 0.2^c	
Proportion of soluble protein (%)	0.3	7.0
Concentration *in vivo* (μM)	0.4–0.8	10–20

[a] Data from Hardie and Cohen.[1,2]
[b] Enzyme prepared by normal procedure.
[c] Enzyme prepared in the presence of 50 mM NaF.

inhibitors listed earlier, without any loss of activity. Fatty acid synthase is reported to dissociate and lose activity if stored at 4°, and the loss of activity may be recovered only partially in rewarming.[7] For prolonged storage, acetyl-CoA carboxylase can be kept in buffer A containing 35% ethanediol at −20°, and fatty acid synthase may be stored frozen in buffer E at −20°. Both enzymes may be stored for at least 1 month under these conditions without significant loss of activity.

Other Properties. The preparations of acetyl-CoA carboxylase and fatty acid synthase are homogeneous by the criteria of polyacrylamide gel electrophoresis in the presence of sodium dodecyl sulfate and by sedimentation velocity experiments in the analytical ultracentrifuge. The properties of the enzymes from lactating rabbit mammary gland are shown in Table III.

Acetyl-CoA carboxylase and fatty acid synthase from rabbit mammary gland and fatty acid synthase from rat mammary gland have identical subunit molecular weights of 250,000. However acetyl-CoA carboxylase isolated from rat mammary gland has a slightly lower molecular weight of 240,000. This 240,000 form can be isolated within 10 min of homogenization using affinity chromatography,[8] which suggests that it represents a genuine species difference rather than an artifact of proteolysis.

[7] S. Smith and S. Abraham, *J. Biol. Chem.* **246**, 6428 (1971).
[8] D. G. Hardie and P. S. Guy, *Eur. J. Biochem.* **110**, 167 (1980).

[4] Acetyl-CoA Carboxylase from Yeast

EC 6.4.1.2 Acetyl-CoA : carbon-dioxide ligase (ADP-forming)

By MANFRED SUMPER

$$CH_3—CO—SCoA + ATP + CO_2 \rightleftharpoons \overset{\displaystyle COOH}{\overset{\displaystyle |}{CH_2}}—CO—SCoA + ADP + P_i$$

Assay Methods

The assay methods for acetyl-CoA carboxylase have been described previously.[1] The optical assay, which measures formation of malonyl-CoA in a combined reaction with fatty acid synthase[2] by the decrease of NADPH,[1,3] was used throughout the purification procedure described here.

Units. One unit of enzyme is defined as that amount of protein which catalyzes the carboxylation of 1 μmol of acetyl-CoA per minute at 25°. Specific activity is expressed as units per milligram of protein.

Purification Procedure

The purification procedure previously described in this series[1] yields an enzyme preparation with a specific activity of about 0.4 unit per milligram of protein. Acetyl-CoA carboxylase purified by the alternative procedure described below has a specific activity of 6.0–9.0 units per milligram of protein and yields a single polypeptide band on sodium dodecyl sulfate gel electrophoresis. All operations were carried out at 4°. Brewer's yeast (4 kg, wet weight), from Löwenbräu (Munich) was washed twice with about 6 liters of 0.1 M potassium phosphate, pH 6.5, and cells were collected by centrifugation.

Step. 1. Disruption of the Cells.[4] The washed cells (4 kg) were sus-

[1] M. Matsuhashi, this series, Vol. 14 [1].

[2] F. Lynen, Vol. 14 [3].

[3] M. Matsuhashi, S. Matsuhashi, S. Numa, and F. Lynen, *Biochem. Z.* **340**, 243 (1964).

[4] Disruption of the cells in a French press yields crude extracts with considerably lower enzymatic activity.

pended in 6 liters of 0.2 M potassium phosphate, pH 6.5, containing 1 mM EDTA, and each 250 ml suspension was agitated with 250 g of glass beads (size 31/10; Dragonwerk Wild, Bayreuth) in the cell homogenizer of Merkenschlager et al.[5] operated at 0° to 5° and run for 40 sec. The combined crude extracts were centrifuged at 15,000 g for 40 min.

Step 2. Ammonium-Sulfate Fractionation. The crude extract from step 1 was taken to 40% saturation by the addition of ammonium sulfate. The precipitate, which contained the acetyl-CoA carboxylase activity, was collected by centrifugation at 15,000 g for 60 min.

Step 3. Ultracentrifugation. The precipitate was dissolved in 1000 ml of 0.1 M potassium phosphate, pH 6.5, containing 1 mM 2-mercaptoethanol. The cloudy liquid was centrifuged at 100,000 g for 90 min, and the supernatant liquid, which contained the acetyl-CoA carboxylase activity, was saved.

Step 4. Ammonium-Sulfate Fractionation (0 to 35%). The supernatant fraction from stage 3 was taken to 35% saturation by the addition of ammonium sulfate. After stirring for 20 min, the precipitate was collected by centrifugation at 15,000 g for 40 min.

Step 5. Polyethylene Glycol Fractionations. The collected precipitate from step 4 was dissolved in 0.1 M potassium phosphate, pH 6.5, containing 1 mM 2-mercaptoethanol. The protein concentration was lowered to 15–20 mg/ml by addition of buffer (about 2000 ml); 230 ml of 50% (w/w) aqueous polyethylene glycol (average mw 1500) per 1000 ml of protein solution were added with stirring, and the solution was stirred for a further 30 min. The resulting precipitate was removed by centrifugation at 15,000 g for 40 min. An additional 220 ml of 50% aqueous polyethylene glycol solution per 1000 ml of initial volume of the protein solution was added to the supernatant. The resulting precipitate was collected by centrifugation at 15,000 g for 40 min. The pellet, which contained the enzyme, was dissolved in 0.05 M potassium phosphate, pH 6.5, to a protein concentration of about 10 mg/ml. For each 100 ml of protein solution, 5 g of solid ammonium sulfate and 10 ml of 50% (w/w) aqueous solution of polyethylene glycol (average mw 6000) was added. The resulting precipitate was removed by centrifugation at 15,000 g for 30 min. The acetyl-CoA carboxylase was then precipitated by further addition of 10 ml of polyethylene glycol solution per 100 ml initial volume. The sample was stirred for 30 min, and the resulting precipitate was collected by centrifugation at 15,000 g for 40 min.

Step 6. DEAE-Cellulose Chromatography. The pellet from step 5 was dissolved in approximately 20 ml of 0.02 M potassium phosphate, pH 7.5,

[5] M. Merkenschlager, K. Schlossmann, and W. Kutz, Biochem. Z. 329, 332 (1957).

PURIFICATION OF ACETYL-CoA CARBOXYLASE FROM YEAST[a]

Fraction	Protein (mg)	Total activity (units)	Specific activity (mU/mg)	Recovery (%)
Crude extract	—	—	—	—
First $(NH_4)_2SO_4$	115,000	(3100)	27	100
Ultracentrifugation	60,000	3000	50	96
Second $(NH_4)_2SO_4$	20,400	2650	130	84
Polyethylene glycol	2,100	2100	1000	68
DEAE-cellulose	400	1450	3600	47
Phosphocellulose	175	1050	6000	34

[a] Yeast (4 kg, wet weight) was used. In the crude extract, reliable results could not be obtained using the spectrophotometric assay.

containing 20% glycerol and 2 mM $MgCl_2$. Water containing 20% glycerol was added to the solution until its conductance was equal to that of the equilibrating buffer. The solution was applied to a DEAE-cellulose column (4.6 × 30 cm) equilibrated with the above buffer. Elution was carried out with a linear concentration gradient established between 1.5 liters of buffer containing 50 mM NaCl and 1.5 liters containing 200 mM NaCl. The fractions of high specific activity (corresponding to those between approximately 1400 to 2100 ml of effluent volume) were taken to 50% saturation with solid ammonium sulfate and centrifuged at 15,000 g for 60 min. The precipitate was dissolved in a minimum volume of 20 mM potassium phosphate, pH 6.5, containing 20% glycerol.

Step 7. Cellulose-Phosphate Chromatography. Water containing 20% glycerol was added to the resulting protein solution until its conductance was equal to that of the equilibrating buffer: 20 mM potassium phosphate, pH 6.5, containing 20% glycerol. This solution was applied to a column (3.6 × 20 cm) of cellulose phosphate. Elution was carried out with a linear concentration gradient established between 1.2 liters of 50 mM KCl in buffer and 1.2 liters of 220 mM KCl in buffer. The acetyl-CoA carboxylase appeared in the eluate after about 1600 ml. The fractions containing acetyl-CoA carboxylase were pooled, and the protein was precipitated by addition of solid ammonium sulfate.

The results of a typical purification are summarized in the table.

Properties

Specificity and kinetic and some other properties of acetyl-CoA carboxylase were described previously in this series.[1]

Stability. The purified enzyme could be kept for at least 1 year without loss of activity when stored in 0.3 M potassium phosphate, pH 6.5, containing 50% glycerol at $-20°$. Low ionic strength and alkaline pH favor the rapid inactivation of the enzyme.

Subunit Structure. The purified enzyme exhibits a single polypeptide band on SDS–polyacrylamide gels with an apparent molecular weight of 189,000.[6] The native enzyme was shown to be a tetramer of identical subunits with a C-terminal amino acid sequence -Leu-Lys-COOH and a sedimentation coefficient of $s_{20,w} = 15.5$ S.[6,7]

Dissociation and Reactivation. A rapid inactivation of acetyl-CoA carboxylase occurs at pH values above 8.0. This loss of enzymatic activity is accompanied by a change of sedimentation behavior, indicating a dissociation of the tetrameric native enzyme in a mixture of monomers, dimers, and trimers of the subunit polypeptide chain.[7] Inactivated acetyl-CoA carboxylase is reactivated upon transfer to concentrated (0.5 M) potassium phosphate buffer, pH 6.5, containing 20% glycerol and 10 mM dithiothreitol.

[6] J. Spiess, Doctoral thesis, University of Munich, 1976.
[7] M. Sumper and C. Riepertinger, *Eur. J. Biochem.* **29**, 237 (1972).

[5] Acetyl-CoA Carboxylase from *Candida lipolytica*

EC 6.4.1.2 Acetyl-CoA : carbon-dioxide ligase (ADP-forming)

By MASAYOSHI MISHINA, TATSUYUKI KAMIRYO, and SHOSAKU NUMA

$$ATP + HCO_3^- + acetyl\text{-}CoA \rightleftharpoons ADP + P_i + malonyl\text{-}CoA$$

Assay Methods

Several methods are available for the assay of acetyl-CoA carboxylase.[1-5] The activity of the *Candida lipolytica* enzyme is routinely determined by the $^{14}CO_2$-fixation assay or by the spectrophotometric assay in combination with the pyruvate kinase and lactate dehydrogenase

[1] M. Matsuhashi, this series, Vol. 14 [1].
[2] S. Numa, this series, Vol. 14 [2].
[3] H. Inoue and J. M. Lowenstein, this series, Vol. 35 [1].
[4] A. L. Miller and H. R. Levy, this series, Vol. 35 [2].
[5] T. Tanabe, S. Nakanishi, T. Hashimoto, H. Ogiwara, J. Nikawa, and S. Numa, this volume [1].

reactions.[6] The latter method is applicable only to enzyme preparations from the polyethylene glycol step and subsequent steps.

$^{14}CO_2$-Fixation Method

Reagents

Tris-HCl buffer, 0.5 M, pH 7.5
Reduced glutathione, 0.1 M, neutralized with KOH
Bovine serum albumin, 3%
Polyethylene glycol (average molecular weight, 7500), 50% (w/v)
Potassium citrate, 0.1 M
$MgCl_2$, 0.1 M
ATP, 0.15 M
Acetyl-CoA, 10 mM
$KH^{14}CO_3$ (0.25 μCi/μmol), 0.2 M
HCl, 5 M
Scintillator solution: 4 g of 2,5-diphenyloxazole and 0.1 g of 1,4-bis[2-(4-methyl-5-phenyloxazolyl)]benzene in 1 liter of toluene plus 0.5 liter of Triton X-100

Procedure. The assay is conducted without preincubation of the enzyme. The reaction mixture contains 40 μmol of Tris-HCl buffer, pH 7.5, 3 μmol of glutathione, 0.6 mg of bovine serum albumin, 80 mg of polyethylene glycol (average MW 7500), 8 μmol of potassium citrate,[7] 8 μmol of $MgCl_2$, 3 μmol of ATP, 0.5 μmol of acetyl-CoA, 10 μmol of $KH^{14}CO_3$ (0.25 μCi/μmol), and enzyme (up to 0.2 mU) in a total volume of 0.8 ml. The reaction is initiated by addition of the enzyme. After incubation at 25° for 10 min, the reaction is terminated by adding 0.2 ml of 5 M HCl. The acid-stable radioactivity is determined in 5 ml of the scintillator solution with a liquid scintillation spectrometer as described elsewhere in this volume.[5] Under the assay conditions described, the reaction follows zero-order kinetics, and the initial rate of reaction is proportional to enzyme concentration.

Spectrophotometric Method

Reagents

$KHCO_3$, 1 M
Potassium phosphoenolpyruvate, 40 mM

---

[6] M. Mishina, T. Kamiryo, A. Tanaka, S. Fukui, and S. Numa, Eur. J. Biochem. **71**, 295 (1976).

[7] Although citrate exhibits essentially no effect on the enzyme activity, it is routinely added to the reaction mixture to obtain a broader optimum with respect to Mg^{2+} concentration (see below).

NADH, 5 mM, pH 8
Pyruvate kinase (rabbit muscle; Boehringer), 10 mg/ml
Lactate dehydrogenase (rabbit muscle; Boehringer), 5 mg/ml
Other reagents, as for the $^{14}CO_2$-fixation method

Procedure. The assay is performed without preincubation of the enzyme. The reaction mixture for the spectrophotometric assay is the same as that for the $^{14}CO_2$-fixation assay, except that $KH^{14}CO_3$ is replaced by 20 μmol of unlabeled $KHCO_3$ and that 0.4 μmol of potassium phosphoenolpyruvate, 0.1 μmol of NADH, 20 μg of pyruvate kinase, and 10 μg of lactate dehydrogenase are added; up to 5 mU of enzyme are added. The reaction is started by addition of acetyl-CoA, and the rate of decrease in absorbance at 340 nm (or at 334 nm) is followed at 25° with a recording spectrophotometer. Under the assay conditions described, the reaction follows zero-order kinetics for at least 3 min, and the initial rate of reaction is proportional to enzyme concentration.

Units. One unit of acetyl-CoA carboxylase activity is defined as that amount which catalyzes the formation of 1 μmol of malonyl-CoA or ADP per minute under the assay conditions described. Essentially identical activities are measured by the $^{14}CO_2$-fixation method and by the spectrophotometric method. Specific activity is expressed as units per milligram of protein. Protein is determined by the method of Lowry *et al.*[8] with crystalline bovine serum albumin as a standard.

Purification Procedure

The lability of *C. lipolytica* acetyl-CoA carboxylase necessitates the use of glycerol as a stabilizer. Unless otherwise specified, all phosphate buffers used are potassium phosphate buffers containing 20% (w/v) glycerol, 5 mM 2-mercaptoethanol, and 1 mM EDTA, and all operations are carried out at 0–4°.

Preparation of Crude Extract. The hydrocarbon-utilizing yeast *Candida lipolytica* NRRL Y-6795 is grown at 25° with vigorous shaking in medium containing 0.5% KH_2PO_4, 0.5% K_2HPO_4, 0.7% Bacto-yeast extract, 0.7% Bacto-peptone, and 2% glucose. Cells are harvested at late logarithmic phase. The yield of wet packed cells is about 26 g per liter of medium. The harvested cells (310 g wet weight) are washed with 2 liters of 0.1 M phosphate buffer, pH 6.5, and suspended in 500 ml of the same buffer. Each 40 ml portion of the cell suspension is agitated with 50 g of glass beads (diameter, 0.45–0.50 mm) in a 75-ml bottle of a Braun cell homogenizer at high speed for 60 sec under cooling with liquid CO_2. The combined

[8] O. H. Lowry, N. J. Rosebrough, A. L. Farr, and R. J. Randall, *J. Biol. Chem.* **193**, 265 (1951).

homogenate is centrifuged at 18,000 g for 40 min, and the supernatant is collected (525 ml).

First Ammonium Sulfate Fractionation (0–40% Saturation). The crude extract is brought to 40% saturation by addition of solid ammonium sulfate. The resulting precipitate is collected by centrifugation at 18,000 g for 40 min and dissolved in 0.1 M phosphate buffer, pH 6.5 (final volume, 240 ml).

Ultracentrifugation and Second Ammonium Sulfate Fractionation (0–33% Saturation). The first ammonium sulfate fraction is centrifuged at 78,000 g for 90 min. The supernatant is filtered through gauze to remove most of the solidified fat and is brought to 33% saturation by addition of solid ammonium sulfate. The precipitate is collected by centrifugation at 18,000 g for 40 min and dissolved in a minimal volume of 0.1 M phosphate buffer, pH 6.5 (final volume, 20 ml). This enzyme preparation can be stored at −70° for periods up to 2 weeks without loss of activity.

Polyethylene Glycol Fractionation (10–16%). Three preparations from the second ammonium sulfate fractionation step are combined and diluted with 0.1 M phosphate buffer, pH 6.5, to give a protein concentration of 10 mg/ml. Seventy-eight milliliters of 50% (w/v) polyethylene glycol (average MW 3000) is added to 312 ml of the diluted enzyme solution. After further stirring for 30 min, the precipitate formed is removed by centrifugation at 18,000 g for 40 min. To the supernatant solution (368 ml) is added 65 ml of 50% (w/v) polyethylene glycol (average MW 3000), and the mixture is further stirred for 30 min. The resulting precipitate is collected by centrifugation at 18,000 g for 40 min and dissolved in 20 mM phosphate buffer, pH 7.5 (final volume, 36 ml).

DEAE-Cellulose Chromatography. A sufficient amount of 20% (w/v) glycerol is added to the polyethylene glycol fraction to reduce the conductance of the enzyme solution to that of 20 mM phosphate buffer, pH 7.5. The diluted enzyme solution (100 ml) is applied to a DEAE-cellulose column (4.1 × 11.4 cm) equilibrated with the same buffer. The column is washed with 100 ml of the equilibrating buffer. Elution is then carried out with a linear concentration gradient established between 600 ml of 20 mM phosphate buffer, pH 7.5, and 600 ml of 0.5 M phosphate buffer, pH 7.5; the flow rate is 1.9 ml/min. Fractions containing more than 1 unit of enzyme activity per milliliter (corresponding to those between 510 ml and 645 ml effluent volume) are combined, and the combined eluate is brought to 50% saturation by addition of a saturated (0°) ammonium sulfate solution (pH adjusted to 7.0 with NH₄OH). The precipitate is collected by centrifugation at 18,000 g for 30 min and dissolved in a minimal volume of 20 mM phosphate buffer, pH 6.5 (final volume, 5.4 ml).

Hydroxyapatite Chromatography. The enzyme solution from the preceding step is dialyzed against 1 liter of 20 mM phosphate buffer, pH

PURIFICATION OF ACETYL-CoA CARBOXYLASE FROM *Candida lipolytica*[a,b]

Fraction	Protein (mg)	Total activity[c] (units)	Specific activity[c] (units/mg)	Yield (%)
Crude extract	37,800	1030	0.027	100
First (NH₄)₂SO₄	11,500	732	0.064	71
Ultracentrifugation and second (NH₄)₂SO₄	3,120	568	0.18	55
Polyethylene glycol	780	541	0.69	53
DEAE-cellulose	169	495	2.9	48
Hydroxyapatite	22	176	8.0	17

[a] From Mishina *et al.*[6]

[b] From 975 g (wet weight) of yeast cells; the purification through the second ammonium sulfate fractionation step was carried out in three portions, and the three preparations of this step were combined for further purification.

[c] Enzyme activity was determined at 25° by the $^{14}CO_2$-fixation method (before the polyethylene glycol fractionation step) or by the spectrophotometric method (after this step). The enzyme from the polyethylene glycol fractionation step was assayed by both methods; the activity given in the table was measured by the $^{14}CO_2$-fixation method; the spectrophotometric method gave a specific activity of 0.76 unit/mg.

6.5, for 2 hr. The dialyzed solution is diluted with an equal volume of 20% (w/v) glycerol and applied to a column (2.2 × 18.2 cm) of granulated hydroxyapatite[9] equilibrated with 20 mM phosphate buffer, pH 6.5. Elution is conducted with a linear concentration gradient established between 660 ml of 50 mM phosphate buffer, pH 6.5, and 660 ml of 0.3 M phosphate buffer, pH 6.5; the flow rate is 0.7 ml/min. Fractions exhibiting specific activities of 7.8–8.3 units/mg (corresponding to those between 450 ml and 495 ml effluent volume) are combined (45 ml).

To concentrate the purified enzyme solution, an equal volume of a saturated (0°) ammonium sulfate solution, pH 7.0, is added to the combined eluate, and the precipitated enzyme is collected by centrifugation at 20,200 g for 30 min and dissolved in 2.5 ml of 0.3 M phosphate buffer, pH 6.5, containing 50% (w/v) glycerol, 5 mM 2-mercaptoethanol and 1 mM EDTA.

The results of the enzyme purification are summarized in the table. The final preparation is purified approximately 300-fold over the crude extract, the yield being 17%.

Properties[6]

Purity. The final enzyme preparation is homogeneous as evidenced by sedimentation velocity experiments in an analytical ultracentrifuge and

[9] A. L. Mazin, G. E. Sulimova, and B. F. Vanyushin, *Anal. Biochem.* **61**, 62 (1974).

sodium dodecyl sulfate (SDS)–polyacrylamide gel electrophoresis as well as by the fact that Ouchterlony double-diffusion analysis with the use of antibody raised against it yields a single precipitation line.

Physical Properties, Biotin Content, and Subunit Structure. The sedimentation coefficient ($s_{20,w}$) of the native enzyme is 18 S. Upon SDS-polyacrylamide gel electrophoresis, the enzyme exhibits only one kind of subunit with a molecular weight of 230,000. The biotin content, determined by the colorimetric method of Green,[10] is 0.93 μg of biotin per milligram of protein, the value being equivalent to 1 mol of biotin per 263,000 g of protein. Thus, the *C. lipolytica* enzyme, like the enzymes from animal tissues[5,11,12] and *Saccharomyces cerevisiae*,[13] has a highly integrated subunit structure, representing a multifunctional polypeptide.[14]

Stability and Storage. The enzyme is extremely labile without a stabilizer. In the presence of 20% (w/v) glycerol, essentially no loss of the enzyme activity occurs at 0° for at least 24 hr. The purified enzyme preparation can be stored at $-70°$ for periods up to 6 months without loss of activity.

Activators and Inhibitors. The enzyme is markedly activated by polyethylene glycols of higher molecular weights (7500 or 3000); the activation is maximal at a concentration of 10% (w/v), whereas at higher concentrations these compounds inhibit the enzyme. The polyethylene glycol activation is due principally to a decrease in the K_m values for substrates. Citrate exhibits essentially no effect on the enzyme activity, although in its presence a larger amount of Mg^{2+} has to be added to the assay mixture to attain maximal enzyme activity. An excess of Mg^{2+} inhibits the enzyme. Glycerol is also inhibitory; 30%, 66%, and 90% inhibition is observed in the presence of 5%, 10%, and 20% (w/v) glycerol, respectively. Although enzyme solutions normally contain glycerol as a stabilizer, it is so diluted in the reaction mixture that the assay is usually not disturbed. All other related compounds tested, including ethanol, ethylene glycol monomethyl ether, ethylene glycol monoethyl ether, acetone, tetrahydrofuran, and dioxane, suppress the enzyme activity almost completely at a concentration of 5% (v/v).

Kinetic Properties. The apparent K_m values for ATP, HCO_3^-, and acetyl-CoA are 0.07 mM, 3.7 mM, and 0.26 mM, respectively, in the presence of 10% (w/v) polyethylene glycol (average MW 7500), and 0.40 mM, 22 mM, and 1.0 mM, respectively, in the absence of the activator.

[10] N. M. Green, this series, Vol. 18A [74].
[11] T. Tanabe, K. Wada, T. Okazaki, and S. Numa, *Eur. J. Biochem.* **57**, 15 (1975).
[12] J. C. Mackall and M. D. Lane, *Biochem. J.* **162**, 635 (1977).
[13] M. Obermeyer and F. Lynen, *Trends Biochem. Sci.* **1**, 169 (1976).
[14] K. Kirschner and H. Bisswanger, *Annu. Rev. Biochem.* **45**, 143 (1976).

Worthy of note is the remarkably high K_m value for acetyl-CoA as compared with those of the enzymes from *S. cerevisiae*[1] and animal tissues.[2-5]

Specificity. The enzyme catalyzes the carboxylation of propionyl-CoA besides acetyl-CoA. The V_{max} value for propionyl-CoA is 13% of that for acetyl-CoA, while the K_m value for propionyl-CoA is essentially the same as that for acetyl-CoA.

pH Optimum. The enzyme exhibits a pH optimum of 7.9–8.1.

Regulation of Cellular Content and Synthesis.[15] The cellular content of the enzyme varies largely with the carbon source on which the yeast is grown. Among the *n*-alkanes and fatty acids tested as carbon sources, *n*-heptadecane, *n*-octadecane, oleic acid, and linoleic acid reduce the enzyme content to the lowest levels, which are 16–18% of that of glucose-grown cells. Isotopic amino acid incorporation studies, combined with the immunoprecipitation technique, have shown that the decrease in the enzyme content is due to diminished synthesis of the enzyme; no appreciable degradation of the enzyme is observed. Thus, *C. lipolytica* represents a suitable eukaryotic model system for studying the mechanism responsible for the regulation of synthesis of the enzyme.

Cell-Free Translation and Regulation of Acetyl-CoA Carboxylase mRNA.[16] Messenger RNA from *C. lipolytica* is translated in the mRNA-dependent reticulocyte lysate cell-free system[17] to synthesize the complete enzyme. The identity of the translation product is evidenced by the following results: first, it is immunoprecipitated with antibody specific to the enzyme and competes with the authentic enzyme for binding to the antibody; second, it comigrates with the authentic enzyme (MW 230,000) upon SDS–polyacrylamide gel electrophoresis; finally, the peptide fragments formed by its partial proteolysis with papain or α-chymotrypsin are identical with those formed from the authentic enzyme. With the use of this assay system, it has been demonstrated that the level of acetyl-CoA carboxylase mRNA activity in *C. lipolytica* cells decreases with increasing concentrations of fatty acid in culture medium and that the changes in the mRNA activity parallel those in the cellular level of the enzyme. This indicates that the diminished synthesis of the enzyme in cells grown in the presence of fatty acid is due to a reduced level of the mRNA coding for the enzyme.

Candida lipolytica mutants defective in acyl-CoA synthetase I,[18] unlike the wild-type and the revertant strains as well as mutants lacking acyl-CoA synthetase II,[18] do not exhibit the repression of acetyl-CoA car-

[15] M. Mishina, T. Kamiryo, A. Tanaka, S. Fukui, and S. Numa, *Eur. J. Biochem.* **71**, 301 (1976).
[16] S. Horikawa, T. Kamiryo, S. Nakanishi, and S. Numa, *Eur. J. Biochem.* **104**, 191 (1980).
[17] H. R. Pelham and R. J. Jackson, *Eur. J. Biochem.* **67**, 247 (1976).
[18] K. Hosaka, M. Mishina, T. Kamiryo, and S. Numa, this volume [39].

boxylase, when cells are grown on fatty acid. Acyl-CoA synthetase I is responsible for the production of long-chain acyl-CoA that is utilized solely for the synthesis of cellular lipids, while acyl-CoA synthetase II provides long-chain acyl-CoA that is exclusively degraded via β-oxidation.[18] Measurement of the two independent long-chain acyl-CoA pools with the aid of appropriate mutants has indicated that the long-chain acyl-CoA to be utilized for lipid synthesis, but not that to be degraded via β-oxidation, is involved in the repression of acetyl-CoA carboxylase.[18,19] It seems probable that the long-chain acyl-CoA for lipid synthesis, or a compound metabolically related to it, acts as a corepressor or its eukaryotic equivalent to suppress the expression of the acetyl-CoA carboxylase gene. Such a regulatory mechanism is evidently of teleological significance in view of the homeostasis of lipid synthesis.

[19] T. Kamiryo, Y. Nishikawa, M. Mishina, M. Terao, and S. Numa, *Proc. Natl. Acad. Sci. U.S.A.* **76**, 4390 (1979).

[6] Acetyl-CoA Carboxylase from Chloroplasts and Cytosol of Higher Plants[1]

EC 6.4.1.2 Acetyl-CoA : carbon-dioxide ligase (ADP-forming)

By NIELS C. NIELSEN

Malonyl-CoA required for long-chain fatty acid and flavinoid biosynthesis in higher plants is produced by acetyl-CoA carboxylase. This enzyme contains covalently bound biotin and catalyzes the two-step reaction shown below. The first step involves an Mg–ATP and bicarbonate-dependent carboxylation of biotin. The second results in transfer of the CO_2 from carboxybiotin to acetyl-CoA to yield the product of the reaction. Two exchange reactions are characteristic of these half reactions, ATP–$^{32}P_i$ and malonyl-CoA–[^{14}C]acetyl-CoA exchanges, respectively. Review articles have been presented that describe the enzymes found in plants, bacteria, and animals in more detail.[2–4]

[1] Mention of a trademark or proprietary product does not constitute a guarantee or warranty of the product by the United States Department of Agriculture, Science and Education Administration, and does not imply its approval to the exclusion of other products that may also be suitable.

[2] P. K. Stumpf, *J. Am. Oil Chem. Soc.* **52**, 484A (1975).

[3] P. R. Vagelos, *Curr. Top. Cell. Regul.* **4**, 119 (1971).

[4] M. D. Lane, J. Moss, E. Ryder, and E. Stoll, *Adv. Enzyme Regul.* **9**, 237 (1971).

$$HCO_3^- + ATP + \text{biotinyl protein} \xrightarrow{Mg^{2+}}$$
$$CO_2\text{-biotinyl protein} + ADP + P_i \quad \text{(1a)}$$

$$CO_2\text{-biotinyl protein} + \text{acetyl-CoA} \rightarrow$$
$$\text{malonyl-CoA} + \text{biotinyl protein} \quad \text{(1b)}$$

Several factors affect the activity of acetyl-CoA carboxylase. Studies with the wheat germ enzyme show that it, like several other biotin-dependent enzymes,[5-7] has a complex dependence on K^+ and Mg^{2+} for activity.[8] Magnesium participates in the reaction in two different ways. First, free Mg^{2+} interacts with the enzyme in a rapid equilibrium ordered manner to activate it. Second, ATP interacts with the catalytic site as an Mg–ATP complex. Monovalent cations such as K^+ and Rb^+ are required to confer maximum activity on the enzyme. Other monovalent cations, such as Na^+, Li^+, NH_4^+, and Cs^+, activate marginally or not at all. Altering the levels of free Mg^{2+}, Mg–ATP, and K^+ in the reaction mixture can result in changes in the activity of the purified enzyme that span several orders in magnitude.

Since the malonyl-CoA produced by acetyl-CoA carboxylase is a key intermediate in long-chain fatty acid and flavinoid biosynthesis, it is conceivable that a regulatory mechanism might be operative that affects the activity of the enzyme. For example, tricarboxylic acids, such as citrate and isocitrate, have long been recognized as activators of acetyl-CoA carboxylase purified from mammalian and avian sources. However, the involvement of these tricarboxylic acids in the activation of the wheat germ enzyme appears to have been ruled out.[8] Furthermore, a substantial number of nucleotides and phosphorylated sugars, as well as dicarboxylic and tricarboxylic acids, have been tested in the chloroplast system and found to have no effect.[9]

Under certain circumstances, the amount of acetyl-CoA carboxylase activity associated with plant tissue responds to light irradiation. Transferring etiolated barley seedlings into white light results in chloroplast development, a process that depends on large amounts of long-chain fatty acids being synthesized de novo.[9a] An increase in total soluble acetyl CoA carboxylase activity occurs during this greening process[10], and is consid-

[5] W. R. McClure, H. A. Lardy, and H. P. Kneifel, J. Biol. Chem. 246, 3569 (1971).
[6] W. R. McClure, H. A. Lardy, and W. W. Cleland, J. Biol. Chem. 246, 3584 (1971).
[7] C. H. Suelter, Science 168, 789 (1970).
[8] N. C. Nielsen, and P. K. Stumpf, Biochem. Biophys. Res. Commun. 68, 205 (1976).
[9] D. Burton, and P. K. Stumpf, Arch. Biochem. Biophys. 117, 604 (1966).
[9a] P. K. Stumpf, in "Biochemistry of Lipids II" (T. W. Goodwin, ed.), Vol. 14, (MTP International Review of Science). p. 215 Butterworth, London, 1977.
[10] L. Reitzel and N. C. Nielsen, Eur. J. Biochem. 65, 131 (1976).

ered to reflect the increased demand for malonyl-CoA by *de novo* synthesis of long-chain fatty acids. In the flavinoid synthetic pathway, malonyl-CoA is necessary both for the formation of aromatic ring A of flavine and malonylation of flavinoid glycosides.[11] When suspension-cultured parsley cells are subjected to ultraviolet irradiation, the specific activity of acetyl-CoA carboxylase increases substantially at the same time as the activity of the other enzymes participating in this pathway goes up. The mechanism underlying the light-dependent increase of carboxylase activity in higher plants is not known.

The instability of acetyl-CoA carboxylase from higher plants has frustrated attempts to purify the enzyme and study its molecular properties. However, two forms of the carboxylases having distinctly different physical properties have been reported. Kannangara and Stumpf[12] described a prokaryotic form of acetyl-CoA carboxylase in spinach chloroplasts that could reversibly be dissociated into subunits. This complex consisted of three components: biotin carboxylase, transcarboxylase, and a biotin carboxyl carrier protein. After disruption and fractionation of chloroplasts, the biotin carboxylase and transcarboxylase proteins were recovered in the stroma fraction, whereas the biotin carboxyl carrier protein was firmly bound to the thylakoid membranes. Biotin carboxylase catalyzed reaction (1a), and transcarboxylase reaction (1b), whereas biotin carboxyl carrier protein participated in both reactions. Biotin carboxylase was stable, but the chloroplast transcarboxylase was very unstable. This frustrated attempts to reassemble the active chloroplast complex. However, it was possible to reassemble an active complex consisting of chloroplast biotin carboxylase, thylakoid membranes containing the biotin carboxyl carrier protein, and transcarboxylase partially purified from *Escherichia coli*.

In contrast to the prokaryotic form of acetyl-CoA carboxylase found in chloroplasts that is dependent upon membrane-bound biotin, the second form of the enzyme is found in the cytosol of higher plants and is completely soluble. This cytosolic form of acetyl-CoA carboxylase was originally described by Hatch and Stumpf,[13] and partially purified from wheat germ. Enzymes with similar properties have subsequently been reported in leaf homogenates of barley seedlings,[10] barley embryos[14] and parsley cell cultures.[15] Attempts to purify the cytosolic enzyme and establish its subunit structure have not resulted in uniform results. The enzyme

[11] J. Ebel, and K. Hahlbrock, *Eur. J. Biochem.* **75**, 201 (1977).
[12] C. G. Kannangara and P. K. Stumpf, *Arch. Biochem. Biophys.* **152**, 83 (1972).
[13] M. D. Hatch and P. K. Stumpf, *J. Biol. Chem.* **236**, 2879 (1961).
[14] K. Brock and C. G. Kannangara, *Carlsberg Res. Commun.* **41**, 121 (1976).
[15] B. Egin-Bühler, R. Loyal, and J. Ebel, *Arch. Biochem. Biophys.* **203**, 90 (1980).

as purified by Heinstein and Stumpf[16] and Nielsen et al.[17] consisted of tightly aggregated polypeptides ranging between 21,500 and 135,000 molecular weight (MW). The enzyme from barley embryos was isolated using a similar purification scheme and also consisted of several different sized polypeptides.[14] However, the purification schemes used by these workers did not make use of protease inhibitors. Recently, Egin-Bühler et al.[15] included the protease inhibitor phenylmethylsulfonyl fluoride in their purification buffers and obtained enzyme preparations composed of one large polypeptide (MW 210,000–240,000) and possibly a smaller one (MW = 98,000–105,000). Since the rat liver and mammary gland acetyl-CoA carboxylases have previously been demonstrated to be particularly sensitive to proteolytic cleavage at specific sites,[18,19] the latter workers concluded that the smaller peptides obtained in earlier preparations were due to proteolysis.

Measurement of Acetyl-CoA Carboxylase Activities

$^{14}CO_2$ Incorporation into Malonyl-CoA

This assay measures the incorporation of $^{14}CO_2$ from bicarbonate into malonyl-CoA. The reaction mixture contains in a final volume of 0.5 ml: 5 μmol of tris(hydroxymethyl)methylaminopropane sulfonic acid (TAPS), pH 8.5, 1.25 μmol of sodium ATP, 2.5 μmol of $MgSO_4$, 0.25 μmol of acetyl-CoA, 0.5 μmol of dithiothreitol (DDT), 30 μmol of KCl, 75 μmol of $NaH^{14}CO_3$ (about 600 cpm per micromole of HCO_3), 0.25 mg of bovine serum albumin, and amounts of protein depending on purity of the carboxylase. All components of the reaction mixture except the acetyl-CoA are incubated at 37° for 3 min, and then acetyl-CoA is added to begin the reaction. After 10 min of incubation, the reaction is stopped by adding 0.1 ml of concentrated glacial acetic acid. Volatile unreacted $H^{14}CO_3$ is removed by spotting aliquots of the reaction mixture onto 2 cm × 8 cm strips of Whatman 1 MM filter paper and drying the paper in a stream of warm air from a hair drier. Nonvolatile radioactivity remaining is counted in a suitable scintillation fluid.

The reaction mixture outlined above will yield maximum rates of acetyl-CoA carboxylase activity. It is important to maintain a sufficient

[16] P. F. Heinstein and P. K. Stumpf, J. Biol. Chem. 244, 5374 (1969).
[17] N. C. Nielsen, A. Adee, and P. K. Stumpf, Arch. Biochem. Biophys. 192, 446 (1979).
[18] T. Tanabe, K. Wada, T. Okazaki, and S. Numa, Eur. J. Biochem. 57, 15 (1975).
[19] K. P. Huang, Anal. Biochem. 37, 98 (1970).

level of free Mg^{2+} in the reaction mixture to obtain optimal activity. Magnesium : ATP ratios of about 2 generally lead to the highest rates. However, the free Mg^{2+} concentration is reduced substantially when compounds that chelate divalent cations strongly are introduced into the reaction mixture. For this reason buffers that bind appreciable amounts of Mg^{2+}, such as Tricine or phosphate, should be avoided.

The method of Huang[19] is a convenient way to monitor the products of the reaction. A sample of the acidified reaction mixture is made alkaline with 2 M KOH, which destroys the thioester bond and produces radioactive malonate. Samples of the acidified and alkaline reaction mixtures are then each applied directly to 1 cm × 20 cm thin-layer chromatography strips covered with Gelman medium (ITLC-SG type 20). Ascending chromatographic separation of the radioactive products using a solvent system of water-saturated ether–formic acid (7 : 1 v/v) requires about 15 min. The chromatographic strips are air dried in a hood, cut into 1 cm × 1 cm segments, and the radioactivity in each segment is determined using an appropriate scintillation fluid. The mobility of the radioactive products in the chromatogram before and after alkaline hydrolysis can be compared with authentic malonyl-CoA and malonate standards. Malonyl-CoA remains at the origin of the chromatogram, whereas malonate moves near the solvent front.

Measurement of Carboxybiotin

The assay system described by Kannangara and Stumpf[12] has been used to measure the $^{14}CO_2$ bound as carboxybiotin following reaction (1a). The $^{14}CO_2$ is bound to the enzyme in the case of the soluble enzyme, whereas it becomes bound to thylakoid membranes in the case of the chloroplast enzyme. The assay mixture contains in a final volume of 0.5 ml: 27.5 μmol of imidazole buffer, pH 7.9, 1.25 μmol of ATP, 2.5 μmol of $MgSO_4$, 5–10 μCi of $NaH^{14}CO_3$ (about 60 μCi/mmol), and enzyme (300 μg of thylakoid chlorophyll or 400 μg of protein of the purified soluble carboxylase). The mixture is incubated 15 min at room temperature and then desalted at 4° using a Sephadex G-50 column (2 cm × 30 cm) equilibrated with 0.01 M Tris-HCl, pH 8.5. Radioactivity bound to protein is recovered in the void volume and counted using a scintillation system suited for aqueous samples.

To identify the radioactive product as carboxybiotin, the bound $^{14}CO_2$–biotin must be stabilized by methylation, the protein hydrolyzed, and the products examined for 1'-N-methoxycarbonylbiocytin. The reaction mixture described above is stopped using 10 μl of 0.1 M EDTA and 30 ml of methanol at $-20°$. This is followed immediately by addition of an ethanolic solution of diazomethane ($-20°$) with rapid stirring until a persisting yellow color is evident. After an additional 10 min of stirring the suspension at $0°$, the excess diazomethane is blown off in a fume hood using a stream of nitrogen, and the protein is recovered by centrifugation. The protein is resuspended in 7 ml of sterile water containing 200 μmol of phosphate (pH 7.5) and 4–10 mg of Pronase. The samples are incubated 48 hr at $37°$; the hydrolyzates are taken to 40% acetone, and the precipitate is removed by centrifugation (neglible radioactivity). The resulting supernatant is taken to 85% acetone to precipitate radioactive material, which is collected by centrifugation. The radioactive pellets are resuspended in a minimal volume of water and subjected to descending chromatography on strips of Whatman 3 MM paper. Either n-butanol–acetic acid–water (80 : 20 : 20) or n-butanol–water–pyridine–acidic acid (30 : 24 : 20 : 6) can be used as a solvent. After chromatography, each strip is cut into segments, and the segments are counted in an appropriate scintillation fluid to locate the radioactive products. The R_f of the radioactive products is compared with that of authentic 1'-N-methoxycarbonylbiocytin, which is 0.35 in the first solvent system and 0.54 in the second.

Transcarboxylation Assay

Transfer of $^{14}CO_2$ from carboxybiotin to acetyl-CoA has been measured for both the chloroplast[14] and barley embryo[14] enzymes using the method of Alberts and Vagilos et al.[20] The reaction mixtures contains: 55 mM imidazole, pH 7.5, protein containing $^{14}CO_2$–carboxybiotin (30,000–60,000 cpm), and 0.15 mM acetyl-CoA. The reaction mixture is incubated at $30°$ for 30 min, and then one-fifth of the reaction volume of concentrated glacial acetic acid is added. Aliquots of the mixture are dried on Whatman 1 MM filter paper strips (2 cm $\times$ 8 cm); the nonvolatile radioactivity remaining is counted in an appropriate liquid scintillation fluid and compared with control samples containing no acetyl-CoA. Evidence that the product obtained is malonyl-CoA can be obtained chromatographically using the method of Huang.[19] (See the section on $^{14}CO_2$ incorporation into malonyl-CoA).

ATP–$^{32}P_i$ Exchange Reaction

The ATP–$^{32}P_i$ exchange reaction of acetyl-CoA carboxylase can be measured as described by Heinstein and Stumpf.[16] The reaction mixture contains in a final volume of 0.5 ml: 30 μmol of Tris-HCl, pH 8.3, 0.5 μmol

[20] A. W. Alberts, and P. R. Vagelos, (1968). *Proc. Natl. Acad. Sci. U.S.A.* **59**, 561 (1968).

of EDTA, 5 μmol of $KHCO_3$, 2 μmol of $MgCl_2$, 0.3 μmol of ATP, 0.2 μmol of ADP, 18.7 μmol of $^{32}P_i$ (2 to 6 × 10^5 cpm/μmol), and 5–25 mg of protein. After incubation at 30° for 15 min, the reaction is stopped by the addition of 3 ml of 5% trichloroacetic acid. Ten micromoles of nonradioactive P_i and then 0.1 ml of 150 mg/ml Norit A are added to adsorb the radioactive ATP. The Norit suspension is centrifuged, and the supernatant is discarded. The pellet is washed twice with 3 ml of 5% cold trichloroacetic acid and finally with 3 ml of cold distilled water. After washing, the pellet is suspended in 2 ml of aqueous 50% ethanol containing 0.1% NH_3, and an aliquot is counted.

Malonyl-CoA–[^{14}C]Acetyl-CoA Exchange Reaction

The exchange reaction can be measured using the procedure of Gregolin et al.[21] as modified by Heinstein and Stumpf.[16] The reaction mixture contains in a final volume of 0.5 ml: 30 μmol of Tris-HCl, pH 8.3, 0.05 μmol of 1-^{14}C-acetyl-CoA (5.9 × 10^6 cpm/μmol), 0.1 μmol of malonyl-CoA, 0.1 μmol of EDTA, and 5–25 mg of purified acetyl-CoA carboxylase. After incubation at 30° for 10 min, the reaction is stopped by heating at 65° for 2 min. Then 25 μmol of potassium arsenate and 2 mg of phosphotransacetylase are added to deacylate acetyl-CoA. After incubation at 37° for 20 min, the reaction is stopped with 0.1 ml of 6 N HCl. An aliquot is taken to dryness at 90°, then 0.4 ml of CCl_4 is added to remove ^{14}C-labeled acetic acid as its azeotropic mixture when taken to dryness again at 90°. Water and an appropriate scintillation fluid are added, and the sample is counted to measure [^{14}C]malonyl-CoA.

Acetyl-CoA Carboxylase Purification

Wheat Germ Enzyme

The source of wheat germ used for purification is important. Red wheat varieties yielded extremely unstable enzyme preparations, whereas the carboxylase preparations from white wheat varieties were considerably more stable.

The purification procedure for wheat germ acetyl-CoA carboxylase is a modification of the one originally described by Hatch and Stumpf,[13] and later modified by Heinstein and Stumpf[16] and Nielsen et al.[17] All procedures are carried out at 0–3°. An acetone powder of fresh raw wheat germ (150 g) is suspended in 1 liter of 0.1 M potassium phosphate, pH 8.3,

[21] C. Gregolin, E. Ryder, and M. D. Lane, J. Biol. Chem. 243, 4227 (1968).

containing 1 mM dithiothretiol and 1 mM EDTA (PED). The suspension is homogenized briefly in a Waring blender to obtain an even suspension, the pH is readjusted to 8.3 using 1 M KOH, and then stirred for 30 min to extract soluble proteins. After extraction, the insoluble residue is removed by centrifugation. The supernatant is brought to 10% polyvinyl-pyrididone to bind flavinoids, stirred for 30 min, and the polyvinyl-polypyrididone is then removed by centrifugation. Subsequently, 0.05 volume of 1 M MnCl$_2$ is added dropwise while stirring the supernatant and maintaining the pH at 8.3. The resulting precipitate is removed by centrifugation, and the supernatant is taken to 40% saturation using crystalline ammonium sulfate. After 30 min of stirring the precipitated carboxylase activity is collected by centrifugation, and the pellet is resuspended in 0.1 M PED, pH 8.3, 20% saturated with ammonium sulfate. The suspension is stirred for 30 min, and the undissolved material is removed by centrifugation. The carboxylase remaining in solution is precipitated by adjusting the solution to 40% saturation with ammonium sulfate and is collected by centrifugation. The pellet is resuspended in sufficient 0.01 M PED, pH 8.3, to give a final concentration of about 20 mg of protein per milliliter and dialyzed overnight against at least 100 volumes of the same buffer. Sometimes a precipitate develops during dialysis, which is removed by centrifugation. At this stage the enzyme generally has a specific activity ranging between 0.02 and 0.04 μmol of CO$_2$ fixed per milligram of protein per minute. When frozen at $-20°$ it is stable for several months. Occasionally during long-term storage a precipitate develops and there are poor recoveries of activity at later stages in purification. This is most pronounced with enzyme prepared from red wheat varieties.

Acetyl-CoA carboxylase freshly prepared as described above can be further purified by chromatography on DEAE-cellulose. A column packed with Whatman DE-52, 20 × 7.5 cm, is washed with 1 M potassium phosphate, pH 7.5, and then equilibrated with 0.01 M PED, pH 7.5. A solution containing approximately 1.5 g of ammonium sulfate-precipitated enzyme is carefully adjusted to pH 7.5 and applied to the column. The column is developed using a 4-liter linear gradient between 0 and 0.4 M KCl in 0.01 M PED, pH 7.5. A typical elution profile is shown in Fig. 1. The carboxylase activity emerges from the column at about 0.2 M KCl as the trailing shoulder of a broad protein peak. The active fractions are pooled, and the enzyme is precipatated by taking the solution to 60% saturation with ammonium sulfate. The precipitated enzyme is resuspended in 0.01 M PED, pH 8.3, to give a protein concentration of approximately 10 mg/ml and then dialyzed against the resuspension buffer. At this stage in purification the enzyme typically has a specific activity rang-

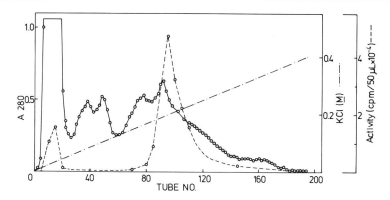

FIG. 1. DEAE-cellulose chromatography of wheat germ acetyl-CoA carboxylase. o——o, Absorbance at 280 nm; o---o, enzyme activity; — · — · , KCl concentration.

ing between 0.15 and 0.35. The enzyme can be stored at $-20°$ for several months with less than 10% loss of activity.

Further purification is achieved by chromatography on hydroxyapatite prepared according to Tiselius et al.[22] Enzyme from the DEAE column is adjusted to pH 6.8, care being taken not to go below this pH because precipitation of the carboxylase then occurs. The enzyme is applied to a column packed with hydroxyapatite (7.5 cm × 7.5 cm) equilibrated with 0.01 M PED, pH 6.8. The column is eluted with a 2-liter gradient between 0.01 and 0.2 M potassium PED, pH 6.8, with the results typical of those shown in Fig. 2. Most of the carboxylase activity appears in the leading shoulder of a major peak of protein eluting at about 0.05 M phosphate. Sometimes a second peak activity of variable size appears shortly after the major breakthrough peak of the column. The active fractions of the peak emerging from the column at about 0.05 mM PED are pooled, and the protein was precipitated by adjusting the solution to 60% saturation with ammonium sulfate. The sedimented precipitate is resuspended in 0.01 M PED, pH 8.3, containing 0.05 M KCl to a final concentration of about 5 mg of protein per milliliter. It is then dialyzed against the same buffer. The specific activity of the carboxylase at this stage of purification generally ranges between 1.5 and 6.5. Recovery is 10–15% of the starting activity. However, after hydroxyapatite chromatography the enzyme is very unstable, losing 50% of its activity after several days. Storage of the enzyme under nitrogen decreases the rate of activity loss somewhat.

Other workers have reported different conditions for purifying acetyl-CoA carboxylase from wheat germ. Heinstein and Stumpf[16] fractionated the enzyme obtained from the DEAE-Sephadex column on a 5 to

[22] A. Tiselius, S. Hjerten, and O. Levin, Arch. Biochem. Biophys. **65,** 132 (1956).

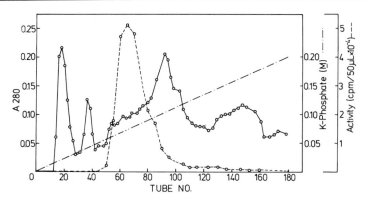

FIG. 2. Hydroxyapatite chromatography of wheat germ acetyl-CoA carboxylase. o——o, Absorbance at 280 nm; o- - -o, enzyme activity; — · — · , phosphate buffer.

20% sucrose gradient. Active fractions obtained from the sucrose gradient had specific activities between 2 and 6 μmol. Egin-Bühler *et al.*[15] introduced 0.2 mmol of phenylmethylsulfonyl fluoride per liter into all buffer solutions to inhibit serine proteases and followed the procedure of Heinstein and Stumpf through the DEAE-cellulose column. The enzyme emerging from the column was fractionated on a 5 to 20% sucrose gradient containing 20 mM Tris, pH 7.5, 50 mM KCl, 14 mM β-mercaptoethanol, and 0.3 mM EDTA. The resulting preparation had a specific activity of 5. The active fractions from the gradient were dialyzed against the gradient buffer solution containing 20% glycerol and 5 mM MgCl$_2$. The resulting solution was applied to a Blue Sepharose CL-6B column equilibrated with the same buffer. The carboxylase was eluted using a linear gradient from 0 to 0.6 M KCl. The enzyme was recovered at between 0.14 and 0.27 M KCl and concentrated by ultrafiltration. It had a specific activity of 24.

Barley Germ Enzyme

Brock and Kannangara[14] purified acetyl-CoA carboxylase from barley embryos. The embryos were homogenized in 55 mM imidazole-HCl, pH 8.3, containing 1 mM dithiothreitol. The homogenate was filtered and then centrifuged for 30 min at 30,000 g in a Sorvall centrifuge using an SS-34 rotor. Manganese chloride (0.025 volume of 1 M MnCl$_2$) was added to the supernatant with rapid stirring, and the precipitate formed was sedimented and discarded. The clarified supernatant was centrifuged in a Beckman ultracentrifuge at 120,000 g for 5 hr to concentrate the enzyme at the bottom of the tube. Fractions having the highest activities were diluted sixfold using the homogenization buffer and were recentrifuged as before. This procedure was repeated two more times to increase the specific activities of the active fractions from 0.9 to 7.4.

Parsley Acetyl-CoA Carboxylase

Egin-Bühler *et al.*[15] purified the carboxylase from parsley cell cultures grown under conditions that were carefully defined to maximize the enzyme activity. All buffers contained 20% glycerol, 14 mM β-mercaptoethanol, and 0.2 mM phenylmethylsulfonyl fluoride (PMSF). Frozen cells (1000 g) were homogenized with 1 liter of 0.1 M Tris-HCl, pH 7.5. Cell debris was removed by centrifugation, and the supernatant was stirred with 100 g of Dowex 1×2. The Dowex resin was removed by filtration through glass wool, and the protein extract was adjusted slowly to contain 0.3% polyethyleneamine. The precipitate formed after 20 min was removed by centrifugation, and the supernatant was slowly brought to 45% saturation with ammonium sulfate. The precipitate was collected by centrifugation and suspended in 0.02 M Tris-HCl pH 7.5. After dialysis against the same buffer, the enzyme was applied to a DEAE-cellulose column (5 × 10 cm). The proteins were eluted from the column stepwise using about 400 ml each of 0.02 M Tris containing 0.08 M and 0.18 M KCl, respectively. Active fractions emerged from the column with the higher salt concentration and were precipitated by making the fractions 50% saturated with ammonium sulfate. After centrifugation, the pellets were suspended in a small volume of 0.1 M Tris-HCl buffer, pH 7.5, and dialyzed against 0.02 M Tris buffer. The enzyme was applied to a Blue Sepharose CL-6B column (1.5 × 6 cm) equilibrated with the same buffer. After washing the column to remove contaminating proteins, 20 mM Tris buffer containing 7 mM ATP was used to elute acetyl-CoA carboxylase from the column. The active fractions were made 80% saturated with ammonium sulfate, and the enzyme was then applied to a Sepharose 6B column (0.9 × 60 cm) that was washed with 0.1 M Tris, pH 7.5, containing 0.1 M KCl. After precipitating the eluted enzyme with 80% saturated ammonium sulfate and redissolving the pellets in 0.1 M Tris buffer, pH 7.5, the enzyme had a specific activity of 6.8.

Chloroplast Enzyme

Although the three components of the prokaryotic chloroplast enzyme have been detected, methods to purify them have not been reported. The methodology used by Kannangara and Stumpf[12] to prepare chloroplast lamellae and stroma fractions is as follows. Fifty grams of washed, deveined spinach leaves are homogenized for 30 sec at full voltage in a precooled Waring blender using 150 ml of cold Honda's medium (2.5% Ficoll, 5% dextran T-40, 0.25 M sucrose. 0.025 M Tricine, 0.01 M $MgCl_2$, and 0.1% sodium ascorbate, pH 7.9). The suspension is filtered through a single layer of Miracloth,[22] and the filtrate is centrifuged at 4° at 1185 g in a Sorvall centrifuge for 5 min. The pellet constituted the "intact" chloroplasts.

To obtain the stroma and lamellar fractions the chloroplast pellets are suspended in a minimal volume of 55 mM imidazole, pH 7.9, and homogenized in a Potter–Elvehjem tissue grinder having a tightly fitting Teflon pestle. Loss of birefringence under phase contract microscopy reveals disrupted chloroplasts. The homogenate is centrifuged at 4° and 110,000 g. The yellow supernatant contains the stroma fraction, where biotin carboxylase and transcarboxylase are located. The pellets are washed several times with 55 mM imidazole to free them of stroma. The pellet contains the biotin carboxyl carrier protein.

Acetyl-CoA carboxylase subunits can be isolated from *E. coli* B using the method of Alberts *et al.*[20] Protein fraction E_a contains biotin carboxylase plus the biotin carboxyl carrier protein, and fraction E_b contains transcarboxylase.

[7] Acyl-CoA Carboxylase from the Nematode *Turbatrix aceti*

By HARUKO MEYER and FRANZ MEYER

$$\text{Acyl-CoA} + CO_2 + \text{ATP} \overset{K^+}{\rightleftharpoons} \alpha\text{-carboxyl acyl-CoA} + \text{ADP} + P_i$$

Assay Method

Principle. *Turbatrix aceti* acyl-CoA carboxylase catalyzes the α-carboxylation of acetyl-CoA, propionyl-CoA, and butyryl-CoA in the presence of ATP, HCO_3^-, Mg^{2+}, and K^+.[1] The rate of carboxylation can be determined by measuring the acetyl-, propionyl-, or butyryl-CoA dependent incorporation of $H^{14}CO_3^-$ into an acid-stable reaction product (malonyl-CoA or its homologs). Under the specified assay conditions, the amount of $H^{14}CO_3^-$ fixed is proportional to the enzyme concentration over

---

[1] H. Meyer and F. Meyer, *Biochemistry* **17**, 1828 (1978).

a limited range (0.5–2.0 μg of enzyme protein) and is nearly linear during the first 5–10 min of the incubation period.[1]

Reagents

Tris-HCl buffer, 0.4 M, pH 8.0
KCl, 0.4 M
MgCl, 0.2 M
ATP, 50 mM, adjusted to pH 7.5 with NaOH
NaH^{14}CO$_3$ (0.2 mCi/mmol), 0.5 M
Acetyl-, propionyl-, or butyryl-CoA, 10 mM

Procedure. The assay mixture has a final volume of 0.4 ml and contains 80 mM Tris-HCl buffer, pH 8.0; 40 mM KCl or 20 mM K$_2$HPO$_4$, pH 8.0; 10 mM MgCl; 1 mM ATP (sodium salt); 1.5 mM acetyl-, propionyl-, or butyryl-CoA (lithium salts); 20 mM NaH^{14}CO$_3$ (0.2 mCi/mmol); and 10–15 mU of enzyme. The reaction is started by the addition of enzyme; it is run for 5 min at 30° and is stopped by the addition of 0.2 ml of 2 N perchloric acid. After centrifugation of the mixture, 0.2 ml of the clear supernatant is placed into a scintillation vial and incubated in a 70° water bath under a gentle stream of air until all unreacted ^{14}CO$_2$ is expelled. The dry residue is then dissolved in 0.1 ml of water, blended with Aquasol (New England Nuclear), and counted in a liquid scintillation counter. One unit of enzyme activity is defined as the amount of enzyme protein (milligrams) required to catalyze the fixation of 1 μmol of H^{14}CO$_3^-$ into malonyl-CoA (or its homologs) per minute under the given assay conditions.

Purification Procedure

All operations are carried out at 0–4°. All buffers contain 1 mM EDTA and 5 mM 2-mercaptoethanol or 3 mM dithiothreitol.[2]

Step 1. Preparation of Crude Homogenate and High-Speed Supernatant. *Turbatrix aceti* can be grown in either a chemically defined medium[3] or a crude medium.[2] It is harvested during the late logarithmic growth phase by centrifugation and washed sequentially in a 50% sucrose solution,[4] distilled water, and 0.1 M phosphate buffer, pH 7.0. The nematodes are then suspended in a minimal volume of the same buffer and used either immediately or after storage at −18°.

About 250 g wet weight of thawed nematodes are diluted to 600 ml with 0.2 M potassium phosphate buffer, pH 7.5. The mixture is passed

[2] H. Meyer, B. Nevaldine, and F. Meyer, *Biochemistry* **17**, 1822 (1978).
[3] M. Rothstein and M. Coppens, *Comp. Biochem. Physiol.* **61B**, 99 (1978).
[4] M. Rothstein, F. Nicholls, and P. Nicholls, *Int. J. Biochem.* **1**, 695 (1970).

through a French pressure cell at 14,000 psi. The homogenate is immediately adjusted to pH 7.0 and centrifuged at 66,000 g for 60 min. The resulting supernatant is set aside while the pellet is washed once with 300 ml of 0.2 M potassium phosphate buffer, pH 7.0. After centrifugation, the two supernatants are combined.

Step 2. First Ammonium Sulfate Fractionation. The high-speed supernatant is fractionated by the stepwise addition of solid ammonium sulfate (Enzyme Grade). During this process, the pH of the mixture is maintained at 7.0. The proteins that precipitate in the range from 0 to 30% ammonium sulfate saturation are removed by centrifugation and discarded. The proteins which precipitate in the range from 30 to 55% saturation are dissolved in a minimal volume of 0.05 M phosphate buffer and dialyzed against the same buffer until free of ammonium sulfate. Steps 1 and 2 are carried out within the same day to minimize the risk of enzyme degradation.

Step 3. Calcium Phosphate Gel Adsorption. The enzyme solution is diluted to a phosphate concentration of 0.02 M and a protein concentration of 20–30 mg/ml. To this solution, calcium phosphate gel suspended in water[5] is added in the ratio of 1.8 g of dry gel per gram of protein. The mixture, totaling 800–1000 ml, is stirred for 10 min, and the calcium phosphate gel, to which the enzyme is bound, is collected by centrifugation at 7000 g for 10 min. The gel is resuspended in 500 ml of 0.05 M phosphate buffer, homogenized gently with a magnetic stirrer, and again collected by centrifugation. After this washing process has been repeated, the enzyme is eluted from the gel with 300-ml portions of 0.25 M phosphate buffer by a repeated homogenization–extraction process.

Step 4. Second Ammonium Sulfate Fractionation. The gel eluate is fractionated by the stepwise addition of solid ammonium sulfate as described in step 2. The enzyme that precipitates in the range from 35 to 55% ammonium sulfate saturation is dissolved in 0.05 M phosphate buffer, pH 7.0, and dialyzed against the same buffer until free of ammonium sulfate.

Step 5. First Cellex E Column Chromatography. The enzyme solution, containing about 2 g of protein, is diluted to 0.02 M phosphate and applied to a Cellex E column (5 × 10 cm) previously equilibrated with 0.02 M phosphate buffer, pH 7.3. The sample is eluted stepwise with three column volumes each of 0.02 M, 0.06 M, and 0.1 M phosphate buffer, pH 7.3. Over 90% of the initial enzyme activity is recovered from the column with 0.06 M phosphate buffer. The fractions containing the enzyme are pooled and concentrated to about 40 ml with an Amicon XM-50 ultrafilter.

Step 6. Second Cellex E Column Chromatography. The enzyme solu-

[5] M. Kunitz and H. Simms, *J. Gen. Physiol.* **11**, 641 (1928).

tion, containing about 400 mg of protein, is diluted to 0.02 M phosphate and placed on a second Cellex E column (2.5 × 20 cm). The enzyme is eluted from the column with a linear phosphate gradient established between 500 ml of 0.02 M potassium phosphate buffer, pH 7.3, and 500 ml of 0.08 M potassium phosphate buffer. A large protein peak without carboxylase activity appears in the eluent at a phosphate concentration of 0.028 M (conductivity: 3.3 × 10^{-3} mho); the enzyme appears at a phosphate concentration of 0.035 M (conductivity: 4.0 × 10^{-3} mho).

Step 7. Third Cellex E Column Chromatography. Chromatography on Cellex E is repeated as described in step 6 to separate fully the carboxylase peak from a nearby contaminating protein peak. The enzyme is then precipitated with ammonium sulfate at 80% saturation, and the precipitate is dissolved in 3 ml of 0.1 M phosphate buffer.

Step 8. Sepharose 4B Gel Filtration. A 2.5 × 60 cm column of Sepharose 4B is equilibrated with 0.2 M potassium phosphate buffer, pH 7.0. The enzyme is applied to the column and eluted with the same buffer. The enzyme-containing fractions, which are eluted with approximately twice the void volume of the column, are combined and concentrated to 20 ml by filtration.

Step 9. Third Ammonium Sulfate Fractionation. Saturated ammonium sulfate, pH 7.0, is added to the enzyme solution until the solution becomes slightly turbid. At this point, the precipitate is removed by centrifugation, and saturated ammonium sulfate is added for a second time until the solution becomes slightly turbid. The sample is then allowed to stand for 15–20 hr at 3°. The precipitate formed during this step is collected by centrifugation at 60,000 g for 30 minutes and redissolved in 2–3 ml of 0.2 M phosphate buffer. After dialysis against the same buffer, the now purified enzyme is stored at −18° in 0.2 M phosphate buffer, pH 7.0, containing 1 mM EDTA, 3 mM dithiothreitol, and 30% glycerol.

Stability. The partially purified enzyme of step 3 retains 80–90% of its initial activity when stored for 3 weeks at 3° in 0.05–0.25 M potassium phosphate buffer, pH 6.5–7.5, containing 1 mM EDTA and 5 mM 2-mercaptoethanol. The fully purified enzyme retains all its activity for at least 3 months when stored at −18°, at a protein concentration of 1–5 mg/ml in 0.2 M phosphate buffer, pH 7.0, containing 1 mM EDTA, 3 mM dithiothreitol, and 30% glycerol.

The results of the purification procedure are summarized in the table.

Properties

Activators and Inhibitors. The *T. aceti* carboxylase is stimulated over 30-fold by the monovalent cations K^+, Rb^+, Cs^+, or NH_4^+. The monova-

PURIFICATION OF ACYL-COA CARBOXYLASE FROM *Turbatrix aceti*[a]

Step	Total protein (mg)[b]	Total activity (units)[c]	Specific activity (units/mg)	Recovery (%)
1. High-speed centrifugation	26,000	549	0.02	100
2. First $(NH_4)_2SO_4$ precipitation	9,200	843	0.1	154
3. Calcium phosphate gel adsorption	3,908	610	0.2	111
4. Second $(NH_4)_2SO_4$ precipitation	2,030	400	0.2	73
5. First Cellex E cellulose chromatography	447	369	0.8	67
6. Second Cellex E cellulose chromatography	188	234	1.3	43
7. Third Cellex E cellulose chromatography	74	229	3.1	42
8. Sepharose 4B gel filtration	34	209	6.1	38
9. Third $(NH_4)_2SO_4$ precipitation	23	146	6.4	27

[a] Starting material: 270 g wet weight of frozen nematodes.
[b] Assayed by the biuret method.
[c] Assayed by measuring the carboxylation of acetyl-CoA. One unit = 1 μmol of HCO_3^- fixed per minute under the assay conditions described in the text.

lent cations primarily increase the rate of the first half of the carboxylation reaction, the ATP-dependent carboxylation of the biotinyl residue.[1]

The enzyme is inhibited by avidin and by sulfhydryl reagents such as N-ethylmaleimide or p-chloromercuribenzoate.

Unlike the acetyl-CoA carboxylases of higher animals,[6] the *T. aceti* enzyme has no requirement for citrate or other tri- or dicarboxylic acids.

Specificity. The enzyme carboxylates a group of straight-chain acyl-CoA esters ranging in chain length from two to four carbon atoms. The relative carboxylation rates for these substrates are: propionyl-CoA, 100%; butyryl-CoA, 23%; and acetyl-CoA, 17%. The enzyme does not carboxylate CoA esters with acyl chains longer than four carbon atoms, and it carboxylates isobutyryl-CoA only at a very low rate.[1]

Kinetics. Under standard assay conditions, the apparent K_m values of the enzyme are: acetyl-CoA, 0.42 mM; propionyl-CoA, 0.23 mM; butyryl-CoA, 0.33 mM; HCO_3^-, 2.3 mM; and ATP, 0.25 mM. The apparent activation constant (K_A) for K^+ is 11.0 mM.

The optimum pH for the enzyme is between 7.8 and 8.6.[1]

Physical Properties. On the basis of sedimentation velocity and sedimentation equilibrium analyses, the enzyme has a sedimentation coefficient ($s_{20,w}$) of 18.0 S and a molecular weight of 667,000. In sodium

[6] J. Moss and M. Lane, *Adv. Enzymol.* **35**, 321 (1971).

dodecyl sulfate, the enzyme dissociates into two distinct polypeptides: a smaller 58,000 MW peptide and a larger 82,000 MW peptide, which alone carries the biotinyl prosthetic group. Since the enzyme contains 1.58 μg of biotin per milligram of protein, it is assumed that four pairs of these polypeptides constitute the oligomeric structure of the native *T. aceti* carboxylase.[1]

[8] Multienzyme Complex Containing Acetyl-CoA Carboxylase from *Euglena gracilis*

EC 6.4.1.2 Acetyl-CoA : carbon-dioxide ligase (ADP-forming)

By MARY LOU ERNST-FONBERG and JACK S. WOLPERT

The acetyl-CoA carboxylase from *Euglena gracilis* is isolated as a component of a multienzyme complex.[1,2] We have suggested previously that the incorporation of CO_2 into acetyl-CoA to form malonyl-CoA for fatty acid biosynthesis is expedited by the existence of the observed complex of enzymes: phosphoenolpyruvate carboxylase, malate dehydrogenase, and acetyl-CoA carboxylase. The ratio of K_m values for HCO_3^- for acetyl-CoA carboxylase and phosphoenolpyruvate carboxylase in the complex from *Euglena* is about 6. Moreover, the catalytic efficiency[3] of the latter enzyme is about 25 times greater than that of the acetyl-CoA carboxylase. Thus, the phosphoenolpyruvate carboxylase captures CO_2 more readily. The resulting oxaloacetate is reduced by malate dehydrogenase, which is specific for NADH, to yield malate, which is then acted upon by malic enzyme (not a constituent of the complex of enzymes) to generate NADPH and the CO_2 substrate for the acetyl-CoA carboxylase (Fig. 1).[4] Malic enzyme in *Euglena* is specific for $NADP^+$.[1] The malonyl-CoA and NADPH resulting from the reactions are then used to form fatty acids.

[1] J. S. Wolpert and M. L. Ernst-Fonberg, *Biochemistry* **14**, 1095 (1975).

[2] J. S. Wolpert and M. L. Ernst-Fonberg, *Biochemistry* **14**, 1103 (1975).

[3] M. Dixon and E. C. Webb, "Enzymes," p. 15. Academic Press, New York, 1958.

[4] L. Drobnica and L. Ebringer, *Folia Microbiol. (Prague)* **8**, 56 (1962).

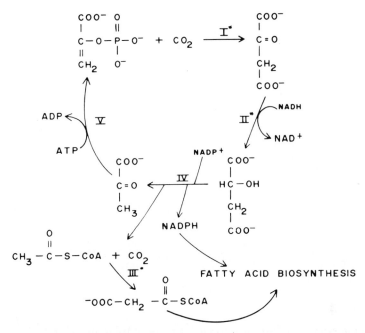

FIG. 1. Metabolic scheme for CO_2 fixation to provide substrates for fatty acid biosynthesis. I*, II*, and III* are the enzymes found within the complex—phosphoenolpyruvate carboxylase, malate dehydrogenase, and acetyl-CoA carboxylase, respectively. IV represents malic enzyme, which is present in *Euglena gracilis*.[1] Pyruvate phosphokinase (V) was first shown to be present in *Euglena gracilis* by L. Drobnica and L. Ebringer.[4]

Assay Methods

Acetyl-CoA Carboxylase

Principle. Acetyl-CoA carboxylase activity is measured by the addition of [^{14}C]bicarbonate to acetyl-CoA to form [3–^{14}C]malonyl-CoA. The rationale for this assay has been described in this series.[5] Spectrophotometric methods for measuring acetyl-CoA carboxylase activity also have been described.[6,7] One of these assays, for use with purified enzyme preparations, measures the formation of ADP in a coupled reaction with pyruvate kinase and lactate dehydrogenase. The decrease in the absorption of NADH is monitored. This linked reaction series which contains acetyl-

[5] H. Inoue and J. M. Lowenstein, this series, Vol. 35 [1].
[6] M. Matsuhashi, this series, Vol. 14 [1].
[7] S. Numa, this series, Vol. 14 [2].

CoA, bicarbonate, $MgCl_2$, NADH, ATP, and phosphoenolpyruvate as substrates cannot be used to measure the activity of the *Euglena* acetyl-CoA carboxylase as isolated in a multienzyme complex. The multienzyme complex contains acetyl-CoA carboxylase, phosphoenolpyruvate carboxylase, and malate dehydrogenase. The phosphoenolpyruvate carboxylase catalyzes the carboxylation of the phosphoenolpyruvate to form oxaloacetate, which is then oxidized in the malate dehydrogenase-catalyzed reaction to form malic acid and NAD. The rate of oxidation of NADH parallels the activity of the radioactive assay of acetyl-CoA carboxylase activity, but it is not a measure of acetyl-CoA carboxylase activity in the multienzyme complex. The oxidation of NADH is independent of acetyl-CoA, ATP, and $MgCl_2$, the substrates for the acetyl-CoA carboxylase-catalyzed reaction; it can be used as a monitor of the presence of the multienzyme complex, since it does measure two of the constituent enzymatic activities.

Reagents

Imidazole buffer, 1.0 M, pH 7.5
ATP, 200 mM
$MgCl_2$, 400 mM
Bovine serum albumin (fatty acid free), 10 mg/ml
Dithiothreitol, 100 mM
Acetyl-CoA, 2.4 mM
$NaH^{14}CO_3$ (0.6 $\mu Ci/\mu mol$), 250 mM
HCl, 3 N
KOH, 40%
Ethanol, absolute
Toluene containing 4.0 g of 2,5-diphenyloxazole per liter

Procedure.[8] All aqueous solutions are prepared in water that was boiled for 30 min to drive off CO_2. Reactions are carried out in glass liquid scintillation counting vials with 15 mm tops sealed with serum stoppers. The serum stoppers are linked in series via two 18-gauge hypodermic needles in each stopper. One needle in each stopper is for air input; the other acts as air outlet. Ten vials is a manageable number to process at one time. The counting vials are secured in a metal rack in a 30° circulating water bath in a fume hood. All reaction components except ATP are added. After a 3–5 min warming period, ATP is added to initiate reaction, and the serum stoppers (with the linkages described above) are secured. After 10 min, 0.1 ml of 3 N HCl is injected, and residual $^{14}CO_2$ is removed by purging with air for 20 min. The effluent air $^{14}CO_2$ is trapped in 40%

[8] A. K. Evers, J, S. Wolpert, and M. L. Ernst-Fonberg, *Anal. Biochem.* **64,** 606 (1975).

KOH. The sealed scintillation vials are placed in a 55° water bath, and the solutions are dried under a current of air that has been passed through Drierite and a Millipore filter holder containing Whatman No. 1 filter paper. After adding (in order) 0.1 ml of H_2O, 3 ml of absolute ethanol, and 15 ml of toluene counting fluid, the vials are counted in a liquid scintillation counter. Quenching is corrected by the channels ratio method.

The enzyme assay in a final volume of 0.25 ml contains: imidazole buffer, 25 μl (0.1 M); ATP, 5 μl (4 mM); $MgCl_2$, 5 μl (8 mM); bovine serum albumin, 10 μl; dithiothreitol, 5 μl (2 mM)) acetyl-CoA, 25 μl (0.24 mM); $NaH^{14}CO_3$, 25 μl (25 mM); and protein. The assay concentrations of reagents are given in parentheses.

Multienzyme Complex

Principle. The multienzyme complex is assayed through the linked activities of two of the three constituent enzymes. The action of phosphoenolpyruvate carboxylase forms oxaloacetate from phosphoenolpyruvate. The oxaloacetate is then reduced by malate dehydrogenase in the presence of NADH to form malate and NAD^+. The rate of NAD^+ formation is measured. This assay is used only for the purified complex.

Reagents

Imidazole buffer, 1.0 M, pH 7.5
$NaHCO_3$, 250 mM
Phosphoenolpyruvate, 75 mM
NADH, 2.25 mM
Dithiothreitol, 100 mM
Bovine serum albumin (fatty acid-free), 10 mg/ml
$MgCl_2$, 400 mM

Procedure. The reaction is carried out at 30° in a spectrophotometer equipped with a recorder. After a 5-min warming period, the reaction is initiated by adding phosphoenolpyruvate. The decrease in absorbance at 340 nm is measured ($\epsilon_{340\,nm}$ = 6.22).

The assay includes in a final volume of 0.25 ml: imidazole buffer, 25 μl (0.1 M); $NaHCO_3$, 25 μl (25 mM); phosphoenolpyruvate, 10 μl (3 mM); NADH, 20 μl (0.18 mM); dithiothreitol, 5 μl (2 mM); bovine serum albumin, 10 μl; $MgCl_2$, 5 μl (8 mM); and enzyme. The numbers in parentheses are the assay concentrations.

Malate Dehydrogenase

Principle. The oxidation of NADH coupled to the conversion of oxaloacetate to malate is measured at 340 nm.

Reagents

Imidazole buffer, 1.0 M, pH 7.5
NADH, 2.25 mM
Dithiothreitol, 100 mM
MgCl$_2$, 400 mM
Oxaloacetate, 20 mM
Bovine serum albumin (fatty acid-free), 10 mg/ml

Procedure. The assay conditions and procedure are similar to those described for the spectrophotometric assay of the multienzyme complex. Reaction is initiated by the addition of oxaloacetate.

The following substances are combined in a final volume of 0.25 ml: imidazole buffer, 25 μl (0.1 M); NADH, 20 μl (0.18 mM); dithiothreitol, 5 μl (2 mM); MgCl$_2$, 5 μl (8 mM); oxaloacetate, 10 μl (0.8 mM); bovine serum albumin, 10 μl; and enzyme sufficient to decrease the absorbance at a rate of 0.05–0.25 per minute. The numbers in parentheses are concentrations in the reaction mixture.

Phosphoenolpyruvate Carboxylase

Principle. The carboxylation of phosphoenolpyruvate with [^{14}C]bicarbonate to oxaloacetate results in the formation of ^{14}C that is nonvolatile in acid. Alternatively, phosphoenolpyruvate carboxylase activity may be measured spectrophotometrically at 340 nm in the presence of NADH and malate dehydrogenase, which catalyzes the reduction of oxaloacetate formed by the action of the carboxylase.

Reagents for Radioactive Assay

Imidazole buffer, 1.0 M, pH 7.5
Phosphoenolpyruvate, 75 mM
Dithiothreitol, 100 mM
NaH^{14}CO$_3$ (0.6 μCi/μmol), 250 mM
Bovine serum albumin (fatty acid-free), 10 mg/ml
MgCl$_2$, 400 mM

Procedure for Radioactive Assay. The apparatus and technique are identical to those described for the assay of acetyl-CoA carboxylase except that freshly prepared oxaloacetate, 0.9 mg in 3 N HCl, is added at the end of the reaction. The reaction is initiated by adding phosphoenolpyruvate.

The following components, combined in a final volume of 0.25 ml, are imidazole buffer, 25 μl (0.1 M); phosphoenolpyruvate, 10 μl (3 mM); dithiothreitol, 5 μl (2 mM); NaH^{14}CO$_3$, 25 μl (25 mM); bovine serum albumin, 10 μl; MgCl$_2$, 5 μl (8 mM); and enzyme.

Reagents for Spectrophotometric Assay

The reagents specified for the radioactive assay except $NaH^{14}CO_3$
$NaHCO_3$, 250 mM
NADH, 2.25 mM
Malate dehydrogenase, 300 units/ml

Procedure for Spectrophotometric Assay. The assay conditions and procedure are similar to those described for the spectrophotometric assay of the multienzyme complex. The reaction is initiated by the adding phosphoenolpyruvate.

The following components are combined in a final volume of 0.25 ml: imidazole buffer, 25 μl (0.1 M); phosphoenolpyruvate, 10 μl (3 mM); dithiothreitol, 5 μl (2 mM); $NaHCO_3$, 25 μl (25 mM); NADH, 20 μl (0.18 mM); $MgCl_2$, 5 μl (8 mM); malate dehydrogenase, 5 μl; bovine serum albumin, 10 μl; and enzyme.

Preparation of Biotin-Affi-Gel 102

Principle. A coupling agent is used to join biotin covalently through the carboxyl moiety of the valeric acid side chain to the ω-amino group of Affi-Gel 102.[9] ω-Aminoalkyl agarose gel (Affi-Gel 102) contains 8–10 μmol of bound 3,3'-diaminodipropylamine side chain, 1.4 nm in length, per milliliter of packed gel.

Procedure. Approximately 10 ml of packed gel is washed with 20 volumes of 0.1 M NaCl, pH 10, followed by distilled water until the absorbance of the effluent at 220 nm reaches zero. The gel is then washed with a few volumes of dimethylformamide–water (1 : 1, v/v) and is transferred to an Erlenmeyer flask with a standard-taper stopper. To the washed gel is added 800 μmol of *d*-biotin in 70 ml of dimethylformamide–H_2O (1 : 1, v/v). The pH is 4.65. 1-Ethyl-3-(3-dimethylaminopropyl)carbodiimide-HCl,   1.0   mmol   dissolved   in   3   ml   of dimethylformamide-H_2O (1 : 1 v/v), is added to the reaction with constant stirring. The reaction mixture is then stirred continuously at room temperature. The carbodiimide (1.0 mmol) is added to the reaction after 27 hr and again after 55 hr. The reaction is allowed to continue and, after 72 hr, the coupled gel is washed with 600 ml of dimethylformamide-H_2O (1 : 1, v/v) followed by about 5 liters of distilled water until the absorbance at 220 nm is zero. The extent of coupling is determined by a hydroxamate assay[10]; typically it is 4 μmol of biotin bound per milliliter of gel, which is about 50% of the theoretical yield.

[9] Bio-Rad Laboratories, Richmond, California 94804.
[10] J. S. Wolpert and M. L. Ernst-Fonberg, *Anal. Biochem.* **52**, 111 (1973).

The coupled gel is chilled to 4°, mixed with an equal volume of water, and poured into a glass column. For use with enzyme, the gel column is equilibrated in 20 mM M imidazole buffer, pH 7.0, 7 mM 2-mercaptoethanol, and 0.04 M NaCl. The method can be scaled up at least 10-fold with similar yields.

The synthesis of biotin-liganded agarose using the water-insoluble dicyclohexylcarbodiimide gives yields similar to those obtained from the method of choice, which is described above.

Purification of the Multienzyme Complex

All procedures are carried out at 4° unless specified otherwise. *Euglena gracilis* strain Z,[11] 50 g, are mixed with 50 ml of 0.1 M imidazole buffer, pH 7.0, 7 mM 2-mercaptoethanol and disrupted with ultrasound. (The use of a nitrogen pressurized bomb for cell disruption gives similar results.) The temperature is kept below 4° with a methanol–ice bath. Multiple short bursts of ultrasound are applied until breakage is complete by microscopic examination. The mixture is centrifuged at 100,000 g for 1 hr. The supernatant solution is collected, and the pellet is washed with one volume of buffer. Following another centrifugation, the supernatant solutions are combined. A small aliquot for assay of protein and enzyme activity is put through a Sephadex G-25 column, 1.3 × 7 cm, in 20 mM imidazole, pH 7.0, 7 mM 2-mercaptoethanol (buffer A). The remainder is brought to 60% saturation with $(NH_4)_2SO_4$. After stirring for 30 min, the precipitate is collected and dissolved in 100 ml of buffer A. A small aliquot for assay of enzyme activity is desalted on Sephadex G-25.

The resulting protein solution is applied to a DEAE-cellulose column (10 mg of protein per milliliter column bed) equilibrated with buffer A. The enzyme activity is washed through the column in the breakthrough peak with the same buffer. Fractions comprising the protein peak are pooled and concentrated by precipitation in 60% saturated $(NH_4)_2SO_4$ solution. The precipitate is collected, dissolved in 15 ml of buffer A, and desalted on Sephadex G-25. The enzyme is then frozen in 1-ml aliquots in liquid N_2 and stored at −60°.

After thawing and centrifuging, about one-fifth of the enzyme solution (120 mg of protein) is applied to a biotin affinity column, 2.5 × 11 cm. The column is eluted with equilibration buffer (20 mM imidazole, pH 7.0, 7 mM 2-mercaptoethanol, and 40 mM NaCl) until no more discrete peaks of protein appear and protein begins to trail from the column. Fractions of 5 ml are collected. At this time, in order to hasten the release of enzyme, the ionic strength of the buffer is raised to 0.29 M with NaCl. A discrete peak

[11] R. K. DiNello and M. L. Ernst-Fonberg, *J. Biol. Chem.* **248**, 1707 (1973).

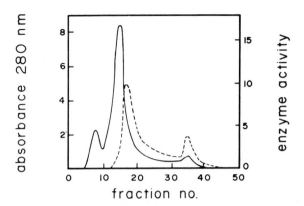

Fig. 2. Biotin affinity chromatography of acetyl-CoA carboxylase. The preparation and use of the column are given in the text. Absorbance at 280 nm is shown by the solid line, and the dashed line gives the profile of enzyme activity determined by the incorporation of $^{14}CO_2$ in the presence of acetyl-CoA (nanomoles per milliliter of enzymes in 15 min). Fractions of 5 ml were collected. The salt increment was started at fraction 23. Fractions 26–42 were pooled and concentrated. The void volume of the column determined with blue dextran was 26–28 ml, and cytochrome *c* appeared at 76 ml of effluent. From Wolpert and Ernst-Fonberg.[1]

of protein is eluted (Fig. 2), the fractions of which are pooled and concentrated in an Amicon Filtration Cell[12] using a UM-20E membrane. A 1.4 ml (10 mg) aliquot of this material is applied to a BioGel A-1.5 column, 2.5 × 30 cm, in buffer A and eluted with the same. The fractions comprising the major protein peak (Fig. 3) are pooled and concentrated by Amicon filtration. A preparation with typical yields is summarized in the table.

Properties of the Multienzyme Complex and Individual Enzymes

Demonstration of Enzymes Present. The preparation catalyzes the carboxylation of acetyl-CoA to form malonyl-CoA. The spectrophotometric assay of the multienzyme complex depends upon the presence of phosphoenolpyruvate and bicarbonate for the oxidation of NADH. Metabolic routes that had to be considered included the presence of the following enzymes: phosphoenolpyruvate carboxylase, phosphoenolpyruvate carboxykinase, pyruvate kinase, pyruvate carboxylase, malate dehydrogenase, and malic enzyme. Product identification and testing the usefulness of the appropriate substrates for all the above metabolic reactions indicate that the multienzyme complex is composed of acetyl-CoA carboxylase, phosphoenolpyruvate carboxylase, and malate dehydrogenase. The activity ratio is 1 : 25 : 500, respectively.

[12] Amicon Corp., Lexington, Massachusetts 02173.

PURIFICATION OF ENZYMES FROM *Euglena gracilis*

		Total activity		
Fraction	Total protein (mg)[a]	Acetyl-CoA carboxylase (units)[b]	Phosphoenolpyruvate carboxylase (units)[b]	Malate dehydrogenase (units $\times$ 10^{-3})[b]
Crude extract[c]	4359	318	—	—
100,000 g 60-min supernatant of crude extract	1568	526	12,834	—
0–60% (NH₄)₂SO₄ fraction	790	516	11,920	7291
DEAE-cellulose	714	578	14,856	7037
0–60% (NH₄)₂SO₄ fraction	636	546	13,213	6716
Biotin affinity chromatography	11	130	3,029	66
BioGel A-1.5	10	127	3,156	46

[a] Protein is measured according to O. H. Lowry, N. D. Rosebrough, A. L. Farr, and R. J. Randall, *J. Biol. Chem.* **193**, 265 (1951).
[b] A unit of activity is the production of 1 nmol of product per minute.
[c] Obtained from the disruption of 50 g of *Euglena*.

Purity. The purified multienzyme complex gives a single band when subjected to discontinuous acrylamide gel electrophoresis in two different systems. Gel filtration chromatography of the purified complex in buffer A on a 2.5 × 30 cm column of BioGel A-1.5 yields a single peak of protein coincident with acetyl-CoA carboxylase activity, phosphoenolpyruvate carboxylase activity, and malate dehydrogenase activity (Fig. 3).

Molecular Weight Estimates. The molecular weight of the complex estimated by gel filtration is 360,000. Acrylamide gel electrophoresis in the presence of sodium dodecyl sulfate and thiol compound shows that the multienzyme complex is composed of subunits of about 12,000 and 51,000 MW. Presumably these include different subunits of the same size.

After dissociation, the following molecular weights are obtained by gel filtration chromatography for the constituent enzymes: malate dehydrogenase, 67,000; acetyl-CoA carboxylase, 127,000; and phosphoenolpyruvate carboxylase 183,000. These values are in agreement with those obtained for highest molecular weight oligomers when the multienzyme complex is subjected to cross-linking with dimethylsuberimidate under conditions where cross-linking between oligomers does not occur.[13] The sum of the molecular weights of the constituent enzymes is 377,000. This

[13] J. S. Wolpert, Ph.D. Thesis, Yale University, New Haven, Connecticut, 1975.

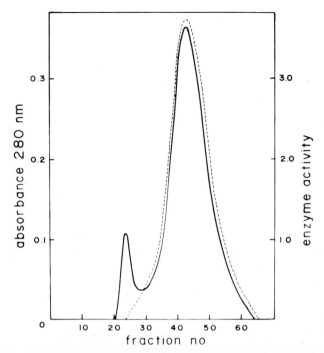

Fig. 3. Gel filtration chromatography of enzyme activity retained by the biotin-liganded gel. Enzyme, 10 mg, retained by the affinity column was concentrated to 1.2 ml and applied to a BioGel A-1.5 column, 2.5 × 30 cm, equilibrated in and eluted with buffer A. The flow rate was 10 ml/cm² per hour. Fractions of 2 ml were collected and measured for absorbance at 280 nm (solid line) and enzyme activity (dashed line, nanomoles per milliliter of enzyme in 15 min). The activity was measured by the incorporation of $H^{14}CO_3$ into acetyl-CoA or by the change in absorbance at 340 nm in the presence of phosphoenolpyruvate and NADH as shown here. Both assays give a similar profile. The void volume was 46 ml. From Wolpert and Ernst-Fonberg.[1]

excellent agreement between the whole and the sum of the parts shows that the multienzyme complex consists of 1 mol each of acetyl-CoA carboxylase, phosphoenolpyruvate carboxylase, and malate dehydrogenase.

Acetyl-CoA Carboxylase Reaction and Stability. The enzyme requires acetyl-CoA, HCO_3^-, ATP, and Mg^{2+} and catalyzes the formation of malonyl-CoA. Propionyl-CoA is carboxylated at 11% of the rate with acetyl-CoA. The enzyme from *Euglena* is unaffected by incubation with citrate or isocitrate prior to starting the reaction. The reaction rate is linear with time and enzyme concentration providing that the total protein in the assay solution is greater than 0.2 mg. This can be achieved by adding bovine serum albumin to the assays. The reaction pH optimum is 7.5 in imidazole buffer.

The presence of phosphate is not important for stability. Omission of ethylenediaminetetraacetic acid considerably improves the stability of the acetyl-CoA carboxylase activity. Furthermore, dialysis overnight results in protein precipitation and large losses of enzyme activity. This is accompanied by dissociation of the multienzyme complex, as can be shown by gel filtration and analytical ultracentrifugation. Although these findings suggest a metal requirement, no specific metal has yet been shown to be important in maintaining the structure of the complex. Repeated freezing and thawing is detrimental to enzyme activity. The best way to store the preparation at all levels of purity is as an ammonium sulfate slurry at 4°.

Acetyl-CoA Carboxylase Kinetics. The effect of changing HCO_3^- concentration on acetyl-CoA carboxylase activity is identical for free enzyme and multienzyme complex. The bicarbonate dependence describes a rectangular hyperbola with a K_m of 4.2–5.2 mM for both physical forms of the enzyme. A sigmoidal relationship is seen between acetyl-CoA concentration and velocity of the reaction catalyzed by the free enzyme.

Phosphoenolpyruvate Carboxylase Reaction. The enzyme requires phosphoenolpyruvate, HCO_3^- and $MgCl_2$. The reaction rate is linear with time (up to 30 min) and enzyme concentration. The activity of the free phosphoenolpyruvate carboxylase is stimulated by acetyl-CoA, NADH, and ATP. (This has not been examined for the enzyme in its multienzyme complex form.) Oxaloacetate is the product of the reaction.

Phosphoenolpyruvate Carboxylase Kinetics. The K_m values for phosphoenolpyruvate are identical (0.9–1.7 mM) for the enzyme in both the free form and the complex. In contrast the K_m for HCO_3^- in the complex is 0.7–1.3 mM, whereas the K_m for the free enzyme is 7.3–9.8 mM. Thus the enzyme in the multienzyme complex is almost an order of magnitude more efficient in incorporating HCO_3^-. This supports the proposed function of the multienzyme complex to capture CO_2 for eventual use by the acetyl-CoA carboxylase component of the complex. NADH increases the V_{max} of the enzyme for phosphoenolpyruvate and HCO_3^- by about 250%, while K_m values for both substrates are unaffected.

Malate Dehydrogenase. The reaction catalyzed by the malate dehydrogenase obtained from the multienzyme complex is specific for NADH. An electrophoretogram at pH 8.4 of *Euglena* soluble extract purified through 0–60% saturation with $(NH_4)_2SO_4$ shows three forms of malate dehydrogenase activity, two positively charged and one negatively charged. The negatively charged isozyme is present in the multienzyme complex.

Dissociation of Multienzyme Complex

Procedure. The enzyme complex obtained from 17.5 g of *Euglena* is purified through DEAE-cellulose chromatography as described earlier

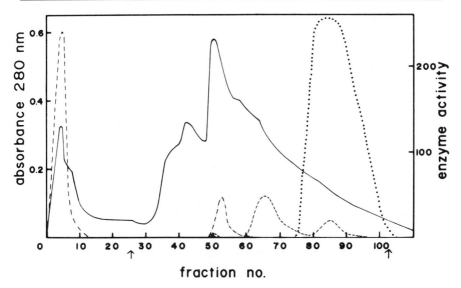

Fig. 4. Dissociation of multienzyme complex by ion exchange chromatography. A DEAE-cellulose column was prepared and operated under the conditions described in the text; the arrows indicate the region of the linear gradient. Fractions of 10 ml were collected and measured for absorbance at 280 nm (solid line), malate dehydrogenase activity (dotted and dashed line, micromoles per milliliter in 15 min), acetyl-CoA carboxylase activity (dashed line, nanomoles per milliliter in 15 min × 10), and phosphoenolpyruvate carboxylase activity (dotted line, nanomoles per milliliter in 15 min). The initial peak of malate dehydrogenase activity contains isozymic forms that are not associated with the multienzyme complex. The second peak, eluted within the gradient, is that from the complex. From Wolpert and Ernst-Fonberg.[2]

and concentrated by precipitation at 60% saturation with $(NH_4)_2SO_4$. The precipitate is dissolved in 10 ml of buffer A and desalted by passage through a Sephadex G-25 column equilibrated and run in the same buffer. The desalted protein solution (22 ml) is clarified by centrifugation (48,000 g, 15 min) and diluted (about four-fold) with 1 mM imidazole, pH 7.0, 7 mM 2-mercaptoethanol until the conductivity equals that of buffer A. The solution, which contains 150 mg of protein, is applied to a DEAE-cellulose column, 2.5 × 10 cm, equilibrated in buffer A. The column is washed with the same buffer until the absorbance at 280 nm is less than 0.05, then an 800 ml linear gradient from an ionic strength of 20 mM (buffer A) to 0.3 M (buffer A which is made 0.28 M in NaCl) is applied. Following the gradient, the column is washed with 100 ml of the higher ionic strength buffer. Fractions of 10 ml are collected at a flow rate of 11 ml/cm² per hour. Aliquots are assayed for malate dehydrogenase, phosphoenolpyruvate carboxylase, and acetyl-CoA carboxylase activities. Fractions containing a given emzyme activity are pooled and precipitated by dialysis against

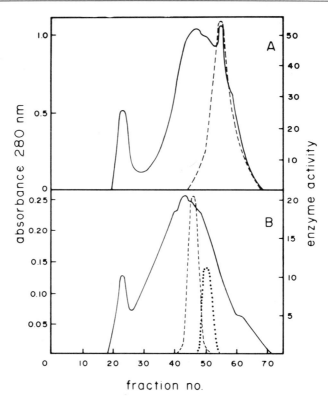

Fig. 5. Gel filtration of the dissociated enzymes of the multienzyme complex. BioGel A-1.5, 2.5 × 30 cm, equilibrated in buffer A which was 0.01 M in NaCl, flow rate 10 ml/cm² per hour. Sample and fraction sizes were 2 ml. The solid line represents absorbance at 280 nm. (A) Malate dehydrogenase from pooled and concentrated DEAE-cellulose fractions 45–60; enzyme activity is shown by dashed line as nanomoles per milliliter in 15 min. (B) Pooled and concentrated fractions 61–80 from DEAE-cellulose chromatography; acetyl-CoA carboxylase is shown by dotted line, and phosphoenolpyruvate carboxylase by dashed line. Both activities are given as nanomoles per milliliter in 15 min. From Wolpert and Ernst-Fonberg.[2]

60% saturated $(NH_4)_2SO_4$ prepared in 0.1 M imidazole, pH 7.0, 7 mM 2-mercaptoethanol. All manipulations are done at 4°, and the enzyme-containing slurries are stored at this temperature.

Properties. Application of the multienzyme complex to DEAE-cellulose under conditions where it is retained leads to its dissociation (Fig. 4). A clean separation of the constituent enzymes is not obtained, but they no longer behave as a single complex of proteins like that seen in Fig. 3. Gel filtration chromatography shows that the conditions of the DEAE-cellulose chromatography result in a separation of the three activities (Fig. 5).

Acknowledgment

Research support is acknowledged from the following sources: Grants-in-Aid from the American Heart Association; National Science Foundation Grants GB-223905, GB-36016X, and PCM-7904062; and Public Health Service Research Career Development Award 5-K04-HD-25169 from the National Institute of Child Health and Human Development (M. L. E.-F.).

[9] Fatty Acid Synthase from Human Liver[1]

By Daniel A. K. Roncari

$$CH_3\text{—}CO\text{—}SCoA + n\ COOH\text{—}CH_2\text{—}CO\text{—}SCoA + 2n\ NADPH$$
$$+ 2n\ H^+ \rightarrow CH_3\text{—}(CH_2CH_2)_n\text{—}COOH + (n + 1)\ CoA\text{—}SH$$
$$+ 2n\ NADP^+ + n\ CO_2 + (n - 1)\ H_2O$$

Assay Methods

Principles. The most convenient method of assaying human liver fatty acid synthase is based on quantifying the decrease in extinction at 340 nm that is associated with the oxidation of NADPH during fatty acid synthesis. This assay is most suitable after the first purification step (ammonium sulfate fractionation). While the spectrophotometric assay is also feasible for crude cell-free fractions, an assay involving the incorporation of $[2-{}^{14}C]$malonyl-CoA into fatty acids extractable with petroleum ether is generally most accurate.

Reagents. Convenient concentrations for the spectrophotometric assay are as follows:

Potassium phosphate buffer, 1.0 M, pH 7.0
Na$_2$EDTA, 0.05 M, pH 7.0
NADPH, 1.1 mM (in Tris-HCl, pH 8.8)
Acetyl-CoA (P-L Biochemicals, Inc.), 1 mM
Malonyl-CoA (P-L Biochemicals, Inc.), 1 mM

---

[1] Supported by grants from the Medical Research Council of Canada (MA-4428) and the Ontario Heart Foundation (1-46).

Convenient concentrations for the radioactive assay are as follows:
NADP$^+$, 0.1 M
Glucose-6-phosphate 0.1 M
Glucose-6-phosphate dehydrogenase (Sigma type XV), 346 units per milligram of protein
Acetyl-CoA, 3 mM
[2–^{14}C]Malonyl-CoA (New England Nuclear Corp.), 3 mM, 1 μCi/μmol

Procedures. For the spectrophotometric assay, potassium phosphate (20 μmol), Na$_2$EDTA (0.2 μmol), DTT^2 (0.2 μmol), NADPH (33 μmol), and acetyl-CoA (12 nmol) are premixed and aliquots are distributed with microsyringes into microcuvettes with a 1-cm light path; the enzyme preparation and water are then added to a volume of 0.18 ml. After 2 min, the reaction is started by adding malonyl-CoA (20 nmol) to a total volume of 0.2 ml. Blanks are devoid of malonyl-CoA. The linear rate of decrease in extinction at 340 nm at 28° is determined using a recording spectrophotometer.

Units. A unit of human liver fatty acid synthase activity is defined as that quantity that catalyzes the malonyl-CoA-dependent oxidation of 1 μmol of NADPH per minute at 28°.

For the radioactive assay, potassium phosphate (100 μmol), Na$_2$EDTA (3 μmol), DTT (1 μmol), acetyl-CoA (19 nmol), and [2 – ^{14}C]malonyl-CoA (40 nmol) are premixed and aliquots are distributed with microsyringes into glass test tubes; the enzyme preparation and water are then added to a volume of 1.97 ml. The reaction is started by adding the "NADPH-generating system," namely, NADP$^+$ (1 μmol), glucose-6-phosphate (2 μmol), and glucose-6-phosphate dehydrogenase (1.8 units) to a total volume of 1.0 ml. Blanks are devoid of the NADPH-generating system. Incubations are conducted in air at 38° for 15 min. The reaction is stopped with 30 μl of 60% perchloric acid, and each mixture is diluted with 1 ml of absolute ethanol. Extraction of fatty acids is carried out with 2 ml of petroleum ether (b.p. 30–60°) three times, the extracts are washed once with water, and after evaporation of the petroleum ether, the residues are dissolved in pentane. Finally, the radioactivity present in aliquots is counted in a liquid scintillation spectrometer.

Units. A unit of enzyme activity is defined as the quantity that catalyzes the incorporation of 1 μmol of radioactive malonyl-CoA into fatty acids extractable with petroleum ether per minute at 38°. For the same amount of enzyme, the value for the unit is one-half of that calculated for the spectrophotometric assay (see equation).

---

^{2} DTT, dithiothreitol.

Purification

Preparation of Cell-Free Fractions. After appropriate legal permission, livers are generally obtained from victims of fatal accidents.[3] The subsequent procedures are carried out at 0–4°. The livers are washed immediately with a solution consisting of 0.1 M potassium phosphate, pH 7.0, 0.25 M sucrose, 1 mM Na$_2$EDTA, and 0.05 M 2-mercaptoethanol. Homogenization with the same solution and a v/w (ml/g) ratio of 4 is carried out with a Waring blender. The homogenate is then centrifuged at either 27,000 g or 100,000 g for 1 hr, the subsequent degree of purification being the same with either supernatant.[3] Purification is generally carried out on the same day, but the supernatant fluid is stable at −75° for several days. However, once purification is begun, it should be completed within the same day.

Ammonium Sulfate Fractionation. This procedure is conducted at 0–4°. The 27,000 g supernatant is stirred gently with a magnet, and solid ammonium sulfate is added slowly to a saturation of 15%. The suspension is stirred for 10 min and then centrifuged at 48,000 g (10,000 g would be adequate) for 10 min. After decanting, solid ammonium sulfate is added slowly to the supernatant to a saturation of 35%. In the case of the 100,000 g supernatant, the 0–15% ammonium sulfate step is not necessary, and one proceeds directly to an ammonium sulfate saturation of 35%. After stirring and centrifuging as described above, the supernatant is decanted and the precipitate is treated as described in the next section.

Calcium Phosphate Gel Adsorption and Elution. This procedure is carried out at 23°. The ammonium sulfate precipitate is dissolved in water containing 0.03 M 2-mercaptoethanol to a conductivity of 2.6 mmho. The subsequent steps are carried out without delay because the synthase is unstable in dilute solutions. Aged calcium phosphate gel (Sigma) is added slowly with gentle stirring, 1 mg of gel per milligram of protein, and stirred further for about 5 min. The slurry is precipitated at 4000 g for 10 min. The synthase is then eluted from the precipitate by gentle stirring with a convenient volume of 0.04 potassium phosphate, pH 7.7, 0.03 M 2-mercaptoethanol.

DEAE-Cellulose Chromatography. This procedure is also carried out at 23°. The calcium phosphate gel eluate is adjusted with a concentrated solution of potassium phosphate, pH 7.0, containing 1 mM DTT, to a conductivity of 5 mmho. Using a peristaltic pump, the solution is then rapidly (for example, at 300 ml/hr) applied to a column of DEAE-cellulose (Whatman DE-52) equilibrated with a solution of potassium phosphate (pH 7.0, conductivity 5 mmho) and 1 mM DTT. About 1 g of total protein

[3] D. A. K. Roncari, *Can. J. Biochem.* **52**, 22 (1974).

is applied to 100 ml of packed DEAE-cellulose. After washing with 3 column volumes of equilibrating solution, proteins including fatty acid synthase, are eluted rapidly with a linear gradient (about 10 column volumes) from 5 to 12 mmho potassium phosphate, pH 7.0, 1 mM DTT. The portion of the eluate containing synthase activity is pooled and, without delay, solid ammonium sulfate is added to 40% saturation. After stirring and centrifugation, the sediment containing purified synthase is suspended in a minimal volume of 0.6 M potassium phosphate, pH 7.0, 0.01 M DTT, and 10% glycerol, and stored under nitrogen at $-75°$. The purification scheme is outlined in Table I.[3]

Purity. After purifying the synthase as described, chromatography on BioGel A-1.5m (Bio-Rad Laboratories) yields a single peak characterized by approximate coincidence of protein concentrations and enzyme activity per unit volume (Fig. 1).[3] Upon analytical ultracentrifugation using schlieren optics, a single symmetrical peak is observed. A single protein band is obtained by polyacrylamide disc gel electrophoresis conducted under alkaline conditions.

On the basis of these findings, the synthase is either homogeneous or highly purified. Its somewhat lower specific activity than that of other mammalian synthases is probably related to the periods of time required to transport the liver to the laboratory.

Properties

Physicochemical Characteristics. On the basis of sedimentation equilibrium studies, the weight-average molecular weight is computed to be 410,000 ± 20,000.[3] Sedimentation velocity studies reveal a sedimentation coefficient ($s_{20,w}$) of 12.0 × 10^{-13} sec.[3]

The content of the prosthetic group, 4′-phosphopantetheine, is determined by analyzing two components of this "acyl-carrying arm," namely, 2-mercaptoethylamine and β-alanine. Performic acid oxidation of the synthase converts 2-mercaptoethylamine to taurine.[3] Amino acid analyses of oxidized synthase reveal 1 mol of taurine per mol of enzyme, whereas analyses of both oxidized and unoxidized enzyme yield 1 mol of β-alanine per mol of enzyme. Thus, the synthase contains one 4′-phosphopantetheine group.

The synthase is quite stable in potassium phosphate solutions of high ionic strength and in the presence of DTT and glycerol. For example, when stored under nitrogen at $-75°$ in 0.6 M potassium phosphate, pH 7.0, 0.01 M DTT, and 10% glycerol, it loses only 30% of its activity after 1 year, and 64% after 7.5 years.

When the enzyme is kept in solutions of relatively low ionic strength

TABLE I

PURIFICATION OF HUMAN LIVER FATTY ACID SYNTHASE[a]

Fraction	Total protein (mg)	Total enzymatic activity (nmol NADPH oxidized/min)	Specific activity (nmol NADPH oxidized/min per milligram protein at 28°)	Purification factor	Yield (%)
Supernatant, 27,000 g	4200	3580	0.9	—	—
$(NH_4)_2SO_4$, 15–35% saturation	1125	3584	3.2	3.6	100
Calcium phosphate gel eluate	120	1470	12.3	14	41
$(NH_4)_2SO_4$, 0–40% saturation, fraction of DEAE-cellulose eluate	2.2	890	404	450	25

[a] Data from Roncari.[3]

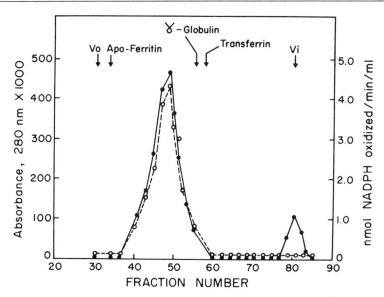

FIG. 1. Gel filtration chromatography of purified fatty acid synthase (FAS) on an agarose column. Four milligrams of FAS in 0.97 ml were applied onto a 0.9 × 90 cm agarose (BioGel A-1.5m) column equilibrated and eluted at 23° with a solution consisting of 0.5 M potassium phosphate, pH 7.0, and 2 mM dithiothreitol. The column was eluted by gravity with a hydrostatic pressure head of 20 cm, and 0.7-ml fractions were collected. The standard proteins were used in separate experiments. ●——●, Absorbance; ○---○, nanomoles of NADPH oxidized per minute per milliliter. From Roncari,[3] with permission.

(for example, less than 0.25 M potassium phosphate), it dissociates into two subunits of very similar, but not identical, molecular weight; as already indicated, only one of the subunits has the 4′-phosphopantetheine group. Dissociation is much more rapid in the absence of DTT and glycerol. Denaturants and detergents also effect dissociation of the synthase. During dissociation, particularly with sodium dodecyl sulfate, great care must be taken to prevent proteolysis; otherwise, numerous fragments of varying molecular weight will appear.

Specificity. The synthase displays considerable preference for NADPH (for which it has a much lower K_m and higher V_{max}), as compared to NADH.[3]

The primer ("starter") substrate specificity of the synthase is outlined in Table II.[3,4] *In vitro,* butyryl-CoA and, to a lesser extent, propionyl-CoA are significantly better substrates than acetyl-CoA. Hexanoyl-CoA is a poor primer.

[4] D. A. K. Roncari and E. Y. W. Mack, *Can. J. Biochem.* **54**, 923 (1976).

TABLE II

PRIMER SPECIFICITY OF HUMAN LIVER FATTY ACID SYNTHASE[a]

Primer[b]	NADPH oxidized (nmol/min)
Acetyl-CoA	1.00 ± 0.02[c]
Propionyl-CoA	1.32 ± 0.11
Butyryl-CoA	1.67 ± 0.04
Hexanoyl-CoA	0.36 ± 0.04

[a] Data are from Roncari and Mack[4] and from Roncari.[3]
[b] The concentration of each thioester was 0.03 mM.
[c] Mean $\pm$ SEM. For any pair of thioesters, $p < 0.005$.

Using the usual assay conditions, which include acetyl-CoA, palmitic acid is virtually the only long-chain fatty acid product.[4]

Activation and Inhibition

In vitro, the synthase is activated by incubation with phosphorylated sugars, the most effective being D-fructose-1,6-diphosphate.[5] It decreases the K_m of the enzyme for NADPH. Malonyl-CoA does not inhibit synthase activity up to the highest concentration tested (0.6 mM).[5] *In vitro*, methylmalonyl-CoA inhibits synthase activity competitively with respect to malonyl-CoA (K_i 8.4 μM).[4]

[5] D. A. K. Roncari, *Can. J. Biochem.* **53**, 135 (1975).

[10] Fatty Acid Synthase from Pig Liver[1]

By I. C. KIM, GARY NEUDAHL, and WILLIAM C. DEAL, JR.

$$CH_3{-}CO{-}SCoA + n\ HOOC{-}CH_2{-}CO{-}SCoA$$
$$+ 2n\ NADPH + 2n\ H^+ \rightarrow CH_3{-}(CH_2{-}CH_2)_n{-}CO{-}SCoA$$
$$+ n\ CoASH + n\ CO_2 + 2n\ NADP^+ + n\ H_2O$$

Assay Methods

Two spectrophotometric assay procedures will be described; both follow the change in absorbance at 340 nm to measure the rate of disappear-

[1] This research was supported in part by grants from the NIH (AM20523) and the Michigan Agricultural Experiment Station (Hatch 1273, Publication No. 8772).

ance of NADPH. In the overall reaction assay, acetyl-CoA and malonyl-CoA are the substrates. In the decalone reductase partial reaction assay,[2] *trans*-1-decalone is the substrate; it is a substrate analog model compound. Dutler *et al.*[3] first reported the reduction of alicyclic ketones by fatty acid synthase. We routinely use the decalone reductase assay because it is at least 15 times more sensitive than the standard overall reaction assay and because it prefers NADPH over NADH, which gives it desirable specificity. In the reagent lists, the final assay concentrations are given in parentheses.

Overall Reaction Assay

Reagents

Potassium phosphate, 0.4 M, pH 6.9 (0.1 M)
Acetyl-CoA, 0.4 mM (50 μM)
Malonyl-CoA, 0.4 mM (50 μM)
NADPH, 4 mM (0.2 mM)
Bovine serum albumin, 8 mg/ml (1 mg/ml)
Enzyme (40–100 μg/ml)

The reaction is initiated by the addition of malonyl-CoA, and the activity is measured from the initial slope of the absorbance curve. A unit of activity is the amount of enzyme that catalyzes the oxidation of 1 μmol of NADPH per minute. Specific activity is expressed as units per milligram of protein.

Decalone Reductase Partial Reaction Assay[2]

Reagents

Potassium phosphate, 0.4 M, pH 6.1 (0.1 M)
trans-1-Decalone, 100 mM, in ethanol (10 mM)
NADPH, 1 mM (50 μM)
Enzyme (4 μg/ml)

The reaction is initiated by addition of *trans*-1-decalone, and the activity is measured from the initial slope of the absorbance curve. Activity units are as described above. For solutions of high ionic strength, it is desirable to have a final concentration of approximately 10% ethanol and 1 mM *trans*-1-decalone, to avoid precipitation.

[2] I. C. Kim, C. J. Unkefer, and W. C. Deal, Jr., *Arch. Biochem. Biophys.* **178**, 475 (1977).
[3] H. Dutler, M. J. Coon, A. Kull, H. Vogel, G. Waldvoge, and V. Prelog, *Eur. J. Biochem.* **22**, 203.

Purification of the Enzyme[2]

Unless indicated otherwise, all purification steps were carried out at 4°
and centrifugation was performed in the Sorvall GSA rotor at 11,000 rpm
for 25 min or in the Sorvall SS-24 rotor at 16,000 rpm for 15 min.

Step 1. Clarified Crude Extract. Frozen pig liver (800 g) was cut into
small pieces with a stainless steel knife and homogenized for 1 min at
maximum speed in a Waring blender in two volumes (1600 ml) of cold 0.05
M Tris-HCl buffer, pH 7.4, containing 1 mM EDTA. The homogenized
solution was centrifuged in a GSA rotor. The supernatant solution was
further centrifuged in polyallomer tubes in a Beckman Model L3-50 preparative ultracentrifuge at 21,000 rpm for 1 hr using a Beckman type 21
rotor. A clear, dark-red supernatant was obtained (1200 ml); this is designated the "clarified crude extract." It was always prepared fresh.

Step 2. Ammonium Sulfate Precipitation. Saturated ammonium sulfate
(25 ml/100 ml of extract), pH 7.4, was added to the clarified crude extract
to 0.2 saturation, gently stirred for 15 min, and centrifuged in a GSA rotor.
The small amount of precipitated protein was discarded. The supernatant
solutions were pooled. Additional saturated ammonium sulfate solution
was added (20 ml/100 ml of supernatant) to bring the supernatant solution
to 0.33 saturation, gently stirred for 15 min, and centrifuged in a GSA
rotor. The pellet was collected and suspended in the same volume of
homogenization buffer as the original clarified crude extract (about 1200
ml). The recovery of decalone reductase activity averaged about 65–75%
up to this point.

Saturated ammonium sulfate solution (50 ml/100 ml of solution) was
added to 0.33 saturation, and the solution was stirred gently for 15 min and
then centrifuged in the GSA rotor. The precipitate was resuspended in 200
ml of the homogenization buffer and dialyzed at 4° overnight against 4
liters of TED[4] buffer with one buffer change. The decalone reductase
activity was stable during overnight dialysis at 4° under these conditions;
also, the enzyme activity was quite stable even at room temperature if the
buffer contained at least 1 mM dithiothreitol. Very little decalone reductase activity is lost in these repeated ammonium sulfate precipitations,
and these steps remove a considerable amount of foreign protein and
reduce the amount of DEAE required in later steps.

The dialyzed enzyme was centrifuged in a SS-34 rotor to remove denatured protein, diluted to 1000 ml with TED buffer at room temperature,
and applied to DEAE-cellulose columns.

---

[4] Abbreviations: TED, buffer containing 0.05 M Tris-HCl, pH 7.4, 1 mM EDTA, and 1 mM
  dithiothreitol; PED, buffer containing 0.2 M potassium phosphate, pH 7.4, 1 mM EDTA,
  and 1 mM dithiothreitol; DTNB, 5,5'-dithiobis(2-nitrobenzoic acid).

Step 3. DEAE-Cellulose Column Chromatography. This step was carried out at 23°. Two columns (4.1 × 33 cm) of DEAE-cellulose were preequilibrated with TED buffer, and their outlet tubes were connected to two separate flow cells of an Isco Model UA-5 absorbance monitor; then 500 ml of diluted dialyzed enzyme were applied to each of the columns. About 300 ml of TED buffer were then applied to each of the columns. A linear KCl gradient ranging from 0 to 0.23 M, in TED buffer, was used to elute the enzyme. The total volume of the gradient (3.1 liters, for each column) was approximately seven times the bed volume (440 ml) of each DEAE-cellulose column. Then 300 ml of 0.5 M KCl in TED was applied to each column.

The eluted fractions were collected using an Isco fraction collector (Model 328). The elution of protein was followed by analysis of absorbance at 280 nm using an Isco UA-5 absorbance monitor and recorder. The flow rate was controlled at about 7 ml/min by keeping the bed head between 55 and 70 cm.

Decalone reductase starts to appear at about 0.09 M KCl, and the peak of enzyme elution is approximately at 0.13 M KCl. This DEAE-cellulose step is the most important step for the removal of catalase. This step yields about 50% pure enzyme. Fresh, unused, acid-base-washed DEAE-cellulose was found to give better capacity, resolution, and reproducibility. The capacity of DEAE-cellulose is at least twofold lower at 4° than at room temperature. At least 1 mM dithiothreitiol should be present in the gradient to maintain enzyme activity. With all other conditions the same, replacing 1 mM dithiothreitol by 10 mM mercaptoethanol markedly decreases (as much as twofold) the capacity of the DEAE-cellulose column.

The 0 to 0.23 M KCl gradient fractions which had decalone reductase activity were pooled, cooled to 4°, and kept there until the next step. The peak of activity eluted with 0.5 M KCl was pooled separately from the 0 to 0.23 M KCl fractions and kept separate through the later steps. Saturated ammonium sulfate solution was added to 0.33 saturation; the solution was stirred gently for 15 min, then centrifuged in a GSA rotor. The resulting yellow pellet was suspended in 5 mM potassium phosphate buffer, pH 7.2, and desalted, in two separate batches, on a Sephadex G-25 column (2.4 × 35 cm) at room temperature. This preparation is bright yellow. This yellow color is due to a hemoprotein; we have purified and characterized this hemoprotein.[5]

Step 4. Calcium Phosphate Gel Treatment. This step was performed at 23°. A suspension of calcium phosphate gel at a concentration of 28 mg/ml was added (2 mg of gel per milligram of protein) to the above desalted

---

[5] I. C. Kim and W. C. Deal, Jr., *Biochemistry* 15, 4925 (1976).

preparation. After stirring gently for about 3 min, this suspension was centrifuged for about 7 min at 16,000 rpm at 4° in a SS-34 rotor. The supernatant solution was brought quickly to 0.5 saturation by adding saturated ammonium sulfate solution, stirred for 15 min, and centrifuged in a SS-34 rotor. This step should be carried out as soon as possible in order to avoid inactivation of the enzyme. The pellet was suspended in a minimal volume of PED[4] buffer and dialyzed against 500 ml of PED buffer for at least 2 hr at 4°.

The calcium phosphate gel step is used solely to remove a yellow substance, a hemoprotein.[5] This step is critical for getting a colorless preparation of fatty acid synthase. Lower ratios of calcium phosphate gel to protein are not effective in removing the yellow substance, and higher ratios decrease the yield.

Step 5. Sucrose Density Gradient Centrifugation. This step was done at 23°. The enzyme solution from the preceding step was diluted to 9 ml. Then 1.5 ml (concentration < 10 mg/ml) was applied to each of six sucrose density gradient tubes (1 × 3.5 in.). The enzyme was then centrifuged at 20° for 18 hr at 27,000 rpm in a Beckman Model L3-50.

There were two protein peaks: a faster moving, 13 S fatty acid synthase, peak; and a slower moving 7 S, peak. The resolution and relative areas under the curves of these two peaks varied from experiment to experiment. We cannot exclude the possibility that some or all of the 7 S

TABLE I

PURIFICATION OF FATTY ACID SYNTHASE (FAS) FROM PIG LIVER (800 g)[a,b]

Fraction	Total volume (ml)	Total protein (ml)	Total units (units)	Specific activity[c] (units/mg)	Yield (%)
Clarified crude extract	1190	70091	1850	0.026	100
(NH₄)₂SO₄-dialyzed enzyme	1000	6300	1290	0.205	69.7
DEAE-cellulose G-25 step (before calcium phosphate step)	100	470	710	1.51	38.4
Enzyme after calcium phosphate treatment	130	152.5	692	4.54	37.4
FAS (after sucrose density gradient[d]	9	89.1	622.7	6.99[d]	33.7

[a] From Kim *et al.*[2]
[b] See text for details.
[c] Based on the *partial reaction* assay of decalone reductase activity, not *overall* fatty acid synthase activity.
[d] The specific activity for the *overall fatty acid synthase reaction* was 0.28 unit per milligram of protein for this pure enzyme.

peak material may be related to fatty acid synthase in some way, such as being a precursor or degradation or dissociation product. Further evaluation of this possibility is needed.

The separation between the two peaks depended on the amount of enzyme applied to each tube. When the resolution between the two peaks was poor, the sucrose density gradient step was repeated a second time to completely remove the 7 S impurity.

The 13 S fractions which had activity were pooled and precipitated at 4° by addition of saturated ammonium sulfate solution to 0.5 saturation. After centrifugation in a SS-34 rotor, the white pellet was suspended in a small volume (about 4 ml) of PED buffer and dialyzed at 4° overnight against 500 ml of PED buffer, with one change of buffer. After centrifugation in a SS-34 rotor to remove a slight amount of precipitated protein, the homogeneous colorless, fatty acid synthase solution was stored at 4°.

The purification steps are summarized in Table I.

Properties of the Synthase

Physical Properties. Schlieren patterns from sedimentation velocity experiments in the analytical Model E ultracentrifuge showed only a

TABLE II
PHYSICOCHEMICAL PROPERTIES OF PIG LIVER FATTY ACID SYNTHASE[a,b]

Parameters	Values
Molecular weight (from $s_{20,w}^{\circ}$ and $D_{20,w}^{\circ}$)	478,000[c]
Molecular weight (from sedimentation equilibrium)	476,000[d]
$s_{20,w}^{\circ}$	13.30 S
$D_{20,w}^{\circ}$	2.60 F[e]
f_s, frictional coefficient (from s)	1.55×10^{-7}
f_D, frictional coefficient (from D)	1.56×10^{-7}
r, Stokes' radius	82.4 Å
f/f_0, frictional ratio	1.58
$[\eta]$, intrinsic viscosity	7.3 ml/g
SH groups per molecule (assuming $MW = 478,000$)	About 90
$E_{280\,nm}^{0.1\%}$, extinction coefficient	1.23
β, Scheraga–Mandelkern shape factor	2.14×10^6

[a] From Kim *et al.*[2]
[b] See text for details.
[c] Calculated from the Svedberg equation using the s and D values listed.
[d] Calculated from meniscus depletion sedimentation equilibrium experiments.
[e] The units of F (Fick) are 10^{-7} cm² s⁻¹.

single Gaussian peak. This is consistent with, but not proof of, homogeneity. Table II summarizes the physicochemical properties of the enzyme.

Absorption Spectra and Fluorescence Spectra. Pig liver fatty acid synthase does not absorb in the visible range from 475 nm down to about 370 nm, but in the ultraviolet range it has a major absorption peak at 280 nm and a shoulder at 290 nm, also noted by Dutler *et al.*[3]

The enzyme showed a typical tryptophan fluorescence peak at 344 nm using various excitation wavelengths ranging from 260 nm to 300 nm. The fluorescence of tyrosine was apparently too weak to show. Maximal fluorescence occurred with an excitation wavelength of 285 nm, followed by 290 nm, and then by 280 nm.

Sulfhydryl Content and Optimum pH. Spectrophotometric titration with DTNB[4] indicated the presence of about 90 SH groups per molecule of enzyme, assuming a molecular weight of 478,000. In 0.1 M potassium phosphate buffer, a pH optimum ranging from 6.5 to 6.8 was observed.

Stability and Storage. In stability tests the decalone reductase activity was completely stable for 11 days in all solutions tested (Table III). Over the first 7 days, the overall fatty acid synthase activity was completely stable in either 0.2 mM NADPH or 25% sucrose; solutions lacking either NADPH or sucrose dropped about 13–16%. After 11 days, full activity was retained only in the solution containing 25% sucrose; the activity dropped 20% in solutions containing 0.2 mM NADPH and about 33% in the solutions containing no NADPH and no sucrose (Table III). Other experiments showed that 0.2 mM NADH had no effect on the stability.

TABLE III
STABILITY OF FATTY ACID SYNTHASE[a]

Addition	Activity[b]	Days					
		0	3	5	7	11	15
No addition	FAS	0.31	0.32	0.32	0.27	0.21	0.24
	DR	4.80	4.80	4.70	4.80	4.30	4.89
NADPH, 0.2 mM	FAS	0.35	0.36	0.36	0.36	0.29	0.30
	DR	4.04	4.13	4.94	4.63	4.32	4.4
Sucrose, 25%	FAS	0.33	0.32	0.35	0.38	0.34	0.32
	DR	4.97	4.18	4.3	4.36	4.67	4.73

[a] Fatty acid synthase solutions (3 mg/ml) were kept at 4° in 0.1 M potassium phosphate, pH 7.4, containing 1 mM EDTA and 1 mM dithiothreitol and assayed at the indicated times. All values are specific activities. See text for further details. Decalone reductase activity was measured at pH 6.9.

[b] FAS, fatty acid synthase; DR, *trans*-1-decalone reductase.

[11] Fatty Acid Synthase from Lactating Bovine Mammary Gland[1]

By Soma Kumar and Peter F. Dodds[2]

Fatty acid synthase of lactating bovine mammary gland is a polyfunctional enzyme complex similar to the enzyme found in other animal tissues and certain microorganisms. It catalyzes the synthesis of varying amounts of fatty acids containing an even number of carbon atoms and between 4 and 18 carbon atoms in length. Using acetyl-CoA as the primer, the predominant products are butyric and palmitic acids. The overall stoichiometry of this reaction can be represented by the equation:

$$CH_3-\underset{O}{\underset{\|}{C}}-SCoA + n\ \underset{O}{\underset{\|}{C}}H_2-\overset{COO^-}{\underset{\|}{C}}-SCoA + 2n\ NADPH + 2nH^+$$

$$\downarrow$$

$$CH_3-CH_2-(CH_2CH_2)_{n-1}-CH_2-COOH + 2n\ NADP^+ + n\ CO_2 + (n-1)\ H_2O$$
$$+ (n+1)\ CoA$$

(1)

This enzyme, unlike the one from yeast[3] and pigeon liver,[4] is able to catalyze two of the partial reactions involved in the synthesis of fatty acids at high rates using normal concentrations of substrates. These reactions are the reduction of acetoacetyl-CoA and of *trans*-crotonyl-CoA. The procedures used for the assay of the overall reaction and of these partial reactions are described.

Substrates. All the substrates required for the assay of the various reactions catalyzed by the bovine mammary fatty acid synthase are currently available commercially.

Fatty Acid Synthesizing Activity

Assay Methods

Spectrophotometric Assay. The overall fatty acid synthesizing activity is most conveniently assayed by following the decrease in the absorbance

[1] Work reported here was carried out with support received from the National Institute of Arthritis, Metabolism and Digestive Diseases, National Institutes of Health, in the form of various grants, including the current AM16086.
[2] Current address: Department of Biochemistry and Soil Science, Wye College, Near Ashford, Kent TN25 5AH, England.
[3] F. Lynen, *Fed. Proc.* **20**, 941 (1961).
[4] S. J. Wakil, *J. Lipid Res.* **2**, 1 (1961).

of NADPH at 340 nm. The assay mixture contains, in 1 ml:

Potassium phosphate buffer, 100 mM pH 7.0
EDTA, 100 μM
Dithiothreitol, 1 mM
NADPH, 125 μM
Acetyl-CoA, 25 μM
Malonyl-CoA, 50 μM
Pure fatty acid synthase, 5 μg

The mixture, without acetyl-CoA and malonyl-CoA, is preincubated at 37° for 8 min. Acetyl-CoA is then added, and a background rate of NADPH oxidation is established. After 2 min, malonyl-CoA is added to start the reaction and the rate of NADPH oxidation is calculated from change in absorbance assuming $\epsilon_{340} = 6.22 \times 10^3$; 25 μM butyryl-CoA may be substituted for acetyl-CoA. One unit of enzymatic activity is the quantity of enzyme that oxidizes 1 nmol of NADPH per minute.

Radioactive Assay. The synthesis of radioactive fatty acids from a radioactive primer, generally acetyl-CoA, or malonyl-CoA can also be used to assay the activity of the enzyme. As we frequently use primers that are not commercially available with a suitable radioactive label, [2−^{14}C]malonyl-CoA is routinely used as the radioactive precursor, although [1−^{14}C]acetyl-CoA of a higher specific radioactivity can also be used without any modification of the procedure. Assays contain, in a volume of 0.5 ml, the following components:

Potassium phosphate buffer, 100 mM pH 7.0
EDTA, 100 μM
Dithiothreitol, 5 mM
NADPH, 5 mM
Purified fatty acid synthase, 12–15 μg
Acetyl-CoA as primer, 0.5 mM
[2−^{14}C]Malonyl-CoA (0.1 μCi/μmol), 1 mM

The mixture with the exception of the acetyl-CoA and malonyl-CoA is allowed to preincubate at 37° for 10 min before the reaction is started by the almost simultaneous addition of both acetyl-CoA and malonyl-CoA. After 12 min the reaction is stopped by the addition of 50 μl of 10 M KOH. It is then heated to 95° for 2 hr. This treatment prevents the binding of the long-chain acids to the protein, thereby reducing the efficiency of their extraction. When crude enzyme fractions are used, this digestion step is extended to at least 6 hr. Since the products of the reaction contain considerable amounts of butyrate using acetyl-CoA a primer, an extraction procedure is used that is designed to extract quantitatively butyric acid as well as the longer-chain acids.

Extraction of Fatty Acids. Ten milligrams of carrier n-butyric acid in 0.1 ml of n-hexane is added to the reaction mixture followed by 100 μl of 5 M

H_2SO_4. The acids liberated are extracted four times, using 5-ml aliquots of diethyl ether. The ether extracts are added to a test tube containing 100 μl of 10 M KOH. The contents are gently evaporated to dryness under a stream of air at 40°. The residue is acidified with 150 μl of 5 M H_2SO_4. Anhydrous Na_2SO_4 is added in small amounts just sufficient to make the contents of the tube into a thin paste. It is then extracted using three 1-ml aliquots of *n*-butanol–hexane (1 : 99, v/v).

Fractionation of the Fatty Acids. The butanol–hexane extract is fractionated into *n*-butyric acid and a fraction containing *n*-hexanoic and longer-chain acids on a silicic acid column originally described for the isolation of butyric acid from butterfat.[5] For this, 5 g of silicic acid (Mallinckrodt AR powder, prepared for chromatography by the method of Ramsey and Patterson) is mixed in a mortar with small aliquots of 2–3 ml of an ammoniacal solution of bromocresol green in ethylene glycol prepared as follows: 700 mg of bromocresol green is dissolved in 700 ml of ethylene glycol by warming on a steam bath. To the cooled solution, 200 ml of H_2O and 40 ml of 0.1 N NH_4OH are added; the volume is made up to 1 liter and stored tightly closed in a dark brown bottle. Silicic acid prepared for chromatography acquires an olive green color and should be saturated with the ethylene glycol solution while retaining a powdery consistency. The quantity of the bromocresol green solution needed for this varies from batch to batch of the silicic acid. The powder is then exposed to NH_3 vapor and mixed with the pestle until the color changes to faint blue-green. Sufficient butanol–hexane (1 : 99) is added to the powder to obtain a darkish green slurry, which is poured into a chromatographic column (1 cm in diameter) plugged with glass wool. After the column is packed, the fatty acid extract is layered carefully at the top of the column, allowed to flow in, and eluted with butanol–hexane. Butyric acid appears as a clear yellow band and moves slowly down the column. The fore-run, the eluate emerging before the yellow band, containing hexanoic and longer acids, and the yellow butyric acid band are collected separately and counted for radioactivity by liquid scintillation spectrometry. (Recovery of butyric acid after extraction and fractionation is 92–98%.) This procedure serves not only to determine the extent of formation of butyric acid, it also removes the radioactive acetic acid formed from acetyl-CoA in the event that this is the labeled compound used. Using this procedure, one observes considerable formation of butyric acid not only by the fatty acid synthase of cow mammary gland, but by the enzyme from lactating mammary gland of goat, rabbit, and rat, as well as the enzyme from rat liver.

[5] M. Keeney, *J. Assoc. Off. Agr. Chem.* **39**, 212 (1956).

Analysis of the Products by Radio-Gas Chromatography. This method is used to determine the amount of radioactivity from $[2-^{14}C]$malonyl-CoA incorporated into individual fatty acid with chain lengths between $C_{4:0}$ and $C_{18:0}$. The reaction is carried out as described above, except that the specific radioactivity of $[2-^{14}C]$malonyl-CoA should be 1.0 $\mu Ci/\mu mol$. Before the alkaline mixture is heated, 10 μl each of solutions of known composition of $C_{4:0}$ and $C_{13:0}$ and of all the even-chain acids between $C_{8:0}$ and $C_{18:0}$ are added to act as carriers. The acids, after acidification, are extracted four times using 2 ml aliquots of *n*-pentane, and the pooled extracts are adjusted to 10 ml. One milliliter of this is used for scintillation counting without prior evaporation. Malonic acid is not extracted in this solvent. Based upon the radioactivity present in the extract a suitable volume is transferred to a conical tube with its bottom drawn out to a fine point. The volume of the extract is reduced very slowly to about 10 μl under a stream of dry nitrogen, the tubes being kept in a salt–ice bath at $-10°$. The extract is drawn into a 50-μl gas-tight syringe. The sides of the tube are washed with about 50 μl of *n*-pentane-diethyl ether (1:1, v/v) which is also reduced to about 10 μl and drawn into the same syringe. The combined extracts of free acids is then used for radio–gas chromatography. The conditions for gas–liquid chromatography are as follows:

Column 2 m × 2 mm (internal diameter) glass

Packing, 10% SP-216-PS on 100/120 mesh Supelcoport (Supelco)

Injector temperature, 190°

Column temperature, 130° for 2 min, increasing at 8°/min and held at 195°

Detector temperature, 275°

Carrier gas, helium at 30 ml/min

The effluent from the column is split roughly in the ratio 10:1, the minor portion going to the hydrogen flame ionization detector and the major portion along a heated (275°) transfer line to a gas flow proportional counter where the fatty acids are oxidized in a CuO furnace at 750° to water and $^{14}CO_2$. The latter is counted in a gas flow proportional counter.[6] The loss of the various short-chain acids relative to $C_{13:0}$ is determined from the mass trace, and appropriate corrections are applied to the radioactive peaks. Using this method, the proportion of ^{14}C found in $C_{4:0}$ is in very close agreement with results obtained using silica gel–ethylene glycol partition chromatography. Fractionation of free acids rather than their methyl esters has the advantage of there being fewer steps in the analysis and less loss due to the evaporation of $C_{4:0}$ and $C_{6:0}$ acids. We recover up to 45% of $C_{4:0}$ and 95% of $C_{6:0}$ acid relative to $C_{13:0}$ acid. The chromatog-

[6] C. R. Strong, I. Forsyth, and R. Dils, *Biochem. J.* **128**, 509 (1972).

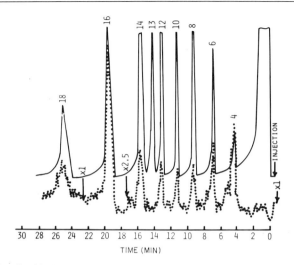

FIG. 1. Gas–liquid chromatography of fatty acids synthesized by fatty acid synthase using acetyl-CoA and [2–^{14}C]malonyl-CoA in the assay mixture described in the text. The column length was 180 cm. The oven was held at 115° for 2 min and was programmed to increase at 8°/min to a final temperature of 195°. The rate of flow of helium was 20 ml/min. The solid line represents the mass of the fatty acids, and the dotted line their ^{14}C content. The numbers refer to fatty acids containing the corresponding number of carbon atoms.

raphy of free acids generally presents problems due to their polarity, which results in their tailing. This can be minimized by the use of a very polar stationary phase, employing phosphoric acid as a tail reducer, and by ensuring that the acids come into contact with only the least polar surfaces possible. For this reason, we use a glass-lined stainless steel transfer line with all the glass surfaces siliconized. The joints at the ends of the column are of Teflon, and those exposed to higher temperature are of nickel-200. Nickel-200 is also used for the constricted tube leading from the splitter to the flame ionization detector. Figure 1 shows a radio–gas chromatogram that was obtained under conditions slightly different from those described above.

Purification of the Enzyme

These methods have been described, in part, in previous publications from our laboratory.[7,8]

[7] S. K. Maitra and S. Kumar, *J. Biol. Chem.* **249**, 111 (1974).
[8] K. A. Strom, W. L. Galeos, L. A. Davidson, and S. Kumar, *J. ∴ol. Chem.* **254**, 8153 (1979).

Preparation of the Particle-Free Supernatant. The tissue is obtained from a slaughterhouse. It has been our experience that the best enzyme preparations result from the mammary tissue of animals carrying the most milk. Immediately after the cow is slaughtered and the udder removed, pieces weighing approximately 100 g are removed from the most secretory parts of the gland and placed in heavy duty polyethylene bags. These are frozen immediately on Dry Ice in lots of 0.5–1 kg and brought to the laboratory, where they are stored at −80°. Storage for as long as 6 months shows no loss of enzyme activity.

All solutions used in the preparation of the enzyme contain 1 mM dithiothreitol and 0.1 mM EDTA; unless specified otherwise, all operations are carried out at 0–4°. The tissue from each lot is cut into small pieces using a sharp knife or Stadie–Riggs blade before it has completely thawed and then is freed from connective tissue. The pieces are rinsed with 0.25 M sucrose until free of milk and then blended with two volumes of 0.25 M sucrose using a top drive Omnimixer (Ivan Sorvall, Inc.). The resulting suspension is homogenized using a Potter–Elvehjem homogenizer with a Teflon pestle. The homogenate is filtered through glass wool and centrifuged at 13,000 g,[9] filtered again through glass wool, and then centrifuged at 125,000 g for 45 min. When this spin is complete, the supernatant is filtered through glass wool once more to obtain the particle-free supernatant (PFS).

Precipitation and Adsorption Steps. The next step is the collection of the protein that precipitates in 33⅓% saturated (by volume) ammonium sulfate. Some workers preparing fatty acid synthase from other sources first discard the protein, containing acetyl-CoA carboxylase, that precipitates from a 25% saturated solution of ammonium sulfate. This procedure was of little benefit in our preparations and is therefore omitted.

With constant stirring of the PFS a saturated solution of ammonium sulfate is added, a little at a time, until a final concentration of 33⅓% saturation is achieved. The stirring is continued for another 20 min. The precipitate is collected by centrifugation at 6000 g for 15 min and stored overnight, if necessary, at 0–5° in an atmosphere of nitrogen.

The next two steps require the use of buffers of low ionic strength in which fatty acid synthase dissociates into its two component subunits. They are, therefore, performed as rapidly as possible. The ammonium sulfate precipitate is dissolved in 2.5 mM potassium phosphate buffer, pH 6.8, and adjusted to a protein concentration of 2–4 mg/ml. Calcium phosphate gel is added with constant mixing to the protein solution to obtain a ratio of 1 : 2 (gel : protein, by weight). They are allowed to mix for 30 sec and immediately centrifuged at 6000 g for 5 min. The fatty acid synthase

[9] All gravitational forces are average values.

remains in the supernatant that is collected. Although the calcium phosphate adsorption step results in little purification of the enzyme per se (Table) it has been found to be necessary for the eventual success of the enzyme purification.

The supernatant from the calcium phosphate gel is added with stirring to the appropriate amount of aged alumina Cγ gel to obtain a gel : protein ratio of 2 : 1, by weight. This is stirred for 20 min before centrifugation at 6000 g for 5 min. Fatty acid synthase, which is adsorbed onto the gel, is collected and the supernatant is discarded. The enzyme is extracted from the gel with 0.25 M potassium phosphate buffer, using a volume equal to the volume of the starting PFS. The mixture is stirred with a magnetic stirrer for 20 min, and the gel is reextracted using half the volume of the same buffer, the extracts are pooled.

The protein is reprecipitated from the extract saturated with ammonium sulfate at $33\frac{1}{3}\%$ saturation as already described. This precipitate (second ammonium sulfate precipitate) may be stored overnight at 0–5° in an atmosphere of N$_2$ or for longer periods in liquid N$_2$. At this stage and in the subsequent stages of purification, the enzyme needs to be handled with great care because it is very susceptible to denaturation. Agitation of its solution and exposure to scratched or sharp glass surfaces are scrupulously avoided, and exposure to oxygen is kept to a minimum by the use of only deaerated solutions.

Gel Filtration Chromatography. We obtained a substantial improvement in the purity of the preparation when we replaced Sephadex G-200 (Pharmacia)[7] with Ultrogel AcA22 (LKB)[8] which has a larger exclusion limit for molecular sieving, gives better flow rates, and is less sensitive to hydrostatic pressure. A column of between 600 ml and 800 ml volume (30 to 40 × 5 cm) of Ultrogel AcA22 is packed and equilibrated with 0.1 M potassium phosphate buffer, pH 7.0, which was previously deaerated under vacuum. Typically, flow rates of 150–180 ml/hr are obtained at room temperature, the condition under which the column is run. The second ammonium sulfate is redissolved in a minimum volume of 0.1 M potassium phosphate buffer, pH 7.0, and centrifuged at 40,000 g for 10 min to remove the undissolved or denatured protein. This solution (10 mg of protein per milliliter) in a volume 3–5% of that of the gel volume is allowed to flow into the top of the gel and is eluted by descending chromatography using the buffer used for equilibration, which is kept under an atmosphere of nitrogen. The eluate is continuously monitored for ultraviolet absorption at 278 nm and is collected in 12-ml fractions in tubes kept in a refrigerated bath. The enzyme is eluted at a volume equal to 0.65 times that of the bed volume of the column. The fractions having the highest fatty acid synthase activity are pooled.

FATTY ACID SYNTHASE AND THE REDUCTASE ACTIVITIES DURING PURIFICATION OF BOVINE MAMMARY ENZYME

Step	Volume (ml)	Protein (mg/ml)	Specific activity (units/mg)	Relative purification	Recovery (%)	Reductase activities (nmol NADPH oxidized min^{-1} mg^{-1})	
						Acetoacetyl-CoA	*trans*-Crotonyl-CoA
1. Particle-free supernatant	1610	13.7	26.1	1.0	100	140	5.6
2. First (NH$_4$)$_2$SO$_4$ precipitate	1150	6.8	61.0	2.3	86	219	43.6
3. Calcium phosphate gel supernatant	1950	3.0	75.4	2.9	76	240	75.4
4. Alumina Cγ extract	2700	0.52	239.0	9.2	67	765	299
5. Second (NH$_4$)$_2$SO$_4$ precipitate	60	16.7	370.0	14.2	64	1332	308
6. Ultrogel	256	1.55	630	24.1	43	2142	450
7. DEAE-BioGel	455	0.60	798	30.6	38	2873	725
8. Fraction 7 after concentration	55	4.32	941	36.1	39	3388	855

Ion-Exchange Chromatography. Although DEAE-cellulose has been used in the preparation of fatty acid synthase from several different tissues including lactating bovine mammary gland,[10] we have had no success with this exchanger obtained from several different commercial sources. Our search for a suitable ion-exchange medium resulted in the finding of the DEAE-agarose, DEAE-BioGel A, 100–200 mesh, (Bio-Rad Laboratories) to be adequate. Accordingly, a 5 cm × 2 cm column of this material is packed and equilibrated with 33 mM potassium phosphate buffer, pH 7.5. The protein solution is diluted with two volumes of a solution of 1 mM dithiothreitol and 0.1 mM EDTA to adjust the concentration of potassium phosphate to 33 mM. The conductivity of the protein solution as well as the buffer emerging from the column should be less than 5000 μmho. After this is ascertained, the protein is loaded onto the column and eluted using a linear gradient from 33 mM to 100 mM potassium phosphate, pH 7.5. A suitable device for obtaining this gradient consists of two 600-ml bottles of identical diameter connected in series to the column. At the outset the bottle farther away from the column contains 500 ml of 100 mM buffer, and the other bottle, the mixing reservoir, contains 500 ml of 33 mM buffer. Both of these solutions are kept under nitrogen by continuously flushing the tops with this gas. The mixing reservoir is gently stirred with a magnetic stirrer throughout the elution of the column. With a freshly packed column the flow rate is in excess of 100 ml/hr. This column is also eluted at room temperature. Continuous monitoring of the protein and the collection of 12-ml fractions are carried out as in gel filtration. The enzymatically active fractions are pooled and concentrated.

Concentration of the Enzyme Solution. The enzyme solution, after diluting to a potassium phosphate concentration of 33 mM, pH 7.5, is adsorbed onto a 2 ml DEAE-BioGel A column, preequilibrated with the same buffer. It is then eluted slowly with 0.1 M potassium phosphate, pH 7.0. The enzyme elutes as a single sharp peak. Alternatively, the enzyme is precipitated at 50% saturation of ammonium sulfate. The precipitate is dissolved in 0.1 M potassium phosphate buffer, pH 7.0, and dialyzed for 4 hr against 100 volumes of the same buffer with one change of the buffer. The enzyme is then centrifuged to remove the denatured protein and the final protein concentration is adjusted to 2 mg/ml.

Storage. Enzyme solution is transferred in small aliquots to nitrocellulose tubes, sealed with Teflon tape, and stored in liquid nitrogen. In the presence of 1 mM dithiothreitol and 100 μM EDTA, the enzyme is stable for over 6 months.

Purity. The enzyme is homogeneous as judged by gel filtration, sedimentation velocity, and sedimentation equilibrium centrifugation and

---

[10] J. Knudsen, *Biochim. Biophys. Acta* **280**, 408 (1972).

sucrose density centrifugation. Sodium dodecyl sulfate (SDS)–gel elec-trophoresis results in one intense protein band, corresponding to a molec-ular weight of 210,000. In addition, one or two very faint bands of smaller molecular weight are visible, when high protein concentrations are used. These bands could be impurities or, more likely, proteolytic products released from the large molecule by the detergent action of SDS.

Properties

Physical. Fatty acid synthase from bovine mammary tissue, like the enzyme from other animal tissues, has an s°_{20w} of 13.5 and a molecular weight of about 530,000. Unlike the pigeon liver and rat mammary en-zyme, the bovine mammary enzyme is not cold labile. It is dissociated into enzymatically inactive half-molecular, hitherto inseparable, 9 S subunits, in Tris (10 mM)–glycine (35 mM) buffer at pH 7.0. These can be reasso-ciated into the 14 S species with the recovery of all the different enzymatic activities (see below).

Chemicals. PRODUCTS. With acetyl-CoA as primer, under standard conditions of assay, the major products are butyric acid, 35%, and palmi-tic acid, 60%; the remaining products are various intermediate-chain acids. Nearly three-fourths of the butyric acid is present as a thioester, the remaining acid being free acid, while over 90% of the palmitic acid occurs as free acid.[11] Butyric acid production can be as high as 60%, with corre-sponding decreases in palmitate production. Unlike the rabbit mammary enzyme[12] there is no formation of butyrate, or of any other acids, in the absence of malonyl-CoA.

Substrate Specificity. This enzyme is able to synthesize fatty acid using butyryl-CoA as primer instead of acetyl-CoA. The kinetic constants for the former substrate are K_m, 5 μM, and V_{max}, at least 800 nmol of NADPH oxidized min^{-1} mg^{-1}, depending upon the preparation. The cor-responding values for acetyl-CoA dependent reaction are 9 μM and 800. It shows a total dependence on malonyl-CoA and a high specificity for NADPH for fatty acid synthesis.

Partial Reactions. In contrast to pigeon liver and yeast enzyme, the bovine mammary enzyme exhibits a very active NADPH-dependent acetoacetyl-CoA and *trans*-crotonyl-CoA reductase activities.[7,13] The kinetic constants for the former are K_m, 8 μM, and V_{max}, 4200 nmol of NADPH oxidized per minute per milligram.[14] The kinetic constants for

[11] Ali Abdinejad and S. Kumar, unpublished observations.
[12] R. Dils and E. M. Carey, this series, Vol. 35 [9].
[13] S. K. Maitra and S. Kumar, *J. Biol. Chem.* **249**, 118 (1974).
[14] A. G. F. Guzman, P. F. Dodds, and S. Kumar, unpublished observations.

the latter are K_m, 33 μM, and V_{max}, 800 nmol. These reactions are assayed as follows: The product of both reactions has been shown to be butyryl-CoA according to Eqs. (2) and (3).

$$CH_3-\overset{O}{\overset{\|}{C}}-CH_2-\overset{O}{\overset{\|}{C}}-SCoA + 2\ NADPH + 2\ H^+ \rightarrow$$

$$CH_3-CH_2-CH_2-\overset{O}{\overset{\|}{C}}-SCoA + 2\ NADP^+ \quad (2)$$

$$CH_3-\overset{H}{\overset{|}{C}}=\underset{H}{C}-\overset{O}{\overset{\|}{C}}-SCoA + NADPH + H^+ \rightarrow$$

$$CH_3-CH_2-CH_2-\overset{O}{\overset{\|}{C}}-SCoA + NADP^+ \quad (3)$$

It appears very likely that in these reactions the acyl groups are transferred to the 4'-phosphopantetheine prosthetic group prior to reduction.[8,15] It has been observed further that both these reactions are lost on dissociation of the dimeric enzyme and are restored on reassociation.[14]

Acetoacetyl-CoA and Crotonyl-CoA Reductases

The methods for assaying the two activities are rather similar and again make use of the decrease in absorbance at 340 nm as NADPH is ozidized. The assay mixture contains, in volumes of 1 ml, the following components:

Acetoacetyl-CoA Reductase[14]

      Potassium phosphate, 100 mM, pH 7.6

      EDTA, 100 μM

      Dithiothreitol, 1 mM

      NADPH, 125 μM

      Acetoacetyl-CoA, 50 μM

      Purified fatty acid synthase, 5 μg

Crotonyl-CoA Reductase[8]

      Bicine-KOH, 100 mM, pH 8.3

      EDTA, 100 μM

      Dithiothreitol, 1 mM

      NADPH, 125 μM

      trans-Crotonyl-CoA, 50 μM

      Purified fatty acid synthase, 5 μg

[15] K. A. Strom and S. Kumar, *J. Biol. Chem.* **254**, 8159 (1979).

The acetoacetyl-CoA reductase reaction is initiated by the addition of enzyme and the crotonyl-CoA reductase assay by the addition of crotonyl-CoA after 10 min of preincubation at 37°, the last 2 min of which are used to observing A_{340} for possible changes in the background. It is necessary to start the acetoacetyl-CoA assay with enzyme rather than substrate in order to prevent a "lag period" of about 2 min that occurs when it is initiated by the addition of the substrate. The rate of *trans*-crotonyl-CoA reduction ceases to be a linear function of time very quickly, so care must be taken to mix the reaction and return the cuvette to the spectrophotometer very quickly. We also find it useful to expand the scale of the chart recorder to obtain 5 μM full scale. The activities obtained during the course of a typical purification are shown in the table.

[12] Fatty Acid Synthase from Red Blood Cells

By ROBERT A. JENIK and JOHN W. PORTER

Reticulocytes are capable of de novo biosynthesis of fatty acids whereas erythrocytes are not.[1-3] In erythrocytes acetyl-CoA carboxylase, a key enzyme in fatty acid synthesis, is lost. However, these cells still possess fatty acid synthase activity.[1] Surprisingly, the fatty acid synthase isolated from rat red blood cells differs immunochemically and in size[4] from the enzyme isolated from other tissues, e.g., liver, adipose tissue, mammary gland, and lung. In this chapter the isolation and some properties of the red blood cell fatty acid synthase are described.

Assay Method

Principle. The red blood cell fatty acid synthase catalyzes the conversion of acetyl-CoA and malonyl-CoA to long-chain fatty acids in the presence of NADPH. The activity of this enzyme may be determined either by radiochemical or spectrophotometric assay procedures. In the radiochemical method, the incorporation of [2-14C]malonyl-CoA into fatty acids in the presence of acetyl-CoA and NADPH is measured. This method is the more reliable of the two for assaying crude and partially purified enzyme preparations. The spectrophotometric method measures the oxidation of

[1] J. G. Pittman and D. B. Martin, *J. Clin. Invest.* **45**, 165 (1966).
[2] G. C. Weir and D. B. Martin, *Diabetes* **17**, 305 (1968).
[3] P. A. Marks, A. Gellhorn, and C. Kidson, *J. Biol. Chem.* **235**, 2579 (1960).
[4] R. A. Jenik and J. W. Porter, *Int. J. Biochem.* **10**, 609 (1979).

METHODS IN ENZYMOLOGY, VOL. 71

NADPH at 340 nm in the presence of acetyl-CoA and malonyl-CoA and should be used only for the DEAE-cellulose purified enzyme. The spectrophotometric method is described elsewhere in this series,[5] and the radiochemical method is described below. For either method, the amount of red blood cell fatty acid synthase protein required is 10–20 times greater than the required amount of liver enzyme.

Reagents

Potassium phosphate buffer, 0.2 M, pH 7.0
EDTA, 0.1 M, pH 7.0
Acetyl-CoA, 1.67 mM
[2-^{14}C]Malonyl-CoA, 3.33 mM, 420 dpm/nmol
NADPH, 2 mM
Liquid scintillator: 4 liters of toluene, 16 g of 2,5-diphenyloxazole (PPO), and 0.4 g of 2,5-bis-2-(5-t-butylbenzoxazolyl)thiophene (BBOT)

Procedure. The reaction mixture (total volume, 1.0 ml) contains 0.2 mmol of potassium phosphate buffer, pH 7.0, 1 μmol of EDTA, 33 nmol of acetyl-CoA, 100 nmol of [2-^{14}C]malonyl-CoA, 100 nmol of NADPH, and enzyme protein (50–200 μg). The reaction mixture is allowed to incubate at 30° for 5 min prior to the addition of fatty acid synthase to start the reaction. After 10 min at 30°, the reaction is terminated by the addition of 30 μl of 60% perchloric acid. One milliliter of ethanol is added to the reaction mixture and the ^{14}C-labeled fatty acids are extracted three times with 2 ml each of petroleum ether. Aliquots are then transferred to vials containing toluene-based scintillation cocktail and assayed for radioactivity in a liquid scintillation spectrometer.

Units. A unit of fatty acid synthase activity is defined as the amount of enzyme required to catalyze the formation of 1 nmol of palmitate per minute under the conditions of assay. Specific activity is expressed as units per milligram of protein (determined by the biuret[6] or Folin[7] method).

Purification Procedure[4]

The procedure described below results in a final purification of the red blood cell fatty acid synthase of about 680-fold (Table I). Unless indicated otherwise, all steps were carried out at room temperature.

[5] R. A. Muesing and J. W. Porter, this series, Vol. 35 [7].
[6] A. G. Gornall, C. J. Bardawill, and M. M. David, *J. Biol. Chem.* **177**, 751 (1949).
[7] O. H. Lowry, N. J. Rosebrough, A. L. Farr, and R. J. Randall, *J. Biol. Chem.* **193**, 265 (1951).

TABLE I
PURIFICATION OF RAT RED BLOOD CELL FATTY ACID SYNTHASE[a]

Step	Total protein (mg)	Total activity (units)[b]	Specific activity (units/mg)	Recovery of activity (%)	Purification (fold)[c]
Hemolyzate[d]	19,052	57.2	0.003	100	—
100,000 g supernatant solution	2,052	36.9	0.018	65	6
$(NH_4)_2SO_4$ fractionation	1,123	23.6	0.021	41	7
DEAE-cellulose chromatography	4.5	9.1	2.03	16	677

[a] Reprinted from Jenik and Porter,[4] with permission from Pergamon Press, Ltd.
[b] A unit is defined as the amount of enzyme required for the formation of 1 nmol of palmitate per minute.
[c] Based on change in specific activity.
[d] From 30 ml of blood.

Animals. Male Holtzman albino rats (150–250 g) were maintained on an ad libitum diet of Purina rat chow and water. Blood was obtained from the rats by cardiac puncture using a plastic syringe containing heparin (200 units per 10 ml of blood).

Separation of Blood Cells. A dextran flotation technique was used to separate the erythrocytes from leukocytes and platelets.[8] Dextran[9] with an average molecular weight of 250,000 was dissolved in 0.9% NaCl to make a 3% solution. For each milliliter of blood, 2 ml of the dextran solution were added, and the blood–dextran solution was mixed by inversion for 2 min in graduated plastic tubes. The erythrocytes were then allowed to sediment under gravity for 15 min at 0–4°. The supernatant solution containing the leukocytes and platelets was carefully pipetted from the sedimented erythrocytes and discarded. The erythrocytes were resuspended in 0.9% NaCl and sedimented at 125 g and 4°. This wash procedure was repeated two more times. After this procedure there was less than one leukocyte per 5000–10,000 erythrocytes.

Preparation of Hemolyzate and High-Speed Supernatant Solution. The volume of the packed erythrocytes[10] was measured, and the erythrocytes were lysed by adding two volumes of deionized water to the cells and mixing by inversion. The hemolyzate was centrifuged for 30 min at 4° and

[8] W. A. Skoog and W. S. Beck, *Blood* 11, 436 (1956).
[9] Dextran T 250 from Pharmacia was used but is no longer available. Dextran with an average molecular weight between 140,000 and 250,000 can be substituted as described by Skoog and Beck.[8]
[10] The volume of packed cells is usually 50% of the volume of the blood obtained from the rats.

30,000 g. The cell debris was discarded, and the ionic strength of the supernatant solution was increased to 0.1 M with 1 M potassium phosphate buffer (pH 7.0) containing 1 mM EDTA and 2 mM dithiothreitol. The supernatant solution was centrifuged at 4° and 100,000 g for 45 min. The 100,000 g supernatant solution was stored at −20° after increasing the dithiothreitol concentration to 10 mM.[11,12]

Ammonium Sulfate Precipitation. After thawing, the 100,000 g supernatant solution was centrifuged at 4° and 30,000 g for 30 min to remove denatured protein. The supernatant solution was then brought to 50% saturation by the addition of ammonium sulfate (313 g/liter) with stirring under a stream of nitrogen gas. After stirring for 10 min, the enzyme solution was centrifuged at 4° and 30,000 g for 15 min. The supernatant solution was discarded, and the pellet was dissolved in 40 mM potassium phosphate buffer (pH 7.0) containing 1 mM EDTA and 5 mM 2-mercaptoethanol. The enzyme solution was subjected to DEAE-cellulose chromatography as soon as the pellet was dissolved because the fatty acid synthase dissociates to half molecular weight species in low ionic strength buffer.[4]

DEAE-Cellulose Chromatography. A column of DEAE-cellulose (2.8 × 10 cm) was packed with material that had been prepared for use by soaking overnight in 1 M potassium phosphate buffer (pH 7.0) and then washing thoroughly with deionized water. The DEAE-cellulose column was equilibrated with approximately 400 ml of 40 mM potassium phosphate buffer (pH 7.0) containing 1 mM EDTA and 5 mM 2-mercaptoethanol. The enzyme solution, dissolved in 40 mM potassium phosphate buffer as described above, was diluted with an equal volume of deionized water and immediately loaded onto the DEAE-cellulose column. After the enzyme protein adsorbed to the ion exchanger (hemoglobin did not bind to DEAE-cellulose under these conditions), the column was washed with 400–500 ml of 40 mM potassium phosphate buffer until the light absorbance at 280 nm of the eluent was less than 0.05 unit. The fatty acid synthase was eluted with 150 ml of 250 mM potassium phosphate buffer (pH 7.0) containing 1 mM EDTA and 5 mM 2-mercaptoethanol. Five-milliliter fractions were collected and those with a light absorbance at 280 nm of greater than 0.1 unit were combined. The fatty acid synthase was concentrated by ammonium sulfate precipitation and dialyzed at 4° against 125 ml of 0.2 M potassium phosphate buffer (pH 7.0) containing 1 mM EDTA and 1 mM dithiothreitol overnight, with two or three changes of dialysis buffer.

---

[11] The supernatant solution may be stored for 1–2 months.
[12] The dithiothreitol should be added as a solution because the addition of solid dithiothreitol results in the precipitation of some protein.

Storage and Stability of Red Blood Cell Fatty Acid Synthase. The enzyme was stored at 4° in 0.2 *M* potassium phosphate buffer (pH 7.0) containing 1 m*M* EDTA and 10 m*M* dithiothreitol at a protein concentration of 2–4 mg/ml. Under these conditions, 75% of the original activity was retained after 5–7 days. However, in order to obtain maximum activity, it was necessary to allow the enzyme to stand at room temperature for 30–60 min prior to assay for fatty acid synthesis.

Properties of Rat Red Blood Cell Fatty Acid Synthase

Purity. The fatty acid synthase obtained by the above procedure gives one major band and two minor bands when subjected to electrophoresis under nondenaturing conditions on 5% polyacrylamide gels. A minor component that sediments more slowly than the fatty acid synthase is also observed during sucrose density gradient centrifugation.

Molecular Weight. The molecular weight of the enzyme is about 300,000 as estimated by gel filtration on BioGel A-1.5m (8% Agarose in gel, exclusion limit 1.5×10^6 daltons, obtainable from Bio-Rad).

Dissociation of Enzyme. The fatty acid synthase is dissociated into two

TABLE II

ENZYME ACTIVITIES OF LIVER AND RED BLOOD CELL FATTY ACID SYNTHASES[a]

Reaction	Source of fatty acid synthase (nmol product formed per mg protein per minute)	
	Rat liver	Rat red blood cell
Fatty acid synthesis	60–80	1.4–2.0
Acetyl-CoA : pantetheine transacylase	590	68
Malonyl-CoA : pantetheine transacylase	222	10
β-Ketoacyl thioester synthase	3.9	1.0
β-Ketoacyl thioester reductase	248	20
β-Hydroxyacyl thioester dehydrase	576	22
Enoylacyl thioester reductase	237	20
Palmitoyl-CoA deacylase	39.2	9.1

[a] Reprinted from Jenik and Porter,[4] with permission of Pergamon Press, Ltd.

TABLE III

CHAIN LENGTHS OF FATTY ACIDS SYNTHESIZED BY RAT RED BLOOD CELL
FATTY ACID SYNTHASE

Chain length	$8:0$	$10:0$	$12:0$	$14:0$	$16:0$	$18:0$	$20:0$	$22:0$
Distribution of radioactivity[a] (%)	1.1	2.2	2.8	38.6	42.6	5.1	1.3	2.6

[a] Determined by gas–liquid chromatography on an SE-30 column.

125,000 MW species when dialyzed against buffer comprised of 5 mM Tris base, 35 mM glycine, pH 8.4, 1 mM EDTA, and 2 mM 2-mercaptoethanol at 0° for 2–4 hr.[4]

pH Optimum. When the pH is varied, with all other assay conditions constant, a pH optimum of 6.6–6.7 is found.

Nucleotide Specificity. The red blood cell fatty acid synthase is a NADPH-dependent enzyme. NADH is only 10–15% as efficient as NADPH for fatty acid synthesis. Flavin mononucleotide (100 μM) does not increase the activity of the enzyme.

Component Activities Exhibited by the Enzyme. The fatty acid synthase from red blood cells catalyzes the same component reaction activities as the liver enzyme.[4] Table II shows that the red cell enzyme catalyzes these reactions much more slowly than the liver enzyme.

Products of the Reaction. The chain lengths of the fatty acids synthesized by the red blood cell enzyme can be determined by using either radioactively labeled acetyl-CoA or malonyl-CoA, followed by derivatization of the products and subjection of the derivatives to gas–liquid chromatography. Table III shows the distribution of radioactivity into fatty acids of chain length C_8–C_{22}. Myristate ($C_{14:0}$) and palmitate ($C_{16:0}$) are formed in approximately equal amounts, and together they account for more than 80% of the fatty acids synthesized.

Immunochemical Cross-Reactivity of the Fatty Acid Synthase. Immunodiffusion of the red blood cell fatty acid synthase did not give a precipitin band with rabbit antiserum to rat liver fatty acid synthase. The failure of the red blood cell fatty acid synthase to cross-react with antisera to the liver enzyme distinguishes it from the enzymes isolated from liver, adipose tissue, mammary gland, and lung. The fatty acid synthases from the latter four tissues have been shown to be immunologically identical.[13,14] No cross-reaction is observed using antisera to pigeon liver fatty acid synthase. It is evident from these results that the gene for the red blood cell fatty acid synthase is different from that for the liver enzyme.

[13] S. Smith, *Arch. Biochem. Biophys.* **156**, 751 (1973).
[14] S. Kumar, *Arch. Biochem. Biophys.* **183**, 625 (1977).

Acknowledgments

This work was supported in part by a grant (AM 01383) from the National Institute of Arthritis and Metabolic Diseases of the National Institutes of Health, United States Public Health Service, and by the Medical Research Service of the Veterans Administration.

[13] Fatty Acid Synthase from the Uropygial Gland of Goose

By P. E. KOLATTUKUDY, A. J. POULOSE, and J. S. BUCKNER

Fatty acid synthase from goose uropygial gland catalyzes the synthesis of *n*-fatty acids from malonyl-CoA and multimethyl branched fatty acids from methylmalonyl-CoA with acetyl-CoA and NADPH as the preferred primer and reductant, respectively.[1,2]

$$CH_3\text{—}\overset{\overset{O}{\|}}{C}\text{—SCoA} + 7\ \overset{HOOC}{\underset{\,}{C}}H_2\text{—}\overset{\overset{O}{\|}}{C}\text{—SCoA} + 14\ NADPH + 14\ H^+$$

$$\downarrow$$

$$CH_3\text{—}(CH_2)_{14}\text{—COOH} + 8\ CoASH + 14\ NADP + 7\ CO_2 + 6\ H_2O$$

(1)

$$CH_3\text{—}\overset{\overset{O}{\|}}{C}\text{—SCoA} + 4\ CH_3\text{—}\overset{HOOC}{\underset{\,}{C}}H\text{—}\overset{\overset{O}{\|}}{C}\text{—SCoA} + 8\ NADPH + 8\ H^+$$

$$\downarrow$$

$$CH_3\text{—}(CH_2\text{—}\underset{\underset{CH_3}{|}}{C}H)_4\text{—COOH} + 5\ CoASH + 8\ NADP + 4\ CO_2 + 3\ H_2O$$

(2)

Assay Methods

Enzyme Assay

Principle. Fatty acid synthase can be assayed either by determining the radioactivity incorporated from malonyl-CoA or methylmalonyl-CoA into fatty acids or by spectrophotometrically following the oxidation of NADPH by measuring the absorbance decrease at 340 nm.

[1] J. S. Buckner and P. E. Kolattukudy, *Biochemistry* 14, 1774 (1975).
[2] J. S. Buckner and P. E. Kolattukudy, *Biochemistry* 15, 1948 (1976).

Reagents

Malonyl-CoA, 2 mM
NADPH, 2.2 mM
Acetyl-CoA, 0.5 mM
Methylmalonyl-CoA, 1 mM
Buffer, 10 mM sodium phosphate, pH 7.0, containing 1.0 mM dithioerythritol (DTE), 1 mM EDTA, 0.2 M KCl. All of the reagents are prepared in this buffer.

Radiochemical Assay

Preparation of Labeled Substrates. Monothiophenyl malonate and monothiophenyl methylmalonate are prepared by the method of Trams and Brady.[3] The products are purified by thin-layer chromatography on silica gel G (1.0 mm) with hexane–ethyl ether–formic acid (60 : 60 : 1, v/v/v) as the solvent. The monothiophenyl esters are eluted from the silica gel with ethyl ether. This method avoids the losses encountered by the charcoal treatment method. Another useful modification that improves the yield is the use of freshly distilled (over LiAlH$_4$) tetrahydrofuran instead of ethyl ether as the solvent for addition of the monothiophenyl ester into the CoA solution. The CoA esters are purified by DEAE-cellulose column chromatography with a linear LiCl gradient.[4] The fractions containing the labeled CoA ester are pooled, lyophilized, and desalted with a BioGel P-2 column using 3 mM HCl as the solvent. The fractions containing the labeled CoA esters are pooled and lyophilized. Labeled methylmalonyl-CoA can be prepared in a similar manner. Labeled malonyl-CoA and methylmalonyl-CoA can be purchased from commercial sources (e.g., New England Nuclear Corp.).

Assay Procedure. The reaction mixture consists of 0.1 μmol of [2-^{14}C]malonyl-CoA (~200,000 cpm) 0.05 μmol of acetyl-CoA, 0.26 μmol of NADPH, 3.9 μmol of glucose 6-phosphate, 1 unit of glucose-6-phosphate dehydrogenase[5] (Sigma Chemical Co.), 0.25 μmol of DTE, 0.5 μmol of EDTA, 5 μmol of sodium phosphate buffer (pH 7.0), 100 μmol of KCl, and 10–20 mU of enzyme in a total volume of 0.5 ml. Assay is initiated by the addition of malonyl-CoA, the reaction mixture is incubated for 10 min at 30°, and the reaction is terminated by the addition of 0.1 ml of 45% NaOH. The reaction mixture is heated for 10 min in a boiling water bath and then acidified with 6 N HCl followed by extraction with chloroform. The

---

[3] E. G. Trams and R. O. Brady, *J. Am. Chem. Soc.* **82,** 2972 (1960).
[4] C. Gregolin, E. Ryder, and M. D. Lane, *J. Biol. Chem.* **243,** 4227 (1968); See also this series, Vol. 34, p. 74.
[5] Glucose 6-phosphate and glucose-6-phosphate dehydrogenase can be omitted when purified enzyme is used.

chloroform extract is washed with acidified water, and the chloroform is removed with a rotary evaporator. The residue is mixed with 50 μg of myristic acid, and the mixture is subjected to thin-layer chromatography on silica gel G with hexane–ethyl ether–formic acid (40 : 10 : 1) as the developing solvent. The chromatogram is sprayed with a 0.1% alcoholic solution of 2′, 7′-dichlorofluorescein, then the fatty acid region is located under ultraviolet light. The silica gel from this region is scraped into a counting vial containing 15 ml of a counting fluid consisting of 30% ethanol in toluene containing 4 g per liter, of Omnifluor (New England Nuclear Corp.) and assayed for radioactivity in a scintillation counter. Assay for methylmalonyl-CoA incorporation is done in an identical manner except that 60 nmol of [methyl-^{3}H]methylmalonyl-CoA are used instead of malonyl-CoA, 250 mU of enzyme are used, and the incubation time is extended to 30 min.

Spectrophotometric Assay

The reaction mixture consists of 0.04 μmol of malonyl-CoA, 0.01 μmol of acetyl-CoA, 0.04 μmol of NADPH, 0.2 μmol of DTE, 2 μmol of sodium phosphate, pH 7.0, 40 μmol of KCl, 0.2 μmol of EDTA, and 1–2 mU of enzyme in a total volume of 0.2 ml. The reaction is carried out in 0.5-ml quartz microcells (1 cm pathlength) at 30°, and malonyl-CoA is added to initiate the reaction. Initial rates are measured by monitoring the absorbance decrease at 340 nm. One unit of enzyme activity is defined as the enzyme that oxidizes 1 μmol of NADPH per minute at 30°. Specific activity is defined as units of enzyme per milligram of protein.

Enzyme Purification

Preparation of 105,000 g Supernatant. The goose is killed by exsanguination and the uropygial gland is excised from the bird.[6] All the operations after excision of the gland are carried out at 0–4°. The fat and muscle tissue adhering to the gland are removed with a razor blade. The two lobes of the gland are sliced and homogenized in 100 mM sodium phosphate buffer at pH 7.0 containing 250 mM sucrose, 1.0 mM DTE, and 1.0 mM EDTA. The crude homogenate is centrifuged at 12,000 g for 20 min. After removal of the fat layer by filtration through cheesecloth, the supernatant is centrifuged at 15,000 g for 20 min. The supernatant is centrifuged at 105,000 g for 90 min, followed by the recentrifugation of the supernatant at the same speed for 60 min. The supernatant (5–7 ml) is filtered through cheesecloth to remove floating fat.

[6] P. E. Kolattukudy, this series, Vol. 72 [60].

Sepharose 4B Gel Filtration. The high-speed supernatant (5–7 ml) is subjected to gel filtration with a Sepharose 4B column (2.7 × 96 cm) equilibrated with 100 mM sodium phosphate buffer, pH 7.0, containing 0.5 mM dithioerythritol, and 1.0 mM EDTA. At a flow rate of 20 ml/hr, 5-ml fractions are collected and the absorbance of the effluent is monitored at 280 nm. The first major protein peak (after the void volume) corresponds to fatty acid synthase. To avoid malonyl-CoA decarboxylase activity[7] which elutes just after fatty acid synthetase, only the fractions with highest activity are pooled. The pooled enzyme solution (2–4 mg/ml) can be stored at 0–4° for at least a week. Further concentration of the enzyme is achieved by ultrafiltration using a PM-30 membrane and extremely slow rate of stirring. One gland from a mature Chinese white goose yields 150–200 mg enzyme during the winter months and 80–100 mg during the summer months.

Purity. The enzyme is homogeneous according to analytical ultracentrifugation, immunodiffusion assays, and polyacrylamide disc gel electrophoresis and has a specific activity of 0.8–1.2.

Properties

Stability. Purified enzyme can be stored in 0.1 M phosphate, pH 7.0, containing 1 mM DTE at a protein concentration of 2–5 mg/ml for at least 1 week at 0–4° C. Lower protein concentrations and lack of DTE causes loss of activity although addition of 1 mM DTE often reverses the inactivation at least partly. Addition of EDTA also tends to prolong the storage life of the enzyme.

Products. Free fatty acids are generated by the enzyme. Palmitic acid is the major product generated from malonyl-CoA and acetyl-CoA, but small amounts of myristic acid and stearic acid are also produced. 2,4,6,8-Tetramethyldecanoic acid is the major product generated from methylmalonyl-CoA with acetyl-CoA as the primer and 2,4,6,8-tetramethylundecanoic acid is the major product, with propionyl-CoA as the primer.[2] A mixture of malonyl-CoA and methylmalonyl-CoA generates longer aliphatic chains containing methyl branches.[8,9] Identical results are obtained with fatty acid synthase from Peking duck, Muscovy duck, mallard, wood duck, and Canadian goose. Ability to catalyze the synthesis of the above branched fatty acids from methylmalonyl-CoA is an inherent property of fatty acid synthase from animals in general. Thus

[7] J. S. Buckner and P. E. Kolattukudy, *Biochemistry* **14**, 1768 (1975).
[8] M. deRenobales, L. Rogers, and P. E. Kolattukudy, *Arch. Biochem. Biophys.*, in press (1980).
[9] J. S. Buckner, P. E. Kolattukudy, and L. Rogers, *Arch. Biochem. Biophys.* **186**, 152 (1978).

the enzyme from goose liver, rat liver, and rat mammary glands generates the same branched fatty acids as those synthesized by the enzyme from the uropygial gland of goose.[9]

Substrate Specificity. Fatty acid synthase from the uropygial gland shows strict requirement and specificity for NADPH.[2] NADH is not utilized by the enzyme at detectable rates. Acetyl-CoA is the preferred primer over propionyl-CoA for the synthesis of methyl branched acids.[1] Fatty acid synthase does not show any stereospecificity for methylmalonyl-CoA incorporation.[1,10] The stereochemical purity of the products synthesized *in vivo*[11] (D,D,D,D-tetramethyldecanoate) is apparently due to the specificity of propionyl-CoA carboxylase of the gland, which gives rise to the *S* isomer of methylmalonyl-CoA.[10]

General Properties. With malonyl-CoA, linear rates are obtained up to 20 μg of protein per milliliter and 15 min of incubation time. With methylmalonyl-CoA, linear rates are obtained up to 250 μg of protein per milliliter and 60 min incubation time. In both cases maximal rates are obtained at pH 6.5–7.5. K_m for malonyl-CoA and NADPH are 40 μM and 20 μM, respectively. Methylmalonyl-CoA is a competitive inhibitor of malonyl-CoA incorporation with a K_i of 16 μM.[9] With the 105,000 g supernatant, typical Michaelis–Menten substrate saturation patterns are observed for methylmalonyl-CoA with K_m of 0.2 mM whereas with the purified enzyme a sigmoidal substrate saturation is observed with this substrate but not with malonyl-CoA. A maximum rate of 1 nmol per milligram of protein is obtained at a methylmalonyl-CoA concentration of 0.1 mM.

Molecular Weight. Fatty acid synthase from the uropygial gland of goose has a molecular weight of 547,000 as determined by sedimentation equilibrium centrifugation and an $s_{20,w}$ of 13.5 S. Native enzyme consists of two apparently identical subunits with $s_{20,w}$ of 9.35 and molecular weight of 269,000 as estimated by sodium dodecyl sulfate polyacrylamide disc gel electrophoresis. Similar results are obtained also with the enzymes from the uropygial glands of wood duck, mallard, Muscovy duck, Peking duck, and Canadian goose.[8]

Dissociation and Association of Subunits. Dimeric enzyme is the active form of fatty acid synthase. In low ionic strength buffers the enzyme dissociates into subunits, resulting in loss of activity. For example, dialysis of the enzyme (5 mg/ml) for 24 hr against 1 mM phosphate buffer (pH 8.0) containing 1 mM DTE results in ~50% dissociation, whereas dialysis for 48 hr against 0.44 mM Tris, 35 mM glycine containing 1 mM DTE results in ~70% dissociation.[2] Complete association of the dis-

---

[10] Y. S. Kim and P. E. Kolattukudy, *J. Biol. Chem.* **255**, 686 (1980).
[11] G. Odham, *Ark. Kemi* **21**, 379 (1963).

sociated enzyme as well as recovery of full activity can be achieved by incubating the dissociated enzyme with $0.2 M$ KCl or 0.2 mM NADPH for 2 hr at room temperature. Also incubation of the enzyme (0.25 mg/ml) with 200 μM palmitoyl-CoA results in complete inactivation of the enzyme. The mechanism of this inactivation is not known. This enzyme is not as readily dissociated by low temperature and/or low ionic strength as the enzyme from rat mammary glands.[12]

Amino Acid Composition. The amino acid composition of the enzyme from the uropygial gland of goose[2] and that from mallards and wood ducks[8] is very similar to the composition of fatty acid synthases from rat mammary gland,[13] chicken liver,[14] rat liver, pigeon liver,[15] and rabbit mammary gland.[16] Partial specific volume calculated from amino acid composition is 0.73 ml/g. Fatty acid synthase from goose uropygial gland contains one covalently attached 4-phosphopantetheine per subunit of the enzyme.[2]

Amino Acids Essential for Activity. Treatment of the enzyme with diisopropylfluorophosphate results in modification of "active serines" at the active sites of the thioesterase domains of fatty acid synthase,[17] thus inactivating the enzyme. Correlation of the number of serine residues modified with the loss of enzymatic activity suggests that there is one thioesterase domain per subunit of fatty acid synthetase. Chemical modification studies with phenylglyoxal and 2,3-butanedione show that there is one essential arginine present at each of the ketoacyl reductase and enoyl-CoA reductase active sites of fatty acid synthetase that specifically interacts with the 2'-phosphate of NADPH.[18] This study also indicates that each subunit of fatty acid synthase contains one ketoacyl reductase domain and one enoyl-CoA reductase domain. Modification of fatty acid synthase with pyridoxal 5'-phosphate results in the inactivation of the enzyme by the specific inhibition of enoyl-CoA reductase activity.[19] Protection experiments with NADPH analogs suggest that the lysine modified by pyridoxal 5'-phosphate interacts with the 5'-pyrophosphate of NADPH. Furthermore, modification of one lysine per subunit results in complete inactivation of the enoyl-CoA reductase and hence the overall activity of fatty acid synthase.

[12] S. Smith and S. Abraham, *J. Biol. Chem.* **246**, 6428 (1971).
[13] S. Smith and S. Abraham, *J. Biol. Chem.* **245**, 3209 (1970).
[14] S. Yun and R. Y. Hsu, *J. Biol. Chem.* **247**, 2689 (1972).
[15] D. N. Burton, A. G. Havik, and J. W. Porter, *Arch. Biochem. Biophys.* **126**, 141 (1968).
[16] E. M. Carey and R. Dils, *Biochim. Biophys. Acta* **210**, 371 (1970).
[17] P. E. Kolattukudy, J. S. Buckner, and C. J. Bedord, *Biochem. Biophys. Res. Commun.* **68**, 379 (1976).
[18] A. J. Poulose and P. E. Kolattukudy, *Arch. Biochem. Biophys.* **199**, 457 (1980).
[19] A. J. Poulose and P. E. Kolattukudy, *Arch. Biochem. Biophys.* **201**, 313 (1980).

Proteolysis of Fatty Acid Synthase. Limited proteolysis of the enzyme with trypsin releases both thioesterase domains from fatty acid synthase, and this activity is contained in a 33,000 MW fragment that has been purified and characterized.[20] Similarly, elastase treatment of the enzyme also releases a fragment similar in molecular weight to that of trypsin-derived fragment that contains thioesterase activity.[21] Since the limited proteolysis treatment removes the thioesterase fragments, and leaves the rest of the protein unaffected, it appears that the thioesterase active site is at one end of the 269,000 MW peptide. Since the isolated thioesterase segment and the intact enzyme have blocked N-termini, the thioesterase domain appears to be located at the N-terminal region of the intact enzyme.

Immunologic Comparison. The antiserum prepared against the uropygial gland fatty acid synthase cross-reacts with the goose liver fatty acid synthase, and complete fusion of the immunoprecipitant lines is observed in Ouchterlony double-diffusion analysis showing that the enzyme from these two tissues are immunologically identical.[9] Fatty acid synthases from the uropygial glands of wood duck, mallards, Peking duck, Muscovy duck, and Canadian goose also cross-react with the rabbit antibodies prepared against the enzyme from the goose, and complete fusion of the immunoprecipitant lines in all cases indicates that the enzyme from all these sources are immunologically identical.[8] Fatty acid synthase from eared grebe (*Podiceps caspicus*) also cross-reacts with the antibody prepared against the enzyme from the goose, but in this case Ouchterlony double diffusion reveals spurs showing immunological nonidentity.[22]

Identity of Subunits. Identity of subunits is shown by the evidence that each peptide of the synthase contains one 4'-phosphopantetheine,[2] one ketoreductase (K_d of NADPH binding, 7 μM),[18,23] one enoyl reductase (K_d of NADPH binding, 1.3 μM),[19,23] one thioesterase,[17] and one acetyl loading site.[21]

Acknowledgment

This work was supported in part by a grant (GM-18278) from the U.S. Public Health Service. Dr. Charles Bedord and Linda Rogers made valuable contributions to this work.

[20] C. J. Bedord, P. E. Kolattukudy, and L. Rogers, *Arch. Biochem. Biophys.* **186,** 139 (1978).
[21] A. J. Poulose and P. E. Kolattukudy, unpublished results, 1978.
[22] Y. S. Kim and P. E. Kolattukudy, unpublished results, 1977.
[23] A. J. Poulose, R. J. Foster, and P. E. Kolattukudy, *J. Biol. Chem.* (in press, 1980).

[14] Fatty Acid Synthase from *Mycobacterium smegmatis* [1]

By WILLIAM I. WOOD and DAVID O. PETERSON

$$CH_3COSCoA + n\ HOOCCH_2COSCoA + CoA + n\ NADPH$$
$$+ n\ NADH + 2n\ H^+ \rightarrow CH_3(CH_2CH_2)_nCOSCoA + (n + 1)\ CoA$$
$$+ n\ NADP^+ + n\ NAD^+ + n\ CO_2 + n\ H_2O$$

Assay[3-5]

Fatty acid synthase is assayed either spectrophotometrically as a malonyl-CoA-dependent oxidation of reduced pyridine nucleotide or radiochemically as [2-^{14}C]malonyl-CoA incorporation into long-chain fatty acids. One unit of activity is defined as 1 μmol of malonyl-CoA incorporated into fatty acids per minute.

Spectrophotometric Assay

Fatty acid synthase is routinely assayed at 340 nm in a Gilford spectrophotometer thermostatted at 37°.

Reagents

> Potassium phosphate buffer, 1.0 M, pH 7.0 (0.636 M KH_2PO_4, 0.364 M K_2HPO_4)
> 2-Mercaptoethanol
> Acetyl-CoA, 10 mM (store frozen)
> Malonyl-CoA, 1.6 mM (store frozen)
> CoA, 10 mM (store frozen)
> NADH and NADPH, 0.5 mM each (made up fresh)
> MCHA[6] or the mycobacterial polysaccharides MMP or MGLP, 5 mM (store frozen). MCHA is prepared as described by authors

---

[1] The organism used for these studies, *M. smegmatis* ATCC 356, was recently found to have been previously misclassified as *M. phlei*.[2] Earlier literature refers to this organism as *M. phlei*.

[2] R. W. Hendren, *Int. J. Sys. Bacteriol.* **24**, 491 (1974).

[3] W. I. Wood, D. O. Peterson, and K. Bloch, *J. Biol. Chem.* **252**, 5745 (1977).

[4] W. I. Wood, Ph.D. Thesis, Harvard University, Cambridge, Massachusetts, 1977.

[5] D. O. Peterson, Ph.D. Thesis, Harvard University, Cambridge, Massachusetts, 1977.

[6] The abbreviations used are MCHA, 2,6-di-*O*-methylcycloheptaamylose; MMP, 3-*O*-methylmannose-containing polysaccharide; MGLP, 6-*O*-methylglucose-containing lipopolysaccharide.

[7] R. Bergeron, Y. Machida, and K. Bloch, *J. Biol. Chem.* **250**, 1223 (1975).

[8] Y. Machida, R. Bergeron, P. Flick, and K. Bloch, *J. Biol. Chem.* **248**, 6246 (1973).

METHODS IN ENZYMOLOGY, VOL. 71   ISBN 0-12-181971-X cited in footnotes 7 and 8; MMP and MGLP, as described in references cited in footnotes 9 and 10.

Assay mixtures contain 0.1 M potassium phosphate buffer, 5 mM 2-mercaptoethanol,[11] 300 μM acetyl-CoA, 80 μM malonyl-CoA, 100 μM CoA, 100 μM NADH, 100 μM NADPH, 200 μM MCHA (or MMP or MGLP), and 0.05–5 mU of enzyme in a final volume of 200 μl. A solution containing 0.5 M potassium phosphate buffer, 25 mM 2-mercaptoethanol, and 500 μM CoA is first incubated at 37° for 30 min to ensure complete reduction of the CoA. Assay components other than malonyl-CoA and enzyme are combined in 1-ml quartz cuvettes, and the mixture is equilibrated at 37° for a few minutes. Enzyme is added, and a background rate of pyridine nucleotide oxidation is determined. (The background rate is the same with or without enzyme when the enzyme is highly purified.) Malonyl-CoA is then added, and the linear rate is determined. The enzyme activity is the difference in rate before and after addition of malonyl-CoA assuming that 2 mol of pyridine nucleotide ($\Delta\epsilon_{340} = 6.22$ mM^{-1} cm^{-1})[13] are oxidized per mole of malonyl-CoA incorporated.

Radiochemical Assay

The radiochemical assay is used for enzyme fractions prior to DEAE-cellulose chromatography and for radio-gas chromatographic product analysis.

Additional Reagents

[2-^{14}C]Malonyl-CoA, 3–30 Ci/mol (New England Nuclear)

KOH, 500 g/liter

Carrier fatty acids: a mixture of C_{12}, C_{14} through C_{24} or C_{32} fatty acids or their methyl esters in hexane or chloroform (about 1.5 g/liter each)

Methanol

Thymol blue, 0.1 g/liter

HCl, 6 M

n-Pentane (99%)

Methanol/HCl: made by passing HCl gas through a fresh bottle of anyhdrous methanol to 25–30 g/liter

Isooctane

[9] D. E. Vance, O. Mitsuhashi, and K. Bloch, *J. Biol. Chem.* **248**, 2303 (1973).
[10] M. Ilton, A. W. Jevans, E. D. McCarthy, D. Vance, H. B. White, III, and K. Bloch, *Proc. Natl. Acad. Sci. U.S.A.* **68**, 87 (1971).
[11] 2-Mercaptoethanol is used because of a chemical reaction of dithiothreitol with thioester substrates to form an O-acyldithiothreitol.[12]
[12] G. B. Stokes and P. K. Stumpf, *Arch. Biochem. Biophys.* **162**, 638 (1974).
[13] B. L. Horecker and A. Kornberg, *J. Biol. Chem.* **175**, 385 (1948).

Radioassays are performed in 15-ml glass conical centrifuge tubes. The assay mixtures are the same as for the spectrophotometric assay, except that the assays are initiated with enzyme rather than with malonyl-CoA. The assays are incubated at 37° for 2–5 min and then terminated by the addition of KOH, 500 g/liter, to a concentration of 115 g/liter. Five microliters of the carrier fatty acid solution are added, and the assay mixtures are saponified in a boiling water bath for 30 min. Methanol is added to 100–200 ml/liter, 1 or 2 drops of 0.1 g/liter thymol blue is added as an acid-base indicator, the assays are acidified with 6 M HCl, and the fatty acids are extracted with three 5-ml portions of pentane. The combined pentane layers are evaporated to dryness in glass scintillation vials at 30–35°, 15 ml of 4 g of 2,5-diphenyloxazole per liter in toluene are added, and the radioactivity is determined by liquid scintillation counting. The radioassay gives 70–100% of the activity found by the spectrophotometric assay.

Radioassays for product chain-length distribution are best followed spectrophotometrically to verify their linearity. After termination and addition of carrier fatty acids, assay mixtures are transferred from cuvettes to conical centrifuge tubes, and the cuvettes are rinsed three times with an equal volume of 115 g of KOH per liter. After saponification and extraction of the fatty acids, the combined pentane layers are evaporated to dryness in 16 × 125-mm screw-cap culture tubes. Methanol–HCl (2.25 ml) is added, and the tubes are sealed with Teflon-lined screw caps. After heating at 80° for 1.5 hr, 0.25 ml of H_2O is added and the methyl esters are extracted with three 5-ml portions of pentane. The pentane is evaporated with a stream of nitrogen without heating, and pentane is used to transfer the methyl esters to a small glass tube (made by sealing off a Pasteur pipette). The methyl esters are taken up in as little as 10 μl of isooctane and analyzed on an F and M 400 gas chromatograph equipped with a Packard 894 radiodetector. Separations are performed with a 0.4 × 180 cm column of 10% SP-2340 on Supelcoport 100/120 (Supelco Inc). The column temperature is programmed to rise from 150° to 225° at 4° per minute. Peaks of radioactivity are integrated by planimeter, and the rate of synthesis of each fatty acid, expressed in terms of malonyl-CoA incorporation, can be calculated assuming that the sum of all the peak areas equals the spectrophotometric rate.

Enzyme Source

Mycobacterium smegmatis ATCC 356 (formerly classified as *M. phlei*)[1] is grown on a medium containing glucose and Tween 80.[4,5,14] Cultures are

[14] W. I. Wood, D. O. Peterson, and K. Bloch, *J. Biol. Chem.* **253**, 2650 (1978).

maintained on agar slants. For slants, a stock solution containing per liter, 10 g of K_2HPO_4, 0.2 g of $FeSO_4$, 1.46 g of $MgSO_4$, 20 g of Tween 80, 6 g of fumaric acid, and 130 g of casamino acids is adjusted to pH 7.0 with KOH, 450 g/liter. Twenty milliliters of this solution are added along with 4 g of agar to 160 ml of water and autoclaved. Twenty milliliters of glucose, 200 g/liter (autoclaved separately) are then added, and, while still hot, the solution is divided into tubes for slants. For large-scale growth a stock solution containing per liter, 10 g of K_2HPO_4, 0.2 g of $FeSO_4$, 4.6 g of $MgSO_4$, 20 g of Tween 80, 7.5 g of fumaric acid, and 50 g of glutamic acid is adjusted to pH 7.0 with KOH, 450 g/liter. Two hundred milliliters of this solution are added to 1.6 liters of water in a 4-liter flask and autoclaved. Two hundred milliliters of 220 g of glucose per liter (autoclaved separately) are then added. Seed flasks (100 ml in a 250-ml flask) are inoculated by loop transfer from slants of *M. smegmatis*. After shaking at 37° for 24 hr, the seed flasks are poured into the 4-liter flasks and grown at 37° with shaking for about 32 hr to log phase ($A_{620} = 2.0$–2.5). Cells are harvested in a Sharples continuous flow centrifuge and stored at $-65°$. The yield of cells is about 2–2.5 g per liter of culture. The amount of fatty acid synthase activity recovered per gram of cells falls dramatically if the cells are allowed to grow to high density (4–6 g/liter).

Purification of Fatty Acid Synthase[4,5,14]

All purification steps are performed at 0–4°. Buffers contain 1 mM dithiothreitol and 1 mM Na_2EDTA. Between steps, the enzyme is frozen in liquid N_2 and stored at $-65°$. Purified enzyme stored in this manner is stable for more than 2 years.

Frozen cells (184 g) are dispersed with the aid of a blender in 553 ml (3 ml per gram of cells) of freshly prepared 0.1 M potassium phosphate buffer (pH 7.1 at 0°) containing 2 mM phenylmethylsulfonyl fluoride and 50 ml of isopropanol per liter. The cells are broken by passage through a French pressure cell (American Instrument Co.) operated at 16,000 psi. This solution is adjusted to pH 7.1 (glass electrode) by the addition of 1 M KOH. The broken cells are removed by centrifugation for 20 min at 9500 rpm in a Sorvall GSA rotor (14,600 g), and the supernatant recentrifuged for 90 min at 30,000 rpm in a Beckman 30 rotor (105,000 g).

The supernatant is brought to 1.4 M by the slow addition of finely ground ammonium sulfate,[15] allowed to stir for 15 min, and centrifuged for 15 min at 9500 rpm in a Sorvall GSA rotor (14,600 g). The resulting supernatant is brought to 2.0 M ammonium sulfate,[15] allowed to stir for 15 min, and centrifuged as above. The precipitate is suspended in 200 ml of

[15] W. I. Wood, *Anal. Biochem.* **73**, 250 (1976).

0.25 M potassium phosphate buffer (pH 7.1), dialyzed[16] for 4 hr against 5 liters of the same buffer, and stored. Overnight dialysis at this stage has resulted in large losses in activity.

The dialyzed fraction is thawed and centrifuged for 15 min at 15,000 rpm in a Sorvall SS-34 rotor (27,000 g). The supernatant conductivity is checked, and the solution is diluted slightly if necessary to the conductivity of 0.25 M potassium phosphate buffer (pH 7.1). This material is applied to a 360-ml bed (6.2 × 12 cm) of DEAE-cellulose (Bio-Rad Cellex D, high capacity) equilibrated with 0.25 M potassium phosphate buffer (pH 7.1). (The DEAE-cellulose is washed with 0.25 M NaOH and 0.25 M HCl, and fines are removed before use.) The column is washed with 800 ml of the same buffer (flow rate 425 ml/hr), eluted with a 1-liter linear gradient of 0.25 M to 0.5 M potassium phosphate buffer (pH 7.1), and washed with the final buffer. The pooled, active fractions (425 ml) are concentrated to 11.5 ml with Amicon concentration cells equipped with XM-50 membranes and stored.

The concentrated DEAE-cellulose fraction is applied to a 430-ml bed (2.5 × 88.5 cm) of BioGel A-15m (200–400 mesh) equilibrated with 0.5 M potassium phosphate (pH 7.1). The column is eluted with the same buffer (flow rate 46 ml/hr). Fractions with a specific activity greater than 1400 units per gram of protein (48.5 ml) are pooled, concentrated as above, and stored.

The affinity column step is performed in several batches. In a typical procedure 4 ml of the concentrated BioGel A-15m fraction is applied to a 1-ml bed (0.5 × 5 cm) of NADP$^+$-Agarose (P-L Biochemicals, AGNADP, type 4) equilibrated with 0.5 M potassium phosphate buffer (pH 7.1). The column is washed with about 2 ml of the buffer, and the enzyme is eluted with 5 ml of 50 mM NADP$^+$ in 0.5 M potassium phosphate buffer (pH 7.1). The enzyme can also be eluted by raising the temperature to 20°. Active fractions (2.8 ml) are pooled. NADP$^+$ is removed by passage through an 80-ml bed (1.6 × 40 cm) of BioGel A-1.5m (100–200 mesh) equilibrated and eluted (flow rate 50 ml/hr) with 0.5 M potassium phosphate buffer (pH 7.1). (This column is first washed with 2 ml of albumin, 5 g/liter, in 0.5 M potassium phosphate buffer, pH 7.1.) Activity elutes at the void volume, and active fractions (8 ml) are concentrated as above to 2 ml and stored.

A summary of the enzyme purification is given in the table. Enzyme prepared by this method is more than 95% pure as judged by SDS–gel electrophoresis.[4,5,14] Polyacrylamide gel electrophoresis in 0.1 M potassium phosphate buffer gives a single protein band that is coincident with a peak of malonyl-CoA-dependent pyridine nucleotide oxidation activity when the gel is soaked in assay solution.[4,5,14] The extinction coefficient of the purified synthase at 280 nm is 1.3 $(g/l)^{-1}$ cm^{-1}.[4,14]

---

[16] P. McPhie, this series, Vol. 22, p. 23.

SUMMARY OF THE PURIFICATION OF FATTY ACID SYNTHASE FROM *Mycobacterium smegmatis*

Fraction	Protein (mg)	Activity (mU/g cell)	Yield (%)	Specific activity (units/g protein)	Purification (fold)
Broken cells	12,400	678	100	10.1	1.0
High-speed supernatant	6,210	494	73	14.6	1.5
$(NH_4)_2SO_4$	2,450	802	118	58.3	5.8
DEAE-cellulose	162	267	39	303	30
BioGel A-15m	21.0	172	25	1510	150
NADP$^+$ affinity[a]	7.83	88.2	12.1	1920	190

[a] Corrected, about one-half of the material was processed through this step.

Properties

Kinetic Properties. Mycobacterium smegmatis fatty acid synthase produces fatty acyl-CoA derivatives ranging in chain length from 14 to as long as 32 carbons.[3,4,17,18] This product distribution is distinctly bimodal with respect to chain length with peaks at C_{16} and C_{24} fatty acids. The bimodal distribution persists throughout a wide variety of conditions although the proportion of the short-chain (C_{14}–C_{20}) and long-chain (C_{22}–C_{32}) peaks varies considerably. Low concentrations of acetyl-CoA or polysaccharide favor the longer-chain products, and high concentrations favor the shorter chains. The product distribution is nearly independent of the malonyl-CoA concentration, higher concentrations slightly favoring the long-chain products. At one extreme the synthase can produce 90% C_{22}–C_{32} fatty acids (at 20 μM acetyl-CoA and in the absence of polysaccharide) and at the other, 90% C_{14}–C_{20} fatty acids (at 3 mM acetyl-CoA and 100 μM polysaccharide).[3,4] The fatty acid distribution synthesized under the standard assay conditions described above is 32% C_{16}, 37% C_{18}, 1% C_{20}, 2% C_{22}, and 29% C_{24}.

The synthase is stimulated by the polysaccharides MMP or MGLP,[10] which are isolated from *M. smegmatis,* or by the synthetic polysaccharide 2,6 di-*O*-methylcycloheptaamylose, MCHA.[8] This initial steady-state stimulation can be as large as 50-fold depending on the assay conditions.[3-5] MMP, MGLP, or MCHA have no effect on fatty acid synthases from other organisms. *Mycobacterium smegmatis* fatty acid synthase is not stimulated by the unmethylated cycloheptaamylose or by amylose, starch, or glycogen.[9] A wide variety of studies have demonstrated that release of the acyl-CoA product from the enzyme is the rate-limiting step

[17] D. N. Brindley, S. Matsumura, and K. Bloch, *Nature (London)* **224,** 666 (1969).
[18] P. K. Flick and K. Bloch, *J. Biol. Chem.* **249,** 1031 (1974).

in the absence of polysaccharide; polysaccharide stimulates the enzyme rate by forming a ternary complex with the enzyme and the product and by markedly accelerating the product release.[3–5,19,20]

In the absence of polysaccharide the V_{max} for the synthase is relatively small. Under these conditions the K_m for acetyl-CoA is unusually large (about 400 μM). Polysaccharide increases the V_{max} 6- to 7-fold with a K_m of 50–100 μM. Saturating concentrations of polysaccharide lower the K_m for acetyl-CoA more than 10-fold to about 30 μM.[3–5]

The synthase utilizes NADPH for the β-ketoacyl reduction step and NADH for the enoyl reduction.[21] The enzyme will also elongate C_{16}-CoA when it is used as a primer instead of acetyl-CoA.[5,17,22] The enzyme is inhibited by the antibiotic cerulenin.[23]

Structural Studies.[4,5,14] Strong evidence has been presented that *M. smegmatis* fatty acid synthase is composed of identical subunits of 290,000 MW. A wide variety of electrophoresis, centrifugation, and stoichiometry data support this conclusion. The enzyme contains 1 mol of pantetheine and 1 mol of FMN per subunit. One mole of fatty acid is synthesized per subunit during the first enzyme turnover for both acetyl-CoA de novo synthesis and for C_{16}-CoA elongation.

The native molecular weight is about 2,000,000, and thus the enzyme contains 6–8 subunits. Active enzyme sedimentation shows that the 2,000,000 MW form of the enzyme is the active form under routine assay conditions.[4,14]

Acknowledgment

We gratefully acknowledge the guidance and support of Dr. Konrad Bloch, in whose laboratory this work was performed.

[19] D. O. Peterson and K. Bloch, *J. Biol. Chem.* **252**, 5735 (1977).
[20] R. J. Banis, D. O. Peterson, and K. Bloch, *J. Biol. Chem.* **252**, 5740 (1977).
[21] H. B. White, III, O. Mitsuhashi, and K. Bloch, *J. Biol. Chem.* **246**, 4751 (1971).
[22] D. E. Vance, T. W. Esders, and K. Bloch, *J. Biol. Chem.* **248**, 2310 (1973).
[23] D. Vance, I. Goldberg, O. Mitsuhashi, K. Bloch, S. Omura, and S. Nomura, *Biochem. Biophys. Res. Commun.* **48**, 649 (1972).

[15] Fatty Acid Synthase from *Cephalosporium caerulens*[1]

By AKIHIKO KAWAGUCHI, HIROSHI TOMODA, SHIGENOBU OKUDA, and SATOSHI ŌMURA

Cerulenin is an antifungal antibiotic isolated from culture filtrates of the fungus *Cephalosporium caerulens*[2] and is a potent inhibitor of fatty acid synthase systems isolated from various microorganisms and from rat liver.[3,4] This antibiotic specifically blocks the activity of β-ketoacyl thioester synthase (condensing enzyme).[3,4] Fatty acid synthase from *C. caerulens* is much less sensitive to cerulenin than fatty acid synthases from other sources.[1]

Assay Method

The enzyme activity is assayed spectrophotometrically by measuring the rate of oxidation of NADPH at 340 nm. Assays are carried out at 37° in a mixture containing 0.1 M potassium phosphate buffer (pH 7.15), 5 mM dithiothreitol (DTT), 0.2 mM NADPH, 50 μM acetyl-CoA, 40 μM malonyl-CoA, 0.1 mg bovine serum albumin, and enzyme protein to produce an absorbance change of 0.05–0.15 per minute in a total volume of 0.5 ml. After measuring NADPH oxidation without malonyl-CoA, the reaction is started by adding malonyl-CoA. The rate of NADPH oxidation prior to malonyl-CoA addition serves as a blank value, which is subtracted from the total rate observed in the presence of malonyl-CoA. One unit of activity is defined as the amount of enzyme required to incorporate 1 nmol of malonyl-CoA into fatty acids per minute.

Purification of the Enzyme

The method is the same as the procedure described previously,[1] but with the last step slightly modified. All steps are performed at 0–4°. All buffers are potassium phosphate, pH 7.0, containing 10 mM

[1] A. Kawaguchi, H. Tomoda, S. Okuda, J. Awaya, and S. Ōmura, *Arch. Biochem. Biophys.* **197**, 30 (1979).

[2] S. Ōmura, M. Katagiri, A. Nakagawa, Y. Sano, S. Nomura, and T. Hata, *J. Antibiot. Ser. A* **20**, 349 (1967).

[3] G. D'Agnolo, I. S. Rosenfeld, J. Awaya, S. Ōmura, and P. R. Vagelos, *Biochim. Biophys. Acta* **326**, 155 (1973).

[4] D. Vance, I. Goldberg, O. Mitsuhashi, K. Bloch, S. Ōmura, and S. Nomura, *Biochem. Biophys. Res. Commun.* **48**, 649 (1972).

2-mercaptoethanol and 1 mM EDTA, unless otherwise stated. Protein is determined by the method that involves the binding of Coomassie Brilliant Blue G-250 to protein.[5] *Cephalosporium caerulens* is grown to the mid-log phase for 24 hr at 27° in a medium containing 30 g of glycerol, 10 g of glucose, 5 g of peptone, and 2 g of NaCl per liter. Cells can be stored frozen ($-20°$) for a few months.

Preparation of Supernatant Solution. Frozen *C. caerulens* cells (135 g) are thawed in 540 ml of 0.1 M potassium phosphate buffer. The cells are broken by passage through a French pressure cell operated at 20,000 psi, and disrupted cells are centrifuged at 25,000 g for 20 min. The volume of the resulting supernatant is adjusted to 1000 ml by adding 0.1 M potassium phosphate buffer.

Ammonium Sulfate Fractionation. To the diluted supernatant solution solid ammonium sulfate is added slowly to 30% saturation. After stirring for 15 min, the precipitate is removed by centrifugation at 25,000 g for 20 min and discarded. The supernatant is brought to 50% saturation with solid ammonium sulfate, stirred for 15 min, and centrifuged at 25,000 g for 20 min. The precipitate is dissolved in 35 ml of 0.05 M potassium phosphate buffer and dialyzed against 0.05 M potassium phosphate buffer for 3 hr.

DEAE-Cellulose Column Chromatography. The dialyzed solution is applied to a DEAE-cellulose column (90 g of Whatman DE-52, 4.5 cm × 12 cm), equilibrated with 0.05 M potassium phosphate buffer. The column is washed with 500 ml of 0.05 M potassium phosphate buffer, and the enzyme is eluted with a linear gradient from 0.05 to 0.25 M potassium phosphate buffer (900 ml). The major fractions containing enzyme activity are pooled.

Ammonium Sulfate Precipitation. Saturated ammonium sulfate solution (pH 7.0) is added dropwise to the pooled fractions to 50% saturation with stirring. Direct addition of solid ammonium sulfate at this step causes changes in pH and leads to loss of enzyme activity. The suspension is stirred for 15 min and centrifuged at 25,000 g for 20 min. The precipitate is dissolved in 4 ml of 0.25 M potassium phosphate buffer (pH 7.0) containing 1 mM EDTA and 1 mM DTT instead of 10 mM 2-mercaptoethanol.

BioGel A-5m Gel Filtration. The solution is applied to a BioGel A-5m column (2.6 cm × 95 cm), equilibrated with 0.25 M potassium phosphate buffer (pH 7.0) containing 1 mM EDTA and 1 mM DTT, and the enzyme is eluted with the same buffer. The fractions containing enzyme activity are combined and concentrated to 6.5 ml using a Diaflo apparatus with an XM-50 membrane. The results of this purification are summarized in the

[5] M. M. Bradford, *Anal. Biochem.* **72**, 248 (1976).

PURIFICATION OF *Cephalosporium caerulens* FATTY ACID SYNTHASE

Fraction	Volume (ml)	Total activity (units)	Total protein (mg)	Specific activity (units/mg)	Yield (%)
Crude extract	1000	64,500	2720	23.7	100
$(NH_4)_2SO_4$, 30–50%	43	47,900	594	80.6	74.3
DEAE-cellulose	400	30,400	70.5	430	47.1
BioGel A-5m	6.5	15,500	7.49	2070	24.0

table. After the final purification step, the synthase is homogeneous, as judged by the appearance of a single band when subjected to electrophoresis on 1% agarose gels.[1] The molecular weight, estimated by chromatography on BioGel A-5m, is 1.2×10^6. The purified enzyme can be stored frozen ($-20°$) in 0.25 M potassium phosphate buffer (pH 7.0) containing 1 mM EDTA and 1 mM DTT for several months with negligible loss of activity.

Cerulenin Resistance of the Enzyme[1]

Under the standard assay conditions, the products are palmitic and stearic acids. NADPH is the specific electron donor for both β-ketoacyl and enoyl reductions. The apparent K_m values for the three substrates are: malonyl-CoA, 9.1 μM; acetyl-CoA, 9.3 μM; and NADPH, 14 μM. Catalytic properties and the product patterns of *C. caerulens* fatty acid synthase

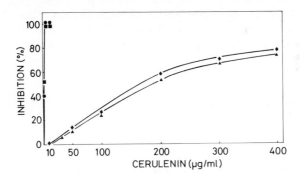

FIG. 1. Percentage of inhibition of fatty acid synthases and yeast alcohol dehydrogenase by cerulenin. ◆, *Cephalosporium caerulens* fatty acid synthase; ●, yeast fatty acid synthase; ■, *Brevibacterium ammoniagenes* fatty acid synthase; ▲, yeast alcohol dehydrogenase. From Kawaguchi *et al.*[1]

are almost the same as those of yeast fatty acid synthase.[6] However, a striking difference between these synthases is their sensitivity to cerulenin (Fig. 1). The sensitivity of *C. caerulens* synthase to cerulenin is much less than that of yeast and *Brevibacterium*[7] synthases. The inhibition of *C. caerulens* synthase with high concentrations of cerulenin might be nonspecific because such high concentrations of cerulenin cause the inhibition of yeast alcohol dehydrogenase to the same degree. The cerulenin resistance of the synthase may be responsible for the protection of *C. caerulens* from inhibitory effects of this antibiotic.

[6] F. Lynen, this series, Vol. 14 [3].
[7] A. Kawaguchi and S. Okuda, *Proc. Natl. Acad. Sci. U.S.A.* **74**, 3180 (1977).

[16] Fatty Acid Synthase from *Brevibacterium ammoniagenes*[1]

By AKIHIKO KAWAGUCHI, YOUSUKE SEYAMA, TAMIO YAMAKAWA, and SHIGENOBU OKUDA

Fatty acid synthesis from acetyl-CoA and malonyl-CoA in *Brevibacterium ammoniagenes* is catalyzed by a fatty acid synthase of the multienzyme complex type (type I). *Brevibacterium ammoniagenes* synthase, like the multienzyme complexes of yeast[2] and *Mycobacterium smegmatis*,[3] produces fatty acyl-CoA derivatives. The prominent feature of this synthase is its ability to synthesize not only saturated fatty acyl-CoAs (palmitoyl- and stearoyl-CoA), but also oleoyl-CoA. Oleoyl-CoA is synthesized by an anaerobic process involving β,γ-dehydration of a β-hydroxyacyl thioester intermediate and subsequent chain elongation of the β,γ-enoate without reduction of the double bond (Fig. 1). The enzyme component responsible for the oleate formation is tightly associated with the various activities for carbon chain elongation.

Assay Method

The enzyme activity can be determined by either a radioactive, spectrophotometric, or mass fragmentographic method. In the radioactive method, incorporation of [2-14C]malonyl-CoA into fatty acids is measured in the presence of acetyl-CoA, NADPH, and NADH. This method is reliable for the assay of crude and purified preparations. The spec-

[1] A. Kawaguchi and S. Okuda, *Proc. Natl. Acad. Sci. U.S.A.* **74**, 3180 (1977).
[2] F. Lynen, this series, Vol. 14 [3].
[3] K. Bloch, this series, Vol. 35 [10].

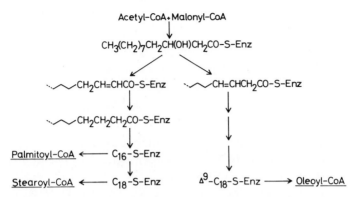

FIG. 1. Proposed mechanism of *Brevibacterium ammoniagenes* fatty acid synthase.

trophotometric method measures the rate of oxidation of NADPH and NADH at 340 nm and is most useful for the purified enzyme. The mass fragmentographic method is best suited for quantitative determination of the amounts of individual fatty acids synthesized by the enzyme reaction and is applicable to crude and purified enzyme preparations.

Reagents

Potassium phosphate buffer, 1 M, pH 7.3
Dithiothreitol (DTT), 0.125 M
NADPH, 2.5 mM, prepared fresh and kept at 0°
NADH, 2.5 mM, prepared fresh and kept at 0°
Acetyl-CoA, 2.5 mM
Malonyl-CoA, 2.5 mM
Heptakis (2,6-di-O-methyl)-β-cyclodextrin (CH$_3$-cyclodextrin), 2.5 mg/ml

Spectrophotometric Assay.[1] Measurement is carried out in a quartz cuvette (d = 1.0 cm) at 340 nm using a recording spectrophotometer at 37°. The reaction mixture contains 0.4 M potassium phosphate buffer (pH 7.3), 5 mM DTT, 0.1 mM NADPH, 0.1 mM NADH, 40 μM malonyl-CoA, 50 μM acetyl-CoA, 50 μg of CH$_3$-cyclodextrin, and enzyme protein to produce an absorbance change of 0.05–0.15 per minute in a total volume of 0.5 ml. After measuring NADPH- and NADH-oxidation without malonyl-CoA, the reaction is started by adding malonyl-CoA. The rate of NADPH- and NADH-oxidation prior to malonyl-CoA addition serves as a blank value that is subtracted from the total rate observed in the presence of malonyl-CoA.

Radioactive Assay.[1] The reaction mixture used in this assay is the same as that used in the spectrophotometric assay except that [2-^{14}C]malonyl-CoA (56,200 dpm per assay) is used. The reaction is started by the addition of [2-^{14}C]malonyl-CoA and carried out at 37° for 15 min. Then 0.15 ml of 50% KOH is added to terminate the reaction, and the mixture is heated in boiling water for 30 min. The solution is acidified to pH 1 with 8–10 drops of 6 N HCl, and the fatty acids are extracted three times with 3 ml of n-hexane. The hexane extracts are combined and washed with water (acidified with acetic acid), then evaporated at 70° under nitrogen. The radioactivity of the residue is measured in a liquid scintillation spectrometer.

Mass Fragmentographic Assay.[4] The individual fatty acids synthesized by the enzyme reaction are assayed according to the following method. The reaction mixture used in this assay is the same as that described for the spectrophotometric assay except for the use of deuterated water instead of water. The reaction is stopped by freezing with Dry Ice–acetone, and heptadecanoic acid (444 ng) is added to the reaction mixture as an internal standard. Fatty acid methyl esters are liberated from the dried residue by methanolysis at 100° for 3 hr with 2 ml of 3% anhydrous methanolic HCl and are extracted with 2 ml of n-hexane (three times). Methyl esters are applied to a 2 m × 3 mm glass column containing 1.5% OV-1 on Chromosorb W (AW-DMCS) and are chromatographed at 160° with helium as the carrier gas at a flow rate of 30 ml/min. Mass fragmentography is performed with a gas chromatography–mass spectrometer combined with a computerized mass fragmentograph. The temperatures of the ion source, line, and separator are 270°, 260°, and 250°, respectively. The amounts of fatty acids synthesized by the enzyme reaction and endogenous material are determined by monitoring the intensities of the m/e 77 and m/e 74 fragment ions, and peak area measurements are carried out with a computer system.

Unit. One unit of enzyme activity is defined as the amount of enzyme required to incorporate 1 nmol of malonyl-CoA into fatty acids per minute.

Purification of the Enzyme

The method is the same as the procedure described previously[1] but with the first step slightly modified. All steps are performed at 0–4°. All buffers are potassium phosphate, pH 7.0, containing 1 mM EDTA and 1 mM DTT unless otherwise stated. Protein is determined by the method that involves the binding of Coomassie Brilliant Blue G-250 to protein.[5] *Brevibacterium ammoniagenes* is grown to the mid-log phase (A_{660} = 9.5–10.0) in a medium containing, per liter, 10 g of peptone, 7 g of meat

[4] Y. Seyama, A. Kawaguchi, S. Okuda, and T. Yamakawa, *J. Biochem.* **84**, 1309 (1978).
[5] M. M. Bradford, *Anal. Biochem.* **72**, 248 (1976).

extract, 3 g of NaCl, and 10 g of dextrose. Cells can be stored frozen ($-20°$) for a few months.

Preparation of Supernatant Solution. Frozen *B. ammoniagenes* cells (110 g) are thawed in 110 ml of 0.1 *M* potassium phosphate buffer. The cells are broken with 400 g of glass beads (0.17–0.18 mm) in a Waring blender for a total of 15 min, three 5-min periods with 5-min intervals. After addition of a 110 ml of the buffer with stirring, the homogenate is filtered through a single layer of cheesecloth and centrifuged at 25,000 *g* for 20 min. The sediment is discarded, and the supernatant solution is recentrifuged at 105,000 *g* for 1 hr. The resulting supernatant solution contains the fatty acid synthase.

Ammonium Sulfate Fractionation. The supernatant solution from the preceding step is stirred gently with a magnet, and solid ammonium sulfate is added slowly to 30% saturation. After stirring for 15 min, the precipitate is removed by centrifugation at 25,000 *g* for 20 min and discarded. The supernatant is slowly brought to 55% saturation with solid ammonium sulfate, stirred, and centrifuged as before. The precipitate is dissolved in 50 ml of 0.1 *M* potassium phosphate buffer. The solution is passed through a Sephadex G-25 column (3.5 cm × 60 cm) which is equilibrated with the same buffer.

DEAE-Cellulose Treatment. The Sephadex G-25 treated solution is diluted five times with 0.1 *M* potassium phosphate buffer, and DEAE-cellulose (160 g of Whatman DE-52) equilibrated with the same buffer is then added. After the mixture has been stirred for 30 min, the DEAE-cellulose containing absorbed enzyme if filtered and washed with 1 liter of 0.25 *M* potassium phosphate buffer for 30 min with stirring. After filtration, the DEAE-cellulose is suspended in 300 ml of 0.25 *M* potassium phosphate buffer and packed into a column. The enzyme is eluted with 0.5 *M* potassium phosphate buffer, and the major fractions containing enzyme activity are pooled.

Ammonium Sulfate Precipitation. Saturated ammonium sulfate solution, pH 7.0, containing 1 m*M* EDTA and 1 m*M* DTT is added dropwise to the pooled fractions to 50% saturation with stirring. Direct addition of solid ammonium sulfate causes changes in pH, and leads to loss of enzyme activity. The suspension is stirred for 15 min and centrifuged at 25,000 *g* for 20 min. The precipitate is dissolved in 8 ml of 0.5 *M* potassium phosphate buffer. The solution is centrifuged at 25,000 *g* for 10 min to remove insoluble materials.

Sepharose 6B Gel Filtration. The clear solution is applied to a Sepharose 6B column (2.6 cm × 95 cm), equilibrated with 0.5 *M* potassium phosphate buffer, and the enzyme is eluted with the same buffer. Enzymatic activity is eluted with the second protein peak. Active frac-

PURIFICATION OF *Brevibacterium ammoniagenes* FATTY ACID SYNTHASE

Fraction	Volume (ml)	Total activity (units)	Total protein (mg)	Specific activity (units/mg)	Yield (%)
105,000 g supernatant	210	52,700	2440	21.6	100
$(NH_4)_2SO_4$ (30–55%)	124	50,700	704	72.0	96.2
DEAE-cellulose	186	45,600	195	234	86.5
$(NH_4)_2SO_4$ (0–50%)	8.5	38,900	142	274	73.8
Sepharose 6B	4.0	15,500	9.3	1670	29.4

tions are combined and concentrated to 4.0 ml using a Diaflo apparatus with a XM-50 membrane. The results of this purification are summarized in the table. After the final purification step the synthase is homogeneous, as judged by gel electrophoresis on 1% agarose gel.[1] The molecular weight, estimated by chromatography on Sepharose 6B, is 1.2×10^6.

Properties of the Enzyme

Storage and Stability of the Enzyme. The purified *B. ammoniagenes* fatty acid synthase can be stored frozen ($-20°$) in 0.5 M potassium phosphate buffer (pH 7.0) containing 1 mM EDTA and 1 mM DTT for at least a month. The synthase can also be stored in the same buffer at 0–4° for a few days with negligible loss of activity, but the enzyme rapidly loses the activity on dialysis against 0.005 M potassium phosphate buffer, pH 7.0, containing 1 mM EDTA and 1 mM DTT (a half-life of less than 1 hr). Reactivation takes place upon dialysis against 0.5 M potassium phosphate buffer, pH 7.0, containing 1 mM EDTA and 1 mM DTT.[6]

Products of the Reaction. Under the standard assay conditions the enzyme produces a mixture of fatty acyl-CoA derivatives. The fatty acid composition of the products is palmitic, stearic, and oleic acids. The relative amounts of the three fatty acids depend on the assay conditions. The ratio of unsaturated to saturated fatty acids is dependent on the temperature of the enzyme reaction (Fig. 2).[7] The proportions of palmitic and stearic acids are subject to the concentrations of acetyl-CoA and malonyl-CoA.[8]

[6] A. Kawaguchi and S. Okuda, unpublished data, 1977.
[7] A. Kawaguchi, Y. Seyama, K. Sasaki, S. Okuda, and T. Yamakawa, *J. Biochem.* **85**, 865 (1979).
[8] A. Kawaguchi, K. Arai, S. Okuda, Y. Seyama, and T. Yamakawa, *J. Biochem.* **88**, 303 (1980).

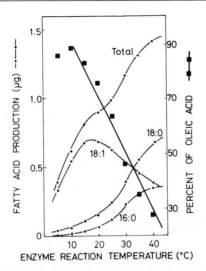

FIG. 2. Temperature-sensitive synthesis of fatty acid species by *Brevibacterium am-moniagenes* fatty acid synthase. From Kawaguchi *et al.*[7]

Specificity and Kinetics of the Reaction. The pH optimum of the enzyme activity is 7.3, and relatively high concentrations of potassium phosphate buffer (optimal at 0.4 M phosphate) are required.[1] NADPH is the specific electron donor for the β-ketoacyl reductase step, and NADH is specifically utilized for enoyl reduction.[9] This synthase is specific for malonyl-CoA, but acetyl-CoA may be replaced by longer acyl-CoAs up to 10 carbon atoms in chain length. Enzymatic activity is highest with acetyl-CoA and decreases with increasing chain length of the acyl-CoA derivatives.[8] The K_m values for substrates are malonyl-CoA, 9 μM; acetyl-CoA, 11 μM; NADPH, 18 μM; and NADH, 11 μM.[1]

Inhibitors and Activators. The enzyme is inhibited by cerulenin (50% inhibition at 1.0 μg/ml).[10] 3-Decynoyl-*N*-acetylcysteamine has inhibitory effects on oleate synthesis, but not on the synthesis of saturated fatty acids.[1] Long-chain acyl-CoA derivatives (C_{12}–C_{20}) inhibit the synthase. Palmitoyl-CoA, stearoyl-CoA, and oleoyl-CoA are the most potent inhibitors (50% inhibition at 3 μM). Bovine serum albumin (0.1 mg/ml) and CH_3-cyclodextrin relieve the long-chain fatty acyl-CoA inhibition. The polysaccharides MMP and MGLP, which are isolated from *Mycobacterium phlei*,[11] also prevent the inhibition by long-chain fatty acyl-CoA.

[9] Y. Seyama, T. Kasama, T. Yamakawa, A. Kawaguchi, and S. Okuda, *J. Biochem.* **81**, 1167 (1977).
[10] A. Kawaguchi, H. Tomoda, S. Okuda, J. Awaya, and S. Omura, *Arch. Biochem. Biophys.* **197**, 30 (1979).
[11] R. Bergeron, Y. Machida, and K. Bloch, *J. Biol. Chem.* **250**, 1223 (1975).

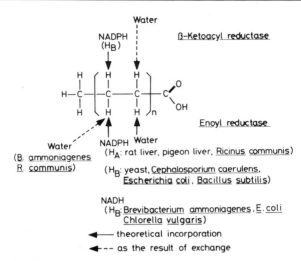

FIG. 3. Origin of hydrogen atoms in fatty acids synthesized by various fatty acid synthases. H_A: A-side hydrogen of reduced pyridine nucleotides; H_B: B-side hydrogen of reduced pyridine nucleotides. From Saito et al.[14]

Stereospecificity for NADPH and NADH. The two hydrogen atoms at C-4 of the dihydropyridine ring of NADPH or NADH are enzymatically nonequivalent, and pyridine nucleotide-linked dehydrogenases stereospecifically transfer the hydrogens to substrates.[12] The A-specific dehydrogenases remove the hydrogen on the A-side (*pro-R*) of the dihydropyridine ring and the B-specific dehydrogenases remove the B-side hydrogen (*pro-S*). This stereospecificity has been recognized as one of the most highly conserved characteristics of dehydrogenases.[13] The β-ketoacyl reductase reaction is B-specific regardless of the source of the fatty acid synthase, and the stereospecificity of the enoyl reductase reaction for NADPH or NADH is dependent on the source of the enzyme (Fig. 3).[14] Fatty acid synthase from *B. ammoniagenes* utilizes the B-side hydrogen of NADH for enoyl reduction but the number of incorporated hydrogen atoms is less than expected.[9] We suspect that there is some exchange of the B-side hydrogen of NADH with water catalyzed by this enoyl reductase, but the detailed mechanism is not clear.

[12] H. F. Fisher, P. Ofner, E. E. Conn, B. Vennesland, and F. H. Westheimer, *J. Biol. Chem.* **202**, 687 (1953).
[13] K. You, L. J. Arnold Jr., W. S. Allison, and N. O. Kaplan, *Trends Biochem. Sci.* **3**, 265 (1978).
[14] K. Saito, A. Kawaguchi, S. Okuda, Y. Seyama, T. Yamakawa, Y. Nakamura, and M. Yamada, *Plant Cell Physiol.* **21**, 9 (1980).

Mechanism of Oleoyl-CoA Formation[1]

The synthesis of oleoyl-CoA does not require molecular oxygen and is inhibited by 3-decynoyl-*N*-acetylcysteamine, which is a potent inhibitor of β-hydroxydecanoyl thioester dehydrase from *Escherichia coli*.[15] Palmitic, stearic, and oleic acids are synthesized at almost a linear rate during the first 15 min of the enzyme reaction, and there are no significant differences in the ratio of saturated to unsaturated acids throughout a 1-hr reaction period. The ratio of saturated to unsaturated fatty acids is essentially constant across the enzyme active peak of a Sepharose 6B profile and also constant for all preparations throughout enzyme purification. These results suggest that oleoyl-CoA is synthesized by the anaerobic process shown in Fig. 1 and that the enzyme component responsible for oleate formation is tightly associated with the various activities for carbon chain elongation as an integral part of the *Brevibacterium* fatty acid synthase complex.

[15] G. M. Helmkamp, Jr., R. R. Rando, D. J. H. Brock, and K. Bloch, *J. Biol. Chem.* 243, 3229 (1968).

[17] Fatty Acid Synthase From *Ceratitis capitata*

By A. M. MUNICIO and J. G. GAVILANES

The de novo synthesis of fatty acids during the development of the insect *Ceratitis capitata* appears to be hardly regulated.[1] *In vitro* and *in vivo* experiments of lipid biosynthesis from [^{14}C]acetyl-CoA as well as measurements of both acetyl-CoA carboxylase and fatty acid synthase activities during development and metamorphosis of *C. capitata* show that fatty acid synthesis reaches maximal activity at the early third-instar larval stage and declines soon afterward.[1] Acetyl-CoA carboxylase and fatty acid synthase activities are very low during the pharate adult stage and exhibit a slow but steady increase after emergence. The very sharp peak of fatty acid synthesis capacity that appears at the larval stage can be easily correlated with the previous formation of the substrate for de novo synthesis. For these reasons, the fatty acid synthase complex has been isolated from the particle-free supernatant fraction of homogenates from the 6-day-old *C. capitata* larvae.[2]

[1] A. M. Municio, J. M. Odriozola, M. Pérez-Albarsanz, J. A. Ramos, and E. Relaño, *Insect Biochem.* 4, 401 (1974).

[2] H. E. Hinton, *Proc. R. Soc. London Ser. B* 35, 55 (1971).

Assay Methods

Principle. Fatty acid synthase activity is determined by a radiochemical assay that monitors the incorporation of [^{14}C]acetyl-CoA into fatty acids. Alternatively, a spectrophotometric assay is used that measures the NADPH oxidation during the synthetic process.

Radiochemical Assay

Reagents

Potassium phosphate buffer, 0.25 M, containing 3 mM EDTA and 3 mM mercaptoethanol, pH 7.5
Malonyl-CoA, 50 nM
NADPH, 300 nM
[1-^{14}C]Acetyl-CoA, approximately 3 $\times$ 10^6 cpm/μmol
Liquid scintillator: 4.0 g of 2,5-diphenyloxazole (PPO), 100 mg of 4-bis-2-(4-methyl-5-phenyloxazolyl)benzene (dimethyl-POPOP), to 1 liter of toluene

Procedure. The complete reaction mixture contains 0.2 ml of the cofactor solution composed of malonyl-CoA, NADPH, and 13.5 nmol of [1-^{14}C]acetyl-CoA (0.05 μCi/ml). The reaction is initiated by addition of 0.3 ml of the enzyme preparation to 0.2 ml of the reaction mixture and is continued at 37° for 10 min in a gently shaken water bath. The reaction is stopped at 0–4 by adding (*a*) 0.15 ml of 50% KOH, (*b*) 1.05 ml of methanol and 0.5 ml of chloroform, according to either the Goldberg and Bloch[3] method or the Bligh and Dyer[4] method for the extraction procedure, respectively. In the procedure of Goldberg and Bloch,[3] the alkalinized incubation mixture is warmed in a boiling water bath for 20 min and, after being acidified with HCl to pH 1.0 at 0–4, fatty acids are extracted with three 3-ml portions of hexane; this method requires a protein concentration of less than 0.3 mg/ml as well as a low concentration of detergents and lipids. The method of Bligh and Dyer[4] uses 2.1 volumes of methanol and 1.0 volume of chloroform. In both methods the organic solvent phase, which contains the labeled fatty acids, is taken to dryness at 60° in 15 $\times$ 45 mm glass vials (flat-bottomed shell vials) in a forced draft oven. Three milliliters of liquid scintillator are added to each vial. The radioactivity is determined with a liquid scintillation spectrometer.

Enzyme activity is expressed as nanomoles of fatty acids synthesized at 37° in 10 min.

---

[3] I. Goldberg and K. Bloch, *J. Biol. Chem.* **247**, 7349 (1972).
[4] E. G. Bligh and W. J. Dyer, *Can. J. Biochem. Physiol.* **39**, 911 (1959).

Identification of Products

The labeled fatty acid residues are methylated with the boron trifluoride–methanol reagent (25%, w/w) in sealed tubes at 100–110° for 90 min.[5] Fatty acid methyl esters are extracted with pentane, washed with water, and dried over sodium sulfate. Radioactivity of the methyl esters is determined by means of a flow-through reactor proportional counter connected to a gas–liquid chromatographic unit through a stream-splitting device (split ratio 16 : 1). The gas–liquid unit has a flame ionization detector and a stainless steel column (6 feet × ⅛ in.) packed with 10% EGS on Chromosorb W (60–80 mesh) and a carrier gas (He) flow of 18.3 ml/min at 180°C. The main effluent stream is driven into the flow-through reactor, where oxidizing degradation of the methyl esters takes place; water is removed with the aid of a drying tube at the end of which methane is continuously fed through a T piece to give a He/CH_4 ratio of 1 : 3. The counting tube has an active volume of 10 ml and is operated at 2700 V. Working temperatures are as follows: proportional counting tube, 100°; reactor furnace, 620°; connection tube between stream-splitter and furnace, 200°. Relative activities are determined by dividing the number of counts under a given peak by the average residence time of components within the active volume of the proportional detector and correcting for background.

Purification Procedure

Insects are cultured under carefully controlled conditions of diet, temperature, and humidity.[6] The results of a typical purification of fatty acid synthase from the cytoplasmic cell fraction of 6-day larvae of *C. capitata* are summarized in Table I. All operations are carried out at 2–6.

Preparation of the Cytoplasmic Fraction. Six-day larvae, 255 g, are homogenized in an Omni-Mixer (Sorvall, USA) with 100 ml of 0.1 M potassium phosphate buffer, pH 7.5, containing 2 mM EDTA and 1 mM mercaptoethanol, for 6 min in 2-min periods. The homogenate is centrifuged at 34,800 g for 15 min. The supernatant is passed through two layers of cheesecloth, and the solution is submitted to a second centrifugation at 95,000 g for 1.5 hr.

Ammonium Sulfate Fractionation. To the supernatant fraction is added solid ammonium sulfate to a 50% concentration; this is done with magne-

---

[5] L. D. Metcalfe, A. A. Schmitz, and J. R. Pelka, *Anal. Chem.* **38**, 514 (1966).
[6] J. M. Fernández-Sousa, A. M. Municio, and A. Ribera, *Biochim. Biophys. Acta* **231**, 527 (1971).

TABLE I
PURIFICATION OF FATTY ACID SYNTHASE FROM *Ceratitis capitata*

Procedure	Volume (ml)	Protein (mg/ml)	Specific activity	Total activity	Yield (%)
Ultracentrifuged solution	120	37.4	0.42	1900	100
(NH₄)₂SO₄ fractionation	100	7.2	2.11	1520	80.3
DEAE-cellulose column	200	0.7	7.15	1020	53.7
Sephadex G-150	76	0.1	76.64	699	37.0

tic stirring over a period of approximately 45 min. After standing for 30–60 min, the precipitated protein is collected by centrifugation at 34,800 g for 30 min. The pellet is brought into solution in 100 ml of the potassium phosphate buffer, pH 7.5, containing 2 mM EDTA and 1 mM mercaptoethanol, with stirring for 45 min. The insoluble residue is removed by centrifugation, and the solution is dialyzed against 3 liters of potassium phosphate buffer in a cellulose acetate ultrafilter (b/HFU-1).

DEAE-Cellulose Fractionation. The solution from the preceding step is stirred with 150 ml of DEAE-cellulose ion exchanger in potassium phosphate buffer, pH 7.5, for 1 hr. The exchanger is filtered and resuspended in 100 ml of buffer; after stirring for 1 hr, the exchanger is recovered by filtration and the process is repeated twice. The enzyme is eluted from the exchanger with 100-ml batches of 0.3 M potassium phosphate buffer, containing 2 mM EDTA and 1 mM mercaptoethanol, pH 7.5, with stirring for 1 hr. The pooled fractions showing fatty acid synthase activity are concentrated through ultrafiltration, using an Amicon filter Model UF-52 with PM-10 membranes.

Gel Filtration with Sephadex G-150. Enzyme solution from the preceding step is applied to a column (2.5 × 50 cm) of Sephadex G-150, equilibrated with the 0.25 M phosphate buffer. The enzyme is eluted with this buffer at a rate of 15–20 ml/hr. Enzyme activity appears in a first sharp peak that accounts for 50–60% of the activity applied to the column. At this stage the preparations show a single band on gel electrophoresis.

Table I shows that the fatty acid synthase complex is purified approximately 180 times in good yield by this procedure.[7,8]

[7] A. M. Municio, M. A. Lizarbe, E. Relaño, and J. A. Ramos, *Biochim. Biophys. Acta* **487**, 175 (1977).
[8] M. A. Lizarbe, Doctoral thesis, Complutensis University, Madrid, 1979.

AMINO ACID COMPOSITION OF FATTY ACID SYNTHASE FROM *Ceratitis capitata*

Amino acid	%	Amino acid	%
Asx	10.92	Met	1.65
Thr	6.39	Ile	5.80
Ser	6.43	Leu	8.27
Glx	13.28	Tyr	3.77
Pro	4.26	Phe	3.26
Gly	7.18	His	1.79
Ala	7.83	Lys	5.72
Cys	1.74	Arg	4.56
Val	5.87	Trp	1.28

Molecular and Catalytic Properties

Amino Acid Composition. The amino acid composition of the purified fatty acid synthase complex has been determined on a Durrum automatic analyzer. Tryptophan has been quantitated by the methods of Beaven and Holiday,[9] Edelhoch,[10] and Bencze and Schmid.[11] Cysteine has been evaluated as carboxymethylcysteine through hydrolysis of the carboxymethylated protein for 36 hr. Serine and threonine contents are corrected by extrapolation to zero time; valine and isoleucine are calculated from 120-hr hydrolyzates. Results are given in Table II.[8]

Major Fatty Acids Synthesized. Radio-gas chromatographic analysis of the end products shows the composition of the main fatty acids synthesized to be as follows: capric acid (2.1%), lauric acid (27.5%), myristic acid (12.4%), palmitic acid (56.4%), and stearic acid (1.6%) at a malonyl-CoA : acetyl-CoA ratio of $1:4^7$.

Kinetic Properties. The optimum pH is 7.4 ± 0.1. At this pH value the observed kinetic parameters are K_m for acetyl-CoA = 10 μM, K_m for malonyl-CoA = 80 μM, and K_m for NADPH = 15 μM.

Molecular Weight. Gel filtration on a Sepharose-6B (1.1 × 60 cm) column equilibrated with 0.25 M sodium phosphate buffer, pH 7.5, shows that fatty acid synthase (0.1–0.2 mg/ml) has a molecular weight of 650,000. At higher enzyme concentrations anomalous results are obtained due to protein polymerization.[8]

Partial Specific Volume. The partial specific volume of the polypeptide component of the complex is $\bar{v}$ = 0.728–0.731 cm³/g as calculated by the

---

[9] G. H. Beaven and E. R. Holiday, *Adv. Protein Chem.* **7**, 319 (1952).
[10] H. Edelhoch, *Biochemistry* **6**, 1948 (1967).
[11] W. L. Bencze and K. Schmid, *Anal. Chem.* **29**, 1193 (1957).

TABLE III

LIPID CONTENT OF FATTY ACID SYNTHASE PREPARATIONS FROM *Ceratitis capitata* AND
LIPID/PROTEIN-ENZYME ACTIVITY RELATIONSHIP

Enzyme activity (nmol FFA)	Lipid/protein ratio	Percentages[a]				
		TG	DG	FFA	PG	CE
0.08	0.16	81.2	10.1	4.8	2.9	1.0
2.36	0.99	81.6	10.1	2.7	4.7	0.9
2.55	1.01	81.2	9.9	3.0	4.9	1.0
2.67	1.04	81.3	10.0	2.8	5.0	1.0
3.26	1.09	81.1	9.9	2.3	5.6	1.1

[a] TG, triacylglycerols; DG, diacylglycerols; FFA, free fatty acids; PG, phosphoglycerides; CE, cholesterol esters.

method of Cohn and Edsall.[12] The variation is related to the degree of amidation of the acidic amino acids.[8]

Stability. The purified enzyme maintains 90% of the activity after 10 days of storage in the presence of 1 mM mercaptoethanol at $-60°$.[7]

Lipid Content–Enzyme Activity Relationship. Purified enzyme preparations exhibit an enzyme activity dependent on the lipid : protein ratio, as shown in Table III. The main lipid classes are also recorded in this table; these lipid classes are triacylglycerols, diacylglycerols, free fatty acids, phosphoglycerides, and cholesterol esters.[13] Delipidated preparations do not show any residual activity. Phospholipase A_2 and C treatments exert a strong inhibitory effect; addition of phosphoglycerides can restore the enzyme activity up to 90% of the control values,[13] phosphatidylethanolamine being the most efficient lipid in this respect.

Conformational Properties

Circular dichroism studies on the native fatty acid synthase from *C. capitata* provide evidence of a secondary structure of the enzyme consisting of 43% α helix, 26% β structure, and 31% random coil.[14] Lipidation and cholate treatment does not modify the structure of the enzyme complex, whereas the sodium dodecyl sulfate treatment changes the native conformation into a structure containing 43% α helix, 8% β structure, and

[12] E. J. Cohn and J. T. Edsall, "Proteins, Amino Acids, and Peptides," p. 370. Reinhold, New York, 1943.
[13] J. G. Gavilanes, M. A. Lizarbe, A. M. Municio, and M. Oñaderra, *Biochem. Biophys. Res. Commun.* **83**, 998 (1978).
[14] J. G. Gavilanes, M. A. Lizarbe, A. M. Municio, J. A. Ramos, and E. Relaño, *J. Biol. Chem.* **254**, 4015 (1979).

19% random coil. The removal of lipids results in the disorganization of the complex as judged from both the total enzyme inactivation and the changes in the helical conformation. Urea induces a conformational transition at the 3 M concentration; cholate and sodium dodecyl sulfate have little effect on the α-helical structure although both agents induce a loss of enzyme activity in a similar manner to that induced by urea. All data[8,13,14] support the interpretation that phospholipids play a fundamental role in the conformation of the active site and the secondary structure, whereas triacylglycerols contribute to the general support of the oligomeric enzyme.

[18] Cyclopropane Fatty Acid Synthase from *Escherichia coli*

By FREDERICK R. TAYLOR, DENNIS W. GROGAN, and
JOHN E. CRONAN, JR.

Cyclopropane fatty acids (CFA)[1] are found in the phospholipids of many eubacteria[2] and have also been reported in a few eukaryotic organisms.[2]

In bacteria, Law[3] and his co-workers showed that CFAs are formed by methylenation of the double bond of unsaturated fatty acids. These workers also demonstrated an enzyme, CFA synthase, in *Clostridium butyricum* that catalyzed methylenation of the unsaturated fatty acid moieties of phospholipids using the methyl carbon of S-adenosyl-L-methionine (SAM) as the methylene donor.

This synthase is one of the few enzymes known to act on the nonpolar portion of phospholipids dispersed in a vesicle. The substrate of the enzyme is the double bond of a phospholipid unsaturated fatty acid residue.[3] This double bond must be 9–11 carbon atoms removed from the glycerol backbone of the phospholipid molecule;[4,5] therefore, the site of action is well within the hydrophobic region of the lipid bilayer. For these reasons, CFA synthase is an unusually interesting system for the study of protein–lipid interactions. It binds to substrate phospholipid vesicles. This binding greatly stabilizes the enzyme and is exploited in its purification.[6]

[1] H. Goldfine, *Adv. Microbiol. Physiol.* **8**, 1 (1972).
[2] W. W. Christie, *Top. Lipid Chem.* **1**, 1 (1970).
[3] J. H. Law, *Acc. Chem. Res.* **4**, 199 (1971).
[4] L. A. Marinari, H. Goldfine, and C. Panos, *Biochemistry* **13**, 1978 (1974).
[5] J. B. Ohlrogge, F. D. Gunstone, I. A. Ismail, and W. E. M. Lands, *Biochim. Biophys. Acta* **431**, 257 (1976).
[6] F. R. Taylor, and J. E. Cronan, *Biochemistry* **15**, 3292 (1979).

METHODS IN ENZYMOLOGY, VOL. 71

Assay

Principle. The assay is based on the incorporation label from [methyl-^{3}H]S-adenosyl-L-methionine (SAM) into phospholipid.[6,7] After incubation the reaction mixture is pipetted onto a filter paper disk. The disks are washed in trichloroacetic acid (which precipitates phospholipids, but not SAM). CFA synthase activity is then quantitated by scintillation counting of the disk.

Reagents

[Methyl-^{3}H]S-adenosylmethionine   (commercially   available)   25
   Ci/μmol; final concentration 0.5 mm
Phospholipid dispersion, 0.1 mg, final concentration 1 mg/ml
Buffer, potassium phosphate (pH 7.5), final concentration 20 mM
S-adenosyl-L-homocysteine hydrolase (SAHase), 1 unit/ml final con-
   centration

Partial Purification of SAH Nucleotidase (SAHase). This enzyme was purified from *E. coli* B by essentially the method of Duerre.[8] The SAHase preparations purified by ammonium sulfate fractionation and DEAE-cellulose chromatography were free of CFA synthase activity. It was found that less purified preparations could be freed of CFA synthase with full retention of SAHase activity by heating to 45° for 15 min. The preparations were stored at $-20°$ in 50 mM Tris-HCl (pH 7.5) buffer and were stable for at least 1 year. SAHase activity is assayed as described by Duerre,[8] except that Nelson's test[9] is used to measure the production of reducing sugar. One unit of SAHase activity is defined as 1 μmol of reducing sugar formed from SAH per minute at 37° and pH 7.5. SAHase activity can also be assayed indirectly by relief of the inhibition of CFA synthase by added SAH.

Preparation of Phospholipid Dispersions. Phospholipids deficient in CFA (hence rich in unsaturates) are extracted either from *E. coli* wild-type cells grown to early log phase or from stationary phase cells of a mutant strain deficient in CFA synthase.[10] The cells are grown in a broth medium, harvested, and the phospholipids extracted as described by Ames.[11] The phospholipids are purified free of neutral lipids by the solvent precipitation method of Law and Essen[12] or by chromatography on a silicic acid column.[13] The resulting phospholipids are dried under a stream of nitrogen

[7] H. Goldfine, *J. Lipid Res.* **7**, 146 (1966).
[8] J. A. Duerre, *J. Biol. Chem.* **237**, 3737 (1962).
[9] N. Nelson, *J. Biol. Chem.* **153**, 375 (1944).
[10] F. R. Taylor, and J. E. Cronan, Jr. *J. Bacteriol.* **125**, 518 (1976).
[11] G. F. Ames, *J. Bacteriol.* **95**, 833 (1968).
[12] J. H. Law, and B. Essen, this series, Vol. 14, p. 665.
[13] J. C. Rittmer, and M. A. Wells, this series, Vol. 14, p. 483.

and then dispersed into 1 mM EDTA in glass-distilled water (1 ml/10 mg of lipid) by sonication for 1 min resulting in monolamellar vesicles or by vigorous agitation (vortex mixer) and homogenization resulting in multilamellar liposomes. After dispersion the phospholipid concentration is determined by either the hydroxamate test for lipid esters[14] or by phosphate analysis.[15]

Hydrogenation of unsaturated phospholipids (50–100 mg of phospholipid) as performed with 5 mg of Adams catalyst (PtO$_2$) in 20 ml of tetrahydrofuran–methanol (1:1) or chloroform–methanol (1:1) in an apparatus made from two 125 ml sidearm Erlenmeyer filter flasks connected via the sidearm with thick-wall rubber tubing. One flask contained the phospholipid solution and the catalyst, and the second flask contained 50 ml of 5 N HCl. The first flask was stoppered with a silicone stopper pierced by a glass tube, to the external end of which was wired a rubber policeman. The second flask was capped with a rubber septum. A stabilized NaBH$_4$ solution[16] was injected into the acid of the second flask; this resulted in the immediate evolution of H$_2$ (monitored by expansion of the rubber policeman). The apparatus was shaken at room temperature until no further uptake of H$_2$ occurred. Hydrogenation under these conditions has no effect on cyclopropane rings.[2] Liposomes of these phospholipids are made as described above except the solution is heated to 60–70° periodically during homogenization.

Assay Procedure. To the buffer is added the [³H]SAM and then the phospholipid dispersion and enzyme fraction. The buffer is added first to neutralize the dilute sulfuric acid in which the SAM is stored and thus prevent acid precipitation of the phospholipids. It is important that the [³H]SAM be mixed with carrier SAM (if needed) just before use. Storage of concentrated [³H]SAM solutions results in formation of a trichloroacetic acid (TCA) insoluble product, which leads to high background values.

After incubation at 37° for 30 min, the entire reaction mixture is pipetted onto a 2.4 cm disk of Whatman No. 3 MM filter paper mounted on a pin. The filter disks are dried in a stream of hot air for 20 sec and immersed in TCA (10% w/v) for 5 min at room temperature. The disks are then placed in a boiling solution of 5% (w/v) TCA solution, washed in two changes of distilled water for 15 min each, then dried, and assayed for radioactivity after addition of scintillation fluid. The assay is linear with protein from 0.02 to 5 mg of crude supernatant protein and is linear with time for at least 1 hr. A unit of CFA synthase activity is defined as 1 pmol of CFA formed per minute at 37°.

[14] B. Shapiro, *Biochem. J.* **53**, 663 (1953).
[15] B. N. Ames, this series, Vol. 8, p. 115.
[16] L. F. Fieser, and M. Fieser, "Reagents for Organic Synthesis," Vol. I, pp. 1045–1055. Wiley, New York.

Purification of CFA Synthase

Cell Disruption and Ammonium Sulfate Fractionation. Cyclopropane fatty acid synthase has been purified from either freshly grown (then frozen) *E. coli* K12 or commercially grown *E. coli* B stationary phase cells. The freshly grown cells have higher activities and give more consistent results. The cell paste (20 g) is thawed and homogenized in 20 ml of 50 m*M* potassium phosphate buffer, pH 7.6, containing MgCl$_2$ (5 m*M*) and about 1 mg of deoxyribonuclease I. This and all subsequent steps are done at 0–4°. The cells were disrupted by two passages through a French pressure cell at 11,000 psi. The resulting lysate is cleared of large particulate material by centrifugation at 10,000 *g* for 10 min, and the supernatant is retained. The centrifugation supernatant was diluted to a protein concentration of 10 mg/ml, and ammonium sulfate is slowly added to 40% of saturation. After equilibration the precipitate is collected by centrifugation at 10,000 *g* for 15 min and dissolved in the phosphate buffer. Residual ammonium sulfate is removed either by dialysis or by gel filtration.

Flotation. The CFA synthase binds to and is stabilized by vesicles of phospholipids containing unsaturated fatty acids.[6] To purify CFA synthase the ammonium sulfate-fractionated enzyme is exposed first to vesicles made of phospholipids containing only saturated fatty acids (hydrogenated phospholipids). The synthase does not bind to such vesicles, but, since other proteins do, this provides a negative purification step; CFA synthase is then bound to vesicles made of phospholipids that contain unsaturated fatty acids and separated from the bulk of the other proteins by equilibrium sucrose gradient centrifugation.[6]

Ammonium sulfate-purified enzyme, 60% sucrose, and a suspension of multilamellar liposomes (made of hydrogenated phospholipids) are mixed to give a solution containing final concentrations of sucrose, protein, and liposomes of 30% (w/v), 10 mg/ml, and 4 mg/ml, respectively, in 50 m*M* potassium phosphate (pH 7.5). After incubation at 37° for 15 min, 4 ml of this mixture are placed in a centrifuge tube and sequentially overlaid with 0.5 ml of phosphate buffer containing 25% (w/v) sucrose (*d* = 1.09 g/ml), 0.5 ml of phosphate buffer containing 20% (w/v) sucrose (*d* = 1.08 g/ml), and 0.1 ml of buffer. The tube is then centrifuged at 80,000 *g* for 1–2 hr. After centrifugation, the lipid is visible as an opalescent band in or on the surface of the 20% (and sometimes in the 25%) sucrose layer (Fig. 1). This layer was removed by puncturing the side of the tube and removing the band with a syringe. The protein remaining in the 30% sucrose layer is removed by puncturing the bottom of the tube. This fraction is mixed with liposomes made of unsaturated phospholipids, overlaid with 25% and 20% sucrose layer, and recentrifuged. The lipid band containing CFA synthase is removed as before.

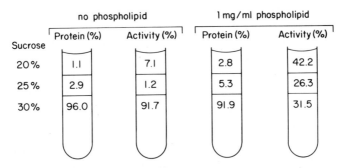

FIG. 1. Distribution of cyclopropane fatty acid synthase activity and protein in sucrose step gradients in the absence (left) or presence (right) of liposomes (1 mg/ml) formed from phospholipids containing unsaturated fatty acids. The enzyme preparation (10 mg/ml ammonium sulfate purified enzyme) was mixed with liposomes in 30% sucrose overlaid first with 25% sucrose and then with 20% sucrose as described. After centrifugation, the 20 and 25% layers were collected through the side of the tube whereas the 30% layer was collected through the bottom of the tube.

The rationale of the double flotation step procedure is that CFA synthase does not bind to liposomes made from saturated phospholipid whereas other proteins do. This step is therefore a negative purification step and leaves CFA synthase in the 30% sucrose layer. The synthase is then bound to liposomes of substrate (unsaturated) phospholipids and resolved from the bulk of the protein by the second centrifugation.

To remove the lipids from the CFA synthase, the vesicle-bound synthase is mixed with sufficient sucrose and KCl to give concentrations of 40% (w/v) and 1 M, respectively. This mixture is centrifuged at 50,000 g for 1 hr. Centrifugation results in the lipid being layered in the surface of the solution, and lipid-free CFA synthase is obtained by puncturing the bottom of the tube. The enzyme is very unstable in the absence of phospholipid and is best stored in the presence of 1–2 mg of phospholipid per milliliter at $-70°$.

Purity of CFA Synthase Preparations. Using the double flotation technique, 500- to 600-fold purification CFA synthase have been obtained.[6] If the negative purification step (the first flotation) is omitted, then the resulting enzyme is purified only 20 to 50-fold. Preparations obtained by double flotation vesicles are not homogeneous, they contain at least 15 proteins of differing molecular weight as estimated by gel electrophoresis in the presence of sodium dodecyl sulfate.[6] The molecular weight of the active CFA synthase is estimated to be 80,000 to 100,000.[6] The *E. coli* gene coding for CFA synthase has been cloned using *in vivo* techniques.[17] Such strains

[17] D. W. Grogan and J. E. Cronan, Jr., unpublished observations. Purified preparations with specific activities of 5 × 10⁵ U/mg protein have been obtained from these strains using the procedure given in the table [modified by omission of the $(NH_4)_2SO_4$ step].

PURIFICATION OF CYCLOPROPANE FATTY ACID SYNTHASE[a]

Step	Protein (mg)	Specific activity (unit/mg)	Yield (%)	Purification (fold)
Cell extract	74,100	24.3	(100)	(1)
Centrifugation	27,500	49.1	76	2
$(NH_4)_2SO_4$	9,400	95.3	50	4
Double flotation	39.1	14,094	6.5	580

[a] The synthase was purified from 500 g of frozen cell paste of commercially grown *Escherichia coli* B.

show elevated CFA synthase levels and should greatly facilitate purification of this enzyme.[17]

A summary of a sample purification is given in the table.

Properties of Purified CFA Synthase

Purified preparations of CFA synthase are very labile. In the absence of lipid, all activity is lost in <30 min at 37°. In the presence of phospholipid vesicles (2 mg/ml) the enzyme is greatly stabilized (no loss in activity in 30 min at 37°) whereas S-adenosylmethionine (1 mM) preserves about 50% of the activity under similar conditions. The purified enzyme has a Michaelis constant for SAM of 90 μM. S-Adenosylhomocysteine is a competitive inhibitor with a K_i of 220 μM.[6] CFA synthase is saturated with phospholipid vesicles at a liposome concentration of about 0.67 mM (0.5 mg/ml) of phospholipid.[6] Phospholipids dispersed by sonication are somewhat more effective on a weight basis. This is probably due to the greater external surface area per weight of lipid of single bilayer vesicles. However, the heterogeneous site of sonicated vesicles made from *E. coli* lipids and the unusual mechanism of action of CFA synthase (see below) precludes calculation of a Michaelis constant. Liposomes made of hydrogenated lipids are inactive as substrates and fail to inhibit the reaction of the enzyme with liposomes made of unsaturated phospholipids.

The CFA synthase is sensitive to sulfhydryl reagents. Dithiobis (nitrobenzoic acid) (DTNB), N-ethylmaleimide, and p-hydroxymercuribenzoate all inhibit the enzyme; however, iodoacetic acid does not. The most specific of these reagents, DTNB, inhibits the enzyme by over 90% at a concentration of 0.5 mM. Inhibition by DTNB is completely reversed by addition of a reducing reagent such as dithiothreitol at 2 mM. The CFA synthase is destabilized and inhibited by high salt concentrations. This is probably due to dissociation of the enzyme from stabilizing lipids. The finding that CFA synthase requires the presence of lipid for maintenance of activity explains the activity losses encountered during standard protein chromatographic fractionation procedures since such methods tend to resolve protein and lipid. Owing to these considerations we developed methods for the purification of CFA synthase that avoided conventional chromatographic steps. The commonly used detergents (Tritons, Tweens, Brijs, etc.) destroy enzyme activity. Two detergents, the monooleate and the monolaurate esters of sorbitol, stabilize the enzyme. Unfortunately, these detergents inhibit the enzyme assay and disperse poorly, thus limiting their usefulness.

In vesicles composed of a mixture of these lipids,[6] CFA synthase reacts equally well with each of the phospholipid species present in *E. coli* (phosphatidylethanolamine, phosphatidylglycerol, and cardiolipin). The enzyme is equally active on vesicles of phospholipid in either the ordered or disordered states of the lipid phase transition[6] and appears able to react with phosphatidylethanolamine molecules of both the outer and inner leaflets of single bilayer phospholipid vesicles.[6]

[19] Isolation, Translation *in Vitro*, and Partial Purification of Messenger RNA for Fatty Acid Synthase from Uropygial Gland[1]

By ALAN G. GOODRIDGE, SIDNEY M. MORRIS, JR., and TAMAR GOLDFLAM

Fatty acid synthase is one of the set of lipogenic enzymes whose activities are inhibited by starvation and stimulated by refeeding starved animals.[2] In both avian and mammalian liver, changes in lipogenic enzyme activities are a function of altered enzyme protein concentration, which, in turn, are primarily functions of selective alterations in the rates of synthesis of these enzymes.[3-6] In maintenance cultures of chick embryo

[1] This work was supported in part by Grant AM 21594 from the National Institute of Arthritis, Metabolism, and Digestive Diseases. We thank Ms. Sally Mansbacher for excellent technical assistance.

[2] J. J. Volpe and P. R. Vagelos, *Physiol. Rev.* **56**, 339 (1976).
[3] J. J. Volpe and J. C. Marasa, *Biochim. Biophys. Acta* **380**, 454 (1975).
[4] M. C. Craig, C. M. Nepokroeff, M. R. Lakshmanan, and J. W. Porter, *Arch. Biochem. Biophys.* **152**, 619 (1972).
[5] Z. E. Zehner, V. C. Joshi, and S. J. Wakil, *J. Biol. Chem.* **252**, 7015 (1977).
[6] P. W. F. Fischer and A. G. Goodridge, *Arch. Biochem. Biophys.* **190**, 332 (1978).

hepatocytes, the synthesis of fatty acid synthase is stimulated by insulin and thyroid hormone and inhibited by glucagon.[6] The next step in the analysis of the nutritional and hormonal regulation of the synthesis of this enzyme involves measurement of the levels of the corresponding messenger RNA.

Assays of the mRNA[7] levels could be accomplished by *in vitro* translation of mRNA or by RNA–DNA hybridization techniques, using a DNA sequence complementary to fatty acid synthase mRNA. Translation assays have recently been used to obtain useful data correlating changes in synthesis of fatty acid synthase with changes in levels of its corresponding polyribosomal mRNA.[8,9] However, the wheat germ cell-free translation system did not produce a full-length product.[8] Another *in vitro* translation system has been reported,[10] but it required substantial enrichment of the fatty acid synthase mRNA fraction before synthesis of the full-length polypeptide could be obtained. The amphibian oocyte translation system, though generating full-length fatty acid synthase,[9] reveals only relative changes in fatty acid synthase mRNA levels without yielding information on the abundance of fatty acid synthase mRNA within a given mRNA population.

RNA–DNA hybridization methods using a specific complementary DNA (cDNA) can provide all the information obtainable by *in vitro* translation assays, and also permit determination of mRNA synthesis and turnover rates. Thus, our efforts have been directed toward obtaining a cDNA for fatty acid synthase mRNA. The goose uropygial gland was selected as the richest probable source of fatty acid synthase mRNA since fatty acid synthase comprises one-sixth to one-third of total soluble protein in this tissue.[11] The enzyme purified from the gland is physically, kinetically, and immunologically identical to the liver enzyme[12] and likely represents the product of the same gene. We describe here optimal conditions for *in vitro* synthesis of full-length fatty acid synthase and partial purification of its corresponding mRNA.[13] The partially purified mRNA will be used for synthesis of cDNA.

[7] Abbreviations used: cDNA, complementary DNA; mRNA, messenger RNA; SDS, sodium dodecyl sulfate; anti-FAS, rabbit antiserum to fatty acid synthase; poly(A)$^+$ RNA, polyadenylated RNA.

[8] P. K. Flick, J. Chen, A. W. Alberts, and P. R. Vagelos, *Proc. Natl. Acad. Sci. U.S.A.* **75**, 730 (1978).

[9] C. M. Nepokroeff and J. W. Porter, *J. Biol. Chem.* **253**, 2279 (1978).

[10] H.-P. Lau, C. M. Nepokroeff, and J. W. Porter, *Biochem. Biophys. Res. Commun.* **89**, 264 (1979).

[11] J. S. Buckner and P. E. Kolattukudy, *Biochemistry* **15**, 1948 (1976).

[12] J. S. Buckner, P. E. Kolattukudy, and L. Rogers, *Arch. Biochem. Biophys.* **186**, 152 (1978).

[13] A. G. Goodridge, S. M. Morris, and T. Goldflam, *Fed. Proc.* **38**, 300 (1979).

Materials

Adult domestic geese were obtained from local suppliers. Sucrose (ribonuclease free) was purchased from Schwarz-Mann, sodium dodecyl sulfate (SDS) from BDH Chemicals Ltd., calf liver tRNA from Boehringer Mannheim, oligo(dT) cellulose (types 2 and 3) from Collaborative Research, and Nonidet P-40 from Shell Oil Co. Emetine hydrochloride, guanidine hydrochloride, and CsCl (optical grade) were obtained from Sigma Chemical Company. Acrylamide, N,N'-methylene bisacrylamide, and unlabeled marker proteins for molecular weight determinations were purchased from Bio-Rad Laboratories. Nuclease-treated reticulocyte lysate, [^{35}S]methionine (>700 Ci/mmol), and ^{14}C-methylated marker proteins for molecular weight determinations were purchased from New England Nuclear. Rabbit antiserum against chicken liver fatty acid synthase was prepared and characterized as described previously.[6]

Sterile Technique

Solid CsCl and all glassware were baked at 180° for 6 hr to minimize ribonuclease contamination. Plastic ware was autoclaved or soaked in 0.05% diethyl pyrocarbonate. Solutions containing nonlabile components were treated with 0.05% diethylpyrocarbonate and then autoclaved or heated at 75° for 1 hr. Other solutions were made up with sterile water and filtered through type GS Millipore filters.

Isolation of RNA

Procedure I. The technique described here is based on a procedure previously developed for chicken liver.[14] This method avoids the use of detergents or inhibitors of ribonuclease until lysosomes have been removed by centrifugation. The use of phenol is also avoided. The success of this procedure depends upon the rapidity of operation and avoidance of lysosomal breakage.

Adult domestic geese (4–5 kg body weight) were killed by injection of 8–10 ml of emetine solution (30 mg/ml in 0.15 M NaCl) into the wing vein. The uropygial glands (6–8 g per animal) were removed, freed of nonglandular tissue, and placed in an ice-cold mixture of 20 mM Tris-HCl, 100 mM KCl, 5.25 mM Mg acetate, 0.25 mM EDTA, 1 mM dithiothreitol, 200 mM sucrose, and 0.3 mM emetine hydrochloride, pH 7.4. Subsequent operations were carried out at 0–4°. The tissue was minced finely with scissors

[14] A. G. Goodridge, O. Civelli, C. C. Yip, and K. Scherrer, *Eur. J. Biochem.* **96,** 1 (1979).

and then, using a pestle, forced through a stainless steel strainer (1 mm² mesh) nested in an evaporating dish containing 3 volumes of the same buffer. The suspension of small pieces of gland was homogenized in a glass Dounce homogenizer (Kontes) with five strokes of the loose piston, followed by five strokes of the tight piston. This homogenate was centrifuged at 4100 g for 8 min. Then, without deceleration, speed was increased to 16,300 g for an additional 15 min. The supernatant was removed, taking care not to include any material from the pellet. Heparin solution (one-ninth volume, 10 mg/ml in homogenization buffer without emetine or sucrose) was added to the supernatant. The preparation was adjusted to pH 5.2 by dropwise addition of 1 M acetic acid with stirring. After 5 min the precipitate that formed was collected by centrifugation at 11,000 g for 10 min and the supernatant was discarded. If the pellet is not to be immediately dissociated for passage over oligo(dT) cellulose as described below, this last centrifugation step should be performed in a polyethylene tube and the pellet immediately frozen in liquid N_2 and stored at $-70°$. When the frozen pellet is needed, it can be easily dislodged by striking the tube sharply against the laboratory bench.

Procedure II. This procedure is essentially a combination of two previously described RNA isolation methods.[15,16] Geese were killed, and the uropygial glands were removed and placed in the same buffer used in Procedure I. After weighing, the glands were coarsely minced with a razor blade and immediately homogenized in 6 volumes of 8 M guanidine hydrochloride—20 mM sodium acetate–2 mM dithiothreitol, pH 5, in an ice-cooled Waring blender using several 10–20 sec bursts to effect complete homogenization. The homogenate was centrifuged at 17,000 g at 4° for 10 min, and the top lipid layer was removed with a spatula and cotton-tipped applicator sticks. The clear supernatant was filtered through two layers of cheesecloth and transferred to sterile polyallomer or cellulose nitrate tubes for the Beckman SW27 rotor. The supernatant was underlaid with 2 ml of 5.7 M CsCl–0.1 M EDTA, and the preparation was centrifuged in the SW27 rotor at 25,000 rpm at 20° overnight. The lipid layer and supernatant were carefully removed and discarded. The pellet resulting from two uropygial glands was resuspended in 5 ml of a mixture of 7 M guanidine hydrochloride, 20 mM sodium acetate, 1 mM dithiothreitol, and 20 mM EDTA, pH 7, and heated at 65° for 5 min with intermittent vortexing. Insoluble material was removed by centrifugation at 17,000 g at 4° for 10 min. The supernatant was vigorously extracted with an equal volume of chloroform–isoamyl alcohol (24 : 1). The phases were separated

[15] R. G. Deeley, J. I. Gordon, A. T. H. Burns, K. P. Mullinix, M. Bina-Stein, and R. F. Goldberger, *J. Biol. Chem.* **252**, 8310 (1977).
[16] V. Glisin, R. Crkvenjakov, and C. Byus, *Biochemistry* **12**, 2633 (1974).

by brief centrifugation, the aqueous phase was carefully removed, and RNA was precipitated from it by addition of an equal volume of ethanol at $-20°$. The mixture was allowed to stand at $-20°$ for at least 2 hr. The pellet was collected by centrifugation and vigorously washed twice with 66% ethanol–0.1 M sodium acetate, pH 5. It was then dissolved in 2 ml of sterile water and stored at $-70°$.

Chromatography on Oligo(dT) Cellulose. The pH 5.2 pellet (procedure I) from two uropygial glands was solubilized by homogenization in 35 ml of buffer I (0.5 M NaCl, 10 mM Tris-HCl, 1 mM EDTA, 0.5% SDS, and 0.2% sodium deoxycholate, pH 7.4). This and subsequent steps were carried out at room temperature. Insoluble material was removed by centrifugation at about 4500 g for 2 min. The resulting clear solution was passed over a column of oligo(dT) cellulose (1 g of type 2 or 0.5 g of type 3). The eluate was passed over the column a second time. The column was washed with 25–30 ml of buffer I, 25–30 ml of buffer II (buffer I minus detergents), and finally with six 2-ml portions of buffer III (10 mM Tris-HCl, 1 mM EDTA, pH 7.4). Polyadenylated RNA generally eluted in the second and third fractions of the buffer III wash. These fractions were pooled and adjusted to 0.1 M NaCl, and RNA was precipitated overnight at $-20°$ with 2 volumes of ethanol. Precipitated RNA was collected by centrifugation at 10,500 g for 1 hr, redissolved in H_2O, and stored at $-70°$. An $E_{1 cm}^{1 \%}$ of 200 at 260 nm was used to determine RNA concentration. The yield of poly(A)$^+$ RNA should be about 120–150 μg.

The solution of RNA obtained from two uropygial glands by procedure II was heated at 70° for 5 min, then immediately diluted into 35 ml of buffer I. The remaining steps of chromatography were carried out exactly as described above. Yield of poly(A)$^+$ RNA should be about 150–250 μg.

In vitro Translation

Incubation Mixture. Cell-free protein synthesis was carried out with a commercial, nuclease-treated rabbit reticulocyte lysate kit (New England Nuclear) prepared according to previously described methods.[17] Assays were routinely carried out in a total volume of 50 μl, containing 20 μl reticulocyte lysate, 4 μl of the translation cocktail included in the kit (containing spermidine, creatine phosphate, and GTP), 40–50 μCi of [^{35}S]methionine, appropriate amounts of RNA sample, and sufficient amounts of supplemental stocks to yield final concentrations of 50 mM potassium acetate, 0.65 mM magnesium acetate, 40 μg/ml calf liver tRNA, and 30 μM each of 19 amino acids (except methionine). Incuba-

---

[17] H. R. B. Pelham and R. J. Jackson, *Eur. J. Biochem.* **67**, 247 (1976).

tions were carried out for variable periods of time at 30°. Aliquots of the incubation mixture (total products) were taken for determination of radioactivity and SDS gel analysis as described below.

Immunoprecipitation. For immunoprecipitation of labeled fatty acid synthase, the incubated reaction mixture was diluted to 1 ml with 85 mM Tris-HCl, 11 mM MgCl$_2$, 0.1 mM EDTA, and 1% Nonidet P-40, pH 7.4. This mixture was centrifuged at 200,000 g at 4° for 1 hr to obtain a postmicrosomal fraction (total released polypeptides). Aliquots were taken for radioactivity measurement, then the supernatant was adjusted to contain 0.15 M NaCl and 0.1 mM phenylmethylsulfonyl fluoride, plus sufficient uropygial gland postmicrosomal supernatant to give 20 mU of fatty acid synthase activity, and 25 μl of rabbit anti-FAS. This amount of antiserum was sufficient to precipitate twice the amount of carrier enzyme added to the immunoprecipitation mixture. The immunoprecipitation reaction mixtures were incubated in 1.5-ml Eppendorf tubes for 1 hr at room temperature and overnight at 4°. Immunoprecipitates were collected by centrifugation for 2 min in a Beckman Microfuge and washed 4 times with 1-ml portions composed of 0.15 M NaCl and 1% Nonidet P-40, with vigorous vortexing for each resuspension. Excess liquid was carefully blotted from the washed pellets by the tip of a tightly twisted strip of laboratory tissue wiper. The washed immunoprecipitate was dissolved in SDS gel sample buffer as described below, and aliquots were taken for determination of radioactivity.

The specificity of rabbit anti-FAS has been demonstrated previously.[6] In addition we have determined that chicken anti-FAS and the goose uropygial gland enzyme are highly cross-reactive.[18] Antiserum to chicken albumin does not immunoprecipitate appreciable radioactivity from a lysate programmed with uropygial gland RNA (not shown), although significant radioactivity is immunoprecipitated when the lysate is programmed with liver RNA. This observation further demonstrates the specificity of the anti-FAS and suggests that little or none of the radioactivity found in the fatty acid synthase immunoprecipitate was due to trapping or nonspecific adsorption.

Determination of Incorporated Radioactivity. Radioactivity in the total products, total released polypeptides, and immunoprecipitate fractions was determined after spotting duplicate aliquots of appropriate volume on Whatman No. 540 disks. The disks were placed in approximately 200 ml of 10% trichloroacetic acid, which was then boiled for 10 min to hydrolyze charged tRNA. After cooling for several minutes in an ice bath, the 10% trichloroacetic acid was discarded and the disks were washed with 200 ml

[18] A. G. Goodridge, unpublished results.

of 5% trichloroacetic acid and two 100-ml washes of methanol. Air-dried filters were counted by liquid scintillation spectrometry.

Electrophoretic Analysis of Translation Products. Sodium dodecyl sulfate gel sample buffer (80 mM Tris-HCl, 4% SDS, 5% β-mercaptoethanol, 12% sucrose, 0.002% bromophenol blue) was added to the washed immunoprecipitate and aliquots (5 or 10 μl) of the original incubation mixture (total products) to give a final volume of 50 μl. Samples were thoroughly dissociated by heating at 100° for 5 min. Total products and immunoprecipitates were subjected to electrophoresis on 7.5% and 10% polyacrylamide slab gels containing SDS, according to Laemmli.[19] For fluorography, gels were fixed in 12.5% trichloroacetic acid and either processed according to Bonner and Laskey[20] or incubated in 3 volumes of EN³HANCE (New England Nuclear) for 1 hr, then in 6 volumes of H$_2$O for 1 hr. In some cases staining of the gel essentially according to Chrambach *et al.*,[21] followed by destaining in 7.5% acetic acid and 15% methanol, was carried out just prior to the EN³HANCE step. Gels were vacuum-dried onto Whatmann No. 3 MM paper and exposed to Kodak XR-1 film at −70°. In some experiments the region of the dried gel containing the stained carrier fatty acid synthase was cut into approximately 2-mm strips, digested with 30% H$_2$O$_2$ at 60° for several hours, and assayed for radioactivity by liquid scintillation counting.

Optimization of Translation Conditions. Because very few polypeptides of molecular weight greater than 200,000 have been synthesized *in vitro,* efforts were directed toward achieving optimal conditions for synthesis of full-length fatty acid synthase (MW 230,000). These conditions were determined with uropygial gland poly(A)⁺ RNA prepared by procedure I. The criteria used for identification of *in vitro* synthesized fatty acid synthase polypeptides were comigration with authentic carrier enzyme and specific immunoprecipitation by anti-FAS. Optimal conditions were determined with reference only to incorporation into full-length enzyme,[13] not into incomplete but immunoprecipitable polypeptides. Similar translation conditions have been reported for *in vitro* synthesis of fatty acid synthase from rat liver.[10]

Added calf liver tRNA markedly stimulated (nine-fold) production of a full-length product, whereas synthesis of total released polypeptides was only slightly stimulated. The unsupplemented reticulocyte lysate may be deficient in one or more specific tRNAs that are rate-limiting for synthesis of fatty acid synthase, resulting in a high degree of premature termination.

[19] U. K. Laemmli, *Nature (London)* **227,** 680 (1970).
[20] W. M. Bonner and R. A. Laskey, *Eur. J. Biochem.* **46,** 83 (1974).
[21] A. Chrambach, R. A. Reisfeld, M. Wyckoff, and J. Zaccari, *Anal. Biochem.* **20,** 150 (1967).

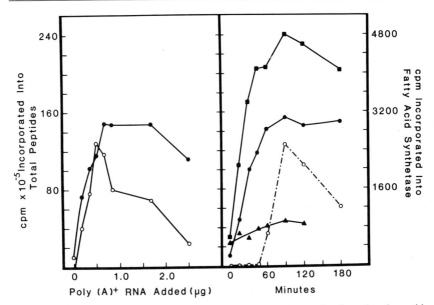

FIG. 1. Cell-free synthesis of total products (■———■), total released polypeptides (●———●), and full-length fatty acid synthase (O-----O, O———O) in the nuclease-treated reticulocyte lysate as a function of poly(A)⁺ RNA concentration and incubation time. *Left panel:* Incorporation of [³⁵S]methionine into designated fractions in a 50-μl assay as a function of added uropygial gland poly(A)⁺ RNA. Incubation time was 120 min. *Right panel:* Incorporation into 50-μl aliquots of an initial 450 μl of reaction mixture containing 10 μg of uropygial gland poly(A)⁺ RNA per milliliter at varying time points. Incorporation into total products in the absence of added RNA is also shown (▲———▲).

Alternatively, the stimulation may have been due to protection of the mRNA from ribonuclease by competing tRNA.

Cation concentrations were also found to be critical. At a concentration of 50 mM added K⁺, significant synthesis of total released polypeptides occurred in the absence of added Mg²⁺ but synthesis of full-length fatty acid synthase was undetectable. The optimal concentration of added Mg²⁺ for synthesis of the full-length enzyme subunit was about 0.65 mM. In the presence of 0.65 mM added Mg²⁺, added K⁺ only slightly stimulated synthesis of total released polypeptides, whereas synthesis of fatty acid synthase was markedly stimulated, with a sharp optimum at 50 mM.

Synthesis of total released polypeptides was maximal at a concentration of about 14 μg of poly(A)⁺ RNA per milliliter, and synthesis of fatty acid synthase was maximal at a lower poly(A)⁺ RNA concentration of about 10 μg/ml (Fig. 1, left panel). Incorporation into both fractions was inhibited at higher poly(A)⁺ RNA concentrations, the inhibition being more marked for synthesis of fatty acid synthase than for total released polypeptides.

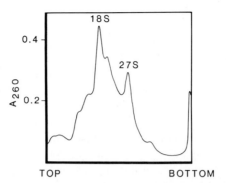

F<small>IG</small>. 2. Fractionation of 65 μg of goose uropygial gland poly(A)$^+$ RNA on an isokinetic 10 to 29.7% sucrose density gradient. Aliquots were taken at various points along the gradient, precipitated by ethanol, and assayed for fatty acid synthase mRNA activity by *in vitro* translation.

Synthesis of total and total released polypeptides increased in an essentially linear fashion from the initiation of the reaction to 45–60 min of incubation (Fig. 1, right panel). In contrast, full-length fatty acid synthase polypeptides did not appear until 60–90 min after initiation of the reaction and then accumulated for an additional 30 min (Fig. 1, right panel). Between 90 and 120 min and the end of the incubation at 180 min, significant loss of radioactivity in full-length enzyme was observed. In contrast, total radioactivity immunoprecipitable by anti-FAS increased slowly during the early part of the incubation period, rapidly when full-length fatty acid synthase polypeptides were appearing, but then leveled off near the end of the incubation (data not shown). There was little or no loss of counts in total products or total released polypeptides during the latter part of the reaction. The pattern of these results could be due to a combination of premature termination and proteolysis. Thus, as has been observed for other high molecular weight peptides synthesized *in vitro,* such as vitellogenin[22] and the multifunctional enzyme involved in eukaryotic pyrimidine biosynthesis,[23] the immunoprecipitate generally contains a spectrum of discrete polypeptides smaller than the full-length product. The ratio of full-length to incomplete products depends greatly on the translation conditions, so for this reason alone determinations of incorporation into full-length product may underestimate the relative concentration of the corresponding mRNA in the poly(A)$^+$ RNA population.

[22] D. J. Shapiro and H. J. Baker, *J. Biol. Chem.* **252**, 5244 (1977).
[23] R. A. Padgett, G. M. Wahl, P. F. Coleman, and G. R. Stark, *J. Biol. Chem.* **254**, 974 (1979).

Sucrose Gradient Fractionation of RNA

Sixty-microgram aliquots of poly(A)$^+$ RNA were fractionated on isokinetic 10 to 29.7% sucrose density gradients[24] (Fig. 2). Samples of RNA were heated at 75° for 5 min, quick-chilled in ice, and immediately loaded onto the gradient. Centrifugation in the Beckman SW65 rotor was at 60,000 rpm for 270 min at 4°. Gradients were scanned at 260 nm with a density gradient fractionator (Isco Model 640) and absorbance monitor (Isco Model UA-5). The tubing and flow cell were flushed with 0.05% diethylpyrocarbonate before use. Carrier calf liver tRNA (10 μg) was added to gradient fractions containing less than 2 μg of RNA. The RNA was recovered from gradient fractions by ethanol precipitation.

Uropygial gland RNA, before and after gradient fractionation, was assayed by *in vitro* translation (Fig. 3). The unfractionated poly(A)$^+$ RNA directed the synthesis of polypeptides as large as 230,000 MW. Depending on the RNA preparation, about 0.5 to 1.5% of the total released counts were immunoprecipitated by anti-FAS. Although fatty acid synthase comprises one-sixth to one-third of total soluble protein in the uropygial gland,[11] this relatively high abundance is not reflected in the poly(A)$^+$ RNA population, either by analysis of sucrose density gradient profiles (Fig. 2) or of *in vitro* translation products (Fig. 3). This suggests that *in vivo* the mature enzyme is highly stable, or that the mRNA for the enzyme is highly stable or very efficient in translation. After gradient fractionation, fatty acid synthase mRNA activity was found exclusively in the small discrete peak migrating at about 36 S. Full-length fatty acid synthase was a major product translated from this mRNA fraction (Fig. 3). Depending on the mRNA preparation, from 30 to 65% of the total released counts were immunoprecipitated by anti-FAS.

Comments

Both procedures for isolation of RNA described here yield uropygial gland mRNA that will direct *in vitro* synthesis of polypeptides up to 230,000 MW. Messenger RNA has also been successfully isolated from avian liver by procedure I.[14] A method very similar to procedure II has been used to isolate mRNA from rat pancreas.[25,26] Procedure I is more rapid. However, without the use of additional RNA extraction techniques,

[24] K. S. McCarty, Jr., R. T. Vollmer, and K. S. McCarty, *Anal. Biochem.* **61**, 165 (1974).
[25] A. Ullrich, J. Shine, J. M. Chirgwin, R. Pictet, E. Tischer, W. J. Rutter, and H. M. Goodman, *Science* **196**, 1313 (1977).
[26] J. M. Chirgwin, A. E. Przybla, R. J. MacDonald, and W. J. Rutter, *Biochemistry* **18**, 5294 (1979).

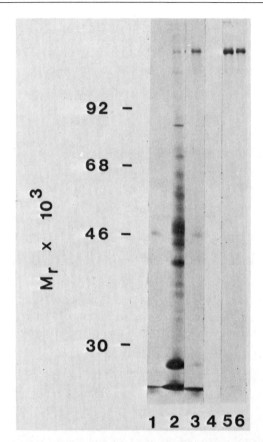

FIG. 3. Fluorogram of *in vitro* translation products directed by uropygial gland poly(A)$^+$ RNA, following electrophoresis on a sodium dodecyl sulfate–10% polyacrylamide gel. The molecular weight (M$_r$, molecular weight ratio) scale was determined by migration of ^{14}C-methylated marker proteins electrophoresed on the same gel (not shown). Total products: (1) endogenous lysate products (no added RNA); (2) unfractionated goose uropygial gland poly(A)$^+$ RNA; (3) enriched fatty acid synthase mRNA fraction from isokinetic sucrose gradient (36 S peak, far right, Fig. 2). Polypeptides immunoprecipitated by anti-FAS: (4) endogenous lysate (from lane 1); (5) unfractionated uropygial gland poly(A)$^+$ RNA (from lane 2); (6) enriched fatty acid synthase mRNA fraction (from lane 3).

procedure I permits analysis only of RNA that binds to oligo(dT) cellulose, whereas analysis of total cellular RNA is possible with procedure II.

The first two centrifugation steps in procedure I remove a significant amount of membrane-bound polyribosomes, so that the final mRNA preparation is more representative of free, rather than total, polyribosomal mRNA. Therefore, this will probably not be the method of choice for isolation of mRNAs for polypeptides synthesized on membrane-bound polyribosomes. However, if the objective is the isolation and purification of a specific mRNA for a soluble cytoplasmic enzyme, this selective enrichment may be more an advantage than a problem.

[20] Malonyl-CoA Decarboxylase from Avian, Mammalian, and Microbial Sources

EC 4.1.1.9 Malonyl-CoA carboxy-lyase

By P. E. KOLATTUKUDY, A. J. POULOSE, and YU SAM KIM

Malonyl-CoA decarboxylase catalyzes the following reaction:

$$\text{Malonyl-CoA} + \text{H}^+ \rightarrow \text{acetyl-CoA} + \text{CO}_2$$

Cell-free preparations from many organisms from both the plant and the animal kingdom catalyze malonyl-CoA decarboxylation.[1-3] This enzyme has been purified from the uropygial glands of waterfowl,[4,5] rat liver mitochondria,[6] rat mammary glands,[7] and *Mycobacterium tuberculosis*.[8]

Assay Method

Radiotracer Assay

Principle. $^{14}\text{CO}_2$ released from [3-^{14}C]malonyl-CoA is absorbed in a base and assayed for radioactivity. Alternatively [1,3-^{14}C]malonyl-CoA can be used, but only one-half of the ^{14}C present in the substrate is released.

Reagents

Tris-HCl buffer, 0.2 *M*, pH 9.0
Dithioerythritol (DTE), 10 m*M*

[1] M. D. Hatch and P. K. Stumpf, *Plant Physiol.* **37**, 121 (1962).
[2] F. Lynen, this series, Vol. 5 [60].
[3] H. I. Nakada, J. B. Wolfe, and A. N. Wick, *J. Biol. Chem.* **226**, 145 (1957).
[4] J. S. Buckner, P. E. Kolattukudy, and A. J. Poulose, *Arch. Biochem. Biophys.* **177**, 539 (1976).
[5] Y. S. Kim and P. E. Kolattukudy, *Arch. Biochem. Biophys.* **190**, 585 (1978).
[6] Y. S. Kim and P. E. Kolattukudy, *Arch. Biochem. Biophys.* **190**, 234 (1978).
[7] Y. S. Kim and P. E. Kolattukudy, *Biochim. Biophys. Acta* **531**, 187 (1978).
[8] Y. S. Kim, P. E. Kolattukudy, and A. Boos, *Arch. Biochem. Biophys.* **196**, 543 (1979).

METHODS IN ENZYMOLOGY, VOL. 71

Malonyl-CoA, 1.5 mM (0.3 Ci/mol)
Hyamine hydroxide in methanol, 1 M
H_3PO_4, 1 M

Preparation of [3-^{14}C]malonyl-CoA. Acetyl-CoA is carboxylated using $H^{14}CO_3^-$ with acetyl-CoA carboxylase obtained from any of the convenient sources.[9] Acetyl-CoA carboxylase present in the extracts (100,000 $g \times$ 90 min) of the uropygial gland emerges from a Sepharose 4B column immediately after the void volume, and this fraction can be used as the source of acetyl-CoA carboxylase.[10] After removal of ATP with hexokinase, the reaction mixture is acidified with $HClO_4$, the precipitated protein is removed by centrifugation, and the product is purified by DEAE-cellulose column chromatography with a linear LiCl gradient.[11,12] The malonyl-CoA fraction is lyophilized, desalted with a BioGel P-2 column (1.5 $\times$ 60 cm) using 0.3 mM HCl as the solvent, and the final product is checked for radiochemical purity by paper chromatography with isobutyric acid : 1 M ammonium acetate : water (57 : 35 : 8 v/v/v).

Assay Procedure. Into a disposable culture tube (15 $\times$ 85 mm) the following reagents are added: 50 μl of Tris-HCl buffer, pH 9.0, 5 μl of DTE, 20 μl of malonyl-CoA. The test tube is sealed with a Thomas serum cap (14 $\times$ 18 mm) through which is fixed a disposable polypropylene center well (Kontes) containing 125 μl of hyamine hydroxide and a 5 $\times$ 1 cm fluted filter paper. The reaction is started by injecting 25 μl of enzyme solution into the test tube, the reaction mixture is thoroughly mixed, and the test tube is incubated at 30° for 10 min. The reaction is stopped by injection of 60 μl of 1 M H_3PO_4 into the reaction mixture, and the tubes are incubated for 30 min at 30° to complete the absorption of CO_2. The contents of the center well are transferred into 10 ml of ScintiVerse (Fisher), and ^{14}C is assayed in a scintillation counter.

Spectrophotometric Assay

Principle. Production of acetyl-CoA is measured spectrophotometrically by coupling with malic dehydrogenase and citrate synthase.

$$\text{Malonyl-CoA} \xrightarrow{\text{decarboxylase}} \text{acetyl-CoA} + CO_2$$
$$\text{Malate} + \text{NAD} \rightleftharpoons \text{oxaloacetate} + \text{NADH}$$
$$\text{Acetyl-CoA} + \text{oxaloacetate} \rightarrow \text{citrate} + \text{CoA}$$

[9] See this volume [1,2].
[10] J. S. Buckner and P. E. Kolattukudy, *Biochemistry* **14**, 1768 (1975).
[11] C. Gregolin, E. Ryder, and M. D. Lane, *J. Biol. Chem.* **243**, 4227 (1968).
[12] S. Smith and S. Abraham, this series, Vol. 35, [8].

Acetyl-CoA generated by malonyl-CoA decarboxylase shifts the equilibrium of the dehydrogenase reaction by condensing with oxaloacetate, resulting in the production of more NADH, which can be followed by increase in absorbance at 340 nm.

Reagents

Tris-HCl, 1 M, pH 8.0
DTE, 10 mM
L-Malate, 0.1 M
NAD, 10 mM
NADH, 1 mM
Malonyl-CoA, 2 mM
Malic dehydrogenase from pig heart (1250 units/mg, Sigma Chemical Co), 1 unit/μl
Citrate synthase from pig heart (type III, 10 units/mg, Sigma Chemical Co), 0.17 unit/μl

Assay Procedure. The following reagents are added into a 0.4-ml quartz cuvette with 1 cm path length: Tris-HCl buffer, 20 μl; DTE, 10 μl; L-malate, 20 μl; NAD, 10 μl; NADH, 25 μl; water, 80 μl; malic dehydrogenase, 5 μl. The contents are mixed and incubated for 2 min followed by the addition of 5 μl of citrate synthase. After one additional minute of incubation, 5 μl of malonyl-CoA decarboxylase preparation and 20 μl of malonyl-CoA are added and absorbance change at 340 nm is measured for 5 min. The initial linear velocity is used to calculate the enzyme activity.

Malonyl-CoA Decarboxylase from the Uropygial Glands of Waterfowl

Uropygial gland of waterfowl, which synthesizes large quantities of multiple methyl-branched fatty acids, is an unusually rich source of malonyl-CoA decarboxylase.[4,5]

Purification

Preparation of 105,000 g Supernatant. Twenty geese are killed by exsanguination, and their uropygial glands are excised.[13] All the following procedures are done at 0–4°. Each gland, after slicing into small pieces with a razor blade, is homogenized in approximately 10 ml of 100 mM sodium phosphate buffer (pH 7.6) containing 250 mM sucrose, 20 mM sodium citrate, 1 mM MgCl$_2$, and 1 mM DTE. The homogenates are combined and centrifuged at 15,000 g for 20 min. The supernatant, filtered through cheesecloth to remove the floating fat pad, is recentrifuged at

---

[13] See this series, Vol. 72 [60].

15,000 *g* for another 20 min, the resulting supernatant is centrifuged at 105,000 *g* for 90 min, and the supernatant is filtered through cheesecloth.

Ammonium Sulfate Fractionation. The 105,000 *g* supernatant is diluted to 400 ml with 100 m*M* phosphate buffer, pH 7.6, containing 0.5 m*M* DTE. Finely powdered ammonium sulfate is slowly added with constant slow stirring until 45% saturation is achieved. After stirring for an additional 45 min, the precipitated protein is removed by centrifugation at 10,000 *g* for 20 min. The supernatant is similarly treated with ammonium sulfate until 70% saturation is achieved, and the precipitated protein is recovered by centrifugation at 10,000 *g* for 20 min. The precipitate is suspended in 20 ml of 100 m*M* sodium phosphate buffer, pH 7.6, containing 0.5 m*M* DTE, and dialyzed overnight against the same buffer. Any insoluble material present is removed by centrifugation, and the supernatant is concentrated by ultrafiltration using a PM-30 membrane to about 15 ml.

Sepharose 4B Column Chromatography. Three Sepharose 4B columns (2.7 × 95 cm) are equilibrated with 100 m*M* sodium phosphate buffer, pH 7.6, containing 0.5 m*M* DTE, and 5-ml portions of the above concentrated enzyme solution are subjected to gel filtration on each column with the same buffer at 0.4 ml/min, collecting 6-ml fractions. Malonyl-CoA decarboxylase is eluted in the fractions consisting of the leading shoulder of the major protein peak. The fractions containing the enzyme activity are pooled, concentrated to 5–10 ml by ultrafiltration, and dialyzed against 20 m*M* Tris-HCl buffer, pH 7.6, containing 0.5 m*M* DTE.

QAE-Sephadex Chromatography. The enzyme solution is applied to a QAE-Sephadex A-25 column (1.9 × 27 cm) equilibrated with 20 m*M* Tris-HCl buffer, pH 7.6, containing 0.5 m*M* DTE. After washing the column with approximately two void volumes of the same buffer, which removes a large portion of the protein, a linear gradient of 0 to 300 m*M* NaCl in a total volume of 500 ml of the starting buffer is applied and 5-ml fractions are collected. Fractions containing the enzyme (0.1 to 0.15 *M* NaCl) are pooled and concentrated by ultrafiltration and dialyzed against 20 m*M* citrate-phosphate buffer, pH 5.8, containing 0.5 m*M* DTE.

CM-Sephadex Chromatography. The protein solution is applied to a CM-Sephadex A-25 column (1.9 × 25 cm), equilibrated with the citrate-phosphate buffer. After washing the column with one void volume of the same buffer, which removes a large portion of the protein, a linear gradient of 0 to 300 m*M* NaCl in a total volume of 500 ml of the citrate-phosphate buffer is applied and 5-ml fractions are collected. Malonyl-CoA decarboxylase is eluted with the major protein peak at a salt concentration of 200 to 250 m*M*. If the CM-Sephadex step is carried out at pH 6.4 the enzyme can be eluted at 100 to 150 m*M* NaCl.[14] Fractions containing the

[14] D. L. Rainwater and P. E. Kolattukudy, unpublished results, 1979.

TABLE I

PURIFICATION OF MALONYL-CoA DECARBOXYLASE FROM THE UROPYGIAL GLANDS OF WATERFOWL

Fraction	Total protein (mg)	Total activity (μmol/min)	Specific activity (μmol/mg) per minute)	Recovery (%)	Purification (fold)
Domestic goose					
105,000 g supernatant	7630	5200	0.7	100	—
$(NH_4)_2SO_4$ (45–70%)	940	3100	3.3	60	4.7
Sepharose 4B	324	2100	6.5	40	9.3
QAE-Sephadex	51	2300	45	44	64.3
CM-Sephadex	23	1700	74	33	106
Muscovy ducks					
105,000 g supernatant	5712	4051	0.7	100	1
$(NH_4)_2SO_4$ (45–70%)	1071	4080	3.8	101	5.4
Sepharose 4B	185	2995	16.2	74	23.1
QAE-Sephadex	46	1738	37.8	43	54
CM-Sephadex	13.8	1109	80.4	27	115
North American wood ducks					
105,000 g supernatant	1176	536	0.5	100	1
Sepharose 4B	322	424	1.3	79	2.6
$(NH_4)_2SO_4$ (45–70%)	115	295	2.6	55	5.2
QAE-Sephadex	28	270	9.6	50	19.2
CM-Sephadex	2	166	83	30	166

enzyme are pooled, the pH is adjusted to 7.6 with dibasic sodium phosphate, and the solution is concentrated by ultrafiltration to a protein concentration of 2–6 mg/ml. The concentrated protein solution is dialyzed against 100 mM sodium phosphate buffer, pH 7.6, containing 0.5 mM DTE and stored at 4°.

Purity. The above procedure results in 100 to 200-fold purification with about 30% yield. A typical purification scheme is shown in Table I. The enzyme obtained from the final step is homogeneous as shown by analytical ultracentrifugation, Ouchterlony double diffusion, and polyacrylamide disc gel electrophoresis.

Properties

Stability. Purified malonyl-CoA decarboxylase is stable for 2 weeks when kept at 4° in 100 mM sodium phosphate buffer, pH 7.6, containing 0.5 mM DTE. At a concentration of <1 μg/ml the enzyme is unstable,

TABLE II
PHYSICOCHEMICAL AND KINETIC PROPERTIES OF MALONYL-CoA DECARBOXYLASE[a,b]

Property	Value
$s_{20,w}$	7.8×10^{-13} S
$\bar{v}$	$0.736 \ cm^3 \times g^{-1}$
Molecular weight	186,000
Subunit molecular weight	47,000
pH optimum	8.5–9.5
K_m for malonyl-CoA	0.1 mM
V_{max}	80 μmol/mg per minute

[a] From J. S. Buckner, P. E. Kolattukudy, and A. J. Poulose, *Arch. Biochem. Biophys.* **177**, 539 (1976).
[b] K_m of malonyl-CoA for the decarboxylase from wood duck and Muscovy duck was 33 μM, and all the other properties were the same as shown in this table.[5]

losing 50% of the activity in 2–3 min at 30°. The substrate protects the enzyme from this inactivation.

Physicochemical and Kinetic Properties. Some of the physicochemical and kinetic properties of malonyl-CoA decarboxylase are given in Table II.

Substrate Specificity. Enzyme is specific for malonyl-CoA; methyl malonyl-CoA is decarboxylated at only a small percentage of the rate observed with malonyl-CoA, and malonic acid and succinyl-CoA are not decarboxylated. The enzyme is specific for (*S*)-methylmalonyl-CoA, and the decarboxylation occurs with retention of configuration.[15]

Amino Acid Composition–Immunological Comparison. Malonyl-CoA decarboxylase from the goose uropygial gland has a very high glutamic acid and leucine content (Table III), and the amino acid compositions of the enzyme from Muscovy duck, wood duck and goose are quite similar. Ouchterlony double-diffusion analyses with rabbit antibodies prepared against the enzyme from domestic goose show fusion of precipitin lines with the enzyme from Muscovy duck, wood duck, and Canadian goose, whereas spurs are observed with the decarboxylase from mallard and Peking ducks.

Inhibitors. p-Hydroxymercuribenzoate severely inhibits (at 0.1 mM, 66% inhibition) malonyl-CoA decarboxylase, and malonyl-CoA protects the enzyme from inactivation, suggesting the presence of essential thiol group(s) in the enzyme. Alkylating agents, such as bromopyruvic acid (at 1 mM, 69% inhibition) and N-ethylmaleimide (at 5 mM, 45% inhibition), also inhibit the enzyme. Arginine modifying reagents, such as

[15] Y. S. Kim and P. E. Kolattukudy, *J. Biol. Chem.* **255**, 686 (1980).

TABLE III

AMINO ACID COMPOSITION OF MALONYL-CoA DECARBOXYLASES FROM DOMESTIC GOOSE, MUSCOVY DUCK, AND WOOD DUCK

	Number of residues per molecule[a]		
Amino acid	Goose	Muscovy duck	Wood duck
Asx	101	120	107
Thr	70	67	64
Ser	123	111	114
Glx	229	236	228
Pro	71	71	75
Gly	110	117	111
Ala	124	130	125
Cys	27	33	30
Val	103	114	99
Met	23	26	26
Ile	63	69	61
Leu	205	193	229
Tyr	49	30	27
Phe	60	56	60
Lys	93	94	100
His	40	41	40
Arg	112	119	129
Trp	22	29	26

[a] Based on a molecular weight value of 186,000.

phenylglyoxal and 2,3-butanedione, also inhibit malonyl-CoA decarboxylase activity, suggesting that arginine residues are essential for enzymatic activity.[14] Both acetyl-CoA (at 0.6 mM, 44% inhibition) and methylmalonyl-CoA (at 1.0 mM, 49% inhibition) inhibit malonyl-CoA decarboxylase activity. Free CoA shows only slight inhibition (at 0.9 mM, 12% inhibition). Chelators, such as EDTA and bipyridyl, do not inhibit the enzyme, suggesting that there is no metal ion requirement for the enzymatic activity. Also it appears that the decarboxylase activity is not biotin dependent, since avidin does not inhibit the enzyme.

Subunit Structure. Polyacrylamide disc gel electrophoresis of the purified enzyme shows two bands and both bands have malonyl-CoA decarboxylase activity. Reelectrophoresis of the band with low mobility gives rise to both bands while the reelectrophoresis of the band with high mobility does not give rise to both bands. Also the intensity of both bands varies from one preparation to the next. Therefore it appears that the band with lower mobility is an aggregated form of the protein with higher mobility. This conclusion is also supported by the fact that sodium dodecyl sulfate polyacrylamide disc gel electrophoresis of the enzyme preparation gives only one band. The enzyme as isolated from the extract is a tetramer of identical molecular weight.

Subcellular Localization. Sucrose density gradient centrifugation with marker enzymes shows that the enzyme is located mainly in the cytoplasm but a small portion (a few percent) is present in the mitochondria. The mitochondrial enzyme is immunologically similar to the cytoplasmic enzyme.[16]

Malonyl-CoA Decarboxylase from Rat Liver Mitochondria

In the rat the highest level of malonyl-CoA decarboxylase appears to be present in the liver, where the enzyme is localized in the mitochondrial matrix.[17]

Purification

Preparation of Mitochondria. Mitochondria are isolated from freshly killed adult male rats by a recently described high yield preparative method.[18] One hundred male albino rats are sacrificed by a blow on the head followed by decapitation. The livers are quickly removed (1400 g), cleansed, minced with scissors, and homogenized in 220 mM mannitol containing 70 mM sucrose, 2 mM HEPES, and 0.5 g of bovine serum albumin per liter, pH 7.4 (4 ml/g tissue), using a glass Thomas homogenizer and a Teflon pestle (~1500 rpm, two strokes). The homogenate is centrifuged at 1100 g for 3 min. The pellet is suspended in the grinding medium by one stroke of the pestle at 1500 rpm, and the suspension is centrifuged. This procedure is repeated twice more on the recovered pellet. The pooled supernatant is centrifuged at 6800 g for 15 min, the pellet is gently suspended in one-half the original volume and centrifuged at 20,000 g for 10 min. The pellet is suspended in one-fourth the original volume and centrifuged at 3000 g for 3 min. The supernatant is saved, and the pellet is resuspended in one-eighth the original volume and centrifuged for 3 min. This supernatant is combined with the saved supernatant from the previous centrifugation and centrifuged at 20,000 g for 15 min. The fluffy layer is removed by carefully rinsing with the grinding medium, and the hard-packed pellet is resuspended in one-eighth the original volume of the grinding buffer and centrifuged at 20,000 g for 20 min. The resulting

[16] Y. S. Kim, P. E. Kolattukudy, and A. Boos, *Comp. Biochem. Physiol.* **62B**, 443 (1979).
[17] C. Landriscina, G. V. Gnoni, and E. Quagliariello, *Eur. J. Biochem.* **19**, 573 (1971).
[18] E. Bustamante, J. W. Soper, and P. L. Pedersen, *Anal. Biochem.* **80**, 401 (1977).

pellet is rinsed with the grinding medium to yield 250 g mitochondrial pellet.

Extraction of the Enzyme from Mitochondria. The freshly prepared mitochondrial pellet is suspended in 10 mM sodium phosphate buffer, pH 7.6, containing 0.5 mM DTE (50 ml/g of wet mitochondrial pellet). The suspension is stirred in an ice bath for 2 hr, then the mixture is centrifuged at 24,000 g for 10 min. The pellet is resuspended and reextracted in the same manner. The first treatment releases 60% of the activity, and second treatment releases 60% of the remaining activity resulting in the recovery of about 85% of the total enzyme activity in the extract.

Ammonium Sulfate Fractionation. The combined extract is 40% saturated with ammonium sulfate by slow addition of the powdered salt with stirring. The precipitated protein is removed by centrifugation at 24,000 g for 10 min and discarded. The supernatant is 55% saturated with ammonium sulfate; the precipitated protein, collected by centrifugation, is dissolved in 0.1 M sodium phosphate buffer, pH 7.6, containing 0.5 mM DTE, dialyzed overnight against the same buffer, and centrifuged to remove any insoluble material.

Sepharose 4B Gel Filtration. The protein solution is subjected to gel filtration on a 95 × 3.2 cm Sepharose 4B column (1–1.5 g of protein per column) with 0.1 M sodium phosphate buffer, pH 7.6, containing 0.5 mM DTE, at ~0.5 ml/min. The enzyme activity appears at the leading shoulder of the unsymmetrical protein peak. The fractions (10 ml each) containing the enzyme activity are pooled, concentrated by ultrafiltration with a PM-30 membrane, and dialyzed overnight against 20 mM Tris-HCl buffer, pH 8, containing 0.5 mM DTE.

QAE-Sephadex Chromatography. The enzyme solution is applied to a 30 × 2 cm QAE-Sephadex column (~600 mg of protein per column) equilibrated with the Tris buffer. After washing the column with two bed volumes of the same buffer (a large proportion of the protein is not retained), 500 ml of a 0 to 0.3 M NaCl gradient prepared in the Tris buffer is passed through the column at 0.3 ml/min. The fractions (7 ml each) containing the enzyme activity, which elutes at about 0.12 M NaCl, are pooled, concentrated by ultrafiltration and dialyzed overnight against 20 mM citrate-phosphate buffer, pH 5.8, containing 0.5 mM DTE.

CM-Sephadex Chromatography. The enzyme solution is applied to a 25 × 1.8 cm column of CM-Sephadex equilibrated with the citrate–phosphate buffer (150 mg of protein per column). After washing the column with two bed volumes of the same buffer, which removed the bulk of the protein but no decarboxylase, a 0 to 0.3 M linear gradient of NaCl prepared in the citrate–phosphate buffer is applied at a flow rate of 16 ml/hr. Fractions (4 ml each) containing the enzyme activity, which

TABLE IV
PURIFICATION OF MALONYL-CoA DECARBOXYLASE FROM RAT LIVER MITOCHONDRIA

Fraction	Protein (mg)	Enzyme activity (nmol/min)	Specific activity (nmol/mg per minute)	Recovery (%)	Purification (fold)
Mitochondrial extract	18,302	147,266	8.0	100	1
$(NH_4)_2SO_4$ (40–55%)	5,548	78,415	14.1	53	1.8
Sepharose 4B	3,824	126,569	33.1	86	4.1
QAE-Sephadex	627	64,125	102.3	44	12.8
CM-Sephadex	180	80,750	448.6	55	56.1
QAE-Sephadex	54	54,340	1,006	37	126
Sephadex G-150	6	42,443	7,074	29	884
NADP-agarose	0.9	14,816	16,462	10	2058

emerges at about $0.12\ M$ NaCl, are pooled, concentrated by ultrafiltration, and dialyzed against 20 mM Tris-HCl buffer, pH 8.0, containing 0.5 mM DTE. This step gives a slight (25%) increase in the total units of enzyme activity.

Second QAE-Sephadex. The enzyme solution (180 mg of protein) is applied to a 25 × 1.8 cm column of QAE-Sephadex equilibrated with the Tris buffer indicated above, and the procedure described under the first QAE step is followed. The fractions (4 ml each) containing enzyme activity eluting at about 56 mM NaCl are pooled, concentrated by ultrafiltration, diluted with $0.1\ M$ sodium phosphate buffer, pH 7.6, containing 0.5 mM DTE, and concentrated. The dilution and concentration is repeated twice more.

Sephadex G-150 Gel Filtration. The enzyme solution is passed through a 2.8 × 90 cm column of Sephadex G-150 with $0.1\ M$ sodium phosphate buffer, pH 7.6, containing 0.5 mM DTE. The fractions containing enzyme activity (eluting at the leading shoulder of the major protein peak) are pooled, then concentrated by ultrafiltration; the buffer is changed to 20 mM sodium phosphate, pH 7.0, containing 0.5 mM DTE by repeated dilution and ultrafiltration.

NADP-Agarose Chromatography. The protein (6 mg) solution is passed through a 10-ml NADP-agarose column containing 24 μmol of bound NADP. The column is washed with 20 ml of buffer, which removes the bulk of the protein but no enzyme activity, then a 0 to $0.2\ M$ NaCl linear gradient is applied. The fractions (2.2 ml each) containing the enzyme activity, which elutes at about 50 mM NaCl, are pooled, then concentrated by ultrafiltration; the buffer is changed to $0.1\ M$ sodium phosphate, pH 7.6, containing 0.5 mM DTE.

This procedure results in over 2000-fold purification giving a 10% yield of enzyme with a specific activity of 15–20 μmol/min per milligram of protein (Table IV).

Purity of the Enzyme. Polyacrylamide electrophoresis shows a single protein band. Ouchterlony double diffusion also shows homogeneity.

Properties of the Enzyme

General Properties. Molecular weight, as estimated by gel filtration, is about 160,000. The enzymatic activity is maximal between pH 8.5 and 10.0. Linear rates are obtained up to 5 μg protein per milliliter and 15 min incubation time. K_m and V_{max} are 50 μM and 19 μmol/mg per minute, respectively.

Substrate Specificity. Malonic acid and succinyl-CoA are not decarboxylated, but (R,S)-methylmalonyl-CoA is decarboxylated at 0.5% of the rate observed with malonyl-CoA.

Inhibitors. Thiol-directed reagents, p-hydroxymercuribenzoate and N-ethylmaleimide at 0.5 mM and 5.0 mM, respectively, nearly completely inhibit the enzyme whereas 5 mM iodoacetamide has no effect. EDTA does not affect the activity. Acetyl-CoA, propionyl-CoA, and methylmalonyl-CoA inhibit the enzyme.

Subcellular Localization. Sucrose density gradient centrifugation with marker enzymes shows that the enzyme is located in the mitochondria, and the release of the activity into the low ionic strength buffer indicates that the enzyme is in the matrix.[6]

Malonyl-CoA Decarboxylase from *Mycobacterium tuberculosis* H37Ra

Purification of the Enzyme

Growth of Bacteria and Extraction of the Enzyme. Mycobacterium tuberculosis H37Ra strain is grown at 37° in either glycerol–alanine–salt medium or Middlebrook 7H9 medium enriched with Tween-80 and ADC (Difco).[19] Bacteria are harvested at stationary phase after about 3 weeks of growth at 37°. About 500 g wet weight of frozen cells are thawed and suspended in 1 liter of 0.1 M sodium phosphate buffer, pH 7.0, containing 0.5 mM DTE. After the suspension reaches 30°, 25 mg of lysozyme are added and incubated for 1 hr with stirring. The cell suspension is cooled to 4° and subjected to ultrasonic treatment with a Biosonik III, and the

[19] K. Takayama, H. K. Schnoes, E. L. Armstrong, and R. W. Boyle, *J. Lipid Res.* **16**, 308 (1975).

mixture is centrifuged at 20,000 g for 10 min. The pellet is suspended in the buffer, the sonication and centrifugation are repeated six times, and the supernatants are pooled.

Ammonium Sulfate Precipitation. The cell-free extract is 35% saturated with ammonium sulfate by slow addition of the powdered salt with stirring. After the solution is stirred for 1 hr the mixture is centrifuged at 20,000 g for 10 min and the pellet is discarded. The supernatant is treated with powdered ammonium sulfate with stirring until 55% saturation is achieved, and the mixture is centrifuged. The pellet is dissolved in 0.1 M sodium phosphate buffer, pH 7, containing 0.5 mM DTE and dialyzed against the same buffer overnight.

Gel Filtration on Sepharose-6B. The enzyme solution (about 700 mg of protein) is passed through a 3.2 × 100 cm column of Sepharose 6B in the same buffer as above. The enzyme activity is contained in the third protein peak. The fractions (8–10 ml) containing malonyl-CoA decarboxylase activity are pooled, concentrated by ultrafiltration using an Amicon PM-10 membrane, and dialyzed against 20 mM Tris-HCl buffer, pH 7.6, containing 0.5 mM DTE.

DEAE-Sephacel Chromatography. The protein solution (about 600 mg of protein) is applied to a 2 × 25 cm column of DEAE-Sephacel, equilibrated with 20 mM Tris-HCl buffer, pH 7.6, containing 0.5 mM DTE. After passing two void volumes of the same buffer, a linear gradient of 0 to 0.3 M NaCl prepared in a total volume of 500 ml of the same Tris buffer is applied. Malonyl-CoA decarboxylase is eluted at about 0.18 M NaCl just prior to the major protein peak. The fractions containing the enzyme activity are pooled, concentrated by ultrafiltration, and dialyzed overnight against 20 mM citrate-phosphate buffer, pH 5.8, containing 0.5 mM DTE.

CM-Sephadex Chromatography. The enzyme solution (~100 mg of protein) is applied to a 1.8 × 25 cm column of CM-Sephadex equilibrated with the citrate–phosphate buffer. After passing two bed volumes of the same buffer, a 0 to 0.3 M linear gradient of NaCl is applied. The bulk of the protein is not retained, and the enzyme is eluted at approximately 0.06 M NaCl. Fractions containing the enzyme activity are pooled, then concentrated by ultrafiltration; the buffer is changed to 10 mM Tris-HCl, pH 7.0, containing 0.5 mM DTE, by repeated dilution and concentration.

NADP-Agarose Chromatography. The enzyme solution is passed through a 1-ml column of NADP-Agarose containing 2.5 μmol of covalently attached NADP, equilibrated with the above Tris-HCl buffer. Even though the enzyme is not retained by the gel, the major portion of the activity is eluted after the major part of the protein is eluted, thus resulting in a twofold increase in specific activity. When purity is a major consideration, the first half of the enzyme activity peak is not included in the fractions pooled.

TABLE V

PURIFICATION OF MALONYL-CoA DECARBOXYLASE FROM *Mycobacterium tuberculosis*

Fraction	Protein (mg)	Enzyme activity (nmol/min)	Specific activity (nmol/mg per minute)	Recovery (%)	Purification (fold)
Crude supernatant	17,457	44,690	2.6	100	1
(NH₄)₂SO₄ (35–55%)	5,523	38,410	7.0	86	2.7
Sepharose 6B	2,260	39,787	17.6	89	6.8
DEAE-Sephacel	294	28,592	97.3	64	37.4
CM-Sephadex	3.8	4,066	1070.0	9	412
NADP-agarose	0.54	1,294	2396.8	3	922

Extent of Purification. The procedures outlined above resulted in about 1000-fold purification with a 3% recovery of the enzyme activity (Table V). Polyacrylamide gel electrophoresis of the preparation shows a major band containing enzymatic activity, and a minor, slower moving band that does not show decarboxylase activity.

Properties of the Enzyme

General Properties. The molecular weight as estimated by gel filtration is about 44,000. The p*I* of this enzyme is 6.7. Linear rates are obtained up to 0.5 mg of protein per milliliter and 10 min incubation time. A relatively sharp pH optimum at around 5.5 is observed. The K_m and V_{max} are 0.2 mM and 3.9 μmol/min per milligram of protein, respectively. This bacterial enzyme does not cross-react with rabbit antibodies prepared against the avian or mammalian malonyl-CoA decarboxylase.

Substrate Specificity

Neither malonic acid nor succinyl-CoA is decarboxylated, but methylmalonyl-CoA is decarboxylated at ~5% of the rate observed with malonyl-CoA.

Inhibitors. Thiol-directed reagents, *p*-hydroxymercuribenzoate and *N*-ethylmaleimide severely inhibit the enzyme, but iodoacetamide, ethylenediaminetetraacetic acid, diisopropylfluorophosphate, avidin, and biotin have no effect on the enzyme. Acetyl-CoA propionyl-CoA, methylmalonyl-CoA, CoA and succinyl-CoA inhibit this decarboxylase.

Another Bacterial Malonyl-CoA Decarboxylase. The decarboxylase from malonate-grown *Pseudomonas fluorescens* is more unstable than the mycobacterial enzyme. The *Pseudomonas* enzyme has been partially purified and characterized. This enzyme has a molecular weight of $\sim56,000$ and a pH optimum of 5.5. The apparent K_m is 1 mM. This enzyme is insensitive to p-hydroxymercuribenzoate, acetyl-CoA, and propionyl-CoA.

Acknowledgment

This work was supported in part by Grant GM-18278 from the National Institutes of Health. Dr. James Buckner contributed to the development of some of the methods described.

[21] Acyl-Acyl Carrier Protein Synthetase from *Escherichia coli**

By CHARLES O. ROCK and JOHN E. CRONAN, JR.

Acyl-ACP synthetase from *Escherichia coli* is localized in the inner membrane[1,2] and functions to ligate free fatty acids to ACP-SH in the presence of ATP and Mg^{2+}.[3] As with other synthetases, AMP was identified as a product of the reaction.[3] The role of acyl-ACP synthetase in *E. coli* lipid metabolism is not clear at present. However, the finding that intracellular free fatty acids can be incorporated into phospholipid in the strains lacking acyl-CoA synthetase activity[4] suggests that acyl-ACP synthetase may function to reintroduce free fatty acids into the biosynthetic pathway. The broad substrate specificity of acyl-ACP synthetase makes purified preparations of this enzyme a useful tool for the synthesis of native acyl-ACP[5] for use in examining other lipid biosynthetic enzymes in *E. coli*.

Assay Method

Principle. Acyl-acyl carrier protein (ACP) synthetase is assayed by measuring the incorporation of [1-^{14}C]palmitic acid into palmitoyl-ACP. The reaction is terminated by applying an aliquot of the incubation mixture to a Whatman 3 MM filter disk and washing the disk in two changes of organic solvent. [1-^{14}C]Acyl-ACP is not soluble in the organic solvent and

* Unknown to the author and Editor articles 21 and 41 were switched. This should be article 41 in Section II.

[1] A. K. Spencer, A. P. Greenspan, and J. E. Cronan, Jr., *J. Biol. Chem.* **253**, 5922 (1978).
[2] C. O. Rock and J. E. Cronan, Jr., *J. Biol. Chem.* **254**, 7116 (1979).
[3] T. K. Ray and J. E. Cronan, Jr., *Proc. Natl. Acad. Sci. U.S.A.* **73**, 4374 (1976).
[4] T. K. Ray and J. E. Cronan, Jr., *Biochem. Biophys. Res. Commun.* **69**, 506 (1976).
[5] C. O. Rock and J. L. Garwin, *J. Biol. Chem.* **254**, 7123 (1979).

TABLE I

REAGENTS AND STOCK SOLUTIONS FOR THE ASSAY OF ACYL–ACYL CARRIER PROTEIN
(ACP) SYNTHETASE

Reagent	Microliters per assay	Final concentration
Tris-HCl, 1 M, pH 8.0	4	0.1 M
LiCl, 4 M	4	0.4 M
ATP, 0.1 M, pH 7.0	2	5 mM
MgCl$_2$, 0.2 M	2	10 mM
DTT, 40 mM	2	2 mM
[1-^{14}C]Palmitic acid,[a] 600 μM	4	60 μM
ACP-SH, 300 μM	2	15 μM
Protein in 2% Triton X-100	20	0–10 mU

[a] [1-^{14}C]Palmitic acid is supplied as a hexane solution. The appropriate volume of the hexane solution is transferred to a glass tube, the solvent is removed under a stream of nitrogen, and the residue is dissolved in 10% Triton X-100, 0.6 mM NaOH. This solution is heated at 55° for 1 hr to ensure a homogeneous solution. The dithiothreitol (DTT) solution should be prepared daily.

is retained on the disk whereas free fatty acids are extracted into the organic phase.[3] After washing, the disks are recovered, dried, and counted to determine the amount of acyl-ACP present.

Reagents. [1-^{14}]Palmitic acid (specific activity 55 Ci/mol) and Triton X-100 are supplied by New England Nuclear. ACP-SH is purified to homogeneity as described in this volume.[6] ATP is purchased from P-L Biochemicals, filter papers (3 MM) are from Whatman, and Tris and dithiothreitol are obtained from Sigma. All other materials should be reagent grade or better. The stock solutions required for the assay are listed in Table I and can be stored frozen for months, with the exception of the DTT solution, which should be prepared daily.

Method. The number of assays to be performed is determined, and the appropriate volumes of the assay reagents (Table I) are combined; 20 μl of the mixture are placed into the 10 × 75 mm glass assay tubes. The reaction is initiated by the addition of 20 μl of the protein solution in 2% Triton X-100, the tubes are briefly vortexed and gently shaken at 37° for 20 min. At the end of this time, 30 μl of the incubation mixture are removed, applied to a Whatman 3 MM filter disk suspended on a steel pin, and allowed to air dry. The filter disks are then washed (20 ml per filter) with two changes of chloroform : methanol : acetic acid (3 : 6 : 1, v/v/v). The filter disks are dried and counted in Aquasol (New England Nuclear) or PCS (Amersham). A unit of acyl-ACP synthetase activity is defined as the

[6] See this volume [41]; see also C. O. Rock and J. E. Cronan, Jr. *Anal. Biochem.* **102**, 362 (1980).

amount of protein required to produce 1 nmol of [1-^{14}C]palmitoyl-ACP per minute.

Purification of Acyl-ACP Synthetase

Materials. Blue Sepharose CL-6B is purchased from Pharmacia, and spheroidal hydroxyapatite is purchased from Clarkson Chemical Co. Deoxyribonuclease II is purchased from Sigma. Frozen *E. coli* B cells (late log) are purchased from Grain Processing Co. The purification procedure has also been successfully accomplished with freshly grown cells, and these preparations have been found to have higher initial activities than do commercially available cell pastes. All other materials should be reagent grade or better.

Step 1. Preparation and Extraction of Membranes. All procedures are performed at 4° unless otherwise indicated. *Escherichia coli* cells (250 g wet weight) are suspended in 50 mM Tris-HCl, pH 8.0, to a final volume of 600 ml, and 3–5 mg of DNase II are added to the cell suspension. The cells are disrupted by passage through a French pressure cell at 16,000 psi. The resulting homogenate is centrifuged at 15,000 g for 20 min to remove debris, and unbroken cells and the pellet is discarded. The supernatant is adjusted to 10 mM MgCl$_2$ by the addition of an appropriate volume of a 1 M MgCl$_2$ stock solution. The Mg^{2+} is added to ensure complete recovery of membranes during the centrifugation steps. The suspension is centrifuged at 80,000 g for 90 min, and the supernatant is discarded. The resulting pellets are thoroughly homogenized in 150 ml of 50 mM Tris-HCl, pH 8.0, and then 150 ml of 50 mM Tris-HCl, pH 8.0, containing 1 M NaCl and 20 mM MgCl$_2$ is mixed with the membrane suspension and the solution sedimented at 80,000 g for 90 min. The salt wash is performed to remove contaminating soluble proteins and extrinsic membrane components. The membrane pellets from the salt wash are thoroughly homogenized in 100 ml of 50 mM Tris-HCl, pH 8.0. To this suspension 100 ml of 50 mM Tris-HCl (pH 8.0) containing 4% Triton X-100 and 20 mM MgCl$_2$ is added, and the solution is slowly stirred for 30 min. The MgCl$_2$ is present at this step to retard solubilization of outer membrane components. The suspension is centrifuged at 80,000 g for 90 min to pellet unsolubilized material, and the Triton X-100 supernatant is saved.

All the above procedures should be executed on the same day. Although membrane-bound acyl-ACP synthetase can be frozen without much loss of activity, 50% inactivation of the Triton X-100 solubilized enzyme is observed upon freeze-thawing. The enzyme is best stored in 50 mM Tris-HCl, pH 7.0, containing 2% Triton X-100, 5 mM MgCl$_2$, 0.02% sodium azide, and 5 mM ATP at 4°. Only a 10–20% loss of activity is found under these conditions after 1 month of storage.

Step 2. Heat Treatment. The Triton X-100 extract is adjusted to 5 mM in ATP by the addition of the appropriate volume of neutralized 0.1 M ATP. Acyl-ACP synthetase was found to become progressively more heat labile as the protein concentration was decreased. Therefore, the protein concentration at this step should be between 2 and 5 mg/ml. Likewise, acyl-ACP synthetase is totally inactivated by heat treatment in the absence of ATP. Heat treatment is performed by first placing the solution in a 37° water bath and stirring with a thermometer until the temperature reaches 33°. The solution is then placed in a 55° water bath and stirred until the temperature reaches 53°, and heating is continued for an additional 5 min. The solution is then cooled with stirring on ice and centrifuged at 15,000 g for 20 min to remove denatured proteins.

Step 3. Blue Sepharose Chromatography. A column (1.2 × 35 cm) is packed with Blue Sepharose CL-6B (bed volume, 39 ml) and equilibrated with 50 mM Tris-HCl, pH 8.0, containing 2%Triton X-100 at 4°. The entire supernatant from the heat step is applied to the Blue-Sepharose column. The column is next rinsed with six column volumes of 50 mM Tris-HCl, pH 8.0, containing 2% Triton X-100 and 0.6 M NaCl. Acyl-ACP synthetase is eluted from the column with 0.5 M KSCN in 50 mM Tris-HCl, pH 8.0, 2% Triton X-100. Fractions containing acyl-ACP synthetase can usually be located visually, since these fractions also contain a bright red cytochrome. Acyl-ACP synthetase activity can be assayed in these fractions provided that the KSCN concentration is reduced to <25 mM in the standard assay system. Higher concentrations of KSCN in the assay system inhibits acyl-ACP synthetase activity, but the enzyme can be exposed to 0.5 M KSCN at 4° for several days without significant loss of activity.

Step 4. Hydroxyapatite Chromatography. The KSCN can be removed by dialysis from the acyl-ACP synthetase-containing Blue Sepharose fractions. However, at protein concentrations typically encountered, a copious precipitate is formed. The precipitate contains approximately 60% of the protein, and only 40% of the acyl-ACP synthetase activity is recovered in the supernatant. The precipitate could not be redissolved under conditions that maintained acyl-ACP synthetase activity. Therefore Blue Sepharose fractions containing acyl-ACP synthetase activity are pooled and diluted with 4 volumes of 50 mM Tris-HCl, pH 8.0, 2% Triton X-100. The sample is then applied to a hydroxyapatite column (1.2 × 25 cm; 28 ml bed volume) at a flow rate of 4 ml/hr. Acyl-ACP synthetase is retained by the column and is eluted using 0.3 M KPO$_4$, pH 7.5, containing 2% Triton X-100. The enzyme is then dialyzed against 50 mM Tris-HCl, pH 8.0 containing 2% Triton X-100 and is stored in the presence of ATP as described in step 1.

Comments on the Purification. The purification procedure results in approximately 3000-fold purification of the enzyme in a 28% yield (Table

TABLE II
PURIFICATION OF ACYL-ACYL CARRIER PROTEIN SYNTHETASE[a]

Step	Activity (units/mg)	Purification (fold)	Yield (%)	Total protein (mg)
15,000 g supernatant	0.6	1	100	30,000
Membranes	2.4	4	98	7,500
Salt-washed membranes	3.6	6	95	5,000
2% Triton X-100 extract	4.8	8	86	3,750
55° supernatant	14.4	24	78	1,250
Blue-Sepharose	960	1500	54	20
Hydroxyapatite	1900	3000	28	10

[a] Enzyme activity was determined using the standard assay system. Protein content was determined by the dye binding assay of McKnight[7] using bovine serum albumin in 2% Triton X-100 as a standard.

II). The purity of acyl-ACP synthetase at this stage is not precisely known. Chromatography on DEAE-cellulose or gel filtration media did not result in appreciable increase in specific activity, although enzyme units are recovered in good yield. Sodium dodecyl sulfate–gel electrophoresis of the final product exhibits an intensely staining band that does not enter the gel even at acylamide concentrations as low as 2.5%. Gel filtration of acyl-ACP synthetase in the presence of Triton X-100 on BioGel A-5.0m indicates a molecular weight of on the order of 500,000 for the protein–detergent complex. Gel filtration experiments with acyl-ACP synthetase need to be performed in the presence of 0.25–0.5 M KSCN to prevent adsorption of enzyme to the chromatographic media.

Properties

Acyl-ACP synthetase activity is not subject to catabolite repression, is not a component of the *fad* (β-oxidation) regulon, and is distinct from acyl-CoA synthetase. The purified enzyme does not require phospholipids for activity, and activity is recovered in good yield following extraction with organic solvents. The enzyme is most active on saturated fatty acids (12:0 to 16:0) having a K_m 5 to 15 μM and is less active on unsaturated fatty acids and shorter chain lengths (K_m 30 to 160 μM). Poorer substrates not only have a higher K_m but also exhibit a lower V_{max}. Care must be taken in determining the specific activity of acyl-ACP synthetase in crude extracts, since these extracts contain a contaminating activity that functions to limit severely the linearity of the time and protein curves.[2,8] This contaminating activity is not present in purified preparations.

[7] G. S. McKnight, *Anal. Biochem.* **78**, 86 (1977).
[8] A. K. Spencer, A. D. Greenspan, and J. E. Cronan, Jr., *FEBS Lett.* **101**, 253 (1979).

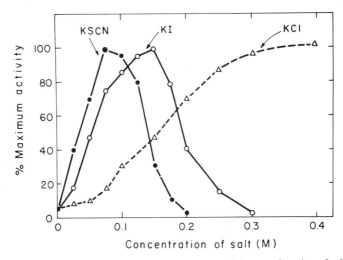

FIG. 1. Acyl–acyl carrier protein (ACP) synthetase activity as a function of salt concentration. The salts tested in this experiment are KSCN (●——●), KI (○——○), and KCl (△——△). Acyl-ACP synthetase activity was assayed as described in the text except that the various concentrations of salt were substituted for LiCl. For a more complete description of salt effects see Rock and Cronan.[2]

One of the most unusual features of acyl-ACP synthetase is the requirement for high concentrations of salts for maximum activity. With the exception of Mg^{2+}, the requirements for ions has been found to be nonspecific. All salts tested were capable of activating acyl-ACP synthetase to comparable extents but were maximally effective only within narrow concentration ranges. Figure 1 illustrates the reaction rate dependence of acyl-ACP synthetase on the concentration of three potassium salts. All salts are capable to activating the enzyme to the same extent, but the salt concentration giving maximal activation of acyl-ACP synthetase varies with the ionic species. Chaotropic salts (e.g., KSCN) are found to be effective at lower concentrations than are less chaotropic species (Fig. 1). Thus activation correlates not with ionic strength per se, but with the disordering influence of the ions on the structure of water. Several other ions have been tested,[7] and the effectiveness of anions ($SCN^- > I > Cl^-$) and cations ($Ca^{2+} > Mg^{2+} > Li^+ > K^+$, Na^+) is the same as in the chaotropic series. LiCl is chosen as the standard assay salt because this salt has a broader plateau region (0.35 to 0.45 M) than other salts tested. Therefore, the enzyme rate is less sensitive to the ionic composition when assayed in the presence of LiCl.

[22] Purification of Acyl Carrier Proteins by Immunoaffinity Chromatography

By MARY LOU ERNST-FONBERG, ANN W. SCHONGALLA, and
THERESA A. WALKER

Similarities in the structure of acyl carrier proteins (ACPs) from different sources[1,2] result in immunologic cross-reactivities that provide the basis of a method[3] for obtaining extensively purified ACPs from different organisms. A procedure for purifying ACPs from *Escherichia coli* and *Euglena gracilis* by immunoaffinity chromatography using anti-*E. coli* ACP bound to Sepharose is described.

Assay of ACP

Activity is measured by a CO_2–malonyl-CoA exchange assay.[2,4,5] The preparation of fraction A and caproylpantetheine needed for this assay are described elsewhere in this series.[4] Under the conditions used, 1 μg of pure, fully reduced *E. coli* ACP fixes 12 nmol of CO_2 in 15 min. Because of its sensitivity and low background, the ACP-dependent fatty acid synthase assay[6] may also be used to detect ACP. In cases where antisera or preparations derived from them are added to assays, controls are run containing equivalent preparations except that normal serum is substituted for antiserum.

Reagents

Imidazole buffer, 1.0 M, pH 6.2
2-Mercaptoethanol, 0.4 M
Caproylpantetheine, 2.0 mM
Malonyl-CoA, 1.0 mM
Fraction A from *E. coli*
$KH^{14}CO_2$ (0.2 μCi/nmol), 0.25 M
Perchloric acid, 10%

---

[1] D. J. Prescott and P. R. Vagelos, *Adv. Enzymol.* **36**, 270 (1972).
[2] R. K. DiNello and M. L. Ernst-Fonberg, *J. Biol. Chem.* **248**, 1707 (1973).
[3] M. L. Ernst-Fonberg, A. W. Schongalla, and T. A. Walker, *Arch. Biochem. Biophys.* **178**, 166 (1977).
[4] P. W. Majerus, A. W. Alberts, and P. R. Vagelos, this series, Vol. 14 [6].
[5] A. K. Evers, J. S. Wolpert, and M. L. Ernst-Fonberg, *Anal. Biochem.* **64**, 606 (1975).
[6] M. L. Ernst-Fonberg, *Biochemistry* **12**, 2449 (1973).

KOH, 40%
Ethanol, absolute
Toluene containing 4.0 g of 2,5-diphenyloxazole per liter

Procedure. Reactions are run in glass liquid scintillation counting vials with 15-mm tops, which are sealed with serum stoppers. Alternatively, small glass vials (Econvials, New England Nuclear) may be used and sealed with high quality No. 000 rubber stoppers. In either case, the stoppers are linked in series via two 18-gauge hypodermic needles in each stopper. One needle is for flushing the air inlet, the other for flushing the air outlet. The outlet needles are all linked in series, and the inlet needles are linked in a separate series. Ten vials is a manageable number to process at one time. The vials are secured in a metal rack in a 30° circulating water bath in a fume hood. Following a 5-min warming period of all components except malonyl-CoA, the reaction is initiated by adding malonyl-CoA, and the stoppers are secured. After 15 min, reactions are stopped by injecting 100 μl of perchloric acid, and residual $^{14}CO_2$ is removed by purging with air for 20 min. The $^{14}CO_2$ in the effluent air is trapped in 40% KOH. The stoppered scintillation vials are placed in a water bath at 55°, and the reaction mixtures are dried under a current of air that has been passed through Drierite and a Millipore filter holder containing Whatman No. 1 filter paper. The following are added in the order shown: 0.1 ml of H_2O, 3 ml of absolute ethanol, and 15 ml of toluene counting fluid. The vials are then counted in a liquid scintillation counter. Quenching is corrected for by the channels ratio method or with internal standards.

The ACP assay contains in a final volume of 0.25 ml: imidazole buffer, 25 μl (0.1 M); 2-mercaptoethanol, 25 μl (40 mM); caproylpantetheine, 25 μl (0.2 mM); malonyl-CoA, 25 μl (0.1 mM); KH$^{14}CO_2$ μl (25 mM); fraction A, an amount that fixes 300–1000 cpm without added ACP; and ACP, about 0.2 μg. The concentrations of reagents are in parentheses.

Preparation of ACPs

ACP for use as an antigen is purified from *E. coli* strain B cells as described in this series.[4]

Crude preparations of ACPs from *Euglena gracilis* strain Z or variety *bacillaris* are obtained by ammonium sulfate fractionation.[2,7] All manipulations are carried out at 4°. *Euglena* cells, grown and stored as described,[2] 361 g, are suspended in 361 ml of 0.05 M imidazole buffer, pH 7.1, 0.07 M 2-mercaptoethanol, 0.02 M potassium citrate, 0.01 M EDTA. The cells are

[7] R. K. DiNello and M. L. Ernst-Fonberg, this series, Vol. 35 [14].

disrupted in a Parr nitrogen bomb or by ultrasound. The cell suspension is centrifuged at 23,000 g for 1.5 hr. The supernatant solution, 550 ml, is adjusted to 70% saturation with ammonium sulfate, 47.2 g/100 ml. After 1 hr of stirring, the precipitate is removed by centrifugation and the supernatant solution is brought to 95% saturation with ammonium sulfate by adding 17.9 g/100 ml. The container is covered with Parafilm, and the mixture is stirred overnight. The precipitate is collected by centrifugation, dissolved in about 35 ml of 0.01 M potassium phosphate buffer, pH 6.2, and desalted on a Sephadex G-25 column. The protein solution, about 200 ml, is stored at 4° until aliquots are applied to the immunoaffinity column.

Preparation of Antibodies against *E. coli* ACP

Mature male rabbits are immunized with 5 mg of pure *E. coli* ACP. The protein, in a volume of 0.5 ml, is mixed with 0.5 ml of Freund's complete adjuvant, and the emulsion is injected intradermally at several sites. A booster of 2 mg of ACP in 0.5 ml, mixed with 0.5 ml of Freund's incomplete adjuvant, is given 3 weeks later. The appearance of antibodies is monitored by immunodiffusion on Ouchterlony plates prepared with a mixture of 2.5% agar in 0.01 M potassium phosphate buffer, pH 7.0, 0.15 M NaCl, and 0.1% NaN$_3$. Boosters of 2 mg of ACP in a volume of 0.5 ml, mixed with 0.5 ml of Freund's incomplete adjuvant, are given whenever the Ouchterlony plates give a negative result. Despite its small size, *E. coli* ACP is a fairly good antigen. Each of three rabbits inoculated developed a positive immunologic response within 5 weeks of the initial immunization.

Escherichia coli ACP, in both crude and pure preparations, gives a single sharp line on double gel diffusion against the antiserum. The antibodies extensively inhibit *E. coli* ACP activity in both the CO$_2$–malonyl-CoA exchange and the ACP-dependent fatty acid synthesis assays.

Although there is no definite precipitin line formed against *Euglena* ACP, cross-reaction between anti-*E. coli* ACP and ACP from *Euglena* does occur. Figure 1 shows antibody inhibition of the activity of *E. gracilis* strain Z ACP in the CO$_2$–malonyl-CoA exchange reaction. The activity of the same ACP monitored in the ACP-dependent fatty acid synthesis system is also inhibited, to 54% of the control value.

Immunologic cross-reaction also exists between anti-*E. coli* ACP and the acyl carrier function of the multienzyme complex fatty acid synthase of *E. gracilis* variety *bacillaris*.[8] The specific inhibition of this fatty acid synthase by anti-*E. coli* ACP is shown in Fig. 2.

[8] T. A. Walker, Ph.D. thesis, Yale University, New Haven, Connecticut, 1980.

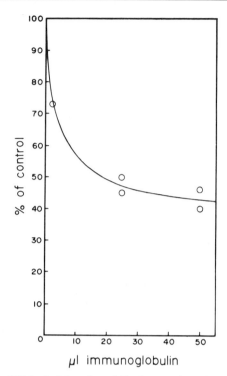

μl immunoglobulin

Fig. 1. Inhibition of biological function of *Euglena* acyl carrier protein (ACP) by antibodies against *Escherichia coli* ACP. Antiserum and normal serum were purified through ammonium sulfate fractionation and were adjusted to equal absorbance at 280 nm (10.0) with 0.01 M potassium phosphate, pH 7.0, and 0.15 M NaCl. The indicated amounts of either normal γ-globulin or immunoglobulin were added to the malonyl-CoA–CO_2 exchange reaction assay of crude ACP from *Euglena gracilis* strain Z. The control assays exchanged 4.5–4.9 nmol of $^{14}CO_2$ per 15 min. There was little or no inhibition by the normal γ-globulin compared to assays that contained no γ-globulin. From Ernst-Fonberg *et al.*[3]

Quantitative Precipitin Test

Optimum concentrations for precipitation of an antigen–antibody complex are determined by varying the amount of antigen available for combination with a fixed amount of antibody.[9]

Reagents

Hyperimmune serum, or 0–40% saturation $(NH_4)_2SO_4$ fraction
Normal serum, or 0–40% saturation $(NH_4)_2SO_4$ fraction
E. coli ACP purified through acid precipitation[3,4]

---

[9] E. A. Kabat and M. M. Mayer, "Experimental Immunochemistry," 2nd ed., p. 22. C Thomas, Springfield, Illinois, 1961.

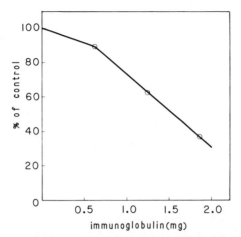

FIG. 2. Inhibition of *Euglena gracilis* var. *bacillaris* multienzyme complex fatty acid synthase by antibodies against *Escherichia coli* ACP. The indicated amounts of either normal γ-globulin or anti-*E. coli* ACP immunoglobulin (40% saturation ammonium sulfate fraction) were added to the fatty acid synthease assay measuring incorporation of [2-^{14}C]malonyl-CoA into long-chain fatty acids. Assays contained 30 μM acetyl-CoA, 60 μM [2-^{14}C]malonyl-CoA (specific activity, 0.5), 0.80 mM NADPH, 0.10 M TES, pH 7.0, 2 mM DTT, enzyme, and γ-globulin, in a final volume of 0.5 ml. After a 15-min incubation at 35°, 0.1 ml of 45% KOH was added; and the reaction mixture was boiled for 30 min. After acidifying with 0.2 ml of 12 N HCl, two 5-ml aliquots of pentane were used to extract the fatty acids from the mixture. The source of fatty acid synthase was a 0 to 35% saturation ammonium sulfate fraction from *E. gracilis* var. *bacillaris*, 7.3 μg of protein per assay. From Walker.[8]

Protein, 0.5 mg/ml
NaCl, 1.0 M
NaCl, 0.15 M
NaOH, 0.1 N

Procedure. Hyperimmune and normal sera are centrifuged just before use to remove any nonspecific precipitated protein and adjusted to equal absorbances at 280 nm. The following components are combined in a final volume of 0.5 ml: serum, 0.25 ml; NaCl, 1.0 M, 0.05 ml; and antigen, 0.01–0.2 ml in increments of 0.02 ml. Controls containing antibody alone and the maximum amount of antigen alone are essential. After incubation at 4° for 48 hr, the immunoprecipitates are isolated by centrifugation, washed twice with 0.5 ml of cold 0.15 M NaCl, and dissolved in 1.0 ml of NaOH; the absorbance at 280 nm is measured.

Purification of Antibodies against *E. coli* ACP

Antiserum can be used directly in the purification of specific antibodies, or the immunoglobulins can be isolated from the antiserum by precipitation at 40% saturation with ammonium sulfate. The ammonium sulfate-precipitated fraction is dissolved in 0.01 M potassium phosphate buffer, pH 7.0, 0.15 M NaCl and dialyzed against several changes of 100 volumes of the same buffer. The precipitation and dialysis are repeated twice. After the final precipitation, the immunoglobulin fraction is dissolved in 0.2 M sodium citrate, pH 6.5, and dialyzed against 100 volumes of the same solution.

Purification of the antibodies is based on the isolation, dissolution, and separation of the components of an immunoprecipitate.[10] In the precipitin test described above, a sharp equivalence point is seen at 0.05 ml of antigen solution. At equivalence, in this instance, 0.6 mg of antibody is precipitated from 1 ml of antiserum; this amount will vary somewhat among various antisera.

Based on the equivalence point, the antiserum pool, 53 ml, is precipitated by the addition, in this case, of 10.5 ml of the $E.$ $coli$ antigen solution. After 4 days at 4°, the precipitate is collected and washed with 0.15 M NaCl, then dissolved in 6.1 ml of 0.02 N HCl, 0.15 M NaCl, and centrifuged to remove any insoluble protein. The solution is applied to a column of Sephadex G-75 (4 × 43 cm) equilibrated and eluted with the same solvent. Fractions of 10 ml are collected and brought to about pH 6.0 with 0.2 ml of 1 N NaOH. The initial peak of absorbance at 280 nm, fractions 16 through 21, contains purified anti-$E.$ $coli$ ACP immunoglobulin. A small peak containing ACP elutes later. The pure immunoglobulin fractions are pooled and concentrated by ultrafiltration at 50 psi in an Amicon stirred cell with a UM-10 membrane. About 24 mg of pure anti-$E.$ $coli$ ACP are obtained from 50 ml of antiserum.

The preparation of pure specific antibodies is not necessary for immunoaffinity chromatography, but it does theoretically allow the construction of smaller, but higher capacity, affinity columns.

Binding of Immunoglobulin to CNBr-Sepharose

The ammonium sulfate-fractionated immunoglobulin, 66 ml, 776 absorbance units at 280 nm, is added to 60 g of swollen and washed CNBr-Sepharose 4B (Pharmacia) at a concentration of about 2.5 mg of protein per milliliter of gel. Both the gel and the protein solution are in 0.2 M sodium citrate, pH 6.5. The mixture is shaken gently in a glass-stoppered Erlenmeyer flask on a wrist-action shaker at 4° for 2 days. The gel is filtered with gentle suction in a sintered-glass funnel at room temperature. It is washed with several 50-ml portions of the citrate buffer, poured into a column, and left to sit overnight at 4° in the same buffer. The washes and

[10] D. Givol, S. Fuchs, and M. Sela, $Biochim.$ $Biophys.$ $Acta$ **63**, 222 (1962).

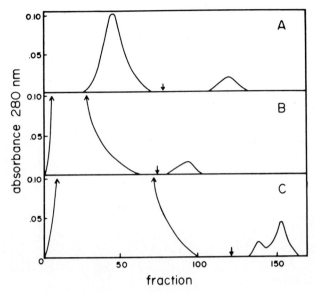

FIG. 3. Elution profiles of various acyl carrier proteins (ACPs) from the anti-*Escherichia coli* ACP immunoaffinity column. The amount of protein applied varied among chromatograms, as did fraction size, as indicated below. In all instances, elution with 0.01 M potassium phosphate, pH 6.2, 0.5 M NaCl was carried out until absorbance at 280 nm was consistently zero; elution was then begun (arrows) with 0.2 M glycine buffer, pH 2.8, 0.5 M NaCl. (A) *Escherichia coli* ACP purified through acid precipitation, 4.11 absorbance units at 280 nm, was applied in a volume of 2.3 ml. The sample was washed on with 0.01 M potassium phosphate buffer, pH 6.2, 0.1 M NaCl and eluted with the same buffer, except that the NaCl concentration was raised to 0.5 M. Fractions of 2 ml were collected. Glycine buffer elution was started at fraction 78. The pH of the fractions began to decrease at fraction 110. Fractions 110–130 were pooled and concentrated by ultrafiltration in an Amicon pressure cell with a UM-2 membrane at 50 psi. (B) *Euglena gracilis* strain Z, a 70–95% saturation ammonium sulfate fraction, 831 absorbance units at 280 nm, was applied in a volume of 36 ml. The sample was washed and eluted as in (A), except that 5-ml fractions were collected and elution with glycine buffer was begun at fraction 74. Fractions 80–100 were pooled and concentrated as above. (C) *Euglena gracilis* var. *bacillaris*, a 70–95% saturation ammonium sulfate fraction, 3500 absorbance units at 280 nm, was applied in a volume of 30 ml. The sample was washed and eluted as in (A), except that 3-ml fractions were collected and elution with glycine buffer was begun at fraction 121. Fractions 142–159 and 160–194 were pooled and concentrated separately. From Ernst-Fonberg *et al.*[3]

effluent are monitored for absorbance at 280 nm. Ethanolamine, 1 M, pH 7.0, 100 ml, is washed through the column until the pH of the effluent is 7.0. The gel is left in this medium at room temperature for 1 hr. The ethanolamine wash ensures blockage of any residual active groups on the gel. Noncovalently bound protein is removed by cyclic washing with high and low pH buffers. The gel is washed extensively at room temperature

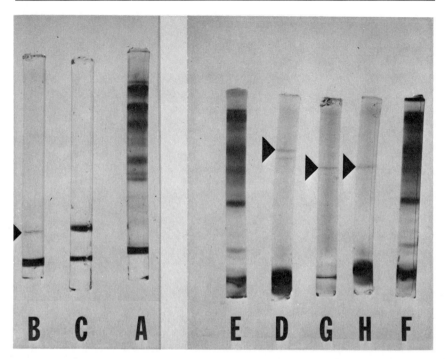

FIG. 4. Acrylamide gel disc electrophoresis of various ACP preparations. Dye fronts appear at the bottom of gels. *Escherichia coli* preparations: A, acid precipitate, 200 μg; B, ACP purified by affinity chromatography 30 μg; C, ACP purified according to Majerus *et al*,[4] 90 μg. *Euglena gracilis* strain Z preparations: D, ACP purified by immunoaffinity chromatography, 35 μg; E, a 70–95% saturation ammonium sulfate fraction, 200 μg. *Euglena gracilis* var. *bacillaris* preparations: F, a 70–95% saturation ammonium sulfate fraction, 200 μg; G, pooled and concentrated immunoaffinity fractions 142–159, 27 μg; H, pooled and concentrated immunoaffinity fractions 160–194, 17 μg. The arrows designate the protein bands of the respective ACPs purified by immunoaffinity chromatography. From Ernst-Fonberg *et al*.[3]

with 0.1 *M* acetate buffer, pH 4.0, 1.0 *M* NaCl, and 0.1 *M* borate buffer, pH 8.0, 1.0 *M* NaCl, alternately through three cycles. Finally, the column is put through two cycles of the buffers to be used in isolating the ACP.

Operation of the Immunoaffinity Column

All operations are carried out at 4°. A column, 4 cm in diameter and 6 cm in height, is prepared from thoroughly washed antibody-liganded Sepharose. The column is equilibrated in 0.01 *M* potassium phosphate buffer, pH 6.2, 0.1 *M* NaCl, and the sample is run into the gel. After about 30 min (to allow time for antigen interaction with antibody), nonspecifically bound protein is eluted by washing with 0.01 *M* potassium phos-

YIELDS OF ACPs RETAINED BY THE ANTI-*Escherichia coli* ACP IMMUNOAFFINITY COLUMN[3]

Source of ACP purified by affinity chromatography	Protein (μg)[11]	Units of ACP activity[a]	ACP-dependent fatty acid synthase activity[b]
E. coli	160	203	+
Euglena gracilis strain Z	151	162	+
Euglena gracilis var. *bacillaris*	168	NM[c]	+

[a] One unit of ACP is 1 nmol of $^{14}C_2$ exchanged per 15 min in the malonyl-CoA–CO_2 exchange reaction. The specific activities of the ACPs purified are lower than those reported for freshly reduced, desalted ACP. This is probably due to the absence of a thiol reagent during the chromatographic procedures and to the high salt concentration of the eluents.
[b] ACP is a substrate for the ACP-dependent fatty acid synthase from *Euglena gracilis*. There is no fatty acid biosynthesis in the absence of ACP; therefore, fatty acid biosynthesis is a sensitive indication of the presence of functional ACP.
[c] NM, not measured.

phate, pH 6.2, 0.5 M NaCl, until the absorbance (280 nm) of the effluent is zero. Finally, specifically bound ACP is removed by elution with 0.2 M glycine, pH 2.8, 0.5 M NaCl. After elution of antigen, the immunoadsorbent is equilibrated and stored in 0.01 M potassium phosphate, pH 6.2, 0.1 M NaCl. It should be washed with several column volumes of the same mixture every 2 weeks.

Properties of the Protein Retained on the Immunoaffinity Column

The elution profiles of *E. coli*, *E. gracilis* strain Z and variety *bacillaris* ACPs from the immunoaffinity column are shown in Fig. 3. In each case, upon application of a crude preparation to the column, protein is specifically retained and later released under acidic conditions. When an excess of pure *E. coli* ACP is processed through the column, a similar result is obtained. The yield achieved in the immunoaffinity chromatography step is a function of the binding capacity of the immunoadsorbent. The yields are identical from one column run to the next (Table) regardless of the excessive amount and stage of purity of the ACP applied to the column. Discontinuous electrophoresis, in 14% acrylamide gels, of the material applied to the affinity column and of the selectively retained protein demonstrates the extent of purification achieved by the single step (Fig. 4).

The material retained from the crude *E. coli* ACP preparation shows a single major band at R_f 0.85 (Fig. 4B), identical to *E. coli* ACP purified according to Majerus *et al.*[4] (Fig. 4C). The single peak retained from *E.*

gracilis strain Z 70–95% saturation ammonium sulfate fraction exhibits a major band at R_f 0.30, a lesser band at R_f 0.35, and one or two other very faint bands (Fig. 4D). Since the immunoaffinity chromatography is done in the absence of thiol reagents, it is possible that some of the ACP is present as disulfide bridge-linked dimer.[2] The elution pattern of *E. gracilis* var. *bacillaris* ACP from the immunoaffinity column is biphasic (Fig. 3C), but electrophoresis of material in each peak shows a single protein band at R_f 0.39 (Fig. 4G,H). In all cases, upon neutralization and concentration, the retained protein is biologically active in the ACP assays (Table).[11]

The small-scale purifications described here illustrate the potential usefulness of immunoaffinity chromatography in obtaining ACPs from diverse sources. Under the conditions described, stable and reproducible results are obtained through more then 25 runs on a single column.

Acknowledgment

Research support from the following sources is acknowledged: Grants-in-Aid from the American Heart Association; National Science Foundation Grants GB-23905, GB-36016X, and PCM-7904062; and Public Health Service Research Career Development Award 5-KO4-HD-25169 from the National Institute of Child Health and Human Development (M. L. E.-F.).

[1] O. H. Lowry, N. J. Rosebrough, A. L. Farr, and R. J. Randall, *J. Biol. Chem.* **193**, 265 (1951).

[23] Acyl-Acyl Carrier Protein Thioesterase from Safflower

By TOM MCKEON and PAUL K. STUMPF

The acyl-ACP[1] thioesterase catalyzes the hydrolysis of acyl-ACP to free fatty acid and ACP-SH.

$$\text{Acyl-S-ACP} + H_2O \rightarrow \text{acyl-OH} + \text{ACP-SH}$$

The thioesterase is of interest because it terminates the set of biosynthetic reactions that take place on ACP, a water-soluble and lipid-insoluble acyl carrier. Further metabolism of fatty acids appears to occur in membrane systems. Thus, acyl-ACP thioesterase may play an important role in regulating the fatty acid composition of plant tissue.[2]

[1] The abbreviations are ACP, acyl carrier protein; BSA, bovine serum albumin.
[2] W. E. Shine, M. Mancha, and P. K. Stumpf, *Arch. Biochem. Biophys.* **172**, 110 (1976).

Assay Method

Principle. The acyl-ACP thioesterase assay involves the measurement of labeled fatty acid released from labeled acyl-ACP. The free fatty acids are extracted into petroleum ether and counted in a liquid scintillation counter.

Reagents

Glycine, 0.20 M, pH 9.0
Bovine serum albumin, 10 mg/ml in water
[^{14}C]Stearoyl-ACP, 10 μM in 0.02 M potassium phosphate, pH 6.8 (synthesis described in this volume[3]
Acetic acid, 1 M, in isopropanol with 5 mg/ml each of palmitic and stearic acid
Petroleum ether, reagent grade, saturated with isopropanol–water, 1 : 1 (v/v)

Procedure. The reaction mixture in a 13 × 100 mm screw-cap tube contains 100 μl of glycine buffer, 70 μl of water, 10 μl of BSA and 10 μl of thioesterase preparation appropriately diluted. The reaction is started by the addition of 10 μl of [^{14}C]stearoyl-ACP, and the reaction is stopped after 10 min at room temperature (20–23°) by the addition of 0.2 ml of the 1 M acetic acid reagent. After 10 min, the free fatty acids are extracted with two 2-ml portions of the petroleum ether and the extract is counted.

The assay is linear with respect to time and enzyme concentration up to 40% hydrolysis of substrate.[4] One unit of activity is equal to a rate of hydrolysis of 1 μmol per minute per milligram of protein.

Purification

Acetone Powder Extract. This material is obtained from acetone powder of safflower by the method described for stearoyl-ACP desaturase.[3]

Acid Precipitate. The acetone powder extract is cooled on ice and acidified to pH 5.2 with glacial acetic acid. After 1 hr, the precipitate is centrifuged at 10,000 g for 10 min, and the supernatant is adjusted to pH 4.3 with acetic acid. After 1 hr, the precipitate is pelleted and resuspended in one-half the starting volume of 0.02 M potassium phosphate buffer, pH 6.8. Insoluble debris is centrifuged out, and the supernatant retains 60% to 80% of the acyl-ACP thioesterase activity (see the table) and less than 5% of the stearoyl-ACP desaturase activity.[4]

ACP-Sepharose 4-B column. This column is run exactly as described

[3] T. McKeon and P. K. Stumpf, this volume [34].
[4] T. McKeon, unpublished data, 1979.

PURIFICATION OF ACYL-ACP THIOESTERASE

Fraction	Total protein[a] (mg)	Total activity (mU)	Specific activity (mU/mg)	Yield (%)	Purification (fold)
Acetone powder extract	400	86	0.22	—	—
Acid precipitate	57	64	1.12	74	5
ACP-Sepharose 4B	.13	23	170	27	770

[a] Protein was determined by the method of O. H. Lowry, N. J. Rosebrough, A. L. Farr, and R. J. Randall, *J. Biol. Chem.* **193**, 265 (1951), using bovine serum albumin as the standard.

for the purification of stearoyl-ACP desaturase. The thioesterase elutes with the 0.30 M phosphate wash, with the early fractions containing proportionally more thioesterase and the later fractions more of the desaturase.[4]

Purity. As seen in the table, the acyl-ACP thioesterase is purified 770-fold by this procedure. The stearoyl-ACP desaturase is present as approximately 5% of the bulk protein in the purified preparations of the thioesterase.[4]

Properties

Specificity. Acyl-ACP thioesterase from safflower has a strong preference for oleoyl-ACP as substrate. The preference for substrates under routine assay conditions is oleoyl-ACP > stearoyl-ACP > palmitoyl-ACP with relative rates of 10 : 2 : 1, respectively. The rates of hydrolysis of oleoyl-CoA and stearoyl-CoA are less than 2% of the rate of hydrolysis of the corresponding acyl-ACP.[4]

Stability. Preparations purified through the ACP-Sepharose 4B column step are stable for 3 weeks at 4° when maintained in 1 mM DTT.[4]

pH Activity Profile. The thioesterase is half-maximally active at pH 8.5 and pH 10.0 with optimum activity at pH 9.5. The thioesterase has less than 2% maximal activity at pH 6.5 and below, where the stearoyl-ACP desaturase is maximally active.[4]

[24] Long-Chain Fatty Acyl-S-4'-Phosphopantetheine-Fatty Acid Synthase Thioester Hydrolase from Rat

By STUART SMITH

Long-chain fatty acyl-S-4'-phosphopantetheine-fatty acid synthase +
H_2O → long-chain fatty acid +
HS-4'-phosphopantetheine-fatty acid synthase

In animals, several of the enzymes involved in fatty acid biosynthesis de novo are arranged as two polyfunctional polypeptides that constitute the fatty acid synthase multienzyme complex. This multienzyme effects the synthesis of fatty acid by catalyzing the sequential addition of two-carbon fragments, derived from malonyl-CoA, to an acyl group, derived initially from acetyl-CoA. The growing acyl chain remains covalently attached to the multienzyme via a thioester linkage. Hydrolysis of the acyl-4'-phosphopantetheine thioester bond by one of the components of the multienzyme results in the termination of growth of the acyl chain and release of a free fatty acid product. The component responsible for the termination step, a long-chain fatty acyl thioesterase (thioesterase I), is specific for long-chain acyl thioesters, and the fatty acid synthases isolated from a variety of tissues and species all synthesize long-chain fatty acid, predominantly palmitic acid. Both subunits of the rat fatty acid synthase contain a thioesterase I domain located at one end of the polyfunctional polypeptide chain. The thioesterase I component can be released from the multienzyme complex by limited proteolysis and recovered in active form.[1] The following procedures have been applied, in our laboratory, for the isolation and characterization of the thioesterase I component from fatty acid synthases purified from rat liver and lactating rat mammary gland.

Substrates

Malonyl-CoA, acetyl-CoA, palmityl-CoA, and [1-¹⁴C]palmityl-CoA are available commercially. Alternatively, they can easily be synthesized and purified in the laboratory. The procedures for the preparation of acetyl-CoA and malonyl-CoA were described in an earlier volume.[2]

---

[1] S. Smith, E. Agradi, L. Libertini, and K. N. Dileepan, *Proc. Natl. Acad. Sci. U.S.A.* **73**, 1184 (1976).
[2] S. Smith and S. Abraham, this series, Vol. 35 [8].

Palmityl-CoA can be prepared either via the N-hydroxysuccinimide derivative[3] or via palmityl chloride[4] and purified by acid precipitation.

Assay of Fatty Acid Synthase Activity

The enzyme is conveniently assayed by monitoring the decrease in absorbance at 340 nm that accompanies the oxidation of NADPH during the reaction.

Reagents

Potassium phosphate buffer, 1.0 M, pH 6.6
NADPH, 7.5 mM
Acetyl-CoA, 1.25 mM
Malonyl-CoA, 1.25 mM

Procedure. The reagents, potassium phosphate (50 μl), NADPH (10 μl), acetyl-CoA (20 μl), and malonyl-CoA (20 μl), are usually premixed and dispensed as a "cocktail," 100-μl portions per assay. Reaction mixtures, final volume 0.5 ml, are contained in semimicro cuvettes having a 1-cm light path. Enzyme (usually 1–10 mU) is added to start the reaction. We have routinely measured fatty acid synthase activities at 30°. A decrease in absorbance at 340 nm of 1.0 corresponds to the oxidation of 80.5 nmol of NADPH.

Units. A unit of fatty acid synthase activity is defined as that amount catalyzing the malonyl-CoA-dependent oxidation of 1 μmol of NADPH per minute at 30°.

Assay of Thioesterase I Activity

The enzyme is assayed by measuring the release of radioactive fatty acid from [1-^{14}C]palmityl-CoA.

Reagents

Potassium phosphate buffer, 1.0 M, pH 8.0
[1-^{14}C]Palmityl-CoA, 0.1 mM, 4 Ci/mol

Procedure. Assays are performed in plastic tubes at 30° with a final volume of 0.5 ml. Buffer (25 μl), enzyme (usually 0.3–1.5 mU), and water are equilibrated to 30°; the reaction is started by the addition of substrate (25 μl). Incubation is continued for 1–2 min, then the reactions are terminated by the addition of 0.1 ml of 21% (w/v) HClO$_4$, followed by 1.5 ml of hexane/isopropanol (3 : 2, v/v). The two phases are mixed by vortexing

[3] A. Al-Arif and M. Blecher, *J. Lipid Res.* **10**, 344 (1969).
[4] W. Seubert, *Biochem. Prep.* **7**, 80 (1960).

vigorously for 1 min and separated by centrifugation. The upper phase (hexane) is carefully removed with a Pasteur pipette and transferred to a scintillation vial. The lower phase is mixed with 2 ml of hexane, and again the upper phase is transferred to the scintillation vial. The last step is repeated once more, and the combined hexane phases from the three extractions are evaporated to dryness in a stream of air. Three milliliters of toluene/2-ethoxyethanol (2 : 1, v/v) containing 15 mg of Omnifluor are added to each vial, and the radioactivity is measured by liquid scintillation spectrometry. We routinely determine counting efficiencies using either the internal standard or channels-ratio method.

Units. A unit of thioesterase I activity is defined as that amount catalyzing the hydrolysis of 1 μmol of palmityl-CoA per minute at 30°.

Assay of Protein Concentrations

Protein concentrations are determined from the absorbance at 280 nm using the specific absorption coefficients ($A_{1\,cm}^{1\%}$) of 8.9 (Libertini and Smith[5]) and 7.14 (Lin and Smith[6]) for fatty acid synthase and thioesterase I, respectively.

Purification of Fatty Acid Synthase

Fatty acid synthase is purified from cytosol prepared from either rat liver or lactating rat mammary gland as described in an earlier volume.[2] In recent years we have preferred to use Sepharose 6B rather than Sephadex G-200 in the final purification step. We have also increased the scale of the operation and can obtain up to 0.8 g of fatty acid synthase in a single purification. The enzyme purified in this way is at least 95% pure as judged by sodium dodecyl sulfate (SDS)-polyacrylamide gel electrophoresis; the major impurities are probably nicked fragments of fatty acid synthase formed by proteolysis. Specific activities vary somewhat from one preparation to another (1–1.4 units/mg) irrespective of the purity of the enzyme.

Isolation of Thioesterase I by Limited Trypsinization of Fatty Acid Synthase

Reagents

Fatty acid synthase, 2 mg/ml in 0.25 *M* potassium phosphate buffer, (pH 7.0), 1 m*M* EDTA, and 1 m*M* dithiothreitol

[5] L. Libertini and S. Smith, *Arch. Biochem. Biophys.* **192**, 47 (1979).
[6] C. Y. Lin and S. Smith, *J. Biol. Chem.* **253**, 1954 (1978).

Bovine pancreatic trypsin, 1 mg/ml, 210 units/mg, in 0.25 M potassium phosphate buffer (pH 7.0), 1 mM EDTA, and 1 mM dithiothreitol

Soybean trypsin inhibitor, 1 mg/ml in the above solution

Procedure. Thioesterase I is released from covalent attachment to the fatty acid synthase by digestion with trypsin under nondenaturing conditions.[1,6] Fatty acid synthase is equilibrated to 30° in a plastic vessel. A 10-μl portion is assayed for fatty acid synthase activity to establish a zero-time value. Proteolysis is initiated by the addition of trypsin (3.8 μl per milliliter of fatty acid synthase solution), and the contents of the vessel are mixed gently with a magnetic stirrer. At 5-min intervals 10-μl portions of the reaction mixture are removed and assayed for fatty acid synthase activity. When the fatty acid synthase activity is reduced to about 50% of the initial level, usually in about 30 min, proteolysis is arrested by the addition of trypsin inhibitor (3.8 μl per milliliter of fatty acid synthase solution). Solid ammonium sulfate, 243 g/liter, is added at 0°, and the mixture is stirred magnetically for 20 min. The precipitated protein, containing trypsinized and residual fatty acid synthase, is removed by centrifugation at 10,000 g for 20 min and discarded. Usually 30–40% of the thioesterase I activity remains attached to the fatty acid synthase; this material can be used to prepare the fatty acid synthase core required for the assay of thioesterase II.[7] Further solid ammonium sulfate, 318 g per liter of starting material, is added at 0°, and the mixture is stirred and centrifuged as before. The precipitated protein, mainly thioesterase I, is dissolved in a minimum volume of 0.25 M potassium phosphate buffer (pH 7.0), 1 mM EDTA, and 1 mM dithiothreitol and applied, at 0–4°, to a column (93 × 2.5 cm) of Sephadex G-75 that has been equilibrated with the same solution. Thioesterase I is eluted as a symmetrical zone of constant specific activity. A small amount of trypsinized fatty acid synthase is commonly eluted near the void volume. See the table for results of a typical isolation.

Homogeneity of Thioesterase I and Reproducibility of the Procedure

Thioesterase I preparations obtained by this procedure appear to be homogeneous when examined, under nondenaturing conditions, by analytical ultracentrifugation and gel filtration.[6] The preparations contain no detectable amounts of any of the other partial activities associated with the native fatty acid synthase. However, thioesterase I contains a site near the middle of the polypeptide, which is susceptible to tryptic attack, and the purified thioesterase consists of a mixture of 35,000 and 17,500

[7] S. Smith, this volume [25].

ISOLATION OF LONG-CHAIN FATTY ACYL THIOESTERASE (THIOESTERASE I) FROM FATTY
ACID SYNTHASE BY LIMITED DIGESTION WITH TRYPSIN

| | | Thioesterase activity | | |
Fraction	Protein (mg)	Total activity (units)	Specific activity (units/mg)	Yield (%)
Fatty acid synthase	620	358	0.577	100
$(NH_4)_2SO_4$				
0–243 g/liter (discarded)	570	122	0.214	(34)
243–561 g/liter	38	110	2.89	31
Sephadex G-75	25.6	87	3.40	24

MW species. The two 17,500 MW halves of the enzyme, formed through
nicking by trypsin, are held together by noncovalent forces and copurify
with intact 35,000 MW polypeptides. The proportion of nicked 17,500
MW polypeptides formed depends on the severity of the limited proteolysis procedure. For this reason, we do not attempt to strip all of the
thioesterase I from the multienzyme, but terminate the digestion when
fatty acid synthase activity is reduced by 50%. Under these conditions the
proportion of nicked polypeptides in the thioesterase I preparations is
minimized. The specific activity of the preparation depends on the relative
proportion of intact and nicked thioesterase I polypeptides present, since
the activity of the nicked thioesterase I is considerably lower than that of
the intact thioesterase I polypeptide. The proportion of 35,000 MW
species in the final preparation ranges from 40 to 60%, and the specific
activity ranges from 2 to 4 units/mg. A procedure has been devised for
separating the intact and nicked thioesterase I polypeptides with partial
retention of enzyme activity.

Separation of Intact (MW 35,000) and Nicked (MW 17,500) Polypeptides
of Thioesterase I

Thioesterase I (2 mg/ml) is incubated for 1 hr, at 20–22°, in 50 mM
sodium phosphate buffer (pH 7.0) and 1% (w/v) SDS. Usually 5–10% of
the enzyme activity is retained after this treatment. The solution is then
applied, at 20–22°, to a column (112 × 1 cm) of Sephadex G-100 (superfine
grade) that has been equilibrated with 50 mM sodium phosphate buffer
(pH 7.0), 0.1% (w/v) SDS, 1 mM EDTA, and 1 mM dithiothreitol. Gel
filtration resolves the thioesterase I into components with elution zone
maxima characterized by relative elution volumes (V_e/V_o) of 1.12 and
1.26; the two components can be identified, by SDS–polyacrylamide gel

electrophoresis, as the intact MW 35,000 and nicked MW 17,500 polypeptides, respectively.[8]

Properties

Location of the Thioesterase I Domains on the Fatty Acid Synthase. Immunochemical studies and kinetic analysis of the limited trypsinization procedure indicate that a thioesterase I domain occupies a terminal locus at one end of each of the two polyfunctional polypeptides that comprise fatty acid synthase.[9] At present it is not known whether this site is at the carboxyl or amino terminus.

Immunochemical Studies. Antibodies raised specifically against the thioesterase I component will precipitate both subunits of the dissociated fatty acid synthase; approximately three antibody molecules combine with each subunit molecule at equivalence.[9] At antibody : antigen ratios below that required for lattice formation, fatty acid synthase is not precipitated by anti-thioesterase I but is completely inhibited.[10] The inhibition appears to be an indirect result of the interaction of multienzyme with antibody, since thioesterase I activity is not inhibited. Most likely, reaction of the fatty acid synthase dimer with anti-thioesterase I induces dissociation into monomers, which are unable to catalyze the overall fatty acid synthase reaction.

Physicochemical Properties. The molecular weight of thioesterase I has been determined by equilibrium sedimentation (32,300), gel permeation chromatography (32,000), and SDS–polyacrylamide gel electrophoresis (35,000).[8]

The sedimentation and diffusion coefficients, determined at a single protein concentration (4.0 mg/ml), are 2.9×10^{-13} sec and 5.0×10^{-7} cm^2 sec^{-1}, respectively.[6]

A value for the partial specific volume of 0.738 ml/g has been computed from the amino acid composition.[6]

Kinetic Properties. Interpretation of the kinetic data obtained for thioesterase I using long-chain acyl-CoAs as substrates is complicated by the fact that these compounds are detergents. The concentration of detergents above which micelles are formed is influenced by solvent conditions. In interpreting kinetic data obtained using such substrates, it is necessary to take into account the effect of reaction conditions on the physical form of both the enzyme and its substrate.[6] Thioesterase I appears to utilize the long-chain acyl-CoA monomers as substrates but is inhibited by micellar

[8] K. N. Dileepan, C. Y. Lin, and S. Smith, *Biochem. J.* **175**, 199 (1978).
[9] S. Smith and A. Stern, *Arch. Biochem. Biophys.* **197**, 379 (1979).
[10] A. Stern and S. Smith, unpublished results.

forms. With palmityl-CoA as substrate the pH affects both V_{max} and K_m ($K_m = 0.5$ μM at pH 6.6, 2.5 μM at pH 8.0), whereas ionic strength affects V_{max} but not K_m.

The effect of pH on reaction velocity has been studied at constant ionic strength ($\mu = 0.143$) employing palmityl-CoA concentrations optimal for each pH. The pH–activity profile resembles that of a simple curve for a single ionizable group; the pH optimum is 8.0, and half-maximum velocity occurs at pH 6.3.

Substrate Specificity. When saturated fatty acyl-CoA derivatives of various chain lengths are used as substrates, thioesterase I displays a marked preference for long-chain acyl moieties.[6] The maximum rates of hydrolysis observed with stearoyl-CoA, palmityl-CoA, and myristyl-CoA were 5.0, 5.0, and 0.6 μmol per minute per milligram of protein.

Long-chain fatty acyl thioesters of a 4'-phosphopantetheine-containing pentapeptide, derived from the fatty acid synthase by peptic digestion, are also hydrolyzed rapidly by thioesterase I.[11]

The ability of thioesterase I to hydrolyze long-chain fatty acyl thioesters of 4'-phosphopantetheine covalently linked to a polypeptide region of the fatty acid synthase indicates that this component of the multienzyme may be responsible for termination of growth of the fatty acyl chain during fatty acid synthesis.

The specificity of thioesterase I is not limited to aliphatic acyl chains. With phenylacetyl-CoA as primer in the fatty acid synthase reaction, the rate of formation of ω-phenyl-fatty acids is about 25% of that of aliphatic fatty acids with acetyl-CoA as primer.[10] Thioesterase I terminates growth of the acyl chain preferentially at the ω-phenyl-C_{12} stage; 12-phenyldodecanoic acid is the sole product.[10]

Activators and Inhibitors. Albumin stimulates thioesterase I activity and also relieves the inhibitory effect of high concentrations of acyl-CoA substrates.[6] These effects may be attributable to the ability of ablumin to bind both fatty acyl-CoAs and fatty acids.

Thioesterase I is inactivated by the serine esterase inhibitors diisopropylphosphofluoridate and phenylmethanesulfonyl fluoride. Experiments with radioactively labeled diisopropylphosphofluoridate have shown that the enzyme contains a single active site per 35,000 MW polypeptide.[8]

Thioesterase I activity is also sensitive to reagents that attack sulfhydryl groups. For example, p-chloromercuribenzoate produces 50% inhibition below 0.1 μM, whereas iodoacetamide produces 50% inhibition at 100 μM. It is not known whether the affected residues are in close proximity to the active site.[6]

---

[11] S. Smith, unpublished results.

Comparison of the Properties of the Thioesterase I Components Obtained
from Rat Liver and Mammary Gland Fatty Acid Synthases

The amino acid compositions of liver and mammary gland thioesterase
I appear to be identical; the enzymes have identical physicochemical and
kinetic properties and the same substrate specificity and are immunologically indistinguishable.[6] These results imply that the mechanism for termination of growth of the acyl chain on the fatty acid synthase is the same
for the liver and mammary gland multienzymes. The ability of the mammary gland to synthesize medium-chain fatty acids results from the presence in this tissue of a unique chain-terminating thioesterase that is not
part of the fatty acid synthase multienzyme.

Acknowledgments

This work was supported by Grants from the National Institutes of Health (AM 16073),
the National Science Foundation (BMS 7412723), and the American Heart Association (EI
73-140).

[25] Medium-Chain Fatty
Acyl-S-4'-Phosphopantetheine-Fatty Acid Synthase Thioester
Hydrolase from Lactating Mammary Gland of Rat

By STUART SMITH

Medium-chain fatty acyl-S-4'-phosphopantetheine-fatty acid synthase +
$H_2O \rightarrow$ Medium-chain fatty acid +
HS-4'-phosphopantetheine-fatty acid synthase

The mammary glands of some species contain an enzyme that can alter
the product specificity of fatty acid synthase, inducing the multienzyme
complex to synthesize medium chain, rather than the usual long-chain,
fatty acids. This enzyme, thioesterase II, is capable of hydrolyzing the
thioester bond linking the growing acyl chain to the 4'-phospho-
pantetheine of fatty acid synthase. Unlike thioesterase I, thioesterase II is
not covalently associated with the fatty acid synthase and is readily isolated by conventional methods of protein purification.[1]

[1] S. Smith, *J. Dairy Sci.* **63**, 337 (1980).

Substrates

Acetyl-CoA and malonyl-CoA are available commercially. We have used decanoyl-CoA and decanoylpantetheine as model substrates for the assay of thioesterase II. Decanoyl-CoA is available commercially. However, decanoylpantetheine can be synthesized easily, and its use results in a considerable reduction in costs. Decanoylpantetheine is synthesized from decanoyl chloride and pantethine, both of which are available from commercial sources. Pantethine is first reduced to pantetheine with sodium amalgam. A 2% sodium amalgam can be prepared by placing 1.38 g of freshly cut sodium in a flask, which is continuously flushed with dry nitrogen. Five milliliters of filtered mercury are added slowly to the sodium, the mixture is warmed gently with a Bunsen flame to start the reaction, and the remaining mercury is added slowly. The amalgam is crushed in a mortar and stored in a tightly closed, dry vessel. The amalgam is added slowly to a solution of pantethine (60 mM) in water, while nitrogen is bubbled through the mixture and an acid pH is maintained by the addition of Dowex 50, H$^+$ form. The reaction is monitored by measuring the pantetheine thiol formed. Ten microliter portions of the solution are transferred to a solution containing 2.5 mM (final concentration) 5,5′-dithiobisnitrobenzoic acid in 2 ml of 50 mM potassium phosphate buffer, pH 7.0. The absorbance of the solutions is determined at 412 nm, and the thiol content is calculated, assuming a molecular extinction coefficient of 13,600 for 5-thio-2-nitrobenzoic acid. The solution of panthetheine is stable for several days when stored under nitrogen at $-20°$. For the acylation reaction, a twofold excess of decanoyl chloride is added stepwise to a solution of pantetheine in tetrahydrofuran/water (1 : 1, v/v), and the pH is maintained between 7.5 and 8.0 by addition of 40% (w/v) KOH. When the reaction is complete, as indicated by a stable pH, the mixture is acidified to pH 1.5–2.5 using 12 N HCl; two phases are formed. The upper phase is collected, and the lower phase washed with an equal volume of diethyl ether. The original upper phase is combined with the ethereal phase, and solvent is evaporated in a stream of nitrogen. Residual decanoic acid is readily removed from the decanoylpantetheine by washing with petroleum ether (b.p. 35–60°). The thioester content of the product is most conveniently assayed by measuring the free thiol released by treatment with hydroxylamine. Decanoylpantetheine is first dissolved in ethanol (0.3 mg/ml), and a 0.8 ml-portion is mixed with 0.2 ml of freshly prepared, neutralized hydroxylamine (14%, w/v). Portions of the reaction mixture are removed at intervals of several minutes, and the sulfhydryl content is determined with 5,5′-dithiobisnitrobenzoic acid, as described above; the maximum value attained represents the thiol ester content of the decanoylpantetheine preparation. The product is a clear, extremely hygroscopic liquid. The major impurity appears to be water, since no contaminants are revealed by thin-layer chromatography. A stock solution of decanoylpantetheine (2 mM) can be conveniently prepared by sonicating the thioester briefly in 3 mM HCl. The solution can be frozen and thawed repeatedly and is stable for several weeks at $-20°$.

Preparation of Trypsinized Fatty Acid Synthase Core

The fatty acid synthase core is prepared by completely stripping the two thioesterase I domains from the native multienzyme, using trypsin under nondenaturing conditions. Rather more rigorous conditions of trypsin treatment are required than those used when the main objective is to isolate intact thioesterase I. The fatty acid synthase discarded during preparative-scale isolation of thioesterase I provides a convenient starting material.[2] This material is a mixture of native fatty acid synthase and fatty acid synthase that has been partially stripped of its thioesterase domains. It is isolated from the original tryptic digest by ammonium sulfate precipitation. The protein pellet is redissolved in 0.25 M potassium phosphate buffer, pH 7.0, 1 mM EDTA, and 1 mM dithiothreitol to give a protein concentration of about 4 mg/ml. A freshly prepared solution of trypsin (5 mg/ml in 0.25 M potassium phosphate buffer, pH 7.0, 1 mM EDTA, and 1 mM dithiothreitol) is added, 4 μl for each milliliter of fatty acid synthase solution. The mixture is incubated at 30° for 1 hr, a second portion of trypsin is added (4 μl per milliliter of fatty acid synthase solution), and the incubation is continued for 1 additional hour. It is not necessary to inactivate the remaining trypsin, since the trypsinized fatty acid synthase core is resistant to further attack, under nondenaturing conditions. The preparation is assayed for fatty acid synthase activity at this point; usually none can be detected. Solid ammonium sulfate (196 g/liter) is added at 0°, and the suspension is stirred gently for 20 min. The precipitated protein, which consists mainly of the trypsinized fatty acid synthase core, is collected by centrifugation at 10,000 g for 20 min and dissolved in a minimum volume of 0.05 M potassium phosphate buffer, pH 7.0, 1 mM EDTA, and 1 mM dithiothreitol. The solution is centrifuged at 10,000 g for 10 min to remove any agglutinated protein and then is applied, at 20–22°, to a column (2.5 × 60 cm) of Sepharose 6B, which has been previously equilibrated with 0.25 M potassium phosphate buffer, pH 7.0, 1 mM EDTA, and 1 mM dithiothreitol. A flow rate of 20–30 ml/hr is normally used, and fractions of 4 ml are collected. The trypsinized fatty acid synthase core emerges from the column with an elution zone maximum characterized by a relative elution volume (V_e/V_o) of 2.0. Fractions from the leading edge

---

[2] S. Smith, this volume [24].

and the peak of the elution zone are combined and retained; those from the trailing edge are discarded. The solvent and solute concentrations are adjusted to give a protein concentration of 3–5 mg/ml in 0.25 M potassium phosphate buffer, pH 7.0, 1 mM EDTA, 10 mM dithiothreitol, and 20% glycerol. The specific absorption coefficient ($A_{1\,cm}^{1\%}$) of the trypsinized fatty acid synthase core is 8.4. The core (molecular weight approximately 440,000) consists of two copies each of polypeptides with molecular weights 125,000 and 95,000. The two pairs of 125,000 + 95,000 MW polypeptides appear to be derived one from each of the two identical subunits of the native multienzyme.[3] The core is completely lacking in thioesterase I activity and is unable to catalyze the formation of fatty acids. However, it retains all other partial activities required for the assembly of a long-chain acyl moiety from acetyl-CoA, malonyl-CoA, and NADPH. A single, enzyme-bound, long-chain, acyl-4'-phosphopantetheine thioester is formed by each molecule of trypsinized fatty acid synthase core (i.e., the 440,000 MW, "nicked dimer" species).[4] This acyl-enzyme thioester can be used effectively as a substrate for thioesterase II.

Preparation of Anti-Rat (Fatty Acid Synthase) Immunoglobulins

Homogeneous preparations of fatty acid synthase, isolated from rat liver or lactating mammary gland[5] and dissolved in 0.25 M potassium phosphate buffer, pH 7.0, 1 mM EDTA, 10 mM dithiothreitol, and 20% glycerol, are centrifuged at 100,000 g for 10 min to remove any agglutinated protein. Approximately 2 mg of enzyme are emulsified with an equal volume (1–2 ml) of incomplete Freund's adjuvant and injected subcutaneously, at monthly intervals, into young male New Zealand White rabbits (2–3 kg body weight). Usually 40–50 ml of blood are collected weekly, once anti-fatty acid synthase antibodies have appeared. Serum is prepared and treated, at 0–4°, with solid ammonium sulfate (243 g/liter) to precipitate immunoglobulins. The precipitated protein is dissolved in 10 mM sodium phosphate buffer, pH 7.5, 10 mM NaCl and dialyzed exhaustively against the same solvent at 0–4°. Agglutinated protein is removed by centrifugation, and the clear supernatant, containing up to 3 g of protein, is applied, at 20–22°, to a column (2.5 × 34 cm) of DEAE-cellulose (Whatman DE-52) that has been equilibrated with 10 mM sodium phosphate buffer, pH 7.5, 10 mM NaCl. Immunoglobulins are eluted from the

[3] S. Smith and A. Stern, *Arch. Biochem. Biophys.* **197**, 379 (1979).
[4] L. Libertini and S. Smith, *Arch. Biochem. Biophys.* **192**, 47 (1979).
[5] S. Smith and S. Abraham, this series, Vol. 35 [8].

column with the equilibration solvent. Fractions from the leading edge and the peak of the elution zone, which consist of highly pure immunoglobulins (according to sodium dodecyl sulfate (SDS)–polyacrylamide electrophoresis), are pooled and stored at $-20°$. Overall recovery of anti-fatty acid synthase immunoglobulins is usually about 50%.

Assay of Thioesterase II Activity

The activity of this enzyme can be measured either with model substrates, such as decanoylpantetheine or decanoyl-CoA, or with the natural substrates, acyl-fatty acid synthase thioesters. The assay system utilizing a model substrate is simple, inexpensive, and convenient to perform. However, it suffers from the disadvantage that it is nonspecific and does not discriminate between thioesterase II and other hydrolases capable of attacking decanoyl thioesters of pantetheine or CoA. The assay system utilizing the natural substrate for thioesterase II is more complex; it cannot be used in the presence of fatty acid synthase and is relatively expensive to perform. However, it has the advantage of being remarkably specific for thioesterase II. Cytosols prepared from a variety of tissues other than mammary gland, which have varying abilities to hydrolyze decanoylpantetheine, do not catalyze the release of the acyl moiety from thioester linkage to the trypsinized fatty acid synthase core. We have used both assay systems routinely. In isolating thioesterase II for the first time it is advisable to use the natural substrate. Once the purification scheme is well established, the model substrate can be used confidently.

Use of the Natural Substrate: the "Reactivation" Assay

Since the coupling of thioesterase II with the trypsinized fatty acid synthase core effectively restores the ability to catalyze the overall reaction of fatty acid synthesis, we refer to this assay as the reactivation assay. The coupled reaction is monitored spectrophotometrically at 340 nm.

Reagents

Potassium phosphate buffer, 1.0 M, pH 8.0
NADPH, 7.5 mM
Acetyl-CoA, 1.25 mM
Malonyl-CoA, 1.25 mM
Trypsinized fatty acid synthase core, 6.8 μM

Procedure. Samples containing fatty acid synthase must be treated with anti-fatty acid synthase immunoglobulins. Since the first step in the purification scheme effects complete separation of fatty acid synthase and thioesterase II, in practice only the cytosol need be treated with antibody. A portion of the cytosol is treated, at 20–22° for 30 min, with sufficient antibody to eliminate all of the endogenous fatty acid synthase activity. The antibody–antigen precipitate is removed by centrifugation, and portions of the supernatant are taken for assay of thioesterase II activity.

The reagents, potassium phosphate (50 μl), NADPH (20 μl), acetyl-CoA (20 μl), and malonyl-CoA (40 μl), are usually premixed and dispensed as a "cocktail," 130-μl portions per assay. A calculated amount of water is added so as to anticipate a final volume for the complete assay of 0.5 ml. The trypsinized fatty acid synthase core (80 μl) is then added, and incubation is continued for 3 min. During this loading period, a fatty acyl moiety is assembled on the multienzyme core, with an accompanying short burst of NADPH oxidation; the rate of NADPH oxidation falls to near zero as growth of the enzyme-bound acyl thioester is completed. At this point the 1–8 mU of thioesterase II are added, and the rate of NADPH oxidation is recorded. Thioesterase II present in the added sample catalyzes the release of the acyl moiety from the 4'-phosphopantetheine of the multienzyme core, permitting synthesis of new acyl moieties with accompanying oxidation of NADPH. Controls lacking thioesterase II, which would monitor any residual fatty acid synthase in the trypsinized fatty acid synthase core, are usually unnecessary provided it has been established that the trypsin treatment was carried out effectively. Controls lacking the trypsinized fatty acid synthase core should be performed to confirm the effectiveness of the antibody-treatment in removing endogenous fatty acid synthase from the cytosol.

Units. A unit of thioesterase II is defined as the amount promoting the oxidation, by the trypsinized fatty acid synthase core, of 1 μmol of NADPH per minute at 37°.

Use of Model Substrates

The assay depends on the detection of the free thiol, liberated from a decanoyl thioester by thioesterase II, with the reagent 5,5'-dithiobisnitrobenzoic acid. Reactions are monitored spectrophotometrically at 412 nm.

Reagents

Tris-HCl buffer, 1 M, pH 8.0
Albumin, 5 mg/ml
5,5'-dithiobisnitrobenzoic acid, 1 mM
Decanoylpantetheine, 2 mM
Procedure. Reactions are performed in glass microcuvettes at 37°, with

a final volume of 0.5 ml. If plastic cuvettes are used, albumin may be omitted from the assay system. The reagents, Tris-HCl (50 μl), albumin (5 μl), and 5,5'-dithiobisnitrobenzoic acid (50 μl), are usually premixed and dispensed as a "cocktail," 105-μl portions per assay. Water and sample (usually containing 0.5–5.0 mU of thioesterase II) are added to a volume of 480 μl, and the absorbance at 412 nm, due to the presence of dithiothreitol and protein thiols in the sample, is monitored until it reaches a stable level, usually in 3–4 min. Decanoylpantetheine (20 μl) is added, and the rate of increase in absorbance at 412 nm is recorded. E_{mM} at 412 nm is 13.6.

Units. A unit of thioesterase II activity is defined as that amount catalyzing the hydrolysis of 1 μmol of decanoylpantetheine per minute at 37°.

Assay of Protein Concentrations

The protein concentrations of partially purified preparations of thioesterase II are calculated from the absorbance at 280 nm and 260 nm, according to the formula developed by Warburg and Christian.[6] Protein concentration of the purified enzyme is calculated from the absorbance at 280 nm, using the specific absorption coefficient $(A_{1\,cm}^{1\%})$ of 8.5.[4]

Purification of Thioesterase II

The starting material is cytosol obtained from lactating rat mammary gland.[5] The enzyme is stable, for at least several months in cytosol stored at $-70°$. Our original purification scheme[7] has been modified, with a resulting improvement in yield.

Cytosol is thawed and mixed with 1 M potassium phosphate buffer, pH 7.0, 0.1 M EDTA, 25 mM dithiothreitol, 1 volume per 100 volumes of cytosol. The solution is applied directly (at 0–4°) to a column containing DEAE-cellulose equilibrated with a solution containing 50 mM potassium phosphate buffer, pH 7.0, 1 mM EDTA, 0.25 mM dithiothreitol, and 20% glycerol. A column bed of 75 volumes per 100 volumes of cytosol is used. The column is washed with a solution containing 10 mM potassium phosphate buffer, pH 7.0, 1 mM EDTA, 0.25 mM dithiothreitol, and 20% glycerol until the absorbance at 280 nm of the eluent no longer continues to decrease; usually this requires about 3 volumes per volume of cytosol. Proteins are then eluted with a linear gradient of NaCl, 0 to 0.2 M, in a solution containing 10 mM potassium phosphate buffer, pH 7.0, 1 mM EDTA, 0.25 mM dithiothreitol, and 20% glycerol. A gradient volume of 1

[6] E. Layne, this series, Vol. 3, p. 447.
[7] L. Libertini and S. Smith, *J. Biol. Chem.* **253,** 1393 (1978).

liter is used for each 100 ml of cytosol applied; the flow rate is usually about 150 ml/hr. The elution profile of a typical DEAE-cellulose chromatography step is shown in Fig. 1. Decanoylpantetheine hydrolase activity is eluted in two zones—one in the unabsorbed fraction, and the other with approximately 70 mM NaCl. The reactivation assay indicates that only the second zone of activity corresponds to thioesterase II. Fatty acid synthase activity is eluted in a single, broad zone beyond thioesterase II. Thioesterase II is eluted quantitatively from the ion-exchange column; selective pooling of fractions usually limits the recovery to 85–90%. In the experiment shown in Fig. 1, fractions 88–100 were pooled and dialyzed overnight against a solution containing 10 mM potassium phosphte buffer, pH 7.0, 1 mM EDTA, 0.25 mM dithiothreitol, and 0.1 M NaCl, to remove

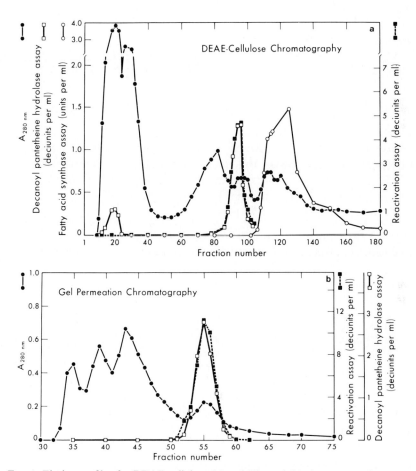

FIG. 1. Elution profiles for DEAE-cellulose (a) and Ultrogel (b) chromatography steps.

glycerol. The presence of glycerol limits the solubility of ammonium sulfate, used in the next step to precipitate the enzyme. Yields of thioesterase II in the precipitation step are reduced to about 75% if the glycerol is not removed. Solid ammonium sulfate (209 g/liter) is added to the dialyzed preparation at 0°, the pH is maintained between 6.9 and 7.1 by the addition of 0.5 M NaOH and the mixture is stirred magnetically for 20 min. A small amount of protein precipitates and is removed by centrifugation at 10,000 g for 20 min and discarded. Solid ammonium sulfate (263 g per liter of starting material) is added to the supernatant, the pH is maintained between 6.9 and 7.1, and the mixture is stirred and centrifuged as before. The precipitated protein, which includes thioesterase II, is dissolved in a minimum volume of 50 mM potassium phosphate buffer, pH 7.0, 1 mM EDTA, and 0.25 mM dithiothreitol. Recovery in the ammonium sulfate precipitation step is normally greater than 90%. The redissolved protein is applied, at 0–4°, to a column (145 × 1.5 cm) of Ultrogel AcA54, which has previously been equilibrated with 50 mM potassium phosphate buffer, pH 7.0, 1 mM EDTA, and 0.25 mM dithiothreitol. Proteins are eluted from the column with the same solution at a flow rate of 5–7 ml/hr. Thioesterase II activity elutes as a single zone with a relative elution volume (V_e/V_o) of 1.6. The elution profiles determined with the decanoylpantetheine hydrolase and reactivation assays coincide (Fig. 1). Elution of thioesterase II activity is quantitative although the actual yield is normally 60–70%, owing to selective pooling of fractions with highest specific activities. A second passage through the Ultrogel column is recommended to isolate the enzyme in highly pure form.

Results of a representative purification are shown in the table.

Homogeneity: Reproducibility of the Procedure

The enzyme is usually about 90% pure, as judged by SDS–polyacrylamide gel electrophoresis after the first Ultrogel step. The second passage through the Ultrogel column reduces the level of impurities to less than 5% with little reduction in yield. The enzyme sediments as a single, symmetrical boundary during velocity sedimentation; it migrates as a single, symmetrical zone of constant specific activity on gel permeation chromatography and migrates as a single zone on polyacrylamide gel electrophoresis at pH 8.9 and 3.6.[7]

In the early stages of purification recoveries of decanoylpantetheine hydrolase and acyl-fatty acid synthase hydrolase activities do not occur in parallel. This appears to be due to the presence of other enzymes capable of hydrolyzing decanoylpantetheine and to the presence of proteins that may modulate decanoylpantetheine hydrolase activity by bind-

PURIFICATION OF MEDIUM-CHAIN FATTY ACYL THIOESTERASE (THIOESTERASE II)

Step	Volume (ml)	Protein (mg/ml)	Total protein (mg)	Reactivation assay					Decanoylpantetheine hydrolase assay			
				Units/ ml	Units/ mg	Total units	Recovery (%)	Purification (fold)	Units/ ml	Units/ mg	Total units	Percent recovery (%)
Cytosol	94.0	14.8	1391	0.30	0.020	28.2	100	1	0.187	0.013	17.6	100
DEAE-cellulose	99.5	0.9	89.6	0.24	0.267	23.9	85	13	0.087	0.097	8.66	49
(NH$_4$)$_2$SO$_4$	1.6	30.6	49.0	12.5	0.407	19.9	71	20	3.80	0.124	6.08	35
Ultrogel I	15.0	0.20	3.0	0.83	4.17	12.5	44	205	0.180	0.900	2.70	15
Ultrogel II	12.9	0.17	2.1	0.89	5.40	11.5	41	266	0.190	1.15	2.45	14

ing the substrate or by protecting the enzyme from inactivation or inhibition. The reactivation assay is not subject to these limitations and provides a more accurate representation of the recovery of thioesterase II through the purification scheme. Overall yields determined by the reactivation assay are usually 40–50%. The procedure offers an increase in yield over our original method. The improvement is attributable to increased recovery in the ion-exchange step when this is carried out directly on the cytosol, rather than after ammonium sulfate precipitation, and a better recovery is obtained in ammonium sulfate precipitation when the pH is maintained at 7.0.

Properties

Immunochemical Properties. Thioesterase II is not recognized by antibodies raised against fatty acid synthase or the thioesterase I component derived from the multienzyme; conversely, antibodies raised against thioesterase II do not recognize fatty acid synthase or thioesterase I.[7] Thioesterase II isolated from rat mammary gland gives a reaction of partial identity with mouse mammary gland thioesterase II.[8] Antithioesterase II antibodies inhibit the ability of thioesterase II to catalyze the hydrolysis of decanoylpantetheine and acyl-fatty acid synthase thioesters.

Physicochemical Properties. The molecular weights for thioesterase II determined by gel permeation chromatography (33,900) and SDS–polyacrylamide gel electrophoresis (32,400) indicate that the enzyme is a monomer.[7] The sedimentation coefficient, determined at a protein concentration of 7 mg/ml, is 2.4×10^{-13} S. A value of 0.74 ml/g for a partial specific volume has been computed from the amino acid composition.

Amino Acid Composition. The complete amino acid composition has been reported; significant differences are found in the compositions of thioesterase I and II.[7]

Substrate Specificity. Oxygen esters are rather poor substrates for thioesterase II; decanoyl-CoA and decanoylpantetheine are hydrolyzed three times faster than *p*-nitrophenyldecanoate. Model compounds containing the cysteamine moiety are the best substrates: decanoylpantetheine > decanoyl-CoA > decanoyl *N*-acetylcysteamine.[9] Thioesters containing the cysteine moiety are very poor substrates. When a homologous series of acyl-CoAs is used as model substrate, maximum velocities are observed close to the critical micelle concentration for each thioester, indicating that the monomeric, but not the micellar, form of the

---

[8] L. Libertini, C. Y. Lin, S. Abraham, R. Hilf, and S. Smith, *Biochim. Biophys. Acta* **618**, 185 (1980).

[9] S. Smith and L. Libertini, *Arch. Biochem. Biophys.* **196**, 88 (1979).

acyl thioester is the substrate.[7] Acyl-CoA thioesters of chain length C_8–C_{18} are acceptable substrates. Longer acyl moieties have not been tested; C_2 and C_4 thioesters are ineffective as substrates, and only a trace of activity is observed with the C_6 thioester.

Thioesterase II will hydrolyze the acyl moiety from thioester linkage to the 4'-phosphopantheteine of the fatty acid synthase. An acyl-4'-phosphopantetheine-containing pentapeptide, derived by peptic digestion of the acyl-fatty acid synthase thioester, is also hydrolyzed by thioesterase II.[9] It is likely, therefore, that this region of the fatty acid synthase provides the natural substrate for thioesterase II. Thioesterase II will hydrolyze fatty acid synthase-linked 4'-phosphopantetheine thioesters of a broad range of acyl chain lengths, C_6–C_{20}.[4] The hydrolysis of the C_6–C_{12} thioesters is likely to be of the most importance physiologically, since these acyl-thioesters are not hydrolyzed efficiently by the thioesterase I component of the fatty acid synthase. When native fatty acid synthase is incubated with substrates in the presence of thioesterase II, the products are shifted from predominantly palmitate toward the medium-chain fatty acids characteristic of milk fats.[7]

The specificity of thioesterase II is not limited to aliphatic acyl chains. Acyl moieties containing ω-phenyl groups, with aliphatic chain lengths of 4–16 carbons that are synthesized by the fatty acid synthase from phenylacetyl-CoA and malonyl-CoA, are released from thioester linkage to the multienzyme by thioesterase II.[10]

Kinetic Properties. The pH–activity profiles for hydrolysis of decanoyl-CoA, decanoylpantetheine, and acyl-fatty acid synthase substrates by thioesterase II are very similar; maximum rates are observed at about pH 8.0, half-maximum velocities occur at pH 6.5–6.6.[4,7] Manganous and magnesium ions exert a small stimulative effect on enzyme activity. The ascending portions of some of the substrate–velocity curves obtained with acyl-CoAs are sigmoidal, and different substrate saturation curves are obtained at different thioesterase II concentrations;[7] thus kinetic data cannot be treated in the conventional manner to obtain values for kinetic parameters. This problem is averted when the natural substrates, acyl-fatty acid synthase-4'-phosphopantetheine thioesters, are used. The K_m for acyl-fatty acid synthase is between 2 and 3 μM, and the V_{max} corresponds to 0.1 μmol of acyl-fatty acid synthase hydrolyzed per milligram of thioesterase II per minute.[7]

Inhibitors. Thioesterase II is inhibited by the serine esterase inhibitors diisopropylphosphofluoridate and phenylmethanesulfonyl fluoride. The kinetics are pseudo-first order, and 1 mol of inhibitor becomes covalently bound to each mole of enzyme.[7] The inhibited enzyme cannot release acyl

[10] A. Stern and S. Smith, unpublished results.

moieties from the fatty acid synthase 4'-phosphopantetheine and affects neither the rate of the fatty acid synthase reaction nor the product chain length.[10]

Acknowledgments

This work was supported by grants from the National Institutes of Health (AM 16073), the National Science Foundation (BMS 7412723), and the American Heart Association (EI 73-140). The author is indebted to Dr. Louis Libertini and Mr. Alan Stern for their contribution to this work.

[26] Medium-Chain Fatty Acyl-S-4'-Phosphopantetheine-Fatty Acid Synthase Thioester Hydrolase from Lactating Rabbit and Goat Mammary Glands

By JENS KNUDSEN, INGER GRUNNET, and RAYMOND DILS

Medium-chain fatty acyl thioesterase is a mammary gland enzyme that interacts with fatty acid synthase to release medium-chain ($C_{8:0}$ and $C_{10:0}$) fatty acids that are characteristic of the milk triacylglycerols of a number of species. The enzyme isolated from lactating-rat mammary gland has been called thioesterase II[1] to distinguish it from the long-chain acyl-thioesterase moiety (thioesterase I) of fatty acid synthase.

Assay Methods

Spectrophotometric Assay of Fatty Acyl Thioesterases Using 5,5'-Dithiobis(2-nitrobenzoic Acid) (DTNB)

Although this method is unspecific in that it will assay any fatty acyl thioesterase, it is convenient to use during purification of the enzyme. The release of free thiol groups from dodecanoyl-CoA is measured by spectrophotometric assay in the presence of DTNB.

The assay mixture contains 0.4 M Tris-HCl buffer, pH 7.4, 1 mM EDTA, 0.2 mM DTNB, and enzyme protein. The mixture is preincubated for 3 min at 37°. The reaction is started by adding dodecanoyl-CoA (final concentration 56 μM) and is followed spectrophotometrically at 412 nm. The amount of thiol released is calculated[2] from the molar extinction coefficient ($\epsilon = 1.36 \times 10^4$ liter mol^{-1} cm^{-1}).

[1] S. Smith, this volume [25].
[2] G. E. Means and R. E. Feeney, "Chemical Modification of Proteins." Holden-Day, San Francisco, California, 1971.

Radiochemical Assay of Fatty Acyl Thioesterases

The specificity of the purified enzyme toward acyl-CoA esters of different chain lengths as model substrates can be determined by measuring the release of $1\text{-}^{14}\text{C}$-labeled fatty acids from the radioactive acyl-CoA esters. Again, the method is unspecific in that it can be used with any fatty acyl thioesterase.

The assay mixture contains $0.1\ M$ potassium phosphate buffer, pH 7.5, 1 mM EDTA, 1 mM dithiothreitol, and 3 or 8 μM $[1\text{-}^{14}\text{C}]$acyl-CoA ester (0.4–1.6 nCi/nmol). The mixture is incubated at 37° for 5 min, and the reaction is started by adding 1–4 μg of enzyme protein that has been preincubated at 37° for 2 min in 14 mM Tris-HCl buffer, pH 7.8. The final volume of the reaction mixture is 0.5 ml. The reaction is stopped after 0.5–3.0 min with 2.0 ml of Dole's reagent.[3] Unesterified ^{14}C-labeled fatty acids of chain lengths $C_{8:0}$–$C_{16:0}$ are extracted by the method of Bar-Tana et al.[4] The top phase, which contains the unesterified ^{14}C-labeled fatty acids, is mixed with 10 ml of xylene that contains 4.0 g of 2,5-diphenyloxazole per liter and 0.1 g of 1,4-bis-(5-phenyloxazol-2-yl)benzene per liter; the radioactivity is determined by liquid scintillation. Portions (0.5–1.0 ml) of the bottom phase are mixed with 10 ml of Triton X-100/xylene (1:2, v/v), which contains 5.5 g of 2,5-diphenyloxazole per liter and 0.1 g of 1,4-bis-(5-phenyloxazol-2-yl)benzene per liter. Water (0.1–0.5 ml) is added, and the radioactivity is determined by liquid scintillation. When butyryl-CoA or hexanoyl-CoA is used as substrate, the reaction is stopped by adding 0.5 ml of 0.1 M HCl. The unesterified fatty acids are extracted four times with 1.0-ml portions of diethyl ether, and their radioactivity is determined by liquid scintillation. The recovery of $[^{14}\text{C}]$butyric acid is 93%. Control incubations without added enzyme should always be used. All assays are performed in triplicate.

Since these assays are unspecific, the following methods are used to measure specifically the activity of medium-chain acyl thioesterase.

Specific Assay of Medium-Chain Fatty Acyl Thioesterase Activity by
    Chain Termination in Fatty Acid Synthesis

The purified enzyme is assayed by its ability to terminate fatty acid synthesis at medium-chain fatty acids. The incubation system (final volume 0.5 ml) contains $0.1\ M$ potassium phosphate buffer, pH 7.0, 1 mM EDTA, 100 μM $[1\text{-}^{14}\text{C}]$acetyl-CoA (5.75 μCi/μmol), 10 mM KHCO$_3$, 5 mM ATP, 8 mM MgCl$_2$, 5 mM tripotassium citrate, 0.24 mM NADPH,

[3] V. P. Dole, *J. Clin. Invest.* **35**, 150 (1956).
[4] J. Bar-Tana, G. Rose, and B. Shapiro, *Biochem. J.* **122**, 353 (1971).

fatty acid synthase (218 μg) purified from lactating rabbit mammary gland, and purified medium-chain acyl thioesterase. [1-^{14}C]Malonyl-CoA is generated *in situ* by adding rate-limiting amounts of acetyl-CoA carboxylase purified from lactating rabbit mammary gland, or by the infusion of 2–3 μl of malonyl-CoA solution per minute (to give rate-limiting malonyl-CoA concentrations) using a Harvard infusion pump, Model 971, fitted with eight 1.0-ml disposable syringes. To ensure efficient mixing of the infused malonyl-CoA in the incubation medium, the infusion tube is fixed near the bottom of the incubation vessel, which is shaken in a rotary motion by an Evapomix (Buckler Instruments, Fort Lee, New Jersey). The mixture is incubated at 37° for 10 min, and the reaction is stopped by adding aqueous NaOH to a final concentration of 2.5 M. The radioactive fatty acids are extracted and are analyzed by radio gas–liquid chromatography.[5] In the absence of medium-chain fatty acyl thioesterase, this system synthesizes negligible proportions of medium-chain fatty acids.

Specific Assay of Medium-Chain Fatty Acyl-Thioesterase Activity by
    Reactivation of Fatty Acid Synthase Inhibited with
    Phenylmethanesulfonyl Fluoride (PMSF)

Medium-chain fatty acyl thioesterase from lactating-rat mammary gland (thioesterase II) can reactivate PMSF-inhibited fatty acid synthase from this tissue.[6] Similarly, medium-chain fatty acyl thioesterase from lactating rabbit mammary gland reactivates PMSF-inhibited fatty acid synthase from both lactating rabbit and goat mammary gland. This effect can be used to assay specifically medium-chain fatty acyl thioesterase that is involved in chain termination.

Inhibition of Fatty Acid Synthase by PMSF. Fatty acid synthase (1–3 mg) in 1.0 ml of 0.25 M potassium phosphate buffer, pH 7.0, is mixed with 40 μl of 25 mM PMSF per milliliter in propan-2-ol and incubated for 1 hr at 30°. The enzyme is precipitated at 4° by adding 200 g of solid $(NH_4)_2SO_4$ per liter and is then dialyzed against 0.25 M potassium phosphate buffer, pH 7.0, which contains 1 mM EDTA and 1 mM dithiothreitol, for 2 hr at 4°.

Reactivation of PMSF-Inhibited Fatty Acid Synthase. Reactivation is measured spectrophotometrically by recording the oxidation of NADPH as the decrease in absorption at 340 nm during incubation at 37°. The assay system consists of 0.1 M potassium phosphate buffer, pH 7.0, 1 mM EDTA, 0.24 mM NADPH, 65 μM malonyl-CoA, 35 μM acetyl-CoA, and 260 μg of PMSF-inhibited fatty acid synthase in a total volume of 1.0 ml. The reaction is started by adding medium-chain fatty acyl thioesterase.

[5] J. Knudsen, *Comp. Biochem. Physiol. B* **53**, 3 (1976).
[6] J. J. Libertini and S. Smith, *J. Biol. Chem.* **253**, 1393 (1978).

This method is specific for the chain terminating medium-chain fatty acyl thioesterase, but is much less sensitive than the method for inhibition of fatty acid synthase described above.

Preparation of Substrates

Acetyl-CoA is prepared from acetic anhydride as described by Stadtman.[7] Malonyl-CoA is synthesized by the method of Eggerer and Lynen.[8] Radioactive and nonradioactive long-chain acyl-CoA esters are synthesized by the method of Sánchez et al.[9] To synthesize the CoA esters of butyric acid, hexanoic acid, or octanoic acid, the method is modified by converting the sodium salts of these acids into the tetraethylammonium salts before the reaction with ethyl chloroformate as follows: Tetraethylammonium hydroxide is prepared by shaking tetraethylammonium bromide (2.1 g) with Ag_2O (3.5 g) in 10 ml of water for 10 min. The AgBr is removed by centrifugation. A column (5 cm × 0.5 cm) of Dowex 50-W (H^+ form) is converted into the tetraethylammonium form by applying 3 ml of the aqueous solution of tetraethylammonium hydroxide and then washing the column with water until the eluate is pH 7.0. The sodium salt of the butyric acid, hexanoic acid, or octanoic acid (in each case 20 μmol in 0.4 ml of water) is then applied to the column. The eluate, which contains the tetraethylammonium salts, is evaporated to dryness under N_2, and the salt is dissolved in methylene chloride.

All acyl-CoA esters are analyzed and purified as described by Pullman.[10]

Materials

New Zealand white rabbits and Wistar rats at 14 days postpartum, red Danish dairy cows at 6–7 months postpartum, and goats and sheep of mixed breeds at 14 days postpartum are used. Coenzyme A was obtained from Boehringer, Mannheim, West Germany, and [1-^{14}C]acetic anhydride was from the Radiochemical Centre, Amersham, Bucks., U.K. Dithiothreitol (DTT) and bovine serum albumin (fraction V, fatty acid-poor) were from Sigma Chemical Co., St. Louis, Missouri. All other reagents were of analytical purity and were obtained from E. Merck, Darmstadt, West Germany.

[7] E. R. Stadtman, this series, Vol. 3, 931.
[8] H. Eggerer and F. Lynen, *Biochem. J.* **335**, 540 (1962).
[9] M. Sánchez, D. G. Nicholls, and D. N. Brindley, *Biochem. J.* **132**, 697 (1973).
[10] M. E. Pullman, *Anal. Biochem.* **54**, 188 (1973).

General Methods

Purification of Fatty Acid Synthase

Fatty acid synthase is purified from the lactating mammary glands of cow, goat, rabbit, and sheep and from rabbit liver as described by Knudsen.[11] Fatty acid synthase from adipose tissue is purified by the same method, except that the adipose tissue is homogenized at room temperature in 2 volumes of 100 mM potassium phosphate buffer, pH 7.0, containing 250 mM sucrose, 4 mM EDTA, and 1.0 mM dithiothreitol, in a Waring blender at full speed for 30 sec.

Protein Determination

Protein is precipitated with 15% (w/v) trichloroacetic acid and determined by the method of Lowry et al.[12] with bovine serum albumin as the standard.

Lactating Rabbit Mammary Gland

Purification of Medium-Chain Fatty Acyl Thioesterase

Animals. New Zealand white rabbits (14 days postpartum) are killed by cervical dislocation and the entire mammary gland is excised and placed on ice. Care should be taken to remove as much of the surrounding tissue as possible. About 180 g wet weight of tissue is obtained from one animal. Mammary gland tissue from two animals is used in each purification. All procedures are at 4° unless otherwise stated.

Preparation of Particle-Free Supernatant. The mammary tissue is cut into thin slices and passed through a kitchen parsley cutter. It is then diluted with two volumes of 0.1 M potassium phosphate buffer, pH 7.0, containing 1 mM EDTA and 1 mM dithiothreitol, and homogenized in a Potter–Elvehjem homogenizer with a loose Teflon pestle. The homogenate is centrifuged at 105,000 g_{av} for 1 hr, and the particle-free supernatant is decanted through a loose plug of glass wool. The particle-free supernatant can be stored several months at −60° without loss of enzyme activity.

Ammonium Sulfate Fractionation. Solid $(NH_4)_2SO_4$ is added continuously to the particle-free supernatant. The protein precipitating between 245 and 390 g of $(NH_4)_2SO_4$ per liter initial volume is collected by centrifu-

---

[11] J. Knudsen, *Biochim. Biophys. Acta* **280,** 408 (1972).

[12] O. H. Lowry, N. J. Rosebrough, A. L. Farr, and R. J. Randall, *J. Biol. Chem.* **193,** 265 (1951).

gation at 10,000 g_{av} for 10 min. The precipitate is dissolved in 100 ml of 1.0 mM potassium phosphate buffer, pH 7.0, which contains 1.0 mM EDTA and 0.5 mM dithiothreitol, and then dialyzed against several 2-liter portions of distilled water to decrease the ionic strength to about 1.0 mohm^{-1} at 4°.

DEAE-Cellulose Chromatography. The dialyzed sample is loaded onto a column (2.5 cm × 25 cm) of DEAE-cellulose that has been previously equilibrated with 1.0 mM potassium phosphate buffer, pH 7.0, containing 1.0 mM EDTA and 0.5 mM dithiothreitol. The absorbed protein is eluted with a linear gradient consisting of 500 ml each of 10 mM and 175 mM potassium phosphate buffer, pH 7.0, which contains 1 mM EDTA, 0.5 mM dithiothreitol, and 20% (v/v) glycerol. The fractions that contain enzyme activity are pooled, and the protein is precipitated by the addition of 470 g of (NH$_4$)$_2$SO$_4$ per liter. The precipitate is suspended in 5 ml of 100 mM potassium phosphate buffer, pH 7.0, which contains 1.0 mM EDTA and 0.5 mM dithiothreitol, and then dialyzed against the same buffer until it is completely dissolved. The elution pattern from a typical DEAE ion-exchange chromatography separation is shown in Fig. 1. Three peaks with thioesterase activity are obtained, but only one (peak III) is able to terminate fatty acid synthesis. The glycerol in the eluation buffer is necessary to stabilize the enzyme during DEAE ion-exchange chromatography. At-

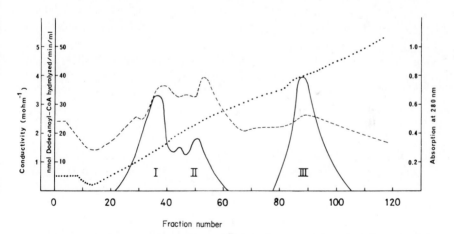

Fraction number

FIG. 1. Purification of medium-chain acyl thioesterase from lactating rabbit mammary gland. DEAE-cellulose ion-exchange chromatography of the (NH$_4$)$_2$SO$_4$ precipitate (245–390 g/liter) of the particle-free supernatant fraction. For details, see the text; enzyme activity is measured as described in Assay Methods, section on spectrophotometric assay. ---, Absorption at 280 nm; ——, fatty acyl thioesterase activity; · · · ·, conductivity.

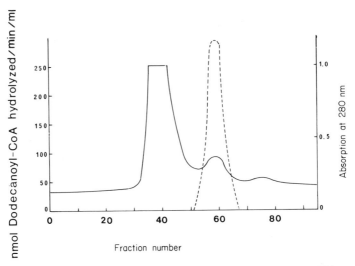

FIG. 2. Purification of medium-chain acyl thioesterase from lactating rabbit mammary gland. Sephadex G-75 filtration of the concentrated eluate from DEAE-cellulose ion-exchange chromatography. For details, see the text; enzyme activity is measured as described in Assay Methods, section on spectrophotometric assay. ——, Absorption at 280 nm; ---, fatty acyl-thioesterase activity.

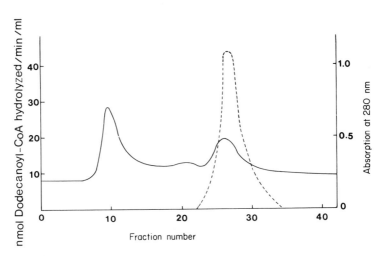

FIG. 3. Purification of medium-chain fatty acyl thioesterase from lactating rabbit mammary gland. Preparative polyacrylamide gel electrophoresis of the concentrated eluate from Sephadex G-75. For details, see the text; enzyme activity is measured as described in Assay Methods, section on spectrophotometric assay. ——, Absorption at 280 nm; ---, fatty acyl-thioesterase activity.

tempts to elute the enzyme with buffers without glycerol result in complete loss of enzyme activity.

Sephadex G-75 Gel Filtration. The dialyzed sample is then applied to a column (2.5 cm × 80 cm) of Sephadex G-75. The column is eluted with 100 mM potassium phosphate buffer, pH 7.0, which contains 1.0 mM EDTA and 0.5 mM dithiothreitol (25 ml/hr), and fractions (7 ml) containing enzyme activity are pooled. A typical separation is shown in Fig. 2. At this stage of the purification, the enzyme is a major component of the pooled fraction and occasionally no other component can be detected by SDS–polyacrylamide gel electrophoresis. In most purifications, however, a second diffuse band with half the molecular weight of the main band is detected by SDS–polyacrylamide gel electrophoresis. This second band cannot be removed by refiltration on Sephadex G-75. The pooled fraction from the Sephadex G-75 column is therefore routinely purified by preparative electrophoresis.

Preparative Electrophoresis. The enzyme in the pooled fraction is concentrated by adding 470 g of $(NH_4)_2SO_4$ per liter. The precipitate is dissolved in 5 ml of 60 mM Tris buffer adjusted to pH 6.9 with concentrated H_3PO_4 and then dialyzed at 4° against 1 liter of the same buffer for 4 hr. The dialyzed sample is made to 15% (v/v) with glycerol, and applied to a preparative electrophoresis gel using an Ultraphore Colora instrument (Ultraphore Colora, Messtechnik G.m.b.H., 7073 Lorch/Wurtt. 1, West Germany). The gel size is 7.5 cm × 13.7 cm and 8 cm high and consists of 2 cm of 2.5% (w/v) polyacrylamide (2.5% cross-linked) stacking gel in 60 mM Tris buffer adjusted to pH 6.9 with concentrated H_3PO_4, and 6 cm of 15% (w/v) polyacrylamide (2.7% cross-linked) separation gel in 280 mM Tris-HCl buffer, pH 8.9.

The upper and lower buffer compartments contain 50 mM Tris buffer, which contains 400 mM glycine, pH 8.5. Electrophoresis is at 2° with a constant power of 25 W during the stacking period and 50 W during the separation period. The protein bands are eluted continuously with 124 mM Tris-HCl buffer, pH 8.1, at a flow rate of 40 ml/hr. A typical chromatogram from a preparative electrophoresis separation is shown in Fig. 3. Fractions (5 ml) containing enzyme activity are pooled and are stored at −60°.

The enzyme purified by this procedure is homogeneous as judged by SDS–polyacrylamide gel electrophoresis. The overall recovery and the purification achieved are difficult to calculate owing to the presence of other thioesterases in the initial particle-free supernatant fraction. It can be seen from Table I that the apparent values for recovery and purification are 6% and 67-fold, respectively.

TABLE I

PURIFICATION OF MEDIUM-CHAIN ACYL-THIOESTERASE FROM LACTATING RABBIT MAMMARY GLAND

Procedure[a]	Protein (mg)	Activity (units)	Specific activity[b] (units/mg of protein)	Purification (fold)	Yield (%)
(2) Particle-free supernatant	3133	30552	9.7	1.0	100
(3) $(NH_4)_2SO_4$ fraction (245–390 g/liter)	1316	24120	18.3	1.9	79
(4) DEAE-cellulose ion-exchange chromatography, peak III	154	2816	18.3	1.9	9.2
(5) Sephadex G-75 gel filtration eluate	16.2	1955	121	12.4	6.4
(6) Preparative electrophoresis eluate	2.8	1835	651	67.1	6.0

[a] Numbers in parentheses refer to the text section Lactating Rabbit Mammary Gland: Purification of Medium-Chain Acyl Thioesterase.
[b] One unit of enzyme activity is defined as 1 nmol of dodecanoyl-CoA hydrolyzed per minute at a substrate concentration of 56 μM in the spectrophotometric assay; see under Assay Methods, the section on spectrophotometric assay.

Properties of Medium-Chain Fatty Acyl Thioesterase

Molecular Weight

The molecular weight of the partially purified enzyme precipitated by 245–390 g of $(NH_4)_2SO_4$ per liter from the particle-free supernatant is approximately 28,000 as estimated by Sephadex G-100 gel filtration using albumin (MW 68,000), peroxidase (MW 40,000), pepsin (MW 35,000), and cytochrome c (MW 11,700) as reference proteins.[13]

The molecular weight of the purified enzyme as established by 10% (w/v) polyacrylamide disc gel electrophoresis in the presence of SDS is $29,000 \pm 500$ (mean value $\pm$ SD) using albumin, ovalbumin (MW 43,000), pepsin, trypsin (MW 23,300), myoglobin (MW 17,200), and cytochrome c as reference proteins.

Substrate Specificity toward Acyl-CoA Esters

The critical micellar concentrations of acyl-CoA esters of different chain lengths are not known for the incubation conditions used to assay the purified enzyme. Hence, meaningful values for K_m and V_{max} cannot be obtained from these assays. To avoid this problem of critical micellar concentrations, the specificity of the purified enzyme toward acyl-CoA esters of different chain lengths in the absence of albumin is measured at low substrate concentrations i.e., 8 μM for acyl-CoA esters of chain lengths $C_{4:0}$–$C_{12:0}$, and 3 μM for acyl-CoA esters of chain lengths $C_{8:0}$–$C_{16:0}$ (Fig. 4).

At both 3 and 8 μM substrate, the rate of hydrolysis is greatest with dodecanoyl-CoA (Fig. 4). The enzyme does not hydrolyze butyryl-CoA or hexanoyl-CoA at measurable rates. The rates of hydrolysis of octanoyl-CoA and of decanoyl-CoA are about 5% and 50%, respectively, of that of dodecanoyl-CoA. Tetradecanoyl-CoA and hexadecanoyl-CoA are hydrolyzed at rates about 60% and 55%, respectively, of that of dodecanoyl-CoA. The enzyme cannot hydrolyze acyl-carnitine esters of chain lengths $C_{4:0}$–$C_{16:0}$ at concentrations of 3 and 10 μM.

Properties of the Enzyme When Dodecanoyl-CoA Is Used as a Model Substrate

Since the purified enzyme shows the highest rate of hydrolysis with dodecanoyl-CoA, this is used as the substrate in the following experiments.

[13] P. Andrews, *Methods Biochem. Anal.* **18**, 1 (1970).

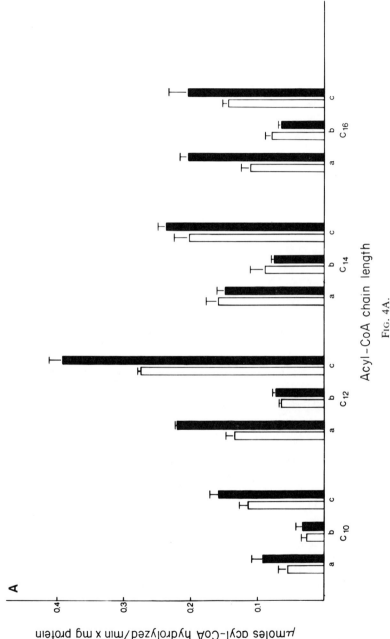

FIG. 4A.

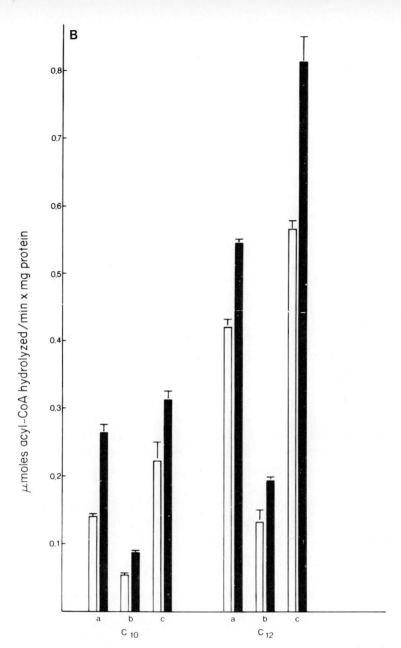

FIG. 4. Substrate specificity of purified medium-chain fatty acyl thioesterase from lactating rabbit mammary gland using acyl-CoA esters as model substrates. The radiochemical assay (see the text section on properties of the enzyme) is used. In panels A and B, substrate concentrations are 3 and 8 μM, respectively. Three purified enzyme preparations have been used; preparations a, b, and c had specific activities, respectively, of 1130, 305, and 2101 nmol of dodecanoyl-CoA hydrolyzed per minute per milligram of protein as determined by spectrophotometric assay (see Assay Methods, section on spectrophotometric assay) using 56 μM dodecanoyl-CoA as substrate. Filled bars: enzyme activity in the presence of 30 μg of albumin per milliliter of incubation mixture. Open bars: enzyme activity in the absence of albumin. The error brackets are the standard deviation of triplicate incubations.

pH-Activity Curve. The enzyme shows the highest activity toward 8 μM [1-^{14}C]dodecanoyl-CoA (in 0.1 M potassium phosphate buffer that contains 1 mM EDTA and 1 mM dithiothreitol) at pH 7.0–7.5.

Substrate Concentration. The rate of hydrolysis of dodecanoyl-CoA increases linearly with substrate concentration up to 45 μM. The rate decreases at higher concentrations, and a slight inhibition is observed at 212 μM substrate, although it is difficult to obtain reproducible results at high substrate concentrations. This may be due to the differing extent of micelle formation, even though the substrate is preincubated with buffer for 5 min before the enzyme is added.

Effects of Albumin. Fatty acid-poor albumin increases the rate of hydrolysis of dodecanoyl-CoA by the purified enzyme. The maximum increase occurs with 30 μg of albumin per milliliter of incubation mixture; higher concentrations decrease the rate, which may be due to the binding of substrate. This optimum concentration of albumin increases the rate of hydrolysis of all fatty acyl-CoA esters of chain lengths $C_{8:0}$–$C_{16:0}$ although it does not affect the overall pattern of chain-length specificity.

This stimulatory effect of albumin could be due to the removal of inhibitory products. However, if increasing concentrations of unesterified dodecanoic acid are added to the incubation system (except that albumin is omitted from the incubations that contain 8 μM dodecanoyl-CoA), no decrease in the rate of hydrolysis of dodecanoyl-CoA is observed even when 3 mM dodecanoic acid is added (data not shown). This indicates that the effect of albumin is unlikely to be due to the removal of inhibitory products.

Concentration of Medium-Chain Fatty Acyl Thioesterase and Fatty Acid Synthase in Lactating-Rabbit Mammary Gland Determined by Rocket Immunoelectrophoretic Estimations

Antibodies to fatty acid synthase and to medium-chain fatty acyl thioesterase are raised in goats. The antiserum is then processed[14] to give fractions that contain immunoglobulins. The immunoglobulin preparation should be shown to be monospecific by cross immunoelectrophoresis in the Svendsen buffer system.[15] Rocket immunoelectrophoresis is carried out as described by Veeke,[15] except that 1.0% (v/v) Triton X-100 is added to the gel solution when the medium-chain fatty acyl thioesterase is to be assayed. All enzyme solutions and dilution buffers used in the assay of the thioesterase are made 1 mg/ml with respect to bovine serum albumin to

[14] N. Harboe and A. Ingild, *Scand. J. Immunol.*, Suppl. 1, 161 (1973).
[15] B. Veeke, *in* "A Manual of Quantitative Immunoelectrophoresis" (N. H. Axelsen, J. Kroll, and B. Veeke, eds.), p. 38. Universitetsforlaget, Oslo, 1973.

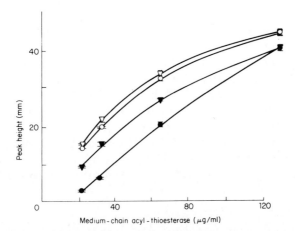

Fig. 5. Effect of dilution method and of bovine serum albumin on the assay of purified medium-chain fatty acyl thioesterase from lactating rabbit mammary gland by rocket immunoelectrophoresis. Dilution method a: peak height with (▽) and without (▼) albumin (1 mg/ml). Dilution method b, peak height with (○) and without (●) albumin (1 mg/ml). Duplicate samples of four standard solutions are used on each plate. Error brackets represent the average value ± half the difference between duplicate determinations.

prevent the adsorption of the enzyme onto glassware. The final concentration of potassium phosphate in all the thioesterase dilutions is adjusted to 50 mM to standardize the effect of potassium phosphate on peak height. All assays are in duplicate.

Concentration of Medium-Chain Fatty Acyl Thioesterase. The addition of 0.2 mg of albumin per milliliter to two different particle-free supernatant fractions that contain 54 and 44 μg of enzyme per milliliter increases the peak height by about 20 and 25%, respectively. Increasing the albumin concentration to 1 mg/ml further increases the peak height slightly, but no further effect on peak height is seen above this concentration of albumin. The results in Fig. 5 show that the effect of albumin can be explained by the decreased adsorption of the enzyme onto glassware. In these experiments the purified enzyme is diluted in two different ways both in the presence and in the absence of 1.0 mg of albumin per milliliter. In dilution method a, each dilution step is obtained by diluting 100 μl of enzyme stock solution with a calculated amount of buffer. Dilution method b uses successive dilution of the previous dilution, and so on. With albumin present, both dilution methods give similar results. Without albumin present, both methods give a lower value, but this is most pronounced for dilution method b at high dilution. The effect of albumin on the adsorption of the enzyme onto glassware might also explain the stimulatory effect of albumin on chain-length termination.

The concentration of the thioesterase in nine different preparations of the particle-free supernatant fraction from lactating rabbit mammary glands has been found to vary from 86 to 480 μg/ml.

Concentration of Fatty Acid Synthase. The concentration of fatty acid synthase in the nine particle-free supernatant fractions used to assay the thioesterase varied between 0.95 and 2.97 mg/ml. This large variation is partly due to the different amount of milk in the glands when the animals were killed.

Assuming a molecular weight of 475,000 for fatty acid synthase[16] and 29,000 for that of the thioesterase, the molar ratio of thioesterase to fatty acid synthetase is 1.99 ± 0.66 (mean $\pm$ SD for the nine preparations of the particle-free supernatant fraction).

Interaction of Medium-Chain Fatty Acyl Thioesterase with Fatty Acid Synthase

Effects of Albumin on Fatty Acid Chain Termination by Medium-Chain Fatty Acyl Thioesterase. The effect of albumin on the chain lengths of the fatty acids synthesized by lactating-rabbit mammary gland fatty acid synthetase in the presence and in the absence of the thioesterase is shown in Table II. Low concentrations of bovine serum albumin (1 mg/ml) have only a slight effect on the pattern of fatty acid synthesized by purified fatty acid synthase alone. By contrast, high concentrations of albumin (10 mg/ml) increase the proportion of short-chain fatty acid synthesized ($C_{4:0}$ plus $C_{6:0}$).

Malonyl-CoA utilization can be calculated from the incorporation of [1-^{14}C]acetate by the following formula, where n represents the number of carbon atoms in the individual fatty acids.

$$\sum_{n=4}^{n=16} [(\text{molar }\% \text{ of } C_n \times n)] - 2 \times \text{nmol of } [1\text{-}^{14}\text{C}]\text{acetate incorporated}$$

The calculated rate of incorporation of malonyl-CoA remains unchanged when bovine serum albumin is added. Therefore the increased rate of incorporation of [1-^{14}C]acetate in the presence of albumin only represents a large number of chain initiations caused by the change of chain length.

The addition of 1.36 mol of thioesterase per mole of fatty acid synthase results in high proportions of medium-chain fatty acids being synthesized without changing the calculated amount of malonyl-CoA utilized (Table II). The addition of 52 μg of albumin per milliliter greatly increases both the rate of [1-^{14}C]acetate incorporation and the calculated rate of malonyl-CoA incorporation. This results in an increased synthesis of me-

[16] I. Grunnet and J. Knudsen, *Biochem. J.* **178**, 929 (1978).

TABLE II

EFFECT OF BOVINE SERUM ALBUMIN ON FATTY ACIDS SYNTHESIZED BY LACTATING RABBIT MAMMARY GLAND FATTY ACID SYNTHASE IN ABSENCE AND PRESENCE OF MEDIUM-CHAIN FATTY ACYL THIOESTERASE

Additions to 0.5 ml incubation mixture[a]	Calculated malonyl-CoA utilization (nmol)	Total acetate incorporation from [1-^{14}C]acetyl-CoA[b] (nmol)	Percentage distribution of radioactivity in fatty acids[c] (mol/100 mol)						
			$C_{4:0}$	$C_{6:0}$	$C_{8:0}$	$C_{10:0}$	$C_{12:0}$	$C_{14:0}$	$C_{16:0}$
None	11.4	2.95 ± 0.01	34	8	2	3	6	41	6
Albumin, 0.52 mg	10.8	2.93 ± 0.04	33	12	2	6	7	37	3
Albumin, 5.2 mg	11.1	3.36 ± 0.02	39	15	4	2	3	31	6
Thioesterase	10.3	3.17 ± 0.08	33	10	6	21	13	17	ND
Thioesterase + 26 µg of albumin	22.4	5.47 ± 0.04	17	6	7	23	15	24	8
Thioesterase + 52 µg of albumin	19.3	5.05 ± 0.04	16	5	11	31	17	17	3
Thioesterase + 130 µg of albumin	16.8	4.77 ± 0.05	21	4	14	36	16	9	ND
Thioesterase + 502 µg of albumin	20.3	5.21 ± 0.04	14	5	11	35	17	16	2
Thioesterase + 5.02 mg of albumin	20.7	5.57 ± 0.18	15	7	14	36	15	12	1

[a] The incubation system is described in Assay Methods, section on specific assay by chain termination. The incubation mixture (0.50 ml) contains 240 µg of fatty acid synthase from lactating-rabbit mammary gland (specific activity 743 nmol of NADPH oxidized per minute per milligram of protein), 20 µg of medium-chain fatty acyl-thioesterase (specific activity 293 nmol of dodecanoyl-CoA hydrolyzed per minute per milligram of protein), bovine serum albumin as indicated, and 42 µM [1-^{14}C]acetyl-CoA (specific radioactivity 3.8 µCi/nmol). Malonyl-CoA is infused at a rate of 2.64 nmol/min for 15 min. The molar ratio of thioesterase to synthase is 1.36.

[b] Values are means ± half the difference between duplicate incubations.

[c] Values are the average of two determinations; ND, not detected.

dium- and long-chain fatty acids with only a slight change in the synthesis of $C_{4:0}$. Increasing the concentration of albumin from 52 to 104 or 260 $\mu g/ml$ does not increase the rate of incorporation of malonyl-CoA, but it further alters the pattern of fatty acids synthesized so that $C_{10:0}$ becomes the predominant fatty acid. Even up to 10 mg/ml of albumin does not further increase the incorporation of malonyl-CoA or increase the synthesis of medium-chain fatty acids. It is interesting that the effect of albumin on both the rate of malonyl-CoA utilization and the pattern of fatty acids synthesized (Table II) occurs in the same concentration range as when albumin affects rocket height in the rocket immunoelectrophoretic assay.

In these experiments, only 27–56% of the infused malonyl-CoA is used for fatty acid synthesis, although the capacity of the added fatty acid synthase to utilize malonyl-CoA is 34 times that of the rate of infusion of malonyl-CoA. However, the calculation of malonyl-CoA concentration at the end of the incubation period is complicated because of a possible, but unknown, rate of malonyl-CoA decarboxylation during the incubation period.

Effects of Rate of Malonyl-CoA Infusion and Amount of Medium-Chain Fatty Acyl Thioesterase on Fatty Acid Synthesis. Increasing the rate of malonyl-CoA infusion from 0.31 to 8.35 nmol/min gradually increases the chain length of fatty acids synthesized by fatty acid synthase in the presence of 1.36 mol of the thioesterase per mole of synthase (Table III). At the highest rate of infusion (where the capacity of fatty acid synthase to convert malonyl-CoA to fatty acids is still three times higher than the infusion rate), $C_{14:0}$ and $C_{16:0}$ become the predominant fatty acids synthesized. At the two lowest infusion rates, no long-chain fatty acids are synthesized. As in the experiment shown in Table II, only a fraction (at most 25 nmol) of the malonyl-CoA infused is converted to fatty acids. This indicates a feedback inhibition of the fatty acid synthase, possibly by CoA. In the experiment described in Table III, the final concentration of CoA in the 0.5-ml incubations would be twice the nanomoles of acetyl-CoA plus malonyl-CoA used, i.e., about 60 μM at most. Hsu *et al.*[17] have shown that 20 μM CoA inhibits pigeon liver fatty acid synthase by 35%.

Increasing the amount of thioesterase relative to fatty acid synthase gradually changes the pattern of fatty acids synthesized from long- to medium-chain without any change in the calculated rate of utilization of malonyl-CoA (Table IV). With 2 mol of thioesterase per mole of fatty acid synthase (which is the average ratio of these enzymes found in the cytosol), the pattern of fatty acids synthesized is very similar to that synthesized by lactating rabbit mammary gland *in vivo*.[18] No inhibitory effect of the thioesterase on the rate of fatty acid synthesis is observed.

[17] R. Y. Hsu, G. Warson, and J. W. Porter, *J. Biol. Chem.* **240,** 3736 (1965).
[18] E. M. Carey and R. Dils, *Biochem. J.* **126,** 1005 (1972).

TABLE III

EFFECTS OF DIFFERENT RATES OF MALONYL-CoA INFUSION ON THE PATTERN OF FATTY ACIDS SYNTHESIZED BY LACTATING RABBIT MAMMARY GLAND FATTY ACID SYNTHASE IN THE PRESENCE OF MEDIUM-CHAIN FATTY ACYL THIOESTERASE

Malonyl-CoA infused/min[a] (nmol)	Calculated malonyl-CoA utilization[b] (nmol)	Total acetate incorporation from [1-^{14}C]acetyl-CoA[b] (nmol)	Percentage distribution of radioactivity in fatty acids[c] (mol/100 mol)						
			$C_{4:0}$	$C_{6:0}$	$C_{8:0}$	$C_{10:0}$	$C_{12:0}$	$C_{14:0}$	$C_{16:0}$
0.31	2.7	1.31 ± 0.01	49	12	20	19	ND	ND	ND
0.68	6.8	2.25 ± 0.08	43	8	23	19	7	ND	ND
1.36	9.5	3.15 ± 0.02	20	4	13	34	16	11	2
2.72	11.8	4.55 ± 0.03	7	ND	6	26	19	29	13
5.43	11.5	4.92 ± 0.04	14	1	3	24	13	27	18
8.35	12.8	4.64 ± 0.03	8	2	4	13	10	24	39

[a] The incubation system is described in Assay Methods, section on specific assay by chain termination. The incubation mixture (0.5 ml) contains 240 μg of lactating rabbit mammary gland fatty acid synthase (specific activity 743 nmol of NADPH oxidized per minute per milligram of protein), 260 μg of bovine serum albumin, 20 μg of medium-chain fatty acyl thioesterase (specific activity 255 nmol of dodecanoyl-CoA hydrolyzed per minute per milligram of protein), and 42 μM [1-^{14}C]acetyl-CoA (specific radioactivity 3.8 μCi/μmol). Malonyl-CoA is infused at the indicated rates for 15 min. The molar ratio of thioesterase to synthase is 1.36.

[b] Values are means ± half the difference between duplicate incubations.

[c] Values are the average of two determinations; ND, not detected.

TABLE IV

EFFECT OF INCREASING THE RATIO OF MEDIUM-CHAIN FATTY ACYL THIOESTERASE TO FATTY ACID SYNTHASE

Molar ratio of medium-chain fatty acyl thioesterase to fatty acid synthase[a]	Calculated malonyl-CoA utilization (nmol)	Total acetate incorporation from [1-^{14}C]acetyl-CoA[b] (nmol)	Percentage distribution of radioactivity in fatty acids[c] (mol/100 mol)						
			$C_{4:0}$	$C_{6:0}$	$C_{8:0}$	$C_{10:0}$	$C_{12:0}$	$C_{14:0}$	$C_{16:0}$
No thioesterase added	17.3	3.50 ± 0.09	19	6	2	2	3	43	25
0.44	16.0	3.80 ± 0.16	16	6	8	23	14	24	9
1.0	16.5	4.12 ± 0.06	13	3	12	36	18	14	4
2.0	16.6	4.56 ± 0.06	10	6	20	40	16	7	1
3.1	18.5	5.10 ± 0.06	9	5	24	43	14	5	ND
5.0	18.4	5.36 ± 0.15	7	4	35	44	8	2	ND

[a] The incubation system is described in Assay Methods, section on specific assay by chain termination. The incubation mixture (0.5 ml) contains 240 µg of lactating rabbit mammary gland fatty acid synthase (specific activity 743 nmol of NADPH oxidized per minute per milligram). Bovine serum albumin (260 µg), medium-chain fatty acyl thioesterase (specific activity 225 nmol of dodecanoyl-CoA hydrolyzed per minute per milligram) as indicated, and [1-^{14}C]acetyl-CoA (43.4 µM; specific radioactivity 5.95 µCi/µmol). Malonyl-CoA is infused at a rate of 2.6 nmol/min for 15 min.

[b] Values are means ± half the difference between duplicate incubations.

[c] Values are the average of two determinations; ND, not detected.

The concentration-dependent effect observed on increasing the molar ratio of thioesterase to fatty acid synthase could indicate a weak and nonspecific interaction between the two enzymes. This was further investigated in the following experiment.

Specificity of Medium-Chain Fatty Acyl Thioesterase toward Fatty Acid Synthases from a Number of Different Species. In comparative experiments where fatty acid synthases from lactating rabbit, cow, goat, and sheep mammary gland, and from lactating rabbit liver and cow adipose tissue are used, the addition of thioesterase in all cases changes the pattern of fatty acids synthesized from long- to medium-chain. The "specificity" of the thioesterase to terminate chain elongation is very broad with respect to fatty acid synthases from different species and tissues.

Lactating Goat Mammary Gland

Goat milk triacylglycerol fatty acids consist of about 10 mol % decanoic acid. This acid is synthesized *de novo* within the gland.[19] Attempts to show the presence of a separate medium-chain fatty acyl thioesterase, similar to that in rabbit mammary gland, in the particle-free supernatant fraction of lactating goat mammary gland have been unsuccessful.[20]

The investigations described below show that the fatty acid synthase complex from lactating goat mammary gland possesses medium-chain fatty acyl thioesterase activity and is able to synthesize high proportions of medium-chain fatty acids under appropriate conditions.

Properties of Lactating Goat Mammary Gland Fatty Acid Synthase

Molecular Weight

The molecular weight, as estimated by sucrose gradient centrifugation,[21] is 484,000 ± 17,000 (mean ± SD) when alcohol dehydrogenase is used as the reference protein. The molecular weight of the subunit, as estimated by SDS–polyacrylamide gel electrophoresis, is 226,000 ± 7000 (mean ± SD of four determinations) when myosin (MW 220,000)[22], phosphorylase *a* (MW 94,000)[22], the β-subunit of RNA polymerase (MW 115,000)[21] and α-macroglobulin (MW 185,000)[23] are used as reference proteins.

[19] E. F. Annison, J. L. Linzell, S. Fazakerley, and B. W. Nicholls, *Biochem. J.* **102**, 637 (1967).
[20] I. Grunnet and J. Knudsen, *Eur. J. Biochem.* **95**, 497 (1979).
[21] R. G. Martin and B. N. Ames, *J. Biol. Chem.* **236**, 1372 (1961).
[22] K. Weber and M. Osborn, *J. Biol. Chem.* **244**, 4406 (1969).
[23] R. R. Burges, *J. Biol. Chem.* **244**, 6168 (1969).

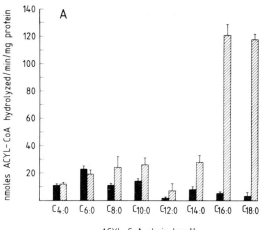

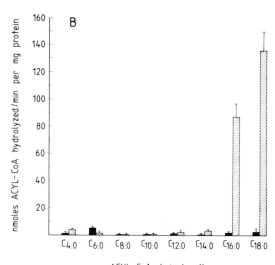

FIG. 6. Effect of phenylmethanesulfonyl fluoride (PMSF) on the fatty acyl thioesterase activity of lactating goat and rabbit mammary gland fatty acid synthases. The synthases are treated with 1.0 mM PMSF (filled bars), or with 4% (v/v) propan-2-ol as control (hatched bars) (see Assay Methods, section on specific assay: inhibition of fatty acid synthase), and their activities are determined as described in Assay Methods, section on radiochemical assay. (A) Lactating goat mammary synthase; (B) lactating rabbit mammary synthase. Error brackets represent the mean value ± standard deviation.

Substrate Specificity of the Terminating Thioesterase in Goat Mammary Gland Fatty Acid Synthase toward Acyl-CoA Esters

The chain length specificity of the terminating thioesterase activity of this synthase toward acyl-CoA esters as model substrates is measured using the radiochemical assay (see Assay Methods). Acyl-CoA esters are used at 3 μM final concentration; the incubation time is 3 min for $C_{4:0}$ to $C_{14:0}$ acyl-CoA esters and 1 min for $C_{16:0}$ and $C_{18:0}$ acyl-CoA esters. Fatty acid synthase is treated with PMSF as described in Assay Methods (section on specific assay inhibited by PMSF).

Fatty acid synthase from goat mammary gland shows significant hydrolytic activity toward both medium- and long-chain acyl-CoA esters as model substrates (Fig. 6A, hatched bars). By contrast, the rabbit mammary gland synthase hydrolyzes only long-chain acyl-CoA esters (Fig. 6B, hatched bars). This indicates that the goat mammary synthase contains either one acyl thioesterase with a chain length specificity different from that of the rabbit mammary synthase, or that it contains two acyl thioesterases as reported for pigeon liver fatty acid synthase.[24,25]

Evidence for the presence of two different fatty acyl-thioesterases in goat mammary synthase is obtained from inhibitor studies using PMSF. This specific inhibitor of serine-containing esterases completely inhibits the terminating fatty acyl thioesterases of fatty acid synthase from pigeon liver.[25] Figure 6A (black bars) shows that 1 mM PMSF almost completely inhibits the long-chain fatty acyl thioesterase activity of goat mammary fatty acid synthase, but inhibits hydrolytic activity toward medium-chain acyl-CoA esters by only 50%. By contrast, when rabbit mammary synthase is incubated under the same conditions, 1 mM PMSF strongly inhibits all the fatty acyl thioesterase activity (Fig. 6B, black bars).

Pattern of Fatty Acids Synthesized by Goat Mammary Gland Fatty Acid Synthase

Incubation Conditions

Two incubation systems are used with the purified fatty acid synthase: (*a*) incubations with a fixed initial concentration of malonyl-CoA and based on the conditions described in Assay Methods section on specific assay by reactivation inhibited by PMSF; (*b*) incubations where rate-limiting concentrations of malonyl-CoA are added by continuous infusion

[24] F. A. Lornitzo, A. A. Qureshi, and J. W. Porter, *J. Biol. Chem.* **250**, 4520 (1975).
[25] S. Kumar, *Biochem. Biophys. Res. Commun.* **53**, 334 (1973).

as described in Assay Methods section on specific assay by chain termination, except that the indicated concentration of [1-^{14}C]acetyl-CoA was
used. Radioactive fatty acids are extracted and analyzed by radio-gas–
liquid chromatography as described by Knudsen.[5]

Fatty Acids Synthesized by Fatty Acid Synthase from Goat Mammary Gland

The purified fatty acid synthase synthesizes predominantly short-
chain and long-chain fatty acids when incubated with 12.5–100 μM
malonyl-CoA. This result is similar to that found for other mammalian
fatty acid synthases.[11,26] In the absence of malonyl-CoA, no measurable
amount of fatty acids are synthesized (this is contrary to previous reports[11,26] that were due to incomplete separation of products by the gas–
liquid chromatography techniques used). Increasing the concentration of
malonyl-CoA increases both the total amount of fatty acids and the relative proportion of long-chain fatty acids synthesized. However, the synthesis of medium-chain fatty acids is low regardless of the malonyl-CoA
concentration used. There is also insignificant synthesis of medium-chain
fatty acids ($C_{8:0}$ and $C_{10:0}$) by purified fatty acid synthase alone when
rate-limiting concentrations of malonyl-CoA (i.e., when the ratio of the
rate of malonyl-CoA infusion to the maximum rate of malonyl-CoA utilization by the synthase is about 1 : 100) are infused, or when rate-limiting
concentrations of malonyl-CoA are generated *in situ* (as in Table V).

According to the model proposed by Sumper *et al.*[27] the chain lengths
of the fatty acids synthesized by fatty acid synthase depend on the relative
rates of chain elongation and of termination. The model predicts that the
shortest possible chain length should be obtained with rate-limiting concentrations of malonyl-CoA. The results described above fully agree with
this model. Even with the lowest concentration of malonyl-CoA used
(12.5 μM), palmitic acid is the predominant long-chain fatty acid synthesized. When the concentration of malonyl-CoA infused or generated *in situ*
is decreased, the predominant long-chain fatty acid synthesized changes
from $C_{16:0}$ to $C_{14:0}$ plus $C_{12:0}$. This occurs without a significant increase in
the relative proportion of $C_{8:0}$ plus $C_{10:0}$ synthesized. Hence, although the
fatty acid chain length can be shortened by lowering the malonyl-CoA
concentration, the fatty acid synthase itself cannot synthesize $C_{10:0}$ in
proportions similar to those found in goat milk triacylglycerols.

[26] E. M. Carey and R. Dils, *Biochim. Biophys. Acta* **210**, 388 (1970).

[27] M. Sumper, O. Osterhelt, C. Riepertinger, and F. Lynen, *Eur. J. Biochem.* **10**, 377 (1969).

TABLE V

EFFECT OF GOAT MAMMARY MICROSOMAL FRACTION ON THE CHAIN LENGTH OF FATTY ACIDS SYNTHESIZED BY PURIFIED GOAT MAMMARY GLAND FATTY ACID SYNTHASE

Experiment[a]	Microsomal fraction (mg protein)	Total acetate incorporated from [1-^{14}C]acetyl-CoA[b] (nmol)	Percentage distribution of radioactivity in fatty acids[c] (mol/100 mol)							
			$C_{4:0}$	$C_{6:0}$	$C_{8:0}$	$C_{10:0}$	$C_{12:0}$	$C_{14:0}$	$C_{16:0}$
1	0	28.27 ± 0.43	16	4	2	3	9	34	32
2	0.54	20.58 ± 0.21	25	12	7	21	10	17	8
3	1.08	18.68 ± 0.23	42	10	7	18	7	11	5
4	1.62	16.45 ± 0.04	49	11	8	16	6	8	2

[a] The incubation system is described in Assay Methods, section on specific assay by chain termination. Incubations (15 min) contain 207 μg of fatty acid synthase (specific activity 850 nmol of NADPH oxidized per minute per milligram of protein), 35 μg of acetyl-CoA carboxylase (specific activity 260 nmol of malonyl-CoA formed per minute per milligram of protein), 100 μM [1-^{14}C]acetyl-CoA (5.52 nCi/nmol) and lactating goat mammary microsomal fraction as indicated, in a total volume of 1 ml.

[b] Values are means ±SD.

[c] Values are the average of two determinations.

Fatty Acids Synthesized by Goat Mammary Gland Fatty Acid Synthase in the Presence of Goat Mammary Microsomal Fraction

Medium-chain fatty acids are synthesized when purified fatty acid synthase from lactating-goat mammary gland and the microsomal fraction from this tissue are incubated with a rate-limiting concentration of malonyl-CoA plus the other substrates required for fatty acid and triacylglycerol synthesis (Table V). Compared with incubations without the microsomal fraction, the total and the relative amount of $C_{10:0}$ synthesized is increased whereas that of longer-chain fatty acids is decreased. Increasing the amount of added microsomal protein decreases total fatty acid synthesis and alters the chain length toward short-chain fatty acids. The microsomal fraction itself synthesizes only small amounts of short-chain and long-chain fatty acids, but no medium-chain fatty acids (Table VI). Boiling the microsomal fraction destroys its ability to induce medium-chain fatty acid synthesis (Table VI).

These results indicate that the microsomal fraction is involved in the control of fatty acid chain termination in goat mammary gland, and that the controlling factor is heat-labile and may therefore be a protein. The following experiments distinguish between a medium-chain terminating fatty acyl thioesterase in the microsomal fraction, which is similar to that in the cytosol of lactating rabbit mammary gland, and the activation of medium-chain fatty acyl thioesterase present in the goat mammary fatty acid synthase complex.

Fatty Acids Synthesized by Goat Mammary Fatty Acid Synthase in the Presence of Mammary Microsomal Fractions from Different Species

Medium-chain fatty acids are synthesized when fatty acid synthase from lactating goat mammary gland is incubated with the microsomal fraction from lactating goat, cow, rat, or rabbit mammary gland (Table VII). By contrast, the microsomal fraction from lactating rat or rabbit mammary gland does not induce the synthesis of medium-chain fatty acids by their corresponding fatty acid synthases. The pattern of fatty acids synthesized is, in fact, similar to that synthesized by the purified synthases alone (Table VIII). Hence the induction by lactating rabbit and rat mammary microsomal fractions of $C_{10:0}$ synthesis by goat mammary synthase is unspecific; that is, it cannot be explained by a contamination of these microsomal fractions with medium-chain fatty acyl thioesterase.

Since lactating cow mammary gland fatty acid synthase also synthesizes $C_{10:0}$ when it is incubated with the microsomal fraction from this tissue (Table VII), these ruminant mammary synthases differ from the lactating rat and rabbit mammary synthases in their ability to terminate

TABLE VI

FATTY ACID SYNTHESIS BY MICROSOMAL FRACTION OF LACTATING GOAT MAMMARY GLAND, AND HEAT SENSITIVITY OF MICROSOMAL INDUCTION OF MEDIUM-CHAIN FATTY ACID SYNTHESIS BY GOAT MAMMARY GLAND FATTY ACID SYNTHASE

Experiment[a]	Fatty acid synthase (mg protein)	Microsomal fraction (mg protein)	Total acetate incorporation from [1-^{14}C]acetyl-CoA[b] (nmol)	Percentage distribution of radioactivity in fatty acids[c] (mol/100 mol)							
				$C_{4:0}$	$C_{6:0}$	$C_{8:0}$	$C_{10:0}$	$C_{12:0}$	$C_{14:0}$	$C_{16:0}$	$C_{18:0}$
1	0	0.41	1.7 ± 0.3	64	ND	ND	ND	ND	10	25	ND
2	0.20	0.41	5.1 ± 0.2	35	6	4	15	7	16	14	3
3	0.20	0.41[d]	5.0 ± 2.5	33	10	3	4	6	24	20	ND

[a] The incubation system is described in Assay Methods, section on specific assay by chain termination. Incubations (15 min) contain fatty acid synthase (specific activity 1100 nmol of NADPH oxidized per minute per milligram of protein) as shown, 40 μM [1-^{14}C]acetyl-CoA (5.95 nCi/nmol), and lactating goat mammary microsomal fraction as indicated, in a total volume of 1 ml.
[b] Values are means ± SD.
[c] Values are the average of two determinations; ND, not detected.
[d] Heated at 100° before being added to the incubation system.

TABLE VII

FATTY ACID SYNTHESIS BY GOAT MAMMARY GLAND FATTY ACID SYNTHASE INCUBATED WITH MAMMARY GLAND MICROSOMAL FRACTIONS FROM VARIOUS SPECIES[a]

Source of mammary fatty acid synthase	Source of mammary microsomal fraction	Total acetate incorporation from [1-^{14}C]acetyl-CoA[b] (nmol)	Percentage distribution of radioactivity in fatty acids[c] (mol/100 mol)						
			$C_{4:0}$	$C_{6:0}$	$C_{8:0}$	$C_{10:0}$	$C_{12:0}$	$C_{14:0}$	$C_{16:0}$
Goat	None	7.5 ± 0.7	18	6	3	3	11	40	19
Goat	Goat	8.5 ± 0.4	18	7	7	21	11	21	15
Goat	Cow	7.1 ± 0.2	23	7	4	14	16	30	6
Goat	Rabbit	7.2 ± 0.6	25	9	8	24	10	19	5
Goat	Rat	8.3 ± 0.1	22	6	5	17	14	22	14
Cow	Cow	5.2 ± 0.3	31	10	3	8	12	25	11
Rabbit	Rabbit	7.4 ± 0.0	26	9	1	1	1	37	25
Rat	Rat	8.0 ± 0.3	26	7	4	4	9	36	14

[a] The incubation system is described in Assay Methods, section on specific assay by chain termination. Incubations (15 min) contain 100 μg of fatty acid synthase, 200 μg of microsomal protein, and 40 μM [1-^{14}C]acetyl-CoA (3.8 nCi/nmol) in a total volume of 0.5 ml. Malonyl-CoA is infused at a rate of 2.1 nmol/min. The specific activities (nanomoles of NADPH oxidized per minute per milligram of protein) of the fatty acid synthases are as follows: goat, 980; cow, 260; rabbit, 740; rat, 1340.

[b] Values are means ± SD.

[c] Values are the average of two determinations.

TABLE VIII

FATTY ACID SYNTHESIS BY MAMMARY GLAND FATTY ACID SYNTHASES FROM VARIOUS SPECIES[a]

Source of mammary fatty acid synthase	Total acetate incorporation from [1-^{14}C]acetyl-CoA[b] (nmol)	Percentage distribution of radioactivity in fatty acids[c] (mol/100 mol)						
		$C_{4:0}$	$C_{6:0}$	$C_{8:0}$	$C_{10:0}$	$C_{12:0}$	$C_{14:0}$	$C_{16:0}$
Goat	7.5 ± 0.7	18	6	3	3	11	40	19
Cow	7.9 ± 0.1	28	6	1	5	7	41	12
Rabbit	5.9 ± 0.1	39	6	ND	ND	ND	35	20
Rat	8.6 ± 0.2	26	4	ND	ND	7	38	25

[a] The incubation conditions are described in the legend to Table VII except that no microsomal protein is added.
[b] Values are means ± SD.
[c] Values are the average of two determinations; ND, not detected.

fatty acid synthesis at milk-specific fatty acids. Further evidence for this is provided by the following results.

Fatty Acids Synthesized by Lactating Goat, Cow, Rat, and Rabbit Mammary Fatty Acid Synthases in the Presence of Lactating Goat Mammary Microsomal Fraction

Fatty acid synthases from lactating goat, cow, rat, and rabbit mammary gland are incubated with the substrates necessary for fatty acid and triacylglycerol synthesis in the presence of lactating goat mammary microsomal fraction. As in the experiments described above, malonyl-CoA is infused continuously at a low rate.

The goat mammary microsomal fraction induces medium-chain fatty acid synthesis by both goat and cow mammary gland fatty acid synthases, but not with the rat and rabbit mammary gland synthases (Table IX). Control experiments show that under these conditions none of the purified fatty acid synthases can themselves synthesize significant amounts of medium-chain fatty acids (Table VIII). This strongly supports the conclusion that these ruminant mammary synthases differ from rat and rabbit mammary synthases in their inherent ability to terminate synthesis at medium-chain fatty acids.

TABLE IX

Fatty Acid Synthesis by Mammary Gland Fatty Acid Synthases from Different Species Incubated with Goat Mammary Gland Microsomal Fraction[a]

Source of mammary fatty acid synthase	Source of mammary microsomal fraction	Total acetate incorporation from [1-^{14}C]acetyl-CoA[b] (nmol)	Percentage distribution of radioactivity into fatty acids[c] (mol/100 mol)						
			$C_{4:0}$	$C_{6:0}$	$C_{8:0}$	$C_{10:0}$	$C_{12:0}$	$C_{14:0}$	$C_{16:0}$
Goat	Goat	8.5 ± 0.4	18	7	7	21	11	21	15
Cow	Goat	8.8 ± 0.3	21	8	6	15	13	25	11
Rabbit	Goat	6.1 ± 1.8	36	10	1	ND	ND	30	23
Rat	Goat	8.5 ± 0.1	26	8	4	3	7	38	14

[a] The incubation conditions are described in the legend to Table VII.
[b] Values are means ± SD.
[c] Values are the average of two determinations; ND, not detected.

Substrate Specificity of the Terminating Thioesterase in Lactating Cow
Mammary Fatty Acid Synthase toward Acyl-CoA Esters

Further evidence that lactating-cow mammary synthetase has the inherent ability to terminate fatty acid synthesis at medium-chain acids is provided by the ability of this synthase to hydrolyze medium-chain acyl-CoA esters with 18% of the activity shown toward palmitoyl-CoA.

Conclusions

Lactating goat and cow mammary fatty acid synthase, unlike lactating rat and rabbit mammary synthases, have the inherent ability to terminate synthesis at milk-specific medium-chain fatty acids. However, interaction with an unknown heat-labile microsomal factor is required to ensure this chain termination. It is unlikely that this microsomal factor is a medium-chain terminating fatty acyl thioesterase similar to the enzyme in the cytosol of lactating rabbit and rat mammary gland.[6] Table VII shows that the microsomal fraction from rat, rabbit or goat mammary gland can induce $C_{10:0}$ synthesis by goat mammary synthase, but not by rat and rabbit mammary synthetases.

Both goat and cow mammary synthases hydrolyze medium-chain acyl-CoA esters, and it has not been possible to detect medium-chain terminating fatty acyl thioesterase in the cytosol of lactating goat mammary gland. We therefore suggest that the synthesis of medium-chain fatty acids by lactating-goat mammary gland occurs by activation of medium-chain fatty acyl-thioesterase within the fatty acid synthase by an unknown microsomal factor. Rabbit and rat microsomal fractions, which also contain this factor, synthesize *in vivo* triacylglycerol, which contains a high proportion of medium chain fatty acids. The factor might therefore be a general component of microsomal systems involved in the esterification of medium-chain fatty acids.

This is a novel mechanism for the synthesis of medium-chain acids, since rat and rabbit mammary glands contain a separate, low molecular weight cytosolic chain terminating thioesterase. Our unpublished results with lactating sheep mammary gland support the view that ruminant mammary tissue synthases have a common mechanism for terminating fatty acid synthesis at milk-specific fatty acids that is distinct from that found in nonruminant mammary tissue.

Acknowledgments

We thank the Wellcome Trust for a travel grant, and the Royal Society and the Agricultural Research Council of Great Britain for financial support.

[27] Thioesterase Segment from the Fatty Acid Synthase of the Uropygial Gland of Goose

By P. E. KOLATTUKUDY, C. J. BEDORD, and LINDA ROGERS

The thioesterase segment of fatty acid synthase from the uropygial gland of goose catalyzed hydrolysis of the acyl thioesters generated by the synthase as well as CoA esters of fatty acids.[1,2]

$$R\overset{\overset{\displaystyle O}{\parallel}}{-}C-SCoA + H_2O \rightarrow RCOOH + CoSH$$

Enzyme Assay

Principle. Thioester hydrolysis catalyzed by purified enzyme can be measured spectrophotometrically by following decrease in absorption at 232 nm ($E = 4000$ M^{-1}).[3] With crude enzyme preparations hydrolysis can be assayed by measuring the production of radioactive fatty acids. The free fatty acids can be easily separated from the substrate by extraction with an organic solvent.

Reagents

Tris-HCl, 0.1 M, pH 7.5, containing 0.5 M dithioerythritol (DTE)
Bovine serum albumin, 2.2 mg/ml in the buffer
Palmitoyl-CoA, 0.3 mM in the buffer
[1-^{14}C]Palmitoyl-CoA, 0.3 mM in the buffer (1000 cpm/nmol)

Spectrophotometric Assay. The reaction mixture containing 20 μmol of Tris-HCl, 0.1 μmol of DTE, 12 nmol of palmitoyl-CoA, 44 μg of bovine serum albumin, and enzyme in a total volume of 0.2 ml is incubated in a 1-cm path length cuvette (0.4 ml capacity) at 30° in a recording spectrophotometer; absorbance is measured at 232 nm. The initial linear rate is used to calculate the rate of hydrolysis.

Radiotracer Assay. Reaction mixture containing 50 μmol of Tris-HCl, 0.25 μmol of DTE, 110 μg of bovine serum albumin, 30 nmol of [1-^{14}C]palmitoyl-CoA (~30,000 cpm), and enzyme is incubated in a small test tube at 30° for 2 min. The reaction is stopped by the addition of 0.1 ml 6 N HCl, and the mixture is extracted twice with 2 ml each of chloroform. The

---

[1] P. E. Kolattukudy, J. S. Buckner, and C. J. Bedord, *Biochem. Biophys. Res. Commun.* **68**, 379 (1976).

[2] C. J. Bedord, P. E. Kolattukudy, and L. Rogers, *Arch. Biochem. Biophys.* **186**, 139 (1978).

[3] W. M. Bonner and K. Bloch, *J. Biol. Chem.* **247**, 3123 (1972).

pooled chloroform extract is washed with 2 ml of acidified (pH <3) water, and the chloroform layer is transferred to a counting vial and evaporated to dryness with a stream of N_2. (All the chloroform should be removed before the addition of the counting fluid because chloroform is a potent quencher.) After the addition of 10 ml of counting fluid, consisting of a 0.4% (w/v) solution of Omnifluor (New England Nuclear) in toluene containing 30% ethanol, ^{14}C is assayed in a scintillation spectrometer. A control reaction mixture containing no enzyme or boiled enzyme is used, and the control value is subtracted from the experimental value. Since CoA esters undergo nonenzymatic hydrolysis, the substrate solutions should not be stored for long periods.

Enzyme Purification

Preparation of Fatty Acid Synthase. Fatty acid synthase is purified to apparent homogeneity as described elsewhere.[4]

Trypsin Treatment of Fatty Acid Synthase. To 500 mg of purified fatty acid synthase in 100 ml of 0.1 M sodium phosphate, pH 7.6, containing 0.5 mM DTE, 1 mg of trypsin (TPCK treated, Worthington Biochemical Corp.) in 1 ml of the same buffer is added, and the mixture is incubated at 30° for 20 min. The reaction is stopped by the addition of 50 mg of soybean trypsin inhibitor (chromatographically prepared, Sigma Chemical Co.) in 1.0 ml of the above buffer. After stirring the mixture for 5 min at 30°, the preparation is cooled in an ice bath to 0–4° and all further procedures are carried out at this temperature.

Ammonium Sulfate Fractionation. To the above solution granulated ammonium sulfate is slowly added until 40% saturation is achieved. The mixture is stirred for an additional 30-min period, then the precipitated protein is removed by centrifugation at 20,000 g for 20 min and discarded. The supernatant is treated with a further quantity of ammonium sulfate until 70% saturation is achieved; the precipitated protein, collected by centrifugation, can be stored at 0–4° for at least a week under N_2. When substantial amounts of the thioesterase segment are needed, it is convenient to store the ammonium sulfate precipitate until three or four batches of fatty acid synthase are processed. The protein pellets are pooled and suspended in about 25 ml of 0.1 M Tris-HCl, pH 7.5, containing 0.5 mM DTE; the suspension is dialyzed overnight against the same buffer. The resulting protein solution is concentrated to 5–7 ml by ultrafiltration with an Amicon PM-10 membrane, and any insoluble material is removed by centrifugation. This process yields about 280 mg of protein from about 2 g of fatty acid synthase.

---

[4] P. E. Kolattukudy, A. J. Poulose, and J. S. Buckner, this volume [13].

Sephadex G-100 Gel Filtration. The enzyme solution is subjected to gel filtration with a 2.5 × 87 cm column of Sephadex G-100 using 0.1 M Tris-HCl, pH 7.5, containing 0.5 mM DTE at a flow rate of 0.4 ml/min. Two protein peaks are observed, the second of which represents thioesterase whereas the first, at the void volume, shows no thioesterase activity. Fractions (5–6 ml each) containing thioesterase activity are pooled and concentrated to 5–7 ml by ultrafiltration as indicated above, yielding about 220 mg of protein. This protein solution is diluted with four volumes of ice-cold water to decrease the buffer concentration to 0.02 M.

QAE-Sephadex Chromatography. The protein solution is passed through a 2.5 × 38 cm column of QAE-Sephadex, equilibrated with 0.02 M Tris-HCl, pH 7.5, containing 0.5 mM DTE. After washing the column with two void volumes of the same buffer, a 0–0.4 M linear gradient of NaCl in a total volume of 650 ml of the buffer is applied, and 5–6 ml fractions are collected. Sodium dodecyl sulfate (SDS)–polyacrylamide gel electrophoresis shows that the leading half of the protein peak is more enriched in thioesterase (MW 33,000) than the trailing half of the protein, which contains a higher proportion of two lower molecular weight fragments (MW 14,000 and 13,000). Therefore the two halves are separately pooled, concentrated by ultrafiltration as above, and dialyzed against 0.02 M citrate-phosphate buffer, pH 6.0, containing 0.5 mM DTE. Any insoluble material is removed by centrifugation, yielding about 70 mg of protein.

Cellulose Phosphate Chromatography. Complete removal of the two smaller fragments with a minimum loss of the thioesterase segment results when 600 μg of protein per milliliter of bed volume are applied. (Higher protein load results in incomplete removal of the small fragments, and lower protein load results in excessive loss of thioesterase.) After the protein solution fully penetrates the cellulose phosphate gel (1.6 × 16 cm), previously equilibrated with the above citrate–phosphate buffer, the flow is turned off and the protein is allowed to remain with the gel at 4° overnight. The following morning the protein is eluted with the same citrate–phosphate buffer at about 0.15 ml/min. The fractions (2–3 ml each) containing thioesterase activity are pooled, concentrated by ultrafiltration and dialyzed against 0.1 M Tris-HCl, pH 7.5, containing 0.5 mM DTE. This procedure does not completely remove the smaller fragments from the second half of the enzyme preparation obtained from the QAE step. In this case the protein solution obtained from the first pass through the cellulose phosphate is concentrated by ultrafiltration and passed through another cellulose phosphate column in an identical manner. The resulting protein can be pooled with the thioesterase obtained from the first half. It is advisable to subject the two protein solutions to SDS–gel electrophoresis prior to pooling them. This step yields about 30–40 mg of protein.

Purity. The protein resulting from the above steps is homogeneous according to the following criteria: Polyacrylamide gel electrophoresis shows a single band containing all the thioesterase activity, and SDS–gel electrophoresis shows a single protein band. Analytical ultracentrifugation shows a single component with a $s_{20,w}$ of about 3.0. Ouchterlony double diffusion with rabbit antibodies prepared against the enzyme shows a single immunoprecipitant line.

Properties

Molecular Weight and Amino Acid Composition. According to SDS–polyacrylamide gel electrophoresis the molecular weight is about 33,000. The amino acid composition is similar to that of fatty acid synthase.

Optimum pH. Maximal rate of hydrolysis of palmitoyl-CoA is at about 7.5, and the rate drops drastically below pH 7.0 and above pH 8.0.

Effect of Bovine Serum Albumin and Concentrations of the Enzyme and Substrate. When the molar ratio of palmitoyl-CoA to thioesterase is higher than 33 no measurable activity is detected, presumably because the substrate binds the enzyme and inactivates it. This probem can be overcome by the inclusion of bovine serum albumin in the reaction mixture. Increasing albumin concentration increases the rate of hydrolysis (up to 220 μg/ml), and when the optimum albumin concentration is exceeded an inhibition is observed. With a 0.5-ml reaction mixture containing 110 μg of albumin, linear rates are obtained up to 30 μg of thioesterase with 30 nmol of palmitoyl-CoA. In the presence of albumin, increasing concentrations of palmitoyl-CoA give a sigmoidal saturation pattern, most probably because of competition of the albumin for palmitoyl-CoA at low substrate concentrations. In the absence of albumin a typical substrate saturation pattern is observed with inhibition at higher concentrations of the substrate. Ignoring this inhibition a K_m of 50 μM and V of 340 nmol/mg per minute are obtained.

Substrate Specificity. With CoA esters as the substrates the enzyme shows a high degree of preference for C_{16} and longer acyl chains, although measurable rates are obtained with acyl chains as short as C_{10}. Since the interaction of the acyl-CoA with the enzyme and bovine serum albumin depends on the chain length, comparative rates of hydrolysis should be viewed with caution.

Inhibitors. The thioesterase is inhibited by the "active serine"-directed reagents, diisopropylfluorophosphate and phenylmethanesulfonyl fluoride as well as by SH-directed reagents, N-ethylmaleimide, and p-chloromercuribenzoate.

Immunological Comparison. Rabbit antibodies prepared against the thioesterase segment cross-reacts with intact fatty acid synthase from the gland as well as the intact synthase from the liver of goose. Intact fatty acid synthase from the uropygial glands of other waterfowl, such as wood duck, mallard, Canadian goose, Muscovy duck, and Peking duck, also cross-reacts with the rabbit antibodies prepared against the thioesterase segment from the uropygial gland of the goose.

Acknowledgment

This work was supported by Grant GM-18278 from the National Institutes of Health.

[28] Long-Chain Fatty Acyl-CoA Hydrolase from Rat Liver Mitochondria

EC 3.1.2.2 Palmitoyl-CoA Hydrolase

By ROLF KRISTIAN BERGE and MIKAEL FARSTAD

Palmitoyl-SCoA + H_2O → palmitate + CoASH

At least two enzymes are present in homogenates of rat liver that catalyze the hydrolytic cleavage of fatty acyl-CoA thioesters.[1] The enzymes are localized in the mitochondrial matrix and in the microsomes[1] and seem to be different proteins.[2-4] The mitochondrial matrix enzyme is specific for long-chain (C_{10}–C_{18}) fatty acyl thioesters with maximal activity toward palmitoyl-CoA.

Assay Methods

Two methods for the assay of long-chain fatty acyl-CoA hydrolase are used, the spectrophotometric method and the radiochemical method.

Spectrophotometric Method

Principle. The spectrophotometric assay method is based on the release from acyl-CoA thioesters of CoASH, which reduces 5,5′-dithiobis(2-nitrobenzoic acid) (DTNB) to 5-thio-2-nitrobenzoic acid, which is yellow. The reaction is measured by the increase in absorbance at 412 nm.

[1] R. K. Berge and M. Farstad, *Eur. J. Biochem.* **95**, 89 (1979).
[2] R. K. Berge and B. Døssland, *Biochem. J.* **181**, 119 (1979).
[3] R. K. Berge and M. Farstad, *Eur. J. Biochem.* **96**, 393 (1979).
[4] R. K. Berge, *Biochim. Biophys. Acta* **574**, 321 (1979).

Reagents

N-2-Hydroxyethylpiperazine-N'-2-ethanesulfonic acid (HEPES), 75
mM, pH 7.5
EDTA, 10 mM
Bovine serum albumin, 10 mg/ml
Palmitoyl-CoA, 2.5 mM (in 30 mM HEPES buffer, pH 7.5)
DTNB, 3 mM (in 30 mM HEPES buffer, pH 7.5)

Procedure. The complete assay mixture contains 30 mM HEPES, pH
7.5, 1.0 mM EDTA, 0.3 mM DTNB, 120–130 nmol of palmitoyl-CoA per
milligram of protein (usually 40 μM palmitoyl-CoA and 0.25 mg/ml of
bovine serum albumin), and 1–4 μg of purified enzyme in a final volume of
1.0 ml. All assay mixtures are incubated for at least 3 min at 35° before the
reaction is started by the addition of enzyme at 35°. The spectrophotometric measurements are performed in a recording spectrophotometer. The
amount of thiol released is calculated from the molar absorption coefficient,[5] $E_{412 nm} = 13,600 \ M^{-1} \ cm^{-1}$, after correction for unspecific reaction
of the sulfhydryl groups of the enzyme with DTNB and the nonspecific
hydrolysis of palmitoyl-CoA. A concentration of DTNB of 0.3 mM is
chosen in order to reduce the blank absorbance and the inhibitory effect
on the hydrolase activity itself. The main disadvantage of the spectrophotometric method is that unspecific interaction of CoASH and other
SH groups is detected. Using di(4-dipyridyl) disulfide (Aldrithiol), which
usually is dissolved in ethanol, the enzyme activity will be overestimated
owing to release of CoASH in the reaction[6]: palmitoyl-CoA + ethanol →
ethylpalmitate + CoASH. Aldrithiol in ethanol is therefore not recommended as a CoASH trapping reagent.

Radiochemical Assay Method

Principle. The radiochemical assay method for the long-chain fatty
acyl-CoA hydrolase is based on the release of 1-^{14}C-labeled fatty acids
from [1-^{14}C]acyl-CoA.

Reagents

HEPES buffer, 100 mM, pH 7.5
Sucrose, 1 M
Triton X-100, 1% (v/v)
Palmitoyl-CoA, 2.5 mM (10–20 nCi) (in 10 mM HEPES buffer, pH
7.5)

[5] G. E. Means and R. E. Feeney, "Chemical Modification of Proteins." Holden-Day, San
Francisco, California, 1971.
[6] L. E. Hagen, R. K. Berge, A. Bakken, and M. Farstad, *FEBS Lett.* **110**, 205 (1980).

Bovine serum albumin, 10 mg/ml

Termination medium: propan-2-ol, heptane, 1 N H_2SO_4 (20 : 5 : 1)

Procedure. The complete assay system contains 0.25 mM sucrose in 10 mM HEPES buffer, pH 7.5, enzyme (usually 1–4 μg of purified long-chain fatty acyl-CoA hydrolase), and bovine serum albumin and [1-^{14}C]palmitoyl-CoA adjusted to a ratio of 120–130 nmol per milligram of protein, in a final volume of 0.5 ml. The reaction is started by adding the substrate. The incubation is carried out at 35° for 1–2 min. The reaction is stopped by adding 1 ml of termination medium followed by 0.5 ml of heptane and 0.5 ml of distilled water. After 10 min the mixture is vigorously shaken for 5–10 min. After centrifugation, the upper heptane layer, which contains the radioactive palmitic acid, is washed with 0.5 ml of heptane-saturated water and the lower aqueous phase is washed with 0.5 ml of water-saturated heptane. A portion of the combined heptane phases is counted in a liquid scintillation counter.

The substrate for long-chain fatty acyl-CoA hydrolase, palmitoyl-CoA, forms micelles in solutions.[7] The critical micelle concentration depends on the buffer composition, pH, salt concentration, and protein concentration.[7] The enzyme shows different reaction behavior toward the substrate in monomeric and micelle form.[3,4,7] It is important to establish that there is linearity of long-chain fatty acyl-CoA hydrolase activity with respect to time, protein, and substrate concentration.[8]

Units. Similar results are obtained by the two methods at a substrate concentration of 120 nmol of palmitoyl-CoA per milligram of protein. An enzyme unit is defined as the amount of activity necessary to catalyze the cleavage of 1 nmol of palmitoyl-CoA per minute at 35° under the stated conditions. Specific activity is expressed as units per milligram of protein.

Purification Procedures

Step 1. Animals and Preparation of the Mitochondria. Male albino rats of 200–300 g body weight (Wistar, Møll), fasted overnight are stunned and decapitated. The livers are cut into small pieces and rinsed in a medium consisting of 0.25 M sucrose and 5 mM HEPES buffer, pH 7.4. The tissue is then homogenized in a Potter–Elvehjem homogenizer at 720 rpm with two strokes of a loose-fitting Teflon pestle. The 10% (w/v) homogenate is centrifuged at 1500 g for 10 min, and the nuclear fraction, which also contains cell debris, is washed three times. The combined supernatants are centrifuged at 12,000 g for 10 min, and the mitochondrial pellet thus obtained is washed twice using the same centrifugation force and time.

[7] R. K. Berge, E. Slinde, and M. Farstad, *FEBS Lett.* **109**, 194 (1980).

[8] R. K. Berge, E. Slinde, and M. Farstad, *Biochem. J.* **182**, 347 (1979).

The centrifugation conditions provide maximal recovery of mitochondria with minimal contamination of other subcellular particles.[9] The swinging-bucket HB-4 rotor of the Sorvall RC 5 refrigerated centrifuge (R_{min} 6.2 cm and R_{max} 14.4 cm) is used at a temperature of 2°. The final mitochondrial pellet is resuspended in the homogenization medium at a concentration of 40–50 mg of mitochondrial protein per milliliter.

Step 2. Preparation of Soluble Fraction. Soluble proteins from the mitochondrial matrix are isolated after disruption of the mitochondrial outer membrane with digitonin.[1] Digitonin dissolved in 0.25 M sucrose containing 5 mM Tris-HCl buffer, pH 7.4, is added to isolated mitochondria to give a final concentration of 0.12 mg of digitonin per milligram of mitochondrial protein. Bovine serum albumin is then added to a final concentration of 0.5 mg/ml. Ten minutes after the addition of digitonin, the mixture is diluted with 3 volumes of 0.25 M sucrose containing 5 mM Tris-HCl buffer, pH 7.4, and 0.5 mg of bovine serum albumin per milliliter. The mixture is centrifuged at 10,000 g for 20 min. The pellet, which contains mitoplasts, is resuspended in 10 mM HEPES buffer and sonicated at 20 μm for 30 sec in an MSE 150 W sonifier. The suspension is centrifuged at 100,000 g for 75 min. The resulting supernatant (fraction II) is little contaminated with other subcellular fractions, as less than 0.4% of the lactate dehydrogenase (a cytosol marker), 0.2% of rotenone-insensitive NADPH–cytochrome c reductase (a microsomal marker), and 5% of the adenylate kinase (a marker for the mitochondrial intermembrane space) of the whole homogenate are found in fraction II.

Normally the purification is continued on the same day, but the supernatant solution can be stored at −80° for several days without significant loss in activity.

Step 3. Ammonium Sulfate Fractionation. Unless otherwise noted, the following steps are carried out at 0–4° and all centrifugations are performed at 27,000 g for 30 min with a SS-34 rotor in a Sorvall RC-5 centrifuge.

Fraction II is gently stirred with a magnetic stirrer, and saturated ammonium sulfate solution (4.1 M, prepared from enzyme grade ammonium sulfate adjusted to pH 7.4 with 0.5 N NaOH) is added over a period of 5 min to give a final concentration of 2.5 M. After 15 min the precipitate is collected by centrifugation and successively extracted with 2.3, 2.1, 1.8, 1.6, and 1.5 M solutions of ammonium sulfate containing 15 mM HEPES, pH 7.4, 1.5 mM MgCl$_2$, 0.1 mM dithiothreitol, 0.1 mM EDTA, and 5% (v/v) glycerol (buffer A). Each extraction is allowed to stand for 15 min before the precipitate is collected by centrifugation for 20 min at 27,000 g. Fractions containing a high specific activity of palmitoyl-CoA hydrolase (usually the 2.3 and 2.1 M fractions) are pooled,

[9] E. Slinde and T. Flatmark, *Anal. Biochem.* **56**, 324 (1973).

and the enzyme is precipitated by adjusting the ammonium sulfate concentration to 2.7 M with crystalline ammonium sulfate. The precipitate is collected by centrifugation, dissolved in buffer A, and dialyzed for 2 hr against buffer A (fraction III). The dialysis tubing used has a molecular weight cutoff of about 10,000.

Step 4. DEAE-Cellulose Chromatography. The DEAE-cellulose (10–12 g of DE-52) is suspended and equilibrated in buffer A. It is then packed into a glass column (1.5 × 30 cm) and washed with 1 liter of buffer A. Fraction III is diluted with buffer A to a protein concentration of 1 mg/ml, and passed through the column at a flow rate of 0.8 ml/min. The enzyme is then eluted from the column using a linear gradient from 0.08 M to 0.5 M $(NH_4)_2SO_4$ in buffer A (total volume of 400 ml) at the above flow rate. Fractions of 5 ml are collected and their absorbance is recorded at 280 and 260 nm. Palmitoyl-CoA hydrolase activity elutes as three peaks between 0.06 and 0.15 M ammonium sulfate. The fractions with the highest specific activity (peaks I and II) (Fig. 1) are pooled and dialyzed against a mixture containing 10 mM potassium phosphate, pH 7.4, 1.0 mM MgCl$_2$, 0.1 mM dithiothreitol, 0.1 mM EDTA, and 5% (v/v) glycerol (buffer B) for 3 hr (fraction IV).

Step 5. Hydroxyapatite Chromatography. Fraction IV is applied to a hydroxyapatite column (30 × 1.5 cm) equilibrated with buffer B. The enzyme is eluted from the column using a linear gradient from 0.08 to 0.5 M KH$_2$PO$_4$, adjusted to pH 7.4, in buffer B. Palmitoyl-CoA hydrolase activity elutes as a sharp peak after approximately 115 ml of buffer at a

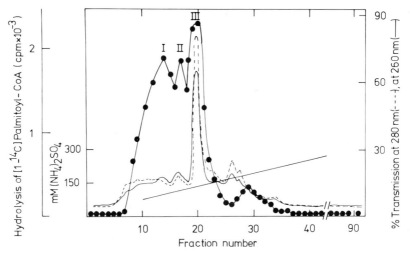

Fig. 1. Elution profile of the long-chain fatty acyl-CoA hydrolase activity of the DEAE-cellulose chromatography.

concentration of 0.18 M. The fractions containing about 75% of the eluted hydrolase activity are collected and dialyzed for 1 hr against 4 liters of a buffer containing 15 mM HEPES, pH 7.4, 0.1 mM dithiothreitol and 50 mM KCl (fraction V).

Step 6. Isoelectric Focusing. This is carried out in a column with a bed volume of 110 ml (LKB Instruments). The amount of carrier ampholyte used is 3% (w/v) in a 0% to 47% (w/v) sucrose gradient with a pH range of 3 to 10. Fraction V is diluted in the middle three-fifths of the sucrose gradient. The electrofocusing is initiated at 200 V. As the current decreases, the voltage is increased to and maintained at 800 V for 48 hr. After completion of the focusing, 1-ml fractions are collected and assayed for palmitoyl-CoA hydrolase activity. The activity elutes as a sharp peak; it is pooled, and concentrated by ammonium sulfate precipitation. The precipitate is dissolved in buffer A and desalted by passage through a Sephadex G-25 column (0.9 × 9 cm) equilibrated with buffer A. Fractions of 1 ml are collected. The palmitoyl-CoA hydrolase activity elutes as a sharp peak, and the fractions containing the major part of the activity (75%) are pooled. The final product is stabilized by dialyzing the enzyme against buffer A with 60% (v/v) glycerol (fraction VI).

Notes on Recovery and Purification. Details of the purification of the enzyme are shown in the table. The ammonium sulfate precipitation and the DEAE-cellulose chromatography show greater variation in yield of enzyme than other steps of the procedure. The distribution of long-chain fatty acyl-CoA hydrolase activity in the peaks I, II, and III (Fig. 1) can vary considerably from one experiment to another, depending on the ammonium sulfate concentration in fraction III. Although the activity of peak I and peak II represents only a portion of the total palmitoyl-CoA hydrolase eluted from the column, the specific activity of these particular fractions is considerably higher than that of the remaining peak III. Effective steps in the final stage of the purification are hydroxyapatite chromatography and isoelectric focusing (see the table). From 680 mg of rat liver mitochondria about 0.4 mg of very active enzyme protein is obtained.

Properties of the Enzyme

Stability. The enzyme retains 60–70% catalytic activity over a 1-month period when stored at −20° in buffer A containing 60% glycerol (v/v). If left for longer periods a precipitate forms, accompanied by loss of enzyme activity. Repeated freezing and thawing inactivates the enzyme. Without glycerol present, 50% of the enzyme activity is lost in 1 day at 4°.

Homogeneity. The enzyme is homogeneous as judged by the following

PURIFICATION OF LONG-CHAIN FATTY ACYL-CoA HYDROLASE FROM RAT LIVER MITOCHONDRIA[a]

Fraction	Volume (ml)	Total protein (mg)	Substrate concentration (nmol/mg protein)	Total activity (units)	Specific activity (units/mg protein)			Recovery (%)	Purification factor
					No BSA	+BSA	+Triton X-100		
I. Mitochondria	13.5	680	150[b](120)[c]	1904(1632)[d]	2.8	2.4	2.3	100(100)[d]	1 (1)[d]
II. Soluble	67.0	320	148 (120)	2208(1888)	6.9	5.9	5.1	116(116)	2.5(2.5)
III. (NH₄)₂SO₄ extract	57	74	317 (120)	800(784)	10.8	10.6	7.9	42(48)	3.9(4.4)
IV. DEAE-cellulose	56	34	385 (120)	697(870)	20.5	25.6	18.0	37(53)	7.3(10.6)
V. Hydroxy-apatite	22	2.2	676 (120)	228(818)	103.5	372.0	270.0	12(50)	37.0(155)
VI. Isoelectric focusing	4.0	0.4	2778 (120)	59(678)	147.3	1695.0	1410.0	3(41)	52.6(706)

[a] Long-chain fatty acyl-CoA hydrolase activity was assayed radiochemically as described in Assay Methods with various amounts of enzyme protein. Assays were also performed in the presence and in the absence of bovine serum albumin (BSA) and in the presence of 0.04% Triton X-100. The concentration of $[1-^{14}C]$palmitoyl-CoA was 40 μM.

[b] Protein refers to the enzyme.

[c] Protein refers to the sum of enzyme and BSA.

[d] Data in parentheses were obtained in the presence of BSA.

criteria: Polyacrylamide gel electrophoresis using 7.5, 9, and 10% gels with 3% cross-linking in the absence and in the presence of SDS reveals one major component,[3] the long-chain fatty acyl-CoA hydrolase activity coincides with the protein component in the absence of SDS,[3] and a single symmetrical peak with constant specific activity is obtained after re-chromatography on a column of Sephadex G-75.[3]

Kinetic Properties. The activity of the purified enzyme is linear with time up to 10 min and with the amount of protein up to 30 μg. Optimal activity is found at pH 7.5–8.0. At the last steps of purification (fractions V and VI) addition of bovine serum albumin increases the activity of the enzyme (see the table). This effect is due to a reduction of the substrate to protein ratio from 2778 nmol of palmitoyl-CoA per milligram of protein in the absence of serum albumin to 120 nmol per milligram of protein in the presence of serum albumin (see the table). We have shown[3] that the activity of the hydrolase is maximal when the substrate to protein ratio (serum albumin + enzyme protein) is about 120 nmol of palmitoyl-CoA per milligram of protein and that a higher as well as a lower ratio results in a lower activity. When all fractions (I to VI) are assayed at this substrate : protein ratio, the specific activity of the hydrolase (1695 units/mg of protein) represents a 706-fold enrichment of the purified enzyme over the initial extract. The table also shows that at the last steps of purification addition of Triton X-100 influences the activity of the enzyme.

Triton X-100 and bovine serum albumin appear to prevent inhibition of the enzyme by binding the inhibitory micellar form of palmitoyl-CoA, as the substrate concentration can be as high as 40 μM in the presence of a few micrograms of protein. Interpretation of the kinetic data for the hydrolase is complicated by the fact that palmitoyl-CoA is hydrophobic and detergent-like and forms micelles in solution[7] and mixed micelles in the presence of a detergent above the critical micelle concentration.[7] The kinetics of long-chain fatty acyl-CoA hydrolase is influenced both by the availability of the substrate and formation of micelles.[7,10] Monomeric palmitoyl-CoA is the substrate for the enzyme, and enzyme activity appears to be influenced by the equilibrium: micelles $\rightleftarrows$ monomer forms.[7,10]

Michaelis–Menten kinetics are observed at palmitoyl-CoA concentrations below the critical micellar concentration. Thus, with palmitoyl-CoA in a monomer form, an apparent K_m of 57 nmol per milligram of protein, or 0.8 μM, is observed. In view of the complications caused by micelle formation, the K_m does not have the usual meaning, but is mainly of practical value for the enzyme assay.

Physicochemical Properties. The isoelectric point of the enzyme is 6.0. The enzyme sediments at 2.1 S in sucrose gradients. As sedimentation of

[10] R. K. Berge, E. Slinde, and M. Farstad (submitted for publication in *J. Biol. Chem.*).

the enzyme protein after ammonium sulfate fractionation gives the same value, it is likely that the 2.1 S form is structurally stable and is not a result of proteolysis during purification. The purified enzyme is composed of one polypeptide with a molecular weight of 19,000. The Stokes' radius of the enzyme is 19 Å.[3]

Substrate Specificity. At saturating levels of substrates and in the presence of serum albumin, maximal rate of hydrolysis is found with palmitoyl-CoA, the activity decreasing upon either increasing or decreasing the length of the saturated fatty acyl-CoA chains.[3] Lauroyl-CoA, myristoyl-CoA, and stearoyl-CoA are hydrolyzed at about 20, 55, and 30% the rate, respectively, of that of palmitoyl-CoA.[3] The purified enzyme also hydrolyzes unsaturated acyl-CoA esters. The rate of oleoyl-CoA hydrolysis is only 15% of the rate of palmitoyl-CoA.[3] No hydrolysis is detectable with decanoyl-CoA or acyl-CoA of shorter chain length. No hydrolysis of palmitoyl-L-carnitine, cholesteryl oleate, tripalmitoyl glycerol, and dipalmitoylphosphatidylcholine is detectable with the purified enzyme.[3] The enzyme has neither palmitoyl-CoA synthase activity, nor carnitine palmitoyltransferase activity, nor acid and alkaline phosphatase activity.[3] The enzyme thus seems to be specific for long-chain fatty acyl-CoA esters.

[29] Long-Chain Fatty Acyl-CoA Thioesterase from *Mycobacterium smegmatis*

By KENICHI K. YABUSAKI and CLINTON E. BALLOU

A thioesterase, with a specificity for long-chain (C_{12}–C_{18}) acyl-CoA derivatives is present in extracts of *Mycobacterium smegmatis* (ATCC 356). The action of this enzyme on palmitoyl-CoA is inhibited by the polymethylpolysaccharides found in mycobacteria.[1]

Assay Procedure

Principle. The assay is based on the liberation of the thiol group of CoA, reaction (1), as measured by the increase in absorbance at 412 nm due to the reaction of free CoA with 5,5'-dithiobis(2-nitrobenzoic acid) (DTNB).[2]

[1] K. K. Yabusaki and C. E. Ballou, *J. Biol. Chem.* **254**, 12,314 (1979).
[2] G. L. Ellman, *Arch. Biochem. Biophys.* **82**, 70 (1959).

METHODS IN ENZYMOLOGY, VOL. 71

$$\text{Acyl-SCoA} + H_2O \rightleftharpoons \text{fatty acid} + \text{CoASH} \qquad (1)$$

Reagents

Tris-HCl buffer, 0.1 *M*, pH 8.0

Palmitoyl-CoA, 5 μM

Ovalbumin, 7.0 mg/ml in 0.1 *M* Tris-HCl buffer, pH 8.0

DTNB, 0.01 *M* in 0.1 *M* potassium phosphate buffer, pH 7.0

Procedure. The complete assay mixture contains 10 μmol of Tris-HCl buffer, pH 8.0, 0.05 μmol of DTNB, 0.01 μmol of palmitoyl-CoA, and 1–1.5 mU of the thioesterase in a final volume of 2.0 ml. To observe initial rates of hydrolysis of palmitoyl-CoA, it was necessary to dilute the enzyme (final step of purification) severalfold. Typically, a 5-μl aliquot of the stock thioesterase solution containing 0.125 mg of protein per milliliter was diluted to 0.2 ml with a solution of ovalbumin at 7.0 mg/ml in 0.1 *M* Tris-HCl buffer, pH 8.0. After temperature equilibration and recording a base line, the sample cuvette was removed, diluted enzyme was added from a 4-μl microcapillary pipette, and a piece of parafilm was placed over the top of the cuvette, which was mixed by several inversions and then replaced in the sample holder. This total operation could be performed within 30 sec; the initial rates of hydrolysis were linear and were recorded over a period of 5 min or longer. A split-beam recording spectrophotometer with 0.05 or 0.1 full scale absorbance slide wire capability is convenient for the above measurements. A molar extinction coefficient of 13,600 is used to quantitate the DTNB reduction.[2] Since DTNB also reacts slowly with the thiol groups of proteins, the use of a split beam spectrophotometer is advantageous in that an aliquot of the protein solution can be added to the reference cuvette, which contains all assay components except the substrate, palmitoyl-CoA. Otherwise, it is necessary to correct the initial rates of reduction of DTNB by proteins by subtraction of controls in which palmitoyl-CoA is omitted.

Units. An enzyme unit is defined as the amount of enzyme activity necessary to catalyze the hydrolysis of 1 μmol of palmitoyl-CoA per minute under the above conditions. Specific activity is expressed as units per milligram of protein.

Purification Procedure

The following procedure, summarized in the table, yields an 8000-fold purification of the acyl-CoA thioesterase.[1] All operations were performed at 4° unless otherwise stated.

Step 1. Preparation of Crude Extract. Frozen *M. smegmatis* wet cells (200 g) were thawed, suspended in 800 ml of 0.1 *M* potassium phosphate buffer, pH 7.0, and broken by two passes through a Manton–Gaulin

PURIFICATION OF *Mycobacterium smegmatis* THIOESTERASE

Step	Total activity[a] (units)	Specific activity (units/mg protein)	Purification (fold)
Extract, 37,000 g[b]	536	0.03	1
Protamine sulfate extract	421	0.05	1.7
Dialyzed $(NH_4)_2SO_4$ pellet	228	0.08	3
DEAE-cellulose chromatography	80	0.6	21
Hexylagarose chromatography	39	86	3007
BioGel A-0.5m chromatography	27	230	8040

[a] Expressed as micromoles of CoA released per minute with 5 μM palmitoyl-CoA as substrate.
[b] From 200 g of wet cell paste.

pressure cell operating at 10,000 psi. The broken cell suspension was centrifuged at 37,000 g for 1 hr.

Step 2. Protamine Sulfate Extract. To the supernatant extract was added, slowly and with stirring, a 1% solution of protamine sulfate in 0.01 M potassium phosphate buffer, pH 7.0, to give a final ratio of 0.12 mg of protamine sulfate per milligram of protein. The solution was stirred for 30 min then centrifuged at 37,000 g for 15 min. The supernatant liquid containing the enzyme activity was removed, the pellet was washed by resuspension in 0.01 M potassium phosphate buffer, pH 7.0, and the suspension was centrifuged at 37,000 g for 15 min. This second supernatant was combined with the first.

Step 3. Ammonium Sulfate Precipitation. The enzyme was precipitated from the above extract by adding ammonium sulfate to 55% of saturation. The mixture was stirred for 3 hr and then centrifuged at 16,000 g. The pellet was suspended in 0.01 M Tris-HCl buffer, pH 8.0, and dialyzed for 24 hr against 7 liters of 0.01 M Tris-HCl buffer, pH 8.0, with two changes of the buffer during that time. The protein solution was concentrated to 50 ml with an Amicon Diaflo apparatus fitted with a PM-10 membrane.

Step 4. DEAE-Cellulose Chromatography. The concentrated protein solution from step 3 was applied to a DEAE-cellulose column (2.8 × 18 cm) equilibrated with 0.01 M Tris-HCl buffer, pH 8.0, and fractions were collected by elution with the same buffer. After the absorbance at 280 nm dropped to a low value, a gradient of 0.01 to 0.5 M Tris-HCl, pH 8.0, was applied and fractions were collected. The major peak of thioesterase activity was eluted at a conductivity of 3–5 mmho. The active fractions were combined and concentrated to 5 ml, and glycerol was added to give a concentration of 10%. This solution was dialyzed against 4 liters of 0.01 M Tris-HCl, pH 8.0, containing 10% glycerol.

Step 5. Hexylagarose Chromatography. The protein solution from step 4 was applied to a hexylagarose column (1.8 × 33 cm) equilibrated with 0.01 *M* Tris-HCl, pH 8.0, containing 10% glycerol, and the column was washed with the same buffer. After the absorbance at 280 nm dropped to a low value, the column was eluted with a gradient of 0.01 to 0.4 *M* Tris-HCl, pH 8.0, containing 10% glycerol. The fractions that contained thioesterase activity were combined and concentrated to 1.5 ml.

Step 6. BioGel A-0.5m Chromatography. The concentrated protein solution from step 5 was applied to a BioGel A-0.5m column (1.8 × 114 cm) equilibrated with 0.1 *M* Tris-HCl, pH 8.0, containing 10% glycerol. Fractions of 2.5 ml were collected by elution with the same buffer, and the active fractions were combined, concentrated, and stored at 4° in the final buffer.

Properties

Purity. Preparations of thioesterase with a specific activity of 230 units per milligram of protein reveal two major components of about 22,000 and 20,000 MW and a few minor components by acrylamide gel electrophoresis under denaturing conditions.[1]

Molecular Weight. The enzyme has an apparent molecular weight by gel permeation chromatography on Sephadex G-100 of about 40,000[1] and is probably composed of two subunits of about MW 20,000 from its properties during acrylamide gel electrophoresis under denaturing conditions.

Requirements and Extent of Hydrolysis. Because of the high activity at the final stage of purification, in order to observe the initial rates of hydrolysis of palmitoyl-CoA, the enzyme has to be assayed in very dilute protein solutions. Dilution of the final stage purified enzyme with buffer alone causes a substantial loss in activity. Therefore, to stabilize the enzyme, a solution of ovalbumin[1] is used to dilute the enzyme instead of serum albumin, which is noted for its ability to bind palmitoyl-CoA.[3] The inactivation of enzyme activity by dilution with buffer alone is not observed at stages of purification prior to step 6. Thioesterase preparations from step 4, or at the final stage of purification in the presence of ovalbumin, will catalyze the complete conversion of palmitoyl-CoA to stoichiometric amounts of CoA and palmitate.

Kinetic Parameters and pH Optimum. The thioesterase has an apparent K_m and V_{max} for monomeric palmitoyl-CoA of 9 μM and 107 μmol of CoA released per minute per milligram of protein, respectively. With micellar palmitoyl-CoA as substrate, the thioesterase has an apparent K_m

[3] H. Knoche, T. W. Esders, and K. Bloch, *J. Biol. Chem.* **248**, 2317 (1973).

and V_{max} of 5 μM and 77 μmol of CoA released per minute per milligram of protein, respectively. It should be noted that the above V_{max} data were determined on the purified enzyme after about 10 months of storage at 4° in the presence of 10% glycerol. These results indicate that the enzyme has good stability under these conditions. The optimal pH for thioesterase activity is 8.0.

Specificity. At saturating levels of substrate, maximal initial rates of hydrolysis are found for palmitoyl-CoA and stearoyl-CoA. Activity decreases with decreasing length of the saturated fatty acyl chain. Approximately 5 times lower activity is observed with lauroyl-CoA over palmitoyl-CoA at equal concentrations. Very low rates of hydrolysis are observed for acyl-CoA thioesters of C_{10} chain length, and no detectable activity is observed for acyl chain lengths of C_8 or less.

Inhibitors. The mycobacterial polymethylpolysaccharides,[4] 6-*O*-methylglucose polysaccharide (MGP) and 3-*O*-methylmannose polysaccharide, which form very stable complexes with palmitoyl-CoA,[5,6] are potent inhibitors of the thioesterase-catalyzed hydrolysis of both monomeric and micellar palmitoyl-CoA. The polysaccharides appear to have no effect on the V_{max} but appear to increase the apparent K_m with both monomeric and micellar palmitoyl-CoA by altering the free substrate concentration.[1] The effect of the polymethylpolysaccharides on the thioesterase-catalyzed hydrolysis of palmitoyl-CoA appears to result solely from the binding of the substrate, not from an interaction with the enzyme itself. This is suggested from the fact that MGP had no effect on the initial rate of hydrolysis of lauroyl-CoA a substrate that, because of its short lipid chain, forms only weak complexes with polymethylpolysaccharides.[1] The same conclusion is supported by the observation that the modified MGP,[1] which binds palmitoyl-CoA poorly,[1,5] also fails to inhibit. However, the possibility that the acyl-CoA–polymethylpolysaccharide complex itself is a very poor substrate has not been ruled out.

[4] G. R. Gray and C. E. Ballou, this series, Vol. 35 [11].
[5] K. Bloch, *Adv. Enzymol.* **45**, 1 (1977).
[6] K. K. Yabusaki and C. E. Ballou, *Proc. Natl. Acad. Sci. U.S.A.* **75**, 691 (1978).

[30] β-Hydroxyacyl-CoA Dehydrase from Rat Liver Microsomes

By JOHN T. BERNERT, JR. and HOWARD SPRECHER

$$R—CHOH—CH_2—CO—SCoA \rightleftharpoons R—CH\!=\!CH—CO—SCoA + H_2O$$

The penultimate step in microsomal fatty acid chain elongation involves the dehydration of β-hydroxyacyl-CoA to the corresponding *trans*-α,β-unsaturated enoyl-CoA. The resulting intermediate is then reduced to the overall chain elongation product at the expense of NADPH. In the above reaction, R may represent a variety of alkyl or alkenyl radicals.

Assay Method

Principle. Two assay procedures have been used to monitor this reaction. With solubilized, optically clear preparations, the reaction may be followed continuously[1] by measuring the increase in ultraviolet absorption (λ_{max} = 263) resulting from the introduction of an α,β-unsaturation in conjugation with the thioester carbonyl group.[2,3] Background absorption contributed by the substrate adenine group may be minimized by the inclusion of ATP in the reference cell and by monitoring the reaction at 280 nm.[4] At the latter wavelength, $\Delta\epsilon$ = 4400.[1,5] With crude microsomal preparations a 3-^{14}C-labeled substrate is used and the reaction is monitored by thin-layer chromatography.[6]

Reagents

Potassium phosphate, 0.2 M, pH 6.5
β-Hydroxystearoyl-CoA, 10 μM
ATP, 20 μM
Triton X-100, 100 mg/ml

Procedure. The incubation mixture contains 1.0 ml of potassium phosphate buffer, pH 6.5, 0.018 ml of DL-β-hydroxystearoyl-CoA, 0.005 ml of Triton X-100, and the enzyme (0.005–0.01 unit) in a total volume of 2.0 ml. The composition of the reference mixture is identical except that ATP is

[1] J. T. Bernert, Jr., and H. Sprecher, *J. Biol. Chem.* **254**, 11584 (1979).
[2] W. Seubert and F. Lynen, *J. Am. Chem. Soc.* **75**, 2787 (1953).
[3] F. Lynen and S. Ochoa, *Biochim. Biophys. Acta* **12**, 299 (1953).
[4] H. M. Steinman and R. L. Hill, this series, Vol. 35 [18].
[5] R. M. Waterson and R. L. Hill, *J. Biol. Chem.* **247**, 5258 (1972).
[6] J. T. Bernert, Jr., and H. Sprecher, *J. Biol. Chem.* **252**, 6736 (1977).

substituted for the acyl-CoA substrate. Mixtures containing all ingredients except the enzyme are preincubated in cells with a 1-cm path length at 25° and monitored at 280 nm to assure the presence of a stable base line. The reaction is then started by addition of the enzyme aliquot to both reference and sample cells and is followed by recording the initial, linear increase in absorption.

Units. One unit of enzyme is defined as the amount required to catalyze the formation of 1 μmol of α,β-enoyl-CoA per minute under the specified conditions. The specific activity is expressed as units per milligram of protein. Protein is measured by the method of Lowry *et al.*[7]

Alternative Assay Procedure. When crude preparations are assayed prior to enzyme solubilization, activity may be determined by incubating 180 nmol of DL-β-hydroxy-[3-^{14}C]stearoyl-CoA with 0.5 mg of microsomal protein in 1.5 ml of 0.1 M potassium phosphate buffer, pH 6.5. After 1.5 min, the reaction is terminated and analyzed by thin-layer chromatography as previously described in detail.[6] The synthesis of both labeled and unlabeled substrates has also been described.[6]

Purification Procedure

All steps are carried out at 0–4°.

Step 1. Preparation of Microsomes. The pooled livers from 10 rats (ca 250 g each) are washed with 0.25 M sucrose, 0.01 M potassium phosphate, pH 7.4, and collected in three volumes of the same medium. The preparation is minced, then homogenized with a loose-fitting Teflon pestle in a Potter–Elvehjem vessel. The homogenate is centrifuged at 500 g for 10 min, and the resulting supernatant is centrifuged at 17,300 g for 15 min. The second supernatant is then centrifuged at 100,000 g for 1 hr. The drained microsomal pellet is washed free of glycogen with sucrose buffer, homogenized briefly, and resedimented at 100,000 g for 50 min. The washed microsomal pellets are suspended in sucrose buffer at 25 mg of protein per milliliter and lyophilized.

Step 2. Solubilization. After determination of the protein content of the lyophilized powder, the powder is suspended in ice-cold solubilization medium to yield a final concentration of 10 mg of protein per milliliter. The solubilization medium consists of 0.5% (w/v) sodium deoxycholate in 0.5 M KCl. The mixture is allowed to stand for 10–20 min with occasional homogenization, then centrifuged at 100,000 g for 1 hr. The clear, amber supernatants are pooled, and a sufficient volume of 20% sodium cholate is added to make the resulting solution 0.5% in cholate.

[7] O. H. Lowry, N. J. Rosebrough, A. L. Farr, and R. J. Randall, *J. Biol. Chem.* **193**, 265 (1951).

Step 3. Ammonium Sulfate Fractionation. Solid enzyme grade ammonium sulfate (0.104 g per milliliter of solution) is slowly added with continuous mixing to make the solution approximately 20% saturated. After incubation with occasional mixing for 15 min, the suspension is centrifuged at 17,300 g for 15 min and the pellet is discarded. The resulting supernatant is then made 33% saturated in ammonium sulfate by the addition of 0.065 g of salt per milliliter of solution. After incubation and centrifugation as before, the supernatant is discarded and the pellet is homogenized in 5 ml of 0.02 *M* Tris-Cl buffer, pH 7.8, containing 2% Triton X-100. This material is applied to a G-50 column (1.6 × 26 cm) and eluted with the same detergent–buffer mixture. The void volume is collected and clarified by centrifugation at 100,000 g for 1 hr.

Step 4. DEAE-Cellulose Chromatography. The supernatant from step 3 is applied to a DE-52 column (1.6 × 15 cm) that has been equilibrated with 0.02 *M* Tris-Cl, pH 7.8, 2% Triton X-100 and eluted with the same buffer, fractions of 2.8 ml each being collected. Under these conditions the enzyme is not retained and elutes in the void volume, generally in fractions 5 through 12. Recovery of total activity at this step is approximately 65–70% of that applied to the column. Attempts to recover the remaining activity with an ionic strength gradient did not yield any additional active fractions.

Step 5. CM-Cellulose Chromatography. The active fractions from the preceding step are pooled (total volume approximately 22 ml), and the pH is adjusted to 6.0 by the dropwise addition of 0.2 *N* HCl. This material is then applied to a CM-52 column (1.6 × 28 cm) that has been equilibrated with 0.01 *M* potassium phosphate buffer, pH 6.0, con-

PURIFICATION OF β-HYDROXYACYL-CoA DEHYDRASE FROM RAT LIVER MICROSOMES

Step	Total protein (mg)	Specific activity[a] (nmol min^{-1} mg^{-1})	Total activity (units)	Purification (fold)	Yield (%)
Total soluble	649.4	44	28.6	1	100
20–33% (NH$_4$)$_2$SO$_4$ pellet	195	149	29.1	3.4	100
G-50/100,000 g supernatant	159.6	135	21.6	3.1	76[b]
DEAE-cellulose	16.0	889	14.2	20.2	50
CM-cellulose	2.9	4164	12.1	94.6	42

[a] All assays were carried out with β-hydroxystearoyl-CoA using the spectrophotometric method.

[b] Although some of the activity was lost in the pellet at this stage, most of the activity lost was sacrificed during the G-50 step owing to tailing.

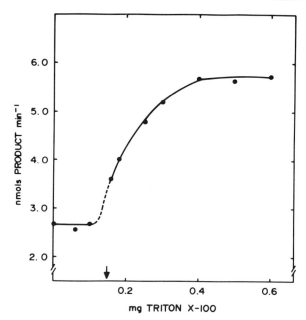

Fig. 1. Effect of Triton X-100 on the dehydrase assay. The reaction was conducted with β-hydroxystearoyl-CoA and 1.45 μg of protein by the standard assay procedure. The enzyme aliquot contained 160 μg of the detergent. The arrow in the figure represents the critical micelle concentration for 2 ml, assuming a mean molecular weight of 645 for detergent monomers.

taining 0.3% Triton X-100. A linear gradient is initiated consisting of 100 ml each of application and limit buffer (0.2 M potassium phosphate, pH 6.0, 0.3% Triton X-100). The column is developed at a flow rate of 9 ml/hr, and fractions are collected as before. Fractions possessing dehydrase activity (generally numbers 30–42) are pooled and concentrated by ultrafiltration using a PM-10 membrane (Amicon). Since the fractions contain Triton X-100 concentrations above the critical micelle concentration,[8] most of the detergent is retained in the enzyme preparation. The final detergent concentration may be estimated by determination of the absorbance of aliquots in water at 276 nm, relative to a Triton standard curve. The concentration of enzyme protein in the final preparation is quite low, and its contribution to the absorbance at 276 is insignificant.

A summary of the purification procedures is given in the table.

[8] A. Helenius and K. Simons, *Biochim. Biophys. Acta* **415**, 29 (1975).

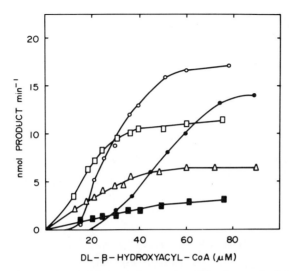

Fig. 2. Kinetics of the dehydrase reaction conducted by the standard assay procedure with 1.9 μg of enzyme protein in the presence of 650 μg of Triton X-100; β-hydroxy-12:0-CoA (●), β-hydroxy-14:0-CoA (○), β-hydroxy-16:0-CoA (□), β-hydroxy-18:0-CoA (△), and β-hydroxy-20:0-CoA (■).

Properties

Stability. The enzyme is stable in the high detergent preparation, maintaining activity for several weeks when stored at 0°.

pH Optimum. The activity range for this enzyme is fairly broad, extending from pH 5.5 to 11.0. Maximum activity occurs at pH 6.5, but additional relative maxima occur at pH 8.5 and 10.0.

Activators. The enzyme is strongly activated by Triton X-100 when the detergent is present in the assay medium at concentrations above its critical micelle concentrations. Activation is concentration dependent up to about 560 μg of total detergent concentration, after which a plateau exists (Fig. 1). By contrast, sodium deoxycholate does not activate the reaction and may be slightly inhibitory in the presence of Triton.

Substrate Specificity. Relatively little specificity is observed for saturated β-hydroxyacyl-CoA esters with chain lengths of 12–20 carbon atoms. However, the substrates of shorter chain length exhibit greatly reduced activity at low substrate concentrations as indicated in Fig. 2. This may reflect the higher critical micelle concentrations of the latter substrates.[1] When the data are linearized[9] and kinetic constants are estimated, the maximum reaction velocity is essentially constant for

[9] S. Gatt and T. Bartfai, *Biochim. Biophys. Acta* **488**, 1 (1977).

β-hydroxyl-14:0-CoA to β-hydroxy-20:0-CoA at about 20 nmol per minute, while the K_m values increase from 24 to 262 μM as the chain length is extended.[1]

The enzyme has also been found to be active with a representative polyunsaturated substrate (β-hydroxy-8,11-eicosadienoyl-CoA), and with 2-*trans*-enoyl-CoA substrates in a reverse reaction.[1] These activities copurify with β-hydroxystearoyl-CoA dehydrase throughout the procedure.

Reversibility. The reaction is readily reversible, and may be assayed with the appropriate enoyl-CoA substrate by either the spectrophotometric or radiochemical method.

[31] Terminal Enzyme of Stearoyl-CoA Desaturation from Chicken Liver

By Vasudev C. Joshi, M. Renuka Prasad, and K. Sreekrishna

The stearoyl-CoA desaturase system of liver microsomes is membrane bound and consists of NADH-cytochrome b_5 reductase (a flavoprotein), cytochrome b_5 (a hemoprotein), and a terminal cyanide-sensitive factor, the Δ^9 terminal desaturase.[1-4] The overall conversion of stearoyl-CoA to oleoyl-CoA utilizes a molecule of O_2 as an acceptor for two pairs of electrons, one pair derived from the substrate and the other pair from the reduced pyridine nucleotide, which is a required coreductant in the reaction. The Δ^9 terminal desaturase binds the fatty acyl-CoA and is also believed to be the site of oxygen activation.[4-6] The Δ^9 terminal desaturase was first purified by Strittmatter *et al.*[3] from rat liver. However, this method of purification is not applicable to the enzyme from chicken liver, possibly owing to the differences in properties of chicken and rat enzymes.[4] The purification of Δ^9 terminal desaturase from chicken liver microsomes is achieved by sequential extraction of other proteins with anionic and nonionic detergents followed by preferential solubilization of the enzyme with Triton X-100 and subsequent column chromatography on DEAE-cellulose, hydroxyapatite, and 8-aminooctyl-Sepharose.[4]

[1] P. W. Holloway and S. J. Wakil, *J. Biol. Chem.* **245**, 1862 (1970).

[2] N. Oshino, Y. Imai, and R. Sato, *J. Biochem.* (*Tokyo*) **69**, 155 (1971).

[3] P. Strittmatter, L. Spatz, D. Corcoran, M. J. Rogers, B. Setlow, and R. Redline, *Proc. Natl. Acad. Sci. U.S.A.* **71**, 4565 (1974).

[4] M. R. Prasad and V. C. Joshi, *J. Biol. Chem.* **254**, 6362 (1979).

[5] H. G. Enoch, A. Catala, and P. Strittmatter, *J. Biol. Chem.* **251**, 5095 (1976).

[6] K. Sreekrishna and V. C. Joshi, *Biochim. Biophys. Acta* **619**, 267 (1980).

Assay Method

The assay of the Δ^9 terminal desaturase activity is based on the observation that hepatic microsomes from 20-day-old chick embryos or 1-day-old chicks lack specifically the terminal component of the stearoyl-CoA desaturation system.[7,8] Such microsomes are substituted for the purified components, cytochrome b_5 reductase, cytochrome b_5, and liposomes used in the existing assays.[3,9] This assay method[8] offers considerable savings in time and effort and avoids the complexity of the assay using purified components.

The activity of Δ^9 terminal desaturase was measured at various stages of purification procedure by preincubating 5–20 μl of the fraction with 400 μg of chick embryo microsomes in the presence of 1% Triton X-100 in a final volume of 50 μl and assaying the reconstituted stearoyl-CoA desaturation system using 100 nmol of NADH, 10 nmol of [1-^{14}C]stearoyl-CoA (5 μCi/μmol), and 30 μmol of potassium phosphate, pH 7.2, in a final volume of 0.5 ml as described by Joshi et al.[8] The amount of [^{14}C]oleate formed is linear with time until about 40% of the substrate is converted to the product. The protein content of sample containing Triton X-100 is measured by the method of Wang and Smith.[10]

Purification Procedure

The content of the hepatic Δ^9 terminal desaturase is increased by dietary induction involving initial starvation for 72 hr followed by refeeding a high-carbohydrate, low-fat diet for 48 hr. Routinely, 6–8 adult chicken are used. The dietary regimen, homogenization of the liver, and subsequent centrifugation for the isolation of high-speed pellet are the same as described for the purification of fatty acid synthase.[11] The pellet is suspended in 400 ml of 0.25 M sucrose and homogenized for 30 sec using a Polytron PT-20 homogenizer (Brinkmann Instruments). The resulting homogenate is centrifuged at 14,000 g for 20 min. The supernatant solution is further centrifuged at 100,000 g for 90 min. The microsomal pellet is frozen at $-70°$ until used for isolation of the enzyme. All centrifugations during the purification are performed at 105,000 g. The various buffers used are supplemented with 10% glycerol (v/v). The entire purification procedure is carried out at 0–5°.

Microsomes are suspended in 0.05 M sodium phosphate, pH 7.2, con-

[7] A. C. Wilson, S. J. Wakil, and V. C. Joshi, *Arch. Biochem. Biophys.* **173**, 154 (1976).
[8] V. C. Joshi, A. C. Wilson, and S. J. Wakil, *J. Lipid. Res.* **18**, 32 (1977).
[9] P. Strittmatter and H. G. Enoch, this series, Vol. 52, p. 188.
[10] C. S. Wang and R. L. Smith, *Anal. Biochem.* **63**, 414 (1975).
[11] M. J. Arslanian and S. J. Wakil, this series, Vol. 35 [7a].

taining 20 mM EDTA and 0.6% sodium deoxycholate at a protein concentration of 35–40 mg/ml. The suspension is sonicated for 45 sec at 60 mW using a microprobe. After centrifugation for 90 min, the clear supernatant solution is removed with a syringe and discarded. This treatment removes all the peripheral proteins and some of the integral proteins of endoplasmic reticulum (microsomes). The loosely packed material above the glycogen pellet is suspended in 0.02 M sodium phosphate, pH 7.2, to yield a final volume equal to 60–80% of the original microsomal suspension. The suspension is again sonicated and centrifuged for 60 min. The tightly packed pellet is homogenized in 0.02 M sodium phosphate, pH 7.2, to yield a final volume equal to 40 to 50% of the starting microsomal suspension. The suspension is adjusted to a final concentration of 1.5% Triton X-100 by the addition of 10% Triton X-100 solution, sonicated for 45 sec, and allowed to stand for 30 min. The sample is centrifuged for 30 min, and the resulting pellet can be stored overnight at $-70°$ without loss of enzyme activity. The pellet is suspended in 0.02 M sodium phosphate (pH 8.0) containing 3% Triton X-100 in a final volume that is about 0.3 times the starting microsomal suspension. The sample is adjusted to a final concentration of 0.2% sodium deoxycholate and 15 mM CaCl$_2$ by the addition of sufficient 10% sodium deoxycholate and 0.8 M CaCl$_2$, respectively. The sample is sonicated for 60 sec, allowed to stand for 30 min, and then centrifuged for 30 min. The clear supernatant solution is brought to a final concentration of 30% ethylene glycol monomethyl ether and again centrifuged for 30 min. The supernatant fraction is diluted four times with 0.02 M sodium phosphate (pH 8.0) and layered on a DEAE-cellulose column (10 × 2 cm) equilibrated with the same buffer. Most of the Δ^9 terminal desaturase activity emerged unadsorbed to the column. The active fraction from DEAE-cellulose column gave at least 10 bands on polyacrylamide gel electrophoresis in sodium dodecyl sulfate (SDS), unlike the preparation from rat enzyme, which is supposed to be pure at this stage.[3,9] The enzyme fraction from the DEAE-cellulose column is loaded on a hydroxyapatite (Clarkson Chemical Co.) column (10 × 1.2 cm) equilibrated with 0.01 M sodium phosphate, pH 8, containing 0.5% Triton X-100. After adsorption on the column, stepwise elution is carried out with four bed volumes each of 0.04 M and 0.08 M sodium phosphate, pH 8, containing 0.5% Triton X-100. The enzyme is eluted with 0.12 M sodium phosphate, pH 8, containing 0.5% Triton X-100. The active fractions are pooled. Generally the enzyme preparation at this stage is nearly 90% pure as demonstrated by polyacrylamide gel electrophoresis in SDS. However, occasionally the enzyme preparation at this stage is slightly contaminated, as indicated by the presence of two protein bands with mobilities faster and slower than the major enzyme band on the polyacrylamide gel. Such preparations are further purified by passing the sample over an 8-aminooctyl-Sepharose (P-L Biochemicals) column (bed volume, 10 ml) equilibrated with 0.12 M sodium phosphate, pH 8, containing 0.5% Triton X-100 and eluted with the same buffer. The void volume contains all the desaturase activity. The enzyme is at least 90–95% pure as judged by the density of the stained band on SDS–polyacrylamide gel. The purification procedure described here results in about 120-fold increase in the specific activity of the enzyme with an overall yield of 17% from microsomes. A typical protocol for the purification procedure is shown in the table. The enzyme is stored in small aliquots at $-70°$. The enzyme preparation is stable for at least 6 months but loses 50% of its activity on repeated freezing and thawing. The enzyme precipitates if the Triton X-100 concentration is reduced below the critical micellar concentration using Bio-Beads SM$_2$ (Bio-Rad laboratories).[4]

Properties

Molecular Weight, Composition, and Immunological Properties. The molecular weight determination by SDS-polyacrylamide gel electrophoresis

PURIFICATION OF TERMINAL DESATURASE ENZYME FROM CHICKEN LIVER

Procedure	Protein (mg)	Total terminal desaturase activity (nmol/min)	Enzyme activity[a] (nmol/mg min^{-1})	Yield (%)
Microsomes	2433	2304	0.95	100
0.6% Deoxycholate pellet	819	2156	2.63	93
Buffer pellet	521	1300	2.49	56
3.0% Triton X-100 supernatant	137	1097	8.00	47
Ethylene glycol monomethyl ether supernatant	95	997	10.50	43
DEAE-Cellulose column	24	550	22.90	24
Hydroxyapatite column	05	400	80.00	17
8-Aminooctyl-Sepharose column	04	440	110.00	17

[a] The presence of Triton X-100 in the assay is required for reconstitution of the desaturation system using chick embryo microsomes. However, this detergent at the concentration present in the assay inhibits the reaction by 66%. Thus, the actual specific activities will be three times those reported in this column.

under denaturing and reducing conditions, using appropriate protein standards, revealed that the avian Δ^9 terminal desaturase has an apparent molecular weight of 33,600 ± 1200.[4] Thus, the chicken enzyme is smaller than the rat enzyme, which has a molecular weight of 55,000.[3] The purified enzyme contains 305 amino acid residues (excluding tryptophan and cysteine), and about 53% of the residues are contributed by nonpolar amino acids.[12] The purified enzyme has no detectable carbohydrate.[4] The enzyme is inhibited by both cyanide and azide, suggesting the involvement of nonheme iron in catalysis.[4] Addition of bathophenanthroline sulfonate to the enzyme in 10% SDS results in the formation of a pinkish complex with λ_{max} at 535 nm only after the addition of sodium dithionite. The iron content was calculated to be 1.25 atom per mole of enzyme. X-Ray fluorescence analysis of the enzyme indicated the presence of iron, copper, and zinc.[13] However, no interference of Cu in the estimation of Fe^{2+}–bathophenanthroline complex was noted, since the Cu^+–bathophenanthroline complex has 25-fold lower sensitivity than the iron complex. The Δ^9 terminal desaturase from rat[14] and chicken[15] is reversibly inactivated in borate buffer by the arginine-specific reagent 2,3-butanedione, and the enzyme is protected against this inactivation by the substrate stearoyl-CoA. The terminal enzyme from rat[14] and chicken[13] is also inhibited by nitration of tyrosyl residues with tetranitromethane. These results implicate the participation of both arginyl and tyrosyl residues in the active site of terminal desaturase. The antibody raised to the Δ^9 terminal enzyme from chicken inhibited the cytochrome b_5 oxidase activity of the chicken terminal desaturase but failed to inhibit the rat terminal enzyme.[4] The inhibition of the Δ^9 terminal desaturase activity in intact microsomal vesicles by antidesaturase antibody and the susceptibility of the Δ^9 terminal desaturase in intact microsomal vesicles to trypsin indicate that the active site and antigenic determinants of the Δ^9 terminal desaturase are exposed on the cytoplasmic surface of the endoplasmic reticulum.[4,12] The differences in susceptibility of the enzyme to proteolysis in native microsomal vesicles and in artificial micelles and liposomes suggest that the orientation of the Δ^9 terminal desaturase is different in the two systems.[12] Immunotitration of the Δ^9 terminal desaturase in induced liver[4] and cultured liver explants[16] indicates that the increase in

[12] M. R. Prasad, K. Sreekrishna, and V. C. Joshi, *J. Biol. Chem.* **255**, 2583 (1980).
[13] K. Sreekrishna and V. C. Joshi, unpublished results.
[14] H. G. Enoch and P. Strittmatter, *Biochemistry* **17**, 4927 (1978).
[15] V. C. Joshi, unpublished results.
[16] V. C. Joshi and L. P. Aranda, *J. Biol. Chem.* **254**, 11779 (1979).

hepatic stearoyl-CoA desaturation in neonatal development[4] and during dietary[4] and hormonal[16] induction is due to an increase in the amount of the terminal protein.

Substrate Specificity. The enzyme catalyzes desaturation of myristoyl-CoA and palmitoyl-CoA at 59 and 62%, respectively, of the rate of stearoyl-CoA desaturation and introduces a double bond at position 9 of the fatty acyl chain, irrespective of the chain length.[4] Although decanoyl-CoA and dodecanoyl-CoA are not desaturated by the enzyme, they inhibit the enzyme activity competitively. On the other hand, oleoyl-CoA, unlike decanoyl-CoA, is a poor inhibitor of the enzyme activity, and the inhibition is not reversed by increasing the substrate concentration.[4] The desaturation of myristoyl-CoA, palmitoyl-CoA, and stearoyl-CoA by chicken liver microsomes is completely inhibited by antibody to the Δ^9 terminal desaturase indicating that there is only one Δ^9 terminal desaturase in chicken liver microsomes.[4] The formation of a stable covalent acyl-enzyme intermediate during catalysis is not detected.[4] The results of substrate specificity and the competitive inhibition of desaturase by decanoyl-CoA are consistent with the hypothesis that the primary substrate recognition on the enzyme is provided by the CoA and 10 carbon atoms from the carboxyl end of the acyl chain, and methylene units beyond carbon 10 provide additional hydrophobic interaction. In the reconstituted system with chick embryo microsomes[4] or with purified components,[6] NADH and NADH–cytochrome b_5 reductase can be replace by ascorbate, although this results in reduced efficiency of desaturation.

Inhibitors. The Δ^9 terminal desaturase is inhibited by p-chloromercuribenzene sulfonate.[4] This inhibition can be reversed by β-mercaptoethanol or prevented by preincubating the enzyme with the substrate, stearoyl-CoA. These results indicate that the enzyme has active sulfhydryl groups. It is particularly noteworthy that N-ethylmaleimide and iodoacetamide, which lack the negatively charged group and the bulkier hydrophobic aromatic ring of p-chloromercuribenzene sulfonate, do not inhibit the terminal enzyme.[4] Divalent copper and its amino acid chelates at low concentration (micromolar) inhibit Δ^9 terminal desaturase activity in both chicken liver microsomes and purified system consisting of the terminal enzyme, cytochrome b_5, ascorbate, and liposomes.[6] Addition of EDTA does not relieve the inhibition, nor does preincubation of the enzyme with the copper chelate increase inhibition, suggesting that inhibition is not due to the oxidation of essential sulfhydryl groups. The presence of copper chelates during catalysis is essential for the observed inhibition, and the possibility that copper and its chelates are acting as superoxide scavengers appears plausible in view of the initial single electron reduction of desaturase iron and involvement of oxygen.[6]

Acknowledgments

This work was supported by United States Public Health Service grant HD 07516 and by a grant from Juvenile Diabetes Fondation. V. C. J. is a recipient of United States Public Health Service Research Career Development Award AM 00397.

[32] A Non-Substrate-Binding Protein That Stimulates Microsomal Stearyl-CoA Desaturase

By DEAN P. JONES and JAMES L. GAYLOR

Stearyl-CoA desaturase is a microsomal multienzyme system that catalyzes the insertion of a double bond in the 9,10 position of stearyl-CoA and other fatty acyl-CoA substrates:

$$NADH + H^+ + O_2 + \text{—}CH_2\text{—}CH_2\text{—} \rightarrow NAD^+ + 2H_2O + \text{—}CH\text{=}CH\text{—}$$

An active desaturase system has been reconstituted from solubilized and purified cytochrome b_5, cytochrome b_5 reductase, phospholipid, and a terminal oxidase.[1,2]

Enzymatic activities of microsomal oxidases are known to be affected in three ways[3]: (a) *assay-specific* stimulation, i.e., stimulation by addition of components such as catalase or serum albumin,[4-6] which occurs only under certain requisite *in vitro* assay conditions and may be of questionable physiological importance; (b) stimulation by *constitutive components* of the enzyme system,[7-10] i.e., stimulation by substrate-binding components such as ligandin, fatty acid-binding protein, and sterol carrier protein(s)

[1] H. G. Enoch, A. Catala, and P. Strittmatter, *J. Biol. Chem.* **251**, 5095 (1976).

[2] P. Strittmatter, L. Spatz, D. Corcoran, M. J. Rogers, B. Setlow, and R. Redline, *Proc. Natl. Acad. Sci. U.S.A.* **71**, 4565 (1974).

[3] D. P. Jones and J. L. Gaylor, *Biochem. J.* **183**, 405 (1979).

[4] R. C. Baker, R. L. Wykle, J. S. Lockmiller, and F. Snyder, *Arch. Biochem. Biophys.* **177**, 299 (1976).

[5] R. Jeffcoat, P. R. Brawn, and A. T. James, *Biochem. Biophys. Acta,* **431**, 33 (1976).

[6] R. Jeffcoat, P. R. Brawn, R. Safford, and A. T. James, *Biochem. J.* **161**, 431 (1977).

[7] I. Jansson and J. B. Schenkman, *Mol. Pharmacol.* **11**, 450 (1975).

[8] A. Catala, A. M. Nervi, and R. R. Brenner, *J. Biol. Chem.* **250**, 7481 (1975).

[9] T. J. Scallen, M. W. Schuster, and A. K. Dhar, *J. Biol. Chem.* **246**, 224 (1971).

[10] M. C. Ritter and M. E. Dempsey, *J. Biol. Chem.* **246**, 1536 (1971).

that may promote catalytic efficiency by improving the access of substrates to the catalytic center in the membrane; and (c) stimulation by *regulatory proteins* that neither bind substrate nor affect the K_m of the enzyme.[3,11] The purification scheme described here was developed for enrichment of a soluble, cytosolic protein that appears to function as a regulatory protein for stearyl-CoA desaturase from rat liver microsomes.

Preparation of Microsomes

Control rats (male, Sprague–Dawley, 150–250 g) are fed on rat chow ad libitum. Rats are decapitated (at 9–10 AM), and blood is removed from livers by perfusion *in situ* with 50 ml of cold 0.25 M sucrose. Livers are removed, minced in three volumes of a mixture of 50 mM potassium phosphate buffer, pH 7.4, 250 mM NaCl, 5 mM EDTA, and 1 mM glutathione, and homogenized with a TenBroeck homogenizer (8–10 strokes). Mitochondria, cell debris, and more dense organelles are removed from the suspension by centrifugation at 12,000 g for 20 min. Microsomes are sedimented from the resulting postmitochondrial supernatant fraction by centrifugation at 105,000 g for 1 hr. The resulting microsomal pellet is suspended in fresh buffer by homogenization, microsomes are sedimented again at 105,000 g for 1 hr, and the resulting washed microsomes are suspended in fresh 0.1 M potassium phosphate buffer to a final concentration of 10–20 mg of protein per milliliter. Protein is determined by the method of Lowry *et al.*[12] with bovine serum albumin as standard.

Fresh, washed microsomes should be used because unwashed microsomes exhibit diminished stimulation by proteins in the cytosolic fraction and microsomes that have been frozen become refractory to the stimulatory effect.[3] Furthermore, even microsomes that have been stored at 4° for only 24 hr show inhibition of desaturase by the cytosolic fraction rather than stimulation.

Assay of Stearyl-CoA Desaturation

Several indirect assays of desaturase are available based upon either oxidation–reduction change or oxygen consumption.[3,13,14] Since the stoichiometry of the desaturase reaction is not affected by the cytosolic

[11] J. L. Gaylor and C. V. Delwiche, *J. Biol. Chem.* **251**, 6638 (1976).
[12] O. H. Lowry, N. J. Rosebrough, A. L. Farr, and R. J. Randall, *J. Biol. Chem.* **193**, 265 (1951).
[13] N. Oshino, Y. Imai, and R. Sato, *J. Biochem.* (*Tokyo*) **69**, 155 (1971).
[14] P. Strittmatter, M. J. Rogers, and L. Spatz, *J. Biol. Chem.* **247**, 7188 (1972).

stimulatory protein,[3] all the documented assays should, in principle, be satisfactory. However, because of the many possible ambiguities of using indirect assays to measure stimulation of a specific reaction in a complex microsomal system, a direct assay is recommended. Measurement of conversion of [^{14}C]stearyl-CoA to [^{14}C]oleyl-CoA is laborious, but quite reliable. Incubations are performed at 37° in 1 ml total volume of 100 mM potassium phosphate buffer under air for 15 min. Initial conditions are as follows: 60 μM [1-^{14}C]stearyl-CoA (4 mCi/mmol), 1.2 mM NADH, and 0.7 mg of microsomal protein per milliliter. Reactions are stopped by addition of 1 ml of 10% KOH in methanol (w/v). Samples are placed on ice until addition of 0.2 ml of benzene containing, per milliliter, 3 mg of stearic acid and 3 mg of oleic acid. Saponification is performed by heating samples under reflux for 30 min at 70–80° on a steam bath. Samples are acidified with 1 ml of 3 N HCl and extracted three times with 15 ml of petroleum ether. The combined extracts are dried over sodium sulfate, and the solvent is evaporated under dry N_2. After addition of 3 ml of 14% of boron trifluoride in methanol, the samples are placed on a steam bath for 5 min. Samples are then extracted into 30 ml of petroleum ether after the addition of 10 ml of H_2O. Water is removed. The petroleum ether extract is washed once with water by a second addition of 10 ml of H_2O. After shaking, the petroleum ether phase is decanted into an extraction tube and the petroleum ether is evaporated on a steam bath. The residue is dissolved in 100 μl of benzene and 20–25 μl are spotted on an argentation thin-layer chromatography plate (250-μm plates, prepared by mixing 20 ml of H_2O, 1 g of AgNO$_3$, and 10 g of type H silica gel, spreading, and drying for 1 hr at 100°). Plates are developed in hexane : diethyl ether (7 : 1), dried, and sprayed with Rhodamine B (50 mg per 100 ml of methanol) for visualization of the esters of stearic and oleic acids. Samples are removed by scraping, then counted in a scintillation counter.

Purification of the Stimulatory Protein

Step 1. Postmicrosomal supernatant fraction is prepared (see section on Preparation of Microsomes) from 10–12 rats. This fraction is termed "crude cytosol" and gives about 100% stimulation of desaturase activity when added to microsomes at a concentration of 4–10 mg of protein per milliliter. This fraction contains several stimulatory fractions as well as inhibitory components.[3]

Step 2. Crude cytosol is fractionated by ammonium sulfate precipitation. Solid ammonium sulfate is slowly added at 0° to give 65% saturation. The mixture is stirred for 15 min and the precipitate is removed by centrifugation at 10,000 g for 10 min. Ammonium sulfate is then added to the resulting supernatant fraction to give 100% saturation. The mixture is stirred, and the protein precipitate is collected by centrifugation. The pellet is solubilized in 10 mM potassium phosphate buffer (pH 7.4) and dialyzed against the same buffer to remove residual ammonium sulfate. There is no apparent enrichment in specific activity at this step, and there is a loss of more than 80% of the cytosolic stimulatory activity. However, these observations are ascribed primarily to the removal of other stimulatory components in the crude cytosolic fraction. Therefore, yield is related to the stimulatory activity of the 65–100% ammonium sulfate fraction rather than to that in the crude cytosol; accordingly, the removal of 85% of the total protein with no change in stimulatory activity may equal as much as a six-fold purification by this fractionation step.

Step 3. The solution from step 2 is stirred for 20 min at 0° with a suspension of calcium phosphate gel (Calbiochem, A-grade, 10 mg solids per ml of suspension) at a ratio of 10 mg of protein per milliliter of gel suspension. The gel is collected by centrifugation (3000 g for 5 min). The adsorbed protein is eluted by suspending the resulting pellet in 250 mM potassium phosphate buffer (pH 7.4; approx. 1 ml for each mg of adsorbed protein) for 20 min. The mixture is then centrifuged. The supernatant fraction resulting from the second centrifugation is decanted from the pellet and dialyzed overnight at 4° against a mixture of 10 mM potassium phosphate buffer, pH 7.4, and 500 mM NaCl.

Step 4. The resulting solution is fractionated by gel filtration on a Sephadex G-100 column (2 cm × 73 cm) that has been equilibrated with a mixture of 10 mM phosphate buffer, pH 7.4, and 0.5 M NaCl. Three stimulatory peaks are observed. The most active of these elutes after about 125 ml, a volume corresponding to a molecular weight of about 26,500. The most active fractions of this peak are combined and dialyzed overnight against 10 mM Tris-HCl, pH 8.0, at 20°.

Step 5. The dialyzed solution is filtered through a PM-30 membrane (Diaflow ultrafilter; Amicon Corp.) to remove hemoglobin.

Step 6. The resultant clear, colorless solution is applied to a DEAE-Sephadex column (1.5 cm × 50 cm) and developed with 420 ml of a linear NaCl (0 to 500 mM) gradient in 10 mM Tris-HCl buffer (pH 8.0 when measured at 20°). Two stimulatory fractions are present that elute at 250 mM NaCl and 310 mM NaCl. The second peak is nearly homogeneous as judged by polyacrylamide gel electrophoresis in the presence of sodium dodecyl sulfate.

Characteristics

The procedure results in an 1100-fold purification (Fig. 1). At all stages of purification the extent of stimulation is proportional to protein concentration. Half-maximum stimulation observed with purified cytosolic pro-

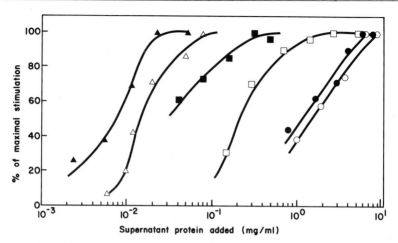

FIG. 1. Stimulation of desaturase activity by the cytosolic protein(s) at various stages of resolution. Fractions are crude cytosol (○), 65–100% saturated $(NH_4)_2SO_4$ (●), calcium phosphate gel (□), Sephadex G-100 (■), PM-30 (△), and DEAE-Sephadex (▲). Values are normalized to percentage of maximal stimulation for individual fractions by dividing observed stimulation at specified protein concentrations by maximal stimulation for that fraction. Reproduced from Jones and Gaylor,[3] with permission of the *Biochemical Journal*.

tein occurs at 8 μg/ml, and maximal stimulation occurs at about 20 μg/ml when incubated with 0.7 mg of microsomal protein per milliliter.

The stimulatory fraction has an absorption spectrum characteristic of protein with no unusual properties. No absorption bands are present in the visible range when measured at concentrations up to 0.113 mg of protein per milliliter. Chromatography on a calibrated Sephadex G-100 column confirmed initial data on the crude fraction that the active fraction is eluted with an approximate molecular weight of 26,500. Examination by polyacrylamide gel electrophoresis in the presence of sodium dodecyl sulfate revealed a major band of approximately 27,000 MW.

The mechanism of the stimulation of activity is not known. The protein exhibits neither catalase nor glutathione peroxidase activity. Addition of the cytosolic protein produces no effect on the desaturase reaction stoichiometry; the proportion of O_2 consumed/NADH oxidized/stearyl-CoA desaturated remains 1 : 1 : 1. Because the K_m for stearyl-CoA is unchanged by addition of the cytosolic protein, no change in substrate affinity is suggested. This, as well as direct measurements, suggests that the protein does not bind substrate. Addition of the cytosolic protein produces no effect on desaturase inhibition by oleyl-CoA; this indicates that the protein does not act by binding acyl-CoA and thus relieving apparent product inhibition.

Comments

The degree of stimulation depends on both the presence of the stimulatory protein and the condition of the microsomes. Freshly prepared, washed microsomes give maximal stimulation. Treatments that disrupt membrane structure, such as freezing or inclusion of detergents, eliminate the stimulatory effect. Thus, as reported for other microsomal oxidases that are affected by cytosolic proteins,[15] an intact membrane is needed for action of the regulatory protein.

Several assay-specific factors can modulate the desaturase activity. These must be carefully considered for interpretation of results associated with apparent stimulation. In addition, the presence of inhibitory factors in the crude cytosol suggests that true regulation of desaturase may be dependent upon several interacting components including other proteins and endogenous metabolites.

Acknowledgment

Supported in part by Research Grant AM21336 from the National Institute of Arthritis, Metabolism, and Digestive Diseases, U.S. Public Health Service. A detailed report can be found in Jones and Gaylor.[3]

[15] T. Ono and K. Bloch, *J. Biol. Chem.* **250**, 1571 (1975).

[33] Enzymatic Reduction of Fatty Acids and α-Hydroxy Fatty Acids

By P. E. KOLATTUKUDY, LINDA ROGERS, and J. D. LARSON

Biological reduction of fatty acids provides fatty alcohol for the synthesis of ether lipids and wax esters. Even though this reduction must be a two-step process involving an aldehyde intermediate, such an intermediate does not accumulate in free form in cases involving microsomal reductases,[1-3] but in cases where soluble enzyme preparations catalyze the

[1] P. E. Kolattukudy, *Biochemistry* **9**, 1095 (1970).
[2] L. Wang, K. Takayama, D. S. Goldman, and H. K. Schnoes, *Biochim. Biophys. Acta* **260**, 41 (1972).
[3] F. Snyder and B. Malone, *Biochem. Biophys. Res. Commun.* **41**, 1382 (1970).

reduction, free aldehyde can be detected and the enzyme preparation[4-6] can be resolved into two reductases catalyzing the following reactions:

acyl-CoA reductase:

$$R-C\overset{O}{\underset{SCoA}{}} + \begin{matrix} NADH \\ (NADPH) \end{matrix} + H^+ \longrightarrow R-C\overset{O}{\underset{H}{}} + \begin{matrix} NAD \\ (NADP) \end{matrix}$$

aldehyde reductase

$$R-C\overset{O}{\underset{H}{}} + NADPH + H^+ \longrightarrow RCH_2OH + NADP$$

Assay Method

Principle. With soluble enzyme preparations the acyl-CoA reductase, when resolved from aldehyde reductase, can be measured spectrophotometrically by following acyl-CoA-dependent decrease in absorbance at 340 nm with either NADH or NADPH, depending upon the source of the enzyme. Similarly, aldehyde-dependent decrease in absorbance at 340 nm is used to measure aldehyde reductase activity. These spectrophotometric assays are not suitable for measuring the activity in crude preparations. In such cases tracer assays with radioactive substrates are used and the radioactivity in the product isolated by thin-layer chromatography is measured.

In cases where the reduction of fatty acyl-CoA is catalyzed by membranes, the two steps cannot be measured separately. Instead the amount of the final product, fatty alcohol, generated from radioactive acyl-CoA is measured after thin-layer chromatographic isolation of the product. Since the membrane preparations also catalyze esterification of the fatty alcohol with acyl-CoA, the resulting esters have to be cleaved prior to isolation of the alcohols.

Uropygial Glands of White-Crowned Sparrow (*Zonotrichis leucophrys*)

Reagents

NADPH, 110 mM in buffer
Bovine serum albumin (BSA), 20 mg/ml of the buffer
[1-^{14}C]Palmitoyl-CoA, 0.5 mM

[4] J. I. E. Day, H. G. Goldfine, and P. O. Hagen, *Biochim. Biophys. Acta* **218**, 179 (1970).
[5] P. E. Kolattukudy, *Arch. Biochem. Biophys.* **142**, 701 (1971).
[6] R. C. Johnson and J. R. Gilbertson, *J. Biol. Chem.* **247**, 6991 (1972).

Enzyme preparation, 0.2–0.8 mg of protein per milliliter
Citrate-phosphate buffer, pH 5.0, 0.1 M containing 0.5 mM
dithioerythritol (DTE) and 1 mM MgCl$_2$

Assay Procedure

To a small test tube the following reagents are added so that the indicated final concentrations are obtained: BSA, 1 mg; NADPH, 11 mM; [1-^{14}C]palmitoyl-CoA, 100 μM; enzyme (10–40 μg of protein) in a final volume of 0.5 ml of citrate (20 mM)–phosphate (40 mM) buffer, pH 5.0. As a control, a test tube containing 0.1 ml of enzyme is heated in a steam bath for 10 min; after cooling the tube in an ice bath, the other reagents are added. The test tubes are incubated at 30° for 10 min. in a water bath shaker, and the reaction is stopped by the addition of a few drops of 6 N HCl followed by 25 ml of a 2 : 1 mixture of chloroform (CHCl$_3$) and CH$_3$OH. The mixture is thoroughly shaken with 6 ml of water in a separatory funnel, and the bottom chloroform layer is collected and evaporated to dryness in a 50-ml round-bottom flask with an evaporator. To the dry residue 10 ml of 14% BF$_3$ in absolute methanol is added and refluxed for 2 hr. The reaction mixture is transferred to a 125 ml separatory funnel containing 20 ml of water. The round-bottom flask is rinsed twice with 5 ml of CHCl$_3$, and the CHCl$_3$ wash is transferred into the separatory funnel. The mixture is extracted three times with 40 ml of CHCl$_3$, and the pooled CHCl$_3$ extract is evaporated to dryness with a rotary evaporator. The residue is transferred to a test tube with CHCl$_3$, and the solvent is removed with a stream of N$_2$. The residue is dissolved in 0.25 ml of CHCl$_3$, and 50 μl of the solution is applied to a thin-layer plate (20 $\times$ 20 cm, 0.5 mm-thick silica gel G) on a 1.5-cm band to which 50 μg of hexadecanol are applied prior to the application of the enzymatic product. After developing the chromatogram in hexane : ethyl ether : formic acid (40 : 10 : 1 v/v/v), the plate is air dried and sprayed lightly with a 0.1% ethanolic solution of 2′, 7′-dichlorofluorescein. Under ultraviolet light the hexadecanol region (with a yellow fluorescence) is marked, and the silica gel from this region is scraped into a counting vial. After thoroughly mixing the gel with 15 ml of counting fluid (0.4% Omnifluor dissolved in toluene containing 30% ethanol), ^{14}C is assayed with a scintillation spectrometer. From the specific activity of the [1-^{14}C]palmitoyl-CoA, the number of nanomoles of hexadecanol generated can be calculated after subtracting the control value obtained with boiled enzyme.

Enzyme Preparation

Uropygial glands are excised from freshly killed (decapitation) white-crowned sparrows. After removal of the adhering skin, fat, and muscle with a razor blade, the glands are sliced and the excess oil is removed by gently blotting with filter paper.[7] The gland slices are ground in a 2-ml Ten Broeck homogenizer in 0.05 M sodium phosphate buffer, pH 6.5, containing 0.25 M sucrose, 1 mM MgCl$_2$ and 0.5 mM DTE (0–4°). Upon centrifugation of the homogenate at 1000 g for 5 min, the cell debris sediments and excess fat floats to form a fat pad. The supernatant is decanted through four layers of cheesecloth and centrifuged at 27,000 g for 20 min. The pellet is suspended in 0.1 M citrate–phosphate buffer, pH 5.0, containing 1 mM DTE and 1 mM MgCl$_2$, and centrifuged at 27,000 g for 20 min; the resulting pellet is suspended in the same buffer (1.25 ml/gland) and used as the enzyme source. About 80% of the acyl-CoA reductase of the gland is contained in this fraction, which upon electron microscopy appears to consist of clusters of membranes similar to those found around the oil drops in the gland.[7,8] Sucrose density gradient centrifugation showed that the reductase is located in the endoplasmic reticulum fraction.

Properties of the Particulate Acyl-CoA Reductase

Specificity. NADPH is highly preferred over NADH, and the reductase is specific for the hydride from the B-side of the nicotinamide ring.[9] This cofactor specificity is also observed with particulate preparations from the uropygial glands of goose and pheasant,[8] mouse preputial gland tumors,[3] and *Mycobacterium*.[2] The chain length specificity for fatty acyl-CoA is not known.

General Properties. The acyl-CoA reductase from the sparrow shows a rather sharp pH optimum at about 5.0.[9] The rate of palmitoyl-CoA reduction is constant for about 15 min of incubation. The protein concentration dependence of the rate of reduction is sigmoidal, presumably because at low protein concentration palmitoyl-CoA binds the membrane and partially inactivates the enzyme (detergent effect). However, this problem can be overcome and linear dependence on protein concentration can be obtained by the inclusion of 1 mg of BSA in the reaction mixture at least up to 60 μg of microsomal protein per milliliter. Apparent K_m values for palmitoyl-CoA (neglecting the fact that higher concentrations of palmitoyl-CoA inhibit the enzyme) and NADPH are 0.3 mM and 3 mM, respectively.

Inhibitors and Activators. p-Chloromercuribenzoate and N-ethyl-

[7] P. E. Kolattukudy, this series, Vol. 72 [60].
[8] J. S. Buckner and P. E. Kolattukudy, *in* "Chemistry and Biochemistry of Natural Waxes" (P. E. Kolattukudy, ed.), p. 147. Elsevier, Amsterdam, 1976.
[9] P. E. Kolattukudy and L. Rogers, *Arch. Biochem. Biophys.* **191**, 244 (1978).

maleimide almost completely inhibit the enzyme at 0.5 mM, and iodoacetamide is a less potent inhibitor. Inclusion of ATP and CoA gives some stimulation, and NADH with ATP severely inhibits the reaction.

Euglena gracilis

Wax esters constitute about 50% of the total lipids of dark-grown *E. gracilis*.[10,11] Therefore this organism is a rich source of acyl-CoA reductase.

Reagents

Potassium phosphate buffer, 0.1 M, pH 6.5, containing 1.0 mM MgCl$_2$ and 0.5 mM DTE
[1-^{14}C]Palmitoyl-CoA, 1.0 mM
NADH, 20 mM in the buffer
Bovine serum albumin, 10 mg/ml in the buffer
Enzyme preparation adjusted to 2 mg of protein per milliliter with the buffer.

Assay Procedure

Into a small test tube the following reagents are transferred: 0.77 ml of potassium phosphate buffer, 0.05 ml of enzyme, 0.1 ml of bovine serum albumin, 0.05 ml of NADH. The reaction is initiated by the addition of 0.03 ml of [1-^{14}C]palmitoyl-CoA, and the reaction mixture is incubated at 30° for 15 min with shaking. The reaction is terminated by the addition of 25-fold excess of a 2 : 1 mixture of CHCl$_3$ and CH$_3$OH, and the subsequent steps are identical to those described above for the uropygial gland.

Enzyme Preparation

Growth of the Organism. Euglena gracilis is maintained on slants composed of 0.5% proteose peptone, 0.2% yeast extract, 0.1% beef extract, 0.3% glucose, and 2% agar-agar at 26° with weekly transfers. Cells are transferred from the slants to a liquid medium containing 50 ml of the same ingredients, but without the agar. The 50-ml cultures are grown at 26° for 1 week in the dark and 2-ml aliquots of this culture are transferred

[10] A. Rosenberg, *Biochemistry* **2**, 1148 (1963).
[11] P. F. Guehler, L. Peterson, H. M. Tsuchiya, and R. M. Dodson, *Arch. Biochem. Biophys.* **106**, 294 (1964).

to flasks containing 1.0 liter of the above medium and grown at 26° for 12–14 days in the dark with shaking.

Isolation of Microsomes. Unless otherwise specified, the remaining steps are carried out at 0–4°. Six liters of *E. gracilis* Z cells are harvested by centrifugation at 4000 g for 10 min. The sediment is resuspended in the homogenizing buffer containing 0.3 M sucrose, 0.1 M potassium phosphate, 1.0 mM magnesium chloride, and 0.5 mM dithioerythritol, pH 7.0, and centrifuged at 4000 g for 10 min. The sediment is resuspended in 100 ml of homogenizing buffer and passed through a French press twice under 4500 psi. The resulting homogenate is centrifuged at 17,000 g for 10 min, and the sediment is discarded. The supernatant is centrifuged again at 17,000 g for 10 min. The supernatant fraction is centrifuged at 105,000 g for 60 min. The sediment is resuspended in 16 ml of homogenizing buffer and centrifuged at 105,000 g for 60 min. The supernatant is discarded and the microsomal pellet is resuspended in 6–8 ml of the homogenizing buffer. These microsomal preparations, containing the bulk of the acyl-CoA reductase activity of the cell,[12] can be stored at −20°.

Properties of the Enzyme

Stability. The enzyme preparation is stable for several months when stored in the homogenizing buffer at −20°.

General Properties. Optimal pH with palmitoyl-CoA is 6.0 whereas with fatty acid, ATP, and CoA the pH optimum is at around 6.5. Reduction of palmitoyl-CoA is linear for 15 min of incubation and from 20 to 100 μg of protein per milliliter. Below 20 μg of protein per milliliter, the rate is not linear, presumably because the substrate, owing to its detergent properties, inactivates the enzyme. A sigmoidal substrate saturation pattern is observed for palmitoyl-CoA, but typical Michaelis-Menten kinetics is observed when 1 mg of bovine serum albumin is included in the assay. Even with 1 mg of albumin, higher concentrations of palmitoyl-CoA (>30 μM) cause inhibition. Ignoring the inhibitory concentration, an apparent K_{m} of 130 μM is calculated. K_{m} for NADH is about 0.25 mM.

Products. The microsomal preparation catalyzes reduction of fatty acyl-CoA to alcohol and esterification of the fatty alcohol with acyl-CoA. Therefore the fatty alcohols generated by the reductase are found both in free and esterified form.[13] Free fatty aldehyde does not accumulate but aldehyde intermediate can be trapped using carbonyl reagents such as phenylhydrazine.[1]

Substrate Specificity. This reductase is quite specific for NADH. With fatty acids as substrates (with ATP and CoA) it shows maximum rates

[12] A. A. Khan and P. E. Kolattukudy, *Biochemistry* **12**, 1939 (1973).
[13] A. A. Khan and P. E. Kolattukudy, *Arch. Biochem. Biophys.* **158**, 411 (1973).

with C_{14} and C_{16} acids and much less activity with shorter and longer acids.[1] The enzyme preparation also catalyzes reduction of hexadecanal with NADH as the preferred reductant, although in this case NADPH also shows considerable ($\sim$75%) activity.

Inhibitors. Thiol-directed reagents such as N-ethylmaleimide (4×10^{-4} M) and p-chloromercuribenzoate (2.5×10^{-4} M) almost completely inhibit the enzyme. Triton X-100, Tween-20, and deoxycholate inhibit the reductase.[14]

Solubilization. Sodium chloride, 3 M, solubilizes the reductase.[14] Removal of the salt results in precipitation of the enzyme, but the precipitated enzyme catalyzes acyl-CoA reduction.

Brassica oleracea

The two steps involved in acyl-CoA reduction to fatty alcohol are catalyzed by two protein fractions, an acyl-CoA reductase and an aldehyde reductase.

Assay for Acyl-CoA Reductase

Reagents

Citrate–phosphate buffer, 0.1 M, pH 6.5
NADH, 3 mM
Palmitoyl-CoA, 0.5mM

Assay Procedure. Procedures similar to that described for avian enzyme are used with NADH in place of NADPH. In addition the following spectrophotometric assay can be used. To a cuvette (1.4 ml capacity) the following reagents are added: 0.1 M citrate–phosphate buffer, pH 6.5, 0.9 ml; NADH, 0.1 ml (0.3 μmol); enzyme, 0.1 ml; palmitoyl-CoA, 0.1 ml (50 nmol). Control reaction mixture contains all reagents except palmitoyl-CoA. Absorbance change at 340 nm is recorded. In crude preparations NADH oxidation is observed without the addition of palmitoyl-CoA, and this oxidation can be inhibited by the CoA ester. In such cases the tracer assay should be used.

Assay for Aldehyde Reductase

Reagents

Citrate-phosphate buffer, 0.1 M, pH 5.6, containing 5 mM β-mercaptoethanol and 1 mM $MgCl_2$

[14] A. A. Khan and P. E. Kolattukudy, *Arch. Biochem. Biophys.* **170**, 400 (1975).

[1-^{14}C]Palmitaldehyde, 1 mM
NADPH, 10 mM

Synthesis of Labeled Fatty Aldehyde. In a 10-ml round-bottom flask [1-^{14}C]palmitic acid (0.5 mCi, 25 μmol) is refluxed with 1 ml of thionyl chloride for 1 hr and the excess reagent and volatile products are removed by evaporation under reduced pressure with a rotary evaporator. The resulting palmitoyl chloride is dissolved in 1 ml of acetone freshly distilled from Drierite. A 50-ml Erlenmeyer flask containing 200 mg of 5% Pd on BaSO$_4$, 10 ml of freshly distilled acetone, and a magnetic stirring bar is tightly closed with a serum cap through which two hypodermic needles are inserted. A Teflon tube is attached to the inlet needle in such a way that H$_2$ bubbles through the reaction mixture. Upon bubbling H$_2$ for 30 min with stirring (magnetic) the catalyst turns black, indicating that it is activated. The palmitoyl chloride solution is injected into the reaction vessel, and hydrogen is continuously bubbled through the mixture for 30 min with stirring. The reaction mixture is transferred to a 125-ml separatory funnel, 30 ml of water are added, and the products are extracted twice with 40 ml of chloroform. The pooled chloroform extract is evaporated to dryness with a rotary evaporator, the residue is applied to a thin-layer plate (20 × 20 cm, 1 mm-thick layer of silica gel G), and the chromatogram is developed with hexane : ethyl ether : formic acid (40 : 10 : 1 v/v/v). The labeled aldehyde ($R_f \simeq 0.7$) is recovered from the silica gel with ethyl ether. If all the operations are done with as little exposure to O$_2$ as possible, >80% radiochemical yield of pure aldehyde is obtained. If the product is to be stored for more than a few weeks the thin-layer chromatographic purification should be repeated just prior to using the stored aldehyde. An appropriate amount of hexadecanal is dissolved in a few milliliters of ethyl ether containing Tween-20 (0.5 mg/μmol of substrate is required to get a good emulsion that would not interfere with the spectrophotometric assay). After evaporating the solvent off with a stream of N$_2$ an appropriate amount of the citrate–phosphate buffer is added and subjected to ultrasonic treatment.

Assay Procedure

Tracer Assay. To a small test tube the following reagents are added: 0.1 M citrate-phosphate buffer, pH 6.5, containing 5 mM β-mercaptoethanol and 1 mM MgCl$_2$, 0.5 ml; 1 mM labeled palmitaldehyde, 0.3 ml; 10 mM NADPH, 0.1 ml; enzyme, 0.1 ml. The test tube is immediately stoppered with a serum cap, and the air is replaced with oxygen-free nitrogen using hypodermic needles as inlet and outlet. After thorough mixing of the reaction mixture, the test tube is incubated at 30° for 30 min in a water bath shaker. The reaction is stopped by mixing the reaction mixture with

25 ml of a 2 : 1 mixture of chloroform and methanol in a separatory funnel. (When a series of assays are done simultaneously, a few milliliters of the solvent mixture can be quickly introduced into each reaction mixture with a squeeze bottle; after mixing well, the test tubes are transferred to an ice bath until the extraction of products.) After the addition of 6 ml acidified (a few drops of dilute HCl) water, the mixture is thoroughly mixed and the bottom layer is transferred to a test tube. The solvent is removed with a stream of N_2 while the test tube is kept in warm (~50°) water. The residue is dissolved in 0.25 ml of $CHCl_3$, a 0.05-ml aliquot is subjected to thin-layer chromatography, and the radioactivity in hexadecanol is assayed as described above for the uropygial gland enzyme.

Spectrophotometric Assay. The following reagents are added into a spectrophotometric cuvette: 0.1 M citrate-phosphate buffer, pH 5.6, 0.5 ml; 1 mM hexadecanal, 0.3 ml; 3 mM NADPH, 0.1 ml; enzyme, 0.1 ml. Absorbance change at 340 nm is recorded with a spectrophotometer. The initial linear rate is used to calculate the enzyme activity.

Enzyme Preparation

Acetone Powder. Leaf blade tissue is collected by removing the midribs of leaves (second and third leaf from the apex) from greenhouse-grown broccoli plants. After washing the tissue with cold (−20°) acetone the tissue is homogenized with 10 volumes of cold acetone for 1 min in a explosion-proof Waring blender. The powder collected by filtration with a Büchner funnel is rehomogenized twice for 30 sec each with 60% of the original volume of cold acetone. The final powder collected in a Büchner funnel is rinsed with cold acetone and transferred to 3 MM filter paper. The powder is stirred with a spatula until dry and stored at −20° in sealed jars until required. The yield of acetone powder is about 10% of the fresh weight of the leaf tissue.

Partial Purification. Acetone powder (2 g) is stirred with a magnetic stirrer for 1 hr with 75 ml of 0.1 M potassium phosphate buffer, pH 7.0, containing 5 mM β-mercaptoethanol and 1 mM $MgCl_2$ in an ice bath. The mixture is centrifuged at 15,000 g for 15 min, and the supernatant is decanted through 4 layers of cheesecloth. To the supernatant powdered ammonium sulfate is slowly added with stirring to reach 25% saturation; after an additional 30 min of stirring, the precipitate, collected by centrifugation at 12,000 g for 10 min, is discarded. The supernatant is 60% saturated with ammonium sulfate in a similar manner, and the precipitate collected by centrifugation is suspended in 0.1 M citrate–phosphate buffer, pH 5.6, containing 5 mM β-mercaptoethanol and 1 mM $MgCl_2$. A 3–4 ml aliquot containing about 250 mg of protein is applied to a 2 × 45 cm Sephadex G-100 column equilibrated with the same buffer. Fractions (4 ml) are collected at 0.5 ml/min flow rate, and assayed for acyl-CoA reductase and aldehyde reductase; the fractions containing the enzyme activities are pooled so that the fractions representing the overlap between the larger acyl-CoA reductase ($\sim$70,000 MW) and the smaller aldehyde reductase ($\sim$35,000 MW) are excluded.

Properties of the Enzyme

Nucleotide Specificity. Acyl-CoA reductase prefers NADH whereas aldehyde reductase prefers NADPH.[15]

General Properties of Aldehyde Reductase. The enzyme shows a rather sharp pH optimum near 5.6. Rate of reduction of hexadecanal is linear to 2 mg of protein per milliliter and 90 min incubation time. In the spectrophotometric assay a linear rate is observed for at least 15 min. Apparent K_m values are 70 μM and 50 μM for hexadecanal and NADPH, respectively.

Inhibitors. SH-directed reagents, such as iodoacetic acid, iodoacetamide, N-ethylmaleimide, and p-chloromercuribenzoate, inhibit the reductase.

Other Sources. Acyl-CoA reductase and aldehyde reductase, similar to those originally described in the plant systems, have been obtained from bovine cardiac muscle[6] and *Clostridium butyricum.*[16]

Reduction of α-Hydroxy Fatty Acid by Microsomes from the Uropygial Gland of White-Crowned Sparrow

Alkane-1,2-diols are produced by avian and mammalian sebaceous glands.[17,18] Such diols are generated by reduction of α-hydroxy acids by endoplasmic reticulum.[9,19] Presumably because the clusters of the membranes around the oil droplets catalyze this reduction a substantial proportion (40%) of the enzymatic activity sediments at 27,000 g.

Enzyme Assay

Principle. α-Hydroxy[1-^{14}C]palmitate is incubated with the appropriate cofactors, and the diol generated is chromatographically isolated

[15] P. E. Kolattukudy, *Arch. Biochem. Biophys.* **142**, 702 (1971).
[16] J. I. E. Day and H. Goldfine, *Arch. Biochem. Biophys.* **190**, 322 (1978).
[17] J. Jacob, *in* "Chemistry and Biochemistry of Natural Waxes"(P. E. Kolattukudy, ed.), p. 93. Elsevier, Amsterdam, 1976
[18] D. T. Downing, *in* "Chemistry and Biochemistry of Natural Waxes" (P. E. Kolattukudy, ed.), p. 17. Elsevier, Amsterdam, 1976.
[19] P. E. Kolattukudy, *Biochem. Biophys. Res. Commun.* **49**, 1376 (1972).

and assayed for ^{14}C. It is advisable to cleave any esters generated during the assay, prior to the isolation of the diols.

Reagents

α-Hydroxy[1-^{14}C]palmitate, 50 μM containing 1.5 mg of Tween-20 per milliliter

Glucose-6-phosphate dehydrogenase (type IX, Sigma) 250 units/ml

Cofactor solution containing 60 mM ATP, 5 mMCoA, 10 mM NADPH, and 170 mM glucose-6-phosphate

Sodium phosphate buffer, 0.05 M, pH 6.5, containing 1 mM MgCl$_2$ and 0.5 mM DTE

Synthesis of α-Hydroxy[1-^{14}C]palmitate. [1-^{14}C]Palmitic acid (50 μCi, 55 Ci/mol) is transferred to a Pyrex tube ($\sim$4 mm i.d.) in 200 μl of diethyl ether, and the solvent is removed with a stream of N$_2$. About 250 μl of an 8% solution of Br$_2$ in thionyl chloride are added, and the tube is sealed with a flame after the solution is thoroughly cooled in a Dry Ice–propanol bath. After heating the reaction tube in the dark at 85° for 12–14 hr, the tube is opened and the excess Br$_2$ and thionyl chloride are evaporated off with a stream of N$_2$. The residue is transferred into a 50-ml two-necked round-bottom flask in 0.5 ml of tetrahydrofuran, 10 ml of 1N KOH are added and refluxed for 16 hr with constant but slow bubbling of N$_2$ through the reaction mixture. The reaction mixture is transferred to a separatory funnel acidified with 6 N HCl, and the products are extracted three times with 50 ml of CHCl$_3$. The pooled chloroform extract is evaporated to dryness in a rotary evaporator and the residue subjected to thin-layer chromatography on silica gel G with diethyl ether: hexane: formic acid (50: 50: 2 v/v/v) as the solvent. The labeled component, which shows an R_f (0.32) identical to that of authentic α-hydroxypalmitate, is recovered and rechromatographed with the same solvent. The final product contains 50–60% of the ^{14}C contained in the starting material.

Assay Procedure. Into a small test tube the following reagents are added: sodium phosphate buffer, 0.05 M, pH 6.5, containing 0.5 mM DTE and 1 mM MgCl$_2$, 0.7 ml; α-hydroxy[1-^{14}C]palmitate solution, 0.1 ml; cofactor solution, 0.1 ml; glucose-6-phosphate dehydrogenase, 1 unit; enzyme, 0.1 ml. The reaction mixture is incubated for 60 min at 30° in a water bath shaker. Termination of the reaction, extraction of the products, treatment of the products with 14% BF$_3$ in methanol, and isolation of the resulting products are done exactly as described above for acyl-CoA reductase from the same tissue. The final residue is applied to thin-layer plates (20 × 20 cm, 0.5 m-thick layer of silica gel G) together with 25 μg of authentic hexadecane-1,2-diol (produced by LiAlH$_4$ reduction of α-hydroxyhexadecanoic acid). The chromatogram is developed with hexane: diethyl ether: formic acid (25:25:1 v/v/v) and sprayed with a 0.1% ethanolic solution of 2',7'-dichlorofluorescein. The silica gel from the alkane-1,2-diol region (located under ultraviolet light) is scraped into a counting vial, mixed with 10 ml of 0.4% solution of Omnifluor (New England Nuclear) in toluene containing 30% ethanol and assayed for ^{14}C in a scintillation counter. A boiled enzyme control is run and the control value is subtracted from the experimental value. The number of nanomoles of diol is calculated from the specific activity of the labeled α-hydroxypalmitate.

Enzyme Preparation

The uropygial gland of white-crowned sparrow is excised and homogenized; a 27,000 g (20 min) pellet is prepared as described above for acyl-CoA reductase. The pellet is suspended in 0.05 M sodium phosphate buffer, pH 6.5, containing 0.5 mM DTE and 1 mM $MgCl_2$ and centrifuged at 27,000 g for 20 min. The washed pellet is resuspended in the same buffer (0.5–0.75 ml/gland).

Properties of the Particulate Enzyme Preparation

Cofactors. NADPH is the preferred reductant, whereas NADH is ineffective; inclusion of an NADPH regenerating system in the reaction mixture gives a higher yield of reduction product.[9] Hydride from the B-side of NADPH is preferentially used for α-hydroxy acid reduction; ATP and CoA are essential for reduction of the α-hydroxy acid, strongly suggesting that the CoA ester is the true substrate for the reductase. However, α-hydroxyacyl-CoA has not been tested as a substrate.

General Properties. The pH optimum for α-hydroxy acid reduction is about 6.5 with a sharp decline in activity within 0.5 pH unit on either side of this optimum. Since reduction involves activation of the carboxyl group, it is possible that the optimal pH observed represents a compromise between the pH optimum for activation and that for reduction. Therefore, with α-hydroxyacyl-CoA as the substrate, the pH optimum should be determined. The reduction is linear up to 60 min incubation time and 1 mg of protein per milliliter.

Subcellular Localization. Sucrose density gradient centrifugation of the gland homogenate and assays for marker enzymes indicate that α-hydroxy acid reduction is catalyzed by enzyme(s) located in the endoplasmic reticulum fraction.[9]

Acknowledgment

This work was supported in part by Grant GM-18278 from the National Institutes of Health. Dr. A. A. Khan contributed to some of the methods described.

[34] Stearoyl-Acyl Carrier Protein Desaturase from Safflower Seeds

By TOM MCKEON and PAUL K. STUMPF

Stearoyl-ACP (acyl carrier protein) desaturase is the enzyme responsible for the synthesis of oleic acid in plants. Nagai and Bloch, who first characterized the activity, found that the enzyme requires stearoyl-ACP, reduced ferredoxin, and molecular oxygen.[1,2]

Stearoyl-ACP + ferredoxin(II) + O_2 + 2H$^+$ →
$$\text{oleoyl-ACP} + \text{ferredoxin(III)} + 2H_2O$$

The stearoyl-ACP desaturase is easily extracted into buffer without the use of detergents, has no requirement for added lipid, and has a lipid-insoluble substrate,[1] all in marked contrast to the stearoyl-CoA desaturase of animal systems.[3] However, because the plant and animal desaturases both require oxygen and an electron transfer system to carry out the same chemical reaction, it is thought that the mechanism of the reaction may be the same for both types of enzyme.

Nagai and Bloch found the stearoyl-ACP desaturase in photosynthetic tissue—*Euglena gracilis* and spinach chloroplasts.[1,2] Subsequently, Jaworski and Stumpf characterized the activity in immature safflower (*Carthamus tinctorius*) seed,[4] a nonphotosynthetic tissue. The activity is also present in avocado mesocarp,[5] immature soybean cotyledons,[6] immature jojoba nuts,[7] and immature coconut.[8] However, this report describes only the stearoyl-ACP desaturase from safflower.

[1] J. Nagai and K. Bloch, *J. Biol. Chem.* **241**, 1925 (1966).
[2] J. Nagai and K. Bloch, *J. Biol. Chem.* **243**, 4626 (1968).
[3] P. W. Holloway, this series, Vol. 35 [31].
[4] J. G. Jaworski and P. K. Stumpf, *Arch. Biochem. Biophys.* **162**, 158 (1974).
[5] B. S. Jacobson, J. G. Jaworski, and P. K. Stumpf, *Plant Physiol.* **54**, 484 (1974).
[6] P. K. Stumpf and R. J. Porra, *Arch. Biochem. Biophys.* **176**, 63 (1976).
[7] M. R. Pollard, T. McKeon, L. M. Gupta, and P. K. Stumpf, *Lipids* 651 (1979).
[8] T. McKeon, unpublished data.

Assay Method

Principle. The assay for stearoyl-ACP desaturase is based on the measurement of ^{14}C-labeled oleic acid produced by desaturation of ^{14}C-labeled stearoyl-ACP. Separation and quantitation of the ^{14}C-labeled fatty acids are carried out by thin-layer chromatography and scintillation counting or by gas–liquid chromatography and radioactive counting in a proportional counter.

Reagents

PIPES, 0.10 M, pH 6.0
NADPH, 25 mM, freshly prepared in 0.1 M Tricine, pH 8.2
Bovine serum albumin (BSA), lipid free, 10 mg/ml in water
Dithiothreitol (DTT), 0.10 M, freshly prepared
Ferredoxin, 2 mg/ml (Sigma, spinach, type III)
NADPH: ferredoxin oxidoreductase (Sigma), 2.5 units/ml
Catalase (Sigma, bovine liver, 800,000 units/ml)
[14C]Stearoyl-ACP, 10 μM in 0.1 M PIPES, pH 5.8; its synthesis is described after the assay procedure
NaOH, 8 M
H_2SO_4, 4 M
Stearic acid and oleic acid, 1 mg/ml each in acetone
Petroleum ether
Diazomethane (20 mg/ml in diethyl ether)
AgNO$_3$–silica gel G thin-layer plates, 0.25 mm thick (Redi-Coat AG, Supelco)
2,7-Dichlorofluorescein, 0.1% in methanol

Procedure. The following reagents are added for each assay: water, 150 μl; DTT 5 μl; BSA, 10 μl; NADPH, 15 μl; ferredoxin, 25 μl; NADPH : ferredoxin oxidoreductase, 3 μl; and catalase, 1 μl. This mixture is kept at room temperature for 10 min and is then added to a 13 × 100 mm screw-cap test tube containing 250 μl of PIPES buffer. The stearoyl-ACP desaturase preparation is added in a volume of 10 μl, and the reaction is started by adding 30 μl of stearoyl-ACP and incubating at 23° with shaking for 10 min. The reaction is stopped by adding 125 μl of 8 M NaOH and 0.1 ml of the fatty acid solution. The tubes are capped and incubated for 1 hr at 80°. The mixture is acidified with 160 μl of 4 M H_2SO_4 and vigorously extracted three times with 2-ml portions of petroleum ether. The extract is evaporated under nitrogen, methylated with 0.5 ml of diazomethane solution for 30 min on ice, and then evaporated to dryness. The methyl esters of stearate and oleate are then separated and quantitated by either of two methods: thin-layer chromatography on AgNO$_3$–silica gel plates as described by Holloway[3] or gas chromatography (10%

DEGS-PS on Supelcoport 80/10; 6 ft. × ¼ in. column at 180°) followed by counting of the radioactivity in a gas proportional counter. Radio–gas chromatography avoids the slight complication of correcting for the [^{14}C]palmitate contaminant present in most stearoyl-ACP preparations, but, for accuracy and sensitivity, thin-layer chromatography is the method of choice. One unit of activity is defined as 1 μmol of oleate produced per milligram of protein per minute.

An alternative method for reducing ferredoxin uses a chloroplast grana suspension, ascorbic acid, 2,6-dichlorophenolindophenol, and light. This system has been described in detail by Jaworski and Stumpf.[4]

Stearoyl-ACP Synthesis

Procedure. Stearoyl-ACP is made with a safflower fatty acid synthase system, [^{14}C]malonyl-CoA, and *Escherichia coli* ACP. The method described herein differs only slightly from that described by Jaworski and Stumpf.[9]

Immature safflower seeds are suspended in an equal volume of 0.10 M potassium phosphate, 5 mM sodium ascorbate, pH 6.8, and are homogenized with a Polytron instrument for three half-minute periods at half speed, centrifuged at 12,000 g for 20 min, and filtered through four layers of cheesecloth and one layer of Miracloth. The safflower supernatant is used as a source of fatty acid synthase with no further purification. It is stable when frozen for 6 weeks.[8]

The incubation medium contains the following components in a total volume of 5 ml: water 2.2 ml; 25 mM NADH, 100 μl; 25 mM NADPH, 100 μl; 1.0 M Tricine (K$^+$), pH 7.9, 250 μl; 0.10 M DTT 25 μl; 200 mM MgCl$_2$, 25 μl; 1.0 mM malonyl-CoA, 1.0 ml; [1,3-^{14}C]malonyl-CoA (50–60 mCi/mmol), 10 μCi in 500 μl; and ACP (4 mg/ml), 175 μl. The ACP used is purified from *E. coli* by the method of Majerus *et al.*[10] to 90% purity, and is reduced with 1 mM DTT for 15 min just prior to use. The reaction mixture is carefully bubbled with nitrogen for 5 min; 625 μl of safflower supernatant are added, then the mixture is again bubbled with nitrogen for a minute, stoppered, and placed in a 23° water bath. The reaction is stopped after 45 min by the addition of 0.55 ml of 50% trichloroacetic acid (TCA) in the hood and bubbled with nitrogen to displace ^{14}CO$_2$; it is held on ice for 30 min and centrifuged at 5000 g for 5 min. The pellet is redissolved in 2.5 ml of 0.10 M PIPES, pH 5.8, titrating with 1 M KOH if necessary; debris is removed by centrifugation, and solid ammonium sulfate is added

[9] J. G. Jaworski and P. K. Stumpf, *Arch. Biochem. Biophys.* **162,** 166 (1974).
[10] P. W. Majerus, A. W. Alberts, and P. R. Vagelos, this series, Vol. 14 [6].

to 70% saturation (0°). The precipitate is centrifuged at 12,000 g for 10 min, and the supernatant is acidified with 50% TCA to 10%. The TCA precipitate is dissolved in 1 ml of PIPES buffer as before; insoluble material is removed by centrifugation, and the concentration of the stearoyl-ACP is adjusted to 10 μM. This preparation provides 25–40% of the theoretical yield of ^{14}C in acyl-ACP. The product, as analyzed by radio-gas chromatography and AgNO$_3$–silica gel TLC, contains 80–90% stearoyl-ACP, 10–20% palmitoyl-ACP, and less than 0.5% oleoyl-ACP. Frozen solutions of acyl-ACP are stable for over 2 months.

An alternative method for making acyl-ACP of specific chain length is the acyl-ACP synthetase reaction described by Spencer et al.[11] Since a specific fatty acid may be ligated to the ACP with this sytem, it has been used to make the substrates employed in specificity studies. However, this system does not efficiently ligate stearic acid to ACP (2–4%) in our hands; therefore, the fatty acid synthase reaction is routinely used to produce substrate for desaturase assays. Another method for making acyl-ACP is described in this volume [21].

Purification

Materials

Immature safflower seed, Gila variety, harvested at approximately 14–18 days after flowering, as indicated by a charcoal gray seed coat

Acetone, reagent grade, $-20°$

Diethyl ether, anhydrous

DEAE-cellulose, equilibrated with 0.02 M potassium phosphate, pH 6.8

ACP-Sepharose 4B, 2 mg of ACP per milliliter of wet gel; the column material was made with purified E. coli ACP, and cyanogen bromide activated Sepharose 4B by the method of March et al.[12] The reaction was carried out at pH 6.5 in 0.1 M NaHCO$_3$ for 1 day at 4°, and 70% of the ACP was covalently bound to the Sepharose

Potassium phosphate buffers, 0.02 M, 0.10 M, and 0.30 M, all pH 6.8, sterilized and degassed

Procedure

Acetone Powder. Immature safflower seeds (stored at $-20°$) are ground with an equal volume of acetone at high speed in a blender. The suspen-

[11] A. K. Spencer, A. D. Greenspan, and J. E. Cronan, Jr., *FEBS Lett.* **101,** 253 (1979).
[12] S. C. March, I. Parikh, and P. Cuatrecasas, *Anal. Biochem.* **60,** 149 (1974).

sion is suction-filtered, and the retained material is repeatedly extracted with acetone as above until the filtrate is clear and colorless. After the third extraction, the acetone suspension is passed through a coarse sieve to remove fragments of seed coat. Generally, five extractions are required to remove the lipids and phenolics. After the final filtration, the retained material is rinsed several times with a small volume of ether at $-20°$ to remove acetone, and then is kept under suction or in a vacuum desiccator to remove the last trace of ether. Stearoyl-ACP desaturase is stable in frozen seeds for at least 2 years and in the acetone powder for at least 3 months.

The following steps are carried out at 0–4°.

Acetone Powder Extract. Acetone powder from a given weight of seed is triturated with twice that weight of 0.02 M phosphate buffer and gently agitated for 1 hr. The suspension is then centrifuged at 12,000 g for 20 min and filtered through Miracloth, the supernatant, which contains the desaturase, is immediately applied to DEAE-cellulose or frozen. The activity in this preparation is stable for 3–4 weeks at $-20°$ or for 1 week at 4°.

DEAE-Cellulose Pass-through. Acetone powder extract is passed through a column of DEAE-cellulose (1 ml bed volume/3 ml extract), and the column is washed with one bed volume of 0.02 M phosphate buffer. The pass-through and effluent from the wash are collected. While this step does afford some purification (see the table), its principal purpose is to eliminate an acyl-ACP thioesterase present in the extract. Approximately 80% of the thioesterase is thus eliminated.[8]

ACP-Sepharose 4B column. The capacity of the ACP-Sepharose is 5 ml of DEAE-cellulose pass-through per milliliter of column material. Decreasing this ratio does not improve the percentage yield, and increasing the ratio decreases the percentage yield.

A column with a 20-ml bed volume is loaded at a flow rate of 0.5 ml per

PURIFICATION OF STEAROYL-ACYL CARRIER PROTEIN (ACP) DESATURASE

Step	Total protein[a] (mg)	Total activity (mU)	Specific activity (mU/mg)	Yield (%)	Purification factor
Acetone powder extract	380	205	0.55	—	—
DEAE-cellulose pass-through	170	162	0.95	79	1.7
ACP-Sepharose 4-B	0.34	38	110	19	200

[a] Protein was determined by the method of O. H. Lowry, N. J. Rosebrough, A. L. Farr, and R. J. Randall, *J. Biol. Chem.* **193**, 265 (1951), using bovine serum albumin as a standard.

minute, washed with two bed volumes of 0.02 M buffer and three bed volumes of 0.10 M buffer. The stearoyl-ACP desaturase is eluted with 0.30 M buffer and collected in fractions of 1.5 ml. The early fractions contain most of the contaminating acyl-ACP thioesterase; the most purified fractions of desaturase contain acyl-ACP thioesterase as 5–10% of the bulk protein.[8] The desaturase activity from this preparation is stable for 1 week at 4°.

Purity. The most purified preparations of stearoyl-ACP desaturase display one prominent band and several minor bands on SDS–gel electrophoresis. By comparing samples containing various amounts of desaturase and thioesterase, it appears that the prominent band corresponds to the stearoyl-ACP desaturase.

Properties

Specificity. At substrate concentrations of 0.3 μM, the stearoyl-ACP desaturase is 40 times more active on stearoyl-ACP than on stearoyl-CoA and 80 times more active than on palmitoyl-ACP.[8] This high specificity for stearoyl-ACP contrasts with the promiscuous activity of the analogous stearoyl-CoA desaturase from animal systems, which is quite active on acyl-CoA containing 13–19 carbon atoms in the acyl chain.[13]

pH Activity Profile. The desaturase is half-maximally active at pH 5.5 and pH 8.5, with the maximum activity at pH 5.5 in acetate buffer. However, activity at a given pH is dependent on the type of buffer, even at constant ionic strength.[8]

Stability. The stearoyl-ACP desaturase appears to be fairly unstable. It is sensitive to pH, losing 50% or more activity irreversibly when incubated at a pH outside the range pH 6.0 to pH 7.5. It is inactivated on heating at 50° for 1 min. It is unstable to dialysis, irreversibly losing 50% to 100% activity. Finally, further attempts to purify or concentrate the eluent from the ACP-Sepharose column result in nearly total loss of activity.[8]

Miscellaneous Properties. The concentration of oxygen required for maximum activity is 320 μM, which is slightly higher than the oxygen concentration in air-saturated incubation medium, namely 280 μM at 23°; half-maximum activity occurs at a concentration of 60 μM.[8]

Catalase is not required for the desaturase reaction to occur; however, it does stimulate the reaction fivefold. Presumably, catalase protects the desaturase system by scavenging H_2O_2. Both the desaturase and the ferredoxin, NADPH : ferredoxin oxidoreductase system are partially inac-

[13] H. G. Enoch, A. Catala, and P. Strittmatter, *J. Biol. Chem.* **251**, 5095 (1976).

tivated by 0.1 mM H_2O_2, and catalase partially reverses this inactivation.[8] However, two other enzymatic H_2O_2 scavengers do not. Neither horseradish peroxidase nor glutathione peroxidase can replace catalase, and horseradish peroxidase inhibits the desaturation reaction.[8]

[35] Acyl Chain Elongation in Developing Oilseeds

By Michael R. Pollard

The lipids of most plant tissues contain a narrow spectrum of fatty acids: palmitate, oleate, linoleate, and α-linolenate. The neutral lipids (triacylglycerols)[1] of oilseeds, however, contain a diverse range of fatty acids.[2] One structural variation found is that of acids with chain lengths greater than the usual 16 or 18 carbon atoms. This chapter describes approaches to studying the biosynthesis of acids of chain length C_{20} or greater in developing oilseeds. Some of the considerations noted for the investigation of acyl chain elongation are valid for the investigation of other types of acyl metabolism found in developing oilseeds. Ideally both *in vivo* and *in vitro* experiments are required to demonstrate chain elongation. A radio-gas chromatography machine is useful for detection of ^{14}C-labeled fatty ester.

Supply of Maturing Seed Tissue

The choice of a suitable plant will greatly facilitate the investigation. An ideal plant will exhibit the following features.

1. It should be able to produce a steady supply of developing seeds. That is, a plant is preferred that can be grown and induced to flower all year round, probably in the controlled environment of a growth chamber or greenhouse. A short growth period and early flowering will give maximum experimental flexibility.
2. Larger seeds will help reduce the considerable labor of hand pollination, picking, and removal of the seed coat or pod.
3. For studies on chain elongation, a seed is required that has a high percentage of its fatty acids with a chain length of C_{20} or greater. More than 10% of the dry weight of the mature seed should be lipid

---

[1] The single exception found in higher plants is the wax esters of jojoba (*Simmondsia chinensis*) seeds [T. W. Miwa, *J. Am. Oil Chem. Soc.* **48**, 259 (1971)].

[2] C. Hitchcock and B. W. Nichols, "Plant Lipid Biochemistry," Chapter 1. Academic Press, New York, 1971.

in order to measure accurately lipogenic activities *in vivo* and *in vitro*.

Plants that have been used to study the biochemistry of chain elongation in maturing oilseeds are high erucate strains of *Brassica napus* (rape),[3] *Brassica campestris* (turnip rape),[4] and *Brassica juncea* (mustard rape),[5] as well as *Limnanthes alba* (meadowfoam),[6] *Tropaeolum majus* (nasturtium),[7] and *Crambé abyssinica*.[8] They are all annuals. Sometimes, the choice of an oilseed that can be harvested only at a particular time is unavoidable, as in the study of wax ester biosynthesis, which is unique to *Simmondsia chinensis* (jojoba).[9,10] In this case the project becomes distinctly seasonal.

An important preliminary step is to monitor the development of the seeds (Fig. 1). This will ensure that maturing seeds are harvested at the time of maximum lipid biosynthesis. Lipid content (expressed as mass of total or neutral lipid per seed or per gram of fresh or dry seed weight) should be measured as a function of days after flowering (field grown plants) or days after pollination (greenhouse plants). Appelquist has reviewed the topic of lipid accumulation during seed maturation.[11] Several extraction procedures for lipids are suitable. Soxhlet extraction of the dried, ground seeds with petroleum ether will yield neutral lipids.[12] Alternatively, total lipids can be extracted from fresh tissue by homogenizing in petroleum ether–isopropanol, 3 : 2 (v/v)[13] or in chloroform–methanol, 2 : 1 (v/v),[14] followed by the appropriate aqueous salt wash. Extraction of the seed residues should be exhaustive. Developing seeds are ideal for biochemical studies when about 10–50% of the eventual neutral lipid mass has been deposited. Over this period the *in vivo* incorporation of [1-^{14}C]acetate into lipids is generally at a maximum (Fig. 1). Seeds picked later have much endogenous lipid. This can cause severe mass overloading during radio-chromatographic analysis.

[3] R. K. Downey and B. M. Craig, *J. Am. Oil Chem. Soc.* **41**, 475 (1964).
[4] L. A. Appelquist, *J. Am. Oil Chem. Soc.* **50**(2) (1973).
[5] A. Benzioni and M. R. Pollard, unpublished observations, 1979.
[6] M. R. Pollard and P. K. Stumpf, *Plant Physiol.* **66**, 649 (1980).
[7] M. R. Pollard and P. K. Stumpf, *Plant Physiol.* **66**, 641 (1980).
[8] R. S. Appleby, M. I. Gurr, and B. W. Nichols, *Eur. J. Biochem.* **48**, 209 (1974).
[9] J. B. Ohlrogge, M. R. Pollard, and P. K. Stumpf, *Lipids* **13**, 203 (1978).
[10] M. R. Pollard, T. McKeon, L. M. Gupta, and P. K. Stumpf, *Lipids* **14**, 651 (1979).
[11] L. A. Appelquist, *in* "Recent Advances in the Chemistry and Biochemistry of Plant Lipids" (T. Galliard and E. I. Mercer, eds.), pp. 247–286. Academic Press, New York, 1975.
[12] A. Vogel, "A Textbook of Practical Organic Chemistry," 4th ed., p. 137. Longmans Green, New York, 1978.
[13] A. Hara and N. S. Radin, *Anal. Biochem.* **90**, 420 (1978).
[14] J. Folch, M. Lees, and G. H. Sloane-Stanley, *J. Biol. Chem.* **226**, 497 (1957).

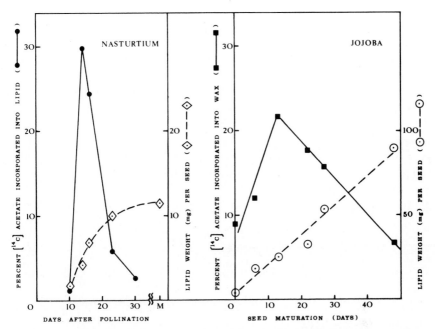

FIG. 1. Profiles for lipid deposition and [1-^{14}C]acetate incorporation into lipids for maturing nasturtium and jojoba seeds. Nasturtiums, grown in the greenhouse, were hand pollinated. M: mature seeds. Labeled lipids were principally triacylglycerols and phospholipids.[7] Lipid weight was measured as milligrams of total acyl residues and was estimated by gas–liquid chromatography. Jojoba seeds were the kind gift of Dr. D. M. Yermanos, University of California, Riverside. Seed maturation was estimated from the date of arrival of seed batches. Lipid (principally wax) was estimated by the weight of material recovered on exhaustive extraction with petroleum ether–isopropanol, 3 : 2 (v/v). Fully mature jojoba seeds contain about 150–200 mg of wax per seed [D. M. Yermanos, *J. Am. Oil Chem. Soc.* **53**, 80 (1976)].

Plants grown in a controlled environment are preferable, as their developing oilseeds can be harvested at the best time. Plants grown in the field may vary considerably in their optimum time for picking, depending on location, time of sowing, etc. Sometimes, however, a change in the appearance of the seed coat or pod (such as browning or yellowing), or of the seed within (turning from green to pale yellow) can be a useful guide to the best time for harvest.

Demonstration of Chain Elongation *in Vivo*

That an active chain elongation step, metabolically separate from *de novo* fatty acid biosynthesis, is occurring *in vivo* in the maturing seed can be demonstrated as follows. Intact tissue is incubated with [^{14}C]acetate or [^{14}C]malonate, which are preferentially incorporated into the chain-elongated portion of the newly synthesized C_{20} or C_{22} fatty acids.[3,5-7,9] Chemical degradation of the ^{14}C-labeled acids isolated after the incubation will highlight any concentration of label at the carboxyl end of the molecule. If a full degradative analysis of a [^{14}C]acyl group is required (Fig. 2), then the *in vivo* experiment should be designed to yield approximately 0.2–1 μCi of the ^{14}C-labeled fatty acyl ester after isolation of the pure material. Incorporation of [^{14}C]acetate into lipids is invariably severalfold greater than incorporation of [^{14}C]malonate,[6,7,9] the latter sometimes being too poor a precursor to produce enough labeled lipid for a full degradative analysis. Although most of the label from [^{14}C]acetate will be found in the carbon atoms added by chain elongation, a little of the [^{14}C]acetate is

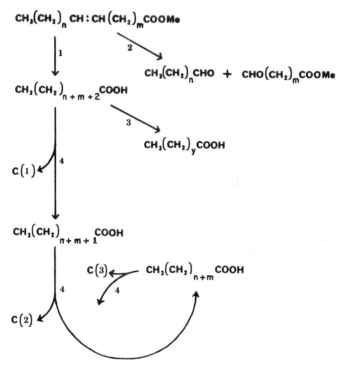

FIG. 2. Summary of the chemical degradation procedures for ^{14}C-labeled unsaturated fatty acids. Reaction arrows: 1, catalytic hydrogenation, followed by saponification; 2, ozonolysis and reduction with triphenyl phosphine; 3, α-oxidation using $KMnO_4$ in acetone and producing a mixture of fatty acids with chain lengths differing by one carbon atom, where $y \leq n + m + 2$; 4, controlled, multistep, and repeatable chemical "decarboxylation" procedure, such as that of Dauben *et al.*[28]

utilized in *de novo* synthesis, resulting in some label in the remainder of the chain for C_{20} and C_{22} acyl groups.

General Method. Finely chopped, maturing seed tissue (2 mm cubes, about 150 mg) is incubated with 5–10 μCi of high specific activity [^{14}C]acetate or [^{14}C]malonate in 0.5 ml of 0.1 M phosphate buffer at pH 6. Incubations in open tubes in a reciprocating water bath at 25° are usually run for 4–6 hr. However, seed slices will remain viable for much longer periods if more precursor incorporation is required.[7] The reaction is terminated by adding 1 ml of isopropanol and heating at 80° for 10 min. On cooling, the tissue is homogenized in 10 ml of either petroleum ether–isopropanol, 3 : 2 (v/v)[13] or chloroform–methanol, 2 : 1 (v/v).[14] The monophasic suspension is left for at least 1 hr in order to facilitate complete lipid extraction. After shaking with the appropriate aqueous phase [about 2–3 ml of 6.7% (w/v) aqueous sodium sulfate[13] or 0.23% NaCl[14]] and centrifugation, the organic phase is evaporated to dryness under nitrogen. For transmethylation, the residual lipids (<30 mg) are refluxed in 3–4 ml of methanol–benzene–concentrated sulfuric acid, 20 : 10 : 1 (v/v/v) for 3 hr. Several other methods are just as suitable.[15,16] After addition of 5 ml of water, the methyl esters are extracted into petroleum ether (3 × 3 ml), and the combined extracts are washed twice with water and evaporated to dryness under nitrogen. Addition of a little methanol helps to remove the last traces of water. Ethanolysis is used for transesterification of wax esters, as in the study of long-chain fatty acid and alcohol biosynthesis in jojoba seeds. The lipid extract is refluxed for 4 hr in anhydrous ethanol–benzene–hydrogen chloride, 20 : 2 : 1 (v/v/w), and worked up as for transmethylation.[17]

Fatty acid esters and fatty alcohols from jojoba wax are analyzed by radio-gas-liquid chromatography.[18] Instruments can vary in design[19] and are commercially available. For studies of acyl chain elongation radio-GLC is a suitable method for routine analysis, but for other types of acyl chain modification occurring in developing oilseeds (e.g., hydroxylation of oleate in the castor bean, Δ^5-desaturation of eicosanoate in meadowfoam seeds) routine analysis by TLC may be just as easy. An inexpensive adaption of a conventional gas chromatograph described by Wels,[20] or even the use of a stream-splitter with the trapping of effluent fractions for

[15] W. W. Christie, "Lipid Analysis." Pergamon, Oxford, 1973.
[16] D. K. McCreary, W. C. Kossa, S. Ramachandran, and R. R. Kurtz, *J. Chromatogr. Sci.* **16**, 329 (1978).
[17] C. C. Duncan, D. M. Yermanos, J. Kumamoto, and C. S. Levesque, *J. Am. Oil Chem. Soc.* **51**, 534 (1974).
[18] Abbreviated to radio-GLC.
[19] A. Karmen, this series, Vol. 14 [52].
[20] C. M. Wels, *J. Chromatogr.* **142**, 459 (1977).

liquid scintillation counting, may suffice for analysis. Typical radio-gas chromatograms have been published for meadowfoam[6] and jojoba.[9] The unlabeled peaks of the endogenous acids in the sample will act as an internal standard for identification of the radioisotope peaks. For initial work on a new oilseed, the identity of any labeled products should be checked by the use of at least two GC columns of widely differing polarity.[21] Suitable stationary phases for this purpose are 10% DEGS and 10% SP-2330 (high polarity), 10% SP-1000 (intermediate polarity), and 14% SE-30 (nonpolar). Chain-elongated products should chromatograph with the fatty ester band on silica gel thin-layer plates. They can be eluted from the silica gel with a polar organic solvent and analyzed by radio-GLC. The TLC methods are described elsewhere.[15,22]

Prior to transesterification the labeled lipid extract can also be examined by TLC, in order to elucidate the lipid classes present and to quantitate the distribution of [^{14}C]acyl groups within each class. Apart from the conventional [^{14}C]acyl lipids, workers should be alert to the possible presence of novel, labeled secondary products.[9,23]

Degradation of Fatty Acids. The crux of the *in vivo* experiment is the measurement of the distribution of label along the acyl or alkyl chain. This will show whether a chain-elongation step, utilizing a separate pool of acetate from that used for *de novo* fatty acid synthesis, has occurred. It is also necessary to determine whether such an elongation has involved palmitate, stearate, oleate, or any other preformed acid as the substrate. Fatty acid degradations by ozonolysis,[24,25] α-oxidation,[26,27] and controlled decarboxylation[28,29] (Fig. 2) can be used in conjunction to give a complete picture of the distribution of label along the acyl or alkyl chain, as has been described for meadowfoam,[6] nasturtium,[7] and jojoba.[9] The table shows a typical set of data obtained by the three methods. A prerequisite to degradative analysis is the separation of individual fatty acid esters by preparative GLC.[19] Since most long-chain acids encountered in oilseeds are unsaturated, ozonolysis is an ideal technique, being quick, easy, and reproducible. If an ozone generator is available, ozonolysis is to be preferred over the von Rudloff periodate–permanganate oxidation[30] as a

[21] R. G. Ackman, this series, Vol. 14 [49].
[22] V. P. Skipsi and M. Barclay, this series, Vol. 14 [54].
[23] M. R. Pollard, J. B. Ohlrogge, and P. K. Stumpf, *Phytochemistry* 17, 1831 (1978).
[24] R. A. Stein and N. Nicolaides, *J. Lipid Res.* 3, 476 (1962).
[25] See this series, Vol. 72 [16].
[26] R. V. Harris, P. Harris, and A. T. James, *Biochim. Biophys. Acta* 106, 465 (1965).
[27] A. G. Netting, *Anal. Biochem.* 86, 580 (1978).
[28] W. G. Dauben, E. Hoerger, and J. W. Peterson, *J. Am. Chem. Soc.* 75, 2347 (1953).
[29] P. Aronsson and J. Gurtler, *Biochim. Biophys. Acta* 248, 21 (1971).
[30] E. von Rudloff, *J. Am. Oil Chem. Soc.* 33, 126 (1956).

EXAMPLE OF DEGRADATION DATA OBTAINED FROM A ^{14}C-LABELED UNSATURATED ACID[a]

$$RO-CO-CH_2-CH_2-CH_2-(CH_2)_8-CH=CH(CH_2)_7CH_3$$

	C$_{13}$ aldehyde ester $\longrightarrow$ 96%				C$_9$ aldehyde $\longleftarrow$ 4%
Distribution of radioactivity in ozonolysis products					
Release of ^{14}C by sequential decarboxylation	C-1 47%	C-2 2%	C-3 45%	C-4 1%	
α-Oxidation: specific activities (cpm/mol) of the chain-shortened acids relative to the parent C$_{22}$ acid	C$_{22}$ 1.0	C$_{21}$ 0.65	C$_{20}$ 0.5	C$_{19}$ 0.1	C$_{18}$ 0.1
Estimated distribution of label along the acyl chain	C-1 45.5%	C-2 0%	C-3 45.5%	C-4 0%	Remaining C atoms: even-numbered, 0% each; odd-numbered, 1% each; total, 9%

[a] The example is [^{14}C]*cis*-13-docosenoic acid derived from [1-^{14}C]acetate.

method of double-bond cleavage. Reduction of the ozonide with triphenyl phosphine gives aldehyde and aldehyde ester fragments (Fig. 2), which can be quantitated directly by radio-GLC. Ozonolysis of [^{14}C]oleate, which is synthesized *de novo* from [1-^{14}C]acetate, produces nonanal and 9-oxononanoate labeled in the ratio 4:5. However, if, for example, [^{14}C]*cis*-5-eicosenoate from the incubation of [1-^{14}C]acetate with developing meadowfoam seeds is subjected to ozonolysis, then 35% of the label is found in pentadecanal and 65% in 5-oxopentanoate. Further degradative studies are required to ascertain whether the preponderance of label associated with the carboxyl end is due to label at C-1 or at C-1 plus C-3, and so on. That is, α-oxidation and/or controlled decarboxylation are required. Both methods require saturated acids as a starting point, so that the ^{14}C-labeled unsaturated ester recovered after preparative GLC must first be reduced[31] and then saponified.[32]

Chemical α-oxidation is suitable for both long-chain saturated acids and alcohols and has been described by Harris *et al.*[26] A modification of the method has been reported by Netting.[27] A problem encountered with long-chain acids ($\geq C_{20}$) is the difficulty in controlling the α-oxidation to produce an approximately equimolar distribution of products. This is apparent in a previously published chromatogram for the α-oxidation of docosanoic acid.[8] Most of the mass and the radioactivity resides with the parent C_{22} acid. Consequently, even with considerable GC overloading, specific activity measurements of the shorter chain acids are not precise. With our instrumentation specific activity values were accurate only to ±0.05 or ±0.1 units, with the specific activity of the parent acid set at unity. Thus chemical α-oxidation of long-chain acids gives a general but not precise picture of the distribution of label.

Controlled chemical decarboxylation is necessary to obtain exact figures for the distribution of label along the chain. A method is required that can be repeated with the chain-shortened product to yield information for successive decarboxylations. A microscale modification[7] of the procedure of Dauben *et al.*,[28] starting with about 1 μCi of [^{14}C]fatty acid plus carrier acid to give 50 μmol, is suitable.

Demonstration of Chain Elongation *in vitro*

Method. Slices of fresh, developing seeds are homogenized to a paste by grinding, using a pestle and mortar, with 4 volumes of buffer (0.3 M sucrose, 80 mM N-2-hydroxyethylpiperazine-N'-2-ethanesulfonic acid, 5

[31] Catalytic hydrogenation, as described by Pollard and Stumpf[7] or on page 135 of Christie[15] is the best method.
[32] The ester is refluxed in 5% methanolic KOH for 30 min.

mM ascorbate, and 1 mM dithiothreitol, at pH 7.5), plus adsorbent material if necessary (see Note 5). The paste is squeezed through several layers of cheesecloth, and the resulting suspension is allowed to drain through Miracloth.[33] This produces a cell-free homogenate, as judged by light microscopy. The above operations need not be performed chilled.

A typical assay for chain elongation will contain [1-^{14}C]oleoyl-CoA (50 μM, 50,000–100,000 cpm), malonyl-CoA (0.2 mM), NADH (0.5 mM), NADPH (0.5 mM), and 0.5 ml of the cell-free homogenate, in a total volume of 0.75 ml. The reaction is started by addition of the enzyme preparation, and the mixture is incubated for 30 min at 25° to 30° in open tubes. The incubation is terminated by addition of 4 ml of 5% (w/v) ethanolic potassium hydroxide, and the mixture is heated at 80° for 1 hr to ensure saponification. Acidification with aqueous sulfuric acid and exhaustive extraction with petroleum ether follow. The [^{14}C]fatty acids recovered are methylated, ethereal diazomethane providing the quickest and easiest method.[15] In the case of incubations with homogenates from jojoba seeds, saponification does not completely hydrolyze the wax. Methylation must be replaced by ethanolysis as described earlier. The distribution of label between the ^{14}C-labeled fatty esters of different chain lengths is measured by radio-GLC, with quantitation of the radioisotope peaks by triangulation or by electronic integration. In quoting the rate of elongation it is preferable to use nanomoles of malonyl-CoA consumed per minute rather than nanomoles of oleoyl-CoA elongated per minute, because [1-^{14}C]oleoyl-CoA can produce both [3-^{14}C]cis-11-eicosenoate and [5-^{14}C]cis-13-docosenoate. The formation of the latter consumes 2 mol of malonyl-CoA.

A complementary activity of lipid biosynthesis, which should be measured along with chain elongation, is de novo fatty acid synthesis. A typical incubation will contain [2-^{14}C]malonyl-CoA (about 50,000 cpm, 0.2 mM), reduced acyl carrier protein (6.5 μM),[34] NADH (1.0 mM), NADPH (1.0 mM) and 0.2 ml of the cell-free homogenate, in a total volume of 0.5 ml. The incubation is carried out at 25° to 30° in open tubes for 30 min. Termination and saponification are as described above for chain elongation. Essentially all the radioactivity recovered in the organic phase after acidification and extraction with petroleum ether is free fatty acid, so liquid scintillation counting of this phase gives a measure of malonyl-CoA incorporation into lipids. Radio-GLC of the ^{14}C-labeled fatty esters shows that palmitate and stearate are the predominantly labeled species.

---

[33] From Chipopee Mills Inc., Milltown, New Jersey.
[34] Acyl carrier protein was isolated from E. coli (P. W. Majerus, A. W. Alberts, and P. R. Vagelos, this series, Vol. 14 [6]), and was reduced by standing with 1 mM dithiothreitol at room temperature for 10 min.

Notes

1. The specific activities of $[1\text{-}^{14}\text{C}]$oleoyl-CoA and $[2\text{-}^{14}\text{C}]$malonyl-CoA used in the above incubations are 1–2 Ci/mol and 0.25–0.5 Ci/mol respectively, but these radiochemicals are commercially available at much higher specific activities. Incubations can therefore be scaled down where necessary while retaining sufficient counts per minute for analysis.

2. The addition of oleoyl-CoA complexed to defatted bovine serum albumin (BSA), to give an incubation concentration of 1–2 mg of protein per milliliter, will give a moderate rate enhancement.[5,10] However, using a cell-free homogenate from developing jojoba seeds, long-chain acyl-CoA reductase and long-chain acyl-CoA : fatty alcohol ligase activities were stimulated severalfold by addition of the $[^{14}\text{C}]$acyl-CoA substrate complexed with BSA.[5,10] Such rate enhancements by BSA may be a general phenomenon for reactions utilizing acyl-CoA thioesters of chain length greater than C_{18}, and they are certainly worth testing for in every oilseed.

3. A large volume of 5% (w/v) ethanolic potassium hydroxide was needed for the satisfactory saponification of incubations with cell-free homogenates from jojoba. With smaller volumes (less than 1.5 ml per 0.5 or 0.75 ml incubation), which still contain more than ample base for complete hydrolysis, gels were sometimes formed during the extraction, and low recoveries of radioactivity were noted.

4. Using cell-free homogenates from developing rapeseed or jojoba beans, the incubations described above for chain elongation and fatty acid synthesis were found to be suitable as assays for enzyme activity. That is, cofactor concentrations were not rate limiting, rates were approximately linear over 30 min and proportional to enzyme concentration, and the substrate concentration was close to a saturating level.[5] However, expression of rates in terms of the amount of enzyme presents a problem. As oilseeds mature, their percentage dry weight increases. Protein and lipid contents, expressed on a per seed basis, also increase sharply. Thus expression of activity on a nanomole per milligram of protein per minute basis is not very helpful. Problems associated with protein determinations in plant tissue have been reviewed by Loomis.[35] Chlorophyll content is not a useful marker either, since seeds are not a photosynthetic tissue and because any greenness disappears over the period of lipid accumulation. Rates should be expressed on a per seed or per gram of seed basis. We have estimated the amount of cell rupture during the homogenization of jojoba seeds by measuring the amount of lipid in the cell-free homogenate and comparing it with the amount of lipid that could be exhaustively extracted from the same batch of seeds. If the seeds were ground to a fine paste, 40–90% of the total lipid was released.[5] Using this simple method,

[35] W. D. Loomis, this series, Vol. 31 [54].

rates measured as nanomoles per milliliter of cell-free extract per minute can be expressed on a per seed basis.

5. Early attempts in our laboratory to obtain active cell-free homogenates from developing nasturtium and jojoba seeds were unsuccessful. When these totally inactive homogenates were added to a 12,000 g supernatant fraction from safflower seeds, which was capable of rapid *de novo* synthesis of fatty acids from malonyl-CoA, severe inhibition of the safflower seed activity was observed. However, jojoba cell-free homogenates prepared in the presence of polyvinylpyrrolidone-40 gave good, reproducible lipogenic activities.[5]

Plant tissues can contain a wide variety of secondary products. When released from their subcellular compartment(s) by homogenization, they may have adverse effects on enzyme activity. The treatment of extracts from plant tissues with various adsorbents and ion exchange resins can remove such troublesome secondary products.[35,36] Recommended materials include polystyrene (Amberlite XAD-4), polyvinylpolypyrrolidone, bovine serum albumin, and Dowex resins. The work of Loomis *et al.* gives an up-to-date summary of this topic.[36] The demonstration of an inhibitory factor in developing jojoba seeds and its removal by the use of an adsorbent material is thought to be the first time such a problem has been reported for the study of fatty acid biochemistry in plants. Therefore for studies on the biochemistry of oilseeds the use of adsorbent materials, included in the homogenization medium or added immediately after homogenization, may be a helpful adjunct or even prerequisite to obtaining good enzyme activities. Suitable adsorbents will vary from plant to plant and can be determined only by experiment. They may not be essential, as in studies on oleate desaturation in maturing safflower seeds,[37,38] Δ^5-desaturation and chain elongation in maturing meadowfoam seeds,[39] or wax degradation in germinating jojoba seeds.[40]

6. The enzyme activities for chain elongation, and indeed for all other reactions required for acyl group biosynthesis in developing oilseeds, appear to be stable to prolonged storage at $-20°$, provided the preparation has been pretreated with adsorbent or no adsorbent is needed.[5,39]

7. For work involving fractionation of the cell-free homogenate it is worthwhile to check rigorously the nature of the ^{14}C-labeled metabolites in each fraction. We encountered ^{14}C-labeled artifacts in various fractions, which were not observed in the cell-free homogenate and which cochromatographed with the expected product under certain conditions.[5]

[36] W. D. Loomis, J. D. Lile, R. P. Sandstrom, and A. J. Burbott, *Phytochemistry* **18**, 1049 (1979).

[37] S. Styme and L. A. Appelquist, *Eur. J. Biochem.* **90**, 223 (1979).

[38] C. R. Slack, P. G. Roughan, and J. Browse, *Biochem. J.* **179**, 649 (1979).

[39] R. A. Moreau, personal communication, 1979.

[40] R. A. Moreau and A. H. C. Huang, *Arch. Biochem. Biophys.* **194**, 422 (1979).

Acknowledgments

This article is based on research funded by the National Science Foundation. I am particularly indebted to my colleagues Dr. P. K. Stumpf and Dr. A. Benzioni for their support and assistance.

[36] Estimation of Active and Inactive Forms of Fatty Acid Synthase and Acetyl-CoA Carboxylase by Immunotitration

By FRANK A. LORNITZO, ROGER F. DRONG, SARVAGYA S. KATIYAR, and JOHN W. PORTER

A determination of the amount of active and inactive species of an enzyme in mammalian tissue homogenates obtained from animals in different nutritional or hormonal states is of critical importance to an understanding of *in vivo* mechanisms of enzyme regulation.[1-4] A direct technique to answer this question would involve the isolation of a homogeneous enzyme and a determination of its specific activity, or a determination of the activity of an enzyme and the amount of immunoprecipitate formed when it is allowed to react with monospecific antibody. Each of these methods has drawbacks. The isolation of pure enzyme from liver homogenates involves several purification steps, and frequently the methods are long and tedious. The total amount of active and inactive enzyme is then inferred from the specific activity of the purified material. The tacit assumption is made that all enzyme species behave in an identical manner during all the purification steps; this may or may not be true. The direct immunoprecipitation of enzyme is probably more reliable than the isolation procedure for determining the ratio of active to inactive enzyme species. However, large amounts of monospecific antibody are required for these determinations, and the assays that are performed are quite time-consuming.

We report here the development of immunotitration methods to estimate the amounts of active and inactive species of the fatty acid synthase (rat or pigeon liver) and acetyl-CoA carboxylase (rat liver). In this method the quantitation of active and inactive species of enzyme is effected by titrating equivalent amounts of enzyme activity in crude homogenates from livers of animals in different nutritional or hormonal states with

[1] D. M. Gibson, *J. Chem. Educ.* **42**, 236 (1965).

[2] D. M. Gibson, S. E. Hicks, and D. W. Allman, *Adv. Enzyme Regul.* **4**, 239 (1966).

[3] M. R. Lakshmanan, C. M. Nepokroeff, and J. W. Porter, *Proc. Natl. Acad. Sci. U.S.A.* **69**, 3516 (1972).

[4] J. J. Volpe and P. R. Vagelos, *Proc. Natl. Acad. Sci. U.S.A.* **71**, 889 (1974).

antibody that may either be partially purified or monospecific. Enzyme that has been partially purified may also be assayed in the same way. Titrations may also be made in the reverse direction, namely by titrating a given quantity of antibody with increasing amounts of enzyme. In these procedures, equivalent units of enzyme activity give the same end point on immunotitration if all enzyme molecules have the same enzyme activity, whereas in the presence of inactive species the end points will be different. Obviously, a standard in which all, or nearly all, of the enzyme molecules are enzymatically active is required for these titrations. In the case of acetyl-CoA carboxylase and fatty acid synthase, such an enzyme is obtained from livers of animals fasted for 48 hr and then refed for 48 hr. The immunotitration method is fast, reproducible, and capable of determining relatively small levels of inactive species of an enzyme.

Two types of immunotitrations are reported. In the first type, fatty acid synthase is titrated directly with monospecific antibody. In this reaction the antibody reacts directly with a site (the condensing activity) required for fatty acid synthesis. In the second type of immunotitration, acetyl-CoA carboxylase is titrated with partially purified or monospecific antibody. In this reaction enzyme activity is not lost. Hence, it is necessary to remove antigen–antibody complex by centrifugation and to assay the supernatant solution for residual acetyl-CoA carboxylase activity. By each of these methods we have found nutritional or hormonal conditions that yield a significant amount of inactive enzyme in rat liver.

Method 1. Rat Liver Fatty Acid Synthase

In this method the binding of antibody to enzyme inhibits fatty acid synthesizing activity.

Preparation of Pure Rat Liver Fatty Acid Synthase. Rats are fasted for 48 hr and then refed a fat-free diet for 48 hr. Fatty acid synthase is isolated from the livers of these animals and purified through the DEAE-cellulose chromatographic step of the method of Burton et al.[5] as modified by Craig et al.[6] The enzyme is further purified to homogeneity by sucrose density gradient centrifugation.[6]

Preparation of Antibody. Sucrose density gradient-purified rat liver fatty acid synthase, 10 mg/ml, is mixed with an equal volume of Freund's adjuvant, and an initial subcutaneous injection of 2.5 mg of fatty acid synthase protein is made into a New Zealand albino rabbit.[7] In subsequent

[5] D. N. Burton, J. M. Collins, A. L. Kennan, and J. W. Porter, *J. Biol. Chem.* **244**, 4510 (1969).
[6] M. C. Craig, C. M. Nepokroeff, M. R. Lakshmanan, and J. W. Porter, *Arch. Biochem. Biophys.* **152**, 619 (1972).
[7] D. M. Livingston, this series, Vol. 34 [91].

injections, incomplete Freund's adjuvant is mixed with the fatty acid synthase, and 2.6 mg doses of enzyme protein of this mixture are injected in two portions into the rabbit every 10–14 days.[7] After 3 or 4 injections, blood is drawn and the antibody titer of the serum is determined by immunotitration against pure fatty acid synthase. The immunoglobulin fraction of the serum is then purified through the DEAE-cellulose chromatography step of the procedure of Livingston.[7] The unbound protein fraction passing through the DEAE-cellulose column is concentrated and further purified by immunoaffinity chromatography by the method of Katiyar et al.[8]

Preparation of Monospecific Antibody by Immunoaffinity Chromatography. A Sepharose-rat liver fatty acid synthase affinity gel is prepared by a procedure similar to those of Katiyar et al.[8] and March et al.[9] One milliliter of Sepharose 4B is suspended in 10 volumes of water and then allowed to settle for 1 hr. The supernatant solution containing fine particles is discarded, and the gel is washed 3 or 4 additional times. Meanwhile, 10 mg of DEAE-cellulose-purified fatty acid synthase protein are dialyzed under nitrogen against a buffer of 0.2 M potassium phosphate and 1 mM EDTA. This step removes dithiothreitol, which would compete with nucleophilic groups on the protein for the activated Sepharose. The dialyzed protein solution is diluted to 4 ml and the solution is buffered at pH 8.5 with 2 M potassium carbonate immediately before activation of the Sepharose by the procedure of March et al.[9] Sepharose, 200 mg, is suspended in 0.2 ml of 2 M potassium carbonate, then 20 μl of 50% cyanogen bromide in acetonitrile are added and the reaction is allowed to proceed for 2 min with stirring. The slurry is poured into a 1 cm coarse sintered-glass funnel, washed with water and with 0.1 M potassium carbonate, pH 8.5. The washed gel is transferred to a capped bottle containing the buffered protein solution, and the coupling of fatty acid synthase to Sepharose is allowed to proceed at room temperature with gentle stirring overnight. The slurry is poured into a sintered-glass funnel and washed with a solution of 0.9% sodium chloride and 1 mM EDTA, pH 7.0. Analysis of the combined filtrate and wash should show a total of not more than 1–2 mg of unbound protein.

The affinity gel is suspended in the saline and EDTA solution and then pipetted into a narrow column. A 1-ml graduated pipette, 2 mm i.d., with about 15 cm cut off the top, is satisfactory for this purpose. A portion of the DEAE-cellulose-purified rabbit anti-rat liver fatty acid synthase immunoglobulin fraction (30 mg of protein per milliliter), with titer equiva-

[8] S. Katiyar, F. A. Lornitzo, R. E. Dugan, and J. W. Porter, *Arch. Biochem. Biophys.* **201,** 199 (1980).
[9] S. C. March, I. Parikh, and P. Cuatrecasas, *Anal. Biochem.* **60,** 149 (1974).

lent to 8–10 mg of fatty acid synthase activity, is loaded onto the column at a flow rate of approximately 1 ml/30 min. The column is then washed with a solution of 0.9% NaCl and 1 mM EDTA until the absorbance at 280 nm is reduced to 0.05. The bound fatty acid synthase specific immunoglobulin is eluted from the column with filtered 4.5 M magnesium chloride that has been neutralized to pH 6.5 with 1 M Tris. (Some lots of magnesium chloride contain a considerable amount of residual HCl.) The flow rate during this step should be 1 ml/hr or less, and the elution pattern should show a sharp peak of light absorption at 280 nm, which returns to background after 5–6 ml. Those fractions showing light absorption at 280 nm are combined for further use.

The eluted antibody protein fractions are poured into dialysis tubing, and one end of the tubing is tied to a glass tube for support. Dialysis of the eluted protein is carried out in 2 liters of a solution of 0.9% sodium chloride and 1 mM EDTA for 24 hr, with two changes of the dialysis solution. A final dialysis is carried out for 3–4 hr in a solution of 0.09% sodium chloride and 0.01 mM EDTA, and the contents of the bag are then lyophilized to a volume of 1–2 ml. The recovery of monospecific immunoglobulin, as measured by immunotitration, should be at least 80% of that in the DEAE-cellulose-purified immunoglobulin fraction. Monospecific anti-rat liver fatty acid synthase antibody constitutes about 3–8% of the DEAE-cellulose purified immunoglobulin fraction. The antibody is stored at $-20°$ at a concentration of approximately 3.5 mg/ml. The fatty acid synthase affinity gel may be washed with 0.9% saline and 1 mM EDTA and used several additional times.

Preparation of Crude Rat Liver Homogenates. Rats are sacrificed by decapitation, and livers are excised and placed in phosphate–bicarbonate buffer (70 mM KHCO$_3$, 85 mM K$_2$HPO$_4$, 9 mM KH$_2$PO$_4$, and 1 mM dithiothreitol) at 0° until weighed. After weighing, the livers are minced with scissors and homogenized in 1.5 volumes of the above buffer with a motor-driven Teflon pestle in a Potter–Elvehjem type glass homogenizer. The homogenate is then centrifuged at 100,000 g and stored as previously described.[5,6]

Immunotitration of Fatty Acid Synthase in the 100,000 g Supernatant Solution of Rat Liver Homogenate. Determination of Inactive and Active Enzyme Species. Active and inactive enzyme protein molecules differ in enzyme activity, but are immunologically identical. Therefore, active and inactive forms of the enzyme will compete equally for immunoglobulin produced to that enzyme. Hence, if a mixture of active and inactive enzyme is titrated with immunoglobulin, there will be less loss of enzyme activity per unit of immunoglobulin than occurs when fully active enzyme is titrated with the same antibody. This decrease in loss of enzyme activity is a direct measure of inactive enzyme present in the sample. Immunotitration may also be made where enzyme is kept constant and antibody is varied. The end point of this titration is determined from measurements of the fatty acid synthase activity remaining after each increase in immunoglobulin. Usually it is necessary to extrapolate the linear portion of this plot of enzyme activity versus quantity of antibody to zero enzyme activity to determine the end point of the titration.

The assays for fatty acid synthase activity remaining after the antibody–antigen reaction are carried out directly with the reaction mixture. In these assays 1 unit of fatty acid synthase activity is defined as equivalent to the oxidation of 1 nmol of NADPH per minute. The specific activity of homogeneous rat liver fatty acid synthase from 48-hr fasted, 48-hr refed rats is 550–660 units per milligram of protein in 0.5 M potassium phosphate, pH 7.0, and 900–1100 units per milligram of protein in 0.2 M potassium phosphate. However, in this procedure the antibody–antigen reactions and the assays are carried out in the higher ionic strength buffer, even though the velocity of the reaction is lower, because fatty acid synthase activity is more stable at the higher ionic strength. Fatty acid synthase activity decreases markedly over a 60- to 90-min period in 0.2 M buffer at 30°.

In an immunotitration such as is shown in Fig. 1, a total of about 12 units of enzyme activity at a concentration of 40 units/ml is needed. A sample of rat liver homogenate, or a partially purified preparation containing fatty acid synthase, is diluted with two volumes of a solution (dilution I) containing 0.67 M potassium phosphate, pH 7.0, 1 mM EDTA, and 10 mM dithiothreitol, and then incubated at 30° for 45 min to ensure that all the fatty acid synthase is in the undissociated form.[10] Dilution I is then assayed for overall fatty acid synthase activity and, if necessary, this sample is diluted a second time (dilution II) to the desired concentration of about 40 units of activity per milliliter in 0.5 M potassium phosphate, pH 7.0, 1 mM EDTA, and 1 mM dithiothreitol. Dilution II is then added with a Hamilton pipette in multiples of 0.4 unit of activity (now equivalent to multiples of 10 μl) to a series of 8 tubes, each containing 100 μg of bovine serum albumin and 1.6 μg of protein of monospecific anti-rat liver fatty acid synthase antibody in 0.5 M potassium phosphate, pH 7.5, and 0.1 M EDTA, in a final volume of 0.45 ml. (The concentration of dithiothreitol in the antibody–antiserum reaction mixture should not exceed 1 mM, as higher concentrations will break the disulfide linkage between the peptide chains of the immunoglobulin.) The tubes are flushed with nitrogen and incubated for 40–60 min at 30°. The contents of each tube are then transferred to a semimicro cuvette and assayed spectrophotometrically at 30°

[10] R. A. Muesing, F. A. Lornitzo, S. Kumar, and J. W. Porter, *J. Biol. Chem.* **250**, 1814 (1975).

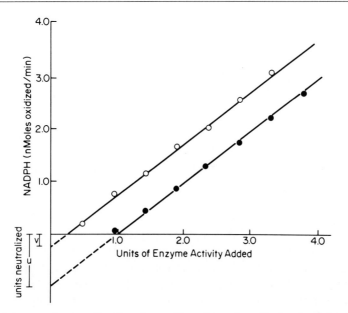

FIG. 1. Immunotitration of rat liver fatty acid synthase: plots of units of activity observed versus units of activity added to a constant amount of antibody. ●——●, DEAE-cellulose-purified fatty acid synthase [67 μg of protein and 40 units/ml, from livers of 48-hr fasted, 48-hr refed (fat-free diet) rats, plus 1.6 μg of monospecific immunoglobulin per tube]. ○——○, the 100,000 g supernatant of homogenate from livers of 48-hr fasted, 4-hr refed (fat-free diet) diabetic rats, plus 1.6 μg of monospecific immunoglobulin per tube. Fatty acid synthase-related protein (calculated from the titration) is 268 μg of protein and 40 units per milliliter of enzyme activity, which is equivalent to 67 μg of protein per milliliter of fully active enzyme per tube. One unit of fatty acid synthase activity is defined as equivalent to the oxidation of 1 nmol of NADPH per minute.

for fatty acid synthase activity.[5] NADPH, 0.05 μmol (25 μl of a 2 mM solution), is added to the sample, and after 3–5 min of incubation the blank rate of disappearance of the nucleotide is determined. Then an acetyl-CoA and malonyl-CoA mixture, 0.0167 and 0.05 μmol, respectively (25 μl of solutions of 0.67 mM and 0.2 mM concentrations), is added, and the rate of the enzymatic reaction is determined by subtracting the rate obtained in the blank from the rate for the complete system.

Calculation of Concentration of Active and Inactive Enzyme in the Sample. When calculations are made, Eq. (1) is used.

$$x = c(u/v - 1) \tag{1}$$

in which the amount of antibody is the same in each titration. x = Concentration of inactive enzyme and c = concentration of active enzyme in the sample; $u = Y$ intercept of the titration of fully active fatty acid synthase with antibody (see Fig. 1), and v = Y intercept of the titration of a mixture of active and inactive enzyme (see Fig. 1).

The probable error of the result x (P.E.$_x$) is calculated from the probable errors of u and v, using Eqs. (5) or (6), which are derived from the relationships of Eqs. (2), (3), and (4) cited by Perrin.[11] In Eq. (2)

$$P.E._x = 0.67 \, \sigma_x \tag{2}$$

and P.E.$_x$ is the probable error in the best estimate of x, and σ_x is the standard deviation in x. In Eqs. (3) and (4), x is stated as a function, f, of two variables, u and v, such as in Eq. (1).

$$x = f(u,v) \tag{3}$$

$$\sigma_x{}^2 = \left[\frac{\partial f(u,v)}{\partial u} \, \sigma_u\right]^2 + \left[\frac{\partial f(u,v)}{\partial v} \, \sigma_v\right]^2 \tag{4}$$

in which $[\partial f(u,v)]/\partial u$ is the partial derivative of $f(u,v)$ with respect to u, and $\partial f(u,v)/\partial v$ is the partial derivative of $f(u,v)$ with respect to v. σ_u is the standard deviation of u, and σ_v is the standard deviation of v.

Carrying out the indicated operations in Eq. (4) on the right-hand side of Eq. (1), and using the relationship of Eq. (2),

$$(P.E._v)^2 = [(c/v) \, P.E._v]^2 + [(cu/v^2) \, P.E._u]^2 \tag{5}$$

in which P.E.$_v$ is the probable error of the best estimate of v, and P.E.$_u$ is the probable error of the best estimate of u. Equation 5 reduces to

$$(P.E._x)^2 = c^2 \left[\frac{1}{v^2} + \frac{u^2}{v^4}\right] P.E._v{}^2 \tag{6}$$

when P.E.$_v$ $\approx$ P.E.$_u$.

Discussion. A plot of a typical immunotitration in which the line is fitted to the points by a linear regression analysis is shown in Fig. 1. Linear extrapolation of the plot to the Y axis gives the units of enzyme activity inactivated by a given amount of antibody to the fatty acid synthase. This value is constant in titrations of enzyme from animals in different nutritional or hormonal states if all the enzyme molecules have the same activity. A different value for the Y axis intercept will be obtained if the population of enzyme molecules contains enzymatically active and inactive species. By these titrations it is possible to establish physiological conditions under which active and inactive fatty acid synthase is present in rat liver.

The greater the percentage of inactive enzyme relative to active en-

[11] C. L. Perrin, "Mathematics for Chemists," p. 149. Wiley, New York, 1970.

zyme in a sample, the smaller will be the resulting value of v for the same amounts of antibody. When enzyme activity is relatively low or absent, the relative error in v will become very large as v vanishes, leading to high uncertainty in the determination of the concentration of enzyme protein. This problem is overcome by adding a specified concentration of fully active enzyme as "carrier" to the unknown sufficient to yield a value of u/v between 1.4 and 3. Then the concentration of inactive enzyme is calculated as for Fig. 1 by Eq. (1), in which c = concentration of active enzyme added plus the concentration of the endogenous active enzyme present in the sample.

In the table are listed the results of two determinations of active and inactive fatty acid synthase on a sample to which differing proportions of "carrier" were added. The variance of the results obtained is relatively small.

Method 2. Rat Liver Acetyl-CoA Carboxylase

This method is employed when the binding of antibody does not inhibit enzyme activity.

Principle. DEAE-cellulose-purified antibody is allowed to react with fully activated acetyl-CoA carboxylase, and then the enzyme–antibody complex is removed by centrifugation and the supernatant solution is assayed for enzyme activity. In other respects the principle involved in carrying out this immunotitration is the same as that for Method 1. Both forward and reverse titrations (constant or variable enzyme concentrations, respectively) may be employed in these titrations, but in this report only the reverse titration will be shown.

Purification of Acetyl-CoA Carboxylase. The purification procedure for this enzyme is a modification of the method of Nakanishi and Numa.[12] Rats are fasted for 48 hr, refed a fat-free diet for 48 hr, sacrificed by decapitation, and the livers then excised. After weighing, the livers are minced and homogenized in 2 volumes (per gram of liver weight) of cold 100 mM potassium phosphate buffer, pH 7.4, 1 mM EDTA, and 5 mM β-mercaptoethanol with a motor-driven Teflon pestle in a Potter–Elvehjem type glass homogenizer. The homogenate is centrifuged at 13,000 g for 20 min and filtered through two layers of cheesecloth, and the filtrate is centrifuged for 75 min at 100,000 g and again filtered.

The 100,000 g supernatant solution is brought to 30% saturation with ammonium sulfate (saturated and neutralized to pH 7.9) at 4°, stirred, and centrifuged. The protein pellet is resuspended in 10 mM potassium phos-

12 S. Nakanishi and S. Numa, *Eur. J. Biochem.* **16**, 161 (1970).

IMMUNOTITRATION OF ACTIVE AND INACTIVE FATTY ACID SYNTHASE IN A LIVER HOMOGENATE OF 48-HR FASTED, 2-HR REFED RATS[a]

Sample titrated	Concentration (μg protein/ml)	Intercept	$(u/v) - 1$	Native homogenate	
				Inactive enzyme (μg/ml)	Active enzyme (μg/ml)
DEAE-cellulose purified fatty acid synthase	150	1.32 ± 0.04	—	—	—
Crude homogenate, 1 ml diluted to 6.5 ml with buffer	177	0.97 ± 0.03	0.36 ± 0.06	414 ± 69	175 ± 10
Crude homogenate, 1 ml diluted to 4 ml with buffer	190	0.81 ± 0.03	0.64 ± 0.06	472 ± 44	160 ± 10

[a] The 100,000 g supernatant solution of liver homogenate (1 ml) was diluted to 6.5 or 4 ml with 0.5 M potassium phosphate, 1 mM EDTA, and 10 mM dithiothreitol, pH 7.0, and then preincubated for 90 min at 30°. A known quantity, 150 μg, of carrier (DEAE-cellulose-purified) rat liver fatty acid synthase was added to the 100,000 g supernatant solution, and the final volume was adjusted to 1 ml. Aliquots of the mixture of diluted 100,000 g supernatant solution plus purified fatty acid synthase protein or purified fatty acid synthase protein alone in increments of 1.5 μg (10 μl) were then added to the buffer in tubes containing a fixed amount of monospecific rabbit anti-rat liver fatty acid synthase antibody, 2.2 μg of protein, and 100 μg of albumin and incubated for 60 min, as described in the text. Assays for enzyme activity were then carried out. For each titration the coordinates for each data point, nanomoles of NADPH oxidized per milliliter per minute versus micrograms of enzymatically active protein, were entered into the linear regression program of the Texas Instruments PL programmable desk computer, and u, v, m, and c were calculated for "carrier" fatty acid synthase alone, 150 μg/ml; for crude sample (1 ml) diluted to 6.5 ml with the addition of 150 μg/ml "carrier"; and for crude sample (1 ml) diluted to 4 ml with the addition of 150 μg of "carrier." The value for m was obtained directly from the slope of the titration line in units of activity per milliliter. Units of activity per milliliter = micrograms of active fatty acid synthase per milliliter × specific activity of active fatty acid synthase; hence, m/specific activity = c. Then the concentration of native active fatty acid synthase in the titrated, diluted samples = c − 150 μg of protein per milliliter. Finally, [c − 150] × dilution factor = micrograms of active fatty acid synthase protein per milliliter in the native homogenate.

phate, pH 7.4, 10 mM potassium citrate, 1 mM EDTA, and 5 mM β-mercaptoethanol and dialyzed against this buffer overnight in the cold. The dialyzed solution is centrifuged and then passed over a DEAE-cellulose column previously washed with the same buffer as that used for dialysis. Unadsorbed protein is washed from the column with the same buffer and then acetyl-CoA carboxylase is eluted from the column with a gradient of equal volumes of the dialysis buffer and 700 mM potassium phosphate, pH 7.4, 10 mM potassium citrate, 1 mM EDTA, and 5 mM β-mercaptoethanol. The most active acetyl-CoA carboxylase fractions are concentrated by ammonium sulfate precipitation, centrifuged, and dialyzed. If pure acetyl-CoA carboxylase is desired, the enzyme is then subjected to gel filtration on Sepharose 4B. Assays for homogeneity of the enzyme are carried out by sodium dodecyl sulfate gel electrophoresis.[13,14]

Production of Antibody. A DEAE-cellulose-purified preparation of acetyl-CoA carboxylase (2.5 mg of protein in 0.5 ml and mixed with an equal volume of Freund's adjuvant) is injected subcutaneously into a white male New Zealand rabbit.[7] At 10-day intervals additional enzyme (2.5 mg of protein) is mixed with incomplete adjuvant and then injected subcutaneously into the same rabbit. After 7 injections blood is collected from an ear vein. Additional bleedings are carried out at 10-day intervals, and the serum immunoglobulin antibody fraction from these bleedings is purified according to the method of Livingston.[7] Antibody may also be prepared against homogeneous acetyl-CoA carboxylase prepared as stated in the preceding paragraph.

Assay of Acetyl-CoA Carboxylase Activity. Acetyl-CoA carboxylase activity is assayed by a modification of the spectrophotometric procedure of Numa.[15] In this procedure acetyl-CoA carboxylase (0.5–10 mU)[16] is incubated in 10 mM potassium citrate, 1 mM dithiothreitol, 50 mM Tris chloride, pH 7.4, for 20–30 min at 30°, and then the following are added consecutively with shaking:

Solution A: bovine serum albumin, 0.72 mg/ml; MgCl$_2$, 10 mM

Solution B: KHCO$_3$, 25 mM; acetyl-CoA, 0.05 mM; NADPH, 0.1 mM

Pigeon liver fatty acid synthase, at least 3.5 mU

ATP, 4 mM

The volume of the incubation mixture is 1 ml, and all concentrations given above are final values.

The change in optical density at 340 nm is followed. A change of 0.1 optical density unit per minute corresponds to an acetyl-CoA carboxylase activity of 8 mu.

It was found in these studies that it is necessary to keep solutions A and B, the fatty acid synthase, and ATP apart until the time of assay in

---

[13] H. Inoue and J. M. Lowenstein, *J. Biol. Chem.* **247**, 4825 (1972).

[14] T. Tanabe, K. Wada, T. Okazaki, and S. Numa, *Eur. J. Biochem.* **57**, 15 (1975).

[15] S. Numa, this series, Vol. 14 [2].

[16] One unit of acetyl-CoA carboxylase activity is defined as the amount of enzyme that catalyzes the carboxylation of 1 μmol of acetyl-CoA per minute. Specific activity of the enzyme is given in units per milligram of protein.

order to obtain maximum activity in all states of purification of acetyl-CoA carboxylase. Also, pigeon liver fatty acid synthase is used instead of the rat liver enzyme in the assay because it is more stable at low ionic strengths and because it does not cross-react with antibody prepared to rat liver proteins.

Protein determinations were made according to the method of Lowry et al.,[17] Gornall et al.,[18] and Bradford[19] as modified by BioRad.[20]

Immunotitration Procedures. The following, in a 300 μl final volume, are added to a 1-ml plastic centrifuge tube: 0.01 M potassium phosphate, pH 7.8, 0.9% NaCl, enzyme (up to 8 mu), and antibody (up to 125 μg of protein). For best results, the enzyme should be in 0.01–0.02 M buffer, pH 7.4, and no more than 100 μl should be added.

The mixture is incubated for 1 hr at 37°, followed by 16 hr at 4°, and then the tubes are centrifuged in a Sorvall centrifuge at 14,000 rpm for 12 min. Two hundred microliters of the supernatant solution are carefully removed with a Hamilton syringe, activated with citrate, and assayed as directed in the preceding section.

In assays for acetyl-CoA carboxylase activity during early stages in the purification of this enzyme, potassium fluorocitrate should be used to activate the enzyme, inasmuch as these preparations contain enzymes that lead to NADPH production from isocitrate. Also, the 100,000 g supernatant solution obtained from homogenates must be passed through a Sephadex G-25 column to remove endogenous substrates that inhibit acetyl-CoA carboxylase activity or interfere in assays of NADPH oxidation.

Types of Immunotitrations. Two types of immunotitrations may be carried out. In one of these (reverse titration) an antibody is titrated with increasing amounts of antigen. In the other (forward titration) a fixed amount of enzyme activity is titrated with increasing amounts of antibody. A linear extrapolation of the decreasing enzyme activity, due to increasing antibody to zero enzyme activity, is defined as the end point or amount of immunoglobulin required to inactivate or titrate the fixed amount (units) of enzyme. The end point relating amount of enzyme neutralized per unit of antibody should be the same for both the forward and reverse immunotitrations. For each of the titrations against enzyme in crude samples, a titration is also made against fully activated acetyl-CoA carboxylase that is purified through the DEAE-cellulose chromatography

---

[17] O. H. Lowry, N. J. Rosebrough, A. L. Farr, and R. J. Randall, *J. Biol. Chem.* **193**, 265 (1951).

[18] A. G. Gornall, C. J. Bardawill, and M. M. David, *J. Biol. Chem.* **177**, 751 (1949).

[19] M. M. Bradford, *Anal. Biochem.* **72**, 248 (1976).

[20] Bulletin 1069, Bio-Rad Laboratories, February 1979.

step. This acetyl-CoA carboxylase is obtained from 48-hr fasted, 48-hr refed animals.

Reverse Immunotitration. The results of a typical set of immunotitrations are shown in Fig. 2A. A sample of DEAE-cellulose-purified acetyl-CoA carboxylase, obtained from 48-hr fasted, 48-hr refed rats, is titrated against a fixed amount of antibody. A linearly increasing activity versus amount of enzyme units added is observed, and the intercept of the plot is displaced from the origin of the axis. A titration of a sample of inactive DEAE-cellulose-purified acetyl-CoA carboxylase obtained from rats starved for 48 hr, to which was added an amount of fully active acetyl-CoA carboxylase equivalent to that used in the first titration, was also made. The plot of the second titration is parallel to that of the first, but with the intercept displaced a much shorter distance from the origin of the axis. This shows a difference in end points and indicates that there is immunotitratable material present in the second sample which is inactive acetyl-CoA carboxylase. Two plots of activity versus amount of enzyme units in the absence of antibody are also shown in this figure. These plots were carried out to establish that the linearity of the coupled assay produces lines that are parallel to the immunotitration plots and pass, as expected, through the origin. Another set of immunotitrations in which the analytical sample consists of partially active acetyl-CoA carboxylase is shown in Fig. 2B. This sample has an inhibitor of acetyl-CoA carboxylase activity present that is lost on further purification of the acetyl-CoA carboxylase of the 100,000 g supernatant solution of livers of 48-hr fasted, 48-hr refed rats. No fully active acetyl-CoA carboxylase was added to this sample when the immunotitrations were carried out.

Forward Immunotitrations. Examples of forward immunotitrations are given by Katiyar et al.[8] and Kleinsek et al.[21] In the latter work[21] plots of immunotitrations of β-hydroxy-β-methylglutaryl-CoA reductase with antibody are presented. The end point of these titrations is established, from a 0 to 50% inactivation of the enzyme, by the intersection of the regression line of the titration points with the x axis. The transition between linearity and saturation kinetics in these titrations is sufficiently sharp so that points beyond the linear range may be excluded from the linear regression plot.

Discussion. The immunotitration of acetyl-CoA carboxylase appears to be a rather special case because of the large size of the polymeric (active) form, 57 S, of this enzyme.[22] Therefore, immunoprecipitates of this

[21] D. A. Kleinsek, A. M. Jabalquinto, and J. W. Porter, *J. Biol. Chem.* **255**, 3918 (1980).
[22] C. Gregolin, E. Ryder, R. C. Warner, A. K. Kleinschmidt, H. C. Chang, and M. D. Lane, *J. Biol. Chem.* **243**, 4236 (1968).

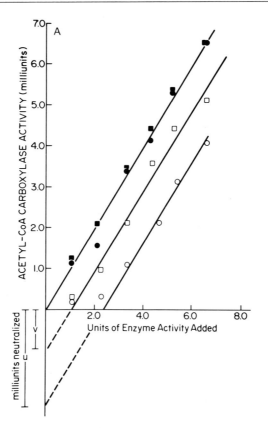

Fig. 2. (A). Immunotitration of DEAE-cellulose-purified acetyl-CoA carboxylase from rats starved for 48 hr: plots of units of activity observed versus units of activity added. ●——●, DEAE-cellulose-purified acetyl-CoA carboxylase from livers of 48-hr fasted, 48-hr refed rats; ■——■, DEAE-cellulose-purified acetyl-CoA carboxylase from 48-hr fasted, 48-hr refed rats, plus DEAE-cellulose-purified acetyl-CoA carboxylase from 48-hr fasted rats; ○——○, same as ●——●, but with 2 units of immunoglobulin added to each assay; □——□, same as ■——■, but with 2 units of immunoglobulin added to each assay.

(B). Immunotitration of acetyl-CoA carboxylase in the 100,000 g supernatant solution of homogenates of rat liver from 48-hr fasted, 48-hr refed animals: plots of units of activity observed versus units of activity added. Both active and inactive enzyme are present. No DEAE-cellulose-purified acetyl-CoA carboxylase was added to the supernatant solution prior to immunotitration. ●——●, DEAE-cellulose-purified acetyl-CoA carboxylase; ■——■, crude extract (100,000 g supernatant solution of homogenate passed through a column of Sephadex G-25); ○——○, same as ●——●, but with 3 units of immunoglobulin added per assay; □——□, same as ■——■, but with 3 units of immunoglobulin added per assay.

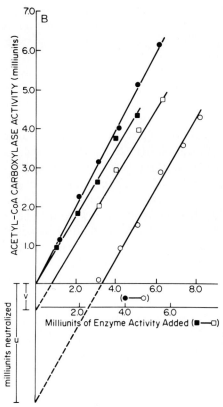

FIG. 2B.

enzyme–antibody complex can be sedimented at low speed because of the size of the complex. The application of this immunoprecipitation technique to smaller enzymes (7 S to 14 S) would require a gravity force sufficient to sediment smaller antigen–antibody complexes without sedimenting the unreacted enzyme. The proper conditions for such separations can be calculated from the clearing times (t) of each species as determined from Eq. (7).[23]

$$t(\text{hr}) = \frac{1}{s} \times \frac{\ln(r_{max}/r_{min})}{(0.10472 \times \text{rpm})^2} \times \frac{10^{14}}{3600} \tag{7}$$

In this equation the clearing time is defined as the time required for the uppermost layer of protein in solution at the start of the centrifugation to

[23] Beckman Instrument Co., "Rotors and Tubes for Preparative Ultracentrifuges." Palo Alto, California, 1972.

reach the bottom of the tube, and r_{max} and r_{min} are the radii measured from the bottom and top of the tube, respectively, that are perpendicular to the axis of rotation. Revolutions per minute is indicated by rpm, and s is the sedimentation coefficient in Svedberg units.

Scope of Application of Methods. Examples of two types of procedures for determining the quantity of active and inactive forms of an enzyme by immunotitration have been presented. Both procedures establish whether differences in the level of activity of enzyme found in animals under differing physiological conditions are related to changes in the amount of protein or to a modification of activity of preexisting enzyme.

In the reporting of our results we have made the assumption that an individual enzyme molecule is either completely active or completely inactive (or of very low activity), as would be the case when the covalent attachment of a prosthetic group is involved. The availability of such results from a series of physiological studies makes it possible to determine which animals have a significant amount of inactive enzyme. This determination then allows for the initiation of the next stage of investigation, namely, the characterization of the mechanism and pathway of modification of the enzyme.

Acknowledgments

The writing of this chapter was supported in part by a grant, AM 01383, from the National Arthritis and Metabolic Diseases Institute, National Institutes of Health, United States Public Health Service, and the Medical Research Service of the Veterans Administration.

Section II

Fatty Acid Activation and Oxidation

PREVIOUSLY PUBLISHED ARTICLES FROM METHODS IN ENZYMOLOGY
RELATED TO SECTION II

Vol. I [85]. Flavin-Linked Aldehyde Oxidase. H. R. Mahler.

Vol. I [89]. Fatty Acid Oxidation in Mitochondria. A. L. Lehninger.

Vol. I [90]. Fatty Acid Oxidation in Higher Plants. P. K. Stumpf.

Vol. I [91]. Butyrate Enzymes of *Clostridium kluyveri*. H. A. Barker.

Vol. I [92]. Butyryl Coenzyme A Dehydrogenase. H. R. Mahler.

Vol. I [93]. Crystalline Crotonase from Ox Liver. J. R. Stern.

Vol. I [96]. Aceto-CoA Kinase. M. E. Jones and F. Lipmann.

Vol. I [97]. Acetate Kinase of Bacteria (Acetokinase). I. A. Rose.

Vol. I [98]. Phosphotransacetylase from *Clostridium kluyveri*. E. R. Stadtman.

Vol. I [99]. Coenzyme A Transphorase from *Clostridium kluyveri*. H. A. Barker, E. R. Stadtman, and A. Kornberg.

Vol. I [112]. Choline Oxidase, J. H. Quastel.

Vol. V [61]. Enzymes of Branched-Chain Amino Acid Metabolism. M. J. Coon and W. G. Robinson.

Vol. V [62]. Assay and Preparation of Yeast Aceto-CoA Kinase. P. Berg.

Vol. V [63]. Activating Enzymes for Higher Fatty Acids. W. P. Jencks.

Vol. V [64]. Activation and Conjugation of Bile Acids. W. H. Elliott.

Vol. V [73]. Acyl Dehydrogenases from Pig and Beef Liver and Heart. H. Beinert.

Vol. V [74]. β-Hydroxybutyryl-CoA Racemase. J. R. Stern.

Vol. V [76]. Choline Dehydrogenase from Rat Liver. T. Kimura and T. P. Singer.

Vol. V [77]. Propionyl-CoA Carboxylase from Pig Heart. A. Tietz and S. Ochoa.

Vol. V [78]. Mitochondrial Propionyl Carboxylase. M. D. Lane and D. R. Halenz.

Vol. V [79]. Methylmalonyl-CoA Isomerase. W. S. Beck.

Vol. IX [67]. Formate Dehydrogenase. J. R. Quayle.

Vol. X [92a]. Microsomal Lipid Peroxidation. L. Ernster and K. Nordenbrand.

Vol. X [114a]. Fatty Acid Activation and Oxidation by Mitochondria. S. G. Van den Bergh.

Vol. XIII [58]. Acetyl-CoA Synthetase. L. T. Webster, Jr.

Vol. XIII [59]. Phosphotransacetylase from *Clostridium kluyberi*. H. R. Klotzsch.

Vol. XIII [60]. Carnitine Acetyltransferase from Pigeon Breast Muscle. J. F. A. Chase.

Vol. XIV [16]. Long-Chain Acyl-CoA Synthetase (GTP-Specific). C. R. Rossi, L. Galzigna, and D. M. Gibson.

Vol. XIV [17]. Δ^3-*cis*, Δ^2-*trans*-Enoyl-CoA Isomerase from Rat Liver Mitochondria. W. Stoffel and W. Ecker.

Vol. XIV [18]. Linoleate Δ^{12}-*cis*, Δ^{11}-*trans*-Isomerase. C. R. Kepler and S. B. Tove.

Vol. XIV [19[. Butyryl-CoA Dehydrogenase from Monkey Liver. D. D. Hoskins.

Vol. XIV [20]. Electron-Transferring Flavoprotein from Monkey Liver. D. D. Hoskins.

Vol. XVIIA [99]. α-Hydroxy-β-keto Acid Reductoisomerase (α-Acetohydroxy Acid Isomeroreductase) (*Neurospora crassa*). K. Kiritani and R. P. Wagner.

Vol. XVIIA [100]. α-Acetohydroxy Acid Isomeroreductase (*Salmonella typhimurium*). S. M. Arfin.

Vol. XVIIA [101]. α,β-Dihydroxy Acid Dehydratase (*Neurospora crassa* and Spinach). K. Kiritani and R. P. Wagner.

[37] Acetate Kinase from *Veillonella alcalescens*[1]

EC 2.7.2.1 ATP: acetate phosphotransferase

By JONATHAN S. NISHIMURA and MICHAEL J. GRIFFITH

$$\text{ATP} + \text{acetate} \xrightleftharpoons{\text{Mg}^{2+}} \text{ADP} + \text{acetyl phosphate}$$

Assay Methods[2,3]

Method I

Principle. The reaction is carried out in the presence of hydroxylamine, which reacts rapidly with acetyl phosphate to form acetylhydroxamic acid. The latter is quantitated as the ferric–acetylhydroxamic acid complex.[4,5] Addition of succinate greatly stimulates the enzymatic reaction.

Reagents

Assay solution in a volume of 0.5 ml,
TEA,[6] 50 mM, pH 8.3
Sodium acetate, 50 mM
Disodium succinate, 50 mM
MgCl$_2$, 5 mM
ATP, sodium salt, 5 mM
Hydroxylamine hydrochloride, neutralized with NaOH, 500 mM
Enzyme, 0.02–0.1 unit
FeCl$_3$–trichloroacetic acid solution (0.37 M FeCl$_3$, 0.2 M trichloroacetic acid, 0.68 N HCl)
Procedure. The assay solution is equilibrated at 37°. The reaction is

---

[1] This work was assisted by Grant AQ-458 from the Robert A. Welch Foundation and Grant GM-17534 from the National Institutes of Health, Institute of General Medical Sciences.
[2] C. M. Bowman, R. O. Valdez, and J. S. Nishimura, *J. Biol. Chem.* **251**, 3117 (1976).
[3] M. J. Griffith and J. S. Nishimura, *J. Biol. Chem.* **254**, 442 (1979).
[4] F. Lippmann and L. C. Tuttle, *J. Biol. Chem.* **159**, 21 (1945).
[5] I. A. Rose, this series, Vol. 1 [97].
[6] Abbreviations used: TEA, triethanolamine adjusted to the desired pH value with HCl; ATP, adenosine triphosphate; ADP, adenosine diphosphate; NADP$^+$ and NADPH, oxidized and reduced species, respectively, of nicotinamide adenine dinucleotide phosphate.

initiated by the introduction of acetate kinase and stopped by the addition of 0.75 ml of $FeCl_3$–trichloroacetic acid solution. The incubation time is usually 5 min. The A_{535} of the reaction solution is measured against a reagent blank. An A_{535} of 0.475 corresponds to the formation of 1 μmol of acetyl phosphate.

Unit. An enzyme unit is defined as that amount of enzyme that catalyzes the formation of 1 μmol of acetylhydroxamic acid per minute under the assay conditions described.

Method II

Principle. In the reverse reaction, ATP formation can be monitored continuously by coupling it to the hexokinase and glucose-6-phosphate dehydrogenase reactions. NADPH that is formed can be determined on the basis of its known extinction coefficient at 340 nm ($6.22 \times 10^3 \ M^{-1}$ cm^{-1}). Addition of succinate is necessary for optimal acetate kinase activity.

Reagents

Assay solution in a volume of 1.0 ml:
  TEA, 100 mM, pH 8.3
  KCl, 50 mM
  Acetyl phosphate, lithium salt, 6.3 mM
  $MgCl_2$, 25 mM
  ADP, sodium salt, 20 mM
  Disodium succinate, 10 mM
  Glucose, 25 mM
  $NADP^+$, 0.5 mM
  Hexokinase, 2.5 units
  Glucose-6-phosphate dehydrogenase, 1 unit
  Enzyme, 0.002–0.025 unit

Procedure. The reaction is carried out at 25° in a silica cuvette of 1-cm light path. Enzyme is added to initiate the reaction. The change in absorbance at 340 nm is monitored over a 5-min period.

Unit. An enzyme unit is defined as that amount of enzyme catalyzing the formation of 1 μmol of ATP per minute, as measured under the conditions described. One unit from this assay method is equal to about 0.5 unit from Method I.

Purification Procedure

The Organism. Strain 221 of *Veillonella alcalescens* (ATCC 17745) was kindly provided by Dr. H. R. Whiteley. The cells are grown for 16 hr at

37° in a static culture. The growth medium contains 2% lactate, 1% yeast extract, and 1% tryptone, pH 7. Kilogram quantities of the organism have been prepared in the past for this laboratory by a commercial firm that no longer offers this service. More recently, the New England Enzyme Center, Tufts University School of Medicine, Boston, Massachusetts, has been contacted for this purpose. The bacterial cell paste is obtained in Dry Ice and stored at $-70°$ until used.

In the purification procedure to be described, all steps are carried out at 4°, unless otherwise indicated.

Preparation of Crude Extract. Extraction of acetate kinase from *V. alcalescens* is facilitated if the bacteria are treated with EDTA and lysozyme prior to sonification. Bladen and Mergenhagen[7] have shown that *Veillonella* produces a thick peptidoglycan wall, uncommon for a gram-negative organism, which is susceptible to lysozyme action after removal of the outer membrane. Electron micrographs of EDTA/lysozyme-treated bacteria show that the thick wall is removed, leaving only the plasma membrane.[7]

Two hundred grams of frozen *V. alcalescens* paste are thawed overnight at 4° and then suspended in 500 ml of 10 mM Na$_2$EDTA. A uniform suspension of the organisms is achieved in a Waring blender. After stirring for 5 min, 500 ml of 10 mM potassium phosphate (pH 6.8) containing 1.0 g of lysozyme are added. Stirring is continued at room temperature for 45 min, and the partially digested bacteria are pelleted by centrifugation at 16,000 g for 10 min. The supernatant fluid is discarded, and the pellet resuspended in 500 ml of 20% (w/v) sucrose. The suspension is stirred for 15 min, then is centrifuged at 16,000 g for 10 min. The cell pellet is resuspended in 300 ml of cold water and sonicated for 10 min in an ice–water bath, at maximum intensity in a Branson sonifier. The supernatant solution is adjusted to contain 100 mM KCl, 50 mM disodium succinate, and 20 mM TEA (pH 8.3). The cell pellet is sonicated again, and the supernatant fluid is treated as described above. The supernatant solutions are combined prior to the next step in the procedure.

Sonication for two 30-min periods without prior treatment with lysozyme results in less than 15% release of acetate kinase.

Heat Treatment. The crude enzyme extract contains dextranase activity that can be eliminated by heat treatment. Acetate kinase is stable to heat treatment,[8] particularly in the presence of succinate.[2]

The crude extract is placed in a covered hot water (5.7 liters) bath maintained at 70°. When the temperature of the extract reaches 50°, the bath is turned off and the cover is removed. The extract is left in the bath

[7] H. A. Bladen and S. E. Mergenhagen, *J. Bacteriol.* **88**, 1482 (1964).
[8] R. A. Pelroy and H. R. Whiteley, *J. Bacteriol.* **105**, 259 (1971).

for 10 min, during which time the temperature reaches 55°. The mixture is cooled rapidly in a salt–ice water bath to 20° and then centrifuged at 16,000 g for 10 min. The supernatant fluid is saved, and the precipitate is discarded.

Protamine Sulfate Treatment. The supernatant fraction from the previous step is stirred briskly as 1.0 g of protamine sulfate is slowly added to it. When addition of the latter is completed, stirring is continued for 15 min. The insoluble matter that is formed is removed by centrifugation at 16,000 g for 10 min and discarded.

Ammonium Sulfate Fractionation. Solid ammonium sulfate (277 g/liter) is added slowly to the solution of partially purified enzyme. The precipitate is removed by centrifugation at 16,000 g for 15 min. More ammonium sulfate (171 g/liter) is added to the supernatant (first supernatant—see the table). The resulting mixture is centrifuged as before. The precipitate (first pellet—see the table) is resuspended in 200 ml of a solution made by dissolving 351 g of ammonium sulfate in 1 liter of TEA buffer (100 mM KCl, 50 mM disodium succinate, 20 mM TEA, pH 8.3). This suspension is stirred at room temperature for 10 min and then centrifuged at 35,000 g for 15 min. The precipitate (second pellet—see the table) is resuspended in 100 ml of a solution prepared by dissolving 277 g of ammonium sulfate in 1 liter of the TEA buffer. The suspension is stirred at room temperature for 10 min and then centrifuged at 35,000 g for 15 min. The precipitate (third pellet—see the table) is dissolved in the TEA buffer. The solution is passed through a 1500-ml Sephadex G-50 column equilibrated with the same buffer, in order to get rid of ammonium sulfate. Fractions containing enzyme activity are pooled and concentrated in an Amicon ultrafiltration apparatus (XM-50 membrane, 50 psi nitrogen).

Sephadex G-200 Gel Filtration. Acetate kinase in 10 ml of TEA buffer is applied to a Sephadex G-200 column (5 × 90 cm) that is equilibrated with the same buffer. A flow rate of 50 ml/hr is established, and 20-ml fractions are collected. The enzyme activity (fractions 32 to 46) is pooled and concentrated by ultrafiltration. The concentrated enzyme is diluted twofold with water.

DEAE-Cellulose Chromatography. The diluted enzyme solution is applied to a DEAE-cellulose column (3.5 × 50 cm) equilibrated with 50 mM KCl, 25 mM disodium succinate, and 10 mM TEA (pH 8.3). The column is then washed with 300 ml of this solution. Contaminating proteins are eluted with a linear 500-ml gradient developed between 250 ml of 50 mM KCl, 25 mM disodium succinate, 10 mM TEA (pH 8.3) and 250 ml of 100 mM KCl, 50 mM disodium succinate, and 20 mM TEA (pH 8.3). The column is then washed with the latter buffer until the enzyme is eluted (400 to 1100 ml). The enzyme activity is concentrated by ultrafiltration.

PURIFICATION OF ACETATE KINASE FROM *Veillonella alcalescens*

Fraction	Volume (ml)	Total units	Protein[a] (mg)	Specific activity (units/mg)	Yield (%)
Crude extract	2000	17,300	40,200	0.43	100
Heat	2000	16,000	34,800	0.46	92
Protamine sulfate	2000	14,700	32,000	0.46	85
Ammonium sulfate					
First supernatant	2300	14,700	28,800	0.51	85
First pellet	200	14,700	6,700	2.2	85
Second pellet	50	14,000	4,200	3.3	81
Third pellet	50	9,000	300	30	52
Sephadex G-50 (concentrated)	20	9,000	300	30	52
Sephadex G-200 (concentrated)	20	8,600	65.7	131	50
DEAE-cellulose (concentrated)	20	6,100	18.6	328	35
Sephadex G-200 (concentrated)	20	6,000	16.9	355	35

[a] Through the Sephadex G-50 step it was assumed that $E_{280}^{0.1\%} = 1.0$.

The protein is about 90% acetate kinase at this point. It can be purified to virtual homogeneity by repetition of either the DEAE-cellulose chromatography or Sephadex G-200 gel filtration. The second DEAE-cellulose chromatography step was generally avoided because of significant losses of enzyme activity.

When kept in 100 mM KCl, 50 mM disodium succinate, and 20 mM TEA (pH 8.3) at 4°, the enzyme is stable for at least 4 weeks.

Properties

Molecular Weight.[2,3] Molecular weight estimation by gel filtration on a Sephadex G-150 column gave a value of 88,000 for acetate kinase. Analysis of the enzyme by sodium dodecyl sulfate–polyacrylamide gel electrophoresis gave a single band of 42,000 MW. These and other criteria[3] indicate that acetate kinase consists of a dimer of identical subunits.

$E_{280}^{0.1\%}$ *of the Enzyme.* A solution of acetate kinase containing 1 mg/ml in 100 mM KCl, 50 mM disodium succinate, and 20 mM TEA (pH 8.3) gave an absorbance of 0.46 at 280 nm. Concentration of the protein was measured by the method of Babul and Stellwagen.[9]

[9] J. Babul and E. Stellwagen, *Anal. Biochem.* **28**, 216 (1969).

Specificity. Adenosine diphosphate, inosine diphosphate, guanosine diphosphate, and thymidine diphosphate are all effective substrates for nucleoside triphosphate synthesis by *V. alcalescens* acetate kinase.[8] Propionate[2] and propionyl phosphate[8] will substitute for acetate and acetyl phosphate, respectively, but with reduced affinity. The divalent metal ion requirement can be met by Mg^{2+} or Co^{2+} and less effectively by Ca^{2+}.[8]

Kinetics and Mechanism of Action. Acetate kinase of *V. alcalescens* is an allosteric enzyme.[2,3,10] It is heterotropically stimulated by succinate. The substrates acetyl phosphate, ADP, acetate, and ATP have been shown to bind cooperatively to the enzyme.[2,3,8,10] Cooperativity has been observed in the binding of ATP and acetyl phosphate to an acetate kinase isolated from another strain of *V. alcalescens* (ATCC 17748).[11] However, it is curious that addition of succinate has no effect on the kinetics of this enzyme.

Allosteric properties have not been observed in acetate kinases isolated from other bacterial sources. However, a phosphoryl enzyme intermediate appears to be indicated, regardless of the origin of the enzyme.[3,12,13] It appears that, with respect to the enzyme described in this chapter, succinate may have an effect on the rate of phosphorylation and dephosphorylation of the active site.[3]

[10] M. J. Griffith and J. S. Nishimura, *J. Biol. Chem.* **254,** 6698 (1979).
[11] F. Yoshimura, *Arch. Biochem. Biophys.* **189,** 424 (1978).
[12] R. S. Anthony and L. B. Spector, *J. Biol. Chem.* **247,** 2120 (1972).
[13] B. C. Webb, J. A. Todhunter, and D. L. Purich, *Arch. Biochem. Biophys.* **173,** 282 (1976).

[38] Acetyl-CoA Synthetase from Bakers' Yeast
(*Saccharomyces cerevisiae*)

EC 6.2.1.1 Acetate : CoA ligase (AMP-forming)

By Eugene P. Frenkel and Richard L. Kitchens

$$\text{Acetate + ATP + CoA} \xrightleftharpoons{\text{Mg}^{2+}} \text{acetyl-CoA + AMP + PP}_i$$

Assay Method[1]

Principle. Enzymatic activity is measured at all stages of the purification by a spectrophotometric technique based on the formation of NADH from a coupled reaction system utilizing citrate synthase and malate dehydrogenase.

$$\text{Acetate + ATP + CoA} \xrightarrow{\text{Mg}^{2+}} \text{acetyl-CoA + AMP + PP}_i \quad (1)$$
$$\text{Acetyl-CoA + oxaloacetate} \rightarrow \text{citrate + CoA} \quad (2)$$
$$\text{Malate + NAD}^+ \rightleftharpoons \text{oxaloacetate + NADH} \quad (3)$$

$$\text{Acetate + ATP + malate + NAD}^+ \rightarrow \text{citrate + AMP + PP}_i + \text{NADH}$$

The equilibrium of the malate dehydrogenase reaction[2] lies strongly in the reverse direction so that only a small amount of oxaloacetate is formed as substrate for the citrate synthase reaction [Eq. (2)]. A slight baseline drift in the direction of NADH formation is observed in the absence of acetate using crude extracts; the drift is subtracted from the value measured with added substrate.

Reagents

Tris, 1 *M*, adjusted to pH 7.6 with HCl
MgCl$_2$, 1 *M*
ATP, 0.3 *M*, adjusted to pH 7–8 with KOH
Malate, 0.5 *M*, adjusted to pH 7–8 with KOH
Dithiothreitol (DTT), 1 *M*
NAD$^+$, 0.1 *M* in 0.1 *M* Tris, pH 7.6
NADH, 0.01 *M* in 0.1 *M* Tris, pH 7.6
CoA, 0.02 *M* (lithium salt in water)
Potassium acetate, 1 *M*, pH 7–8
Citrate synthase, 80 units/mg in (NH$_4$)$_2$SO$_4$ suspension (10 mg/ml)

---

[1] E. P. Frenkel and R. L. Kitchens, *J. Biol. Chem.* **252**, 504 (1977).
[2] W. W. Farrar and K. M. Plowman, *Int. J. Biochem.* **6**, 537 (1975).

METHODS IN ENZYMOLOGY, VOL. 71

Malate dehydrogenase, 1000 units/mg in $(NH_4)_2SO_4$ suspension (10 mg/ml)

Assay. The final reaction volume of 1.0 ml contains the following components (in micromoles): Tris-HCl, pH 7.6, 100; $MgCl_2$, 10; ATP, 6; malate, 5; DTT, 2.5; NAD^+, 1; NADH, 0.1; CoA, 0.2; acetate, 10; as well as citrate synthase, 25 μg, and malate dehydrogenase, 20 μg. The reaction is started by adding the acetate, and the rate of the reaction is measured spectrophotometrically at 340 nm on a recording spectrophotometer at ambient temperature. The extinction coefficient for NADH is 6.22×10^6 cm^2/mol.

For convenience, a reaction mixture may be prepared in the following manner and stored at 0–4° for up to a week. Mix together for 20 assays 14.3 ml of distilled water; 2 ml of Tris-HCl, pH 7.6; 0.2 ml of $MgCl_2$; 0.4 ml of ATP; 0.2 ml of malate; 0.05 ml of DTT; 0.2 ml of NAD^+; 0.2 ml of NADH; 0.2 ml of CoA; 0.2 ml of acetate, 0.05 ml of citrate synthase; and 0.04 ml of malate dehydrogenase. For each assay, 0.9 ml of the above reaction mixture is mixed with 0.1 ml of enzyme. In the assay of crude extracts the acetate is omitted from the reaction mixture and its volume is replaced by water. The blank rate is observed, and 0.01 ml of acetate is then added to start the complete reaction.

Purification Procedure

All operations are performed at 0–4° unless otherwise stated. Protein is determined by the Lowry method.[3]

Step 1. Preparation of Crude Extract. One to two pounds of yeast cake were crumbled into Dewar flasks containing liquid nitrogen; the excess liquid nitrogen was decanted, and the frozen contents were transferred to a 1-gallon capped, stainless steel Waring blender. The yeast was blended to a fine powder, weighed, and slurried with 1.4 volumes of 0.05 M K_2HPO_4, 7 mM 2-mercaptoethanol, and 0.5 mM EDTA. The slurry was permitted to thaw to the point of ice crystal dispersion and was then centrifuged at 13,000 g for 30 min. The extract was decanted through glass wool, the pH was adjusted to 6.8–7.0 with 1 N KOH, and an aliquot was removed for enzyme assay.

Step 2. First Protamine Sulfate Treatment. A 2% solution of protamine sulfate was prepared at 25° by dissolving the compound in 0.02 M potassium phosphate buffer, pH 7.5, to yield a final pH of 6.8. While the crude extract was maintained on ice, 0.12 volume of the protamine sulfate solution was slowly added with constant stirring. After 5 min of stirring, the

[3] O. H. Lowry, N. J. Rosebrough, A. L. Farr, and R. J. Randall, *J. Biol. Chem.* **193**, 265 (1951).

extract was centrifuged at 13,000 g for 5 min. The supernatant solution was assayed for activity, placed in dialysis bags, and dialyzed overnight at 0–4° in 25 liters of 0.02 M potassium phosphate buffer, pH 7.5, 7 mM 2-mercaptoethanol, and 0.5 mM EDTA. The dialysis bags were prepared by boiling for 10 min in 5 mM EDTA, then rinsed with distilled water. After dialysis an aliquot of the solution was removed for assay of enzyme activity.

Step 3. Second Protamine Sulfate Treatment. Protamine sulfate (0.06 volume of the above solution) was added to the dialyzed enzyme. The preparation was mixed for 5 min and centrifuged at 13,000 g for 5 min; the pellet was dissolved with constant stirring in 200 ml of 0.2 M potassium phosphate buffer containing 0.2 M ammonium sulfate, pH 7.0, 7 mM 2-mercaptoethanol, and 0.5 mM EDTA. Enzyme activity was assayed in an aliquot of the solution.

Step 4. Polyethylene Glycol Fractionation. Polyethylene glycol (molecular weight 6000–7500) prepared as a 25% solution in glass-distilled water at room temperature was added slowly to the solution from step 3 to achieve a final concentration of 4%. After 10 min of stirring, the suspension was centrifuged at 13,000 g for 10 min. The pellet was discarded, and polyethylene glycol was again added to the supernatant fraction to achieve a final concentration of 9.5%. The suspension was stirred for 20 min and centrifuged. The resulting pellet was dissolved in 100 ml of 0.2 M potassium phosphate buffer, pH 7.0, 7 mM 2-mercaptoethanol, and 0.5 mM EDTA. The enzyme activity was assayed.

Step 5. Alumina Cγ Gel Fractionation. The protein concentration of the solution from step 4 was determined, and alumina Cγ gel, equivalent to 1 mg dry weight, was added per milligram of protein. The gel suspension was stirred gently for 10 min and centrifuged briefly at 7000 g. The supernatant fraction was saved, and the gel pellet was suspended in 100 ml of 150 mM potassium phosphate buffer, pH 7.5, 14 mM 2-mercaptoethanol, and 0.5 mM EDTA. The pH was adjusted to 7.5 with 1 N KOH. After 10 min of gentle stirring, the preparation was again centrifuged, and the supernatant fraction was pooled with the previous fraction. The gel pellet was washed a total of three times. An aliquot of the resulting solution was removed for assay, and the remainder was dialyzed overnight against 2 liters of 3.6 M ammonium sulfate in 0.1 M potassium phosphate buffer, pH 7.5, 14 mM 2 mercaptoethanol, and 0.5 mM EDTA.

Step 6. BioGel Chromatography. The enzyme precipitated during the dialysis was centrifuged at 20,000 g for 15 min, and the pellet was resuspended in 30 ml of 2.2 M ammonium sulfate and centrifuged again. The resulting pellet was dissolved in 5 ml of 18 mM potassium phosphate buffer, pH 7.5, 14 mM 2-mercaptoethanol, and 0.5 mM EDTA. This solution was placed on a BioGel A-0.5m column (2.6 × 94 cm) equilibrated with the above buffer mixture plus 2.5 mM ATP. The flow rate was 13–15 ml/hr, and 5-ml fractions were collected. Fractions were assayed for enzyme activity, and fractions with high specific activity were pooled.

Step 7. DEAE-Cellulose Chromatography. The pooled enzyme fractions were placed on a DEAE-cellulose column (1 × 40 cm) equilibrated with the above buffer mixture less ATP. The column was then eluted with a 200-ml linear gradient from 20 mM to 150 mM phosphate buffer, pH 7.5. The flow rate was 20 ml/hr, and 4-ml fractions were collected into tubes containing 40 μl of 250 mM ATP, pH 7.5. Fractions were assayed for enzyme activity, and active fractions were pooled.

The pooled fractions were concentrated by dialysis against ammonium sulfate as described above, and the contents of the bags were centrifuged at 20,000 g for 15 min. The pellet was then suspended in 2.4 M ammonium sulfate in 0.1 M potassium phosphate buffer, pH 7.5, 7 mM 2-mercaptoethanol, and 0.5 mM EDTA and stored at 0–4°.

The results of a purification starting with 450 g of bakers' yeast cake are shown in Table I. The final step yielded enzyme with a specific activity of 44 units/mg at 25°, which amounted to a 220-fold purification over the crude extract.

Purity. Sodium dodecyl sulfate (SDS)–polyacrylamide gel electrophoresis of the purified enzyme, previously boiled in 3% SDS and 2% 2-mercaptoethanol and dialyzed at room temperature, revealed a single protein band. When the boiling step was omitted, multiple bands appeared

TABLE I

PURIFICATION OF ACETYL-CoA SYNTHETASE FROM BAKERS' YEAST

Procedure	Volume (ml)	Total activity (units)	Total protein (mg)	Specific activity (units/mg)	Purification (fold)	Yield (%)
Crude extract	785	6,029	30,144	0.20	—	100
First protamine sulfate	842	6,088	19,029	0.32	1.6	101
Dialysis	958	5,729	—	—	—	95
Second protamine sulfate	243	4,811	3,694	1.3	6.5	80
Polyethylene glycol 4–9.5%	101	3,495	1,202	2.9	14.5	58
Alumina Cγ gel	397	2,473	322	7.7	38.5	41
Concentration with (NH$_4$)$_2$SO$_4$	6.5	1,748	282	6.2	31.0	29
BioGel A-0.5m chromatography	21	1,327	37	36	180	22
DEAE-cellulose gradient chromatography	23	543	12.3	44	220	9

on the gels. This is presumably due to the presence of small amounts of proteases that are inactivated by boiling. The enzyme was also pure by the criterion of high-speed sedimentation equilibrium analysis.

The acetyl-CoA synthetase preparation showed no carnitine acetyltransferase, pyruvate carboxylase, NADH oxidase, propionyl-CoA carboxylase, ATPase, acetyl-CoA carboxylase, citrate synthase, alcohol dehydrogenase, myokinase, glutamate dehydrogenase, acetyl-CoA or propionyl-CoA deacylase, ATP-citrate lyase, isocitrate dehydrogenase, or α-ketoglutarate dehydrogenase activity. Malate dehydrogenase activity (0.06–0.08%) was detected in the preparation.

Properties

Stability. The purified enzyme preparation (20–40 units/mg) was stable for at least 2 months when stored suspended in 3 M ammonium sulfate at 0–4°. When the enzyme was maintained in 3 M ammonium sulfate at room temperature for 72 hr, it showed only 10% loss of activity. Storage of the partially purified enzyme following the Cγ gel fractionation procedure in 0.5 M potassium phosphate, pH 7.5, 7 mM 2-mercaptoethanol and 0.5 mM EDTA at −20° resulted in no loss of enzyme activity over a 6-month period as long as the protein concentration was >5 mg/ml. The partially purified enzyme could be stored indefinitely when prepared as an ammonium sulfate pellet and maintained at −20°. Thawing of frozen aliquots of the enzyme purified by BioGel chromatography or at the subsequent purification step resulted in the formation of fibrillar strands and an 80–90% loss of enzyme activity.

Molecular Weight and Sedimentation Analysis. On sedimentation equilibrium analysis a plot of the log of the fringe number versus the radius squared was essentially linear. These studies demonstrated a M_w of 154,000 and a M_n of 148,000 providing a mean molecular weight of 151,000 with a maximum estimated error of 2%. Sedimentation velocity analysis gave a sedimentation coefficient ($S^{\circ}_{20,w}$) of 8.65S. Parallel sedimentation velocity studies in the presence or in the absence of ATP did not result in any differences in the sedimentation behavior of the enzyme. SDS–polyacrylamide gel electrophoresis demonstrated a single band with an estimated molecular weight of 78,000 providing evidence of two identical enzyme subunits.

Enzyme Kinetics: pH Optimum. In the coupled assay employing citrate synthase and malate dehydrogenase, 90% of optimum enzyme activity was obtained in pH range 7.3–8.1. Approximately 50% of maximum values were obtained at pH 6.8 and 8.8.

Kinetic Analysis. Kinetic studies of the bovine heart enzyme by Farrar

TABLE II
APPARENT MICHAELIS CONSTANTS OF SUBSTRATES FOR ACETYL-CoA SYNTHETASE BY
SUBSTRATE PAIR ANALYSIS

Substrate pair	Nonvaried substrate and concentration (mM)	Apparent K_m (mM)		
		ATP	Acetate	CoA
(a) CoA-acetate	ATP, 1.5	—	0.24 ± 0.009^a	0.034 ± 0.001
(b) ATP-CoA	Acetate, 0.5	1.14 ± 0.06	—	0.040 ± 0.002
(c) ATP-acetate	CoA, 0.07	0.74 ± 0.03	0.17 ± 0.009	—

[a] Standard error of the mean.

and Plowman[2] have shown an ordered Bi Uni Uni Bi Ping-Pong mechanism, with ordered substrate addition and release. The kinetics of the yeast-derived enzyme were similarly investigated by varying a substrate pair at a fixed, nonsaturating concentration of the third substrate; the three possible combinations of substrate pairs were studied. Varied and fixed concentrations of ATP, CoA, and acetate were evaluated at room temperature in the coupled citrate synthase–malate dehydrogenase assay system. When the CoA–acetate or ATP–CoA pairs were the varied substrates, a series of parallel lines were obtained. If the ATP–acetate pair was varied, a series of intersecting lines were seen. The kinetic constants from these studies were determined with the Fortran computer program of Cleland[4] using for the intersecting line patterns the equation:

$$V = \frac{VAB}{K_{ia}K_b + K_bA + K_aB + AB}$$

and

$$V = \frac{VAB}{K_bA + K_aB + AB}$$

for the parallel line patterns. The apparent Michaelis constants from this substrate pair analysis are shown in Table II. The initial velocity patterns observed for the yeast-derived enzyme were similar to those obtained for the heart enzyme,[2] suggesting that acetyl-CoA synthetase derived from both yeast and mammalian heart catalyze the reaction by a similar if not identical mechanism. The Michaelis constants, however, for the yeast-derived enzyme are considerably different from those found with bovine heart (Table III).

[4] W. W. Cleland, *Nature* (*London*) **198**, 463 (1963).

TABLE III

COMPARATIVE CHARACTERISTICS OF ACETYL-CoA SYNTHETASE OBTAINED FROM MAMMALIAN[5-9] AND YEAST SOURCES

	Bovine heart	Yeast
Source	Bovine heart mitochondrial preparation	Bakers' yeast autolyzed preparation
Specific activity of purified enzyme	36 units/mg protein (at 37°)	66 units/mg protein (at 37°)
Molecular weight		
Intact enzyme	57,000	151,000
Subunits	?	78,000
pH optimum	8.0	7.6
90% activity range	7.8–8.4	7.3–8.1
59% activity range	6.8–9.2	6.8–8.8
Substrate specificity (fatty acids)	Acetate, propionate, acrylate, butyrate	Acetate, propionate, acrylate
Kinetic constants (K_m)		
Varied substrate pairs		
Acetate	0.82 mM, 0.28 mM	0.24 mM, 0.17 mM
CoA	0.17 mM, 0.18 mM	0.034 mM, 0.040 mM
ATP	0.25 mM, 0.33 mM	1.14 mM, 0.74 mM
Varied individual substrates		
Acetate	0.8 mM, 0.2 mM	0.28 mM
Propionate	10 mM (V_{max} = same as acetate)	10 mM (V_{max} = 46% of acetate)
Acrylate	10 mM (V_{max} = same as acetate)	17 mM (V_{max} = 21% of acetate)
CoA	0.4 mM	0.035 mM
ATP	0.9 mM	1.2 mM
Mg^{2+}	0.7 mM	4.0 mM

For comparison with the data in the literature on the enzyme derived from other sources, kinetic studies were also performed in which individual substrates were varied. These studies utilized the citrate synthase–malate dehydrogenase coupled system except where propionate or acrylate were the varied substrates, in which case the coupled system (described below) for the formation of AMP was used. Acetate was studied in both systems, yielding identical results. The Michaelis constants obtained in this manner are compared with those of the bovine heart enzyme (Table III).[5-9] Propionate and acetate were competitive inhibitors (K_i, propionate = 10 mM). AMP was a competitive inhibitor of ATP (K_i = 0.1 mM). Pyrophosphate showed noncompetitive inhibition of acetate; at approximately 0.8 mM PP_i, the V_{max} was reduced to one-half of that obtained in the absence of PP_i. Finally, NAD^+ added incrementally to a concentration of 2.0 mM did not inhibit the enzyme in the coupled assay system involving citrate synthase and malate dehydrogenase.

Substrate Specificity. Specificity of the enzyme for substrates other than acetate was studied using a coupled assay to determine AMP formation. The assay involved myokinase, pyruvate kinase, and lactate dehydrogenase.[1] At 10 mM concentrations, propionate was 36% as active, acrylate was 12% as active, and formate was 0.5% as active as acetate. Trace contamination of the formate with acetate could not be ruled out.

Effects of Anions and Cations. The inhibitory effects of a variety of monovalent anions and cations on enzyme activity were evaluated.[1] All monovalent cations were inhibitory at 200 mM. Inhibition by sulfate ions was considerably greater than inhibition by chloride or bicarbonate. Tris was the least inhibitory ion, and no differences were noted with the 50 mM and 100 mM concentrations of Tris-HCl.

When 10 mM $MnCl_2$ was used in the enzyme assay, it showed 38% of the rate obtained with $MgCl_2$. Incubation of the enzyme in the presence of 10 mM $MnCl_2$ caused progressive inactivation of the enzyme that was not reversible by the subsequent addition of $MgCl_2$.

Acknowledgment

This work was supported by NIH Grant CA23115, Veterans Administration MRIS 1450, the Meadows Foundation, the Heddens-Good Foundation, and the McDermott Foundation.

[5] L. T. Webster, Jr., *J. Biol. Chem.* **240,** 4158 (1965).
[6] L. T. Webster, Jr., *J. Biol. Chem.* **240,** 4164 (1965).
[7] J. C. Londesborough, S. L. Yuan, and L. T. Webster, Jr., *Biochem. J.* **133,** 23 (1973).
[8] J. R. Williamson and B. E. Corkey, this series, Vol. 13, p. 494.
[9] D. L. De Vincenzi and H. P. Klein, *Fed. Proc.* **29,** 872 (1970).

[39] Long-Chain Acyl-CoA Synthetases I and II from
Candida lipolytica

EC 6.2.1.3 Acid : CoA ligase (AMP-forming)

By Kohei Hosaka, Masayoshi Mishina, Tatsuyuki Kamiryo, and
Shosaku Numa

$$ATP + RCOOH + CoASH \rightleftharpoons PP_i + AMP + RCOSCoA$$

The presence of two distinct long-chain acyl-CoA synthetases has recently been discovered in *Candida lipolytica,* a hydrocarbon-utilizing yeast.[1-5] The two enzymes, designated as acyl-CoA synthetase I and II, play different physiological roles in fatty acid metabolism. The assay methods, separation procedure, properties, and physiological roles of the two acyl-CoA synthetases, as well as the purification procedure of acyl-CoA synthetase I, are described in this chapter.

Assay Methods

Principle. Routinely, the isotopic method or the spectrophotometric method is used. In the isotopic assay,[6] the rate of [14C]palmitoyl-CoA formation from [14C]palmitate is determined by measuring the radioactivity in the aqueous phase after extraction of the unreacted [14C]palmitic acid. In the spectrophotometric assay, the rate of AMP formation is determined by coupling the reaction of acyl-CoA synthetase with those of adenylate kinase, pyruvate kinase and lactate dehydrogenase and following the oxidation of NADH. Occasionally, the hydroxamic acid method,[7] in which the rate of thioester formation is measured, is used.

[1] T. Kamiryo, M. Mishina, S. Tashiro, and S. Numa, *Proc. Natl. Acad. Sci. U.S.A.* **74,** 4947 (1977).

[2] M. Mishina, T. Kamiryo, S. Tashiro, and S. Numa, *Eur. J. Biochem.* **82,** 347 (1978).

[3] M. Mishina, T. Kamiryo, S. Tashiro, T. Hagihara, A. Tanaka, S. Fukui, M. Osumi, and S. Numa, *Eur. J. Biochem.* **89,** 321 (1978).

[4] K. Hosaka, M. Mishina, T. Tanaka, T. Kamiryo, and S. Numa, *Eur. J. Biochem.* **93,** 197 (1979).

[5] T. Kamiryo, Y. Nishikawa, M. Mishina, M. Terao, and S. Numa, *Proc. Natl. Acad. Sci. U.S.A.* **76,** 4390 (1979).

[6] J. Bar-Tana, G. Rose, and B. Shapiro, this series, Vol. 35 [15].

[7] F. Lipmann and L. C. Tuttle, *Biochim. Biophys. Acta* **4,** 301 (1950).

METHODS IN ENZYMOLOGY, VOL. 71

Isotopic Method[2]

This method is used for enzyme preparations from all steps.

Reagents

Tris-HCl buffer, 0.5 M, pH 7.5
Dithiothreitol, 0.1 M
$MgCl_2$, 0.1 M
ATP, 0.15 M
CoA, 20 mM
Potassium [U-^{14}C]- or [1-^{14}C]palmitate (0.2 Ci/mol), 10 mM, in 16 mM
  Triton X-100[8]
Soybean phosphatidylcholine, 20 mM, in 5 mM dithiothreitol, dispersed by sonic oscillation
Isopropanol/n-heptane/1 M H_2SO_4 (40/10/1, v/v/v)
Palmitic acid, 4 mg/ml, in n-heptane
Scintillator solution, 4 g of 2,5-diphenyloxazole and 0.1 g of 1,4-bis[2-(4-methyl-5-phenyloxazolyl)] benzene in 1 liter of toluene plus 0.5 liter of Triton X-100
Enzyme: Dilutions are made with 20 mM potassium phosphate buffer, pH 7.4, containing 2 mM Triton X-100 and 5 mM 2-mercaptoethanol (hereafter referred to as buffer A).

Procedure for the Assay of Acyl-CoA Synthetase I. The standard assay mixture contains in a total volume of 0.2 ml (in micromoles, unless otherwise specified): Tris-HCl buffer, pH 7.5, 20; dithiothreitol, 1; $MgCl_2$, 2; ATP, 3; CoA, 0.2; Triton X-100, 0.32; potassium [^{14}C]palmitate (0.2 Ci/mol), 0.2; enzyme, up to 4 mU. The reaction is started by the addition of enzyme. After incubation at 25° for 5 min or 10 min, the reaction is terminated by adding 2.5 ml of a mixture containing isopropanol, n-heptane, and 1 M H_2SO_4 (40/10/1, v/v/v). Then, 1.5 ml of n-heptane and 1 ml of H_2O are added. After thorough mixing and subsequent centrifugation, the upper layer, containing the unreacted palmitic acid, is removed by suction, and the lower layer is washed twice with 2 ml of n-heptane containing 4 mg of unlabeled palmitic acid per milliliter, each time discarding the upper layer. A 1-ml aliquot of the lower layer is then assayed for radioactivity by scintillation counting with 10 ml of the scintillator solution. Under the assay conditions described, the reaction follows zero-order kinetics, and the initial rate of reaction is proportional to enzyme concentration. The radioactive product in the aqueous phase has been identified as palmitoyl-CoA by chromatographic analysis of the thioester as well as of the hydroxamic acid derived from it.[2,4]

---

[8] The average molecular weight of Triton X-100 is taken to be 628.

Procedure for the Assay of Acyl-CoA Synthetase II. Acyl-CoA synthetase II requires phosphatidylcholine for its activity. The assay procedure is the same as that used for acyl-CoA synthetase I, except that the assay mixture contains 0.5 μmol of soybean phosphatidylcholine. The activity of acyl-CoA synthetase II is measured as phosphatidylcholine-dependent activity, i.e., activity measured in the presence of phosphatidylcholine minus activity measured in the absence of phosphatidylcholine.

Spectrophotometric Method[4]

This method is used for purified acyl-CoA synthetase I preparations that contain no interfering enzymes such as adenosinetriphosphatase and NADH-oxidizing enzymes.

Reagents

Tris-HCl buffer, 0.5 M, pH 7.4
Dithiothreitol, 0.1 M
MgCl$_2$, 0.1 M
ATP, 0.15 M
CoA, 20 mM
Potassium oleate, 10 mM, in 16 mM Triton X-100
Triton X-100, 16 mM
Potassium phosphoenolpyruvate, 50 mM
NADH, 10 mM
Adenylate kinase (Boehringer), 2 mg/ml
Pyruvate kinase (Boehringer), 10 mg/ml
Lactate dehydrogenase (Boehringer), 10 mg/ml
Enzyme: Dilutions are made with buffer A.

Procedure. The standard assay mixture contains in a total volume of 1 ml (in micromoles, unless otherwise specified): Tris-HCl buffer, pH 7.4, 100; dithiothreitol, 5; MgCl$_2$, 10; ATP, 7.5; CoA, 1; potassium oleate, 0.25; Triton X-100, 1.6; potassium phosphoenolpyruvate, 0.2; NADH, 0.15; adenylate kinase, 20 μg; pyruvate kinase, 30 μg; lactate dehydrogenase, 30 μg; enzyme, up to 30 mU. A mixture (0.95 ml) containing all ingredients except CoA is preincubated at 25° for 1 min in a cuvette with 1-cm light path, and the reaction is started by adding 0.05 ml of 20 mM CoA. The oxidation of NADH is followed at 25° with a recording Eppendorf photometer at wavelength 334 nm (340 nm when a Shimadzu spectrophotometer is used). Under the assay conditions described, the reaction follows zero-order kinetics for at least 1 min, and the initial rate of reaction is proportional to enzyme concentration. In this assay, the formation of 1 mol of AMP corresponds to the oxidation of 2 mol of NADH. The ratio of the activity measured by the spectrophotometric method (with the standard assay mixture except that oleate is replaced by palmitate) to that measured by the isotopic method is 1.6.

Hydroxamic Acid Method[2,9,10]

This method is rather insensitive and is used only when the other two methods are not applicable.

Units. One unit of acyl-CoA synthetase activity is defined as that amount which catalyzes the formation of 1 μmol of palmitoyl-CoA or AMP per minute under the assay conditions used. Specific activity is expressed as units per milligram of protein. Protein is determined by the method of Lowry *et al.*[11] with bovine serum albumin as a standard; particulate protein is solubilized with 48 mM sodium deoxycholate prior to the determination.

Separation of Acyl-CoA Synthetases I and II[2]

All operations are carried out at 0–4° unless otherwise specified.

Preparation of Particulate Fraction. Candida lipolytica NRRL Y-6795 is cultivated at 25° with agitation at 72 reciprocations per minute in a 500-ml medium containing 1% oleic acid, 0.1% Bacto-yeast extract, 0.5% $NH_4H_2PO_4$, 0.25% KH_2PO_4, 0.01% NaOH, 0.1% $MgSO_4 \cdot 7 H_2O$ and 0.002% $FeCl_3 \cdot 6 H_2O$. Seven grams of wet yeast cells, harvested at mid-logarithmic phase, are washed with 50 ml of 0.1 M potassium phosphate buffer, pH 6.5, containing 1% Brij 58. The washed cells are suspended in 14 ml of 0.1 M potassium phosphate buffer, pH 6.5, containing 5 mM 2-mercaptoethanol and 1 mM EDTA. The suspended cells are disrupted with 21 g of glass beads 0.45–0.5 mm in diameter (3 g of glass beads per gram of wet cells) in a 50-ml bottle of a Braun cell homogenizer at high speed for 60 sec under cooling with liquid CO_2. The cell homogenate is centrifuged at 2000 g for 10 min to remove cell debris, and the supernatant is further centrifuged at 230,000 g for 1 hr. The resulting precipitate (hereafter referred to as particulate fraction) is suspended in a minimal volume of 0.1 M potassium phosphate buffer, pH 6.5, containing 5 mM 2-mercaptoethanol and 1 mM EDTA to yield a protein concentration of 40–70 mg/ml.

Resolution of Cellular Particles. The suspension of the particulate frac-

[9] A. Kornberg and W. E. Pricer, Jr., *J. Biol. Chem.* **204**, 329 (1953).

[10] S. V. Pande and J. F. Mead, *J. Biol. Chem.* **243**, 352 (1968).

[11] O. H. Lowry, N. J. Rosebrough, A. L. Farr, and R. J. Randall, *J. Biol. Chem.* **193**, 265 (1951).

tion is mixed with a solution containing Triton X-100, Tris-HCl buffer, pH 7.5, dithiothreitol, and glycerol, the final concentration of which are 5 mM, 50 mM, 1 mM, and 20% (w/v), respectively, to yield a protein concentration of 4.5 mg/ml (total volume, 5 ml). The mixture is allowed to stand for 1 hr and then centrifuged at 105,000 g for 1 hr. The resulting supernatant and precipitate (suspended in 50 mM Tris-HCl buffer, pH 7.5, containing 1 mM dithiothreitol and 20% (w/v) glycerol) are assayed for acyl-CoA synthetases I and II. In a typical experiment, 92% of the acyl-CoA synthetase I activity recovered is found in the supernatant, and 87% of the acyl-CoA synthetase II activity recovered is present in the precipitate.

Purification Procedure for Acyl-CoA Synthetase I[4]

All operations are carried out at 0–4° unless otherwise specified.

Preparation of Particulate Fraction. Candida lipolytica NRRL Y-6795 is cultivated as described above, except that the oleic acid in the medium is replaced by 2% glucose and that the total volume of the medium is 12 liters (in twenty-four 500-ml portions). Wet yeast cells (250 g), harvested at mid-logarithmic phase, are washed with 1 liter of 0.1 M potassium phosphate buffer, pH 6.5. The washed cells are suspended in 500 ml of 0.1 M potassium phosphate buffer, pH 6.5, containing 5 mM 2-mercaptoethanol and 1 mM EDTA and then disrupted with glass beads in the same manner as described above, except that 70-ml bottles are used; each bottle contains 45 ml of the cell suspension and 45 g of glass beads. The combined homogenates are centrifuged at 8000 g for 15 min to remove cell debris, and the supernatant is further centrifuged at 230,000 g for 1 hr. The particulate fraction precipitated is suspended in 0.1 M potassium phosphate buffer, pH 6.5, containing 5 mM 2-mercaptoethanol and 1 mM EDTA to yield a protein concentration of 46 mg/ml (72 ml).

Resolution of Cellular Particles. The suspension of the particulate fraction is mixed with a solution containing Triton X-100, potassium phosphate buffer, pH 7.4, 2-mercaptoethanol, and EDTA, the final concentrations of which are 5 mM, 50 mM, 5 mM, and 0.5 mM, respectively, to give a protein concentration of 11 mg/ml. The mixture is allowed to stand for 1 hr and is then centrifuged at 230,000 g for 1 hr. The resulting supernatant is collected (265 ml).

Phosphocellulose Chromatography. The 230,000 g supernatant is diluted with 1.5 volumes of a solution containing 2 mM Triton X-100 and 5 mM 2-mercaptoethanol. The diluted solution is applied to a phosphocellulose (P-11) column (5.2 × 7.5 cm) equilibrated with buffer A. The column is washed with 2 column volumes of buffer A and then eluted with a linear concentration gradient established between 2.5 column volumes of buffer A and the same volume of 0.5 M potassium phosphate buffer, pH 7.4, containing 2 mM Triton X-100 and 5 mM 2-mercaptoethanol; the flow rate is approximately 90 ml/hr, and 15-ml fractions are collected. The enzyme emerges in a single peak at a phosphate concentration range between 0.05 M and 0.15 M. Fractions exhibiting specific activities higher than 0.28 unit/mg are combined (90 ml).

Blue-Sepharose Chromatography. The pooled enzyme solution from the phosphocellulose chromatography step is passed through a Sephadex G-50 column (3.8 × 35 cm) equilibrated with buffer A. All protein-containing fractions are combined and applied to a Blue-Sepharose (CL-6B) column (1.6 × 10 cm) equilibrated with buffer A. The column is washed with 1 column volume of buffer A and then with 2 column volumes of buffer A containing 10 mM ATP. Elution is carried out with a linear concentration gradient made between 3.75 column volumes of buffer A containing 10 mM ATP and the same volume of buffer A containing 1.5 M NaCl and 10 mM ATP at a flow rate of approximately 30 ml/hr; 10-ml fractions are collected. The enzyme appears in a single peak at an NaCl concentration range between 0.1 M and 0.3 M. Fractions exhibiting specific activities higher than 2.6 units/mg are combined (50 ml).

Sephadex G-100 Chromatography. The pooled enzyme solution from the Blue-Sepharose chromatography step is concentrated to 10 ml by ultrafiltration with a Diaflo membrane filter PM-30 and applied to a Sephadex G-100 column (2.7 × 87 cm) equilibrated with buffer A. The column is eluted with buffer A at a flow rate of approximately 30 ml/hr; 10-ml fractions are collected. The enzyme emerges in a single peak in coincidence with the second protein peak. Fractions exhibiting specific activities higher than 9 units/mg are combined (40 ml).

The results of the purification of acyl-CoA synthetase I are summarized in the table. The overall purification is 115-fold with a yield of 12%.

When an enzyme preparation from the Sephadex G-100 chromatography step is not completely homogeneous as judged by dodecyl sulfate-polyacrylamide gel electrophoresis, contaminating proteins are removed by hydroxyapatite (Hypatite C) chromatography as follows. The enzyme preparation (4 mg of protein) is applied to a hydroxyapatite column (1.2 × 3 cm) equilibrated with buffer A. The column is eluted with a linear concentration gradient established between 7.5 column volumes of buffer A and the same volume of 0.5 M potassium phosphate buffer, pH 7.4, containing 2 mM Triton X-100 and 5 mM 2-mercaptoethanol at a flow rate of approximately 5 ml/hr; 1-ml fractions are collected. The enzyme appears in a single peak at a phosphate concentration range between 0.05 M

PURIFICATION OF ACYL-CoA SYNTHETASE I FROM *Candida lipolytica*[a]

Fraction	Volume (ml)	Protein (mg)	Total activity[b] (units)	Specific activity[b] (units/mg)	Yield (%)
Particulate fraction	72	3312	311	0.094	100
230,000 g supernatant	265	1574	238	0.151	77
Phosphocellulose	90	96.0	63.8	0.665	21
Blue-Sepharose	50	10.1	54.8	5.43	18
Sephadex G-100	40	3.5	37.8 (60.5)	10.8 (17.3)	12

[a] From Hosaka *et al.*[4]

[b] Enzyme activity was determined by the isotopic method with the standard assay mixture. The final enzyme preparation was also assayed by the spectrophotometric method with the standard assay mixture, except that oleate was replaced by palmitate; the values in parentheses represent those obtained by the spectrophotometric assay.

and 0.15 *M*. This procedure is also useful for concentrating the enzyme preparation from the Sephadex G-100 chromatography step.

Properties of Acyl-CoA Synthetase I[4]

Purity and Molecular Weight. The purified acyl-CoA synthetase I is essentially homogeneous as evidenced by dodecyl sulfate–polyacrylamide gel electrophoresis[12] as well as by Ouchterlony double-diffusion analysis.[13] The molecular weight of the enzyme, estimated by dodecyl sulfate–polyacrylamide gel electrophoresis, is approximately 84,000.

Specific Activity. The specific activity of the purified enzyme, when measured by the spectrophotometric assay with oleic acid as a substrate, is 20–24 units/mg at 25°, being about 50-fold higher than those of long-chain acyl-CoA synthetases hitherto reported.[14-16]

Storage. The purified enzyme can be stored at −70° for at least 1 month without loss of activity.

Stoichiometry. A stoichiometric relationship is observed between the amount of palmitic acid or CoA added and that of AMP formed as measured by the spectrophotometric method. This, together with the identifi-

[12] K. Weber and M. Osborn, *J. Biol. Chem.* **244**, 4406 (1969).
[13] O. Ouchterlony, *Acta Pathol. Microbiol. Scand.* **26**, 507 (1949).
[14] E. J. Massaro and W. J. Lennarz, *Biochemistry* **4**, 85 (1965).
[15] D. Samuel, J. Estroumza, and G. Alhaud, *Eur. J. Biochem.* **12**, 576 (1970).
[16] J. Bar-Tana, G. Rose, and B. Shapiro, *Biochem. J.* **123**, 353 (1971); see also this series Vol. 35 [15].

cation of palmitoyl-CoA as a reaction product, confirms that acyl-CoA synthetase I catalyzes the stoichiometric conversion of ATP, fatty acid, and CoA to PP_i, AMP and acyl-CoA.[9] The purified acyl-CoA synthetase I can be used for the quantification of fatty acid or CoA.

Specificity and Kinetic Properties. The enzyme specifically utilizes straight-chain fatty acids with 14–18 carbon atoms regardless of the degree of unsaturation. Straight-chain fatty acids containing more than 18 or fewer than 14 carbon atoms (including decanoic acid, octanoic acid, hexanoic acid, *n*-butyric acid, and acetic acid) as well as 16-hydroxy-palmitic acid and hexadecanedioic acid are essentially ineffective.

With respect to acyl acceptor, CoA is utilized most efficiently. The V_{max} values (percentage of that for CoA) and the apparent K_m values for CoA and its effective analogs are as follows: CoA, 100%, 0.042 mM; dephospho-CoA, 70%, 0.11 mM; 1, N^6-etheno-CoA, 40%, 0.50 mM; 4'-phosphopantetheine, 63%, 3.3 mM; pantetheine, 57%, 2.5 mM. *N*-Acetylcysteamine, L-cysteine, glutathione, and dithiothreitol at concentrations up to 5 mM are not utilized. *Escherichia coli* acyl carrier protein at concentrations up to 26 μM is also ineffective.

With respect to nucleoside 5'-triphosphate, only ATP and dATP are effective. At concentrations of 7.5 mM and 15 mM, the activities measured with dATP are 78% and 61%, respectively, of those measured with ATP. At the same concentrations, GTP, UTP, CTP, dTTP, adenylyl (β,γ-methylene)diphosphonate and adenylyl imidodiphosphate are essentially ineffective.

pH Optimum. The enzyme exhibits a broad pH optimum ranging from 7.1 to 9.6.

Immunochemical Properties. Antibody prepared against the purified enzyme inhibits the activity of acyl-CoA synthetase I both from glucose-grown and from oleic acid-grown cells, but does not affect the activity of acyl-CoA synthetase II from oleic acid-grown cells (see below).

Subcellular Localization.[3] Acyl-CoA synthetase I is distributed among different subcellular fractions, including microsomes and mitochondria where glycerolphosphate acyltransferase is located.

Properties of Acyl-CoA Synthetase II[2]

The properties of acyl-CoA synthetase II were studied with the separated but unpurified enzyme preparation (see above).

Activator. Acyl-CoA synthetase II, in contrast to acyl-CoA synthetase I, almost absolutely requires phosphatidylcholine for its activity. Phosphatidylethanolamine and phosphatidylserine exhibit some stimulatory effect, whereas phosphatidic acid, phosphatidylinositol, cardiolipin, and 1-acyl-*sn*-glycero-3-phosphorylcholine are virtually ineffective.

Specificity. Acyl-CoA synthetase II exhibits a broader fatty acid specificity than acyl-CoA synthetase I. The effective substrates include long-chain dicarboxylic acids.

Stability. Acyl-CoA synthetase II is less stable upon treatment with heat or Triton X-100 than acyl-CoA synthetase I.

Subcellular Localization.[3] Acyl-CoA synthetase II is localized in microbodies where the acyl-CoA-oxidizing system is located.

Inducible Nature. Acyl-CoA synthetase II is present in oleic acid-grown cells, but not in glucose-grown cells, whereas acyl-CoA synthetase I is found in both types of cells.

Physiological Roles of Acyl-CoA Synthetases I and II[1,5]

The physiological roles of acyl-CoA synthetases I and II have been disclosed by studies with *Candida lipolytica* mutants defective in the respective enzymes. The mutants lacking acyl-CoA synthetase I cannot incorporate exogenous fatty acid as a whole into cellular lipids, but are able to grow on fatty acid as a sole carbon source by degrading it to yield acetyl-CoA, from which cellular fatty acids are synthesized *de novo*. In contrast, the mutants defective in acyl-CoA synthetase II are unable to grow on fatty acid, but are capable of incorporating exogenous fatty acid into cellular lipids. Thus, acyl-CoA synthetase I is responsible for the production of long-chain acyl-CoA that is utilized solely for the synthesis of cellular lipids, while acyl-CoA synthetase II provides long-chain acyl-CoA that is exclusively degraded via β-oxidation. Consistent with the physiological roles of the two enzymes is the fact that acyl-CoA synthetase II, in contrast to acyl-CoA synthetase I, is induced by fatty acid and exhibits a broad substrate specificity with respect to fatty acid. The subcellular localizations of the two enzymes also support their different functions. The phenotypes of the mutants indicate that there are two functionally distinct long-chain acyl-CoA pools in the cell, i.e., one for lipid synthesis and the other for β-oxidation. The mutants defective in acyl-CoA synthetase I, unlike the wild-type and the revertant strains as well as the mutants lacking acyl-CoA synthetase II, do not exhibit the repression of acetyl-CoA carboxylase, a regulatory enzyme for fatty acid synthesis, when cells are grown on fatty acid. Measurement of the two long-chain acyl-CoA pools with the aid of appropriate mutants has indicated that the long-chain acyl-CoA to be utilized for lipid synthesis, but not that to be degraded via β-oxidation, is involved in the repression of acetyl-CoA carboxylase.

[40] Long-Chain Acyl-CoA Synthetase from Rat Liver

EC 6.2.1.3 Acid : CoA ligase (AMP-forming)

By Takao Tanaka, Kohei Hosaka, and Shosaku Numa

$$ATP + RCOOH + CoASH \rightleftharpoons PP_i + AMP + RCOSCoA$$

The assay method, purification procedures, and properties of long-chain acyl-CoA synthetase from the microsomes as well as from the mitochondrial fraction of rat liver are described in this chapter.

Assay Methods

Principle. Two methods are routinely used to assay long-chain acyl-CoA synthetase from rat liver. In the isotopic assay,[1] the rate of [^{14}C]palmitoyl-CoA formation from [^{14}C]palmitate is determined by measuring the radioactivity in the aqueous phase after extraction of the unreacted [^{14}C]palmitic acid. In the spectrophotometric assay, the rate of AMP formation is determined by coupling the reaction of acyl-CoA synthetase with those of adenylate kinase, pyruvate kinase, and lactate dehydrogenase and following the oxidation of NADH.

Isotopic Method[2]

This method is used for enzyme preparations from all steps.

Reagents

Tris-HCl buffer, 0.5 M, pH 8.0
Dithiothreitol, 0.5 M
KCl, 2 M
MgCl$_2$, 0.15 M
ATP, 0.1 M
CoA, 20 mM
Potassium [U-^{14}C]palmitate (0.2 Ci/mol), 10 mM, in 16 mM Triton X-100[3]
Isopropanol/n-heptane/1 M H$_2$SO$_4$ (40/10/1, v/v/v)
Palmitic acid, 4 mg/ml, in n-heptane

[1] J. Bar-Tana, G. Rose, and B. Shapiro, this series, Vol. 35 [15].
[2] T. Tanaka, K. Hosaka, M. Hoshimaru, and S. Numa, *Eur. J. Biochem.* **98**, 165 (1979).
[3] The average molecular weight of Triton X-100 is taken to be 628.

Scintillator solution: 4 g of 2,5-diphenyloxazole and 0.1 g of 1,4-bis[2-(4-methyl-5-phenyloxazolyl)]benzene in 1 liter of toluene plus 0.5 liter of Triton X-100

Enzyme: Dilutions are made with 20 mM potassium phosphate buffer, pH 7.4, containing 2 mM Triton X-100 and 5 mM 2-mercaptoethanol (hereafter referred to as buffer A).

Procedure. The standard assay mixture contains in a total volume of 0.2 ml (in micromoles, unless otherwise specified): Tris-HCl buffer, pH 8.0, 20; dithiothreitol, 1; KCl, 30; MgCl$_2$, 3; ATP, 2; CoA, 0.2; Triton X-100, 0.32; potassium [^{14}C]palmitate (0.2 Ci/mol), 0.2; enzyme, up to 15 mU. A mixture (0.19 ml) containing all ingredients except CoA is preincubated at 35° for 1 min, and the reaction is started by adding 0.01 ml of 20 mM CoA. After incubation at 35° for 2 min, the reaction is terminated by adding 2.5 ml of a mixture containing isopropanol, n-heptane, and 1 M H$_2$SO$_4$ (40/10/1, v/v/v). The subsequent procedures, including n-heptane extraction of the unreacted [^{14}C]palmitic acid followed by determination of the radioactivity in the aqueous phase, are as described for *Candida lipolytica* acyl-CoA synthetases.[4] Under the assay conditions described, the reaction follows zero-order kinetics, and the initial rate of reaction is proportional to enzyme concentration. The radioactive product in the aqueous phase has been identified as palmitoyl-CoA by chromatographic analysis of the thioester as well as of the hydroxamic acid derived from it.

Spectrophotometric Method[2]

This method is used for purified enzyme preparations that contain no interfering enzymes such as adenosinetriphosphatase and NADH-oxidizing enzymes.

Reagents

Tris-HCl buffer, 0.5 M, pH 8.0
Dithiothreitol, 0.5 M
KCl, 2 M
MgCl$_2$, 0.15 M
ATP, 0.1 M
CoA, 20 mM
Potassium palmitate, 10 mM, in 16 mM Triton X-100
Triton X-100, 16 mM
Potassium phosphoenolpyruvate, 50 mM
NADH, 10 mM

---

[4] K. Hosaka, M. Mishina, T. Kamiryo, and S. Numa, this volume [39].

Adenylate kinase (Boehringer), 5 mg/ml
Pyruvate kinase (Boehringer), 10 mg/ml
Lactate dehydrogenase (Boehringer), 10 mg/ml
Enzyme: Dilutions are made with buffer A.

Procedure. The standard assay mixture contains in a total volume of 1 ml (in micromoles, unless otherwise specified): Tris-HCl buffer, pH 8.0, 100; dithiothreitol, 5; KCl, 150; $MgCl_2$, 15; ATP, 10; CoA, 0.6; potassium palmitate, 0.1; Triton X-100, 1.6; potassium phosphoenolpyruvate, 0.2; NADH, 0.15; adenylate kinase, 45 μg; pyruvate kinase, 30 μg; lactate dehydrogenase, 30 μg; enzyme, up to 25 mU. A mixture (0.97 ml) containing all ingredients except CoA is preincubated at 35° for 1 min in a cuvette with 1-cm light path, and the reaction is started by adding 0.03 ml of 20 mM CoA. The oxidation of NADH is followed at 35° with a recording Eppendorf photometer at wavelength 334 nm. Under the assay conditions described, the reaction follows zero-order kinetics for at least 2 min, and the initial rate of reaction is proportional to enzyme concentration. In this assay, the formation of 1 mol of AMP corresponds to the oxidation of 2 mol of NADH. Essentially identical activities are measured by the isotopic method and by the spectrophotometric method.

Units. One unit of acyl-CoA synthetase activity is defined as that amount which catalyzes the formation of 1 μmol of palmitoyl-CoA or AMP per minute under the assay conditions used. Specific activity is expressed as units per milligram of protein. Protein is determined by the method of Lowry *et al.*[5] with bovine serum albumin as a standard; particulate protein is solubilized with 48 mM sodium deoxycholate prior to the determination.

Preparation of Particulate Fractions[2]

Male Wistar-strain rats weighing 180–200 g, which are maintained on a balanced diet, are used. After being deprived of food for 12 hr, rats are sacrificed by decapitation followed by exsanguination. The livers are quickly removed and homogenized in three volumes of 0.25 M sucrose by three down-and-up strokes of the Teflon pestle in the glass tube. The homogenate is fractionated according to the principle of de Duve *et al.*[6] as follows: The whole homogenate is centrifuged at 600 g for 10 min, and the pellet is washed twice by rehomogenization in three volumes of 0.25 M sucrose and recentrifugation as before. The supernatant and washings are

[5] O. H. Lowry, N. J. Rosebrough, A. L. Farr, and R. J. Randall, *J. Biol. Chem.* **193**, 265 (1951).
[6] C. de Duve, B. C. Pressman, R. Gianetto, R. Wattiaux, and F. Appelmans, *Biochem. J.* **60**, 604 (1955).

combined and subjected to successive centrifugation at 13,000 g for 20 min and at 230,000 g for 60 min to yield the mitochondrial fraction (which includes peroxisomes and lysosomes) and microsomes, respectively; each fraction is washed once with a small volume of 0.25 M sucrose, and the washing is combined with the supernatant at each stage.

Purification Procedure[2]

From Microsomes

All operations are carried out at 0–4° unless otherwise specified.

Resolution of Microsomes. Microsomes derived from 30 livers are suspended in 840 ml of a mixture containing Triton X-100, potassium phosphate buffer, pH 7.4, 2-mercaptoethanol and EDTA, the final concentrations of which are 5 mM, 50 mM, 5 mM, and 1 mM, respectively, to give a protein concentration of 4.5 mg/ml. The mixture is allowed to stand for 1 hr and is then centrifuged at 230,000 g for 1 hr. The resulting supernatant is collected (775 ml).

Blue-Sepharose Chromatography. The 230,000 g supernatant is diluted with five volumes of a solution containing 2 mM Triton X-100 and 5 mM 2-mercaptoethanol. The diluted solution is applied to a Blue-Sepharose (CL-6B) column (2.6 × 18 cm) equilibrated with buffer A. The column is washed with five column volumes of buffer A and then with five column volumes of buffer A containing 10 mM ATP. All the enzyme activity applied is adsorbed to the column, while 94% of the protein applied is found in the effluent. Elution is carried out with buffer A containing 10 mM ATP and 0.8 M NaCl at a flow rate of approximately 15 ml/hr; 15-ml fractions are collected. The enzyme emerges in a single peak. Fractions exhibiting high enzyme activities (more than 1.1 units/ml) are combined (210 ml).

Hydroxyapatite Chromatography. To concentrate the pooled enzyme solution from the Blue-Sepharose chromatography step, 0.7 volume of 50% (w/v) polyethyleneglycol 6000 is added. After gentle stirring for 30 min, the resulting precipitate is collected by centrifugation at 230,000 g for 30 min and then loosely homogenized in 50 ml of 20 mM potassium phosphate buffer, pH 7.4, containing 5 mM Triton X-100 and 5 mM 2-mercaptoethanol. The mixture is gently stirred for 30 min and then freed from insoluble material by centrifugation at 230,000 g for 30 min. This is passed through a Sephadex G-10 column (5 × 25 cm) equilibrated with buffer A. All protein-containing fractions are combined (165 ml) and applied to a hydroxyapatite (Hypatite C) column (2 × 16 cm) equilibrated with buffer A. The column is washed with 2 column volumes of buffer A. All the enzyme activity applied is adsorbed to the column, while 9.8 mg of protein is found in the effluent. Elution is carried out with a linear concentration gradient established between six column volumes of buffer A and the same volume of 0.35 M potassium phosphate buffer, pH 7.4, containing 2 mM Triton X-100 and 5 mM 2-mercaptoethanol at a flow rate of approximately 20 ml/hr; 15-ml fractions are collected. The enzyme appears in a single peak, which coincides with the main protein peak. Fractions having specific activities higher than 8 units/mg are combined (105 ml).

Phosphocellulose Chromatography. The pooled enzyme solution from the hydroxyapatite chromatography step is concentrated to 50 ml by ultrafiltration with a Diaflo membrane filter PM-30. The concentrated enzyme solution is passed through a Sephadex G-10 column (5 × 25 cm) equilibrated with buffer A. The protein-containing fractions combined (165 ml) are applied to a phosphocellulose (P-11) column (2 × 10 cm) equilibrated with buffer A. The column is washed with two column volumes of buffer A. All the enzyme activity applied is adsorbed to the column, while 3.7 mg of protein is found in the effluent. Elution is carried out with a linear concentration gradient made between 10 column volumes of buffer A and the same volume of 0.35 M potassium phosphate buffer, pH 7.4, containing 2 mM Triton X-100 and 5 mM 2-mercaptoethanol at a flow rate of approximately 10 ml/hr; 10-ml fractions are collected. The enzyme emerges in a single peak, which coincides with a nearly single protein peak. Fractions exhibiting maximal and essentially constant specific activities (27–30 units/mg) are combined (50 ml).

The results of the purification of acyl-CoA synthetase from the microsomes are summarized in Table I. The overall purification is 84-fold with a yield of 5%.

From Mitochondrial Fraction

Crude acyl-CoA synthetase in the mitochondrial fraction of rat liver, in contrast to the enzyme in the microsomes, is very unstable. At 0°, approximately 90% of the enzyme activity in the mitochondrial fraction, which is suspended in 50 mM potassium phosphate buffer, pH 7.4, containing 2 mM dithiothreitol and 1 mM EDTA, is lost in 8 hr, while essentially no loss of the activity in the microsomes occurs for at least 24 hr. The addition of dimethyl sulfoxide at a final concentration of 20–35% (v/v) protects the enzyme in the mitochondrial fraction from inactivation; at 0°, 85–95% of the activity is preserved after 8 hr. Therefore, all solutions used for the purification of acyl-CoA synthetase from the mitochondrial fraction contain 25% (v/v) dimethyl sulfoxide.

TABLE I

PURIFICATION OF ACYL-CoA SYNTHETASE FROM RAT LIVER MICROSOMES[a]

Fraction	Volume (ml)	Protein (mg)	Total activity[b] (units)	Specific activity[b] (units/mg)	Yield (%)
Microsomes	95	3780	1290	0.34	100
230,000 g supernatant	775	2460	1230	0.50	95
Blue-Sepharose	210	56.2	349	6.21	27
Hydroxyapatite	105	15.6	162	10.4	13
Phosphocellulose	50	2.2	63.1	28.7	5

[a] From Tanaka et al.[2]
[b] Enzyme activity was determined by the isotopic method with the standard assay mixture.

Prior to the purification, the mitochondrial fraction is treated with dimethyl sulfoxide; 840 ml of a suspension of the particles (protein concentration, 8.2 mg/ml), which are derived from 50 livers, is allowed to stand for 1 hr in the presence of 25% (v/v) dimethyl sulfoxide, 50 mM potassium phosphate buffer, pH 7.4, 5 mM 2-mercaptoethanol, and 1 mM EDTA and is then centrifuged at 230,000 g for 1 hr. The particles precipitated are subjected to Triton X-100 treatment and high-speed centrifugation followed by chromatography on Blue-Sepharose, hydroxyapatite, and phosphocellulose in essentially the same manner as described for the purification of the enzyme from the microsomes; although the starting material contains a larger amount of protein than that used for the purification of the microsomal enzyme, the volume of the enzyme preparation at each step, as well as the sizes of the chromatography columns, is essentially unaltered. The results of a typical purification of acyl-CoA synthetase from the mitochondrial fraction are summarized in Table II. The overall purification is 125-fold with a yield of 6%.

Properties[2]

Purity and Molecular Weight. The purified acyl-CoA synthetase preparation from the microsomes as well as from the mitochondrial fraction is essentially homogeneous as evidenced by dodecyl sulfate–polyacrylamide slab gel electrophoresis,[7] amino terminal analysis,[8] and the elution profile at the final chromatography step. The molecular weight of both enzymes, estimated by dodecyl sulfate–polyacrylamide gel electrophoresis is approximately 76,000.

[7] J. King and U. K. Laemmli, *J. Mol. Biol.* **62,** 465 (1971).
[8] B. S. Hartley, *Biochem. J.* **119,** 805 (1970).

TABLE II
PURIFICATION OF ACYL-CoA SYNTHETASE FROM THE MITOCHONDRIAL FRACTION OF RAT
LIVER[a]

Fraction	Volume (ml)	Protein (mg)	Total activity[b] (units)	Specific activity[b] (units/mg)	Yield (%)
Mitochondrial fraction	132	6890	1450	0.21	100
230,000 g supernatant	760	4030	1050	0.26	72
Blue-Sepharose	180	52.8	528	10.0	36
Hydroxyapatite	105	8.6	157	18.3	11
Phosphocellulose	50	3.3	86.5	26.2	6

[a] From Tanaka et al.[2]
[b] Enzyme activity was determined by the isotopic method with the standard assay mixture.

Specific Activity. The purified enzymes from both sources exhibit a specific activity of 26–29 units/mg at 35°, which is about 60-fold higher than those of long-chain acyl-CoA synthetases hitherto reported.[9–11]

Stability and Storage. When purified, acyl-CoA synthetase from the mitochondrial fraction, like the enzyme from the microsomes, is rather stable even in the absence of dimethyl sulfoxide; both purified enzymes retain their full activity after dialysis at 0° for 24 hr against 50 mM potassium phosphate buffer, pH 7.4, containing 2 mM Triton X-100, 2 mM dithiothreitol, and 1 mM EDTA. When the two dialyzed enzymes are incubated at 35°, no difference is noted in their stability; 80–87%, 63–67%, and 41–42% of the initial activity is preserved after 2.5 min, 5 min, and 15 min, respectively. Both purified enzymes can be stored at −70° for at least 4 months without loss of activity.

Stoichiometry. A stoichiometric relationship is observed between the amount of palmitic acid or CoA added and that of AMP formed as measured by the spectrophotometric method. This, together with the identification of palmitoyl-CoA as a reaction product, confirms that both acyl-CoA synthetases catalyze the stoichiometric conversion of ATP, fatty acid, and CoA to PP$_i$, AMP, and acyl-CoA.[12] The purified enzymes can be used for the quantification of fatty acid or CoA.

Specificity and Kinetic Properties. The purified acyl-CoA synthetases from both sources exhibit essentially identical substrate specificities with respect to fatty acid, utilizing saturated fatty acids with 10–18 carbon

[9] E. J. Massaro and W. J. Lennarz, *Biochemistry* **4**, 85 (1965).
[10] D. Samuel, J. Estroumza, and G. Alhaud, *Eur. J. Biochem.* **12**, 576 (1970).
[11] J. Bar-Tana, G. Rose, and B. Shapiro, *Biochem. J.* **122**, 353 (1971); see also this series, Vol. 35 [15].
[12] A. Kornberg and W. E. Pricer, Jr., *J. Biol. Chem.* **204**, 329 (1953).

atoms and unsaturated fatty acids with 16–20 carbon atoms most efficiently. Saturated fatty acids containing more than 20 or less than 8 carbon atoms (including *n*-butyric acid and acetic acid) as well as hexadecanedioic acid and dodecanedioic acid are essentially ineffective.

The substrate specificities of the two acyl-CoA synthetases with respect to acyl acceptor are likewise indistinguishable. CoA is utilized most efficiently. The V_{max} values (percentage of that for CoA) and the apparent K_m values for CoA and its effective analogs are as follows: CoA, 100%, 0.026–0.028 mM; dephospho-CoA, 45–46%, 0.23–0.29 mM; 4′-phosphopantetheine, 25–26%, 0.41–0.43 mM; pantetheine, 26–27%, 1.1–1.2 mM. *N*-Acetylcysteamine, L-cysteine, glutathione, and dithiothreitol at concentrations up to 2 mM are not utilized.

The substrate specificities of the two acyl-CoA synthetases with respect to nucleoside 5′-triphosphate are also essentially identical. Among the different compounds tested, only ATP and dATP are effective. The V_{max} value for dATP is 27–30% of that for ATP, and the apparent K_m values for ATP and dATP are 2.3–2.4 mM and 12.5–13.3 mM, respectively. GTP, UTP, CTP, ITP, dTTP, adenylyl (β,γ-methylene)diphosphonate, and adenylyl imidodiphosphate at concentrations up to 10 mM are ineffective.

pH Optimum. Both acyl-CoA synthetases exhibit a broad pH optimum ranging from 7.4 to 9.1.

Amino Terminus. Aspartic acid (or asparagine) occupies the amino-terminal position of both acyl-CoA synthetases.

Comparison of Acyl-CoA Synthetases from the Two Sources. The purified acyl-CoA synthetase from the microsomes and that from the mitochondrial fraction, which are obtained by essentially identical procedures, are indistinguishable from each other with respect to all molecular and catalytic properties thus far examined, including molecular weight, amino acid composition, amino-terminal residue, heat stability, specific activity, pH optimum, and substrate specificity regarding fatty acid, acyl acceptor, and nucleoside 5′-triphosphate.

[41] Acyl Carrier Protein from *Escherichia coli**

By CHARLES O. ROCK and JOHN E. CRONAN, JR.

Acyl carrier protein (ACP) is a small, acidic protein that plays a central role in the synthesis of fatty acids in bacteria. The protein and its thioesters interact specifically with at least a dozen different enzymes, and all the acyl intermediates in fatty acid biosynthesis occur as thioesters of

---

* Unknown to the authors and Editor articles 21 and 41 were switched. This should be article 21 in Section I.

ACP (for reviews see footnotes 1–3). The complete primary sequence of *Escherichia coli* ACP is known.[4,5] The protein contains a preponderance of acidic residues (29%) that occur throughout the sequence and a paucity of positively charged residues (8%) that are clustered at the amino terminus.[5] The molecular weight determined from the primary structure is 8847. The protein contains 1 mol of sulfhydryl per mole of protein[6]; this sulfhydryl group has been identified as the 2-mercaptoethylamine[7] portion of the 4′-phosphopantetheine prosthetic group of the molecule.[8] The prosthetic group is attached to the protein via a phosphodiester linkage to serine-36.[5,6,9] Solid-phase chemical synthesis of apo-ACP has been achieved, and the prosthetic group was enzymatically added to the apoprotein utilizing holo-ACP synthase (EC 2.7.8.7).[10,11] Since ACP is not in itself catalytic, large quantities of the protein must be prepared to use as a substrate for preparing acyl-ACP[12] for use in assaying the enzymes of fatty acid and phospholipid synthesis.

Assay for ACP-SH

Principle. ACP-SH is measured enzymatically using purified acyl-ACP synthetase. Reaction conditions are adjusted so that the ligation of ACP-SH with radioactive fatty acid is almost quantitative. 1-^{14}C-Labeled fatty acids are removed, and the amount of [1-^{14}C]acyl-ACP formed is determined.

Materials. ACP-SH is purified as described in this chapter. Acyl-ACP synthetase and the reagents for the assay of this enzyme are prepared as described elsewhere in this volume.[13] Acyl-ACP synthetase purified through the heat step from freshly grown *E. coli* K12 (strain UB1005) is

[1] P. R. Vagelos, *Curr. Top. Cell. Regul.* **3**, 119 (1971).
[2] D. J. Prescott and P. R. Vagelos, *Adv. Enzymol.* **36**, 269 (1972).
[3] P. R. Vagelos, *in* "The Enzymes" (P. Boyer, ed.), 3rd ed., Vol. 8, pp. 155–199. Academic Press, New York.
[4] T. C. Vanaman, S. J. Wakil, and R. L. Hill, *J. Biol. Chem.* **243**, 6409 (1968).
[5] T. C. Vanaman, S. J. Wakil, and R. L. Hill, *J. Biol. Chem.* **243**, 6420 (1968).
[6] P. W. Majerus, A. W. Alberts, and P. R. Vagelos, *Proc. Natl. Acad. Sci. U.S.A.* **51**, 1231 (1964).
[7] F. Sauer, E. L. Pugh, S. J. Wakil, R. Delaney, and R. L. Hill, *Proc. Natl. Acad. Sci. U.S.A.* **52**, 1360 (1964).
[8] P. W. Majerus, A. W. Alberts, and P. R. Vagelos, *Proc. Natl. Acad. Sci. U.S.A.* **53**, 410 (1965).
[9] P. W. Majerus, A. W. Alberts, and P. R. Vagelos, *J. Biol. Chem.* **240**, 4723 (1965).
[10] W. S. Hancock, D. J. Prescott, G. R. Marshall, and P. R. Vagelos, *J. Biol. Chem.* **247**, 6224 (1972).
[11] W. S. Hancock, G. R. Marshall, and P. R. Vagelos, *J. Biol. Chem.* **248**, 2424 (1973).
[12] C. O. Rock and J. L. Garwin, *J. Biol. Chem.* **254**, 7123 (1979).
[13] C. O. Rock and J. E. Cronan, Jr., this volume [21].

REAGENTS FOR THE ASSAY OF ACP-SH

Reagent	Microliters per assay (total volume 40 μl)	Final concentration
Tris-HCl, 1 M, pH 8.0	4	0.1 M
DTT, 40 mM	2	2 mM
MgCl$_2$, 0.1 M	2	5 mM
ATP, 0.1 M	2	5 mM
LiCl, 4 M	4	0.4 M
[1-^{14}C]Palmitic acid, 600 μM	4	60 μM
Acyl-ACP synthetase, 10 units/ml	1	25 mU
Triton X-100, 20%	2	2%
ACP-SH sample ($\leq$1 nmol) in H$_2$O	Up to 19 μl	$\leq$25 μM

suitable for the assay, but the enzyme preparation should be tested to ensure that the acyl-ACP formed is not degraded by a contaminating activity found in less pure preparations.[14,15] If the contaminating activity is present, it can easily be removed by Blue-Sepharose chromatography.[13,14]

Method. Reagents, stock solutions, and volumes to be added to the assay are shown in the table. A standard curve is prepared using ACP purified as described in this chapter. The ACP-SH sample to be measured in a final volume of 10 μl is placed in a 1.5-ml (or smaller) Eppendorf microfuge tube. Next, 30 μl of the reaction mixture (see the table) is added and the solution is briefly vortexed. The tubes are tightly capped to avoid evaporation during the assay and incubated at 37° for 3 hr with gentle shaking. At the end of this time, the tubes are vortexed again and centrifuged for 2 min in a Eppendorf Model 5412 microcentrifuge to force any condensation or droplets on the sides of the tube to the bottom. A 30-μl aliquot of the reaction mixture is removed, deposited on a Whatman 3 MM filter disk suspended with a steel pin, and the filters are washed (20 ml per filter) in two changes of chloroform : methanol : acetic acid (3 : 6 : 1; v/v/v) to remove unreacted fatty acid. The filters are thoroughly dried, placed in scintillation fluid, and counted.

Comments. The assay is specific for ACP-SH. The sensitivity of the assay method is illustrated in Fig. 1. Under these conditions acylation is almost quantitative and as little as 2.5 pmol of ACP-SH can be detected using [1-^{14}C]palmitic acid having a specific activity on the order of 50 Ci/mol. ACP-SH cannot be measured reliably in crude supernatants unless they are first centrifuged at 160,000 g for 2 hr to remove material inhibitory to the acyl-ACP synthetase reaction (presumably localized in the outer membrane).[15] Also the sensitivity of acyl-ACP synthetase to the salt

[14] C. O. Rock and J. E. Cronan, Jr., *J. Biol. Chem.* **254**, 7116 (1979).
[15] A. K. Spencer, A. D. Greenspan, and J. E. Cronan, Jr. *FEBS Lett.* **101**, 253 (1979).

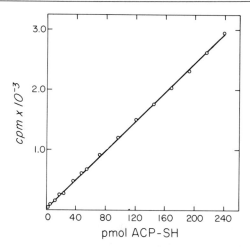

Fɪɢ. 1. Standard curve for the estimation of ACP-SH using the acyl-ACP synthetase reaction.

concentration in the assay[14] should be taken into consideration in preparing the samples. In practice the preparation of ACP is straightforward and the assay need not be set up for the purpose of monitoring the purification of ACP.

The above method for assaying ACP-SH is superior in several respects to the previously available malonylpantetheine–CO_2 exchange assay. First, the malonylpantetheine–CO_2 exchange assay requires the chemical synthesis of several substrates. Second, the assay requires the preparation of an unstable enzyme fraction that is difficult to reproduce. Third, the exact specific activity of the $KH^{14}CO_3$ is difficult to control, and, fourth, the acyl-ACP synthetase method is at least an order of magnitude more sensitive than the malonylpantetheine–CO_2 exchange assay.[15]

Purification of ACP-SH[16]

Materials. Triton X-100, Tris, glycine, PIPES, EDTA, and lysozyme are obtained from Sigma Chemical Co. LiCl and 2-propanol should be reagent grade or better. Fibrous DEAE-cellulose (DE-23) is purchased from Whatman. The purification method has been accomplished using either *E. coli* B cells obtained from Grain Processing, Muscatine, Iowa, or using *E. coli* K12 strain UB1005 grown in a fermentor in a glycerol–tryptone broth.

[16] C. O. Rock and J. E. Cronan, Jr., *Anal. Biochem.* **102,** 362 (1980).

Step 1. Extraction of ACP-SH from Cells. Frozen *E. coli* cells are thawed overnight at 4° and the wet weight of the cell paste is determined. The cells are suspended by addition of 0.1 volume of a stock solution containing 1 *M* Tris, 1 *M* glycine, and 0.25 *M* EDTA, adjusted to pH 8.0 with HCl followed by dilution to a final concentration of 1 g of cells per milliliter with distilled water. This mixture is stirred to make a homogeneous suspension, lysozyme (30 mg per kilogram of cells) is added and the suspension stirred for 2 hr at room temperature. An equal volume of 0.5% Triton X-100 is added to lyse the cells and the suspension is stirred for another hour. The gelatinous mass of DNA extruded by the cells is sheared to obtain a less viscous solution by a brief homogenization (1–2 min) in a Waring blender. An equal volume of 2-propanol is slowly added to this suspension with continual stirring. Homogeneous suspension of the 2-propanol extract is assured by another brief homogenization in the Waring blender. The mixture is stirred for an hour and then centrifuged at 4000 *g* for 20 min: the supernatant is saved.

Step 2. Batch DEAE-Cellulose. The 2-propanol supernatant is titrated to pH 6.1 with acetic acid. De-fined DEAE-cellulose (DE-23, 100 ml per kilogram of cells) is added to the 2-propanol supernatant, and the suspension is stirred for several hours to adsorb the ACP-SH to the DE-23. The DE-23 is then collected on a large filter funnel and washed with five volumes of 10 m*M* PIPES buffer, pH 6.1, 0.25 *M* LiCl, 0.1% Triton X-100 followed by five volumes of 10 m*M* PIPES buffer–0.25 *M* LiCl. The DE-23 is then removed from the funnel and packed into a column. The ACP-SH fraction is eluted from the column with 10 m*M* PIPES [piperazine-*N*,*N*′bis(2-ethane sulfonic acid)] buffer containing 0.6 *M* LiCl.

Step 3. Acid and Ammonium Sulfate Precipitations. The protein peak from the DE-23 column, which is characteristically dark brown, is pooled, the pH is titrated to 3.9 with acetic acid, and the mixture is allowed to stand overnight to precipitate the protein completely. The ACP-SH pellet is recovered by centrifugation 8000 *g* for 20 min. The pellet is resuspended in a minimum volume of water, and the pH is adjusted to 7.0–7.5 by the dropwise addition of 1 *M* Tris base. The entire pellet will redissolve, and the process is facilitated by breaking the pellet by repeatedly drawing it into a Pasteur pipette. At this stage ACP-SH is contaminated by one major protein and about six other minor species. These contaminants are removed by the addition of four volumes of saturated ammonium sulfate. The precipitate is allowed to flocculate for 2 hr, and the suspension is centrifuged at 8000 *g* for 20 min. ACP-SH is recovered from the supernatant by titrating the pH to 3.9 and collecting the ACP precipitate by centrifugation as described earlier.

Comments. In the presence of reducing agents, SDS or SDS–urea gel electrophoresis[16a] of ACP prepared by this method shows only a single band corresponding to ACP. The ACP preparations are also pure as judged by amino acid analysis, gel filtration chromatography, and isoelectric focusing. Native gel electrophoresis (see below) reveals the presence of several ACP derivatives. The ACP purified from freshly grown stationary phase *E. coli* K12 (strain UB1005) shows ACP-SH to be the major component with varying amounts of (ACPS)$_2$ also present. Acyl-ACP and apo-ACP usually comprise less than 10% of the total ACP. When ACP is purified from frozen *E. coli* B cells from Grain Processing Co., similar results are obtained, except that most of the ACP-SH is found in the form of ACP-SSG. However, nonprotein contaminants often are still present (as indicated by an excess of ultraviolet absorption at 260 nm), most likely arising from the coelution of nucleic acid components with ACP from the DEAE-cellulose column. The excess absorbance can be removed by repeated acid precipitation at pH 3.9 as described above.

This procedure results in ACP yields of 120 to 150 mg/kg wet weight of cells. The major loss of ACP occurs during the 2-propanol extraction. If the cells were initially suspended in a smaller volume of buffer than called for in the described method, the losses of ACP were found to be greater. This result is most probably due to the trapping of ACP in the voluminous precipitate formed after the addition of 2-propanol. Resuspension and washing of the 2-propanol precipitate is difficult owing to its pasty consistency. Therefore, the volumes reported in this paper are a compromise between high yields of ACP following 2-propanol extraction and manageable volumes for the subsequent centrifugation steps. This notion was further supported by our experiments on small cultures of *E. coli* cells labeled with D-[3-³H]pantothenic acid. In these experiments recoveries of [³H]ACP have been found to be routinely greater than 90%.

This procedure for isolating ACP offers several advantages over the previous method.[17] First, we have taken advantage of the high solubility of ACP in 2-propanol to design a first step that removes 95% of the protein from the sample. This step dispenses with the need for physically disrupting the cells and the streptomycin sulfate and ammonium sulfate fractionation steps of the previous method.[17] Second, the time required to complete the purification has been reduced from several weeks to several days. The major time savings result from the elimination of the two large ion exchange columns called for in the previous method. Finally, our yield of 120–150 mg/kg cells is considerably better than the 60–80 mg/kg cells usually obtained using the previous method.[17]

[16a] C. O. Rock and J. E. Cronan, Jr., *J. Biol. Chem.* **254**, 9778 (1979).
[17] P. W. Majerus, A. W. Alberts, and P. R. Vagelos, *Biochem. Prep.* **12**, 56 (1968).

Preparation of Labeled ACP-SH

Principle. Pantothenate auxotrophs of *E. coli* (*panB⁻*) cannot synthesize the 4′-phosphopantetheine moiety of CoA or ACP-SH due to a defect
in ketopantoate hydroxymethyltransferase[17a] and are unable to grow unless supplemented with exogenous pantothenate. Similarly, *panD⁻* strains
of *E. coli* are defective in aspartate decarboxylase[17a] and are therefore
auxotrophic for β-alanine. Since ACPSH comprises 95% of the protein-
bound 4′-phosphopantetheine moieties in *E. coli*,[18] the strategy is to label
these auxotrophs with either D-[3-³H]pantothenic acid or β-[1-¹⁴C]alanine
and subsequently isolate the ³H- or ¹⁴C-labeled ACP-SH.

Materials. D-[3-³H]Pantothenic acid, and β-[1-¹⁴C]alanine are purchased from New England Nuclear. β-[1-¹⁴C]Alanine obtained from
commercial sources possesses a relatively low specific activity (ca. 50
Ci/mol), and if higher specific activities are desired a reasonable alternative is to prepare β-[U-¹⁴C]alanine from L-[U-¹⁴C]aspartate (ca. 200 Ci/
mol) using partially purified aspartate decarboxylase from *E. coli*.[19] *Escherichia coli* K12 strains are AB354 (*panD, thr, leu, thi*) Hfr139 (YA139;
panB, thi), CY257 (Hfr P4x, *panB, thr, leu, thi*). DEAE cellulose (DE-52)
is purchased from Whatman. All other materials should be reagent grade
or better.

Method. The *E. coli* strain is grown overnight in the presence of sufficient pantothenate or β-alanine. The cells are harvested by centrifugation
and washed three times in M9 minimal salts to remove all traces of cold
β-alanine or pantothenate from the cells. A small number of cells (ca. 10⁶
cells/ml) are inoculated into 10 ml of M9 minimal salts containing 0.2%
glycerol as a carbon source, the other required nutrients, and 0.25 μM
D-[3-³H]pantothenate in the case of *panB* strains and 28 μM β-
[1-¹⁴C]alanine in the case of *panD* strains. The culture is incubated at 37°
until the cells grow (the usual procedure is overnight growth). The cells
are harvested and washed once with medium E. The cells are suspended
in 0.1 M Tris-HCl, pH 8.0, 25 mM EDTA, lysozyme (0.1 mg/ml) and
incubated at 37° for 10 min; 25 liters of 20% Triton X-100 are added, and
the cells are vortexed. To this suspension, 1 ml of 2-propanol is added to
the mixture, vortexed, and centrifuged at 10,000 g for 10 min. The supernatant containing the ACP-SH and CoA is loaded onto a DE-52 column
(0.5 ml or less will suffice) equilibrated in 10 mM PIPES, pH 6.5. The
2-propanol is cleared from column by washing with 10 mM PIPES containing 2 mM DTT to reduce any CoA dimers that may be present. The
column is next washed with 10 mM PIPES, 0.25 M LiCl to elute the

[17a] J. E. Cronan, Jr., *J. Bacteriol.* **141,** 1291 (1980).
[18] A. W. Alberts and P. R. Vagelos, *J. Biol. Chem.* **241,** 5201 (1966).
[19] J. E. Cronan, Jr., *Anal. Biochem.* **103,** 377 (1980).

labeled CoA. The labeled ACP is collected by eluting the DE-52 column with 10 mM PIPES, 0.5 M LiCl. Greater than 95% of the label in this fraction is in ACP. The remaining 5% is found in another protein component eluting just ahead of ACP on DEAE cellulose columns eluted with a LiCl gradient.[18] Interestingly, almost all the labeled ACP is acyl-ACP and must be cleaved by reaction with 0.1 M hydroxylamine (pH 8.0) for 2 hr to generate ACP-SH. Removal of the acyl moiety can be assessed by Octyl-Sepharose chromatography.[12]

Comments. By far the most efficient way to label ACP-SH is with D-[3-^{3}H]pantothenate. Using the conditions described above half of the label in the medium is incorporated into ACP-SH. In contrast, only a few percent of the β-[1-^{14}C]alanine in the medium is incorporated into ACPSH. Therefore the most cost-effective alternative for preparing [^{14}C]ACP-SH is to use D-[1-^{14}C]pantothenic acid, although it costs three times more per millicurie than β-[1-^{14}C]alanine. Labeled ACPSH prepared by these methods is useful for synthesizing acyl-[^{3}H]ACP and in studying the physical properties of ACP-SH.[20]

Native Gel Electrophoresis of ACP

Principle. ACP undergoes a pH-induced conformational change[16a] characterized by a dramatic loss of α-helical content[21] and an expansion of the Stokes' radius of the protein.[16a] The strategy is to cast the separating gel at a pH and ionic strength that is above the onset of the structural transition. Therefore ACP derivatives that destabilize the protein moiety migrate more slowly than ACPSH due to their expanded Stokes' radius. Correspondingly, ligands that stabilize the protein moiety will migrate faster than ACP-SH.

Materials and Methods. All reagents were purchased from Bio-Rad Laboratories. Stock solutions are:

   Acrylamide stock: 40%, w/v acrylamide; 1.0%, w/v N,N'-methylene-bisacrylamide
   Separating gel buffer: 1.5 M Tris-HCl, pH 9.0
   Stacking gel buffer: 0.5 M Tris-HCl, pH 6.8
   Ammonium persulfate, 10%, w/v
   Running buffer: 14.4 g of glycine, 3.06 g of Tris per liter
The separating gel is composed of 0.5 volume of acylamide stock, 0.25

---

[20] S. Jackowski of this laboratory has recently found that strain SJ16, a *panD* derivative of strain UB1005 (*metB*), accumulates β-alanine very efficiently. A culture of strain SJ16 grown for several generations without β-alanine is added (10^6–10^7 cells/ml) to medium containing labeled β-alanine at 0.5 μM. After growth overnight, about 50% of the added β-alanine is incorporated into ACP-SH.

[21] H. Schulz, *J. Biol. Chem.* **250**, 2299 (1975).

volume of separating gel buffer, 0.25 volume of water, 1.2 μl of *N,N,N′,N′*-tetramethylethylenediamine per milliliter, and polymerization is initiated by adding 4 μl of ammonium persulfate stock per milliliter. The stacking gel is composed of 0.1 volume of acrylamide stock, 0.25 volume of stacking gel buffer, 0.65 volume of water, 1.2 μl of *N,N,N′,N′*-tetramethylethylenediamine per milliliter, and polymerization is initiated by adding 4 μl of ammonium persulfate per milliliter. Samples are loaded in Tris-HCl or Tris-glycine buffer pH 6.8 to 8.3 containing 20% glycerol and bromophenol blue as a tracking dye. Gels are run at 30 mA until the tracking dye reaches the bottom of the gel. Acyl carrier protein stains best using a mixture of 50% methanol, 10% acetic acid, 0.1% Coomassie Blue R-250. The gels are diffusion destained in water containing 10% methanol and 10% acetic acid.

Comments. Native gel electrophoresis is a powerful technique for the analysis of ACP derivatives. Best results are obtained when the separating gel buffer is prepared immediately before pouring the gel. Figure 2 illustrates the resolution of some commonly encountered ACP forms using a 16 cm separating gel. That differences in Stokes' radii account for the observed separations is supported by gel filtration experiments.[16a] In addi-

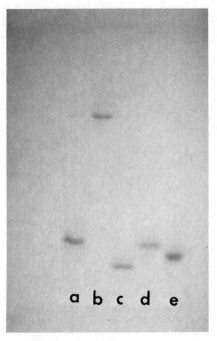

FIG. 2. Native gel electrophoresis of ACP derivatives. Lane a, mixed disulfide of ACP-SH and glutathione (ACP-SSG); lane b, ACP dimer [(ACPS)₂]; lane c, acyl-ACP; lane d, apo-ACP; lane e, ACP-SH.

tion to the five ACP derivatives shown in Fig. 2, any ACP modification that affects the stability of the protein moiety will result in an altered mobility with respect to ACP-SH (e.g., mixed disulfides of ACP-SH and CoASH, thionitrobenzoate, or GSH, acetylated ACP). The sulfhydryl of ACP-SH has been found to react with heavy metal ions that may be present in impure water and to undergo a number of unidentified oxidations that give rise to a variety of bands. To circumvent these difficulties the sample should be reduced by the addition of DTT to the sample buffer. This electrophoresis system also resolves ACP from all other *E. coli* proteins, and ACP is visibly stained if a sufficient quantity of *E. coli* high-speed supernatant is applied to the gel.

Properties of ACP

Criteria for Purity. The best criteria for determining the purity of ACP and its derivatives is native gel electrophoresis. However, this method does not give information on another type of common contaminant in ACP preparations. This contaminant is most likely to consist of nucleic acid fragments that coelute with ACP on the ion exchange columns; it results in ACP preparations that have an absorption at 260 nm far in excess of that calculated from the amino acid composition. This contaminant can be removed by repeated acid precipitation at pH 3.9. The ultraviolet spectrum of pure ACP[16] should show the absence of tryptophan, a λ_{max} of 278 nm, a $\epsilon_{280}^{1\%}$ of 0.2; three peaks are observed between 255 and 262 nm corresponding to the contributions of the two phenylalanine residues. Two other methods for assessing the purity of ACP preparations are SDS–urea gel electrophoresis[16a] and amino acid analysis.[5] The amino acid content of ACP is distinctive, having only one methionine, one histidine, one proline, and one β-alanine arising from the prosthetic group.[5] The protein content of pure ACP solutions can be estimated accurately by the method of Murphy and Kies,[22] but this method underestimates the protein content of acyl-ACP solutions because of the increased absorbance at 225 nm due to the thioester group.

Physical Properties. Acyl carrier protein is an acidic protein having an isoelectric point of pH 4.1, a value very close to the value calculated from the amino acid sequence. The point of minimum solubility of ACP is pH 3.9 to 4.1, and lyotropic salts [e.g., $(NH_4)_2SO_4$] promote precipitation. ACP binds less SDS than typical globular proteins and exhibits anomalous mobility in SDS containing electrophoresis systems.[16a] Gel filtration experiments give a Stokes' radius of 19.6 Å for ACP at neutral pH.[16a] This value is that expected for a globular protein having a molecular weight of ap-

[22] J. B. Murphy and M. W. Kies, *Biochim. Biophys. Acta* **45**, 382 (1960).

proximately 20,000, but sedimentation equilibrium[16a] results and the amino acid sequence[5] give a value of 8847. Therefore, it is concluded that ACP has an asymmetric shape, and a friction ratio of 1.43 is calculated from the gel filtration results.[16a] The range of sedimentation constants reported in the literature (1.23 S to 1.44 S) is probably due to the difficulty of measuring this parameter for small proteins. A predictive algorithm has been used to predict a secondary structure of ACP that provides a rationale for much of the physical and chemical behavior of ACP.[16a]

Stability of ACP. The phosphodiester group that links the prosthetic group to serine-36 of the protein is the most chemically sensitive bond in ACP-SH. The prosthetic group is hydrolyzed by a β-elimination reaction[2] at high pH (e.g., pH >9.5) converting serine-36 to a dehydroalanine residue. The prosthetic group can be removed without modification of the protein by treatment with HF.[23] ACP undergoes a pH-induced conformational change resulting in a dramatic loss of secondary structure[21] and an expansion of the hydrodynamic radius.[16a] Ionic strength and divalent cations increase the stability of ACP.[21] Protein amino groups play a key role in the stability of ACP and chemical modification of ACP by acetylation increases the sensitivity of ACP to pH.[21] The acyl moiety of acyl-ACP stabilizes the protein to pH-induced denaturation.[16a]

ACP dimers, $(ACPS)_2$, are not readily formed in ACP-SH solutions but appear as products of acyl-ACP degradation; $(ACPS)_2$ can also be prepared by allowing ACP-SH to react with the mixed disulfide of ACPSH and thionitrobenzoate at alkaline pH.

[23] D. J. Prescott, J. Elouson, and P. R. Vagelos, *J. Biol. Chem.* **244**, 4517 (1969).

[42] Peroxisomal and Microsomal Carnitine Acetyltransferases

EC 2.3.1.7 Acetyl-CoA : carnitine *O*-acetyltransferase

By L. L. BIEBER and M. A. K. MARKWELL

Assay Methods

Methods for the assay of carnitine acetyltransferase have been reviewed for the enzyme from pigeon breast muscle.[1] The 5,5'-dithiobis(2-nitrobenzoic acid) (DTNB) method is a convenient and sensitive assay for

[1] J. F. A. Chase, see this series, Vol. 13 [60].

detecting carnitine acetyltransferase activity in subcellular fractions[2,3] and for monitoring activity during purification.[4] Because of the high acetyl-CoA hydrolase activity of some enzyme preparations and different kinetic characteristics of the forward and reverse reaction, the reverse reaction has also been used to assay the enzyme activity.

Principle. The enzyme carnitine acetyltransferase catalyzes the reversible transfer of short-chain acyl groups from CoA to carnitine:

$$\text{Acetyl-CoA} + (-)\text{-carnitine} \underset{B}{\overset{A}{\rightleftarrows}} O\text{-acetyl-}(-)\text{-carnitine} + \text{CoASH}$$

Carnitine acetyltransferase can be assayed spectrophotometrically in the forward direction by following the appearance of CoASH using the general thiol reagent DTNB[5] (Ellman's reagent) or in the reverse direction by measuring the appearance of acetyl CoASH.

The reaction shown above leads to the formation of free CoASH, which reacts irreversibly with DTNB to form a mixed disulfide plus 5-thio-2-nitrobenzoate. The latter product has a molar extinction coefficient of 13,600 cm^{-1} at 412 nm.[6]

Reagents for Pathway A (the Forward Direction)

(A) Tris-HCl buffer, 1.50 M, pH 8.0
(B) EDTA, 50 mM, neutralized to pH 8.0
(C) Acetyl-CoA, 2 mM
(D) DTNB, 2.5 mM in 0.01 M NaHCO$_3$, neutralized to pH 8.0
(E) Triton X-100, 1%
(F) 100 mM L-$(-)$-carnitine HCl, 100 mM, neutralized to pH 8.0. L-$(-)$-Carnitine is available from Otsuka Pharmaceutical, Tokushima, Japan, and Sigma-Tau Pharmaceutical, Rome, Italy.

Reagents A, B, E, and F are stored at 4°. The DTNB and acetyl-CoA solutions are stored frozen in aliquots at $-20°$ and thawed only once, on the day of use. A premix for 100 assays is prepared daily by combining 1.55 ml of Tris-HCl, 0.50 ml of EDTA, 1.00 ml of acetyl-CoA, 2.00 ml of DTNB, 2.00 ml of Triton X-100, 0.25 ml of L-$(-)$-carnitine, and 2.70 ml of distilled water. A second premix is prepared to measure acetyl-CoA hydrolase activity. It is identical to the first premix except that 0.25 ml of

[2] M. A. K. Markwell, E. J. McGroarty, L. L. Bieber, and N. E. Tolbert, *J. Biol. Chem.* **248**, 3428 (1973).
[3] M. A. K. Markwell and L. L. Bieber, *Arch. Biochem. Biophys.* **172**, 502 (1976).
[4] M. A. K. Markwell, N. E. Tolbert, and L. L. Bieber, *Arch. Biochem. Biophys.* **176**, 479 (1976).
[5] I. B. Fritz and S. K. Schultz, *J. Biol. Chem.* **240**, 2188 (1965).
[6] G. L. Ellman, *Arch. Biochem. Biophys.* **82**, 70 (1959).

distilled water is substituted for the L-(−)-carnitine. Both premixes are kept on ice.

Procedure. Add 100 μl of premix plus enough water plus sample to bring the final volume to 200 μl in a microcuvette and equilibrate at 25°. Add enzyme to initiate the reaction, mix, and record the initial linear increase in absorbance at 412 nm. The difference between the rates with and without L-(−)-carnitine measures carnitine acetyltransferase activity. One milliunit of enzyme catalyzes the formation of 1 nmol of CoA per minute and causes a carnitine-dependent increase of 0.068 OD unit in a 1-cm light path. The inclusion of Triton X-100 has no effect on the assay of soluble forms of the enzyme, but is necessary for measuring transferase activity associated with subcellular fractions because of latency of the enzyme in mitochondria and peroxisomes and the light scattering at 412 nm by subcellular particles.

Reagents for Pathway B (The Reverse Direction). Carnitine acetyltransferase can be assayed in the reverse direction, pathway B, by measuring the initial rate of acetyl-CoA formation at 232 nm. The assay is less sensitive in this direction, the molar extinction coefficient at 232 nm being 8700 cm^{-1}. In addition, Triton X-100 cannot be used because of its high absorption at 232 nm.

Reagents for the 232 Assay:
   (A) Tris-HCl buffer, 0.4 M; pH 7.5, 0.12 mM CoASH; 2 mM dithio-threitol; 0.4 mM EDTA
   (B) L-(−)Acetylcarnitine, 0.5 mM

Reagents A and B are stored at −20° and reagent A is made fresh weekly and stored in aliquots. After thawing, 100 μl of reagent A and 20 μl of reagent B are added to a quartz microcuvette, and 80 μl of enzyme or enzyme plus distilled water are added; the initial rate of the reaction is monitored at 232 nm. The difference between the rates with and without acetylcarnitine measures the carnitine acetyl-transferase activity. One milliunit of enzyme catalyzes the formation of 1 nmol of acetyl-CoA per minute and causes an acetylcarnitine-dependent increase of 0.048 OD unit in a 1-cm light path. This assay is not satisfactory for measuring activity in intact preparations, but can be used with frozen peroxisome preparations and partially purified enzyme preparations.

An example of data is given in Table 1, which compares the forward and reverse assay described above using frozen preparations of rat liver peroxisomes which had been stored at −20° for several weeks.

Tissue Distribution and Stability. As much as half of the total carnitine acetyltransferase activity of rat liver[2,7,8] is found in peroxisomes and mi-

---

[7] M. T. Kahonen, *Biochim. Biophys. Acta* **428**, 690 (1976).
[8] H. Goldenberg, M. Hutteninger, P. Kampfer, R. Kramer, and M. Pavelka, *Histochemistry* **46**, 189 (1976).

TABLE I

COMPARISON OF THE FORWARD AND REVERSE ASSAY USING FROZEN PEROXISOMES[a]

	Activity (nmol/mg protein per minute)		
Sample	"Reverse assay" at 232 nm	"Forward assay" at 412 nm	Blank "hydrolase" at 412 nm
Peroxisome preparation 1	3.8	5.3	28.6
Peroxisome preparation 2	4.9	7.0	21.0
Peroxisome preparation 3	3.2	5.6	18.7

[a] Dr. N. E. Tolbert, Michigan State University, kindly provided these samples of peroxisomes, and Mr. Kim Valkner performed the assays.

crosomes and the remainder in mitochondria. The high extra mitochondrial activity in rat liver apparently is due to the low specific activity of the enzyme in these mitochondria, i.e., 5 nmol/mg protein per minute for liver versus 250 nmol in skeletal muscle.[9] Although a small amount of enzyme is found in soluble fractions, it has not been determined unequivocally whether this is due to leakage of the enzyme from damaged organelles or to a separate cytoplasmic enzyme. Similar amounts have also been found in heart microsomes. In contrast, in rat and pig kidney[2] and guinea pig small intestine[10] carnitine acetyltransferase appears to be exclusively a mitochondrial enzyme.

Peroxisomal carnitine acetyltransferase is a soluble matrix enzyme[2] that displays a latency that is abolished by disrupting the peroxisomal membrane with detergents, or hypotonic media, or by freeze-thawing, or by prolonged storage at 4°.[2] In contrast, the microsomal enzyme is membrane-bound and sonication, freeze-thawing, or treatment with 0.15 M KCl or 0.15 M KBr solubilizes less than 10% of this activity. The microsomal enzyme occurs to the same extent in both smooth and rough endoplasmic reticulum, but it was not detected in fractions enriched in plasma membrane or in components of the Golgi apparatus.[3]

Some drugs, including the diabetes inducer streptozotocin, increase carnitine acetyltransferase in rat liver.[11] The relative distribution of total carnitine acetyltransferase activity in rat liver changes upon treatment with hypolipidemic drug[7,8,12] clofibrate. The mitochondrial enzyme increased 10-fold in specific activity, which increases the total carnitine

[9] Y. R. Choi, P. J. Fogle, and L. L. Bieber, J. Nutr. 109, 155 (1979).
[10] P. A. Martin, N. J. Temple, and M. J. Connock, Eur. J. Cell Biol. 19, 3 (1979).
[11] P. J. Fogle and L. L. Bieber, Biochem. Med. 22, 119 (1979).
[12] M. A. K. Markwell, L. L. Bieber, and N. E. Tolbert, Biochem. Pharmacol. 26, 1697 (1977).

acetyltransferases in the mitochondria to 90% of the total after clofibrate treatment.[12] In our hands rat liver microsomal carnitine acetyltransferase has been very unstable unless special precautions are taken.

The use of isopycnic sucrose density centrifugation provides both a means of resolving peroxisomes, mitochondria, and microsomes into discrete fractions and of enhancing the stability of the microsomal transferase, probably by lowering contamination by lysosomes and by providing the high osmolarity of the medium. Transferase activity is stable in peroxisomal and microsomal fractions from isopycnic sucrose density gradients during storage at 4° for 10 days and at −20° for at least 3 months. Dilution or dialysis of the fractions reduces the sucrose concentration and destabilizes the transferase.[2]

Carnitine acetyltransferase was also detected in rat liver microsomal fractions isolated by high-speed differential centrifugation or aggregation with Ca^{2+}. It is particularly labile if low ionic strength or low osmolar media are employed.[3] Such instability was not observed for the activity associated with rat heart microsomes.[13] Rat liver microsomes or peroxisomes stored in 0.4 M KCl, 0.02% sodium azide, 150 mM Tris-HCl, pH 8.0, retained at least 95% of their activity when stored at 4° for 18 days and at −20° for 3 months.

Purification Procedure

Partial purification of peroxisomal and microsomal carnitine acetyltransferases was described previously.[4] Zonal isopycnic sucrose density gradients designed to separate the subcellular organelles of rat liver[14] are used to prepare the peroxisomal and microsomal fractions from which the transferase activities are further purified.

Step 1. Solubilization. Rat liver microsomal and peroxisomal fractions from zonal gradients are made 0.4 M in KCl and stirred for 12 hr. This treatment solubilizes the carnitine acetyltransferase from the microsomal membrane, disrupts peroxisomes to release their soluble transferase, and stabilizes the transferases from both organelles.

Step 2. DEAE-Cellulose. The solubilized samples are diluted fourfold with 0.01 M ethanolamine at pH 9.5, applied to DEAE-cellulose columns (4.1 × 35 cm) previously equilibrated in 0.1 M KCl, 0.01 M ethanolamine at pH 9.5, and eluted with the same buffer. The void volume contains >90% of the transferase activity.

Step 3. Cellulose Phosphate. The effluent from the DEAE-cellulose column is concentrated by pressure dialysis using an Amicon PM-10

[13] P. J. Fogle and L. L. Bieber, *Int. J. Biochem.* **9**, 761 (1978).
[14] N. E. Tolbert, this series, Vol. 31, p. 734.

membrane and adjusted to pH 6.0 or 7.5 with 0.015 M phosphate buffer. The microsomal transferase at this stage is more stable at pH 7.5 than at pH 6.0, which is the optimal pH for purification of the peroxisomal enzyme. The dialysis chamber and membrane are washed three times with 0.015 M phosphate buffer at pH 7.5 or 6.0, and the washings are combined with the concentrated enzyme sample to yield a volume of 100 ml. The samples are applied to cellulose phosphate columns[15] (0.6 × 12 cm, obtained from Sigma) and eluted with 50-ml linear KCl gradients containing 0.015 M phosphate buffer. Peroxisomal transferase elutes in a single symmetrical peak at 0.43 M KCl in pH 6.0 buffer and microsomal transferase at 0.34 M KCl in pH 7.5 buffer. The steps using liver peroxisomes and microsomes are summarized in Table II.

Properties

The partially purified peroxisomal and microsomal transferase preparations contain <5% contamination by either acyl-CoA hydrolase or carnitine octanoyltransferase activities, which are also present in peroxisomes and microsomes of rat liver.[3,4] The partially purified carnitine acetyltransferases are stable for several weeks in 0.4 M KCl, 0.02% sodium azide in 100 mM phosphate buffer in a pH range from 5.5 to 7.5 when stored at 4°, or for at least 3 months when frozen at $-20°$. Further attempts to purify the transferases were hindered by their instability when the concentration of KCl was reduced. Inclusion of 50% glycerol, 30% sucrose, 0.05 butylated hydroxytoluene, 2.6 mM EDTA, 2 mg of bovine serum albumin per milliliter, 1 mM L-(−)-carnitine or 1 mM O-acetyl-(−)-carnitine in the presence of 0.4 M KCl failed to stabilize the transferases. The transferases can be concentrated by precipitation with 60% saturated ammonium sulfate but recovery is low.

The general properties of the two partially purified carnitine acetyltransferases from rat liver were compared to those of a commercial preparation of the pigeon breast muscle enzyme, which is the only carnitine acetyltransferase that has been crystallized. The general properties of the peroxisomal and microsomal enzymes appear to be identical except for their native state; i.e., the peroxisomal enzyme is soluble and the microsomal enzyme is membrane-bound. They have similar chromatographic properties, pH optima, isoelectric points, apparent molecular weights, and substrate specificities.[3,4] The molecular weight of 59,000 for the microsomal and peroxisomal enzymes is greater than the 51,000 for the pigeon breast muscle enzyme, but slightly less than the 62,000 found for a highly

---

[15] Whatman cellulose phosphate P-11 completely inactivated microsomal and peroxisomal carnitine acetyltransferase activities whether added batchwise or used as a column.

TABLE II
PURIFICATION PROCEDURE FOR PEROXISOMAL AND MICROSOMAL CARNITINE
ACETYLTRANSFERASE ACTIVITY[a]

Fraction	Carnitine acetyltransferase (units)[b]	Recovery (%)	Protein (mg)	Specific activity (units[b]/mg)	Purification (fold)
Zonal microsomes	1029	100	776	1.3	1.00
DEAE-cellulose	1016	98.7	210	4.8	3.6
Cellulose phosphate (pH 7.5)					
Center fractions (27–33) of peak	450	43.7	1.40	321	242
Remainder of (21–26, 34–39) peak	548	53.3	5.03	109	82.1
	998	97.0			
Zonal peroxisomes	843	100	224	3.76	1.00
DEAE-cellulose	776	92.1	69.8	12.2	3.25
Cellulose phosphate (pH 6.0)					
Center fractions (28–32) of peak	218	25.9	0.26	825	219
Remainder of (24–27, 33–36) peak	582	69.0	3.04	192	50.9
	800	94.9			

[a] Data from Markwell et al.[4] Microsomal and peroxisomal regions from five zonal gradients using two to three female rat livers per gradient were pooled to produce the starting material for purification. Carnitine acetyltransferase activity was assayed by the DTNB method, and protein concentration was calculated from the 210 nm absorbance [M. P. Tombs, F. Souter, and N. F. MacLagan, Biochem. J. 73, 167 (1959)].

[b] A unit of carnitine acetyltransferase activity produces 1 nmol of CoA and L-acetylcarnitine from acetyl-CoA and L-carnitine per milliliter of reaction mixture at 25° and pH 8.0 in the DTNB assay method.

purified preparation from beef heart mitochondria (P. R. H. Clarke, unpublished data).

Substrate Specificity. The rat liver peroxisomal and microsomal acetyltransferases are highly specific for L-(−)-carnitine, CoA, and their short-chain acyl derivatives. O-Acetyl-(+)-carnitine and reduced glutathione will not serve as substrates.[2] The K_m values for L-(−)-carnitine and acetyl-CoA are 150 μM and 69 μM, respectively, for both the microsomal and peroxisomal enzymes, which are similar to Michaelis constants of 120 μM for L-(−)-carnitine and 34 μM for acetyl-CoA for the pigeon breast muscle enzyme.[1]

The microsomal and perixosomal carnitine acetyltransferases show greatest activities with acetyl-CoA and propionyl-CoA as substrates.[4] Relative activities drop off sharply with increasing chain length after butyryl-CoA. The rat liver enzymes can use acetoacetyl-, but not succinyl- or β-hydroxy-β-methylglutaryl-CoA, as substrates. The activity reported for malonyl-CoA could be due to acetyl-CoA contamination of malonyl-CoA. The substrate specificities of the crystallized enzyme from pigeon breast muscle and the partially purified enzyme from pig heart are similar to those of the rat liver microsomal and peroxisomal enzymes.[4] All the carnitine acetyltransferases show very little activity ($<15\%$) with octanoyl-CoA and no activity with palmityl-CoA when assayed in the direction of acylcarnitine formation.

Confirmative Assay for Carnitine Acetyltransferase

With some preparations, for example, fresh microsomes and homogenates, the assays described at the beginning of the chapter can have limitations due to high blank values arising from acetyl-CoA hydrolase for the forward reaction and high 232 nm absorbance for the reverse reactions. When these assays are unsatisfactory, direct formation of acetylcarnitine and other acylcarnitines can be measured by adding 1 to 4×10^6 dpm of radioactive DL-carnitine to assay mixtures. After completion of the reaction, the acetylcarnitine is separated from the free carnitine by paper[16] or thin-layer[17] chromatography, and the radioactivity in the acetylcarnitine spots[17] is determined by scintillation counting. The solvent system for thin-layer chromatography using Brinkman silica gel G plates is $CHCl_3$: MeOH : NH_4OH (50 : 30 : 18); for paper chromatography using Whatman No. 1 paper, the system is isopropyl alcohol : acetic acid : H_2O (8 : 1 : 1).[16] We have found it necessary to purify the radioactive carnitine by chromatography on Dowex 1, OH^- form, to lower blank values. The impurities are removed when the pH of the sample is adjusted to 10.5, and the solution is passed through a small Dowex 1 column (bed volume 2 ml).

[16] L. M. Lewin and L. L. Bieber, *Anal. Biochem.* **96**, 322 (1979).
[17] Y. R. Choi, P. R. H. Clark, and L. L. Bieber, *J. Biol. Chem.* **254**, 5580 (1979).

[43] Butyryl-CoA Dehydrogenase from *Megasphaera elsdenii*[1]

EC 1.3.99.2 Butyryl-CoA:(acceptor) oxidoreductase

By PAUL C. ENGEL

$$R—CH_2CH_2COSCoA + acceptor \rightleftharpoons \underset{H}{\overset{R}{\diagdown}}C=C\underset{COSCoA}{\overset{H}{\diagup}} + reduced\ acceptor$$

Megasphaera elsdenii[1] is an obligate anaerobe that uses the reduction of unsaturated short-chain (C3–C6) acyl-CoA as a means of disposing of excess reducing equivalents.[2,3] Its butyryl-CoA dehydrogenase therefore functions *in vivo* as an enoyl-CoA reductase. The enzyme is nevertheless similar in many properties to the short-chain acyl-CoA dehydrogenases obtained from mammalian mitochondria[4] and, *in vitro*, catalyzes the oxidation of butyryl-CoA and related substrates.[5,6] *Megasphaera elsdenii* is a very convenient source of butyryl-CoA dehydrogenase, because the enzyme constitutes about 2% of the dry weight of the organism and is very readily purified in high yield.

Assay Method

Principle. The physiological reaction involves a second protein donor–acceptor, the "electron-transfer flavoprotein" (ETF).[5,7,8] It is usually more economical and convenient to substitute a coupled dye system for assay purposes. In the assay described below, phenazine methosulfate (PMS) is the intermediate coupling dye, with 2,6-dichlorophenolindophenol (DCPIP) as the terminal acceptor. This has the advantage of sensitivity, but the properties of PMS require that the assay be

[1] M. Rogosa, *Int. J. Syst. Bacteriol.* **21**, 187 (1971). Formerly known as *Peptostreptococcus elsdenii,* but now reclassified.
[2] S. R. Elsden and D. Lewis, *Biochem. J.* **55**, 183 (1953).
[3] H. L. Brockman and W. A. Wood, *J. Bacteriol.* **124**, 1447 (1975).
[4] H. Beinert, *in* "The Enzymes," (P. D. Boyer, H. Lardy, and K. Myrbäck, eds.), 2nd ed., Vol. 7, p. 447. Academic Press, New York, 1963.
[5] R. L. Baldwin and L. P. Milligan, *Biochim. Biophys. Acta* **92**, 421 (1964).
[6] P. C. Engel and V. Massey, *Biochem. J.* **125**, 879 (1971).
[7] C. D. Whitfield and S. G. Mayhew, *J. Biol. Chem.* **249**, 2801 (1974).
[8] H. Beinert, *in* "The Enzymes," (P. D. Boyer, H. Lardy, and K. Myrbäck, eds.), 2nd ed., Vol. 7, p. 467. Academic Press, New York, 1963.

carried out in subdued light. This problem may be circumvented either by using the assay described by Dommes and Kunau,[9] in which PMS and DCPIP are replaced by Meldolablau and a tetrazolium dye, respectively, or by employing phenazine ethosulfate as advocated by Ghosh and Quayle.[10] Both these procedures are satisfactory, but they give lower rates than the assay given here.

It should be noted that, because of variable contamination with ETF and thioesterase, the catalytic assay is not routinely used to monitor the purification. The distinctive visible spectrum of the enzyme provides a more reliable indicator.[6] The assay is, however, a useful check on the state of the purified enzyme.

Reagents

2,6-Dichlorophenolindophenol, 0.005% solution in water

Sodium potassium phosphate buffer, 0.2 M, pH 8

Phenazine methosulfate, 0.4% solution in water, freshly made and stored in ice in a tube protected from light with metal foil

Substrate: butyryl-CoA either prepared from CoA and butyric anhydride[11] or purchased from P-L biochemicals, 2 mM solution in water

Enzyme. The enzyme should be diluted in 0.1 M potassium phosphate buffer at pH 7 to give a concentration of about 10^{-6} M butyryl-CoA dehydrogenase subunits (40–50 μg of pure enzyme per milliliter).

Procedure. Ten milliliter of the DCPIP solution are diluted with 40 ml of the phosphate buffer to give a stock assay solution. Of this solution, 1 ml is taken in a semimicrocuvette (1-cm light path), 50 μl of substrate are added, and the reaction mixture is incubated at 25° for a few minutes. Enzyme solution (10 μl) is then added, and A_{600} is monitored with a thermostatted recording spectrophotometer backed off to allow recording on the 0–0.1 absorbance scale. The rate at this stage with fairly pure enzyme preparations is no more than 1–2% of the final measured rate, since the reduction of DCPIP in the absence of coupling by PMS is very slow. Less pure preparations give a substantial blank rate, as free CoA released by thioesterase action can reduce DCPIP. Finally, in subdued light, 10 μl of PMS solution are added to the cuvette, and the initial rate of dye-coupled reduction of DCPIP is measured.

The rate after correction for the blank reaction is proportional to enzyme concentration, and the purified enzyme should give a specific activity of about 200 μmol per micromole of enzyme subunits per minute.

[9] V. Dommes and W. H. Kunau, *Anal. Biochem.* **71**, 571 (1976).
[10] R. Ghosh and J. R. Quayle, *Anal. Biochem.* **99**, 112 (1979).
[11] E. J. Simon and D. Shemin, *J. Am. Chem. Soc.* **75**, 2520 (1953).

Purification Procedure

Bacterial Growth. Test tube cultures of *M. elsdenii* are maintained in a sloppy agar medium of the following composition: 1.4% sodium lactate, 0.4% Difco yeast extract, 0.2% Davis agar, 0.03% thioglycolic acid, all adjusted to pH 7 with 0.1 N NaOH before autoclaving for 15 min. Plugged test tubes containing 15 ml of this medium are stored in the refrigerator until they are required. For inoculation the culture tubes are heated in a boiling water bath to drive off any oxygen and liquefy the agar and then cooled to 37°. The inoculum of 1–2 ml is taken from a dormant culture with a sterile Pasteur pipette. The new culture tubes are plugged with cotton wool impregnated with alkaline pyrogallol. They are grown overnight at 37–40° and stored in the refrigerator. Fresh subcultures are made every 2–4 weeks.

The medium for larger-scale growth is as for the tube cultures except that it contains no agar and no thioglycolic acid. Sterile medium (2 liters in a 4-liter flask fitted with a Bunsen valve) is thoroughly gassed with nitrogen and then inoculated under sterile conditions with two freshly grown test tube cultures. Solid sodium dithionite is added to a final concentration of 0.02%, still under N_2 bubbling, and the flask is then sealed off. Growth is again at 37–40°. The incubator must be well vented, as the cultures evolve H_2, H_2S, and volatile fatty acids. The 2-liter culture is grown for 12–18 hr and then examined under the microscope. The organism has a highly characteristic appearance[12]; it is much larger than most common contaminants, and the individual cocci tend to be linked in chains of 2, 4, or even 8. The 2-liter culture is used to inoculate the main large-scale culture in a 40-liter carboy. For this growth the final composition of the medium is as for the 2-liter culture, but the nutrients are sterilized in a concentrated solution and diluted into unsterilized tap water just before inoculation. The anaerobic conditions and the use of a vigorously growing 5% inoculum preclude infection.

The 40-liter culture is grown at 37–40° until gas evolution subsides (24–36 hr). As the culture moves into stationary phase the cells tend to clump, forming a thick layer at the bottom of the vessel. Cells are harvested with a Sharples centrifuge, and the cell paste, if not required immediately, is stored in the frozen state. A 40-liter culture yields about 140 g of cell paste, corresponding to about 28 g dry weight.

Cell Breakage. The cell paste is resuspended in 10 ml of water per gram dry weight, and the cells are broken by sonication for 10–15 min in a beaker surrounded by a salt–ice bath. Other methods of breakage (Hughes press, lysozyme-EDTA) are less satisfactory. The efficiency of breakage is

---

[12] S. R. Elsden, B. E. Volcani, F. M. C. Gilchrist, and D. Lewis, *J. Bacteriol.* **72,** 681 (1956).

checked by microscopic examination before centrifuging at 4° to remove cell debris. The supernatant is dark green initially but becomes paler as it is stirred for 6–12 hr in the cold room. This oxidative step is essential for the success of the procedure described below.

Step 1. Acid Treatment. The extract, kept cool in ice, is adjusted to pH 5.6 by gradual addition of 2 N acetic acid. The pale precipitate is removed by centrifugation and discarded. The supernatant is readjusted to pH 7 with 1 N KHCO$_3$ and diluted with an equal volume of 0.2 M potassium phosphate buffer, pH 7. All subsequent steps are carried out in the cold room.

Step 2. Ion Exchange Chromatography. The solution is loaded onto a DE-52 DEAE-cellulose column (30–40 cm $\times$ 3.5 cm) preequilibrated with 0.1 M potassium phosphate buffer, pH 7, and is then washed with about 2 liters of this buffer. This wash removes much of the protein but leaves several colored protein bands overlapping in the top few centimeters of the column, giving a reddish brown color overall. Two factors may result in a different appearance at this stage. If the growth medium has a high iron content, the bacteria produce ferredoxin, which gives a very dark brown band even more tightly bound to the column than the other colored proteins mentioned above. On low-iron media flavodoxin is produced,[13] and this simplifies the purification because flavodoxin is less tightly bound than butyryl-CoA dehydrogenase. Second, if the extract is not allowed to become oxidized before chromatography, the butyryl-CoA dehydrogenase does not bind to the DE-52 column.

Developing the column with 0.17 M potassium phosphate, pH 7, gives rise to a dramatic chromatographic separation of three colored proteins. Pink rubredoxin moves down the column and is gradually eluted under these conditions. Immediately above the pink band is a bright orange-yellow band of flavodoxin, and above this a grass-green band of butyryl-CoA dehydrogenase. Once the rubredoxin has been eluted, the buffer concentration is increased to 0.2 M, and the column is thoroughly washed with this to remove the flavodoxin. Finally, the butyryl-CoA dehydrogenase band, by now broadened to occupy about 40% of the column, can be eluted with a linear gradient of 500 ml of 1 M KCl in 0.2 M potassium phosphate, pH 7, running into 500 ml of the same buffer without the KCl. Merely increasing the phosphate concentration further does not give satisfactory elution. Recent work on salt effects in the chromatography of nucleotides[14] suggested that this anomalous behavior might be the result of enhanced hydrophobic interactions in concentrated phosphate solution.

[13] S. G. Mayhew and V. Massey, *J. Biol. Chem.* **244**, 794 (1969).
[14] P. C. Engel, *Anal. Biochem.* **82**, 512 (1977).

Hence the modification of the original procedure[6] by the use of KCl. The overall procedure described thus allows most other proteins to be eluted from the column while the butyryl-CoA dehydrogenase remains tightly bound. The green enzyme can then be eluted easily with KCl.

Step 3. Salt Fractionation. The concentrated green solution is fractionated at 4° with pulverized ammonium sulfate; 55% ammonium sulfate brings down a muddy yellow-gray precipitate, which is discarded. Butyryl-CoA dehydrogenase is precipitated with 75% ammonium sulfate, leaving an almost colorless supernatant. At this stage the enzyme is sufficiently pure for many purposes. It is conveniently stored as a suspension in 75% ammonium sulfate. When frozen in this state it retains activity for many months.

Step 4. Gel Filtration. If necessary, the enzyme from step 3 can be dialyzed against 0.1 M potassium phosphate, pH 7, and taken through a further purification step by gel filtration on a Sephadex G-200 column. This removes traces of colorless protein. Microcrystals of the enzyme can readily be prepared by the procedure of Jakoby.[15]

The enzyme resulting from this purification procedure gives a single band on polyacrylamide gel electrophoresis with staining either for protein or for activity. There is nevertheless evidence[6] that it may retain traces of enoyl-CoA hydratase activity. This is manifest only in long-term incubations of concentrated solutions of enzyme with substrates.

Properties

Prosthetic Group. The enzyme has FAD as a prosthetic group.[6] This is tightly bound, and it is not necessary to add FAD to reaction mixtures in order to achieve maximal activity. Attempts to obtain an apoprotein that can be reconstituted by adding back FAD have not been successful so far.

Absorption Spectrum. The most distinctive and immediately obvious feature of butyryl-CoA dehydrogenase purified by the above method is its vivid green color, a feature also of butyryl-CoA dehydrogenase from various mammalian sources.[16-18] Greenness is due to a broad absorption band maximal at 710 nm (Fig. 1). The usual flavin absorption bands are blue-shifted to 430 and 365 nm, and there is also a peak at 266 nm. In the course of chromatography or salt fractionation the color reveals clearly which fractions contain the enzyme without the need for assays of catalytic activity. Nevertheless, the color is also a potential source of confusion.

[15] W. B. Jakoby, *Anal. Biochem.* **26**, 295 (1968).
[16] H. R. Mahler, *J. Biol. Chem.* **206**, 13 (1954).
[17] E. P. Steyn-Parvé and H. Beinert, *J. Biol. Chem.* **233**, 853 (1958).
[18] D. D. Hoskins, *J. Biol. Chem.* **241**, 4472 (1966).

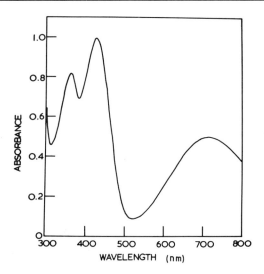

Fig. 1. Visible absorption spectrum of a purified preparation of butyryl-CoA dehydrogenase from *Megasphaera elsdenii*. The preparation shown here is predominantly in the green form.

The green color is due to the presence of an acyl-CoA compound tightly bound at the active site and involved in a charge-transfer interaction with the prosthetic group.[19] Accordingly, the ratio $A_{710}:A_{430}$ depends upon the fractional saturation with this compound and varies from preparation to preparation. Calculations from the anaerobic titrations[6] suggest that for fully green enzyme this ratio would be 0.54. The highest value achieved so far is 0.50,[20] but ratios less than 0.1 have also been obtained. Complete oxidation before chromatography seems to be a prerequisite for achieving very green preparations. Thus low $A_{710}:A_{430}$ ratios do not necessarily imply contamination with other flavoproteins. A further possible source of confusion is the enzyme's capacity for forming complexes with other acyl-CoA compounds: C_4 and C_5 3-oxoacyl-CoA compounds give rise to a long-wavelength band at 580 nm rather than 710 nm.[19] Occasionally bacterial extracts have yielded preparations containing a mixture of butyryl-CoA dehydrogenase forms absorbing at 580 and 710 nm.[20] This depends on the composition of the growth medium at the time of harvesting.

In normal catalytic assays the various spectral variants of the enzyme all perform identically, since the vast molar excess of substrate in such

[19] P. C. Engel and V. Massey, *Biochem. J.* **125**, 889 (1971).
[20] P. C. Engel, *in* "Flavins and Flavoproteins" (K. Yagi and T. Yamano, eds.) (Proceedings of the Sixth International Symposium on Flavins and Flavoproteins, Osaka, 1978), p. 423. Japan Sci. Soc. Press, Tokyo, and University Park Press, Baltimore, Maryland, 1980.

assays displaces the stoichiometrically bound acyl-CoA.[19] For studies either of binding or rapid reaction kinetics, however, it is essential to convert the enzyme to a state of homogeneous liganding. The best method available for this so far is repeated anaerobic dialysis after reduction by sodium dithionite.[19] Reoxidation in air after such treatment yields unliganded yellow enzyme, absorbing maximally at 450 nm.

The absorption spectra of the green and yellow forms of the enzyme are isosbestic very close to 430 nm. Absorbance at this wavelength is thus a useful guide to enzyme concentration even for preparations with low values of $A_{710}:A_{430}$. An extinction coefficient of 9600 liters mol^{-1} cm^{-1}, based on extraction with trichloroacetic acid has been published,[6] but a recent redetermination, based upon release of flavin with guanidine hydrochloride[21] suggests a higher value of 10,400 liters mol^{-1} cm^{-1}. This later estimate is more reliable.

Stability. In the long term, frozen suspensions in ammonium sulfate are very stable, as stated above. In the shorter term, the green form of the enzyme is stable for several days in solution in 0.1 M potassium phosphate buffer, pH 7, at 4°, provided that it is protected against bacterial growth, e.g., with sodium azide (0.02%) or chlorhexidine (0.002%) or by Millipore filtration. It is also resistant to high concentrations of urea. The yellow form of the enzyme described above is much less stable and should be made only in relatively small amounts as and when required.

Substrate Specificity. The chain-length specificity of this enzyme has not been systematically explored in steady-state catalytic assay systems. Spectrophotometric experiments in which various potential substrates are added to concentrated solutions of the enzyme indicate a preference for short-chain (C_3–C_6) substrates.[19] Alicyclic acyl chains are also tolerated; cyclobutanecarboxylic acyl-CoA and cyclopentanecarboxylic acyl-CoA are rapidly oxidized.[22] The enzyme also effects a very slow oxidation of β-hydroxybutyryl-CoA to acetoacetyl-CoA.[19]

Chemical Reducibility. In its green form, butyryl-CoA dehydrogenase from *M. elsdenii* is markedly stabilized against the chemical reductants commonly employed with flavoproteins. Full reduction by EDTA and light[23] takes several days,[7] and even the potent reduction system employing deazaflavin and light[24] is extremely slow with this enzyme.[21] Reduction with excess sodium dithionite is effective, but even this takes 15–20 min to reach completion. The yellow form of the enzyme, by contrast, is instantaneously reduced by an excess of dithionite. Sodium borohydride

[21] P. C. Engel, C. Thorpe, and C. H. Williams, Jr., unpublished results, 1978.
[22] P. C. Engel, *Z. Naturforsch. Teil B* **27**, 1080 (1972).
[23] V. Massey and G. Palmer, *Biochemistry* **5**, 3181 (1966).
[24] V. Massey and P. Hemmerich, *J. Biol. Chem.* **252**, 5612 (1977).

also reduces the yellow form of the enzyme, but in this case the reduced product appears to be the 3,4-dihydroflavin rather than the usual leuco form, the 1,5-dihydroflavin.[6]

[44] Acyl-CoA Dehydrogenase from Pig Kidney

EC 1.3.99.3 Acyl-CoA:(acceptor) oxidoreductase

By Colin Thorpe

Mammalian acyl-CoA dehydrogenases are flavoproteins that catalyze the first step of β-oxidation with the insertion of a *trans* double bond between C-2 and C-3 of their fatty acyl thioester substrates.

$$R-CH_2-CH_2-COSCoA + E \cdot FAD_{ox} \rightleftharpoons$$
$$R-CH=CH-COSCoA + E \cdot FAD_{red}$$
$$E \cdot FAD_{red} + ETF_{ox} \rightleftharpoons E \cdot FAD_{ox} + ETF_{red}$$

The reducing equivalents generated in this reaction enter the respiratory chain via the mediation of a second flavoprotein, electron-transfer flavoprotein (ETF). Three categories of mammalian acyl-CoA dehydrogenases have been described with differing, but overlapping, chain length specificities. Short chain or butyryl-CoA dehydrogenase acts on C_4-C_6 acyl-CoA substrates[1]; general acyl-CoA dehydrogenase acts on C_4-C_{16} acyl-CoA substrates, with maximal activity toward C_{10} derivatives[2-4]; and long-chain or palmitoyl-CoA dehydrogenase acts on C_6-C_{16} substrates with maximal activity toward C_{12} derivatives.[5,6] The three dehydrogenases have similar molecular properties, contributing to the difficulties encountered in separating these enzymes and obtaining them in good yields. It is largely for this reason that knowledge of this important class of flavoproteins is still fragmentary.

Previous purification schemes commence with the isolation of mitochondria from liver or heart, followed by their disruption using organic solvents or, more recently, by sonication. The procedure outlined

[1] D. E. Green, S. Mii, H. R. Mahler, and R. M. Bock, *J. Biol. Chem.* **206**, 1 (1954).
[2] F. L. Crane, S. Mii, J. G. Hauge, D. E. Green, and H. Beinert, *J. Biol. Chem.* **218**, 701 (1956).
[3] C. L. Hall and H. Kamin, *J. Biol. Chem.* **250**, 3476 (1975).
[4] C. Thorpe, R. G. Matthews, and C. H. Williams, *Biochemistry* **18**, 331 (1979).
[5] J. G. Hauge, F. L. Crane, and H. Beinert, *J. Biol. Chem.* **219**, 727 (1956).
[6] C. L. Hall, L. Heijkenskjöld, T. Bartfai, L. Ernster, and H. Kamin, *Arch. Biochem. Biophys.* **177**, 402 (1976).

here yields comparatively large amounts of an acyl-CoA dehydrogenase of general specificity from pig kidney without the prior isolation of mitochondria.

Assay Method

Principle. The enzyme is routinely assayed using phenazine methosulfate (PMS) to mediate the transfer of reducing equivalents to 2,6-dichlorophenolindophenol (DCPIP).

Reagents

Potassium phosphate, 0.25 M, pH 7.6
EDTA, sodium salts, 30 mM, pH 7.6
DCPIP, 1 mM, store at 4°
Octanoyl-CoA, 10 mM, store frozen
PMS, 10 mM, store frozen

Procedure. The assay mixture in a 1-cm light path semimicro cuvette is: 30 μM octanoyl-CoA, 30 μM DCPIP, 1.4 mM PMS, and 0.3 mM EDTA, in a total volume of 0.7 ml of 50 mM phosphate, pH 7.6, at 25°. It is usually convenient to prepare a mixture of all reagents except PMS and dispense 0.595-ml aliquots into cuvettes followed by 0.1 ml of PMS. After temperature equilibration in subdued light, the cuvettes are transferred to the spectrophotometer and the nonenzymatic background rate of approximately $0.004 A_{600}$ per minute is recorded. Enough enzyme is then added to give a decline in absorbance of about 0.06 per minute ($\epsilon_{600,DCPIP} = 21$ mM^{-1} cm^{-1}). The enzyme exhibits a turnover number, calculated on a per flavin basis, of 195 per minute under these conditions.

It should be noted that the PMS concentration used here is not saturating. The K_m for PMS under these conditions is greater than 3 mM.[7] Higher PMS levels lead to increased nonenzymatic background rates.

Alternative assays for acyl-CoA dehydrogenases using the mediator dye medolablau,[8] or the physiological acceptor ETF have been described recently.[3,6,9]

Protein Determination. The biuret method of Gornall *et al.*[10] is used for the crude tissue homogenate. After the DEAE-cellulose batch step, A_{280} and A_{450} readings, together with measurement of visible spectra, provide a convenient means of following the purification. A solution of 1 mg of the pure enzyme per milliliter has an A_{280} of 1.66.

---

[7] C. Thorpe, unpublished results.
[8] V. Dommes and W.-H. Kunau, *Anal. Biochem.* **71**, 571 (1976).
[9] M. C. McKean, F. E. Frerman, and D. M. Mielke, *J. Biol. Chem.* **254**, 2730 (1979).
[10] A. G. Gornall, C. J. Bardawill, and M. M. David, *J. Biol. Chem.* **177**, 751 (1949).

Purification Procedure

A summary of a typical purification procedure is given in Table I. Unless otherwise stated, all buffers contain 0.3 mM EDTA, and all operations are performed at 0–4°.

Homogenization. Pig kidneys are obtained from local sources and stored frozen at $-20°$ or $-70°$. Approximately 1.2-kg batches are allowed to thaw partially, and the cortex layer is chopped into 1–2 cm cubes. The pieces are combined with 2 liters of 50 mM phosphate buffer, pH 5.8, in a 1-gallon Waring blender. A solution of 0.35 g of phenylmethylsulfonyl fluoride (PMSF) in 8 ml of 2-propanol is added, and the mixture is immediately blended for 2.5 min at top speed. The PMSF is added as a precautionary measure to suppress possible proteolysis of the dehydrogenase, but has the further advantage of reducing the decomposition of octanoyl-CoA by enzymes with thioesterase activity in crude kidney homogenates.[7]

The homogenate, covered by a thick layer of foam, is transferred to a 4-liter beaker, and the pH is returned to a value of 5.8 by the addition of about 4 ml of 4 M acetic acid. The liquid is then transferred to 250 ml centrifuge bottles by siphon and centrifuged for 30 min at 25,000 g.

TABLE I
PURIFICATION OF AN ACYL-COA DEHYDROGENASE FROM PIG KIDNEY[a]

Purification step	Total[b] A_{450}	$A_{280}:A_{450}$[b]	Protein (mg)	Total units (μmol/min)	Specific activity (μmol/mg per minute)
1. Supernatant, after pH 5.8 precipitation	—	—	47,000	459	9.8×10^{-3}
2. DEAE-cellulose batch step	127	28	4,400	391	8.9×10^{-2}
3. Calcium phosphate gel–cellulose column	44	15.6	412	—	—
4. 40–80% (NH$_4$)$_2$SO$_4$ fraction	36	10.7	231	—	—
5. DEAE-cellulose column					
a. Pale green fractions	0.9	16	8.4	1.4	0.17
b. Yellow fractions	15.9	6.9	66	171	2.6
c. Yellow side fractions	8.9	11.4	61	79	0.77
6. Sephacryl S-200 (5b)	9.6	5.6	33	114	3.5
7. AH-Sepharose 4B	7.4	5.25	23	92	4.0

[a] From Thorpe *et al.*[4]
[b] This includes the contribution of other chromophores, e.g., hemoproteins and iron-sulfur centers.

DEAE-Cellulose Batch Step. The combined supernatant (approximately 1800 ml) is adjusted to pH 7.2 with about 40 ml of 1 M NH₄OH with vigorous stirring. The supernatant is stirred gently for 90 min with 360 ml of packed wet DEAE-cellulose (DE-52) preequilibrated with 50 mM phosphate, pH 7.2, and then filtered by suction on a large Büchner funnel lined with a Whatman No. 4 filter paper. This grade permits a reasonable flow rate. The cellulose is allowed to just run dry and washed with a total of 3 liters of 50 mM phosphate, pH 7.2. The yellow-brown DE-52 is allowed to suck dry on the funnel, suspended in 300 ml of 50 mM phosphate buffer, pH 7.2, and poured into a column 2.5 cm in diameter. (Although a relatively narrow column was used for the preparation summarized in Table I, columns 4 cm in diameter perform satisfactorily and can be eluted at correspondingly faster flow rates.) The column is then washed with 300 ml of the same buffer followed by 0.3 M phosphate, pH 7.2. The enzyme elutes as a yellow band that gains in intensity as the solvent front moves down the column. Although spectra of the fractions are dominated by hemoprotein contaminants, the presence of a distinct shoulder at 450 nm due to the flavoprotein, and the $A_{410}:A_{450}$ ratio are reliable guides to the selection of fractions to be pooled.

The DE-52 used in this batch step should be cleaned by removing the exchanger from the column and washing with 0.5 M HCl, followed by 0.5 M NaOH following the manufacturer's instructions.

Calcium Phosphate Gel–Cellulose Chromatography. The combined yellow fractions are taken to 85% ammonium sulfate, and the precipitate is redissolved in 100 ml of 100 mM phosphate, pH 7.6, and dialyzed overnight against 2 liters of this buffer to which 35 mg of PMSF in 1.6 ml of 2-propanol have been added. The dialyzed solution is then applied to a 2.5 × 40 cm calcium phosphate gel–cellulose column, prepared as described by Massey,[11] equilibrated with 100 mM phosphate, pH 7.6. The adsorbed protein is washed with 70 ml of this buffer and eluted in 0.2 M phosphate, pH 7.6. The enzyme is preceded by pale brown fractions showing a heme absorption spectrum. The yellow fractions are pooled according to their $A_{280}:A_{450}$ ratios and fractionated with ammonium sulfate.

We experience variability in the capacity of calcium phosphate gel, prepared by the method of Swingle and Tiselius,[12] to bind flavoproteins. The gel used in Table I was worse than average, and other preparations have required up to 0.4 M phosphate, pH 7.6, to elute the enzyme. If the binding capacity of the gel is low, a phosphate buffer of lower concentra-

---

[11] V. Massey, *Biochim. Biophys. Acta* **37**, 310 (1960).
[12] S. M. Swingle and A. Tiselius, *Biochem. J.* **48**, 171 (1951).

tion (e.g., 50 mM) may be used to dialyze the enzyme and equilibrate the column. When loading the enzyme, particular care should be taken to avoid channeling. Either the top of the bed should be lightly compressed with a suitable end fitting, or the bed surface should be allowed to run just dry before the protein is applied.

DEAE-Cellulose Chromatography. The 40–80% ammonium sulfate precipitate is dissolved in 4 ml of 100 mM phosphate buffer, pH 7.6, and dialyzed against 1 liter of 50 mM phosphate, pH 7.2. The clear yellow solution is then applied to a 2.5 × 40 cm DE-52 column and eluted at 58 ml/hr by a gradient formed from 600 ml each of 50 and 250 mM phosphate buffer, pH 7.2. The enzyme binds initially as a narrow band, but broadens considerably in this gradient. (Elution of the enzyme using a 0–0.8 M KCl gradient in 50 mM phosphate, pH 7.2, leads to less broadening but a poorer purification.) The main yellow band is preceded by small amounts of pale green material (λ_{max} 440 nm) with a long wavelength tail of low intensity extending to 700 nm. The yellow fractions across the peak exhibit a constant ratio of activities of 0.1, 1.0, and 0.2 toward butyryl-, octanoyl-, and palmitoyl-CoA, respectively, in the assay system described above. The best fractions are pooled according to their $A_{280}:A_{450}$ ratios (see Table I), and precipitated with 80% ammonium sulfate.

At this stage, the visible spectrum of the yellow fractions should be free of heme interference, and no shoulder at 410 nm should be discernible. During the last two purification steps the visible spectrum of the enzyme undergoes small, but significant, changes. The peak position shifts from 451 nm to 446 nm, and a shoulder at 480 nm becomes much less pronounced (Fig. 1). These spectral perturbations probably reflect the presence of small amounts of tightly bound endogenous thioesters, which are removed during the last two purification steps. Amino acid analysis of the enzyme for step 5 suggests a CoA content (as taurine) of approximately 0.25 mol per mole of FAD,[7] whereas the protein from step 7 does not contain significant amounts of CoA.[4]

Gel Filtration. The precipitate is dissolved in a minimum volume of 20 mM phosphate buffer, pH 7.6, and dialyzed versus 1 liter of this buffer. The solution (less than 3 ml) is applied to a Sephacryl S-200 column (2.5 × 90 cm) and gel filtered at 20 ml/hr. Tubes are combined by absorbance ratio, and the pooled material is divided into batches containing about 300 nmol of enzyme flavin.

AH-Sepharose 4B Column. Each portion is applied in turn to an AH-Sepharose 4B column (1 × 25 cm) equilibrated with 20 mM phosphate, pH 7.6. The enzyme is adsorbed as a narrow band in this buffer and is eluted using 0.16 M phosphate, pH 7.6. Gradients of increasing phosphate concentration do not yield better ratios, but result in considerable smear-

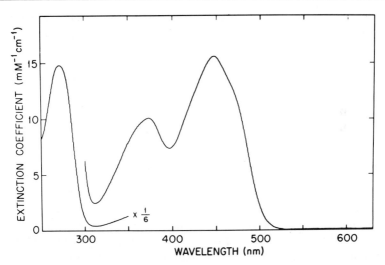

FIG. 1. Visible and ultraviolet spectrum of pig kidney general acyl-CoA dehydrogenase. Spectra were recorded in 160 mM phosphate buffer, pH 7.6, at 25°. From Thorpe *et al.*[4]

ing of the yellow band. The enzyme is then stored either frozen in 100 mM phosphate buffer, pH 7.6, or as an ammonium sulfate suspension at 4°.

Comments on the Purification Procedure. This method has been found to yield about 25 mg protein of high purity per kilogram of kidney cortex. The enzyme shows a sharp band on disc gel electrophoresis with two very minor impurities.[4] Side fractions from steps 5–7 in Table I account for an additional 30 mg of enzyme. This material may be recycled by a combination of these steps or may be used for experiments that do not require enzyme of the highest purity.

The yield of about 500 nmol of highly purified enzyme per kilogram of tissue is roughly 15 times greater than that obtained recently using pig liver mitochondria.[3,8] The procedure using kidney circumvents problems associated with the separation of enzymes of different chain length specificities, since neither the short-chain nor the long-chain dehydrogenases can be recovered in significant amounts. It is conceivable that the comparatively mild disruption procedure (compared to sonication or the use of acetone powders) fails to solubilize the other two enzymes or, alternatively, that general acyl-CoA dehydrogenase predominates in pig kidney.

If desired, the 40–80% ammonium sulfate fractionation (step 4) may be inserted between steps 2 and 3, so that less protein is applied to the calcium phosphate gel–cellulose column.

Pig Kidney Electron-Transfer Flavoprotein. The presumed physiological oxidant for the pig kidney general acyl-CoA dehydrogenase, electron-transfer flavoprotein (ETF), is not retained by DE-52 cellulose in step 2. It may be partially purified using CM-52 cellulose and Sephacryl S-200 columns.[7] The ETF is easily identified by the characteristic fluorescence emission spectrum of the bound FAD chromophore[3] exciting at 440 nm. As would be expected, the oxidized flavin fluorescence of kidney ETF is abolished by incubation with octanoyl-CoA in the presence of the kidney acyl-CoA dehydrogenase.[7]

Properties of the Enzyme

Molecular Weight. The molecular weight of the native enzyme has been estimated as 160,000 by gel filtration on Sephacryl S-200 (Table II). Sodium dodecyl sulfate (SDS)–polyacrylamide gel electrophoresis suggests that the kidney dehydrogenase is composed of four subunits of approximately 42,000 molecular weight.[4]

Isoelectric Focusing. The enzyme precipitates on isoelectric focusing into several bands lying between pH 5 and 5.6 with denaturation and release of free flavin.[7]

Amino Acid Composition. Amino acid analysis indicates a minimum molecular weight per FAD of 47,700.[4] The analysis shows a low content of tryptophan, as is observed with other flavoprotein dehydrogenases,[13] together with relatively few histidine residues. None of the cysteine residues in the native enzyme react rapidly with 5,5'-dithiobis(2-nitrobenzoic acid). High concentrations of guanidine hydrochloride lead to exposure of 5.6 cysteines per FAD.[4]

Crystallization. The pig kidney enzyme may be crystallized by dialysis at 4° against distilled water saturated with toluene to avoid microbial growth. Crystals up to 0.8 mm in their largest dimension have been grown over several weeks by this method; however, the crystals tend to lose flavin to the mother liquor on prolonged storage. Release of flavin is also observed at the isoelectric point of the enzyme during isoelectric focusing (see above).

Inhibitors. The pig kidney enzyme is insensitive to a wide variety of sulfhydryl reagents and thus does not contain rapidly reacting cysteine residues, which are catalytically essential.[7] Iodoacetic acid modifies a single essential methionine residue in the pig kidney enzyme. The resulting S-carboxymethylated inactive flavoenzyme binds octanoyl-CoA, but can no longer be reduced by this substrate.[14] The native enzyme is rapidly

[13] C. H. Williams, Jr., *in* "The Enzymes" (P. Boyer, ed.), 3rd ed., Vol. 13, p. 89. Academic Press, New York, 1976.
[14] C. Thorpe and J. P. Mizzer, *Fed. Proc.* **39**, 1641 (1980).

TABLE II

COMPARISON OF PIG LIVER AND KIDNEY GENERAL ACYL-CoA DEHYDROGENASES

Enzyme	Pig liver general acyl-CoA dehydrogenase		Pig kidney general acyl-CoA dehydrogenase[c]
	Crane et al.[a]	Hall and Kamin[b]	
Absorbance ratios	275, 370, 447	270, 370, 445	272, 280, 373, 447
	6.7 : 0.74 : 1.0	8.0 : 0.6 : 1.0	5.7 : 5.3 : 0.65 : 1.0
Minimum molecular weight/FAD	91,000	40,500[d]	47,700
Subunit molecular weight	ND	42,000	42,000
Native molecular weight	140,000–200,000	ND	160,000
Substrate bleaching; $C_4 : C_8 : C_{16}$ acyl-CoA	55% : 72% : 35% (300 μM)[e]	62% : 82% : 15% (100 μM)	31% : 61% : 26% (13 μM); 70% : 79% : 37% (100 μM)

[a] Data from Crane et al.[2]
[b] Data from Hall and Kamin.[3]
[c] Data from Thorpe et al.[4]
[d] Based on an assumed extinction coefficient; see the text.
[e] Concentration of thioester used in the bleaching experiments.

inactivated by diethylpyrocarbonate at pH 6.5 with consequent release of FAD.[7]

Visible and Ultraviolet Spectrum. The visible and ultraviolet spectra of pig kidney acyl-CoA dehydrogenase from step 7 is shown in Fig. 1. Highly purified preparations show little resolution of the absorption bands in the visible region. No shoulder at 410 nm, indicating heme contamination, should be observed. In addition to $A_{280} : A_{450}$ absorbance ratios (Table II), a useful indicator of purity is the absorbance at 310 nm, which should decline with increasing purification. The 446 nm peak is especially intense for a flavoprotein ($\epsilon_{446} = 15.4$ mM^{-1} cm^{-1}). This value was determined by releasing the flavin using guanidine hydrochloride or trichloroacetic acid.[4] It has previously been assumed that the extinction coefficient of free FAD ($\epsilon_{448} = 11.3$ mM^{-1} cm^{-1}) could be used to calculate the flavin content and concentration of active sites in pig liver general acyl-CoA dehydrogenase. If the extinction coefficient determined for the kidney enzyme applies to other general acyl-CoA dehydrogenases, the flavin content in these preparations has been overestimated by a factor of 1.36.

Correlation of flavin and amino acid analyses for samples of the pig kidney enzyme suggests that the protein, as isolated, contains about 3.5 FAD molecules per tetramer (Table II).[4] The early studies of Crane *et al.*[2] indicate that considerable variation in the flavin content of pure samples of pig liver general acyl-CoA dehydrogenase can occur without loss of tetrameric structure. Attempts to stimulate the activity of the pig kidney enzyme by preincubation with free FAD have not been successful.[7]

Substrate Bleaching Experiments. The extent of bleaching of the flavin chromophore on the addition of substrates of various chain lengths has been widely used to delineate three classes of acyl-CoA dehydrogenases. Usually a single large excess of substrate is used. However, much more information may be obtained from detailed titrations of the enzyme with thioester substrates.[4] These titrations should be performed anaerobically, since E · SH$_2$ complexes of general acyl-CoA dehydrogenase (particularly with butyryl-CoA) are autoxidizable.[4] Techniques for anaerobic titrations have been described earlier in this series.[15]

Acknowledgments

This work was supported in part by U. S. Public Health Service Grant GM 21444 to Dr. Charles H. Williams, Medical Research Service, Veterans Administration; a Michigan Memorial Phoenix Project Grant 534 to Dr. Rowena Matthews; and U. S. Public Health Service Grant GM 26643 to the author.

[15] C. H. Williams, L. D. Arscott, R. G. Matthews, C. Thorpe, and K. D. Wilkinson, this series, Vol. 62 [37].

[45] Acyl-CoA Dehydrogenases from Pig Liver Mitochondria

By CAROLE L. HALL

The overall reaction catalyzed by acyl-CoA dehydrogenases together with electron-transfer flavoprotein (see this volume [46]) is

$$R—CH_2CH_2CO—SCoA + FAD - 2H^+ - 2\ e \rightarrow$$
$$R—CH{=}CHCO—SCoA + FADH_2$$

where R = alkyl group.

The first step of β-oxidation of fatty acids in mitochondria, dehydrogenation of the α,β-bond of the CoA derivatives, is carried out by the action of two types of flavoproteins.[1-7] One of several broadly substrate-specific dehydrogenases binds the appropriate acyl-CoA. However, for electron transfer to the electron transport chain,[6] to dyes (except phenazine methosulfate, PMS), or for turnover of the primary dehydrogenases,[1-7] a second flavoprotein is required. Because of its apparent function in transferring electrons from acyl-CoA dehydrogenases to the electron transport chain it was called electron-transfer flavoprotein (ETF).[2,4] This flavoprotein is discussed in this volume [46].

Three acyl-CoA dehydrogenases acting upon short- (SC-AD), medium- (G-AD), and long- (LC-AD) chain fatty acyl CoAs have been described.[1-7] In addition, a different flavoprotein dehydrogenase, which binds isovaleryl-CoA (IV-D) and interacts with ETF has also been isolated from pig liver mitochondria.[8] This enzyme will not be discussed here except regarding its separation from the fatty acyl-CoA dehydrogenases.

The acyl-CoA dehydrogenases have long been known to form characteristic spectrally distinct air-stable complexes with their substrates as

[1] H. Beinert, this series, Vol. 5 [73].

[2] H. Beinert and J. R. Lee, this series, Vol. 6 [59].

[3] H. Beinert, *in* "The Enzymes" (P. D. Boyer, H. Lardy, and K. Myrbäck, eds.), 2nd ed., Vol. 7, p. 447. Academic Press, New York, 1963.

[4] H. Beinert, *in* "The Enzymes" (P. D. Boyer, H. Lardy, and K. Myrbäck, eds.), 2nd ed., Vol. 7, p. 467. Academic Press, New York, 1963.

[5] C. L. Hall, this series, Vol. 53 [50].

[6] H. Beinert and F. L. Crane, *in* "Inorganic Nitrogen Metabolism" (W. D. McElroy and B. Glass, eds.), p. 601. Johns Hopkins Press, Baltimore, Maryland, 1956.

[7] C. L. Hall and H. Kamin, *J. Biol. Chem.* **250**, 3476 (1975).

well as their products.[3,5,6,8-10] Observation of these spectral changes is the most reliable method of ascertaining which dehydrogenase is the major component in a mixture of dehydrogenases.[5] However, other methods not dependent upon high concentrations of relatively highly purified enzymes can be employed to detect the presence of these enzymes. Some of these methods will be described here, and others in this volume [46].

The initial steps in the separation and purification of these enzymes are as described in detail elsewhere.[5] These steps will be only briefly summarized here, except for recent modifications and further detail regarding some of the later steps in the isolations.

Assay Methods

Catalytic Assay

Principle. The dichlorophenolindophenol (DCPIP) reduction assay has been described,[5] but since rates depend on the concentrations of both acyl-CoA dehydrogenase and ETF, phenazine methosulfate (PMS) was used as ETF to follow purification described in this chapter. However, relative rates with PMS as compared to ETF[see 5] are quite different for the dehydrogenases studied. Specific activities of purified SC-AD, G-AD, and LC-AD using PMS as compared to ETF present at a concentration of around 4 μM flavin are 2.5, 0.1, and 0.005, respectively. Thus the PMS-mediated DCPIP assay overestimates SC-AD and greatly underestimates G-AD and LC-AD. Substrate specificities are broadly overlapping in these enzymes also. However, the assay can be used as a rough indicator of the presence of the different dehydrogenases in various samples, especially if not enough enzyme is present for optical determinations (see later).

Procedure. The procedure and reagents are as described in Vol. 53[5] except that ETF is omitted and, instead, 20 μl of PMS (2 mg/ml) are added just before measurement. The PMS should be freshly prepared each day. Acyl-CoA derivatives (usually obtained from P-L Biochemicals) are prepared in 50 mM NaAc, pH 5, at a concentration of 5 mM and stored frozen.

Units and Specific Activity. Activity is calculated as nanomoles of DCPIP reduced per milligram of protein (or per nanomole of enzyme flavin) per minute using $\epsilon = 21 \times 10^3\ M^{-1}\ cm^{-1}$.[10] The product of the

[8] C. L. Hall, in preparation.
[9] C. L. Hall, J. D. Lambeth, and H. Kamin, *J. Biol. Chem.* **254**, 2023 (1979).
[10] E. P. Steyn-Parvé and H. Beinert, *J. Biol. Chem.* **233**, 843 (1958).

specific activity value so obtained and the total protein (or flavin) is used to express units of activity.

Optical Assays

Principle. Differences in the oxidized spectra of the various acyl-CoA dehydrogenases coupled with spectral changes induced by acyl-CoA substrates, especially palmitoyl-CoA, can be readily perceived and used by the trained eye to identify the various components of a mixture of dehydrogenases. These characteristics were shown and described in detail elsewhere[5] and will not be repeated here. G-AD is bleached about 20% by palmitoyl-CoA and shows a shoulder at 500 nm, LC-AD is bleached about 60%, and the palmitoyl-CoA spectrum is isosbestic with the oxidized spectrum at 500 nm, whereas SC-AD is bleached only 5–10% immediately after addition of palmitoyl-CoA, and a small shoulder appears at 500 nm.[see 5,9] However, since the isovaleryl-CoA dehydrogenase[8] also interacts with palmitoyl-CoA to produce a spectrum that resembles a mixture of SC-AD and G-AD, spectra should also be recorded in the presence of isovaleryl-CoA to check for the presence of that enzyme. Neither highly purified SC-AD, G-AD, or LC-AD show any significant spectral changes in the presence of isovaleryl-CoA. The spectrum of isovaleryl-CoA dehydrogenase (IV-D) in the ultraviolet (UV) resembles that of SC-AD.[5] The visible absorption maximum for IV-D is at 440 nm whereas that for G-AD is at 450 nm; SC-AD is at 430 nm, and LC-AD is also at 440 nm. Using the criteria of these spectral characteristics of the complexed and uncomplexed enzymes, mixtures of dehydrogenases can be roughly assessed for the content of the various enzymes. These assessments are borne out by subsequent separation upon preparative electrophoresis (see Tables I and II).

Procedure. The spectrum of an appropriately diluted sample of dehydrogenase is scanned between 750 nm and 250 nm before and immediately after addition of a three- to fivefold molar excess of the acyl-CoA used. Fresh samples are used for each acyl-CoA tested.

Calculation of Flavin Content and Purity. Flavin content is measured by the absorbance of the oxidized enzyme and calculated as nanomoles of flavin per milliliter using $\epsilon = 11 \times 10^3 \ M^{-1} \ cm^{-1}$ for all dehydrogenases.[5,7,8,11] Purity is assessed by the ratio of the UV maximum absorbance to the visible (Vis) maximum absorbance. The best fractions show

[11] C. L. Hall, L. Heijkensköld, T. Bartfai, L. Ernster, and H. Kamin, *Arch. Biochim. Biophys.* **177**, 402 (1976).
[12] C. L. Hall, in preparation.

UV: Vis absorption ratios of about 6.5 for all dehydrogenases, and these fractions are at least 90% pure as judged by protein staining of polyacrylamide gels.see [5] Absorbance at 280 nm and 260 nm is also used to measure protein content using the formula: 1.42 (A_{280}) − 0.74 (A_{260}) × dilution = milligrams of protein per milliliter.

The content of acyl-CoA dehydrogenase in crude fractions can also be estimated optically. Since the acyl-CoA complexes of the dehydrogenases appear to be tight[3,6,9,10] and ETF is only slowly reoxidizable by air (in the absence of submitochondrial particles),[3,6,12] observation of changes in absorption at 450–460 nm after adding excess acyl-CoA substrate can be used to estimate the dehydrogenase content plus ETF content without including other chromophores, as is the case for dithionite-reducible flavin content.[5] When the ETF content is estimated fluorometrically (see this volume [46]), these methods appear to be fairly reliable and straightforward (see Table I in this chapter and Table I in this volume [46]).

Preparation of the Enzymes from Pig Liver

Isolation of the mitochondria and extraction of the soluble enzymes as well as purification of the enzymes is carried out essentially as described elsewhere[5,7] and is only summarized here, with the exception of alterations now employed and some further detail on the preparative electrophoretic separations. Mitochondria are isolated from pig livers obtained fresh from the kill floor (see Preparation of Mitochondria[5]). Only the "heavy" mitochondria are kept; they are frozen in 250-g (wet weight) aliquots at −90° for up to 1 month. However, extraction and purification usually are carried out within 1 week of isolation of mitochondria. The thawed mitochondria are disrupted by sonic oscillation (see Step A of Hall[5]), followed by high speed centrifugation. The supernatant ("sonic supernatant") is carefully decanted from the heavy pellet and from a small amount of fine suspended particles at the bottom of the tube. The sonic supernatant is diluted and fractionated as described in Step B of Hall.[5] The 40–85% ammonium sulfate precipitate is concentrated, rehydrated, and chromatographed on Whatman DE-52 as described in Step C.[5] The column is eluted with 1 liter of 2.5 mM K_2HPO_4 as described, and fractions containing enoyl hydratase, β-OH acyl-CoA dehydrogenase, and β-ketothiolase activities are eluted from the column just ahead of and overlapping the ETF-containing band, which is readily detected with a hand-held UV lamp as a bright greenish fluorescence.

The purification of ETF is summarized in this volume [46]. Instead of the gradients used for Step D_{AD} of Hall,[5] the column is now washed with a gradient of 1.2 liter each of 50 and 100 mM potassium phosphate, pH 7.6.

During this elution the heme-containing proteins and some brown and colorless proteins are removed, and the greenish dehydrogenase band at the top of the column begins to develop and resolve into various yellow and green bands. The effluent from the 50–100 mM potassium phosphate gradient is usually discarded. The elution continues with 1.2 liters each of 100 and 300 mM potassium phosphate pH 7.6. Fractions containing the various acyl-CoA dehydrogenases are collected (15–20 ml per tube), pooled, and fractionated as described in Step E_{AD} of Hall.[5] The dialyzed fractions are centrifuged at 20,000 g for 15 min to remove any insoluble material and the absorption spectra in the absence and in the presence of palmitoyl-CoA and isovaleryl-CoA are recorded, as described above under Optical Assays. Preparative electrophoresis is carried out as described in Step F_{AD} of Hall.[5] The fractions obtained from Step E_{AD} are concentrated to 2–4 ml using an Amicon ultrafiltration unit (PM-10 membrane). Fractions that are spectrally similar may be pooled, or exceptionally rich fractions may be split so that approximately 100 nmol of enzyme flavin per dehydrogenase will be applied to each column in 2–3 ml. The proteins are focused into a narrow band (a few millimeters) during the first 1–2 hr of electrophoresis and then begin to resolve. After 20–24 hr, SC-AD appears as a green band about 6–8 cm from the bottom (anode) of the column, sometimes with a closely associated yellow band that might be the native, degreened SC-AD (see Beinert[3]). Isovaleryl dehydrogenase appears as a greenish yellow band 2–3 cm above the green SC-AD band; G-AD is just above IV-D when both are present in the same fraction; LC-AD appears at the bottom of the column after about 18 hr of electrophoresis. At the end of the run, the electrodes are disconnected and the column is removed from the electrode baths. The colored bands are collected and fractionated as described in Step F_{AD} of Hall.[5]

Optimal separation during the DEAE column chromatography is essential for good resolution of the enzymes on the preparative electrophoresis column. The methods described herein and in Vol. 53[5] have been used to prepare enzymes from beef heart mitochondria,[5,11] beef adrenals,[13] and pig kidney.[13] It is expected that they would be applicable to a variety of sources. It has been shown that dehydrogenases and ETF from widely disparate sources can interact with each other,[14] suggesting a considerable conservation of structure.[3]

Table I shows recovery of dehydrogenases based on protein, total flavin, acyl-CoA-reducible flavin, and "dehydrogenase" flavin. The latter is obtained by subtracting the "ETF flavin" estimated fluorometrically

---

[13] C. L. Hall, unpublished observations.
[14] H. Beinert and W. R. Frisell, *J. Biol. Chem.* **237**, 2988 (1962).

TABLE I

RECOVERY OF ACYL-CoA DEHYDROGENASES[a]

Line	Fraction	Total Protein (mg)[b]	Total flavin (nmol)			DCPIP reduction (units, μmol/min)[f]			Carrier[g]	Type[h]	Total recovery (nmol)
			Dithionite reducible[c]	Acyl-CoA reducible[d]	Acyl-CoA dehydrogenase[e]	Butyryl-CoA	Octanoyl-CoA	Palmitoyl-CoA			
1.	Sonicated supernatant	21,600	31,800	16,800	13,920	90	210	130	None added	—	—
2.						450	2600	650	ETF	—	—
3.						450	470	170	PMS	—	—
4.	Conc. 40–85% (NH₄)₂SO₄ precipitate	9,724	21,148	11,200	9,190	30	160	100	None added	—	—
5.						240	2600	300	ETF	—	—
6.						230	280	160	PMS	—	—
7.	0–40% Precipitate	10,980	8,433	7,000	7,000	5	20	30	PMS	—	—
8.	A' (1)[i]	10	103[j]	—	—	12	0.24	0.1	PMS	LC-AD	103
9.	A (2)	94	1340	—	—	40	2	1	PMS	SC-AD	1340
10.	B (3)	116	1674	—	—	30	8	3	PMS	SC-AD / (IV-D)	938 / 736
11.	C (2)	105	1406	—	—	15	12	4	PMS	G-AD / (IV-D)	1140 / 266
12.	D (2)	80	1025	—	—	11	13	4	PMS	G-AD / (IV-D)	993 / 90
13.	E, F, G[j]	80	660	—	—	11	12	3	PMS	G-AD / (IV-D)	612 / 48
	Total in A'–G	485	6208			119	47	15			

[a] From 209,700 mg of mitochondrial protein (measured by biuret method[18])

[b] Measured by $A_{280}-A_{260}$ (see Optical Assays).

[c] Measured by dithionite reduction using $\epsilon = 9 \times 10^3 \, M^{-1} \, cm^{-1}$ for the ΔA_{450}.

[d] Measured by acyl CoA-induced bleaching using $\epsilon = 8 \times 10^3 \, M^{-1} \, cm^{-1}$ for the ΔA_{450} (see Optical Assays).

[e] Measured by acyl-CoA-reducible flavin minus ETF flavin (see this volume [46]).

[f] Units = specific activity based on protein content times total protein. DCPIP, dichlorophenolindophenol.

[g] ETF, electron-transfer flavoprotein; PMS, phenazine methosulfate.

[h] LC, long chain; SC, short chain; G, medium chain; AD, acyl-CoA dehydrogenase; IV-D, isovaleryl-CoA dehydrogenase.

[i] Numbers in parentheses denote number of fractions obtained from the pool.

[j] Flavin content of pools A'–G obtained by total absorbance at 430–450 nm using $\epsilon = 11 \times 10^3 \, M^{-1} \, cm^{-1}$ (see Optical Assays).

(see this volume [46]) from acyl CoA-reducible flavin (see Optical Assays). In these experiments a mixture of butyryl-, octanoyl-, palmitoyl-, and isovaleryl-CoA (10 μM each) was used, but the total bleaching accomplished by 25 μM octanoyl-CoA alone was about 80% of that from the mixture. The 0–40% ammonium sulfate precipitate (line 7) did not show any evidence for the presence of ETF in the fluorescence assay, and indeed past attempts to isolate ETF from this fraction have never succeeded.[13] However, this fraction did show some acyl-CoA reducible flavin and served to catalyze reduction of authentic ETF by octanoyl-CoA in the fluorescence assay. An extinction of $8 \times 10^3 \, M^{-1} \, cm^{-1}$ was used to estimate acyl-CoA-reducible flavin, since maximal bleaching by the best substrates is only about 75–80% of the absorbance of the dehydrogenases[3,9,12] as well as the ETF.[6,9,12] Total flavin is estimated by change in absorbance before and after addition of dithionite, using $\epsilon = 9 \times 10^3 M^{-1} cm^{-1}$. Flavin recovered from Step E_{AD} as crude dehydrogenase fractions is shown in lines 8–13. For simplicity, the recovery from all ammonium sulfate fractions of each pool are combined here. The total recovered is about 67% of that estimated in the concentrated 40–85% ammonium sulfate precipitate (line 2). These fractions were judged by their individual spectra to contain a total of 2278 nmol of SC-AD flavin, 2745 nmol of G-AD flavin, 103 nmol of LC-AD flavin, and 1141 nmol of IV-D flavin, suggesting that the dehydrogenase recovered at that step is 35% SC-AD, 42% G-AD, 18% IV-D, and only 1% LC-AD.

In the three columns under DCPIP Reduction are shown units of activity in the DCPIP assay with butyryl-, octanoyl-, and palmitoyl-CoA. Lines 1 and 4 show units obtained from the sonic supernatant and concentrated 40–85% precipitate, respectively, when assayed with no added ETF or PMS. Lines 2 and 5 show units obtained when 3–4 nmol of ETF (70% pH 8.1 ammonium sulfate precipitate, Step E_{ETF} (see Hall[5] and this volume [46]) are added. The pattern of relative activities with the three substrates is very similar to that reported for purified G-AD[5,8] under similar conditions, suggesting that most of the activity being measured is due to that enzyme (see also relative turnover numbers of the different enzymes, below). However, when PMS (100 μM) is used as electron carrier a quite different pattern emerges (lines 3 and 6); activity from butyryl-CoA is equal to activity from octanoyl-CoA, which is about one-tenth of that when assayed with ETF (lines 2 and 5). Overall activity recovery from the dehydrogenase fractions is likewise dominated by butyryl-CoA activity.

The solution to the apparent conflict in the recovery of activities using the PMS-mediated DCPIP reduction assay and the recovery of enzymes judged by spectral characteristics appears upon examination of relative activities of each purified dehydrogenase with PMS and ETF (again at 3–4

μM; note that "$K_{m\ ETF}$" for all dehydrogenases is about 1 μM.[5,7,9,11,13] The relative rates of PMS-mediated and ETF-mediated reduction are as follows: SC-AD, 2.5; G-AD, 0.08; LC-AD, 0.005; IV-D, about 1. When one also considers the overlapping substrate specificities of these enzymes,[1–9,11] it is obvious that activity in the PMS-mediated DCPIP assay cannot be taken as an absolute measurement. However, when large amounts of ETF are not available, the PMS-mediated DCPIP assay can serve as a rough guide for following separation of activities. This would be especially useful if only small amounts of material are available. For example, it can be seen that the A + B fractions contain 63% of the total recovered PMS-mediated reduction of DCPIP by butyryl-CoA, and that these fractions were judged by their spectra to contain most of the recovered SC-AD. Subsequent electrophoresis (see Table II) confirmed this assessment. Fractions C, D, E, F, and G show 75% of the total recovered PMS-mediated reduction of DCPIP by octanoyl-CoA, and inspection of the spectra suggested that all the recovered G-AD was in these fractions. Again, subsequent electrophoresis supported the judgment. Fraction A′ (actually $A_{65-75\%}$ ammonium sulfate precipitate) was the only fraction whose spectra in the absence and in the presence of palmitoyl-CoA unambiguously resembled LC-AD.[15] The PMS-mediated palmitoyl-CoA activities of all the fractions can be accounted for by G-AD activity with this substrate. (Fractions showing activities with isovaleryl-CoA also paralleled appearance of isovaleryl-CoA bleaching in their spectra.) So the PMS-mediated activities would seem to be a useful way of suggesting which dehydrogenases are enriched in which fractions (except for LC-AD) when the optical methods are not feasible and ETF is not readily available.

Table II shows recoveries of dehydrogenases (by flavin content, protein content, and type) after preparative electrophoresis and subsequent ammonium sulfate fractionation (Step F_{AD} above). Columns 2 and 3 show flavin and protein obtained from the dehydrogenase pools (same as columns 4 and 3, lines 8–13, Table I). Columns 3 and 4 show total flavin and protein measured after the various fractions were concentrated prior to preparative electrophoresis. Columns 6 and 7 show the totals of flavin and protein recovered after electrophoresis and subsequent ammonium sulfate fractionation. The percentage recoveries from the electrophoresis step (as percentage of the amount applied, columns 4 and 5) are shown in columns 8 and 9. A total of 17 electrophoresis columns were run, with 100–300 nmol of flavin applied to each. The best recovery is obtained from fractions relatively rich in one dehydrogenase (i.e., A and D). Pools B and C

---

[15] J. G. Hauge, F. L. Crane, and H. Beinert, *J. Biol. Chem.* **219**, 727 (1956).

TABLE II

RECOVERY OF ACYL-CoA DEHYDROGENASES (AD) FROM PREPARATIVE ELECTROPHORESIS[a]

Fraction	Original[b]		Concentrated[c]		Recovered[d]		Percentage recovered[e]	
	Flavin (nmol)	Protein (mg)	Flavin (nmol)	Protein (mg)	Flavin (nmol)	Protein (mg)	Flavin	Protein
A: Long-chain AD	103	10	—	—	15	2	—	—
A: Short-chain AD	1340	94	742	44	440	30	73	68
B: Short-chain AD	938	116	1000	70	246	13[f]	24	20[f]
Isovaleryl AD	736				145	6	14	
C: Medium-chain AD	1140				80		14	
Isovaleryl AD	266	105	570	36	120	—	21	22[f]
Short-chain AD					20	2	3	
D: Medium-chain AD	1025	80	480	30	363	14	75	46
E, F, G: Medium-chain AD	660	80	430	20	100	9	23	45

[a] All flavin contents were obtained by total absorbance at 430–450 nm (see Optical Assays).
[b] From columns 4 and 3, Table I.
[c] After concentration in Amicon prior to loading electrophoresis column(s). See the text.
[d] Total recovered in various fractions.
[e] Percentage recovered from amount loaded.
[f] Total in all fractions recovered.

contain approximately equal amounts of 2 to 3 flavoproteins, while the E, F, and G fractions constitute the G-AD "tail" from the original DEAE column and were perhaps less stable. The purest fractions obtained showed $A_{275}:A_{430-450}$ ratios of 6.5. A total of 892 nmol of SC-AD flavin of varying purities was obtained, 632 nmol of IV-D flavin, 567 nmol of G-AD flavin, and 15 nmol LC-AD flavin after electrophoresis and fractionation. It has been noted previously that LC-AD is obtained rarely if at all from pig liver by this procedure.[5] Inspection of the 0–40% ammonium sulfate precipitate activity patterns in Table I might suggest that fraction contains a significant amount of LC-AD. The sonic supernatant, concentrated 40–85% and 0–40% fractions were all tested for thiolesterase activity using Ellman's reagent.[16] All these fractions showed substantial endogenous rates, but when these rates subsided addition of butyryl-, octanoyl-, or palmitoyl-CoA failed to stimulate any activity. Thus the activity assessments in Table I are probably due mostly to fatty acyl-CoA dehydrogenases. Inspection of the acyl-CoA reducible flavin content of the sonic supernatant as compared to the total flavin content suggests the total of dehydrogenases plus ETF constitutes 53% of the released soluble flavin. This is in agreement with observations made earlier.[3] Proteins shown in Tables I and II were measured by A_{280}, A_{260} (see Optical Assays) and were in good agreement with values obtained from either microbiuret[17] or biuret[18] protein determinations (not shown).

Because ETF is so readily separated from dehydrogenases during the DEAE chromatography step[5,7] (see this volume [46]), it is expected that batch treatments could be effectively used to separate ETF from the dehydrogenases where further purification is not necessary or feasible. The dehydrogenases could be partially purified with preparative electrophoresis following batch elution from the DEAE.

Properties

The structural, spectral, and catalytic properties of the fatty acyl-CoA dehydrogenases have been described in detail, but are summarized here. All the fatty acyl-CoA dehydrogenases appear to be tetramers of 35,000–40,000 MW subunits each containing one FAD.[5,7,11] The fluorescence of the FAD is 90–95% quenched, but the extinction of the flavin is not significantly decreased in the native enzymes.[5,7,11] Although the flavin can be readily detached by boiling or trichloroacetic acid precipitation,[5,7] yellow or green color can be observed on heavily loaded sodium dodecyl

[16] W. E. M. Lands and P. Hunt, *J. Biol. Chem.* **240**, 1905 (1965).
[17] S. Zamenhof, this series, Vol. 3 [103].
[18] A. G. Gornall, C. L. Bardawill, and M. M. David, *J. Biol. Chem.* **177**, 751 (1949).

sulfate–polyacrylamide gels associated with protein bands corresponding to the subunits.[13] The spectral characteristics were summarized under Optical Assays and in references cited in footnotes 5, 7, and 11.

Recent studies with G-AD have shown that binding of substrates is very rapid, tight, but also chain-length dependent.[9] The binding of acyl-CoA and electron transfer to the G-AD flavin appears not to be rate-limiting in the DCPIP assay ("saturating" ETF) for medium-chain substrates.[9] The dehydrogenases cannot turn over enzymatically in the absence of ETF (or PMS),[6,9] and storage for several hours in the presence of substrates or products appears to lead to denaturation.[12] Substrate K_m values for G-AD and LC-AD are virtually impossible to assess by using the DCPIP assay because in addition to the fact that interaction of dehydrogenase and substrate is not rate-limiting (see above), there is evidence of severe substrate–product inhibition at even low levels of substrates having chain lengths longer than four carbon atoms.[7,11,13] The apparent K_m for G-AD with butyryl-CoA is about 40 μM.[9,13] SC-AD tested with butyryl-CoA shows relatively normal hyperbolic plots of ν vs [S], and an apparent K_m of about 1 μM was obtained.[5,13] The K_d values for G-AD for all substrate chain lengths above butyryl-CoA have been estimated to be less than $1 \times 10^{-7} M$.[9] Turnover numbers currently observed from purified dehydrogenases assayed with ETF and DCPIP (V_{max} obtained by extrapolation to saturating ETF[5]) are on the order of 2000 per mole for G-AD (with octanoyl-CoA), 2000 per mole for pig liver LC-AD (with palmitoyl-CoA), and 400 per mole for SC-AD (with butyryl-CoA). The function of the acyl-CoA dehydrogenases appears to be that of tightly and relatively specifically binding their respective substrates prior to interaction with ETF.[9,12]

Acknowledgment

Supported by U.S. Public Health Service Grant GM 25494. The author wishes to thank Eric Juberg for expert technical assistance in obtaining the data presented in Tables I and II.

[46] Electron-Transfer Flavoprotein from Pig Liver Mitochondria

By CAROLE L. HALL

The overall reaction catalyzed by electron-transfer flavoprotein (ETF) with acyl-CoA dehydrogenases is

$$RCH_2CH_2CO—SCoA + FAD - 2\ H^+ - 1\ e \rightarrow$$
$$R—CH{=}CHCO—SCoA + FADH_2$$

where R = alkyl group.

Since its discovery, ETF has been described as the obligate electron acceptor for substrate-reduced acyl-CoA dehydrogenases.[1,2] It could be reduced by saturated fatty acyl-CoA in the presence of catalytic amounts of dehydrogenase and was shown to be required for electron transfer into the electron transport chain as well as for production of substrate for enoyl hydratase.[1-7] Its importance in channeling electrons from fatty acids and from other intermediates of catabolism to the electron transport chain was well recognized.[3] Yet, its status as an enzyme was sometimes questioned because it had no low molecular weight substrate[3] (see discussion to Frisell *et al.*[8]). Recent work now suggests that the ETF is an active partner in the dehydrogenation of saturated acyl-CoA substrates[9] as well as an electron carrier.

Assays

Catalytic Properties

Principle 1. Amounts of ETF in a given sample could be evaluated by assaying with dichlorophenolindophenol (DCPIP) in the presence of a

---

[1] H. Beinert and J. R. Lee, this series, Vol. 6 [59].

[2] H. Beinert, *in* "The Enzymes" (P. D. Boyer, H. Lardy, and K. Myrbäck, eds.), 2nd ed., Vol. 7, p. 467. Academic Press, New York, 1963.

[3] H. Beinert and F. L. Crane, *in* "Inorganic Nitrogen Metabolism" (W. D. McElroy and B. Glass, eds.), p. 601. Johns Hopkins Press, Baltimore, Maryland, 1956.

[4] C. L. Hall, this series, Vol. 53 [50].

[5] C. L. Hall and H. Kamin, *J. Biol. Chem.* **250**, 3476 (1975).

[6] C. L. Hall, J. D. Lambeth, and H. Kamin, *J. Biol. Chem.* **254**, 2023 (1979).

[7] C. L. Hall, unpublished observations.

[8] W. R. Frisell, J. R. Cronin, and C. G. Mackenzie, *in* "Flavins and Flavoproteins" (E. C. Slater, ed.), p. 367 (B.B.A. Library Series 3). Elsevier, Amsterdam, 1966.

[9] C. L. Hall, in preparation.

METHODS IN ENZYMOLOGY, VOL. 71

known amount of an acyl-CoA dehydrogenase and appropriate substrate[4,5] (see also this volume [45]). However, the acyl-CoA dehydrogenase activity would have to be assessed by ETF, so the problem is circular.[10]

Principle 2. The content of ETF can be assessed directly by following the decrease in fluorescence at 495 nm after addition of appropriate acyl-CoA substrate in the presence of acyl-CoA dehydrogenase.[11]

Reagents and Enzymes

1. Buffer, 20 mM potassium phosphate buffer, pH 7.6.
2. Substrate: acyl-CoA derivatives, 5 mM in 50 mM sodium acetate (see this volume [45])
3. Enzymes: 5–20 μl of crude or pure diluted acyl-CoA dehydrogenase
4. ETF-containing sample of appropriate concentration for the sensitivity of the fluorometer (A_{440} ETF about 0.01–0.1) in 1–2 ml

Procedure. The emission of the sample is scanned between 450 and 650 nm (excitation = 380 nm). With the recorder set in time drive, the emission at 495 nm is monitored before and immediately after addition of acyl-CoA substrate. After the emission drop levels off, another scan may be made of the nonfluorescent, reduced ETF. Reoxidation may be monitored if desired at 495 nm. The change in emission at 495 nm is related to the amount of ETF reduced by the substrate chosen. Since octanoyl-CoA substantially reduces G-AD and LC-AD but not IV-D or SC-AD,[4,6,7,12] it is a convenient choice for use in crude mixtures where the dehydrogenase content is not known. The fluorometer must be standardized every day with a stable standard (such as riboflavin, stored in the dark, or quinine) except in models featuring built-in standards. The ETF emission can be quantitated by using the ratio (E/M ETF flavin)/(E/M riboflavin), where E = emission M = mole. Experimentally determined values for this ratio using purified ETF are around 0.38 (see below). The sensitivity of the method is largely dependent on the fluorometer used, but would seem to be acceptable except where mitochondrial content of a sample is very low. The value shown in the table was obtained using 100 μl per milliliter of sonicated supernatant. The method is not affected by the presence of free flavin or other fluorescent flavoproteins because they are not reduced by the saturated acyl-CoA via acyl-CoA dehydrogenase. In

---

[10] H. Beinert, this series, Vol. 5 [73].
[11] C. L. Hall, in preparation.
[12] C. L. Hall, L. Heijkensköld, T. Bartfai, L. Ernster, and H. Kamin, *Arch. Biochim. Biophys.* **177**, 402, 1976.

RECOVERY OF ELECTRON-TRANSFER FLAVOPROTEIN (ETF)

Fraction	Total protein[a] (mg)	Total flavin[b] (nmol)	Total ETF flavin[c] (nmol)
Sonicated supernatant	21,600	31,800	2880
Conc. 40–85% $(NH_4)_2SO_4$ precipitate	9,270	21,150	2012
0–40% $(NH_4)_2SO_4$ precipitate	10,980	7,000	0
Crude ETF (Step C)	1,120	4,720	1300
Partially purified ETF (Step D_{ETF})	277	2,241	1310
Side fractions (Steps C and D_{ETF})	105	290	104
$(NH_4)_2SO_4$ fractions			
60%	110	280	330
70%	50	850	580
75%	20	400	360
80%	5	80	80
85%	1	16	20
Total in $(NH_4)_2SO_4$ fractions	186	1626	1370

[a] Determined by A_{280}-A_{260} (see section Optical Assays in this volume [45]). These values are in good agreement with values obtained by other protein determinations (not shown).

[b] Determined by dithionite reduction, $\epsilon = 9 \times 10^3 \, M^{-1} \, cm^{-1}$ (see this volume [45]) for Sonicated supernatant, Conc. 40–85%, and 0–40% fractions, and by total absorbance at 440 nm for all other samples (see Optical Assays cited in footnote a).

[c] Determined fluorometrically (see Assays, Catalytic Properties) by acyl-CoA reduction of ETF for all samples except 60–85% $(NH_4)_2SO_4$, for which total fluorescence at 495 nm was measured. See the text.

the case of very crude samples, it is sometimes useful to plot the difference spectrum.[11]

Absorbance Properties

Principle. In purified or partially purified fractions where no other flavins are present, the content of ETF can be measured from the absorption spectrum, using $\epsilon = 11 \times 10^3 \, M^{-1} \, cm^{-1}$.[4,5] Results so obtained are in good agreement with results of the enzymatically induced loss of fluorescence. The table shows comparison of two of the above assay methods in following ETF purification.

Preparation of ETF from Pig Liver

Preparation of ETF follows the preparation of acyl-CoA dehydrogenases[4] (see also this volume [45]) through Step C. The purification of the crude ETF is described in detail by Hall.[4] The treatment of crude ETF with CM-cellulose (Step D_{ETF} of Hall[4]) removes hydratase, β-hydroxyacyl-CoA dehydrogenase and β-ketothiolase.[7] Repeated chromatography on DEAE removes some other colorless proteins, and the ammonium sulfate fractionation (Step E_{ETF} of Hall[4]) removes colorless proteins in the 50% and 60% saturation steps, then further fractionation appears to separate ETF containing more flavin from ETF containing less flavin[13] (see the Table).

Columns 2 and 3 of the table show total protein and total flavin (from Dithionite Reducible of Table I this volume [45]). Total ETF was estimated fluorometrically as described under Assays (Catalytic Properties, Principle 2) and is shown in column 4. The total recovered at Step D_{ETF} (both the major pool and the side fractions) represents 45% of the original in the sonicated supernatant and about 65% of that in the concentrated 40–85% ammonium sulfate precipitate. These values were obtained using the ratio $(E/M$ ETF flavin$)/(E/M$ riboflavin$) = 0.38$ to estimate the ETF content. This ratio was the average of experimentally determined values of several samples of ETF from different preparations showing the best $A_{270}:A_{440}$ ratios. Different ammonium sulfate fractions show varying ratios, as also observed by Hall and Kamin.[13] It is not known what are the reasons for these differences. Total fluorescence was measured in the ammonium sulfate fractions rather than octanoyl-CoA reducible fluorescence. Fluorescence measurements are subject to vagaries of the xenon lamps, quenching by unknown materials, or the presence of adventitious flavin,[14] but the fluorescence spectra of the ammonium sulfate fractions of Step E_{ETF} were not suggestive of large amounts of free flavin. In spite of these problems, the fluorescence method of assessing ETF content seems to be useful and is probably fairly reliable.

Properties

Electron-transfer flavoprotein is a dimer of 28,000 MW subunits, each containing one FAD.[4,5] The extinction of the enzyme at 440 nm is not significantly different from the extinction of the detached flavin. The flavin fluorescence is not quenched; in fact, the fluorescence of native ETF is about 3.5 times that of its detached flavin at pH 7, or about equal to its fluorescence at pH 2.75.[4,5,14] Both the absorption and emission spectra are highly resolved. The fluorescence is polarized about equally at both the peak (495 nm) and shoulder (515 nm), indicating that the shoulder at 515 nm is not due to adventitious free flavin. When free flavin is present it

[13] C. L. Hall and H. Kamin, in "Flavins and Flavoproteins" (T. P. Singer, ed.), p. 679. Elsevier, Amsterdam, 1976.
[14] L. M. Siegel, this series, Vol. 53 [43].

appears as an increased emission centered around 525 nm.[7] The spectral and structural properties of ETF are described in detail by Hall.[4]

Electron-transfer flavoprotein rapidly accepts one electron from acyl-CoA dehydrogenase–substrate complex to form a catalytically competent anionic semiquinone.[15] Transfer of a second electron to form the fully reduced ETF occurs more slowly.[9,14] Reoxidation of ETF either slowly by air or more rapidly in the presence of submitochondrial particles is accompanied by a rise in A_{280} indicative of the appearance of free enoyl-CoA. Since the increase in A_{280} is not correlated with transfer of the second electron to ETF, the ETF appears to bind some intermediate form of the acyl-CoA, which remains bound until the ETF is reoxidized. Thus ETF appears to have a role in dehydrogenation beyond its role in transferring electrons from acyl-CoA dehydrogenase to the electron transport chain.[9]

Acknowledgment

Supported by U.S. Public Health Service Grant GM 25494. The author wishes to thank Eric Juberg for expert technical assistance in obtaining the data presented in the table.

[15] C. L. Hall and J. D. Lambeth, *J. Biol. Chem.* **255**, 3591 (1980).

[47] Short-Chain and Long-Chain Enoyl-CoA Hydratases from Pig Heart Muscle

By JIM C. FONG and HORST SCHULZ

$$R—CH{=}CH—\underset{\underset{O}{\|}}{C}—SCoA + H_2O \rightleftharpoons R—\underset{\underset{OH}{|}}{CH}—CH_2—\underset{\underset{O}{\|}}{C}—SCoA$$

The purification of short-chain enoyl-CoA hydratase (crotonase, EC 4.2.1.17) from bovine liver has been reported in previous volumes of this series.[1,2] In this chapter the purification of short-chain enoyl-CoA hydratase (crotonase) and the partial purification of long-chain enoyl-CoA hydratase from pig heart muscle are described. Short-chain enoyl-CoA hydratase is most active with crotonyl-CoA as a substrate, whereas long-chain enoyl-CoA hydratase acts most effectively on Δ^2-octenoyl-CoA but is virtually inactive toward crotonyl-CoA.[3,4]

[1] J. R. Stern, this series, Vol. 1 [93].
[2] H. M. Steinman and R. L. Hill, this series, Vol. 35 [18].
[3] H. Schulz, *J. Biol. Chem.* **249**, 2704 (1974).
[4] J. C. Fong and H. Schulz, *J. Biol. Chem.* **252**, 542 (1977).

Assay Method

Principle. Two spectrophotometric methods for assaying enoyl-CoA hydratase have been described. The direct method is based on the decrease in absorbance at 263 nm due to the hydration of the $\Delta^{2,3}$-double bond of the enoyl-CoA substrate.[1,5] The indirect method relies on a combined assay procedure in which L-3-hydroxyacyl-CoA formed by hydration of the corresponding $\Delta^{2,3}$-enoyl-CoA is oxidized in the presence of NAD and L-3-hydroxyacyl-CoA dehydrogenase.[6,7] The formation of NADH is recorded spectrophotometrically at 340 nm.

Procedure. A single-beam recording spectrophotometer with a full scale absorbance of 0.2 unit and zero offset capability is used for assaying enoyl-CoA hydratase by either the direct or indirect method. Both procedures, as detailed below, have been used to assay short-chain as well as long-chain enoyl-CoA hydratase.[3,4] Assays are performed at 25°. A unit of activity is defined as the amount of enzyme that catalyzes the hydration of 1 μmol of substrate per minute.

DIRECT METHOD. A standard assay mixture contains 0.1 M potassium phosphate, pH 8, bovine serum albumin (0.1 mg/ml), 30 μM Δ^2-enoyl-CoA, and enoyl-CoA hydratase to give an absorbance change at 263 nm of approximately 0.02 unit/min. Under these conditions rates are linear for 2 min. For diluting solutions of short-chain enoyl-CoA hydratase, a buffer containing 0.2 M potassium phosphate, pH 6.4, 20% glycerol (v/v), and 10 mM mercaptoethanol is used, whereas long-chain enoyl-CoA hydratase can be diluted with the assay buffer. The molar extinction coefficient used for calculating rates is 6700 M^{-1} cm^{-1}. The total absorbance, which is primarily due to the adenine residue of CoA should not exceed two absorbance units. If rates need to be determined at high substrate concentrations, measurements should be done at 280 nm.[2]

INDIRECT METHOD. A standard assay mixture contains 0.1 M Tris-HCl, pH 9, 0.1 M KCl, bovine serum albumin (0.1 mg/ml), 120 μM NAD, 30 μM Δ^2-enoyl-CoA, L-3-hydroxyacyl-CoA dehydrogenase (2 units), and enoyl-CoA hydratase to give an absorbance change at 340 nm of approximately 0.02 unit/min. The molar extinction coefficient used for calculating rates is 6220 M^{-1} cm^{-1}.

Preparation of Substrates

Crotonyl-CoA, which is commercially available (P-L Biochemicals, Inc., Sigma Chemical Co.), can also be prepared easily from CoA and

[5] F. Lynen and S. Ochoa, *Biochim. Biophys. Acta* **12**, 299 (1953).
[6] F. Lynen, L. Wessely, O. Wieland, and L. Rueff, *Angew. Chem.* **64**, 687 (1952).
[7] S. J. Wakil and H. R. Mahler, *J. Biol. Chem.* **207**, 125 (1954).

crotonic anhydride.[2] Coenzyme A derivatives of Δ^2-hexenoic acid and its longer-chain homologs are synthesized by the mixed anhydride procedure.[8] In a standard preparation, Δ^2-enoic acid (0.2 mmol) is dissolved in tetrahydrofuran (2 ml), which has been freed of water and ether peroxides by distillation over $LiAlH_4$. The acid is neutralized with triethylamine (0.2 mmol) and allowed to react with ethyl chloroformate (0.2 mmol). After 10 min at 25° the solution is filtered rapidly through a Pasteur pipette containing a glasswool plug and added dropwise over a period of 10 min to a solution of CoA (7 μmol) dissolved in a mixture (5 ml) of water and tetrahydrofuran (3:2, v/v) that has been adjusted to pH 8 by the addition of $NaHCO_3$ (40 mg) and has been equilibrated with N_2. In order to optimize the yield of product, phase separation in the reaction mixture must be prevented by the addition of a few drops of water. After a total reaction time of 20 min, the pH of the reaction mixture is adjusted to approximately 3 and tetrahydrofuran is removed by evaporation under vacuum. Longer-chain enoyl-CoA compounds (C_{10} and longer) are purified by precipitation at pH 1 and by washing the precipitate with pentane. The resulting precipitates are redissolved in water by adjusting the pH to 3. In order to precipitate Δ^2-decenoyl-CoA it is necessary to saturate the solution with NaCl in addition to adjusting the pH to 1. The shorter chain Δ^2-enoyl-CoA compounds are used without further purification after excess enoic acid has been removed by extraction with pentane.

The concentration of Δ^2-enoyl-CoA solutions are determined by the method of Ellman[9] after cleaving the thioester bond with hydroxylamine at pH 7. An aliquot (5–50 μl) of the Δ^2-enoyl-CoA solution containing 0.15 μmol or less is reacted with an equal volume of 2 M hydroxylamine at pH 7 for 5 min or longer when long-chain Δ^2-enoyl-CoA compounds are involved. Ten microliters of 10 mM 5,5'-dithiobis(2-nitrobenzoic acid) are added to the mixture and allowed to react for 30 sec; then 0.1 M potassium phosphate (0.3 ml), pH 8, and water are added to a final volume of 1 ml. The absorbance at 412 nm is recorded until its value has reached a maximum. The resulting absorbance less that of a blank sample, which includes all components except hydroxylamine, is used to calculate the enoyl-CoA concentrations. The molar extinction coefficient is 13,700 M^{-1} cm^{-1}.[9]

Purification of Short-Chain Enoyl-CoA Hydratase

A summary of the purification procedure is presented in Table I. All operations are performed at 4° unless otherwise indicated. All phosphate

[8] P. Goldman and P. R. Vagelos, *J. Biol. Chem.* **236**, 2620 (1961).
[9] G. L. Ellman, *Arch. Biochem. Biophys.* **82**, 70 (1959).

buffers mentioned below contain additionally 10% glycerol (v/v) and 5 mM 2-mercaptoethanol, which were found to protect short-chain enoyl-CoA hydratase from pig heart against inactivation.[4] Crotonyl-CoA is used as a substrate for assaying short-chain enoyl-CoA hydratase.

Step 1. Preparation of an Acetone Powder Extract. Frozen pig hearts (700 g) are cut into pieces and forced through a meat grinder. Batches of one-third of the material are blended together with 500 ml of cold acetone (−20°) in a Waring blender twice for 1 min at top speed. The resulting acetone suspensions are combined and filtered. The solid retentate is washed twice with 500 ml of cold ether (−20°) each. Residual ether is removed by evaporation under vacuum. A total of 230 g of acetone powder is thus obtained. The acetone powder is immediately extracted by stirring it for 8 hr with 800 ml of 0.02 M potassium phosphate, pH 6.3. Insoluble material is removed by centrifugation for 45 min at 30,000 g. The clear supernatant is dialyzed for 8 hr against 7.5 liters of 0.1 M potassium phosphate, pH 6.3, and the precipitate formed is removed by centrifugation at 30,000 g for 30 min and discarded.

Step 2. Phosphocellulose Chromatography (Batch). The acetone powder extract is applied to a phosphocellulose column (4 × 44 cm) equilibrated with 0.01 M potassium phosphate, pH 6.3. The column is washed with the same buffer until UV-absorbing material ceases to be eluted. Short-chain enoyl-CoA hydratase is then eluted with approximately 1200 ml of 0.2 M potassium phosphate, pH 6.3. The eluate is concentrated in an Amicon concentrator (PM-30 membrane) to 400 ml.

Step 3. Phosphocellulose Chromatography (Gradient). The concentrate from step 2 is extensively dialyzed against 0.01 M potassium phosphate, pH 6.3, and applied to a phosphocellulose column (5 × 36 cm) equilibrated with 0.01 M potassium phosphate, pH 6.3. The column is washed with five column volumes of 0.02 M potassium phosphate, pH 6.3, and is then

TABLE I
PURIFICATION OF SHORT-CHAIN ENOYL-CoA HYDRATASE[a]

Step	Total protein (mg)	Total activity[b] (units)	Specific activity (units/mg)	Purification (fold)	Yield (%)
1. Acetone powder extract	19,800	34,400	1.74	1	100
2. Phosphocellulose (batch)	3,358	27,000	8.04	4.6	78
3. Phosphocellulose (gradient)	560	25,600	45.7	26.3	74
4. Hydroxyapatite	11	14,570	1334	767	42

[a] Data for 700 g of pig heart.
[b] Determined with crotonyl-CoA as substrate.

developed with a gradient made up of 2 liters each of 0.02 M potassium phosphate, pH 6.3, and 0.2 M potassium phosphate, pH 6.3. Fractions of 16 ml are collected and assayed for enoyl-CoA hydratase activity. Fractions with high activity are pooled and concentrated in an Amicon concentrator (PM-10 membrane).

Step 4. Chromatography on Hydroxyapatite. The concentrate from step 3 is directly applied to a hydroxyapatite (Bio-Rad) column (2.5 × 15 cm) equilibrated with 0.2 M potassium phosphate, pH 6.3. The column is washed first with four column volumes of 0.2 M potassium phosphate, pH 6.3, then with two column volumes of 0.35 M potassium phosphate, pH 6.3, and is finally developed with a gradient made up of 300 ml each of 0.35 M potassium phosphate, pH 6.3, and 0.7 M potassium phosphate, pH 6.3. Fractions of 6 ml are collected at a flow rate of 60 ml/hr. Fractions with high enoyl-CoA hydratase activity are pooled and concentrated on an Amicon concentrator (PM-10 membrane) to give a solution with 2.4 mg of protein per milliliter. Solid $(NH_4)_2SO_4$ is added to 50% saturation. This preparation when stored at 4° is stable for several months after 70% of the initial activity is lost during the first 3 days.[4] Short-chain enoyl-CoA hydratase purified as outlined above is nearly homogeneous as judged by polyacrylamide gel electrophoresis in the presence of sodium dodecyl sulfate and is free of thiolase and 3-hydroxyacyl-CoA dehydrogenase activities.

Purification of Long-Chain Enoyl-CoA Hydratase

A summary of the purification procedure is presented in Table II. All operations are performed at 4°.

Step 1. Preparation of a Pig Heart Homogenate. Frozen pig heart (225 g) is cut into small pieces and homogenized with 600 ml of 0.05 M potassium phosphate, pH 7, containing 5 mM mercaptoethanol for 5 min in a Waring blender operated at top speed. The resulting suspension is centrifuged for 30 min at 27,000 g, and the precipitate is discarded.

Step 2. Ammonium Sulfate Fractionation. To the stirred supernatant from step 1 is added slowly solid ammonium sulfate to bring the solution to 30% saturation. After stirring the suspension for an additional 20 min, it is centrifuged for 30 min at 27,000 g. The precipitate is discarded, and to the stirred supernatant is added solid ammonium sulfate to bring the solution to 75% saturation. The precipitate formed is isolated by centrifugation at 27,000 g for 60 min and dissolved in a minimal volume of 0.01 M potassium phosphate, pH 7, containing 5 mM mercaptoethanol.

Step 3. DEAE-Cellulose Chromatography. The redissolved ammonium sulfate precipitate from step 2 is extensively dialyzed against 0.01 M

TABLE II

PURIFICATION OF LONG-CHAIN ENOYL-CoA HYDRATASE[a]

Step	Total protein (g)	Total activity (units)			Specific activity (units/mg)			Purification (fold)	Yield[c] (%)
		Δ^2-C$_4$-CoA[b]	Δ^2-C$_{10}$-CoA[b]	Δ^2-C$_{16}$-CoA[b]	Δ^2-C$_4$-CoA[b]	Δ^2-C$_{10}$-CoA[b]	Δ^2-C$_{16}$-CoA[b]		
1. Homogenate	12.66	2457	2961	819	0.194	0.234	0.065	1	100
2. (NH$_4$)$_2$SO$_4$, 30–70% saturation	5.89	814	1322	515	0.138	0.224	0.087	1.34	63
3. DEAE-cellulose	0.184	1.1	114	102	0.006	0.62	0.554	8.5	12.5

[a] Data for 225 g of pig heart.
[b] Δ^2-C$_4$-CoA, crotonyl-CoA; Δ^2-C$_{10}$-CoA, Δ^2-decenoyl-CoA; Δ^2-C$_{16}$-CoA, Δ^2-hexadecenoyl-CoA.
[c] Yield based on Δ^2-hexadecenoyl-CoA hydratase activity.

potassium phosphate, pH 7, containing 5 mM mercaptoethanol and applied to a DEAE-cellulose column (5 × 45 cm) equilibrated with the dialysis buffer.

The column is washed with 0.01 M potassium phosphate, pH 7.0, containing 0.1 M NaCl and 5 mM mercaptoethanol until ultraviolet-absorbing material ceases to be eluted. This forerun contains all short-chain and part of the long-chain enoyl-CoA hydratase activity. The column is developed with a gradient made up of 1.6 liters each of 0.01 M potassium phosphate, pH 7, 0.1 M NaCl, 5 mM mercaptoethanol, and 1.6 liters of 0.01 M potassium phosphate, pH 7, 0.6 M NaCl, 5 mM mercaptoethanol. Fractions of 25 ml are collected and assayed for enoyl-CoA hydratase activity with crotonyl-CoA and longer-chain substrates. Fractions eluted with buffer containing approximately 0.5 M NaCl exhibit high activity with Δ^2-decenoyl-CoA and Δ^2-hexadecenoyl-CoA but are virtually devoid of crotonase, thiolase, as well as 3-hydroxyacyl-CoA dehydrogenase activities. These fractions are pooled and brought to 90% saturation with ammonium sulfate. The resulting precipitate is isolated by centrifugation for 30 min at 27,000 g and dissolved in a minimal volume of 0.01 M potassium phosphate, pH 7, containing 5 mM mercaptoethanol. Preparations of partially purified long-chain enoyl-CoA hydratase remain active for several months when stored at $-20°$.

Properties of Short-Chain Enoyl-CoA Hydratase

The estimated molecular weights of the native enzyme and of its subunits are 155,000 and 27,300, respectively. It appears that the pig heart enzyme is composed of six, possibly identical, subunits, as is the well studied bovine enzyme.[10] The pH optimum is 8.5. Kinetic parameters (K_m, V_{max}) for the hydration of substrates of various chain lengths catalyzed by this enzyme are listed in Table III. The K_m values for all substrates listed are 29–30 μM except for crotonyl-CoA, for which 13 μM has been determined. The V_{max} values decrease nearly linearly from 1670 units/mg with crotonyl-CoA to 40 units/mg for Δ^2-hexadecenoyl-CoA. The short-chain enoyl-CoA hydratase from pig heart is inhibited by p-chloromercuribenzoate and N-methylmaleimide. It is also competitively inhibited by acetoacetyl-CoA (K_i = 14 μM), as is the bovine liver enzyme.[11]

Properties of Long-Chain Enoyl-CoA Hydratase

Kinetic parameters (K_m, V_{max}) for this enzyme are listed in Table III. The enzyme is virtually inactive toward crotonyl-CoA which is hydrated

[10] G. M. Haas and R. L. Hill, *J. Biol. Chem.* **244**, 6080 (1969).
[11] R. M. Waterson and R. L. Hill, *J. Biol. Chem.* **247**, 5258 (1972).

TABLE III

SUBSTRATE SPECIFICITIES OF ENOYL-CoA HYDRATASES FROM PIG HEART MUSCLE

| | Short-chain enoyl-CoA hydratase (crotonase) | | | Long-chain enoyl-CoA hydratase | |
	K_m (μM)	V_{max} (μmol/mg per minute)	Rel. V_{max} (%)	Rel. V_{max} (%)	K_m (μM)
Crotonyl-CoA	13	1,670	100	0	—
Δ^2-Hexenoyl-CoA	29	1,280	77	78	45
Δ^2-Octenoyl-CoA	29	910	54	100	24
Δ^2-Decenoyl-CoA	29	540	32	74	24
Δ^2-Dodecenoyl-CoA	30	160	9.6	42	24
Δ^2-Tetradecenoyl-CoA	—	—	—	34	24
Δ^2-Hexadecenoyl-CoA	30	40	2.4	—	—

at approximately one-hundredth the rate observed with Δ^2-hexenoyl-CoA. The K_m values for substrates of various chain lengths are 24 μM except for Δ^2-hexenoyl-CoA, for which a value of 45 μM is determined. The pH optimum of the enzyme is 8.5. It is inhibited by sulfhydryl inhibitors, for example, p-chloromercuribenzoate and N-methylmaleimide, but not by acetoacetyl-CoA. Inhibition by longer-chain substrates can be prevented by bovine serum albumin. This hydratase tends to aggregate upon storage, but it can be resolubilized with detergents such as Triton X-100.

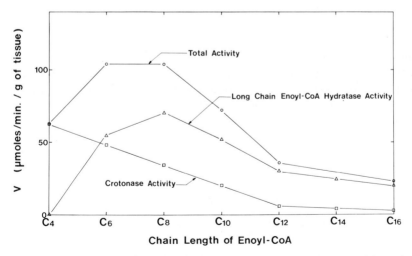

FIG. 1. Contributions of short-chain enoyl-CoA hydratase (crotonase) and long-chain enoyl-CoA hydratase to the total enoyl-CoA hydratase activity in pig heart.

Complementation of Short-Chain and Long-Chain Enoyl-CoA Hydratases

Based on the kinetic parameters listed in Table III, the contributions of the two enzymes to the total enoyl-CoA hydratase activity in pig heart have been calculated (see Fig. 1). Clearly the presence of these two enzymes assures a high rate of hydration of at least 23 units per gram of wet tissue over the whole substrate range. Evidence for the presence of more than one enoyl-CoA hydratase in other tissues and animals has been presented.[4]

Acknowledgment

The studies described were supported by research grants from the National Institutes of Health (AM 15299 and HL 18089) and from the City University of New York Faculty Research Award Program.

[48] 3-Ketoacyl-CoA-Thiolase with Broad Chain Length Specificity from Pig Heart Muscle

EC 2.3.1.16 Acyl-CoA : acetyl-CoA C-acyltransferase

By Horst Schulz and Harold Staack

$$R-\underset{\underset{O}{\parallel}}{C}-CH_2-\underset{\underset{O}{\parallel}}{C}-SCoA + CoASH \rightleftharpoons R-\underset{\underset{O}{\parallel}}{C}-SCoA + CH_3-\underset{\underset{O}{\parallel}}{C}-SCoA$$

Thiolases are ubiquitous enzymes that catalyze essential steps in fatty acid oxidation, ketone body metabolism, and cholesterol biosynthesis. Two types of thiolases have been identified in mitochondria. One is specific for acetoacetyl-CoA as substrate and is referred to as acetoacetyl-CoA thiolase (EC 2.3.1.9), which is believed to function in ketone body metabolism.[1-3] The other thiolase acts on 3-ketoacyl-CoA compounds of various chain lengths and is therefore named 3-ketoacyl-CoA thiolase (EC 2.3.1.16).[2-4] This enzyme catalyzes the last step in the β-oxidation cycle. In heart muscle only the two mitochondrial thiolases mentioned above have been identified[2,3] whereas in liver additionally a peroxisomal thiolase activity[5] and a cytosolic acetoacetyl-CoA

[1] U. Gehring, C. Riepertinger, and F. Lynen, *Eur. J. Biochem.* **6**, 264 (1968).
[2] B. Middleton, *Biochem. J.* **132**, 717 (1973).
[3] H. Staack, J. F. Binstock, and H. Schulz, *J. Biol. Chem.* **253**, 1827 (1978).
[4] W. Seubert, L. Lamberts, R. Kramer, and B. Ohly, *Biochim. Biophys. Acta* **164**, 498 (1968).
[5] P. B. Lazarow, *J. Biol. Chem.* **253**, 1522 (1978).

thiolase[2,6,7] are present. The latter enzyme catalyzes the formation of acetoacetyl-CoA required for cholesterol biosynthesis. The purification of the mitochondrial 3-ketoacyl-CoA thiolase from pig heart muscle is described herein.

Assay Method

Principle. The thiolase assay is based on the CoASH-dependent breakdown of acetoacetyl-CoA and its longer-chain homologs. The rate of the thiolytic cleavage can be measured spectrophotometrically at 303 nm owing to the disappearance of the Mg^{2+}–enolate complex of the 3-ketoacyl-CoA substrates.[8]

Procedure. Spectrophotometric measurements are performed on a recording spectrophotometer with a full-scale absorbance of 0.2 unit. A standard assay mixture contains 0.1 M Tris-HCl, pH 8.3, 25 mM $MgCl_2$, 33 μM acetoacetyl-CoA or 10 μM longer-chain 3-ketoacyl-CoA, 70 μM CoASH, and thiolase to give an absorbance change at 303 nm of approximately 0.04 unit/min. With substrates that have six-carbon or longer acyl chains, bovine serum albumin (0.1 mg/ml) is included in the assay mixture. The reaction is initiated by the addition of CoASH, and the CoASH-dependent stimulation of the decrease in absorbance at 303 nm is used to calculate rates. With purified preparations of thiolase the reaction can be started by the addition of enzyme because the absorbance change in the absence of CoASH is negligible or so small that a separately determined background value can be used for correcting the thiolytic rates. For diluting the enzyme a buffer containing 0.1 M Tris-HCl, pH 8.3, 10% glycerol (v/v), and 10 mM mercaptoethanol is used. Acetoacetyl-CoA, which is commercially available (P-L Biochemicals, Sigma Chemical Co.), can be prepared from CoASH and diketene by the general method developed by Lynen and co-workers[9]; all longer-chain substrates are prepared enzymatically from the corresponding $\Delta^{2,3}$-enoyl-CoA compounds by the method of Seubert *et al.*[4] The concentrations of substrate solutions are determined by measuring spectrophotometrically at 340 nm the decrease in absorbance due to the oxidation of NADH in the presence of L-3-hydroxyacyl-CoA dehydrogenase. An assay mixture contains 0.1 M potassium phosphate, pH 7, 20 μM 3-ketoacyl-CoA, 0.1 mM NADH, and L-3-hydroxyacyl-CoA dehydrogenase (3 mU/ml). An extinction coefficient

[6] K. D. Clinkenbeard, T. Sugiyama, J. Moss, W. D. Reed, and M. D. Lane, *J. Biol. Chem.* **248**, 2275 (1973).

[7] B. Middleton, *Biochem. J.* **139**, 109 (1974).

[8] F. Lynen and S. Ochoa, *Biochim. Biophys. Acta* **12**, 299 (1953).

[9] F. Lynen, L. Wessely, O. Wieland, and L. Rueff, *Angew. Chem.* **64**, 687 (1952).

of 6220 M^{-1} cm^{-1} is used for calculating substrate concentrations. Molar extinction coefficients for the Mg^{2+}–enolate complexes of the various substrates are determined from observed absorbance changes associated with the complete thiolytic cleavage of substrates under the assay conditions given above. The molar extinction coefficients thus obtained are: acetoacetyl-CoA, 21,400 M^{-1} cm^{-1}; 3-ketohexanoyl-CoA, 16,600 M^{-1} cm^{-1}; 3-ketooctanoyl-CoA, 14,400 M^{-1} cm^{-1}; 3-ketodecanoyl-CoA 13,900 M^{-1} cm^{-1}; 3-ketododecanoyl-CoA, 11,000 M^{-1} cm^{-1}; 3-ketohexadecanoyl-CoA, 9900 M^{-1} cm^{-1}. A unit of enzyme activity is defined as the amount of enzyme that catalyzes the hydration of 1 μmol of substrate per minute.

Purification Procedure

A summary of the purification procedure is presented in the table. All operations are performed at 4°. Fractions are assayed for long-chain 3-ketoacyl-CoA thiolase with 3-ketodecanoyl-CoA as substrate.

Step 1. Preparation of a Homogenate. Fresh pig hearts (580 g) are minced with a meat grinder and homogenized together with 2.2 liters of 0.05 M potassium phosphate, pH 6.6, containing 5% glycerol (v/v) and 10 mM mercaptoethanol for 2 min in a Waring blender at top speed. The homogenate is centrifuged for 1 hr at 11,900 g.

Step 2. Phosphocellulose Chromatography (Batch). The centrifuged homogenate from step 1 is applied to a phosphocellulose column (5 × 50 cm) equilibrated with 0.05 M potassium phosphate, pH 6.6, containing 5% glycerol (v/v), and 10 mM mercaptoethanol. The column is washed with the same buffer until material absorbing light at 280 nm ceases to be eluted. At this point another batch of centrifuged homogenate prepared from approximately 150 g of pig heart as described under step 1 can be

PURIFICATION OF 3-KETOACYL-CoA THIOLASE[a]

Step	Total protein (g)	Total activity[b] (units)	Specific activity (units/mg)	Purification (fold)	Yield (%)
Homogenate	20.24	3300	0.163	1	100
Phosphocellulose (batch)	1.468	2935	2.0	12	89
Phosphocellulose (gradient)	0.116	1484	12.8	79	45
CM-cellulose	0.018	1056	59.7	366	32

[a] Data for 715 g of pig heart.
[b] Determined with 3-ketodecanoyl-CoA as substrate.

absorbed to the same column. If so, the column is washed again with 0.05 M potassium phosphate, pH 6.6, containing 5% glycerol (v/v) and 10 mM mercaptoethanol until all material absorbing light at 280 nm has been eluted. The column is then washed with 0.15 M potassium phosphate, pH 6.6, containing 5% glycerol and 10 mM mercaptoethanol, and batches of 200 ml are collected. Two of the batches that contain most of the 3-ketoacyl-CoA thiolase activity are combined and concentrated in an Amicon concentrator (PM-10 membrane) to approximately 30 ml.

Step 3. Phosphocellulose Chromatography (Gradient). The 3-ketoacyl-CoA thiolase concentrate from step 2 is diluted with water containing 5% glycerol and 10 mM mercaptoethanol to a final concentration of 0.05 M potassium phosphate. The resulting solution is applied to a phosphocellulose column (2.5 × 36 cm) equilibrated with 0.05 M potassium phosphate, pH 6.6, containing 10% glycerol and 10 mM mercaptoethanol. The column is washed with three volumes of the same buffer and then developed with a gradient made up of 1 liter of 0.05 M potassium phosphate, pH 6.6, containing 10% glycerol and 10 mM mercaptoethanol and 1 liter of 0.2 M potassium phosphate, pH 6.6, containing 10% glycerol and 10 mM mercaptoethanol. Fractions of 18 ml are collected and assayed for 3-ketoacyl-CoA thiolase activity with acetoacetyl-CoA and 3-ketodecanoyl-CoA. Both activities peak in the same fractions. Fractions with high thiolase activity are pooled and concentrated in an Amicon concentrator (PM-10 membrane).

Step 4. Chromatography on CM-cellulose. The concentrate from step 3 is dialyzed against 4 liters of 0.01 M potassium phosphate, pH 6.6, containing 20% glycerol and 10 mM mercaptoethanol and applied to a CM-cellulose column (1.2 × 46 cm) equilibrated with the dialysis buffer. The column is washed with five column volumes of the dialysis buffer and is then developed with a gradient made up of 200 ml of 0.01 M potassium phosphate, pH 6.6, containing 20% glycerol and 10 mM mercaptoethanol and 200 ml of 0.1 M potassium phosphate containing 20% glycerol and 10 mM mercaptoethanol. Fractions of 3.3 ml are collected and assayed for 3-ketoacyl-CoA thiolase activity. The most active fractions are combined and concentrated in an Amicon concentrator (PM-10 membrane). After the addition of glycerol to a final concentration of 50% (v/v) and dithiothreitol to 5 mM, the preparation can be stored at −75° for more than a year without loss of activity. 3-Ketoacyl-CoA thiolase thus prepared is nearly homogeneous (>90%) and contains only small amounts of crotonase and 3-hydroxyacyl-CoA dehydrogenase activities.

Partial Purification of Acetoacetyl-CoA Thiolase. Acetoacetyl-CoA thiolase, the second thiolase present in pig heart muscle, can be obtained by washing the first phosphocellulose column with 0.4 M potassium phosphate, pH 6.6, containing 5% glycerol and 10 mM mercaptoethanol after 3-ketoacyl-CoA thiolase has been removed with 0.15 M potassium phosphate (see under step 2). Acetoacetyl-CoA thiolase is further purified by following the procedure developed for 3-ketoacyl-CoA thiolase. After a 300-fold purification the enzyme has a specific activity of 43 units/mg and is approximately 50% pure as judged by polyacrylamide electrophoresis in the presence of sodium dodecyl sulfate.

Properties

Stability. 3-Ketoacyl-CoA thiolase when diluted to 1 μg/ml is stable for days at 25° in the presence of 0.75 M Tris-HCl, pH 8.1, containing 25% glycerol (v/v) and 10 mM mercaptoethanol. At much lower Tris-HCl concentrations (0.02 M), or in the absence of either glycerol or mercaptoethanol, inactivation occurs. The remarkable stability of this enzyme contrasts with that of acetoacetyl-CoA thiolase, which in diluted form (1 μg/ml) is stable for days only when kept at 5° in the presence of 50 mM potassium phosphate, pH 6.6, containing 25% glycerol (v/v), bovine serum albumin (1 mg/ml), and 10 mM mercaptoethanol.

Physical Properties. The molecular weight of the native 3-ketoacyl-CoA thiolase was estimated by gel filtration on Sephadex G-200 to be 200,000.[3] A subunit molecular weight of 46,000 was determined by polyacrylamide gel electrophoresis in the presence of sodium dodecyl sulfate.[3] It thus appears that 3-ketoacyl-CoA thiolase is composed of four, possibly identical, subunits, as is acetoacetyl-CoA thiolase from the same tissue.[10]

Kinetic Properties. Pig heart 3-ketoaceyl-CoA thiolase acts almost equally well on CoA derivatives of 3-keto acids having 6 to 14 carbons. Acetoacetyl-CoA is a poorer substrate. This finding is similar to the reported chain length specificity of bovine liver 3-ketoacyl-CoA thiolase.[4] Kinetic constants (apparent K_m and V_{max} values) for several substrates with different chain lengths have been determined.[3] The apparent K_m values for acetoacetyl-CoA, 3-ketohexanoyl-CoA, 3-ketooctanoyl-CoA, and 3-ketodecanoyl-CoA are 17 μM, 8.3 μM, 2.4 μM, and 2.1 μM, respectively. The V_{max} values with the above listed substrates are 65–69 units/mg except for a V_{max} of 29 units/mg observed with acetoacetyl-CoA as substrate. The apparent K_m for CoASH with acetoacetyl-CoA as the substrate is 8.7 μM. 3-Ketoacyl-CoA thiolase is most active at pH 7.6 but still exhibits 75% of its maximal activity at pH 8.3.

[10] U. Gehring and J. I. Harris, *Eur. J. Biochem.* **16**, 487 (1970).

Inhibition. 3-Ketoacyl-CoA thiolase is inhibited by reagents that react with sulfhydryl groups. Incubation of the enzyme (5 μg/ml) with 0.1 M iodoacetamide at pH 8.1 results in the loss of 84% of its activity within 15 min whereas the additional presence of 1 mM acetoacetyl-CoA completely protects the enzyme against inactivation. N-Methylmaleimide (0.1 M) causes activity losses of 30% and 60% within 15 min and 30 min, respectively. Long-chain 3-ketoacyl-CoA compounds also inhibit the enzyme. However, bovine serum albumin partially protects against this inhibition. Antibodies raised in a rabbit against 3-ketoacyl-CoA thiolase inhibit the enzyme partially (60%), but precipitate it completely at a ratio of immunoglobulin to 3-ketoacyl-CoA thiolase of 10 : 1. Antibodies against 3-ketoacyl-CoA thiolase from pig heart did not cross-react with acetoacetyl-CoA thiolase from the same tissue, but did so with 3-ketoacyl-CoA thiolases from beef heart[3] and rat liver.[11]

Acknowledgments

The studies described were supported by research grants from the National Institutes of Health (HL 18089) and from the City University of New York Faculty Research Award Program.

[11] B. M. Raaka and J. M. Lowenstein, *J. Biol. Chem.* **254**, 6755 (1979).

[49] Fatty Acid Oxidation Complex from *Escherichia coli*

By JUDITH F. BINSTOCK and HORST SCHULZ

Fatty acid oxidation in *E. coli* is catalyzed by an enzyme system that is induced when the bacterium is grown on long-chain fatty acids as the sole carbon source.[1,2] Although all enzymes of β oxidation are inducible, three of them, enoyl-CoA hydratase (crotonase, EC 4.2.1.17), L-3-hydroxyacyl-CoA dehydrogenase (EC 1.1.1.35), and 3-ketoacyl-CoA thiolase (EC 2.3.1.16) are induced in a highly coordinate fashion.[2,3] The genes of these three enzymes are closely linked and possibly form an operon.[3] The isolation of the three enzymes from *E. coli* led to the finding that they, in contrast to the same enzymes present in higher organisms,

[1] P. Overath, E. M. Raufuss, W. Stoffel, and W. Ecker, *Biochem. Biophys. Res. Commun.* **29**, 28 (1967).
[2] G. Weeks, M. Shapiro, R. O. Burns, and S. J. Wakil, *J. Bacteriol.* **97**, 827 (1969).
[3] P. Overath, G. Pauli, and H. U. Schairer, *Eur. J. Biochem.* **7**, 559 (1969).

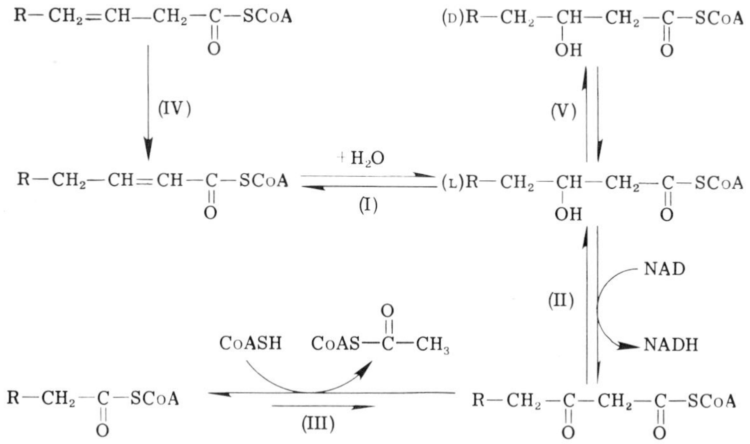

R—CH$_2$=CH—CH$_2$—C—SCoA
‖
O

(IV)

R—CH$_2$—CH=CH—C—SCoA
‖
O

(D)R—CH$_2$—CH—CH$_2$—C—SCoA
| ‖
OH   O

(V)

+ H$_2$O

(I)

(L)R—CH$_2$—CH—CH$_2$—C—SCoA
| ‖
OH   O

NAD

(II)

NADH

O
‖
CoASH   CoAS—C—CH$_3$

R—CH$_2$—C—SCoA
‖
O

(III)

R—CH$_2$—C—CH$_2$—C—SCoA
‖   ‖
O   O

FIG. 1. Reactions catalyzed by the multienzyme complex of fatty acid oxidation from *Escherichia coli*.

are associated with each other and behave during purification like a single protein.[4,5]

In this chapter the purification of the multienzyme complex of fatty acid oxidation from *E. coli* B cells is described. This complex contains, in addition to enoyl-CoA hydratase, L-3-hydroxylacyl-CoA dehydrogenase, and 3-ketoacyl-CoA thiolase mentioned above, 3-hydroxyacyl-CoA epimerase (EC 5.1.2.3) and Δ^3-*cis*-Δ^2-*trans*-enoyl-CoA isomerase (EC 5.3.3.3) activities,[6] which are required for the degradation of unsaturated fatty acids.

Assay Methods

Principle. All five component enzymes of the complex can be assayed spectrophotometrically. Δ^2-Enoyl-CoA hydratase, L-3-hydroxyacyl-CoA dehydrogenase, and 3-ketoacyl-CoA thiolase, which catalyze reactions (I), (II), and (III), respectively (Fig. 1), are assayed spectrophotometrically as described in principle by Lynen and Ochoa.[7] Δ^3-*cis*-Δ^2-*trans*-Enoyl-CoA isomerase and 3-hydroxyacyl-CoA epimerase, which catalyze reactions (IV) and (V), respectively (see Fig. 1) are both measured spectrophotometrically at 340 nm by use of coupled assays similar to those described by Stoffel and co-workers.[8]

[4] J. F. Binstock, A. Pramanik, and H. Schulz, *Proc. Natl. Acad. Sci. U.S.A.* **74**, 492 (1977).
[5] W. J. O'Brien and F. E. Frerman, *J. Bacteriol.* **132**, 532 (1977).
[6] A. Pramanik, S. Pawar, E. Antonian, and H. Schulz, *J. Bacteriol.* **137**, 469 (1979).
[7] F. Lynen and S. Ochoa, *Biochim. Biophys. Acta* **12**, 299 (1953).
[8] W. Stoffel, R. Ditzer, and H. Caesar, *Hoppe Seyler's Z. Physiol. Chem.* **339**, 167 (1964).

Procedure. A single-beam recording spectrophotometer with a full-scale absorbance of 0.2 unit and zero offset capability is used for assaying all five component enzymes. Assays are performed at 25° and the amount of enzyme per assay is adjusted to obtain absorbance changes between 0.02 and 0.06 unit/min. Solutions of the fatty acid oxidation complex are generally diluted with potassium phosphate buffers used for assaying the enzymes and containing bovine serum albumin (1 mg/ml). However, when thiolase is being assayed the dilution buffer contains 1 M potassium N-2-hydroxyethylpiperazine-N'-2-ethanesulfonate (HEPES), pH 8.1, 25% (v/v) glycerol, bovine serum albumin (1 mg/ml), and 10 mM mercapto-ethanol. When L-3-hydroxyacyl-CoA dehydrogenase, 3-hydroxyacyl-CoA epimerase, and Δ^3-*cis*-Δ^2-*trans*-enoyl-CoA isomerase are assayed in crude cell extracts, the preparations are first heated for 1 min at 70° to inactivate NADH dehydrogenase. A unit of enzyme activity is defined as the amount that catalyzes the conversion of 1 μmol of substrate to product per minute.

Reaction (I). Enoyl-CoA hydratase is assayed by following the decrease in absorbance at 263 nm due to the hydration of the $\Delta^{2,3}$-double bond of the substrate. An assay mixture contains 0.2 M potassium phosphate, pH 8, bovine serum albumin (0.2 mg/ml), and 30 μM crotonyl-CoA or 20 μM of longer-chain enoyl-CoA. The reaction is started by the addition of enzyme. An extinction coefficient of 6700 M^{-1} cm^{-1} is used to calculate rates.

Reaction (II). L-3-Hydroxyacyl-CoA dehydrogenase is routinely assayed by measuring the decrease in absorbance at 340 nm due to the dehydrogenation of NADH. An assay mixture contains 0.1 M potassium phosphate, pH 7, bovine serum albumin (0.2 mg/ml), 0.1 mM NADH, 30 μM acetoacetyl-CoA or 20 μM longer-chain 3-ketoacyl-CoA. The assay is started by the addition of enzyme. The rates thus measured are corrected for unspecific NADH dehydrogenase activities. An extinction coefficient of 6220 M^{-1} cm^{-1} is used to calculate rates. Occasionally L-3-hydroxyacyl-CoA dehydrogenase is assayed in the forward direction by recording the increase in absorbance at 340 nm due to the reduction of NAD$^+$. An assay mixture contains 0.1 M potassium phosphate, pH 8, bovine serum albumin (0.3 mg/ml), 2 mM mercaptoethanol, 0.25 mM CoASH, and 17 μM Δ^2-enoyl-CoA. The enzyme complex is added and the hydration of Δ^2-enoyl-CoA is allowed to come to equilibrium. The L-3-hydroxyacyl-CoA dehydrogenase assay is started by the addition of 0.45 mM NAD$^+$. The presence of CoASH in the assay mixture results in the thiolytic cleavage of the formed 3-ketoacyl-CoA, which otherwise would inhibit 3-hydroxyacyl-CoA dehydrogenase.

Reaction (III). 3-Ketoacyl-CoA thiolase is assayed by measuring the decrease in absorbance at 303 nm due to the disappearance of the Mg^{2+}–enolate complex of the substrate. An assay mixture contains 0.1 M Tris-HCl, pH 8.1 (or 0.1 M HEPES, pH 8.1), 25 mM $MgCl_2$, bovine serum albumin (0.2 mg/ml), 2 mM mercaptoethanol, 5% (v/v) glycerol, 0.1 M CoASH, 30 μM acetoacetyl-CoA or 20 μM longer-chain 3-ketoacyl-CoA. The reaction is started by the addition of enzyme. Extinction coefficients for acetoacetyl-CoA and 3-ketodecanoyl-CoA in Tris-HCl are 16,900 M^{-1} cm^{-1} and 13,500 M^{-1} cm^{-1}, respectively, and in HEPES buffer 12,000 M^{-1} cm^{-1} and 6700 M^{-1} cm^{-1}, respectively.

Reaction (IV). Δ^3-*cis*-Δ^2-*trans*-Enoyl-CoA isomerase is assayed by following the increase in absorbance at 340 nm due to the formation of NADH. An assay mixture contains 0.15 M potassium phosphate, pH 8, 0.5 mM NAD^+, 0.1 mM CoASH, 30 μM Δ^3-*cis*-octenoyl-CoA, beef liver crotonase (7.5 μg/ml), pig heart L-3-hydroxyacyl-CoA dehydrogenase (2 μg/ml), and pig heart 3-ketoacyl-CoA thiolase (2.5 μg/ml). The reaction is started by the addition of the fatty acid oxidation complex.

Reaction (V). 3-Hydroxyacyl-CoA epimerase is assayed by measuring the increase in absorbance at 340 nm due to the formation of NADH. An assay mixture contains 0.15 M potassium phosphate, pH 8, bovine serum albumin (0.3 mg/ml), 0.5 mM NAD^+, 0.1 mM CoASH, 60 μM DL-3-hydroxydodecanoyl-CoA, pig heart L-3-hydroxyacyl-CoA dehydrogenase (2 μg/ml), and pig heart 3-ketoacyl-CoA thiolase (1 μg/ml). The reaction is allowed to proceed until the L isomer of 3-hydroxydodecanoyl-CoA is completely degraded. The epimerase assay is then initiated by the addition of the fatty acid oxidation complex.

Preparation of Substrates and Enzymes. DL-3-Hydroxyacyl-CoA as well as Δ^3-*cis*-octenoyl-CoA are prepared from the corresponding carboxylic acids and CoASH by the mixed anhydride procedure of Goldman and Vagelos.[9] Δ^3-*cis*-Octenoic acid is synthesized from 3-octyl-1-ol[10] by following the procedure of Stoffel and Ecker.[11] The preparation of Δ^2-enoyl-CoA compounds is described in this volume.[12] 3-Ketoacyl-CoA derivatives, with the exception of acetoacetyl-CoA,[13] are synthesized from the corresponding Δ^2-enoyl-CoA derivatives by the method of Seubert *et al.*[14]

[9] P. Goldman and P. R. Vagelos, *J. Biol. Chem.* **236**, 2620 (1961).
[10] Obtained from Pfaltz and Bauer, Stamford, Connecticut.
[11] W. Stoffel and W. Ecker, this series, Vol. 14 [17].
[12] J. C. Fong and H. Schulz, this volume [47].
[13] Commercially available from P-L Biochemicals and Sigma Chemical Co.
[14] W. Seubert, L. Lamberts, R. Kramer, and B. Ohly, *Biochim. Biophys. Acta* **164**, 498 (1968).

Growth of *E. coli* Cells

Escherichia coli B cells (ATCC 11775) are grown in a M9 mineral salts medium containing 0.1% (v/v) of oleate as the sole carbon source and 0.4% (w/v) of Triton X-100 to disperse the oleate. Cells are grown at 37° with shaking to the late exponential growth phase (absorbance of 1.7 to 1.8 at 420 nm). The cells are harvested by centrifugation for 15 min at 7500 g, washed once with M9 mineral salts medium, and stored at −20°.

Purification Procedure

A summary of the purification procedure is presented in the table. All operations are performed at 4° unless otherwise stated.

Step 1. Preparation of a Homogenate. *Escherichia coli* B cells (20 g) induced for the enzymes of β oxidation are suspended in 40 ml of 10 mM potassium phosphate, pH 7, containing 25% (v/v) glycerol and 10 mM mercaptoethanol. Phenylmethylsulfonyl fluoride dissolved in isopropanol is added to a final concentration of 1 mM. The cell suspension is sonicated (Branson sonifier, Model W 185, 0.5-in. tip, 60 W output) 12 times for 20 sec each. Between sonic treatments the homogenate is cooled to 0°. After

PURIFICATION OF THE MULTIENZYME COMPLEX OF FATTY ACID OXIDATION FROM
Escherichia coli[a]

		Specific activity (units/mg)		
Enzyme	Substrate	Homogenate[b]	Purified complex[c]	Purification (fold)
Enoyl-CoA hydratase	Crotonyl-CoA	4.1	115	28
L-3-Hydroxyacyl-CoA dehydrogenase	Acetoacetyl-CoA	0.71	20	28
3-Ketoacyl-CoA thiolase	Acetoacetyl-CoA	0.22	5.5	25
3-Hydroxyacyl-CoA epimerase	D-Hydroxydodecanoyl-CoA	0.07	2.4	34
Δ³-cis-Δ²-trans-Enoyl-CoA isomerase	Δ³-cis-Octenoyl-CoA	0.24	8	33

[a] Data for 20 g of *E. coli* B cell paste.

[b] Contained 1.7 g of protein.

[c] After chromatography on phosphocellulose, contained 11.2 mg of protein. Since all assays are performed with 30 μM substrates, these specific activities are significantly lower than the corresponding V_{max} values. The complex is usually obtained in 20% yield.

the first sonication for 20 sec, phenylmethylsulfonyl fluoride is added again to a final concentration of 1 mM. After completion of the sonic treatment, the homogenate is centrifuged 30 min at 31,300 g.

Step 2. Heat Treatment. The supernatant from step 1 is heated to 60° for 10 min and is then centrifuged for 20 min at 31,300 g. The extent of the purification ultimately achieved is hardly or not at all affected by the heat treatment. However, the elimination of approximately two-thirds of the protein before the first column chromatography permits the processing of a larger batch of cells.

Step 3. Phosphocellulose Chromatography. The supernatant from either step 1 or step 2 is extensively dialyzed against 6 liters of 50 mM potassium phosphate, pH 6.6, containing 25% glycerol and 10 mM mercaptoethanol. The dialyzate is applied to a phosphocellulose column (5 × 45 cm) equilibrated with the dialysis buffer. The column is washed with the same buffer until material absorbing light at 280 nm ceases to be eluted and is then developed with a potassium phosphate gradient made up of 2 liters each of 50 mM potassium phosphate, pH 6.6, and 0.5 M potassium phosphate, pH 6.6. Both solutions contain additionally 25% (v/v) glycerol and 10 mM mercaptoethanol. Fractions of 20 ml are collected and routinely assayed for either L-3-hydroxyacyl-CoA dehydrogenase or any of the other component enzymes of the complex. The fractions with high enzyme activities, which are eluted with approximately 0.2 M potassium phosphate, are combined, concentrated by ultrafiltration in an Amicon concentrator (PM-10 membrane) to 2–3 ml, and stored at −76°.

Step 4. Chromatography on Sephacryl S-200. Occasionally when a less than optimal purification is achieved in step 3, chromatography on Sephacryl S-200 (or Sepharose) results in a substantial purification of the complex. Additionally in this step, some inactive polymeric material can be removed. The preparation obtained in step 3 (1 ml containing approximately 5 mg of protein) is placed on a Sephacryl S-200 column (1.2 × 80 cm) equilibrated with 0.2 M potassium phosphae, pH 6.6, containing 10 mM mercaptoethanol. Fractions of 1 ml are collected and assayed for 3-hydroxyacyl-CoA dehydrogenase activity or any of the other component enzymes of the complex. Fractions with high enzyme activity are combined, concentrated in an Amicon concentrator (PM-10 membrane) to 1 ml. After addition of glycerol to a final concentration of 25% (v/v) the preparation is stored at −76°.

Properties

Stability. The purified multienzyme complex of fatty acid oxidation is stable for months when stored in 0.2 M potassium phosphate, pH 6.6, containing 25% (v/v) glycerol and 10 mM mercaptoethanol at −76°. However, the complex is rapidly inactivated in Tris-HCl buffer. For example, when the complex is diluted to a concentration of 27 μg/ml with 1 M Tris-HCl, pH 8, and kept at 0°, 3-ketoacyl-CoA thiolase is inactivated with a half-time of 2 min whereas all other enzyme activities decline with a half-time of 4–4.5 min.[6] Under the same conditions, but in the presence of 0.2 M potassium phosphate, pH 8, or 1 M potassium N-2-hydroxyethyl-piperazine-N'-2-ethanesulfonate, pH 8.1, instead of Tris-HCl, the complex remains fully active. In order to protect the thiolase activity for hours, a sulfhydryl compound such as mercaptoethanol must be present in the enzyme solution. The reaction of sulfhydryl modifying reagents, such as N-ethylmaleimide, with the complex results in the rapid inactivation of 3-ketoacyl-CoA thiolase and the much slower inactivation of Δ^3-cis-Δ^2-trans-enoyl-CoA isomerase, whereas all other enzyme activities are unaffected.

Purity and Molecular Properties. When subjected to electrophoresis on polyacrylamide gels, the purified preparation gives rise to a single protein band if the gels have been preelectrophoresed for 30 min. The purified complex behaves as a single protein when chromatographed on Sephadex G-200 or Sepharose 6B. Electrophoresis on polyacrylamide gradient gels generally gives rise to more than one band. Always observed and most intense is a band corresponding to a protein with a molecular weight of 270,000. Usually an additional band indicative of a protein with a molecular weight of 580,000 is observed. This protein is presumed to be a dimer of the complex.[6] Occasionally a small amount of a putative trimer of the complex can be detected. The assumed monomer and dimer of the complex exhibit all five enzyme activities in the expected ratios.[6] Evidence for polymeric forms of the complex has also been obtained by chromatography of E. coli K12 homogenates on Sepharose 4B.[15] The molecular weight of the complex, which contains phospholipids,[5] has been estimated by several methods and by different groups to be between 245,000[5] and 300,000.[4] When the complex is subjected to electrophoresis on polyacrylamide in the presence of sodium dodecyl sulfate it gives rise to two bands that correspond to proteins with molecular weights of 78,000 and 42,000, respectively. The two types of subunits seem to be present in the complex in equimolar amounts.

Substrate Specificity and Kinetic Properties. The component enzymes of the complex, enoyl-CoA hydratase, L-3-hydroxyacyl-CoA dehydrogenase, and 3-ketoacyl-CoA thiolase, are active with substrates of various

[15] F. R. Beadle, C. C. Gallen, R. S. Conway, and R. M. Waterson, *J. Biol. Chem.* **254**, 4387 (1979).

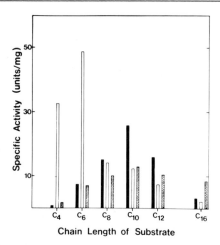

FIG. 2. Chain length specificities of the component enzymes of the complex of fatty acid oxidation from *Escherichia coli*. Solid bars, 3-ketoacyl-CoA thiolase; open bars, enoyl-CoA hydratase; hatched bars, L-3-hydroxyacyl-CoA dehydrogenase. Assays were performed as described under Assay Methods with 17 μM substrates. L-3-Hydroxyacyl-CoA dehydrogenase was assayed in the forward direction.

chain lengths (see Fig. 2).[4] Similar to the corresponding enzyme from higher organisms, enoyl-CoA hydratase (crotonase) is most active with short-chain substrates whereas L-3-hydroxyacyl-CoA dehydrogenase and 3-ketoacyl-CoA thiolase exhibit their highest activities with medium-chain substrates. Only a few of the kinetic parameters (K_m, V_{max}) for the various component enzymes have so far been determined. The following values are preliminary: enoyl-CoA hydratase with crotonyl-CoA, $K_m = 50$ μM, V_{max} 300 units/mg; with Δ^2-decenoyl-CoA, $K_m = 8$ μM, $V_{max} = 35$ units/mg. L-3-Hydroxyacyl-CoA dehydrogenase with acetoacyl-CoA, $K_m = 66$ μM, $V_{max} = 64$ units/mg. 3-Ketoacyl-CoA thiolase with acetoacetyl-CoA, $K_m = 31$ μM,[16] $V_{max} = 10$ units/mg; K_m for CoASH is 20 μM;[16] with 3-ketodecanoyl-CoA, $K_m = 2$ μM,[16] $V_{max} = 60$ units/mg. The V_{max} values for 3-ketoacyl-CoA thiolase have been determined with potassium *N*-2-hydroxyethylpiperazine-*N'*-2-ethanesulfonate as buffer. When Tris-HCl is used, lower values are obtained. In view of the K_m values listed above, especially those for short-chain substrates, and in view of the relatively low substrate concentrations used in the standard assays, it is obvious that the specific activities presented in the table and in Fig. 2 are significantly lower than the V_{max} values and are comparable only to values determined under identical conditions.

[16] H. Staack, J. F. Binstock, and H. Schulz, *J. Biol. Chem.* **253**, 1827 (1978).

Acknowledgments

The studies described were supported by research grants from the National Institutes of Health (HL 18089) and from the City University of New York Faculty Research Award Program. We also would like to acknowledge the important contributions made by Ms. Shashi Pawar and Mr. Ajay Pramanik in determining the properties of the complex.

[50] ω-Hydroxy Fatty Acid: NADP Oxidoreductase from Wound-Healing Potato Tuber Disks

By P. E. KOLATTUKUDY and V. P. AGRAWAL

ω-Oxidation of fatty acids is catalyzed by microsomal enzymes, and the resulting ω-hydroxy fatty acid is converted to a dicarboxylic acid, which is presumably degraded by β-oxidation. Although the biological role of this catabolic process is not known, in plants ω-oxidation is the first step involved in the biosynthesis of the hydroxy and epoxy fatty acids from which plants synthesize the polyester membrane barriers, cutin and suberin.[1] During wound-healing suberin is deposited on the plasma membrane side of the cell wall; this polymer contains alkane-α,ω-dioic acids as major components. Since the synthesis of the dioic acid occurs in response to wounding the following enzymatic steps involved in the conversion of ω-hydroxy fatty acids to dicarboxylic acids were elucidated in cell-free extracts from wound-healing potato tuber disks.[2]

ω-Hydroxy fatty acid + NADP $\rightleftharpoons$ ω-oxo fatty acid + NADPH + H$^+$
ω-Oxo fatty acid + NADP $\rightarrow$ dicarboxylic acid + NADPH + H$^+$

ω-Hydroxy fatty acid: NADP oxidoreductase, the enzyme that catalyzes the first step, is induced during wound healing, whereas the ω-oxoacid dehydrogenase is present in fresh tissue and the latter enzymatic activity level does not change during wound healing. The wound-induced ω-hydroxy fatty acid dehydrogenase from potato tuber has been purified and characterized.[3,4]

[1] P. E. Kolattukudy, R. Croteau, and J. S. Buckner in "Chemistry and Biochemistry of Natural Waxes" (P. E. Kolattukudy, ed.), p. 290. Elsevier, Amsterdam, 1976.
[2] V. P. Agrawal and P. E. Kolattukudy, Plant Physiol. 59, 667 (1977).
[3] V. P. Agrawal and P. E. Kolattukudy, Arch. Biochem. Biophys. 191, 452 (1978).
[4] V. P. Agrawal and P. E. Kolattukudy, Arch. Biochem. Biophys. 191, 466 (1978).

Assay

Three types of assays have been used depending upon the state of purity of the enzyme and availability of appropriate substrates: (*a*) tracer assay with 16-hydroxy[G-³H]palmitate and measure the labeled product isolated by TLC; (*b*) tracer assay with 16-hydroxy[16-³H]palmitate and measure ³H in the water-soluble fraction; (*c*) spectrophotometric assay in which 16-oxohexadecanoic acid-dependent NADPH oxidation is followed by the decrease in absorbance at 340 nm.

Reagents

NADPH, 1 mM
NADP, 10 mM
ω-Hydroxypalmitate, 1 mM
Glycine–NaOH buffer, pH 9.5, 0.4 M
Dithioerythritol (DTE), 10 mM
16-Oxopalmitate, 0.5 mM
Buffer A: 0.05 M Tris-HCl, pH 8.3, containing 10 mM β-mercaptoethanol and 20% by volume glycerol
Buffer B: 0.05 M Tris-HCl, pH 7.3, containing 10 mM β-mercaptoethanol and 20% glycerol
Buffer C: 5 mM sodium phosphate, pH 6.8, containing 10 mM β-mercaptoethanol and 5% glycerol
Buffer D, 0.01 M Tris-HCl, pH 7.5, containing 10 mM β-mercaptoethanol and 20% glycerol

Synthesis of Substrates

16-Hydroxy[G-³H]palmitate. About 100 mg of methyl 16-hydroxypalmitate is exposed to 6 Ci of ³H$_2$ for 15 days according to Wilzbach's technique (any of the radiochemical suppliers performs this service). This method results in the formation of highly radioactive impurities, and therefore rigorous purification of the labeled substrate is absolutely essential. Repeated thin-layer chromatography (TLC) on silica gel G with hexane : ethyl ether : formic acid (65 : 35 : 2 v/v/v) as the developing solvent followed by hydrolysis of the methyl ester in 10% KOH in 95% ethanol (2 hr under reflux) generates labeled ω-hydroxypalmitate. The reaction mixture is acidified, and the product is recovered by extraction with CHCl$_3$ using a separatory funnel (the aqueous layer contains substantial amounts of ³H). The solvent is evaporated off under reduced pressure using a rotary evaporator, and the residue is subjected to TLC on silica gel G with ethyl ether : hexane : formic acid (30 : 20 : 1 v/v/v) as the developing solvent. The specific activity of the product varies from batch to batch (100–400 Ci/mol) and therefore the specific activity of each batch of tritiated material should be determined.

16-Hydroxy[16-³H]palmitate. The specifically labeled substrate is generated by reduction of 16-oxopalmitate with NaB^3H_4. One millimole of methyl 16-hydroxypalmitate in 1.5 ml of CH_2Cl_2 is added dropwise to 1.5 mmol of pyridinium chlorochromate (Aldrich Chemical Co.) suspended in 2 ml of CH_2Cl_2. The reaction mixture is stirred for 1 hr, diluted with 15 ml of ethyl ether, and filtered through a 1.3 × 4 cm column of silicic acid. The ether is removed under a stream of N_2, then the product is purified by TLC on silica gel G with hexane : ethyl ether : formic acid (65 : 32 : 2 v/v/v) as the developing solvent ($R_f = 0.51$). This procedure gives about 80% yield of methyl 16-oxopalmitate. About 200 mg of the methyl ester of the oxoacid is dissolved in 3 ml of tetrahydrofuran (distilled over $LiAlH_4$) and 25 mCi NaB^3H_4 (150–400 Ci/mol) is added. After stirring overnight the reaction mixture is acidified and the product is recovered by three extractions with chloroform ($CHCl_3$). The chloroform is removed with a rotary evaporator under reduced pressure, then the hydroxyacid is purified by repeated TLC as indicated above. The methyl ester is subjected to hydrolysis with ethanolic KOH, and the free acid is recovered and purified as described above.

Assay Procedure

Tracer Assay I. Aliquots of the following reagents are added to a small test tube so that the indicated final concentrations are obtained in a total volume of 0.5 ml of 0.2 M glycine–NaOH buffer, pH 9.5: NADP, 1 mM; ω-hydroxy[G-³H]palmitate, 0.1 mM; DTE, 1 mM; glycine-NaOH buffer; enzyme. The reaction mixture is incubated for 60 min at 30°, and then 0.25 ml of dilute HCl is added. The mixture is transferred to a 60-ml separatory funnel, and the test tube is rinsed twice with 0.25 ml of water and the aqueous solution is extracted three times with 5 ml (each) of $CHCl_3$. The $CHCl_3$ extract collected in a test tube is evaporated off under a stream of N_2 while the test tube is placed in warm (~50°) water; the inner wall of the test tube is washed down with 2 ml of ethyl ether, and the ether is evaporated with a stream of N_2. The residue is dissolved in 0.1 ml of tetrahydrofuran, and 10 μl are assayed for ³H to check for the recovery. Another aliquot, 25 μl, is applied to a thin-layer plate (0.5 mm silica gel G, 20 × 20 cm) as a band. On top of this thin band is applied 25 μl of a solution containing 20 μg each of 16-oxopalmitate and hexadecane-1,16-dioic acid. The chromatogram is developed in ethyl ether : hexane : formic acid

(30:20:1 v/v/v) as the solvent. The plate is air dried and sprayed with a 0.1% ethanolic solution of 2',7'-dichlorofluorescein. Under an ultraviolet lamp the fluorescing areas are marked and the silica gel from the areas corresponding to 16-oxopalmitic and hexadecane-1,16-dioic acids are scraped into counting vials. After the addition of 10 ml of a counting fluid consisting of 30% ethanol in toluene containing 4 g of Omnifluor (New England Nuclear) per liter, the vials are thoroughly shaken with a Vortex mixer and assayed for ^{3}H in a scintillation spectrometer. When the enzyme preparation contains ω-oxo acid dehydrogenase, the amount of ^{3}H present in the oxo acid and dicarboxylic acid together is used to calculate the number of nanomoles of ω-hydroxyacid oxidized. After removal of the oxoacid dehydrogenase only 16-oxoacid needs to be measured.

Tracer Assay II. Aliquots of the following reagents are added to a small test tube so that the indicated final concentrations are obtained in a final volume of 0.2 ml of 0.2 M glycine-NaOH buffer, pH 9.5: NADP, 1 mM; ω-hydroxy[16-^{3}H]palmitate, 0.1 mM ($\sim$1.5 $\times$ 10^6 cpm); enzyme; glycine–NaOH buffer. The mixture is incubated for 30 min at 30°, and the reaction is terminated by the addition of 1.8 ml of 0.05 N HCl. The reaction mixture is extracted four times with 5 ml CHCl$_3$ using a Vortex mixer. One milliliter of the aqueous phase is transferred into a counting vial, and the traces of CHCl$_3$ present are removed with a stream of N$_2$ while the vial is placed in warm (60°) water. After the addition of 10 ml of ScintiVerse (Fisher) ^{3}H is assayed with a scintillation spectrometer. In this assay the number of nanomoles of ω-hydroxyacid oxidized is calculated from the specific activity of the substrate; the resulting number is doubled because the substrate is a DL-mixture.

Spectrophotometric Assay. The following reagents are added into a 1-cm spectrophotometer cuvette (0.5-ml capacity) so that the indicated final concentrations are obtained in a final volume of 0.25 ml of 0.2 M glycine–NaOH buffer, pH 8.5: 16-oxopalmitate, 0.04 mM; NADPH, 0.08 mM; enzyme; in the blank, buffer is used in place of the enzyme. Decrease in absorbance at 340 nm is recorded and initial linear rate is used for calculating the enzyme activity.

Purification Procedure

Preparation of Acetone Powder from Wound-Healing Potato Tuber Disks. Potato tubers (White Rose) purchased from grocery stores can be used as long as they have been stored for less than 6 months after harvest. The tubers are surface sterilized by immersing them in 2% (w/v) hypochlorite (30% v/v Chlorox) solution for 5 min followed by a thorough wash with water. They are broken into halves and tissue cores 1 cm in diameter are punched from the inside with a cork borer. From these cores 2-mm disks are cut with a slicer consisting of double-edged razor blades spaced with washers and held tightly together by two threaded rods. The disks are transferred to layers of rubberized mesh in 1-gallon jars; each jar contains five layers of tissue disks each separated by a rubberized mesh. All materials used are sterilized prior to use, and all operations are done under sterile conditions in a glove box. Humid air is passed at a moderate rate (0.6 liter/hr) through the jars for 120 hr at 22° in the dark. Then the disks are homogenized with cold (−20°) acetone (5 ml/g tissue) for 1 min in an explosion-proof Waring blender. The suspension is filtered under suction, the residue is rehomogenized with cold acetone for 30 sec and filtered again, and the resulting insoluble material is air dried. The yield of acetone powder is 12–18% of the fresh weight of the disks; this powder can be stored in airtight containers at −20° for 4–8 weeks without significant loss in enzymatic activity.

Ammonium Sulfate Precipitation of the Acetone Powder Extract. Acetone powder (400 g) is slowly added with stirring to 4 liters of 0.1 M sodium-phosphate buffer, pH 6.7, containing 50 mM β-mercaptoethanol while the buffer is being stirred in an ice bath. The thick suspension is squeezed through 8 layers of cheesecloth and the filtrate is adjusted to 4 liters with the same buffer. To this solution 704 g of granular ammonium sulfate are slowly added with stirring; the resulting precipitate is removed by centrifugation at 12,000 g for 10 min. To the supernatant 856 g of ammonium sulfate are slowly added with stirring, and the resulting precipitate, collected by centrifugation as above, is dissolved in a minimum volume of buffer A. The solution is dialyzed against the same buffer for 48 hr, and the volume is adjusted to 500 ml with buffer A.

Treatment with DEAE-Cellulose. To the 500 ml of protein solution (usually containing about 11 g of protein), 70 g of DE-52, previously equilibrated with buffer A, are added. The mixture is stirred for 30 min, then centrifuged at 12,000 g for 10 min; the pellet is resuspended in 500 ml of buffer A and centrifuged; the supernatants are discarded. The pellet is stirred with 300 ml of buffer A containing 0.2 M KCl for 1 hr and centrifuged. The pellet is resuspended in 200 ml of the KCl-containing buffer A and centrifuged. The two supernatants are combined and concentrated to about 40 ml by ultrafiltration with UM-10 membrane.

Gel Filtration. The protein solution is divided into four portions each containing about 1.5 g of protein; each 10-ml aliquot is passed through a Sepharose 6B column (3.9 × 100 cm) with buffer B as the eluent at 30–40 ml/hr, collecting 9–10 ml fractions. Aliquots are assayed for ω-hydroxyfatty acid dehydrogenase, and fractions containing enzyme activity greater than half of the maximum activity are pooled. This step yields 2.2 g of protein in a total volume of about 300 ml.

DEAE-Sephadex Anion Exchange Chromatography. The 300-ml protein solution is passed at a flow rate of 40 ml/hr through a DEAE-Sephadex column (bed volume 200 ml) equilibrated with buffer B. After washing the column with 300 ml of buffer B, a linear gradient of 0 to 0.2 M KCl in buffer B is applied in a total volume of 1 liter with a flow rate of 30 ml/hr, collecting 10-ml fractions. The enzyme elutes in the range of 0.09 to 0.14 M KCl and the fractions containing the enzymatic activity are pooled and concentrated to 20–30 ml by ultrafiltration. This solution is diluted to 200 ml with buffer C and concentrated again to the original volume. This process is repeated four times to change the medium, resulting in 40 ml containing 600 mg of protein.

Hydroxylapatite Chromatography. The protein solution is applied to a hydroxyapatite column (25 ml bed volume) equilibrated with buffer C. After washing the column with 50 ml of buffer C, a phosphate gradient is applied at a flow rate of 15–20 ml/hr to elute the enzyme. The mixing chamber contains 125 ml of buffer C, and the reservoir contains 125 ml of 0.2 M sodium phosphate, pH 6.8, containing 5% glycerol and 10 mM β-mercaptoethanol. (Although 20% glycerol protects the enzyme from inactivation, significantly higher than 5% of glycerol drastically reduces the flow rate and makes this step impractical.) The fractions containing the enzyme activity (eluted at around 0.14 M phosphate) are pooled, the medium is changed to buffer D by repeated ultrafiltration and dilution as in the preceding step, and the final volume is adjusted to 20 ml with buffer D; this step yields 115 mg of protein.

NADP–Sepharose Chromatography. The enzyme solution is passed through 1 × 25 cm NADP Sepharose column[5] equilibrated with buffer D. After washing the column with 40 ml of buffer D, which removes the bulk of the protein and some of the dehydrogenase, a linear gradient of 0 to 2 M KCl in a total volume of 200 ml of buffer D is applied at a flow rate of 10–12 ml/hr. The fractions containing enzyme activity (eluting as a rather sharp peak at around 0.3 M KCl) are pooled, KCl is removed by repeated ultrafiltration and dilution with buffer D, and the final volume of the protein solution containing 2 mg of protein is adjusted to 5 ml with buffer D. The protein not retained by the first pass through the column can be recycled to recover more dehydrogenase.

---

[5] NADP-Sepharose is prepared by coupling periodate-oxidized NADP with the dihydrazide–Sepharose 4B prepared by treating CNBr-activated Sepharose 4B with the dihydrazide of adipic acid (R. Lamed, Y. Levin, and M. Wilchek, *Biochim. Biophys. Acta* **304**, 231 (1973)).

The protein solution is passed through an NADP–Sepharose column (1 × 6.4 cm) equilibrated with buffer D, and the column is washed with 15 ml of buffer D. A 0 to 0.1 mM linear gradient of NADP is applied in a total volume of 40 ml of buffer D at 5–6 ml/hr, collecting 1.4-ml fractions. The enzyme emerges at around 60 μM NADP, and the fractions containing enzymatic activity are pooled and concentrated to 1 ml by ultrafiltration. The solution is diluted to 10 ml with buffer D and concentrated; this process is repeated four times to remove NADP. The above procedure yields only 0.2 mg of enzyme (Table I); the major reason for the low (1%) yield is the instability of the enzyme. Furthermore, at every step, particularly the DEAE-Sephadex chromatography step, a large portion of the enzyme is sacrificed to ensure purity of the final product.

Properties of the Enzyme

Purity and Stability. Polyacrylamide gel electrophoresis shows one band that contains the enzymatic activity. Sodium dodecyl sulfate (SDS)–gel electrophoresis also shows one band. The enzyme is quite unstable, but 20% glycerol stabilizes it. After the final purification step it is stable at 0° for a few weeks.

General Properties. This dehydrogenase has a molecular weight of 60,000 and a Stokes' radius of 3.1 nm; SDS gel electrophoresis shows that the enzyme is a dimer of equal-sized protomers. The reaction is reversible with an equilibrium constant of 1.4 × 10^{-9} at pH 9.5 and 30°. The rate of the forward reaction is linear up to 5 μg of protein per milliliter and 30 min of incubation at 30°. The rate of reverse reaction is linear up to 1 μg of protein per milliliter and 5 min of incubation. The optimum pH for the forward reaction and reverse reaction are 9.5 and 8.5, respectively. In both directions activity is much less on either side of the optimal pH.

Substrate Specificity and Kinetic Parameters. NADP is preferred over NAD; the rate obtained with the latter is only 25–33% of that with NADP. Apparent K_m values for NADP, 16-hydroxyhexadecanoate, NADPH, and 16-oxohexadecanoate are 100, 20, 5, and 7 μM, respectively. The K_m for 16-oxohexadecanoate does not change from pH 7.5 to 9.0 but increases about 10-fold from pH 9.0 to pH 10.0. This change is most probably attributable to deprotonation of an ε-amino group of lysine, which is involved in binding this substrate to the enzyme via an ionic interaction with the distal carboxyl group. Analogs lacking the distal carboxyl group bind the enzyme with a higher K_m and serve as substrates (Table II). The hydrophobic aliphatic chain also participates in the binding of the substrates as indicated by the K_m and V values shown in Table II.

A surprising finding is that hydride from the B side of the nicotinamide

TABLE I

Purification of Wound-Induced ω-Hydroxy Fatty Acid: NADP Oxidoreductase from Suberizing Potato Tuber Disks

Step	Total protein (mg)	Total enzyme activity[a] (nmol/min)	Specific activity (nmol/mg per minute)	Yield	Purification (fold)	Specific activity in reverse[a]/specific activity in forward direction[b]
1. Extraction of acetone powder	12,150	64,400	5.3	100	1	—
2. 30–60% (NH4)2SO4	11,100	59,250	5.3	92	1	—
3. DEAE-cellulose	6,520	45,080	6.9	70	1.3	—
4. Sepharose 6B	2,210	32,250	14.6	50	2.8	17.2
5. DEAE–Sephadex chromatography	610	9,670	15.9	15	3.0	21.6
6. Hydroxylapatite chromatography	115	2,260	19.7	3.5	3.7	23
7. NADP–Sepharose affinity chromatography I (KCl as eluent)	2	1,290	645	2	122	23
8. NADP–Sepharose affinity chromatography II (NADP as eluent)	0.2	635	3175	1	599	23

[a] Assayed in reverse direction spectrophotometrically.
[b] Assayed in forward direction by tracer assay II.

TABLE II
K_m AND V OF ω-HYDROXY FATTY ACID : NADP OXIDOREDUCTASE FOR OXO SUBSTRATES

Substrate	K_m (μM)	V (nmol/mg per minute)	V/K_m
A. Alkanals			
C_2	ND^a	ND^a	ND
C_3	7000	287	0.04
C_4	1800	906	0.50
C_6	292	1724	5.9
C_8	88	5525	63.14
C_{10}	65	3536	54.40
C_{12}	45	1879	41.64
C_{14}	38	1724	45.37
C_{16}	50	1768	35.36
C_{18}	62	1503	24.24
C_{20}	71	641	9.03
C_{22}	ND	ND	ND
C_{24}	ND	ND	ND
B. ω-Oxo acid			
8-Oxo C_8 acid	310	928	2.79
10-Oxo C_{10} acid	55	8376	152.3
16-Oxo C_{16} acid	12	2210	184.2
C. Other substrates			
Benzaldehyde	400	1658	4.14
Neral	70	354	5.06

[a] ND = K_m and V could not be determined due to very low rates with these substrates.

ring is nearly as efficiently transferred as that from the A side by this enzyme. This apparent lack of stereospecificity is also manifested by the enzyme contained in a single 1-mm gel slice obtained after polyacrylamide gel electrophoresis.

Kinetic Mechanism. The kinetic parameters shown in Table III indicate an ordered sequential mechanism, where NADPH is added first, followed by 16-oxohexadecanoate, and NADP is released after 16-hydroxy-hexadecanoate.

Inhibitors. Substrate inhibition of ω-oxoacid reduction is observed with >2 mM NADPH. Thiol-directed reagents, such as *p*-hydroxymercuribenzoate, severely inhibit the enzyme. Diethyl-pyrocarbonate (histidine modifier) inhibits the enzyme, and hydroxylamine reverses this inhibition. Phenylglyoxal (arginine modifier) and pyridoxal phosphate (lysine modifier) inhibit the enzyme, and substrates protect the enzyme from such chemical modification.

TABLE III

KINETIC PARAMETERS OF ω-HYDROXY FATTY ACID : NADP OXIDOREDUCTASE FROM
SUBERIZING POTATO DISKS

A. Kinetic constants obtained from initial velocity experiments[a]

Forward reaction (μM)		Backward reaction (μM)	
K_{NADP}	90.2 ± 2	K_{NADPH}	5.9 ± 0.2
$K_{\omega-hydroxy\ C_{16}\ acid}$	27.2 ± 0.2	$K_{\omega-oxo\ C_{16}\ acid}$	8.3 ± 0.4
K_{iNADP}	56.8 ± 0.1	K_{iNADPH}	3.5 ± 0.1

B. Inhibition constants obtained from product inhibition experiments[a]

Inhibitor	Variable substrate	Pattern	K_{ii}	K_{is}
NADP	NADPH	Competitive		14.4
NADP	ω-Oxo C_{16} acid	Noncompetitive	200	120
ω-Hydroxy C_{16} acid	NADPH	Noncompetitive	274^{b}	900^{b}
ω-Hydroxy C_{16} acid	ω-Oxo C_{16} acid	Noncompetitive	300^{b}	260^{b}

[a] K_{ii} and K_{is} are inhibition constants calculated from intercept and slope replots, respectively.

[b] Equations $[intercept]_I = [intercept]_0 (1 + [I]/K_{ii})$ and $[Slope]_I = [Slope]_0 (1 + [I]/K_{is})$ are used to calculate K_{ii} and K_{is}, respectively, where subscripts I and 0 denote, respectively, the presence and the absence of inhibitor.

Other Sources. Oxidation of ω-hydroxyfatty acids to dicarboxylic acids occurs in animals,[6-8] microbes,[9] and plants[10]; therefore the dehydrogenase could be purified from any such source.

Acknowledgment

This research was supported in part by Grant PCM77-00927 from the National Science Foundation.

[6] B. Preiss and K. Bloch, *J. Biol. Chem.* **239,** 85 (1964).

[7] M. A. Mitz and R. L. Heinrikson, *Biochim. Biophys. Acta* **46,** 45 (1961).

[8] I. Bjorkhem, *Biochim. Biophys. Acta* **260,** 178 (1972).

[9] M. Kusunose, E. Kusunose, and M. J. Coon, *J. Biol. Chem.* **239,** 1374 (1964).

[10] P. E. Kolattukudy, R. Croteau, and T. J. Walton, *Plant Physiol.* **55,** 875 (1975).

[51] Enoyl-CoA Hydratases from *Clostridium acetobutylicum* and *Escherichia coli*

EC 4.2.1.17 L-3-Hydroxyacyl-CoA hydro-lyase

By ROBERT M. WATERSON and ROBERT S. CONWAY

$$RCH{=}CH{-}\overset{\overset{\displaystyle O}{\|}}{C}{-}SCoA + H_2O \rightleftharpoons L\text{-}(+)\text{-}R{-}\overset{\overset{\displaystyle OH}{|}}{CH}{-}CH_2\overset{\overset{\displaystyle O}{\|}}{C}{-}SCoA$$

Enoyl-CoA hydratase participates in the fatty acid β-oxidation pathway, catalyzing the stereospecific hydration of $\Delta^{2,3}$-*trans*-enoyl-CoA thioesters to the corresponding L-(+)-β-hydroxyacyl-CoA derivatives. Bovine liver crotonase is the most thoroughly characterized hydratase; descriptions of this mitochondrial enzyme of broad substrate chain length specificity have been presented previously in this series.[1,2] Crotonase has been purified from pig heart[3] and found closely to resemble the bovine enzyme in a number of its physicochemical and enzymatic properties. Hydratases active upon only longer-chain length substrates have been partially purified from several animal tissues[3,4] and are discussed in this volume. In this paper we describe the preparation, assay, and properties of enoyl-CoA hydratases from *Clostridium acetobutylicum*[5] and *Escherichia coli*.[6]

Assay Method

Principle. The activity of this enzyme is directly measured[7,8] by continuous monitoring of the decrease in ultraviolet absorbance in the region

[1] J. R. Stern, this series, Vol. 1 [93].

[2] H. M. Steinman and R. L. Hill, Vol. 35 [18].

[3] J. M. Fong and H. Schulz, *J. Biol. Chem.* **252**, 542 (1977).

[4] H. Schulz, *J. Biol. Chem.* **249**, 2704 (1974).

[5] R. M. Waterson, F. J. Castellino, G. M. Hass, and R. L. Hill, *J. Biol. Chem.* **247**, 5266 (1972).

[6] F. R. Beadle, C. C. Gallen, R. S. Conway, and R. M. Waterson, *J. Biol. Chem.* **254**, 4387 (1979).

[7] J. R. Stern and A. del Campillo, *J. Am. Chem. Soc.* **75**, 2277 (1953).

[8] S. J. Wakil and H. R. Mahler, *J. Biol. Chem.* **207**, 125 (1954).

of 250–290 nm due to hydration of the enoyl-CoA conjugated double bond.[9] In addition, hydration may be determined indirectly by coupled assay in the presence of β-hydroxyacyl-CoA dehydrogenase; in this instance, one measures the reduction of NAD (absorbance increase at 340 nm) that accompanies the NAD-dependent oxidation of L-(+)-β-hydroxyacyl-CoA produced by enoyl-CoA hydratase.[8] The relative merits of these two methods have been discussed in an earlier volume of this series.[2]

Reagents

Potassium phosphate,[10] 0.3 M, pH 8.0, containing ovalbumin (0.05 mg/ml)

Crotonyl-CoA or longer chain $\Delta^{2,3}$-*trans*-enoyl-CoA, 50–250 μM

Crotonyl-CoA is prepared by the reaction of CoA with crotonic anhydride,[12] the longer-chain enoyl-CoA substrates are prepared by the mixed anhydride synthetic method of Goldman and Vagelos.[13]

Procedure. The direct enzyme assay is carried out using a split-beam recording spectrophotometer equipped with a temperature control unit equilibrated at 30°. An aliquot of buffer (0.5 ml) is pipetted into a 1-cm quartz cuvette, and 0.005–0.040 ml of enoyl-CoA solution is added. When using high substrate concentrations the reference cuvette may be optically balanced by addition of a small volume of a solution of AMP (about 10 mg/ml) in water. The reaction is initiated with the addition of 0.005–0.020 ml of enzyme, appropriately diluted prior to assay in 0.02 M potassium phosphate, pH 8.0, containing ovalbumin (1 mg/ml). The absorbance trace at 0.1–0.25 full-scale sensitivity is measured at 263 nm or 280 nm and is usually linear for 2–4 min. The decreasing slopes of the tracings are determined and expressed as units of enzymatic activity. One unit of hydratase activity is equivalent to the hydration of 1 μmol of substrate per minute, based upon molar extinction coefficients of 6700 and 3600 at 263 nm and 280 nm, respectively.[9] Specific activity is expressed as units of activity per milligram of protein.

[9] F. Lynen and S. Ochoa, *Biochim. Biophys. Acta* **12**, 299 (1953).
[10] The buffer employed previously[5] in the assay mixture for crotonase from *C. acetobutylicum* was 0.05–0.10 M Tris-HCl. The use of potassium phosphate is required for assay of the hydratases from *E. coli* because these activities are inactivated by Tris-HCl.[6,11]
[11] A. Pramanik, S. Pawar, E. Antonian, and H. Schulz, *J. Bacteriol.* **137**, 469 (1979).
[12] E. J. Simon and D. Shemin, *J. Am. Chem. Soc.* **75**, 2520 (1953).
[13] P. Goldman and P. R. Vagelos, *J. Biol. Chem.* **247**, 5258 (1972).

Crotonase from *C. acetobutylicum*

Purification Procedure

A summary of the purification procedure[5] is given in the table.

Step 1. Preparation of Acetone Powder and Cell Extracts. Clostridium acetobutylicum is grown anaerobically as previously described,[14] and cells are harvested (225 g from 50 liters) after reaching a growth plateau. Material sufficient for several enzyme preparations may be stored at $-20°$ in the form of an acetone powder. Approximately 100 g (wet weight) of cells are mixed with 800 ml of acetone chilled with Dry Ice–ethanol; the mixture is homogenized in a Waring blender for approximately 2 min at high speed. After blending, the cell suspension is poured from the blender; the residue in the blender after two or three 100-g batches is suspended in 700 ml of acetone and treated identically. All fractions are combined, stirred for 10 min in a beaker, and filtered through a large Büchner funnel. The residue is then sequentially washed with 2 liters each of chilled acetone $(-5°)$ and ether $(-5°)$ and dried under vacuum to remove residual ether. In a representative experiment, a total of 56.5 g of powder was obtained from 225 g of wet cells. Extracts are prepared by suspending 10 g of the acetone powder in 500 ml of 0.02 M potassium phosphate, 0.003 M potassium EDTA, pH 7.5, and stirring at $4°$ for 24 hr. Insoluble material is removed by centrifugation for 1 hr at 9000 rpm (Sorvall RC-2, GSA rotor).

Step 2. Acid–Heat Treatment. The clear supernatant is adjusted to pH 5 by the slow addition of 1 M acetic acid. The acidified solution is placed in a $65°$ water bath and heated to $60°$ with overhead stirring for 4 min, then cooled by transfer of the beaker to an ice bath. When sufficiently cooled $(15°$ or below) the acidified and heated extract is adjusted to pH 7 by the slow addition of 1 M potassium bicarbonate. Denatured protein is removed by centrifugation for 45 min at 9000 rpm.

Step 3. Acetone Precipitation. This step is performed in a $4°$ room. The supernatant solution from the preceding step is cooled to $-1°$ to $-2°$ by hand swirling in a 2-liter flask immersed in a Dry Ice–ethanol bath or a circulating low-temperature cooling unit. When a thin layer of ice has formed on the walls of the flask, an equal volume of cold acetone $(-15°)$ is gradually added while maintaining the temperature between $0°$ and $-5°$. The enzyme is rapidly and irreversibly inactivated at temperatures much above $0°$ in the presence of acetone. Therefore, to minimize the heat of mixing accompanying addition of acetone at this stage, the first 150 ml of acetone are added in 10–15-ml portions over a 10-min period. Precipitation

[14] R. Davies, *Biochem. J.* **37**, 230 (1943).

PURIFICATION OF *Clostridium acetobutylicum* CROTONASE

Step	Volume (ml)	Protein concentration[a] (mg/ml)	Total protein (mg)	Activity[b] (units/mg)	Total activity (units)	Yield (%)	Purification (fold)
1. Extraction from 10 g of acetone powder[c]	401	2.20	882	47	41,464	100	1
2. Acid–heat	421	1.75	737	61.6	45,377	100	1.31
3. 0 to 50% acetone precipitate	50	3.30	165	195.9	32,324	78	4.17
4. Sephadex G-200 chromatography[d]	66	0.19	12.5	2,310.0	28,877	69.6	49.15
5. DEAE pool and concentrate	8	0.56	4.48	6,155	27,575	66.5	131.0

[a] Protein determined by biuret.
[b] One unit = 1 µmol of enzyme hydrated per minute per milligram of protein.
[c] Obtained from approximately 40 g of wet cells.
[d] Using $E_{280}^{1\%} = 0.89$.

is evident by this time. The remaining acetone is then added slowly, with continuous mixing, and the temperature is maintained between $-5°$ and $-10°$. The acetone suspension is added to previously cooled centrifuge bottles and spun at 9000 rpm for 40 min at $-5°$. The pellet is resuspended in about 50 ml of the potassium phosphate–EDTA extraction buffer, concentrated fivefold by vacuum dialysis, and dialyzed against 4 liters of the same buffer at $4°$ for 18 hr.

Step 4. Gel Filtration Chromatography. The dialyzed solution (10 ml) from the preceding step is applied to a column (2.5×85 cm) of Sephadex G-200 equilibrated at $4°$ in 0.02 M potassium phosphate, 0.003 M EDTA, pH 7.4. The column is developed at a flow rate of 24 ml/hr, and 6-ml fractions are collected. Assays for crotonyl-CoA hydratase activity reveal that the enzyme is eluted from the column between 190 and 250 ml. The hydratase is obtained after this step in 90% yield and possesses a specific activity of about 2300 at this stage.

Step 5. DEAE–Sephadex Chromatography. The pooled fractions (66 ml) from step 4 are dialyzed exhaustively at $4°$ against 0.05 M Tris-HCl, pH 8.0, and applied to a column (0.9×15 cm) of DEAE-Sephadex equilibrated with the same buffer. Buffer is pumped through the column at 18 ml/hr for approximately 5 hr to wash through material that fails to stick to the resin. The column is then eluted with a linear gradient composed of 250 ml of 0.05 M Tris-HCl, pH 8.0, and 250 ml of 0.4 M Tris-HCl, pH 8.0. The activity peak emerges in the gradient between 0.23 and 0.27 M Tris-HCl and is coincident with the third of four well separated protein peaks as determined by ultraviolet absorbance measurements. The pool material is concentrated by vacuum dialysis. In the purification shown in the table, 4.48 mg of the enzyme was obtained from 40 g of wet cells; this represents a 131-fold purification and a 66.5% yield based upon the crude extract.

Properties of the Enzyme from *Clostridium acetobutylicum*

Purity and Stability. Enzyme activity is rapidly lost when stored in buffer solution at $4°$ or $-20°$ but can be recovered nearly quantitatively from lyophilized samples. Thus, the protein is normally stored in the dehydrated form. Preparations obtained as described above appear to be homogeneous as judged by sedimentation equilibrium analyses and polyacrylamide gel electrophoresis using three different polyacrylamide gel systems involving the presence or the absence of 6.25 M urea or 1% sodium dodecyl sulfate (SDS).

Physical Properties. The native molecular weight of the enzyme has been estimated to be $158,000 \pm 3000$ as determined by sedimentation equilibrium analysis and gel filtration chromatography in dilute salt solutions. A (weight average) molecular weight of 43,700 ± 200 was obtained by sedimentation equilibrium studies performed in 6 M guanidine hydrochloride, both in the absence and in the presence of β-mercaptoethanol, and a subunit molecular weight of 40,000 was estimated by polyacrylamide gel electrophoresis in the presence of SDS. These results suggest that crotonase from C. *acetobutylicum* is composed of four subunit polypeptide chains that are combined through noncovalent interactions in the native enzyme.

The polypeptide chains of the bacterial hydratase each are composed of approximately 370 amino acid residues and contain 3 residues of histidine and 8 residues of cysteine (measured as cysteic acid). Spectral measurements[15] indicate the presence of four residues of tryptophan per subunit of polypeptide, a value that is consistent with the demonstrated extinction coefficient ($E_{280}^{0.1\%} = 0.89$) for the enzyme.

Catalytic Properties. Crotonase from *C. acetobutylicum* displays a pronounced specificity for short (C_4–C_6) chain length substrates and is inactive toward C_8–C_{16} enoyl-CoA thioesters. The enzyme exhibits an extremely high turnover number for crotonyl-CoA ($V_{max} = 6.5 \times 10^6$ mol per minute per mole of enzyme; $K_m = 30 \ \mu M$ at 25°) and is markedly sensitive to inhibition by crotonyl-CoA at concentrations greater than about 75 μM. In contrast, the rate of hydration of hexenoyl-CoA obeys normal Michaelis–Menten kinetics ($V_{max} = 3.9 \times 10^4$ mol per minute per mole of enzyme; $K_m = 130 \ \mu M$) and no substrate inhibition is observed with hexenoyl-CoA, even at levels at which the enzyme is totally inactivated by crotonyl-CoA. The bacterial enzyme resembles bovine crotonase[16] in requiring a complete CoA thioester structure for efficient catalysis. The rate of hydration of crotonylpantetheine is very low and can be stimulated about 10-fold by the addition of free CoA or ATP. Studies with the bovine enzyme have shown that this enhancement is due to a substrate complementation phenomenon,[17] similar to the stimulation by alkyl ammonium or alkyl guanidinium ions of the tryptic hydrolysis of N-acetylglycine methyl ester.[18]

Enoyl-CoA Hydratases from *E. coli*

Procedure for Fractionation and Partial Purifications[6]

Step 1. Preparation of Extracts. Escherichia coli Kl2 is grown at 37° in a modified minimal salts medium containing 0.03% oleate, 0.7% Na_2HPO_4,

[15] H. Edelhoch, *Biochemistry* **6**, 1948 (1963).
[16] J. R. Stern, *in* "The Enzymes" (P. D. Boyer, H. Lardy, and K. Myrbäck, eds.), 2nd ed., Vol. 5, p. 511. Academic Press, New York, 1961.
[17] R. M. Waterson, G. M. Hass, and R. L. Hill, *J. Biol. Chem.* **247**, 5258 (1972).
[18] T. Inagami and S. S. York, *Biochemistry* **7**, 4045 (1968).

0.3% KH_2PO_4, 0.01% $MgSO_4 \cdot 7 \ H_2O$, 0.1% $(NH_4)_2SO_4$, 0.0014% $CaCl_2 \cdot 2 \ H_2O$, 0.5% trypticase, and 0.1% Brij-35. The cells are harvested in late logarithmic growth by centrifugation at 3000 g and washed twice with sterile 0.01 M potassium phosphate, pH 8.0. Soluble extracts are prepared by sonic disruption of cell suspensions (0.125 g/ml) in 0.02 M potassium phosphate, pH 8.0, 0.01 M β-mercaptoethanol, 0.1 mM phenylmethylsulfonyl fluoride (PMSF), 25% glycerol. Sonication using a Branson Model W-140 sonifier is performed by exposure to three bursts for 45 sec each with cooling for 30 sec between bursts in an ice–salt bath. The sonicate is then centrifuged at 3° and 27,000 g for 50 min in a Sorvall RC-2 centrifuge, and the supernatant fraction thus obtained is dialyzed against 4 liters of 0.025 M sodium phosphate, pH 6.85, 0.001 M β-mercaptoethanol at 4°.

Step 2. Aminohexyl–Sepharose Chromatography. The dialyzed supernatant prepared from 5 g of cells is applied to a column (2.5 × 25 cm) of aminohexyl–Sepharose equilibrated in the dialysis buffer. The flow rate is 50 ml/hr, and fractions of 9.5 ml are collected. The column is washed with four volumes of starting buffer and subsequently developed with a 0 to 0.5 M linear gradient of sodium chloride in the same buffer (2 × 600 ml). Fractions are assayed as soon as possible with crotonyl-CoA and dodecenoyl-CoA substrates. Three peaks of hydratase activity are normally resolved by this procedure (Fig. 1). Activity toward only substrates of longer chain length is detected in breakthrough fractions, whereas a peak of hydratase activity for crotonyl-CoA (but not dodecenoyl-CoA) is observed in early gradient fractions. A third heterogeneous peak containing activities toward both substrates is found to elute later in the gradient. Each of the activity peaks are separately pooled, concentrated by Amicon filtration, and stored at 4°.

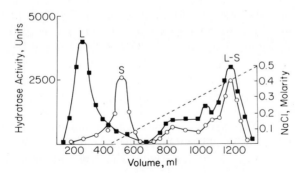

FIG. 1. Elution profile for aminohexyl-Sepharose chromatography. Enzyme activity is shown in units per milliliter for crotonyl-CoA (o) and $\Delta^{2,3}$-dodecenoyl-CoA hydration ($\blacksquare$; multiplied fivefold). L, long chain; S, short chain.

Comments about the Purification Procedure. The fractionation scheme described here has been employed for maximal separation of enoyl-CoA hydratase activity into different hydratases of limited chain length specificity. For this reason glycerol is omitted from the buffer system, in spite of its stabilizing effect on hydratase activity. When glycerol is included (25%, v/v), essentially all the hydratase activity toward substrates of both short and long chain length is recovered in a relatively homogeneous peak that elutes in the gradient between 0.1 and 0.25 M NaCl.[19] The example shown in Fig. 1 represents the highest recovery of separate hydratase activities we have observed to date. The relative yields of the separated activities vary considerably among experiments, and in some instances little or no short-chain activity peak is obtained. This variability is primarily due to the marked lability of the individual components following their fractionation, especially the short-chain "crotonase" activity. Thus, it has not been possible to purify either of the individual hydratase components to homogeneity. Long-chain specific activity also has been observed in membrane fractions[6,20] and following chromatography on phosphocellulose.[21] The existence of at least two hydratase components has also been suggested by *in vivo* studies[6]; the induction rate of short (C_4–C_6) chain length hydratase activity is different from that for longer (C_8–C_{16}) chain length-dependent activity.

Properties

Stability. The associated enoyl-CoA hydratase activities are stable at 4° for several months when samples at concentrations of 1–5 mg/ml are stored in 10 mM potassium phosphate, pH 7.5, 10 mM β-mercaptoethanol, 0.1 mM PMSF, and 25% glycerol. The activities are less stable when stored at $-20°$ or $-80°$. Nonidet P-40 (0.1–0.5%) is nearly as protective as glycerol but presents a disadvantage due to its ultraviolet absorbance properties. In the absence of glycerol, severe and irreversible inactivations result from exposure to Tris-HCl.[6,11] These losses are reduced to 30–40% by the presence of 25% glycerol. Activity in the absence of glycerol is also labilized by EDTA and EGTA (ethyleneglycol bis(β-aminoethyl ether) N,N'-tetraacetic acid).[6]

Molecular and Catalytic Properties. The substrate specificities for each of the three aminohexyl-Sepharose chromatography components were determined using the series of homologous enoyl-CoA substrates ranging in chain length from 4 to 16 carbon atoms. The first column peak is active

[19] C. C. Gallen and R. M. Waterson, unpublished observation, 1977.
[20] W. J. O'Brien and F. Frerman, *J. Bacteriol.* **132**, 532 (1977).
[21] J. F. Binstock, A. Pramanik, and H. Schulz, *Proc. Natl. Acad. Sci. U.S.A.* **74**, 492 (1977).

upon substrates from C_8 to at least C_{16} and is inactive toward crotonyl-CoA and $\Delta^{2,3}$-hexenoyl-CoA. The second component, on the other hand, hydrates only the two shorter substrates. The third activity peak is found to hydrate both short and longer chain length substrates and to exhibit maximal activity toward $\Delta^{2,3}$-hexenoyl-CoA.[6] This aggregate, which contains both hydratase activities, also contains β-hydroxyacyl-CoA dehydrogenase and thiolase activities in the form of multienzyme complexes.[6,11,20,21] The reader is referred to this volume [49] for more information concerning the associated β-oxidation enzyme activities.

When tested at 2×10^{-4} M substrate concentration the stable, associated hydratase activities exhibit relative rates of hydration for $C_4 : C_6 : C_8 : C_{10} : C_{12} : C_{16}$ enoyl-CoA substrates of $1.0 : 1.6 : 0.3 : 0.26 : 0.18 : 0.11$, respectively. Different pH dependencies of hydration are observed with the *E. coli* enzymes, when assayed either before or after their fractionation.[6] Activity toward all substrates is maximal and essentially constant in the pH 7 to 9 range. However, the hydration of short-chain substrates is stimulated to a greater degree by increases in pH from 6.0 to 7.0 than is hydration of long-chain substrates; for example, the fold increases in activity over this pH increment are 2.0 for crotonyl-CoA hydration and 1.1 for $\Delta^{2,3}$-dodecenoyl-CoA hydration.

The associated hydratase activities exhibit size heterogeneity when examined by gel filtration chromatography.[6,20,21] Molecular weights of 180,000–200,000 and 40,000 are estimated for the partially purified short-chain and long-chain hydratases, respectively, by their elution volumes from Sephadex G-200. The third, associated component is heterogeneous in size and chromatographs on Sepharose 6B in the 250,000–400,000 molecular weight range. Prior to aminohexyl–Sepharose chromatography the short-chain and long-chain hydratase activities cochromagraph on Sepharose 4B as a series of high molecular weight aggregates, having estimated molecular weights from 750,000 to 2×10^6 or greater.[6] The larger species are dissociated to aggregates of approximately 200,000–400,000 MW by exposure to solutions containing either 1 M potassium phosphate[22] or 6 mM deoxycholate.[23] Moreover, O'Brien and Frerman[20] have reported that partially purified aggregates of 250,000 MW contain more phospholipid per milligram of protein than is observed in crude lysates. This suggestion of associated phospholipid has been confirmed[24] by direct extraction and identification of cardiolipin from 220,000 and 330,000 MW aggregates purified by several distinct procedures, including

[22] R. S. Conway, C. C. Gallen, and R. M. Waterson, *Fed. Proc.* **38**, 2207 (1979).
[23] C. E. Giarratana and R. M. Waterson, unpublished observation, 1979.
[24] R. S. Conway, C. E. Giarratana, and R. M. Waterson, *Fed. Proc.* **39**, 2115 (1980).

that of Feigenbaum and Schulz.[25] Neither phosphatidylethanolamine nor phosphatidylglycerol is detected. It thus appears possible that phospholipids play a role in the association process.

[25] J. Feigenbaum and H. Schulz, *J. Bacteriol.* **407** (1975).

[52] Lipoxygenase from Rabbit Reticulocytes

EC 1.13.11.12 Linoleate : oxygen oxidoreductase

By TANKRED SCHEWE, RAINER WIESNER, and SAMUEL M. RAPOPORT

Linoleate + O_2 → hydroperoxylinoleate

The lipoxygenase of the reticulocyte cytosol, like that of soybeans, is a dioxygenase containing nonheme iron. It catalyzes the oxidation by molecular oxygen of polyunsaturated free and esterified fatty acids that possess a 1,4-*cis,cis*-pentadiene system. The reticulocyte lipoxygenase shows high reactivity toward phospholipids both in liposomes and in biological membranes. Its action on the mitochondrial inner membrane leads to inactivation of the electron transfer system. It plays an important role in the inactivation and degradation of mitochondria during the maturation process of red cells.[1]

Assay Methods

Principles. All methods used for the assay of activity of other lipoxygenases can also be applied for the reticulocyte lipoxygenase. The following substrates may be used: free polyunsaturated fatty acids (e.g., linoleate), natural sonified phospholipids (e.g., from soybeans of beef heart mitochondria), submitochondrial particles, intact mitochondria. If pure lipids are used, a low assay temperature is recommended in order to minimize disturbances by side reactions and by self-inactivation of the enzyme.

The measurement of the oxygen consumption by oxygen electrode polarography is applicable for all substrates. Another convenient assay is spectrophotometric monitoring, at 234 nm, of the formation of the conjugated diene system from linoleic acid. According to the equation

$$R_1\text{—CH}\text{=}\text{CH—CH}_2\text{—CH}\text{=}\text{CH—R}_2 + O_2$$
$$\rightarrow R_1\text{—CH—CH}\text{=}\text{CH—CH}\text{=}\text{CH—R}_2$$
$$\overset{|}{O}\text{—OH}$$

the dioxygenation of the fatty acid chain to the hydroperoxide is accompanied by the isomerization of the 1,4-pentadiene structure to a butadiene-like system showing an absorbance maximum in the ultraviolet region.

With more complex substrates, such as membranes, the colorimetric determination of thiobarbituric acid (TBA) reactive substances may be employed.[1,2] Thiobarbituric acid condenses with lipohydroperoxides or with their secondary product malonyl dialdehyde to yield a dye with an absorbance maximum at 532 nm. A rather sensitive assay for the action of reticulocyte lipoxygenase on submitochondrial particles is the determination of the inhibition of the respiratory chain. The site of the inhibitory action on the electron transfer system is located between the two iron–sulfur regions of complexes I and II and ubiquinone.[3] It seems to be related to the lipohydroperoxidase activity of reticulocyte lipoxygenase. Soybean lipoxygenase shows respiration-inhibitory activity only in the presence of hemoglobin.

The various assay methods for reticulocyte lipoxygenase activity used by us are compiled in Table I. None of these methods affords reliable results if the *amount* rather than the activity of the lipoxygenase is to be determined, since this enzyme tends to form oligomers that are enzymatically inactive. In this case immunoprecipitation has to be applied, which also implies the oligomers.

Assay Method A. Conjugated Diene Formation from Linoleic Acid

Reagents

Potassium phosphate buffer, 0.1 *M*, pH 7.4
Sodium cholate, 10%
Ethanol, reagent grade
Linoleic acid, >99%, stored under nitrogen, 20 vol % in methanol

Procedure. The assay medium is prepared by mixing 93 volumes of phosphate buffer, 5 volumes of ethanol, and 2 volumes of sodium cholate. The substrate solution is freshly obtained by mixing 1 volume of an anaerobic linoleic acid–methanol mixture 1 : 5 (v/v) and 100 volumes of assay medium.

Assay medium and substrate solution are precooled in the ice-bath.

[1] S. M. Rapoport, T. Schewe, R. Wiesner, W. Halangk, P. Ludwig, M. Janicke-Höhne, C. Tannert, C. Hiebsch, and D. Klatt, *Eur. J. Biochem.* **96,** 545 (1979).
[2] W. Halangk, T. Schewe, C. Hiebsch, and S. M. Rapoport, *Acta Biol. Med. Germ.* **36,** 405 (1977).
[3] T. Schewe and S. P. J. Albracht, *Biochim. Biophys. Acta.,* in press.

TABLE I
ASSAY METHODS FOR RETICULOCYTE LIPOXYGENASE

Substrate	Methods
Free fatty acids	Oxygen electrode; conjugated diene formation
Sonified phospholipids	Oxygen electrode; TBA test
Submitochondrial particles	Oxygen electrode; inhibition of respiratory chain
Rat liver mitochondria	Oxygen electrode; TBA test

Then 1.8 ml of assay medium and 0.2 ml of substrate solution (final concentration 0.53 mM) are put into a 1.0-cm light path cuvette; after a constant absorbance is reached, the enzyme solution is added and the increase in absorbance is monitored at 234 nm using a recording spectrophotometer with cryostat at 2°. After a short lag period due to product activation of the enzyme, the reaction proceeds in a linear manner for several minutes up to an increase in absorbance of about 0.7.

Assay Method B. Oxygen Consumption with Sonified Phospholipids

Reagents

Potassium phosphate buffer, 0.1 M, pH 7.4, air-saturated at assay temperature
Tris-HCl buffer, 10 mM, pH 8.0, containing 1 mM EDTA
Phospholipid from soybeans (asolectin)

Procedure. The substrate is prepared in the following manner: A suspension of 50 mg of phospholipid (e.g., commercial lecithin) in 5 ml of Tris-HCl–EDTA buffer is sonified 4 times for 1 min in a nitrogen atmosphere and under ice-cooling by means of a tip-type sonifier (e.g., Branson B-12, Micro-tip) at medium output.

For the assay 30 μl of the sonified substrate suspension are added per milliliter of potassium phosphate buffer. The reaction is started by adding enzyme. Measurement of the oxygen consumption is carried out by means of a Clark-type oxygen electrode with a polyethylene or polypropylene membrane. Teflon membranes are not recommended, since lipoxygenases tend to adsorb on this material.

Assay Method C. Determination of the Respiration-Inhibitory Activity

Reagents

Tris–H$_2$SO$_4$ buffer, 0.05 M, pH 8.0, containing human serum albumin (HSA), 1 mg/ml
Potassium phosphate buffer, 0.1 M, pH 7.4

Sucrose, 0.25 M, containing 10.6 mM K_2HPO_4
Potassium phosphate buffer, 0.01 M, pH 6.0
NADH, 20 mg/ml in Tris-HCl buffer, 0.1 M, pH 8.6
Potassium succinate, 0.33 M
Ferricytochrome c, 1 mM
Potassium cyanide, 0.15 M, neutralized immediately before use
Beef heart electron-transfer particles (ETP) suspended in 0.25 M sucrose[4] (see also this series, Vol. X [47]

Preincubation Samples. Electron transfer particles suspension (50 μl, diluted by sucrose–phosphate medium to 1–2 mg/ml) is mixed with 5, 10, or 20 μl of lipoxygenase solution (diluted with 0.01 M phosphate buffer, pH 6.0). A final volume of 70 μl is adjusted by addition of phosphate buffer. Control samples without enzyme are prepared in parallel. The samples are preincubated for 15 min at 37°. Thereafter either the NADH oxidase or the succinate–cytochrome c oxidoreductase activity is determined. The amount of lipoxygenase should be adjusted to give an inhibition in the range from 25 to 70%, where approximate proportionality between the lipoxygenase amount and the percentage of inhibition is observed.

Measurement of NADH Oxidase Activity. Tris–H$_2$SO$_4$HSA buffer, 1.0 ml, tempered at 37° is added to the preincubation sample. The reaction is started by addition of 10 μl of NADH solution; the mixture is quickly transferred into a cuvette tempered at 37°, and the NADH oxidase activity is measured at 340 nm by means of a recording spectrophotometer at 37°. After a lag period of about 1 min, a linear reaction rate can be observed up to the exhaustion of NADH. The control activity of the ETP should be in the range of 0.200–0.400 ΔA_{340}/min. The inhibition produced by lipoxygenase treatment is expressed as a percentage, referred to the control activity. An inhibitory unit is defined as that amount of lipoxygenase which causes an inhibition of 50%.

Measurement of the Succinate-Cytochrome c Oxidoreductase Activity. For the activation of the succinate dehydrogenase in the ETP, 5 μl of 0.1 M potassium succinate are added to the preincubation sample and the mixture is incubated for another 5 min at 37°. Thereafter the reaction is started by addition of 0.90 ml of 0.1 M phosphate buffer, pH 7.4, 50 μl of 0.33 M succinate, 50 μl of 1 mM ferricytochrome c, and 10 μl of 0.15 M neutralized potassium cyanide. The samples are quickly transferred into prewarmed cuvettes, and the reduction of ferricytochrome c is measured, with a recording spectrophotometer, by the increase in absorbance at 550 nm at 37°. The initial rate of the reaction is evaluated. The inhibitory activity is calculated in a manner analogous to NADH oxidase activity.

[4] F. L. Crane, J. L. Glenn, and D. E. Green, *Biochim. Biophys. Acta* **22**, 475 (1956).

Precautions

1. Hemoglobin disturbs all assays of lipoxygenase activity. Whereas the formation of hydroperoxylinoleate is partly inhibited, the respiration-inhibitory activity and the formation of TBA-reacting substances are enhanced. Moreover, pseudolipoxygenase activity of hemoglobin may appear under special conditions. The effects of hemoglobin are presumably in a great part due to heme-catalyzed lipohydroperoxidase activity; i.e., the product of the lipoxygenase, hydroperoxylinoleate, is converted to various consecutive products presumably via the linoleate-oxy-radical. The disturbances by hemoglobin are largely excluded after removal of the bulk of the hemoglobin (preparation step 2). Disturbances by catalase were not observed by us. If the lipoxygenase activity is to be determined in stroma-free hemolyzate, 1.1 volume of saturated ammonium sulfate solution is added, the precipitate is centrifuged after standing 1 hr, washed with 55% saturated ammonium sulfate, and dissolved in 0.01 M potassium phosphate buffer equal to the original volume. This solution is used for the assay.

2. With assay method A a complete inactivation of the lipoxygenase during the assay may appear at temperatures higher than 20° owing to a special impurity present in some commercial cholate preparations. Method A can also be applied with a detergent-free assay medium; in this case 0.2 M potassium phosphate buffer, pH 7.8, containing 5% ethanol is recommended. Here again, reproducible measurements can be performed at 2°, but not at 37°. At 37° there is interference by changes in turbidity.

3. In the measurement of oxygen consumption care should be taken that no other oxygen-consuming processes interfere. Mitochondrial respiration can be simply excluded by the addition of antimycin A. The oxygen uptake produced by lipoxygenase is inhibited by 1 mM salicylhydroxamic acid.

Units. One unit of lipoxygenase (1 μkat) is defined as that quantity of enzyme which catalyzes the formation of 1 μmol of hydroperoxylinoleate per second at 2°. An ϵ_{234} of 25,000 mol^{-1}/liter cm^{-1} is used for the calculations.

Purification Procedure

Reticulocytosis of the rabbit is provoked by extensive bleeding. This causes a "superinduction" of the lipoxygenase, and the concentration of this enzyme reaches values up to about 4 mg of protein per milliliter of packed cells. In the "stress reticulocytes" poured out upon continuous strong bleeding, the lipoxygenase constitutes one of the main nonhemoglobin proteins of the red cells, and an approximately 80-fold enrichment suffices to achieve purity. All preparative steps are carried out between 0 and 5°. The enzyme is rather oxygen-sensitive, and traces of heavy metals enhance the oxygen sensitivity. It is also labile toward freezing and thawing and storage in the range of $-1°$ to $-30°$. Anaerobic conditions are useful in step 3, even though it makes the procedure more cumbersome. The enzyme is also protected by 20 mM 2-mercaptoethanol and 1 mM Mg^{2+}.

Step 1. Stroma-Free Supernatant Fluid of Reticulocyte-Rich Rabbit Blood Cells. Young lean rabbits of 3–4 kg body weight are most suitable for the preparation. Reticulocytosis is provoked by withdrawal of 17 ml of blood per kilogram of body weight from the ear vein daily for 3 days. The hematocrit is checked every day. If the hematocrit decreases below 16%, the amount of blood withdrawn has to be reduced. The regeneration of blood in the animals can be improved, especially in the winter period by daily oral administration to each animal of a mixture that contains 15 g of soybean grits, 5 g of egg albumin, 150 mg of iron, e.g., as powdered ferrous salt, ascorbate tablets, 3 g of partially hydrolyzed chlorophyll, and 0.5 g of sodium chloride as well as subcutaneous administration of 10 μg of vitamin B_{12} and 3 mg of folic acid.[5] After an interval of 2 days, i.e., beginning on day 6, a high reticulocytosis with large amounts of lipoxygenase is achieved, so that blood for preparation of the enzyme can be collected up to about day 25. The amount of blood withdrawn daily is chosen to maintain a hematocrit of 18–20%. It ranges between 50 and 75 ml.

The blood is allowed to drop into a glass vessel containing glass beads under continuous shaking in order to precipitate the fibrin. Thereafter the blood is filtered through wide-meshed Perlon or nylon gauze. The blood is cooled in an ice-bath and is centrifuged at 2000 g for 12 min. The cell mass is resuspended in at least three volumes of 0.9% NaCl and centrifuged at 800 g for 5 min. The supernatant and the fluffy layer containing leukocytes and thrombocytes are removed by suction. This washing step is repeated twice. After the last washing the cells are centrifuged at 2000 g for 10 min. This procedure removes a large part of the leukocytes and thrombocytes. A complete separation of these cells by the use of a cotton-wool column is unnecessary.

One volume of packed cells (to be calculated from the original volume of blood and the hematocrit value) are mixed with two volumes of cold-

[5] T. Schewe, H. Scharfschwerdt, and E. Rumpel, *Z. Med. Labortechnik* **13**, 366 (1972).

distilled water. The hemolyzate is allowed to stand for 10 min and is adjusted to pH 6.0 with a glass electrode by slow continuous addition of 1 N HCl from a polyethylene capillary under gentle stirring. Immediately thereafter the hemolyzate is centrifuged at 20,000 g for 20 min. The stroma-free supernatant fluid is rapidly frozen in small portions in an acetone–Dry Ice bath and stored at $-30°$ or less. The precipitate is discarded. The processing of mixed bloods from more than one animal is not recommended.

Step 2. Ammonium Sulfate Precipitation. The lipoxygenase is completely precipitated by addition of ammonium sulfate to 55% saturation, whereas the bulk of hemoglobin remains in the supernatant fluid. Ammonium sulfate (analytical grade) is twice recrystallized, at first in the presence of 1 mM EDTA. Metal-free ammonium sulfate (333 g/liter) is slowly added with stirring to the supernatant of step 1. The precipitate is allowed to stand for 1 hr and is then sedimented by centrifugation for 10 min at 4000 g. The precipitate is resuspended in ammonium sulfate solution of 55% saturation and resedimented. The washed precipitate is dissolved in twice the volume of distilled water and dialyzed against the hundredfold volume of 0.01 M potassium phosphate buffer, pH 6.0, overnight. During dialysis a precipitate appears that is sedimented by centrifugation and discarded. The clear supernatant is slightly red-brown in color; it may be diluted 1 : 100 for assay of lipoxygenase activity (methods A or C). After rapid freezing the supernatant is stored at less than $-30°$.

Step 3. Chromatography on DEAE–Sephadex A-50. Chromatography is carried out on a column (5 × 60 cm) with a filled volume of 900 ml. The gel is equilibrated with 0.03 M Tris-HCl buffer, pH 7.2, freed from heavy metals by passing through a Dowex resin A-1 column containing 1 mM MgCl$_2$ and 20 mM 2-mercaptoethanol. Nearly anaerobic conditions are maintained by several alternate applications of vacuum and gassing with nitrogen of all solutions used. The pooled dialyzates from step 2 containing 1–2 g of total protein are loaded onto the DEAE-Sephadex A-50 column, which is then thoroughly washed with the equilibrating buffer to remove the residual amounts of hemoglobin and other unbound material. The lipoxygenase is eluted by means of a two-reservoir linear gradient system with 400 ml of starting buffer in the mixing reservoir and 400 ml of 0.4 M sodium chloride in starting buffer in the other reservoir. Fractions of 5 ml are collected in a nitrogen atmosphere. The lipoxygenase is eluted in the range of 0.06–0.12 M sodium chloride. The active fractions, which also contain catalase as detectable by the absorbance at 410 nm, are pooled and made anaerobic, as described above, before rapid freezing in small portions in acetone–Dry Ice and storage at $-80°$.

Step 4. Isoelectric Focusing. Isoelectric focusing is carried out by means of a 440-ml focusing column (LKB, Sweden). Oxygen-free 1%

(w/v) Ampholine (pH range 5.0–7.0) in a sucrose gradient serves as separation medium. Protein (100 mg) of the peak fractions from step 3 is applied in the middle of the column. The separation is performed at an initial current of 4 mA and a constant voltage of 500 V under cooling (2°) and is completed after 72 hr. Lipoxygenase is focused at pH 5.5. If larger amounts of proteins are applied, precipitation of the focused enzyme occurs. However, with an application of 200 mg of protein, satisfactory separations can be still achieved despite precipitation. The separation medium seems to stabilize the lipoxygenase and is also used for storage. Sucrose and Ampholine do not interfere with the enzymatic assays of lipoxygenase.

The purification of reticulocyte lipoxygenase is summarized in Table II. The procedure yields about 300 mg of purified enzyme from 100 ml of packed cells. The final product contains only minor impurities as judged by sodium dodecyl sulfate(SDS)-polyacrylamide gel electrophoresis.

Alternative Methods of Purification. If the dialyzate of step 2 is submitted to gel chromatography on Sephadex G-200, the lipoxygenase behaves like a protein of a molecular weight of about 100,000 and is completely separated from catalase and hemoglobin. This preparation rapidly loses its enzymatic activity unless special measures are taken.

The dialyzate of step 2 can also be loaded onto a CM-cellulose column previously equilibrated with 0.01 M potassium phosphate buffer, pH 6.0. The lipoxygenase is eluted then by an ionic strength gradient of the same buffer. Here several well defined activity peaks with different specific activity and stability appear, indicating heterogeneity. A batch of CM-cellulose equilibrated with 0.01 M potassium phosphate buffer, pH 6.0, is convenient for the complete removal of hemoglobin from the dialyzate of step 2; under batch conditions the lipoxygenase is not adsorbed on the cation exchanger. This hemoglobin-free preparation is relatively stable.

The lipoxygenase protein can also be fractionated by means of a hydroxyapatite column and thereby completely separated from hemoglobin and catalase. Unfortunately, the enzyme loses much of its activity.

Affinity chromatography on the linoleic acid derivative of aminohexyl–Sepharose 4B under strictly anaerobic conditions is suitable for further purification of the dialyzate from step 2. The adsorbed enzyme is eluted with a linear gradient of increasing ionic strength. This method is described in detail elsewhere.[1]

Properties

Purity and Association Behavior. Only traces of impurities can be detected both by the analytical ultracentrifuge and by polyacrylamide gel electrophoresis with SDS containing 2-mercaptoethanol. The enzyme be-

TABLE II

PREPARATION OF LIPOXYGENASE FROM RABBIT RETICULOCYTES

Step	Volume (ml)	Protein absorbance (A_{280} ml^{-1})	Activity (μkat ml^{-1})	Total activity (μkat)	Specific activity (μkat A_{280}^{-1})	Yield (%)
1. Stroma-free hemolyzate	1200	160	0.78	960	0.005	53[a]
2. 55% (NH$_4$)$_2$SO$_4$ fraction	95	77	19.00	1805	0.25	100
3. Chromatography on DEAE-Sephadex A-50						
Total peak	940	2.9	1.37	1290	0.47	71
Top peak	330	3.7	2.33	770	0.63	43
4. Isoelectric focusing with top peak ($n = 9$)	270	2.7	1.88	507	0.70	28

[a] The lipoxygenase is partly inhibited by hemoglobin in the hemolyzate.

haves in a homogeneous manner in immunological tests. However, the enzyme shows an open association behavior, i.e., the enzymatically active monomer is in equilibrium with dimers, trimers, tetramers and possibly also higher species that are immunologically reactive but enzymatically inactive.[6]

If the monomer isolated by means of sucrose density gradient centrifugation is subjected to recentrifugation, it shows the same distribution pattern, the oligomers constituting about 40% of the total amount of lipoxygenase protein. The enzyme also shows microheterogeneity with respect to the C-terminal amino acid (see below), which may result from endogenous carboxypeptidase-like actions in the reticulocyte.

Stability. Enzyme purified through step 2 exhibits the highest stability and is therefore a suitable source of lipoxygenase for a variety of experiments. The purified enzyme is most stable in the separation medium of the isoelectric focusing. If this medium interferes with special investigations, a rapid desalting of the sample by means of a Sephadex G-50 minicolumn immediately before the experiment is recommended. Concentration of enzyme solutions is best performed by putting them into a dialysis tubing, which is immersed in dry Sephadex G-25. Ultrafiltration leads to considerable losses owing to adsorption of the enzyme on the common commercial filter materials. Reticulocyte lipoxygenase is inactivated below pH 5.5 and above pH 8. As mentioned before, the enzyme is labile toward oxygen and heavy metals as well as freezing and thawing.

Molecular Properties. A molecular weight between 76,500 and 80,000 has been determined by sucrose density gradient centrifugation,[1] by means of analytical ultracentrifugation,[7] and by polyacrylamide gel electrophoresis in SDS containing 2-mercaptoethanol.[8] The enzyme consists of a single polypeptide chain. The isoelectric point is 5.5 as judged by isoelectric focusing.[1] The amino acid composition[8] is shown in Table III. The high percentage of leucine and tryptophan is remarkable. The N-terminal amino acid is glycine.[8] There is heterogeneity at the C-terminal end. Asparagine, isoleucine, and histidine have been identified with the aid of tritium labeling and carboxypeptidase A. The enzyme contains 1 mol of iron per mole. The value of 1.8 mol per mol of enzyme reported elsewhere[1] is an overestimate. Moreover, the enzyme contains $4.6 \pm 1.2\%$ neutral sugars[1]; sialic acid is absent.

Immunological Characterization. Antiserum against reticulocyte lipoxygenase can be produced by immunization of guinea pigs four times

[6] M. Höhne and H. Andree, *Acta Biol. Med. Germ.* in press (1980).
[7] J. Behlke and R. Wiesner, *Eur. J. Biochem.* **96**, 561 (1971).
[8] R. Wiesner, C. Tannert, G. Hausdorf, T. Schewe, and S. M. Rapoport, *Acta Biol. Med. Germ.* **36**, 393 (1977).

TABLE III
PARTIAL AMINO ACID COMPOSITION OF LIPOXYGENASE
FROM RABBIT RETICULOCYTES

Amino acid	Amino acid/enzyme (mol/mol)
Asp	48
Thr	22
Ser	29
Glu	74
Pro	40
Gly	43
Ala	46
Cys[a]	12
Val	40
Met	12
Ile	26
Leu	78
Phe	28
Trp	21
Lys	35
His	12
Arg	32

[a] Half-cystine + cysteine.

with 100–170 μg of lipoxygenase each (for details, see Wiesner et al.[8]). A single precipitating line is observed both in the immunodouble-diffusion test according to Ouchterlony and in the radial immunodiffusion according to Mancini during all steps of purification. Immunoprecipitation cannot be performed in bone marrow cell lysates owing to unspecific precipitations.

Substrate Specificity. All polyene fatty acids having a 1,4-pentadiene structure are attacked. The V_{max} values show the following trend: $C_{20:3} \gtrsim C_{18:2} > C_{18:3} \gtrsim C_{20:4} >$ methyl-$C_{18:2}$.[2] K_m values decrease with increasing number of double bonds.[2] Even phospholipids of various sources are good substrates for the reticulocyte lipoxygenase. All classes of phospholipids are attacked.[9] The enzyme attacks intact mitochondria, which leads to the formation of TBA-reactive substances and to the release of matrix enzymes.[2] Erythrocyte ghosts are much more resistant than mitochondrial membranes.[10] This may be due both to their high content of cholesterol and to specific protein–lipid interactions. Reticulocyte lipoxygenase causes respiratory inhibition of submitochondrial particles of mitochon-

[9] T. Schewe, W. Halangk, C. Hiebsch, and S. M. Rapoport, *FEBS Lett.* **60**, 149 (1975).
[10] W. Fritsch, D. Maretzki, C. Hiebsch, and T. Schewe, *Acta Biol. Med. Germ.* **38**, 1315 (1979).

dria of various origin. In contrast, respiring membranes of *Escherichia coli* are resistant.

Kinetic Properties. The pH optimum of the dioxygenation of linoleate is in the range of 7.4–7.8. The apparent K_m has been tentatively estimated to be 12 μM.[11] The affinity toward oxygen is rather high. The turnover number seems to be by one order of magnitude lower than that of soybean lipoxygenase. Like the enzyme of soybeans, that of reticulocytes is activated by hydroperoxylinoleate. The amount of hydroperoxylinoleate needed for maximal activation is proportional to the concentrations of both linoleate and oxygen.[11] The reticulocyte lipoxygenase shows a lipohydroperoxidase activity like the soybean enzyme in the presence of linoleate plus hydroperoxylinoleate as judged by the appearance under anaerobic conditions of an absorbance maximum at 295 nm due to oxodienes. A special feature of the reticulocyte lipoxygenase is its suicidal character, i.e., a self-inactivation during reaction with substrates. This self-inactivation appears at temperatures above 20° and is possibly due to products formed via the lipohydroperoxidase activity. With soybean phospholipid the self-inactivation at 37° shows a half-life of 60–90 sec.

Inhibitors. Salicylhydroxamic acid is the best inhibitor of the reticulocyte lipoxygenase; 1 mM suppresses all reactions of the enzyme by more than 90%. Ferric ions counteract the inhibition. Some antioxidants, e.g., propyl gallate or α-naphthol, are also inhibitory. α-Tocopherol inhibits the reaction with lipids, but not with mitochondrial membranes.[2] EDTA, 1 mM, prevents the respiratory inhibition of submitochondrial particles, but is without effect on the linoleate lipoxygenase activity.

[11] P. Ludwig and P. Fasella, preliminary results, 1979.

[53] Lipoxygenase from Soybeans

EC 1.13.11.12 Linoleate : oxygen oxidoreductase

By Bernard Axelrod, Thomas M. Cheesbrough, and Simo Laakso

$$-CH \overset{c}{=} CH-CH_2-CH \overset{c}{=} CH- \ +O_2$$
$$\rightarrow \ -CH(OOH)-CH \overset{t}{=} CH-CH \overset{c}{=} CH-$$

Lipoxygenase is a dioxygenase that catalyzes, as a primary reaction, the hydroperoxidation, by molecular oxygen, of linoleic acid and other polyunsaturated lipids that contain a *cis,cis*-1,4-pentadiene moiety. The reaction can be followed by observing the increase in absorption at 234 nm arising from the conjugated double bonds formed during the reaction, or by measuring the utilization of O_2 with the aid of a recording O_2 electrode. One may also measure the formation of the lipid hydroperoxide by various colorimetric methods. Unfortunately, the hydroperoxide measurements, which are troublesome and less accurate, require the rigorous exclusion of O_2 during the chromogen formation and are not adaptable to continuous measurement. Soybean seeds are the richest known source of lipoxygenase. Four isozymes have been isolated: lipoxygenase-1, -2, -3a, and -3b, hereafter designated as L-1, L-2, L-3a, and L-3b, respectively. L-3a and L-3b are very similar in their properties and, for assay purposes, may be considered identical; they are referred to as L-3. Under appropriate conditions, L-1 exhibits ideal kinetic behavior, and forms linoleic acid hydroperoxide and conjugated diene, and takes up O_2 with perfect stoichiometry. Moreover, the reaction rate is directly proportional to the amount of enzyme over a reasonable range. In contrast, L-3 kinetics are complicated by the fact that the primary hydroperoxidation product is in part transformed into keto-diene products that do not absorb at 234 nm, but instead have a peak near 280 nm.[1] Although material absorbing with a maximum at 234 nm, is formed simultaneously, the progress curve for the 280 nm absorbing product is reasonably linear with time and provides the basis for a selective assay. Utilization of O_2 and formation of products absorbing at 234 nm and at 280 nm are not stoichiometric. For L-2, absorption at 234 nm provides a satisfactory assay, but since it is far more active against arachidonic acid than linoleic acid, compared to the other isozymes, the former is the substrate of choice. The apparent specific activity of L-3 and, to a considerably lesser degree, that of L-2, shows an anomalous inverse dependence on the concentration of enzyme. The effect of substrate concentration, especially on the L-2 and L-3 reactions, which are assayed under slightly acid conditions is complicated by the biphasic nature of the substrate in the pH range of assay, and by the micellar structure assumed by a portion of the substrate. The sensitivity of the physical structure to ionic strength, pH, Ca^{2+}, and temperature adds to the complexity. Comprehensive reviews on plant lipoxygenases are available[2,3] as well as a recent one on assay procedures.[4]

[1] B. Axelrod, E. K. Pistorius, and J. P. Christopher, *Abstracts, 11th World Congress International Society for Fat Research,* p. 117 Goteborg (1972).

[2] G. A. Veldink, J. F. G. Vliegenthart, and J. Boldingh, *Prog. Chem. Fats Other Lipids* **15,** 131 (1977).

[3] B. Axelrod, *in* "Food Related Enzymes" (J. Whitaker, ed.), *Adv. Chem. Ser.* **136,** 324 (1974).

[4] S. Grossman and R. Zakut, *Methods Biochem. Anal.* **25,** 303 (1979).

Assay Method

Principle. Considering the difficulties noted above, it is obvious why rigorously controlled empirical procedures are required. Relatively rapid and sensitive spectrophotometric assay procedures are described. The substrate concentrations have been made as low as possible to minimize problems arising from the physical instability of the substrate mixture. The extent of reaction during the course of the assay is sufficiently limited to give linear results. It is essential to adhere stringently to the procedures described below to obtain uniform results. Assays based on O_2 uptake and employing the O_2 electrode are useful, especially for rough monitoring of the fractions obtained in the initial column chromatographic purification. However, such assays must be followed up by the spectrophotometric assays described below, especially when it is necessary to discriminate between L-2 and L-3. Although we and others have long used the oxygen electrode routinely in lipoxygenase assays we find that the method is not highly reliable with L-2 on arachidonic acid and with L-3 on linoleic acid.

Reagents

Sodium phosphate buffer, 0.2 M, pH 6.1
Sodium phosphate buffer, 0.2 M, pH 6.5
Sodium borate buffer, 0.2 M, pH 9.0
Calcium chloride, 0.1 M
Sodium linoleate, 10 mM. Weigh out 70 mg of linoleic acid. Add an equal weight of Tween 20 plus 4 ml of O_2-free water. Homogenize by drawing back and forth in a Pasteur pipette, avoiding air bubbles. Add sufficient 0.5 N NaOH to yield a clear solution (0.55 ml). Make up to 25 ml total volume. Divide into 1–2 ml portions in small screw-cap vials, flush with N_2 before closing, and keep frozen until needed.
Sodium arachidonate, 10 mM. This is prepared in the manner described above, using arachidonic acid instead of linoleic acid.

Procedure

LIPOXYGENASE-1. The reaction is carried out at 25° in a quartz cuvette with a 1.0-cm light path. The assay mixture contains $(2.975 - x)$ ml of borate buffer, pH 9.0, 0.025 ml of sodium linoleate substrate, and x ml of enzyme. After each addition the mixture is stirred with a few strokes of a plastic puddler. The mixture should be at 25° when the enzyme is added. The reference cuvette contains no enzyme. Absorption at 234 nm is recorded, and the reaction rate is determined from the slope of the straight-line portion of the curve. Because of the lag period that occurs at extremely low concentrations of linoleic acid hydroperoxide,[5] care must be taken in determining the true reaction velocity when using a highly pure substrate.

LIPOXYGENASE-2. The assay is carried out with 0.2 M phosphate buffer (pH 6.1) and sodium arachidonate. The hydroperoxide product has an absorption maximum at 238 nm instead of 234 nm for the linoleic acid product, but the extinction coefficient is substantially the same, 2.5×10^4 M^{-1} cm^{-1}. The reason for using pH 6.1 buffer instead of pH 6.5 buffer, as employed in the L-3 assay below, is that the activity of L-2 against arachidonate is markedly increased relative to L-3. It is essential to prepare a fresh reference cuvette as close as possible to the time of the assay to control for the tendency toward increasing turbidity. Since some L-2 may absorb to the cuvette walls at this pH, the cuvette must be cleaned with Triton X-100 or other detergent between runs. If the enzyme is known to be free of L-3, the assay may be carried out at pH 6.8, where immobilization of the enzyme on the wall does not occur, and where the stability of the emulsion is improved. Here Ca^{2+} should be present at a final concentration of 0.001 M, since it doubles the activity at this pH; Ca^{2+} is unnecessary at pH 6.1. The rate observed at pH 6.8 should be multiplied by 1.5 to convert to the value found at pH 6.1, in order to calculate activity in the defined units.

LIPOXYGENASE-3. The assay reaction mixture contains $(2.800 - x)$ ml of 0.2 M phosphate buffer, pH 6.5, 0.2 ml of sodium linoleate, and x ml of enzyme. The reference solution from which enzyme is omitted is prepared anew for each assay. After a brief lag the rate as measured by absorption at 280 nm becomes linear and is proportional to enzyme concentration if the rate of change of absorbance is ≤ 0.15 per minute.

Units. One unit of L-1 or L-2 is defined as the quantity of enzyme that generates 1 μmol of conjugated diene per minute under standard assay conditions. The extinction coefficient for the diene is taken to be 2.5×10^4 M^{-1} cm^{-1} for the linoleic acid product at 234 nm and the arachidonic acid product at 238 nm. For optimal reproducibility and proportionality, the quantity of enzyme tested must be in the range of 0.1 ± 0.05 unit. One unit of L-3 is defined as the amount of enzyme that generates 1 μmol of ketodiene per minute under standard assay conditions ($E = 2.2 \times 10^4 M^{-1}$ cm^{-1}). The amount of enzyme to be assayed should be ≤ 0.15 unit. The assays for L-1, L-2, and L-3 may be scaled down to smaller volumes if the enzyme range is adjusted proportionately.

Assay with Oxygen Electrode. The O_2 electrode procedure is used in many laboratories, especially in the assay of L-1 at pH 9.0 with linoleic acid as substrate, and is entirely satisfactory. It is not, as noted, well

[5] J. L. Haining and B. Axelrod, *J. Biol. Chem.* **232**, 193 (1958).

suited for measuring L-3 with linoleic acid as substrate. It fails to give good stoichiometry, and the apparent specific activity varies inversely with enzyme concentration over much of its range.[6] While the results with L-2 are reasonably good when linoleic acid is employed as substrate, the stoichiometry is not strictly linear and is strongly influenced by substrate concentration. If arachidonic acid is used, as is desirable when differentiating between L-2 and L-3, the stoichiometric relationship between O_2 consumed and product(s) formed is most complicated. Nonetheless the O_2 uptake procedure is useful for locating elution peaks in chromatography. The procedure described below for L-1 is typical of those commonly employed.[6]

Lipoxygenase activities can be determined using a Clark oxygen electrode in a Gilson Medical Electronics Oxygraph, Model KM.[6] The reaction vessel has a volume of 1.5 ml and is thermostatted at 15°. The substrates are prepared in a manner similar to that described for the spectrophotometric assay. The final concentrations of borate buffer and linoleic acid in the substrate used at pH 9.0 for L-1 are 165 and 1.23 mM, respectively. Phosphate buffer, pH 6.8, replaces the borate buffer in the substrate used for L-3.[6] For L-2, the pH 6.8 buffer is used, and arachidonic acid replaces the linoleate.

Purification of Lipoxygenase Isoenzymes

Soybean seeds are finely ground in a burr mill, taking care to avoid significant heating, and then exhaustively extracted with hexane. The extraction is continued until the eluate becomes colorless.

Isolation of L-1. The defatted soybean meal (200 g) is extracted with 7.5 volumes of 0.2 M sodium acetate buffer (pH 4.5) by slow mechanical stirring for 1 hr. The suspension is filtered through cheesecloth and centrifuged at 16,300 g for 10 min at 4°. The resulting supernatant is adjusted to pH 6.8 with strong NaOH. All subsequent steps are performed at 0–5° and centrifugations are carried out at 16,300 g for 10 min. The clarified extract is taken to 30% saturation with ammonium sulfate and stirred for 1 hr. The suspension is centrifuged and the pellet discarded. Solid ammonium sulfate is added to bring the concentration to 60% of saturation. In both steps 0.2 N NaOH is added during the procedure to keep the pH at 6.8. The pellet obtained by centrifugation is resuspended in 20 mM sodium phosphate buffer, pH 6.8, and dialyzed against three changes of 6000 ml of 20 mM sodium phosphate, overnight. The resulting enzyme solution is loaded on a 5 × 60 cm column of DEAE–Sephadex, which has been preequilibrated with 20 mM sodium phosphate, pH 6.8. The enzyme is eluted with 3000 ml of a linear gradient formed from 1500 ml each of 20

[6] J. P. Christopher, E. K. Pistorius, and B. Axelrod, *Biochim. Biophys. Acta* **284**, 54 (1972).

and 220 mM sodium phosphate, at a flow rate of 30 ml/hr. L-1 is the last isozyme to appear, eluting at a concentration of 200–220 mM phosphate.

Remaining contaminants are removed by a second fractionation on DEAE–Sephadex. L-1, recovered from the first column, is dialyzed against two changes, 6000 ml each, of 20 mM sodium phosphate, pH 6.8, for 15 hr. The enzyme is centrifuged to remove any precipitate before loading on the second column. The second elution is done on a 2.6 × 14 cm column of DEAE–Sephadex prepared as before, employing a phosphate gradient formed from 300 ml each of 20 and 220 mM sodium phosphate buffer, at a flow rate of 10 ml/hr. The purified L-1 can be lyophilized after extensive dialysis against glass-distilled water, to give a stable product. The details presented above are taken from one of many preparations. Column volumes and flow rates given are not highly critical, since L-1 is ordinarily well separated from the other isozymes. A typical purification is summarized in Table I.

Purification of L-2 and L-3. Freshly defatted soybean meal (70 g) is extracted with 10 volumes of 0.05 M, pH 6.8, sodium phosphate buffer at 0–5° for 30 min while stirring. The extract is filtered through one layer of cheesecloth at room temperature, and the resulting filtrate is centrifuged for 10 min at 16,300 g at 4°. The precipitate is discarded, and the supernatant is further purified as described below. All the subsequent steps are performed in the cold at 0–5° with centrifugations at 16,300 g for 10 min. Although extraction of the soybean meal at pH 4.5 offers the advantage of excluding large amounts of unwanted protein, without any serious loss of L-1, the extraction should be done at pH 6.8 when L-2 and L-3 are desired.

The supernatant is made 30% saturated with solid ammonium sulfate, and the pH is adjusted to 6.8 ± 0.2 with 2 N NaOH. After 30 min the suspension is centrifuged and the resulting precipitate is collected and

TABLE I

SUMMARY OF PURIFICATION OF LIPOXYGENASE-1 FROM SOYBEANS

Step	Volume (ml)	Protein[a] (mg)	Total activity[b] (units)	Specific activity (units/mg)
Crude extract	1160	20,800	95,900	4.72
$(NH_4)_2SO_4$, 0.3–0.6 saturation	170	3,424	73,400	21.4
DEAE–Sephadex chromatography	172	229	24,900	101
Rechromatography on DEAE–Sephadex	106	81	14,600	180

[a] Protein values are based on $E_{280 \text{ nm}}^{1\%}$ = 14.0.

[b] Assays were performed using an O_2 electrode.

discarded. The supernatant is brought to 60% saturation with solid ammonium sulfate, while adjusting the pH to 6.8 ± 0.2 as before. After 30 min, the 30–60% saturation fraction is collected by centrifugation and dissolved in 25–50 ml of 0.02 M phosphate buffer, pH 6.8. The protein solution is dialyzed for 10 hr each time against two changes of 50-volumes each of 0.02 M, pH 6.8, phosphate buffer containing 0.5 mM calcium ion.

The dialyzed and centrifuged protein solution is placed on a Sephadex DEAE-A50 column (5.3 × 70 cm) equilibrated as described above for the L-1 purification. Elution is carried out with one column volume of 0.02 M sodium phosphate, pH 6.8, and then with an increasing linear gradient formed from 1500 ml of each of 0.02 M and 0.22 M phosphate buffer, pH 6.8. The flow rate is adjusted to 40 ± 5 ml/hr and fractions of 15 ml are collected. L-2 appears immediately after L-3. Although the L-2 and L-3 isozymes often overlap, L-2 can be located by using the assay method described in this paper. L-1 is always sharply separated from L-2 and L-3. Fractions containing 5 units/ml (oxygraph units) or more of either L-2 or L-3 activity are pooled, and the solution is made up to 70% saturation by slow addition of solid ammonium sulfate while maintaining the pH at 6.8 ± 0.2 with 0.5 N NaOH. An additional 45 min is allowed for complete precipitation of L-2 and L-3. After centrifugation, the isolated precipitate is dissolved in 25 ml of 0.02 M, pH 6.8, sodium phosphate buffer and dialyzed against two changes of 50 volumes each of 0.02 M, pH 6.8, sodium phosphate buffer containing 0.5 mM calcium ion.

The dialyzed protein solution is loaded on a second Sephadex DEAE-A50 column (2.8 × 20 cm) equilibrated as above. Elution is carried out with an increasing linear gradient formed from 750 ml each of 0.02 M and 0.22 M sodium phosphate buffer, pH 6.8. The flow rate is adjusted to 25 ml/hr, and fractions of 10 ml are collected (see Fig. 1).

A typical purification of L-2 and L-3 is summarized in Table II.

Properties

Purity. Employing the above procedures it is possible to obtain preparations of L-1, L-2, and L-3 that appear homogeneous on disc electrophoresis[7] in Tris-glycine at pH 9.5. The respective R_f values employing bromophenol blue as a standard are 0.30, 0.25, and 0.28. L-3 can be resolved into L-3a and L-3b by procedures described elsewhere.[7] As previously noted, the composition and behavior of these isozymes are somewhat alike, and the mixture can be treated as a single enzyme.

[7] J. P. Christopher, "Isoenzymes of Soybean Lipoxygenase: Isolation and Partial Characterization," Ph.D. thesis, Purdue Univ., West Lafayette, Indiana, 1972.

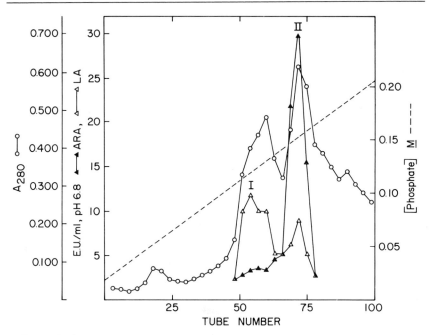

FIG. 1. Rechromatography of fractions rich in lipoxygenases-2 and -3 (L-2, L-3) obtained from first DEAE chromatography. Assays were conducted by the O_2-electrode method. Activities shown are in micromoles of O_2 taken up per minute and are not necessarily equivalent to the spectrophotometric units. This figure serves to illustrate the differential behavior of L-2 (peak II) and L-3 (peak I) with respect to linoleic (LA) and arachidonic (ARA) acids.

Molecular Weight. All four isoenzymes coelute from Sephadex G-150 with an apparent Stokes' radius corresponding to a globular protein of molecular weight 100,000 ± 5000.[7] The isoenzymes are not dissociable on disc gel electrophoresis in 0.1% sodium dodecyl sulfate (SDS). Prior reduction of L-1 with mercaptoethanol under denaturing conditions, followed by carboxymethylation of free sulfhydryl groups to prevent formation of disulfide linkages, does not alter the behavior of the protein in SDS–gel electrophoresis.[8]

Iron Content. Each isoenzyme contains one covalently bound atom of Fe per molecule. In the case of L-1, which has been extensively studied, the Fe atom cannot be removed by Fe chelators such as bathophenanthroline, unless the enzyme is treated with a sulfhydryl-containing re-

[8] E. K. Pistorius, "Studies on Isoenzymes of Soybean Lipoxygenase, Ph.D. thesis, Purdue Univ., West Lafayette, Indiana, 1974.

TABLE II
SUMMARY OF PURIFICATION OF LIPOXYGENASE-2 AND -3 (L-2, L-3) FROM SOYBEANS

Step	Volume (ml)	Protein (mg)	L-2 Total activity[a] (units)	L-2 Specific activity[b] (units/mg)	L-3 Total activity[a] (units)	L-3 Specific activity[b] (units/mg)
Crude extract	510	20800	42000	2.02	39100	1.8
Ammonium sulfate, 0.3–0.6 saturation	122	5190	38600	14.5	19800	3.8
DEAE–Sephadex chromatography	150	851	18400	21	15600	18.2
Rechromatography on DEAE–Sephadex of combined fractions rich in L-2 and L-3						
Peak I	86	50	2949	59	—	—
Peak II	80	31	—	—	1990	64

[a] Activities are based on O_2 uptake and are only approximate, especially for first three steps.
[b] Specific activity is based on λmoles of O_2 utilized min^{-1} mg^{-1} protein. Protein values are based on $E^{1\%}_{280\,nm} = 14.0$.

agent. Chelators that contain reduced S are able to remove the metal from L-1.[9]

Regiospecificity. Under ideal conditions, L-1 forms the 13-hydroperoxy derivative of linoleic acid virtually quantitatively; L-2 forms almost equal amounts of the 9- and 13- products, while L-3 yields about 65% and 35% of these products.[10,11] When L-1 acts on arachidonic acid, the first major product is the 15-hydroperoxide; this in turn undergoes hydroperoxidation mainly at the 8 position, and to a lesser extent at the 5 position.[12] Although the pH optimum for the primary hydroperoxidation of arachidonic acid is 9.0, the subsequent hydroperoxidation occurs best at pH 7.5.

Substrate Specificity. Lipid-like substances which carry one or more *cis,cis*-1,4-pentadiene groups can serve as substrates. Straight-chain fatty acids, esters, phospholipids, alcohols, glycerides, alcohol sulfates, hydroxamates, and even polyunsaturated alkyl-1-halides are utilized.[2,3]

This isozymes L-1, L-2, L-3a, and L-3b contain 4, 4.2, 5.6 (6), and 5.9 (5) sulfhydryl residues per molecule, respectively. After denaturation in urea and reduction with $NaBH_4$, the number of sulfhydryl groups increased by 8, 18, 6, and 6 in L-1, L-2, L-3a and L-3b, respectively, indicating 4, 9, 3, and 3 disulfide bonds present in the corresponding native enzymes.[8]

pH Optimum. The apparent pH optima for the various isozymes depends on a number of factors such as substrate concentration, nature of substrate, ionic strength, temperature, presence of detergent such as Tween 20, Ca^{2+}, etc. Isozyme L-1 acting on linoleic acid is most effective at pH 9.0–9.5; L-2 and L-3 are most active from pH 6 to 7, but activity can be observed above 7 with nonpolar substrates. The importance of the ionic nature of the substrate is well illustrated by the behavior of 9,12-octadecadienol 1-sulfate, which is an excellent substrate for L-1, even at pH 6.8 as well as at pH 9.[13] In contrast, linoleic acid is peroxidized by L-1 at pH 6.8 at about 3% of the rate at pH 9. The sulfate, which is fully ionized over the entire range studied, is untouched by L-2 and L-3.[14]

Secondary Reactions. The lipoxygenases all catalyze secondary reactions yielding products other than the simple hydroperoxides. Thus L-1 under anaerobic conditions catalyzes the interaction of linoleic acid and

[9] E. K. Pistorius and B. Axelrod, *J. Biol. Chem.* **249**, 3183 (1974).
[10] J. P. Christopher, E. K. Pistorius, F. E. Regnier, and B. Axelrod, *Biochim. Biophys. Acta* **289**, 82 (1972).
[11] J. P. Christopher and B. Axelrod, *Biochem. Biophys. Res. Commun.* **44**, 731 (1971).
[12] G. S. Bild, C. S. Ramadoss, and B. Axelrod, *Arch. Biochem. Biophys.* **184**, 36 (1977).
[13] J. C. Allen, *Chem. Commun.* **16**, 906 (1969).
[14] G. S. Bild, C. S. Ramadoss, and B. Axelrod, *Lipids* **12**, 732 (1977).

the primary oxygenation product, 13-hydroperoxy-9,11-octadecadienoic acid (13-hydroperoxylinoleic acid) to form 13-oxo-9,11-octadecadienoic acid, as well as 13-oxo-9,11-tridecadienoic acid, pentane and some C_{18}–C_{18} dimers.[2] Isozyme L-3 catalyzes *inter alia,* the formation of 13-oxo-9,11-octadecadienoic acid, 9-oxo-10,13-octadecadienoic acid, as well as some dimeric compounds and other products.[15] Unlike L-1, L-3 catalyzes these reactions in the presence or the absence of O_2, and thus these products can appear as soon as the primary oxygenation begins; L-2 generates very little ketodiene in the absence or the presence of oxygen. The secondary reaction is not prevented by low temperature, in contrast to that of L-1, which is suppressed at 4°.[16]

Isozyme L-2 catalyzes the formation of secondary products other than ketodienes. Thus substantial amounts of prostanoic acid derivatives[17] are formed, as well as the cyclic ether 9(12)-oxy-8,11,15-trihydroxyeicosa-5,13-dienoic acid, from arachidonic acid.[18] Analogous compounds are formed from 8,11,14-eicosatrienoic acid. This behavior is also useful in distinguishing L-2 from L-3 in mixtures.

K_m *Values.* The following K_m values have been determined using the standard spectrophotometric procedures described with varying concentrations of substrate: L-1, 0.012 mM linoleic acid; L-3, 0.34 mM linoleic acid; L-2, 0.016 mM (pH 6.1), 0.0096 mM, pH 6.8 (Ca^{2+} present), arachidonic acid.

The value for L-3 must be considered an apparent K_m, since there is evidence of serious substrate inhibition based on initial rate as measured at 234 nm. The concentration specified in the assay gives a linear rate of production of 280 nm of absorbing product. An O_2 concentration ≥ 0.025 mM (approximately 10% of the O_2 concentration in water in equilibrium with air at 25°) is sufficient to maintain maximum rates.

[15] S. Lim, "Studies on the Behavior of Soybean Lipoxygenase-3," M.S. thesis, Purdue Univ., West Lafayette, Indiana, 1978.
[16] E. K. Pistorius, J. F. Christopher, and B. Axelrod, *Int. Cong. Biochem. 10th, Abstr.,* p. 393 (1976).
[17] G. S. Bild, S. G. Bhat, C. S. Ramadoss, and B. Axelrod, *J. Biol. Chem.* **253,** 21 (1978).
[18] G. S. Bild, S. G. Bhat, C. S. Ramadoss, B. Axelrod, and C. C. Sweeley, *Biochem. Biophys. Res. Commun.* **81,** 486 (1978).

Section III

Hydroxymethylglutaryl-CoA Enzymes

[54] 3-Hydroxy-3-Methylglutaryl-CoA Reductase from Yeast

By NILOFER QURESHI, SUKANYA NIMMANNIT, and JOHN W. PORTER

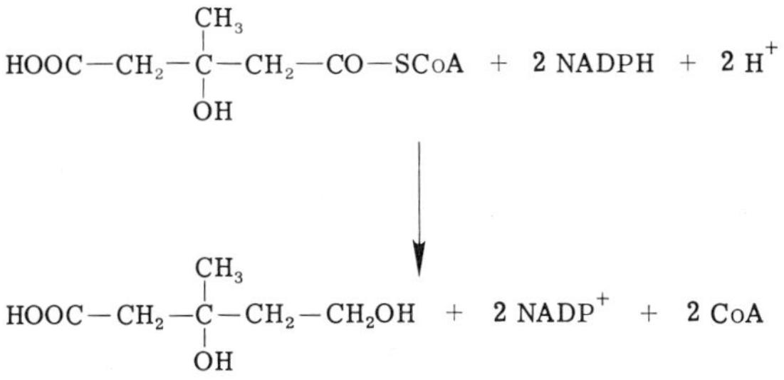

$$\text{HOOC}-\text{CH}_2-\overset{\overset{\displaystyle \text{CH}_3}{|}}{\underset{\underset{\displaystyle \text{OH}}{|}}{\text{C}}}-\text{CH}_2-\text{CO}-\text{SCoA} \ + \ 2\,\text{NADPH} \ + \ 2\,\text{H}^+$$

$$\downarrow$$

$$\text{HOOC}-\text{CH}_2-\overset{\overset{\displaystyle \text{CH}_3}{|}}{\underset{\underset{\displaystyle \text{OH}}{|}}{\text{C}}}-\text{CH}_2-\text{CH}_2\text{OH} \ + \ 2\,\text{NADP}^+ \ + \ 2\,\text{CoA}$$

3-Hydroxy-3-methylglutaryl-CoA (HMG-CoA) reductase catalyzes the reduction by NADPH of D-HMG-CoA to mevalonic acid. The stoichiometry of this reaction is shown above. In yeast, this enzyme is present in the mitochondrial matrix.[1,2] Originally, this enzyme was solubilized by prolonged autolysis[3] and then purified approximately 200-fold to a specific activity of 1.4 μmol of NADPH oxidized per minute per milligram of protein.[4] Later, this enzyme was purified to a specific activity of approximately 20 μmol of NADPH oxidized per minute per milligram of protein.[5] This procedure yields a homogeneous preparation of yeast HMG-CoA reductase.[5]

Assay Method

Principle. The HMG-CoA reductase activity can be conveniently determined by a modification of the spectrophotometric method of Kirtley and Rudney,[4] which measures the oxidation of NADPH at 340 nm.

Reagents

Potassium phosphate buffer, 1 M, pH 7.0
Dithiothreitol, 0.2 M

[1] I. Shimizu, J. Nagai, H. Hatanaka, and H. Katsuki, *Biochim. Biophys. Acta* **296,** 310 (1973).
[2] P. T. Trocha and D. B. Sprinson, *Arch. Biochem. Biophys.* **174,** 45 (1976).
[3] I. F. Durr and H. Rudney, *J. Biol. Chem.* **235,** 2572 (1960).
[4] M. E. Kirtley and H. Rudney, *Biochemistry* **6,** 230 (1967).
[5] N. Qureshi, R. E. Dugan, S. Nimmannit, W.-H. Wu, and J. W. Porter, *Biochemistry* **15,** 4185 (1976).

METHODS IN ENZYMOLOGY, VOL. 71

HMG-CoA,[6] 18 mM
NADPH, 1.6 mM

Procedure. The complete assay mixture[5] contains 0.1 mmol of potassium phosphate buffer, pH 7.0, 5 μmol of dithiothreitol, 300 nmol of HMG-CoA, 0.16 μmol of NADPH, and HMG-CoA reductase, in a total volume of 1.0 ml. Following a 10-min preliminary incubation at 30° to establish a base line rate of absorbance change at 340 nm, the reaction is initiated by the addition of HMG-CoA. The rate of the reaction is monitored by following the oxidation of NADPH at 340 nm ($\epsilon_{340 \text{ nm}}^{1 \text{ cm}}$ = 6.22 × 10³).

Assay for Protein. Protein assays are carried out by the biuret method[7] and by the method of Lowry *et al.*[8] after the protein solution is dialyzed against water to remove dithiothreitol.

Units. One unit of HMG-CoA reductase is defined as the amount of enzyme that catalyzes the oxidation of 1 nmol of NADPH per minute per milliliter of incubation mixture. Specific activity is expressed as units per milligram of protein.

Purification Procedure

The results of a typical purification of yeast HMG-CoA reductase are summarized in the table. Unless otherwise indicated, all steps are carried out at 4°.

Solubilization of Enzyme. HMG-CoA reductase is solubilized from Fleischmann's active dry yeast for bakers by a modification of the autolysis procedure of Kirtley and Rudney.[4] Twelve pounds of dry yeast are suspended in 12 liters of dibasic phosphate, 0.3 *M,* containing 1 mM dithiothreitol and 1 mM EDTA and then stirred for 12 hr at 4°. This suspension is centrifuged for 15 min at 20,000 *g* or by continuous flow centrifugation. The supernatant is discarded, and the gummy precipitate is resuspended in 9 liters of the same buffer and stirred another 48 hr at 4°. The suspension is centrifuged as before, and the precipitate is resuspended in 9 liters of the same buffer and stirred for 40 hr. The supernatant solution obtained after centrifugation is retained. This contains the major portion of the solubilized enzyme.

Heat Treatment. The crude supernatant solution (7130 ml) is heated (in

[6] HMG-CoA was prepared as described by S. Goldfarb and H. C. Pitot, *J. Lipid Res.* **12,** 512 (1971). It was then purified by paper chromatography as described by J. Brodie and J. W. Porter, *Biochem. Biophys. Res. Commun.* **3,** 173 (1960).

[7] A. G. Gornall, C. J. Bardawill, and M. M. David, *J. Biol. Chem.* **177,** 751 (1949).

[8] O. H. Lowry, N. J. Rosebrough, A. L. Farr, and R. J. Randall, *J. Biol. Chem.* **193,** 265 (1951).

PURIFICATION OF HMG-CoA REDUCTASE FROM YEAST[a]

Fraction	Volume (ml)	Protein (mg/ml)	Total units ($\times 10^{-4}$)	Specific activity[b]	Recovery (%)	Purification (fold)
Crude	7130	23.2	66	4	100	1
Heated	6620	19.7	65	5	98	1.25
First $(NH_4)_2SO_4$ precipitate	345	47.2	42	26	63	6.5
$Ca_3(PO_4)_2$ gel	2000	0.5	39	386	58	96
Second $(NH_4)_2SO_4$ precipitate	82	11.2	32	345	47	86
DEAE-cellulose	270	0.46	23	1,860	35	465
Affinity chromatography	3	1.16	6.9	19,800	10	4950

[a] Reprinted, with permission, from Qureshi et al.[5]; Copyright 1976, American Chemical Society.
[b] Nanomoles of NADPH oxidized per minute per milligram of protein.

centrifuge cups) in a water bath (65°) until the temperature of the enzyme solution is 65°, which takes approximately 30 min, and then for an additional 5 min. This mixture is immediately cooled on ice and centrifuged at 20,000 g for 20 min.

Ammonium Sulfate Fractionation. The supernatant solution (6620 ml) is brought to 60% saturation with solid ammonium sulfate (390 g/liter); after stirring for 30 min at 4°, the precipitated protein is collected by centrifugation. The pellet is resuspended in a minimum volume of 0.1 M potassium phosphate buffer, pH 7.0, containing 1 mM dithiothreitol and 1 mM EDTA. (The latter components are present in all buffers used in subsequent steps.) The enzyme is then dialyzed against the same buffer for 3 hr, with a change of buffer after 1.5 hr.

Calcium Phosphate Gel Adsorption and Ammonium Sulfate Precipitation. The dialyzed protein (345 ml) is diluted four times with water containing 1 mM dithiothreitol and 1 mM EDTA to a final concentration of 0.025 M potassium phosphate, and then the protein concentration is adjusted to 7 mg/ml with 0.025 M potassium phosphate buffer. This solution is mixed with calcium phosphate gel prepared as described by Tsuboi and Hudson[9] at a ratio of protein (in milligrams) to gel (dry weight) of 1 : 1. After stirring for 15 min, the gel is centrifuged at 8000 g for 5 min and the pellet is washed twice with one-tenth the original volume of 0.1 M potassium phosphate buffer. The HMG-CoA reductase is eluted from the gel by washing three or four times with 0.4 M potassium phosphate buffer. The enzymatically active eluate fractions are then combined (2000 ml), and the enzyme protein is concentrated by precipitation with solid ammonium sulfate (0–65% saturation, 430 g/liter). The precipitate is dissolved in a minimum volume of 0.1 M potassium phosphate buffer and then dialyzed against the same buffer for 3 hr at 4°, with a change of buffer at 1.5 hr.

DEAE-Cellulose Column Chromatography. Standard DEAE-cellulose ion exchanger is purchased from Schwarz-Mann and then prepared for enzyme purification by washing with acid, followed by base and water, according to the method of Peterson and Sober.[10] The dialyzed protein (82 ml) is diluted to a concentration of 0.01 M potassium phosphate buffer, and about 190 mg of protein are applied to a DEAE-cellulose column (3.5 × 28 cm) previously equilibrated by washing overnight with 0.01 M monobasic potassium phosphate and then with 250 ml of 0.01 M potassium phosphate buffer, pH 7.0. Protein is applied to the column, and the latter is washed with 0.01 M potassium phosphate buffer. The HMG-CoA reductase is eluted from the column with 450 ml of a linear concentration gradient of 0.01–0.50 M potassium phosphate buffer, pH 7.0, at room

[9] K. K. Tsuboi and P. B. Hudson, *J. Biol. Chem.* **224**, 879 (1957).
[10] E. A. Peterson and H. A. Sober, this series, Vol. 5, p. 6.

temperature. Most of the HMG-CoA reductase is eluted when the concentration of potassium phosphate is approximately 0.125 M. Ten-milliliter fractions are collected, and enzymatically active fractions are pooled and concentrated by ultrafiltration in an Amicon cell with a PM-10 membrane.

Affinity Chromatography. DEAE-cellulose-purified enzyme (15 × 10⁴ units and 72 mg of protein) in 20 ml of 0.125 M potassium phosphate buffer, pH 7.0, is diluted fivefold with water containing EDTA, 1 mM, and dithiothreitol, 5 mM, and then applied to a 0.75 × 3.4 cm column (1.5 ml of gel) of agarose-hexane-CoA (type V) gel supplied by P-L Biochemicals. The gel is previously washed with 0.025 M potassium phosphate buffer, pH 7.0, containing 1 mM EDTA and 5 mM dithiothreitol. Nonadsorbed protein is washed from the column at 22° with 10 ml of the pH 7.0 buffer used to equilibrate the column, and then the column is washed with 50 ml of 0.075 M potassium phosphate buffer containing 5 mM dithiothreitol. The HMG-CoA reductase is eluted from the column with a KCl gradient (0 to 2 M, 14 ml) in 0.025 M potassium phosphate buffer, pH 7.0, containing 5 mM dithiothreitol. The rate of elution is 1 ml/min. Ten-milliliter fractions are collected before the start of the gradient, and 1-ml fractions are collected thereafter. Most of the protein passes directly through the column without binding, and then the bound reductase is removed on gradient elution with KCl (Fig. 1). A range in specific activity of 15,000 to 22,000 nmol of NADPH oxidized per minute per milligram of protein is normally obtained in the purifications of this enzyme.

The recovery of enzyme from the gel is usually 40–45%, but it may vary from 20 to 70%. In order to obtain a good yield, careful regulation of the ratio of gel to protein is necessary. Maximum recovery of the active enzyme is obtained when the column is almost saturated with enzyme. After elution, the protein concentration is very low and the enzyme loses activity unless it is concentrated immediately. Dialysis with collodion tubes (Schleicher and Schuell) is used to concentrate the enzyme and to remove salt from the protein.

Properties of Yeast HMG-CoA Reductase

As shown in the table, yeast HMG-CoA reductase is purified approximately 5000-fold by the procedure outlined above. This preparation is stable for at least 3 months when stored at −20°.

Purity. The enzyme purified as described is homogeneous, or nearly so, since it migrates as a single component on BioGel filtration and on a DEAE-cellulose column. When electrophoresed on 5% acrylamide gel, one major band of protein is observed, which coincides exactly with HMG-CoA reductase activity.

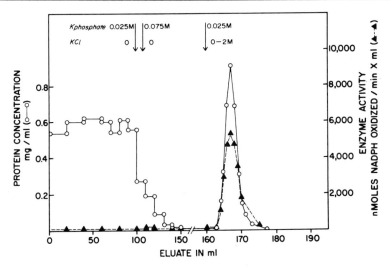

FIG. 1. Affinity chromatography of DEAE-cellulose-purified HMG-CoA reductase obtained from autolyzed yeast. The details of this chromatographic separation are given in the text. Reductase activity (▲----▲) and protein (o——o) were assayed in the effluent. Reprinted, with permission, from Qureshi et al.[5]

Molecular Weight. The molecular weight of the yeast HMG-CoA reductase is estimated to be approximately 2.6×10^5, based on its elution profile on a BioGel column.

Electrophoresis of the dissociated enzyme on a sodium dodecyl sulfate–polyacrylamide (10%) disc gel showed one major staining protein band that corresponded to a subunit weight of approximately 60,000. This indicated that the native yeast HMG-CoA reductase is a tetramer that consists of subunits of the same molecular weight.

Substrate Specificity. This enzyme catalyzes the reduction of HMG-CoA, mevaldic acid,[11] and mevaldic hemithioacetal[12] to mevalonic acid in the presence of NADPH. It also catalyzes the oxidation of mevaldic acid to HMG-CoA in the presence of NADP+ and CoA.

Kinetic Properties. Yeast HMG-CoA reductase is inactivated upon preincubation with CoA,[4] but this inhibition can be reversed by dialysis.[13] Without prior incubation, CoA is an activator for the reduction of mevaldate to mevalonic acid and it reduces the K_m for mevaldate 20- to 30-fold.[11] The K_m values for the substrates for this enzyme are as follows: for the overall reaction, HMG-CoA = 2.4 μM[4] and NADPH = 89 μM[4]; for

[11] N. Qureshi, R. E. Dugan, W. W. Cleland, and J. W. Porter, *Biochemistry* **15**, 4191 (1976).
[12] J. Rétey, E. von Stetten, U. Coy, and F. Lynen, *Eur. J. Biochem.* **15**, 72 (1970).
[13] A. Tan-Wilson and G. B. Kohlhaw, *Biochem. Biophys. Res. Commun.* **85**, 70 (1978).

the second reductive step, mevaldate $= 8$ mM and 0.4 mM,[11] respectively, in the absence and in the presence of CoA, and NADPH $= 24$ μM.[11]

Irreversible Inhibition. Yeast HMG-CoA reductase is very susceptible to inhibition by low concentrations of certain thiol group reagents, e.g., iodoacetamide,[4] p-hydroxymercuribenzoate,[4] and fungal products such as citrinin,[14] ML236A, and ML 236B.[15,16] It is also inhibited by o-phenanthroline.[4]

Mechanism of Reaction. Data from kinetic studies suggest that the oxidation of mevaldic acid to HMG-CoA and the reduction of mevaldic acid to mevalonic acid, catalyzed by the yeast HMG-CoA reductase in the presence or the absence of CoA, proceeds by a sequential mechanism. In terms of the overall reaction, HMG-CoA and NADPH bind to the enzyme in an ordered or random fashion. Reduction of HMG-CoA to mevaldate hemithioacetal then takes place, and NADPH replaces NADP+. The hemithioacetal then reverts to its components, mevaldate and CoA, and mevaldate is reduced to mevalonate. The products mevalonate and CoA, followed by NADP+, then leave the enzyme.[11] The hydrogen used in this reduction is transferred from the A side of the pyridine ring of NADPH in both of the reactions in the conversion of HMG-CoA to mevalonate.[17]

Acknowledgments

This work was supported in part by a research grant, HL 16364, from the National Heart and Lung Institute, National Institutes of Health, United States Public Health Service, and by the Medical Research Service of the Veterans Administration.

[14] M. Kuroda, Y. Hazama-Shimada, and A. Endo, *Biochim. Biophys. Acta* **486**, 254 (1977).
[15] K. Tanzawa, M. Kuroda, and E. Endo, *Biochim. Biophys. Acta* **488**, 97 (1977).
[16] A. Endo, M. Kuroda, and K. Tanzawa, *FEBS Lett.* **72**, 323 (1976).
[17] R. E. Dugan and J. W. Porter, *J. Biol. Chem.* **246**, 5361 (1971).

[55] 3-Hydroxy-3-Methylglutaryl-CoA Reductase from Rat Liver

By DON A. KLEINSEK, RICHARD E. DUGAN, TERRY A. BAKER, and JOHN W. PORTER

$$\text{HOOC}-\text{CH}_2-\underset{\underset{\text{OH}}{|}}{\overset{\overset{\text{CH}_3}{|}}{\text{C}}}-\text{CH}_2-\text{CO}-\text{SCoA} \quad + \quad 2\,\text{NADPH} \quad + \quad 2\,\text{H}^+$$

$$\downarrow$$

$$\text{HOOC}-\text{CH}_2-\underset{\underset{\text{OH}}{|}}{\overset{\overset{\text{CH}_3}{|}}{\text{C}}}-\text{CH}_2-\text{CH}_2\text{OH} \quad + \quad 2\,\text{NADP}^+ \quad + \quad \text{CoASH}$$

The two-step reduction of 3-hydroxy-3-methylglutaryl-CoA (HMG-CoA) to mevalonate, the rate-determining step in hepatic cholesterol biosynthesis, is catalyzed by microsomal HMG-CoA reductase[1-3] [EC 1.1.1.34 mevalonate : NADP$^+$ oxidoreductase (CoA-acylating)]. As expected from the central role of this enzyme in cholesterogenesis, its activity varies markedly as a function of the nutritional or hormonal state of the animal.[4-13] As a first step to a better understanding of the molecular mechanism(s) of control of this activity, it is necessary to obtain homogeneous HMG-CoA reductase. In this chapter we describe purification procedures for the enzyme that result in homogeneous or nearly homogeneous preparations of high specific activity.

A number of purification procedures have been reported for rat liver

[1] M. D. Siperstein and V. M. Fagan, J. Biol. Chem. 241, 602 (1966).
[2] L. W. White and H. Rudney, Biochemistry 9, 2725 (1970).
[3] D. Shapiro and V. W. Rodwell, Biochemistry 11, 1042 (1972).
[4] M. E. Dempsey, Annu. Rev. Biochem. 43, 967 (1974).
[5] M. D. Siperstein, Curr. Top. Cell. Regul. 2, 65 (1970).
[6] V. W. Rodwell, J. L. Nordstrom, and J. J. Mitscheleń, Adv. Lipid Res. 14, 2 (1976).
[7] A. A. Kandutsch, H. W. Chen, and H. J. Heiniger, Science 201, 498 (1978).
[8] R. E. Dugan and J. W. Porter, in "The Biochemical Actions of Hormones" (G. Litwack, ed.), Vol. 4, p. 197. Academic Press, New York, 1977.
[9] J. M. Dietschy and J. D. Wilson, N. Engl. J. Med. 282, 1128, 1179, 1241 (1970).
[10] V. W. Rodwell, D. J. McNamara, and D. J. Shapiro, Adv. Enzymol. 38, 373 (1973).
[11] W. M. Bortz, Metab. Clin. Exp. 22, 1507 (1973).
[12] M. S. Brown and J. L. Goldstein, Science 191, 150 (1976).
[13] J. L. Goldstein and M. S. Brown, Annu. Rev. Biochem. 46, 897 (1977).

HMG-CoA reductase.[14-21] However, these procedures do not yield homogeneous preparations.[22] Recently, homogeneous enzyme of high specific activity has been obtained by combining conventional protein purification steps with affinity chromatography. A description of these techniques, presented below, is divided into two sections. The first section describes two purification procedures used to produce homogeneous enzyme of high specific activity that employ standard purification steps and an affinity chromatographic separation. A description of the assay procedures and the properties of the enzyme is included in this section. The second section describes three purification procedures which produce enzyme of high specific activity by employing similar standard purification steps plus two successive affinity chromatographic fractionation steps.

Selected Materials and Reagents

Chemicals and reagents were obtained from the following sources: 3-hydroxy-3-methylglutaric acid (Methods 1 and 2), Schwarz/Mann and (Method 4) Sigma Chemical Co.; 3-hydroxy-3-methyl [3-^{14}C]glutaryl-CoA (Method 4), 3-hydroxy-3-methyl [3-^{14}C]glutaric acid (Methods 1–3, and 5), New England Nuclear; BioGel A-0.5m, 200–400 mesh (Method 1), Affi-Gel Blue (Methods 4 and 5), Bio-Rad protein dye concentrate (Method 4), Bio-Rad Laboratories; cholestyramine (Questran), Mead Johnson Laboratories; blue Sepharose CL-6B (Method 3), Pharmacia; agarose–hexane–CoA, type 5 (Methods 1–3), agarose-hexane-HMG-CoA, type 5 (Methods 4 and 5), P-L Biochemicals, Inc.

Assay Procedures and Properties; Purification by Methods 1 and 2

Procedure of Assay for Enzyme Activity

Methods 1, 2, and 5

Radiochemical Method of Assay. The activities of microsomal and solubilized HMG-CoA reductase are determined by measurement of the

[14] T. Kawachi and H. Rudney, *Biochemistry* 9, 1700 (1970).
[15] M. Higgins, T. Kawachi, and H. Rudney, *Biochem. Biophys. Res. Commun.* 45, 138 (1971).
[16] M. J. P. Higgins, D. Brady, and H. Rudney, *Arch. Biochem. Biophys.* 163, 271 (1974).
[17] M. S. Brown, S. E. Dana, J. M. Dietschy, and M. D. Siperstein, *J. Biol. Chem.* 248, 4731 (1973).
[18] R. A. Heller and R. G. Gould, *Biochem. Biophys. Res. Commun.* 50, 859 (1973).
[19] R. A. Heller and M. A. Shrewsbury, *J. Biol. Chem.* 251, 3815 (1976).
[20] C. D. Tormanen, W. L. Redd, M. V. Srikantaiah, and T. J. Scallen, *Biochem. Biophys. Res. Commun.* 68, 754 (1976).
[21] M. V. Srikantaiah, C. D. Tormanen, W. L. Redd, J. E. Hardgrave, and T. J. Scallen, *J. Biol. Chem.* 252, 6145 (1977).
[22] C. B. Berde, R. A. Heller, and R. D. Simoni, *Biochim. Biophys. Acta* 488, 112 (1977).

conversion of [3-^{14}C]HMG-CoA to mevalonate in an assay mixture that employs a NADPH-generating system.[23] The reaction mixture in this assay contains potassium phosphate buffer, pH 7.0, 50 μmol; dithiothreitol, 2 μmol; glucose 6-phosphate, 2 μmol; NADP$^+$, 0.5 μmol; DL-[3-^{14}C]HMG-CoA, 0.15 μmol; glucose-6-phosphate dehydrogenase, 1.25 units; and protein, 300–1200 μg, in a total volume of 0.5 ml. This reaction mixture is preincubated at 37° for 10 min with all components except HMG-CoA reductase. The reaction is then carried out for 10 min at 37° following the addition of microsomes. The assay for enzyme activity is linear at protein concentrations of 600–2400 μg/ml; however, the measurement of the activity of solubilized enzyme is linear with lesser amounts. The reaction is terminated by the addition of 50 μl of 2.4 M HCl, and the sample is incubated at 37° for 30 min to allow for the formation of mevalonolactone. To determine background radioactivity, acid is added to control tubes prior to the addition of enzyme. The reaction mixture is then centrifuged (tabletop centrifuge) to remove denatured protein. An aliquot of 200 μl of the reaction mixture is spotted on activated (110°, 30 min) silica gel G thin-layer plates and the chromatogram is developed in benzene–acetone (1 : 1, v/v). The chromatogram is then divided into 2 × 2 cm sections, and the appropriate sections scraped into vials for assay of radioactivity.

Spectrophotometric Method of Assay.[24] The HMG-CoA reductase activity in solubilized fractions is assayed spectrophotometrically by measuring the rate of decrease in absorbance at 340 nm due to the oxidation of NADPH. The reaction mixture in a volume of 0.5 ml contains potassium phosphate buffer, pH 7.0, 50 μmol; dithiothreitol, 2 μmol; NADPH, 0.3 μmol; DL-HMG-CoA, 0.15 μmol; and enzyme, 0.2–400 μg of protein.

The reaction mixture is preincubated in a 2-mm light path glass cuvette without HMG-CoA present for 5 min at 37°. The assay for enzyme activity is then carried out by the addition of HMG-CoA to the reaction mixture at 37° in a recording spectrophotometer. The initial velocity of the reaction is measured, and the net rate of NADPH oxidation is determined by subtracting the rate of its oxidation in the absence of HMG-CoA from the rate observed with both substrates present.

Microsomal activity cannot be assayed by this procedure owing to the low HMG-CoA reductase activity and the turbidity produced by the particulate membrane fraction in the incubation medium. However, the

[23] C. M. Nepokroeff, M. R. Lakshmanan, G. C. Ness, R. E. Dugan, and J. W. Porter, *Arch. Biochem. Biophys.* **160**, 387 (1974).
[24] D. A. Kleinsek, S. Ranganathan, and J. W. Porter, *Proc. Natl. Acad. Sci. U.S.A.* **74**, 1431 (1977).

spectral method is a more rapid, convenient, and economical (though less sensitive) means of measuring soluble enzyme activity than the radiochemical procedure.

Units. One unit of enzyme activity is that quantity of enzyme that oxidizes 1 nmol of NADPH per minute at 37° under the above assay conditions. Specific activity is expressed in activity units per milligram of protein.

Protein Determination. All protein solutions are treated with 10% (w/v) trichloroacetic acid. The precipitate obtained with microsomes is assayed for protein by a modification of the biuret method[25] that employs deoxycholate (40 mM). Solubilized enzyme preparations are assayed for protein content by the method of Lowry et al.[26]

Preparation of Enzyme

Method 1[24]

Treatment of Animals. Male albino Holtzman rats are acclimated to an alternate 12-hr light–dark cycle for a period of 2–3 weeks. The animals, weighing 180–200 g, are fed ad libitum a 2% cholestyramine Wayne Lab Blox powdered diet for a minimum of 4 days prior to sacrifice at the mid-dark period, which is the diurnal high point of enzyme activity.

Preparation of Microsomes. After the animals are decapitated livers are excised and immediately placed in an ice-cold homogenization buffer. All solutions used in the purification scheme contain deionized water, and all further operations for the isolation of the microsomes are carried out at 4°. The homogenization medium contains 50 mM potassium phosphate buffer, pH 7.0, 0.2 M sucrose, and 2 mM dithiothreitol (buffer A). The livers are homogenized in buffer A (2 ml per gram of liver) in a Waring blender at full speed for 15 sec, followed by three strokes with a motor-driven Teflon pestle in a Potter–Elvehjem type glass homogenizer. The homogenate is centrifuged at 15,000 g for 10 min, and the postmitochondrial supernatant solution is retained and centrifuged at 100,000 g for 75 min in 30-ml ultracentrifuge tubes filled to near capacity. The supernatant solution is decanted, and the white lipid-like material remaining on the ultracentrifugation tube wall is removed with cotton-tipped applicators. The microsomes are washed by resuspension of the pellet in buffer A containing 50 mM EDTA (1 ml per gram of liver) and homogenized as before. The

[25] A. G. Gornall, C. J. Bardawill, and M. M. David, *J. Biol. Chem.* **177**, 751 (1949).
[26] O. H. Lowry, N. J. Rosebrough, A. L. Farr, and R. J. Randall, *J. Biol. Chem.* **193**, 265 (1951).

homogenate is centrifuged at 100,000 g for 60 min in tubes of the same size, and the washed microsomal pellets are slow-frozen by placing the ultracentrifuge tubes in a $-20°$ freezer. Each tube should contain the material obtained from approximately 1.5 rat livers.

Solubilization of Enzyme. Microsomal pellets are kept frozen at $-20°$ for a minimum of 2 hr or may be stored at this temperature for several weeks without loss of enzyme activity. The microsomal pellets are thawed at room temperature, and a slight modification of the method of Heller and Gould[18] is used to liberate the enzyme from the membrane matrix. Three milliliters of buffer B (50 mM potassium phosphate, pH 7.0, 0.1 M sucrose, 2 mM dithiothreitol, 50 mM KCl, and 30 mM EDTA) are added to each microsomal pellet and homogenization is effected with three strokes of a tight-fitting Teflon pestle in a Potter–Elvehjem type glass homogenizer. This procedure is followed by an additional three strokes of the microsomal suspension, after dilution with 7 ml of buffer. Excessive heat produced during the homogenization is avoided by surrounding the glass homogenizer with cold water. After standing for 15–30 min at room temperature, the suspension is centrifuged at 100,000 g for 60 min at 20°. The supernatant solution, containing soluble HMG-CoA reductase, is saved and used for purification of the enzyme. All further operations are carried out at room temperature unless otherwise indicated.

Salt Fractionation. A filtered saturated solution of ammonium sulfate is added dropwise at room temperature under a stream of nitrogen to a slow-stirring solution of soluble extract. The protein fraction precipitating between 35 and 50% saturation with ammonium sulfate is collected by low-speed centrifugation.

Heat Treatment. The protein pellet obtained above is dissolved in heat-treatment buffer [50 mM potassium phosphate, pH 7.0, 3 mM dithiothreitol, 1.0 M KCl, and 30% (v/v) glycerol]. The concentration of protein before the heat treatment should be between 4 and 10 mg/ml, since protein concentrations higher than this result in a viscous solution that is difficult to centrifuge. The enzyme solution, 4–5 ml, is placed in a Pyrex glass tube (16 × 125 mm). After heating at 65° for 6 min in a water bath, the enzyme solution is rapidly cooled on ice to room temperature and then centrifuged at 100,000 g for 30 min to remove denatured protein.

Enzyme Concentration. The supernatant solution is diluted 1:1 with buffer B and subjected to a fractionation by 0 to 50% saturation with ammonium sulfate. After centrifugation at 10,000 g for 5 min, the protein pellet is dissolved in a minimal volume of either buffer C or D, depending on whether a BioGel filtration or a sucrose density gradient ultracentrifugation step is used.

BioGel Filtration. The protein pellet is dissolved in a volume of 1–2 ml of buffer C (50 mM potassium phosphate, pH 7.0, 2 mM dithiothreitol, 30 mM EDTA, 50 mM KCl, and 10% sucrose) and then 4000 units of enzyme activity are applied to a BioGel A-0.5m gel filtration column (2 × 44 cm) having a particle size of 200–400 mesh. The enzyme is eluted at a flow rate of 6–9 ml/hr with the same buffer. About 80 eluate fractions (each 1 ml) are collected and analyzed for protein and enzyme activity. Those fractions that have minimal light absorbance at 280 nm and contain high enzyme activity are pooled for affinity chromatography. The gel may be used several times by washing the column between preparations with a buffer containing 50 mM potassium phosphate, pH 7.0, and 0.5 M KCl. The success of BioGel filtration is dependent upon the batch of agarose gel and the ionic strength of the eluent buffer.[27] Use of a high ionic strength eluent buffer, or a gel in which an interaction with the enzyme is not present, results in a poorly resolved enzyme fraction. However, the use of low ionic strength buffer in combination with a "pseudoaffinity" type of agarose gel provides excellent resolution of enzyme from the majority of protein contaminants.

The choice of the proper type of agarose gel is determined by establishing the molecular weight of eluted reductase through a comparison with protein standards of known molecular weight. Agarose gel in which the reductase elutes at an apparent molecular weight of less than 10,000, is suitable for proper enzyme purification. Unfortunately, gels of this type are difficult to obtain. Thus, an alternative fractionation (sucrose density gradient ultracentrifugation) has been developed to replace the BioGel filtration.

Method 2[27]

Sucrose Density Gradient Ultracentrifugation. The 5 to 20% sucrose density gradient contains 50 mM potassium phosphate buffer, pH 7.0, 50 mM KCl, 30 mM EDTA, and 2 mM dithiothreitol (buffer D), in a total volume of 38 ml. The concentrated heat-treated enzyme pellet is dissolved in the above buffer at a sucrose concentration of 2% and a protein concentration of approximately 1.5 mg/ml. (It is important to apply a minimum of 500 units of enzyme activity to the gradient if a good recovery of enzyme activity is to be obtained. However, overloading of the gradient with more than 2000 units of enzyme activity will result in a poor purification of HMG-CoA reductase.) The ultracentrifugation is carried out at 14° for 41 hr at 25,000 rpm in a Beckman Model L3-50 ultracentrifuge, with a SW-27 swinging-bucket rotor. The gradient is then tapped at room temperature and collected in 35 fractions (each 1 ml) with an Isco Model 184 tube

[27] D. A. Kleinsek and J. W. Porter, *J. Biol. Chem.* **254**, 7591 (1979).

holder and piercing mechanism. The fractions that contain the peak enzyme activity and minimal protein (normally tubes 16 to 20) are pooled.

Enzyme Concentration. Before application to the affinity column it is necessary to concentrate the dilute enzyme solution from the BioGel filtration or sucrose density gradient ultracentrifugation steps. These preparations are concentrated in collodion bags at room temperature to a protein concentration of 1 mg/ml or greater. After concentration, the enzyme solution is dialyzed against the low ionic strength buffer used for affinity chromatography. The normal recovery of enzyme activity after concentrating and dialyzing is about 50%. Diluting or dialyzing the enzyme preparation before concentrating results in a greater loss of enzyme activity.

Affinity Chromatography. Affinity chromatography is carried out at 4° with a thioester-linked agarose–hexane–coenzyme A gel. The wet gel, 0.2 ml, is packed into a Pasteur pipette and then washed with approximately 20 volumes of a solution containing 25 mM potassium phosphate, pH 7.0, 1 mM EDTA, and 10% sucrose. After equilibration of the gel with this buffer, enzyme (1400 units) is rapidly cooled in ice to 4° and applied to the gel. A flow rate of 1 ml per 10 min is maintained during the binding process. One-milliliter eluate fractions are collected, and light absorbance at 280 nm is measured. The majority of protein, with negligible reductase activity, appears in the first two 1-ml fractions. When light absorbance decreases to the base-line level, reductase is eluted with binding buffer containing 0.5 M KCl. The elution flow rate is 1 ml/min, and fractions of 1 ml are collected. Of the applied enzyme, 95–100% is recovered in the first and second fractions.

Yield, Purity and Properties of Enzyme (Methods 1 and 2)

Yield. The above procedure normally yields (from 25 rat livers) about 10% of the total enzyme activity present in the solubilized enzyme state, or 3–4% of that in microsomes. Extraction of enzyme by a repeated freeze-thawing of the microsomes increases this yield somewhat. About 2 μg of pure enzyme are obtained from 10 g of rat liver. Thus, per 10 g of rat liver tissue, approximately 60 μg of active microsomal HMG-CoA reductase are available for purification. Therefore, large amounts of starting material are required for a substantial yield of enzyme. A typical purification by Methods 1 and 2 is summarized in Table I. The final specific activity is approximately 19,000–20,000 units per milligram of protein when the BioGel agarose column is used. A slightly lower value of around 17,000 is obtained when sucrose density gradient ultracentrifugation is substituted. The overall purification-fold from microsomes varies between 3700 and 4100, and both methods yield homogeneous enzyme.

TABLE I

PURIFICATION OF HMG-CoA REDUCTASE FROM RAT LIVER BY METHODS 1 AND 2

Step	Volume (ml)	Total protein (mg)	Protein (mg/ml)	Total activity (units)	Specific activity (units/mg protein)	Yield (%)	Purification (fold)
Microsomal suspension	510	4,800	9.4	22,510	4.6	100	1
Soluble extract	165	198	1.2	7,054	36	31	8
(NH₄)₂SO₄, 35–50% saturation	17	65	3.8	6,378	98	28	22
Heat treatment	9	13	1.5	5,012	368	22	82
(NH₄)₂SO₄, 0–50% saturation	1.9	9	4.7	4,100	456	18	97
BioGel filtration	10	0.42	0.042	1,554	3,700	7	787
CoA affinity column	1	0.036	0.036	690	19,166	3	4078
Sucrose density gradient fractionation[a]	9	0.81	0.090	2,620	3,231	12	702
CoA affinity column[a]	1.6	0.053	0.033	890	16,885	4.0	3671

[a] Data normalized to 25 rats are reported for the sucrose density gradient and subsequent CoA affinity fractionations that were performed after precipitation of the enzyme at 0 to 50% saturation with ammonium sulfate. These data were obtained by Method 2. The data obtained by Method 1 are also normalized to 25 rats.

Enzyme Purity.[24] The enzyme is judged to be homogeneous by a number of physicochemical criteria. The purified enzyme (10–40 μg) migrates as a single band of protein when electrophoresis is carried out on SDS–5% polyacrylamide disc gels. Furthermore, it migrates as a single band of protein (10–40 μg) with an R_f of 0.52 on 5% nondenaturing polyacrylamide disc gels when electrophoresed in a pH 8.9 system. This single band of protein comigrates with HMG-CoA reductase activity. Purified enzyme reacts against crude antiserum to form a single precipitin band on Ouchterlony double diffusion plates.

Size.[27] The molecular weight of HMG-CoA reductase is 200,000, as determined by molecular sieve chromatography on two different gel matrices, agarose gel (BioGel) and dextran gel (Sephadex), and it consists of four subunits of 51,000 MW each, as determined by SDS–disc gel electrophoresis. Upon cold inactivation, the enzyme remains as a 200,000 MW entity. Thus, the molecular basis of cold inactivation is probably due to discrete conformational changes that affect the active site of HMG-CoA reductase.

Apparent K_m Values.[27] Spectrophotometric assays of the initial velocity of oxidation of NADPH have been used to determine HMG-CoA reductase reaction rates. From double reciprocal plots of the data in which one substrate concentration was present in saturating amounts and the other concentration was varied, the apparent K_m values were found to be $1.7 \times 10^{-5}\ M$ for D-HMG-CoA and $3.0 \times 10^{-5}\ M$ for NADPH. The saturating concentrations of DL-HMG-CoA and NADPH used were $3.6 \times 10^{-4}\ M$ and $7.8 \times 10^{-4}\ M$, respectively. The concentration of DL-HMG-CoA was determined by light absorbance at 260 nm and by sulfhydryl analysis with Ellman reagent after complete hydrolysis of HMG-CoA with base. The concentration of NADPH was determined from light absorbance at 340 nm.

Reversible Cold Lability.[27] Both soluble and microsomal HMG-CoA reductase exhibit reversible cold lability in low ionic strength phosphate buffer (0.1 M potassium phosphate, pH 7.0, 2 mM dithiothreitol). At low protein concentrations, partially purified enzyme (0.11 mg/ml) loses 50% of its activity within 90 sec. After 70% of the activity is lost, a slower rate of decline is observed until approximately 70 min, at which time a complete cessation of enzyme activity occurs. Rapid and complete recovery of enzyme activity occurs within 7.5 min when the solution is warmed at 37°. Microsomal enzyme exhibits a similar but slower loss at 4° and a rapid recovery of enzyme activity at 37°.

pH and Temperature Optima.[27] The enzyme has a pH optimum of 6.25–7.25 and a temperature optimum in 0.1 M phosphate buffer, pH 7.0, and 2 mM dithiothreitol of approximately 47°.

Isoelectric Point.[27] The enzyme has a p*I* of 6.2, as determined from isoelectric focusing on a 5% polyacrylamide disc gel.

Stability and Storage.[28] Soluble enzyme, at a protein concentration of 1.2 mg/ml in buffer B, exhibits a significant loss of activity at room temperature, i.e., 25% and 50% after 5 and 19 hr, respectively. Slow freezing at −20°, or fast freezing of the soluble enzyme in a Dry Ice–ethanol bath results in a 20% and 10% loss in activity, respectively. Heat-treated enzyme, at a concentration of 2 mg/ml, is stable at room temperature for 30 hr, and at 4°, only a 10% loss of activity of heat-treated enzyme occurs in 5 days. Slow or fast freezing and subsequent storage at −20° for 5 days results in a negligible loss of enzyme activity. Activity of the purified enzyme is rapidly lost in low ionic strength buffer, but in the presence of the affinity eluent buffer the enzyme is stable at room temperature. Fast freezing of the enzyme results in variable losses of 10–30%, which are dependent on the protein concentration. Enzyme stored at 4° or −20° is preincubated for 5–10 min at 37° before assays are carried out for enzyme activity. Both the soluble and microsomal enzyme require a sulfhydryl reagent for stability.

Immunogenicity of the Enzyme.[24] Enzyme from the heat treatment step is effective in generating antiserum to HMG-CoA reductase. The enzyme is dispersed 1 : 1 with complete Freund's adjuvant for the initial injection, and 1 : 1 with incomplete Freund's adjuvant for booster injections. The titer of antibody is raised by subcutaneous multisite injections of the enzyme preparation into a rabbit. Additional injections are made at 2-week intervals, and blood is withdrawn and assayed for antibody titer at periodic intervals.

Enzyme Orientation.[27] HMG-CoA reductase is solubilized from microsomal membranes by mild treatment, and it remains soluble in aqueous solution. The soluble enzyme and the membrane-bound microsomal enzyme have similar neutralization end points per unit of enzyme activity when titrated with antiserum produced against the soluble species. These physical and immunochemical properties of the enzyme support the conclusion that it is an extrinsic protein of the endoplasmic reticulum in which the active site of the enzyme is positioned toward the cytosol. This positioning is consistent with the location of HMG-CoA synthesis and mevalonate phosphorylation within the cell.

Other Enzyme Fractionation Steps.[28] DEAE-cellulose and NADP+ affinity chromatography also have been used successfully as fractionation steps, with respectable yields, in the purification of the enzyme. Both

[28] D. A. Kleinsek, Ph.D. thesis, University of Wisconsin-Madison, 1979.

columns require a low ionic strength buffer for binding of the enzyme, and elution is effected by increasing the ionic strength of the buffer with KCl.

Inactive HMG-CoA reductase.[24] The presence of an inactive species of HMG-CoA reductase which is formed in the liver or during the purification procedure and copurifies with the active species has been demonstrated by immunochemical and physical tests. On Ouchterlony double diffusion plates, the unbound protein fraction from the CoA affinity gel shows a partial identity of antigenic determinants to the active enzyme fraction that binds to, and is eluted from, the gel. Also, the majority of the protein in the unbound fraction shows an identical R_f with that of the purified active enzyme on SDS–polyacrylamide disc gel electrophoresis.

Other HMG-CoA Reductase Purification Procedures

This section describes purification procedures that employ two affinity chromatographic steps in conjunction with other purification techniques described in Method 1. These procedures also yield high specific activity enzyme (Table II).

Method 3[29]

Preparation of Microsomes[30] *and Solubilization of the Enzyme.*[31] Microsomes are obtained from the livers of male Sprague–Dawley rats weighing 200–300 g and fed *ad libitum* a 5% cholestyramine powdered rat chow for 4 days. The animals are sacrificed at the middle of the 12-hr dark period. Each rat liver is placed in 25 ml of ice-cold buffer A (40 mM potassium phosphate, 50 mM KCl, 30 mM potassium EDTA, and 0.1 M sucrose, pH 7.2). Homogenization of the liver is effected by six passes with a motor-driven, tight-fitting glass–Teflon Potter–Elvehjem homogenizer. The homogenate is then subjected to two successive 15-min centrifugations at 10,000 g at 4°, and the postmitochondrial supernatant solution is centrifuged at 100,000 g for 60 min. The microsomal pellet obtained is resuspended in buffer A, and then centrifuged at 100,000 g for 45 min. The resulting pellet is resuspended in buffer A to a concentration of about 82 mg/ml, and solid dithiothreitol is added to a final concentration of 10 mM. Homogenization of this suspension is carried out with a hand-driven, all-glass Potter–Elvehjem homogenizer (Kontes; clearance 0.004–0.006 inch), and the suspension is divided into 3-ml aliquots in glass tubes and

[29] P. A. Edwards, D. Lemongello, and A. M. Fogelman, *Biochim. Biophys. Acta* **574,** 123 (1979).
[30] P. A. Edwards and R. G. Gould, *J. Biol. Chem.* **247,** 1520 (1972).
[31] P. A. Edwards, D. Lemongello, and A. M. Fogelman, *J. Lipid Res.* **20,** 40 (1979).

TABLE II
PURIFICATION OF HMG-CoA REDUCTASE FROM RAT LIVER BY METHODS 3, 4, AND 5

Step[a]	Total protein (mg)	Total activity (units)[b]	Specific activity (units/mg protein)	Yield (%)	Purification (fold)
Microsomal suspension	11,554[c]	79,721	6.9	100	1
	5,179[d]	28,608	6.0	100	1
	3,202[e]	12,807	4.0	100	1
Soluble extract	932	71,421	76	90	11
	—	—	—	—	—
	291	10,501	36	82	9
35–50% (NH$_4$)$_2$SO$_4$	307	70,650	230	89	33
30–50% (NH$_4$)$_2$SO$_4$	330	21,822	66	76	12
35–50% (NH$_4$)$_2$SO$_4$	79	8,581	112	67	30
Heat treatment	54	66,786	1,239	84	180
	84	21,678	258	76	46
	20	10,246	500	80	127
Blue Sepharose	2.93	30,364	10,311	38	1494
Affi-Gel Blue	3.95	12,680	3,220	44	576
	2.13	6,275	2,897	49	724
Agarose-CoA	0.790	13,425	17,405	17	2522
HMG-CoA agarose	0.164	6,322	38,400	22	6860
	0.193	1,129	5,842	9	1461

[a] For each purification step, values are given in the order of Methods 3, 4, and 5.
[b] The data in this table are normalized to a rat kill of 25.
[c] Data derived from studies of Edwards et al.[31] (Method 3) in which livers from 7 rats were used.
[d] Data derived from studies of Ness et al.[33] (Method 4) in which 14 rats were used.
[e] Data derived from unpublished results[35] (Method 5) of studies in which 106 rats were used.

frozen at a rate of 6–8° per min. The microsomes can be stored at −20° for up to 2 months.

Solubilization of the reductase is initiated by thawing the microsomal suspension at room temperature or at 37° and then adding an equal volume of 50% glycerol in buffer B (buffer A plus 10 mM dithiothreitol) preheated to 37°. After homogenization of the suspension with ten downward passes of a hand-driven, all-glass Potter–Elvehjem homogenizer, the suspension is incubated at 37° for 60 min. A threefold dilution with buffer B warmed to 37° is then made to yield a final glycerol concentration of 8.3%. The microsomal suspension is again homogenized with ten downward passes of the homogenizer pestle. After centrifugation at 100,000 g for 60 min at 25°, the supernatant solution is removed and used immediately for enzyme purification. All subsequent steps of enzyme purification are performed at room temperature.

Preparation of Enzyme for Affinity Chromatography. The protein precipitating between 35 and 50% saturation with ammonium sulfate is dissolved in buffer B containing 30% glycerol (v/v) and 1.0 M KCl. Four-milliliter aliquots of this solution, at a protein concentration of 3.6–4.8 mg/ml, are heated at 65° for 8 min and then centrifuged at 100,000 g for 30 min. The supernatant solution is diluted 1:1 with buffer B, and the enzyme is precipitated with ammonium sulfate (0 to 50% saturation). The solution is centrifuged and the pellet is dissolved in a small volume of buffer B and stored overnight at room temperature under nitrogen.

Affinity Chromatography. Two types of affinity gel are used to purify HMG-CoA reductase. The first type, Blue Sepharose, contains the chromophore of Blue Dextran, Cibracron Blue F3GA, a sulfonated polyaromatic blue dye. It interacts with HMG-CoA reductase via the binding sites for both NADPH and (S)-HMG-CoA.[29] The dye also interacts with proteins that possess a supersecondary structure called the dinucleotide fold.[32]

The enzyme suspension is incubated at 37° for 30 min and then centrifuged at 100,000 g for 30 min. Enzyme present in the supernatant solution, which contains less than 19.0 mg of protein, is applied to a Blue Sepharose affinity column, 5.5 × 1.0 cm. The gel is then eluted sequentially with two column volumes of buffer C (buffer A plus 2.5 mM dithiothreitol), which also contains the following sets of components, listed in the order of application: (a) 4 mM NAD$^+$, 0.2 M KCl, (b) 4 mM NADH, 0.2 M KCl; (c) 4 mM NADPH, 0.2 M KCl; (d) 2 mM NADPH, 0.2 M KCl, 0.75 mM CoA. The reductase is then eluted with 10 column volumes of buffer C plus 0.5 M KCl. The enzyme is concentrated and diluted to an ionic strength of 0.09 before application to an agarose-CoA column (0.7 × 1.3 cm). The gel is washed with 10 column volumes of buffer C, diluted 1:1 with a solution containing 0.1 M sucrose and 2.5 mM dithiothreitol, and then with 6 volumes of buffer C. The reductase is eluted with 3 column volumes of buffer C containing an additional 0.4 M KCl.

Protein Determination. Soluble protein is determined by the method of Lowry *et al.*[26] and by a modification of the Bradford procedure with Coomassie Brilliant Blue G-250.[31]

Enzyme Purity. The enzyme, specific activity of 17,500 units per milligram of protein, migrates as a single band of protein (15 μg) on SDS and nondenaturing 5% polyacrylamide disc gel electrophoresis, and enzyme activity comigrates with the single band of protein. As determined by SDS–disc gel electrophoresis, the enzyme has a subunit molecular weight of 52,000. The high specific activity obtained for the Method 3 preparation

---

[32] S. T. Thompson, K. H. Cass, and E. Stellwagen, *Proc. Natl. Acad. Sci. U.S.A.* **72,** 669 (1975).

is similar to that for the enzyme demonstrated to be homogeneous by Method 1. However, when antisera were raised to four preparations of enzyme varying in specific activity from 15,000 to 22,500, Ouchterlony double diffusion patterns against impure enzyme preparations yielded three precipitin lines.

Method 4[33]

Preparation of Microsomes and Solubilization of Enzyme. Male Sprague–Dawley rats, weighing 125–150 g at time of purchase, are fed a 2% cholestyramine powdered Wayne Lab Blox diet for 6–8 days prior to sacrifice. The animals are then decapitated at 3 hr into the 12-hr dark period, and the livers are placed in cold PESK buffer (40 mM potassium phosphate, 30 mM EDTA, 0.1 M sucrose, 50 mM KCl, and 1 mM dithiothreitol, pH 7.2). The livers are minced with scissors and then homogenized with 4 volumes of cold PESK buffer with a motor-driven Potter–Elvehjem glass–Teflon homogenizer. The homogenate is centrifuged at 10,500 g for 15 min, and the postmitochondrial supernatant solution is centrifuged at 99,500 g for 60 min. The microsomal pellets are frozen and stored at −20° for up to 10 days before solubilization of the enzyme.

The solubilization of the enzyme is based on the method of Heller and Gould.[18] The frozen microsomal pellets are thawed at room temperature and then homogenized in cold PESK buffer (12 ml per each microsomal pellet). Each pellet is derived from 7 g of liver. The homogenate is centrifuged at 100,000 g for 60 min at 4°, and the supernatant solution containing the soluble enzyme activity is extracted with a Pasteur pipette and saved. The microsomal pellet is frozen, thawed, and extracted four additional times as stated above. The five extracts are combined and used for further enzyme purification.

Preparation of Enzyme for Affinity Chromatography. The extract is incubated at 37° for 90 min and then centrifuged at 20,000 g for 10 min. The supernatant solution is fractionated between 30 and 50% saturation with ammonium sulfate, and the precipitate is removed by centrifugation. The sedimented pellet is dissolved in PESK buffer and frozen at −20°. The concentrated extract is then thawed at 37°, and glycerol and potassium chloride are added to the extract to final concentrations of 33% and 1 M, respectively. This enzyme solution is added to prewarmed tubes and heated at 64° for 10 min. The turbid solution is rapidly cooled to room temperature by placement in an ice-water bath and then centrifuged at

[33] G. C. Ness, C. D. Spindler, and M. H. Moffler, *Arch. Biochem. Biophys.* **197,** 493 (1979).

144,000 g for 15 min. The supernatant solution containing the enzyme activity is saved for subsequent affinity chromatography.

Affinity Chromatography. The first affinity chromatographic step utilizes an Affi-Gel Blue column (1.5 × 4.0 cm) containing the ligand Cibacron Blue F3GA. The column is equilibrated at room temperature with PESK buffer containing 0.2 M KCl, and the enzyme solution is diluted fivefold with PESK buffer to reduce the concentration of KCl to 0.2 M. After approximately 4800 units of heat-treated enzyme in a volume of 300 ml is applied, the column is washed with 100 ml of PESK buffer containing 0.2 M KCl. The enzyme is then eluted in PESK buffer with a linear gradient of 0.2 to 2.0 M KCl. Fractions of 7.5 ml are collected, and those containing more than 100 units of activity are saved and concentrated with an Amicon ultrafiltration cell containing a YM-10 filter. Repeated ultrafiltration with additions of PESK buffer removes the KCl. Glycerol is then added to a final concentration of 50%, and the enzyme is stored at −70°.

The final step of purification of the enzyme requires a thioester-linked agarose–hexane–HMG-CoA column (0.6 × 2.0 cm). The enzyme is thawed and water is added to dilute the glycerol concentration to 25%. Thirty-eight milliliters of an enzyme solution containing 2050 units are applied to a column which has been equilibrated at room temperature with buffer A (PESK : glycerol : water mixed in ratios of 1 : 1 : 2). The column is then eluted sequentially with the following solutions, and 4-ml fractions are collected: 40 ml of buffer A containing 70 mM KCl; 10 ml of buffer A containing 70 mM KCl and 200 μM RS-HMG-CoA; and buffer A containing 0.5 M KCl. Most of the HMG-CoA reductase is eluted by the addition of HMG-CoA to buffer A.

Protein Determination. The protein concentrations of soluble fractions are determined by the Coomassie dye-binding method.[34]

Enzyme Purity. Small amounts of protein are analyzed by SDS and nondenaturing polyacrylamide disc gel electrophoresis. Purified enzyme (14 μg) migrates as a single band corresponding to a molecular weight of 50,000 on SDS–7.5% polyacrylamide disc gel electrophoresis. On native 6% polyacrylamide disc gels, electrophoresis of 8 μg of protein yields a single staining band, which comigrates with enzyme activity. The specific activity of purified enzyme varies from 32,000 to 46,000 nmol of NADPH oxidized per minute per milligram of protein. However, when protein determinations were carried out by the Lowry method[26] the specific activity value of 46,800 decreased to 27,000. This value is approximately 40% higher than the value reported for the homogeneous enzyme preparation obtained by Method 1.

[34] M. M. Bradford, *Anal. Biochem.* **72**, 248 (1976).

Method 5[35]

Preparation of Microsomes and Solubilization of the Enzyme. The treatment of animals and the preparation of microsomes are carried out as described for Method 1, but microsomal pellets are stored at −60°.

The solubilization of the reductase is performed by thawing the microsomes at room temperature and then resuspending the pellets at a protein concentration of 70–85 mg/ml. This procedure is similar to that of Method 3, except that the solubilization buffer is at pH 7.0 and the homogenization is carried out with a motor-driven Teflon pestle in a Potter–Elvehjem type glass homogenizer.

Preparation of Enzyme for Affinity Chromatography. The salt fractionation of enzyme is performed with a saturated solution of enzyme grade ammonium sulfate, pH 7.0, as described in Method 1. The protein pellet derived from this step is dissolved in buffer containing 50 mM potassium phosphate, pH 7.0, 30 mM EDTA, 1.0 M KCl, 30% (v/v) glycerol, and 10 mM dithiothreitol, and the heat treatment is carried out at 65° for 8 min as described for Method 1.

Affinity Chromatography. Purification of the reductase with Affi-Gel Blue and by HMG-CoA affinity chromatography is accomplished by a minor modification of the procedure of Ness *et al.*[33] Heat-treated enzyme is diluted 1:1 with PESK buffer and precipitated between 0 and 50% saturation with ammonium sulfate as described in Method 1. The pellet is dissolved in a minimum amount of buffer containing 50 mM potassium phosphate, pH 7.0, 2 mM dithiothreitol, 30 mM EDTA, and 10% (v/v) glycerol, and dialyzed for 1 hr. After dialysis, solid dithiothreitol is added to the enzyme solution to give a final concentration of 5 mM. The Affi-Gel Blue column (1.2 × 5.0 cm) is then equilibrated with buffer A [50 mM potassium phosphate, pH 7.0, 30 mM EDTA, 10% glycerol (v/v), and 5 mM dithiothreitol]. Enzyme (approximately 22,000 units) is applied to the column at a flow rate of 1 ml per 2 min. Protein on the column is eluted sequentially with buffer A and buffer A plus 0.5 M KCl. Enzyme is eluted at a flow rate of 1 ml/min with buffer A containing 2.0 M KCl, and eluates are collected in 2-ml fractions. Those tubes with more than 300 units of enzyme activity are pooled and stored overnight at room temperature, when necessary.

The thioester-linked agarose–hexane–HMG-CoA column (0.6 × 1.8 cm) is equilibrated with buffer B [50 mM potassium phosphate, pH 7.0, 2 mM dithiothreitol, 30 mM EDTA, and 15% (v/v) glycerol]. Enzyme is dialyzed against the above buffer, and 10,543 units of activity are applied to the affinity column at a flow rate of 1 ml in 8–10 min. The column is

[35] T. A. Baker, R. E. Dugan, and J. W. Porter, unpublished results.

sequentially washed with buffer B, buffer B containing 70 mM KCl, buffer B containing 70 mM KCl and 0.2 mM DL-HMG-CoA, and buffer B containing 1.0 M KCl. Enzyme is eluted at the fastest flow rate possible with buffer B containing 70 mM KCl and 0.2 mM DL-HMG-CoA, and eluates are collected in 1-ml fractions. The fractions containing enzyme activity are pooled, and KCl is added to a final concentration of 2.0 M before assaying for total enzyme activity. Glycerol is added to a final concentration of 50% before storage at $-60°$.

Protein Determination. Protein concentrations of soluble enzyme preparations are determined by the method of Lowry *et al.*[26]

Enzyme Purity. The purified enzyme has a specific activity of approximately 5800 nmol of NADPH oxidized per minute per milligram of protein. However, 40 μg of the enzyme migrates as one major band corresponding to a molecular weight of 50,000 on SDS–10% polyacrylamide disc gel electrophoresis.

Comparison of Methods

A comparison of the five methods of purification of HMG-CoA reductase from microsomes shows that each includes five major fractionation steps and that there are many similarities in these methods. All methods incorporate a freeze-thaw treatment of microsomes to solubilize enzyme. All the methods also include an ammonium sulfate fractionation, a heat treatment step, and at least one specific affinity chromatographic separation.

The major differences among these methods are the following. Methods 1 and 2 use a freeze–thaw treatment of microsomes to solubilize enzyme, whereas Method 4 uses a repeated freeze–thaw extraction procedure. Methods 3 and 5 solubilize enzyme by incubation with glycerol after a freeze-thaw treatment of microsomes. Methods 3, 4, and 5 substitute the affinity gels Affi-Gel Blue or Blue Sepharose in place of the BioGel filtration or sucrose density gradient ultracentrifugation step presented in Methods 1 and 2, respectively. Methods 4 and 5 use an agarose–hexane–HMG-CoA affinity gel as the final step of purification, whereas Methods 1, 2, and 3 use an agarose–hexane–CoA gel for the affinity chromatographic separation.

Methods 3 and 4 show a recovery of enzyme units in the 17–22% range, which is greater than the 9% yield for Method 5 or the 3–4% yield for Methods 1 and 2. The increased yield of enzyme obtained in Methods 3, 4, and 5 results from an increase in the amount of enzyme solubilized from the microsomes. In Methods 3, 4, and 5, yields of 90, 76, and 82%, respectively, are observed, as compared to 31% for Methods 1 and 2.

Additional modest gains are also observed in the heat treatment or ammonium sulfate fractionation steps for Methods 3, 4, and 5.

The amount of protein recovered for Methods 1 and 2 is 53 μg, whereas Methods 3, 4, and 5 yield 790, 164, and 193 μg, respectively. These figures are based on a normalized rat kill of 25, and they appear to correlate with the final enzyme activity yield, except for Method 3 where much larger amounts of protein are obtained. The increase in protein by this method appears to be due to a three- to fivefold increase in the amount of starting enzyme units over that in other methods. This high microsomal activity may be a result of the treatment of the animals (i.e., 5% instead of 2% cholestyramine in the diet) or to the preincubation of microsomes before solubilization of the enzyme. The degree of purification for enzyme derived from microsomes were 4100, 3700, 6900, 2500, and 1500, respectively, for Methods 1–5.

Methods 1 and 2 produce enzyme of high specific activity (17,000 to 19,000 units per milligram of protein). Enzyme obtained by these methods has undergone a number of stringent physicochemical tests for homogeneity. By all the criteria used (including electrophoresis of a range of protein concentrations on SDS and native polyacrylamide disc gels and immunochemical experiments with several levels of protein) the enzyme is judged to be homogeneous. Small amounts of enzyme from Method 3 migrate as a single band on SDS and nondenaturing polyacrylamide disc gel electrophoresis. Although the specific activity is similar to enzyme purified by Method 1, Method 3 produces enzyme that is immunochemically impure, and thus it would appear not to be homogeneous. Method 4 yields enzyme of a specific activity greater than other reported values. Small amounts of this protein act as a homogeneous species on SDS and native polyacrylamide disc gel electrophoresis. Method 5 produces enzyme of lower specific activity than that obtained by Methods 1 through 4. However, large amounts of this enzyme migrate as a single band of protein that corresponds with the subunit molecular weight of HMG-CoA reductase when subjected to SDS–polyacrylamide disc gel electrophoresis. In addition, rocket immunoelectrophoresis of a crude HMG-CoA reductase preparation into antibody that was produced in response to the purified enzyme showed one major and one trace precipitin arc.

Acknowledgments

This work was supported in part by a Research Grant, HL 16364, from the National Heart and Lung Institute, National Institutes of Health, United States Public Health Service, and by the Medical Research Service of the Veterans Administration.

[56] S-3-Hydroxy-3-Methylglutaryl-CoA Reductase from Pseudomonas

By Victor W. Rodwell and William R. Bensch

$$\text{HMG-CoA} + 2\,\text{NADH} + 2\,\text{H}^+ \rightarrow \text{mevalonate} + \text{CoA} + 2\,\text{NAD}^+ \qquad (1)$$
$$\text{Mevalonate} + \text{CoA} + 2\,\text{NAD}^+ \rightarrow \text{HMG-CoA} + 2\,\text{NADH} + 2\,\text{H}^+ \qquad (-1)$$
$$\text{Mevaldate} + \text{NADH} + \text{H}^+ \rightarrow \text{mevalonate} + \text{NAD}^+ \qquad (1/2)$$
$$\text{Mevaldate} + \text{CoA} + \text{NAD}^+ \rightarrow \text{HMG-CoA} + \text{NADH} + \text{H}^+ \qquad (-1/2)$$

The mevalonate-inducible, NADH-dependent HMG-CoA[1] reductase of *Pseudomonas* catalyzes all of the above reactions. For specialized growth conditions (mevalonate as sole source of carbon), HMG-CoA reductase thus functions as a biodegradative enzyme (Reaction −1). In all other instances, the reaction (1) is of interest . Reactions are therefore numbered so as to emphasize both the biosynthetic role of HMG-CoA reductase and analogies to HMG-CoA reductases from other sources. These reactions are: reductive deacylation of HMG-CoA (1), the reverse reaction, oxidative acylation of mevalonate (−1), and the two half-reactions, reduction of mevaldate to mevalonate ($\frac{1}{2}$) and reductive acylation of mevaldate to HMG-CoA ($-\frac{1}{2}$). If not otherwise stated, all information is from Bensch and Rodwell.[2]

Assay Methods

Principle. All four reactions may be followed kinetically by observing the change in absorbancy at 340 nm consequent to the oxidation–reduction of NADH/NAD$^+$.

Bacteria. Pseudomonas sp. M1 is not in ATCC or presently available. The organism may be isolated from soil by aerobic elective culture on R,S-mevalonate as the sole source of carbon, essentially as described for the isolation of actinomycete S4[2,3] but employing a shaken rather than stationary elective culture.

Reagents

Tris-HCl, 200 mM, pH 7.1
CoA, 5.0 mM, pH 7.1
R,S-Mevalonate, 30 mM pH 7.1

---

[1] The abbreviation used is HMG-CoA, S-3-hydroxy-3-methylglutaryl-coenzyme A.
[2] W. R. Bensch and V. W. Rodwell, *J. Biol. Chem.* **245**, 3755 (1970).
[3] M. A. Siddiqi and V. W. Rodwell, *J. Bacteriol.* **93**, 207 (1967).

R,S-HMG-CoA, 10 m*M* pH 7.1

R,S-Mevaldate, 30 m*M*, pH 7.1, prepared just prior to use from the dibenzylethylenediammonium salt of *R,S*-mevaldate acetal by treatment with NH$_4$OH at 4°. The liberated base is removed by ether extraction. The aqueous phase is adjusted to 0.5 *N* in HCl and incubated 15 min at room temperature to cleave the acetal. Liberated methanol and residual ether are then removed under a nitrogen stream, and the pH of the solution is adjusted to 7.1 with KOH and Na$_2$CO$_3$.[4,5]

Procedures

Spectrophotometric Assays of Enzymatic Activities. The change in absorbance at 340 nm due to NADH formation or utilization is measured at 30° in a double-beam spectrophotometer equipped with a constant temperature cell compartment and a scale expander. Originally, Reaction (−1) was employed to assay fractions during purification. It may, however, be preferable to follow Reaction (1). Assays are conveniently conducted in 700 μl of 100 m*M* Tris-HCl, pH 7.1. Enzyme is added last to initiate the reaction. Standard assay conditions are as follows: Reaction (1): 0.57 m*M R,S*-HMG-CoA, 0.3 m*M* NADH; Reaction (−1): 3.0 m*M R,S*-mevalonate, 3.0 m*M* NAD$^+$, 0.6 m*M* CoA; Reaction ($\frac{1}{2}$): 3.0 m*M R,S*-mevaldate, 0.25 m*M* NADH; Reaction (−$\frac{1}{2}$): 5.7 m*M R,S*-mevaldate, 3.0 m*M* NAD$^+$, 0.6 m*M* CoA.

Other Assays of Enzymatic Activity. Reaction (1) may be assayed also by measuring incorporation of isotope from (R,S)-[3-^{14}C]HMG-CoA into mevalonate and Reaction (−1) by the incorporation of isotope from (R,S)[2-^{14}C]mevalonate into HMG-CoA.[6] HMG-CoA lyase is assayed as described by Bachhawat *et al.*[7] and HMG-CoA hydrolase as described by Dekker *et al.*[8]

Protein. Protein may be determined by the spectrophotometric method of Waddell[9] and Murphy and Kies[10] employing 0.9% NaCl as diluent and serum albumin as standard.

Disc Gel Electrophoresis. Samples (100–200 μg of protein) in 40% su-

[4] F. Lynen and M. Grassl, *Hoppe-Seyler's Z. Physiol. Chem.* **313,** 291 (1958).

[5] M. A. Siddiqi, Ph.D. thesis, p. 13, Univ. of California, San Francisco Medical Center, 1962.

[6] D. J. Shapiro, J. L. Nordstrom, J. J. Mitschelen, V. W. Rodwell, and R. T. Schimke, *Biochim. Biophys. Acta* **370,** 369 (1975).

[7] B. K. Bachhawat, W. G. Robinson, and M. J. Coon, *J. Biol. Chem.* **216,** 727 (1955).

[8] E. E. Dekker, M. J. Schlesinger, and M. J. Coon, *J. Biol. Chem.* **233,** 434 (1958).

[9] W. J. Waddell, *J. Lab. Clin. Med.* **48,** 311 (1956).

[10] J. B. Murphy and M. W. Kies, *Biochim. Biophys. Acta* **45,** 382 (1960).

crose, 50 mM in Tris-HCl, pH 7.1, are applied to 10% acrylamide–0.03% N,N'-methylenebisacrylamide gels[11] at pH 9.5. Electrophoresis is for 15 min at 1 mA per tube, then at 3 mA per tube until the tracking dye is 1 cm from the bottom of the gel. Gels are stained for protein using Coomassie blue dye.[12] To detect HMG-CoA reductase activity, gels in aluminum foil-covered tubes are incubated with a minimum volume of staining mixture containing 3.5 mM R,S-mevalonate, 3.5 mM NAD$^+$, 1.0 mM nitro blue tetrazolium, 0.34 mM phenazine methosulfate, and 100 mM Tris-HCl, pH 7.1, until purple bands become visible (5–15 min). When 0.72 mM CoA is included, staining mixtures rapidly turn brown; omission of CoA does not adversely affect visualization.

Units. One enzyme unit is the amount catalyzing turnover of 1 μmol of substrate (mevalonate, mevaldate, or HMG-CoA) per minute at 30°.

Growth of Cells. Cells grown on mevalonate contain 2000-fold higher levels of HMG-CoA reductase than cells grown on neutral carbon sources, such as malate. Cells grown on leucine, which is degraded to HMG-CoA, are induced only sevenfold above basal levels. Cells are therefore grown on mevalonate. Since specific activity varies little with phase of growth, cells grown aerobically at 30° at pH 7.0 in ionic medium[3,13] 12 mM in ammonium R,S-mevalonate are harvested in stationary phase.

Enzyme Purification

All manipulations are at 0–5°. Centrifugations are at 12,000 g for 30 min unless otherwise stated. Cells in stationary phase are collected by centrifugation and washed twice by resuspension in a mixture of 60 mM K$_x$PO$_4$, pH 7.0, and 20 mM Na$_x$PO$_4$, pH 7.0. Frozen cells retain full HMG-CoA reductase activity for several months.

A 7.5% (w/v) suspension of fresh or frozen cells in 1.2 mM EDTA–50 mM Tris-HCl, pH 7.1, is disrupted by sonic oscillation. Alternatively, packed frozen cells may be ground with alumina in a chilled mortar and buffer added after lysis occurs. The broken-cell suspension is centrifuged to sediment unbroken cells and debris, which are discarded. The turbid supernatant liquid (crude extract) is decanted, centrifuged (60,000 g, 2 hr), and the precipitate is again discarded. Saturated (NH$_4$)$_2$SO$_4$ solution (adjusted to pH 8.0 when diluted 1 : 20) is added to the clear, amber supernatant liquid to 25% saturation (33.3 ml/100 ml of 60,000 g supernatant

[11] S. Hjertén, S. Jerstedt, and A. Tiselius, *Anal. Biochem.* **11**, 219 (1965).
[12] A. Chrambach, R. A. Reisfeld, M. Wyckoff, and J. Zuccari, *Anal. Biochem.* **20**, 150 (1967).
[13] G. M. Fimognari and V. W. Rodwell, *Biochemistry* **4**, 2086 (1965).

liquid). The precipitate is removed by centrifugation and discarded. The supernatant liquid is adjusted to 35% saturation by adding 15.4 ml of saturated $(NH_4)_2SO_4$ solution per 100 ml of 25%-saturated supernatant solution. The 25–35% $(NH_4)_2SO_4$ precipitate, which may be stored frozen for several weeks without loss of activity, is dissolved in 10 μM dithiothreitol–10 mM K_xPO_4, pH 6.8, dialyzed overnight against this buffer in tubing previously boiled in 1 mM EDTA, and applied to a column (2.5 × 6.0 cm) of hydroxyapatite equilibrated with 10 μM dithiothreitol–10 mM K_xPO_4, pH 6.8. The column is washed with one column volume (30 ml) of the above buffer, then eluted, successively, with 60-ml portions of 10 μM dithiothreitol in 20, 28, and 400 mM K_xPO_4, pH 6.8. Fractions, 3 ml, are collected. Activity emerges in high specific activity (HT fractions 42–48) and lower specific activity fractions (HT fractions 49–70) that contain about 15% and 75% of the applied activity, respectively. The HT fractions 49–70 may be combined, dialyzed, and refractionated on hydroxyapatite. Gradient elution gives an improved yield but lower final specific activity. High specific activity fractions are pooled, made 2.5 mM in EDTA, and brought to 40% saturation in $(NH_4)_2SO_4$ by adding 66.7 ml of saturated $(NH_4)_2SO_4$ solution, pH 8.0, per 100 ml. The precipitate is dissolved in 2.5 mM EDTA–10 μM dithiothreitol–100 mM Tris-HCl, pH 7.1, and dialyzed against this buffer. The purified enzyme loses activity rapidly in the absence of EDTA or dithiothreitol. Table I summarizes a typical purification.

Properties

Homogeneity and Molecular Weight. The high activity fractions 42–48 are over 95% homogeneous as judged by polyacrylamide gel electrophoresis at pH 9.5. One major and two minor bands are observed when

TABLE I
SUMMARY OF TYPICAL PURIFICATION OF HMG-CoA REDUCTASE FROM *Pseudomonas*

Fraction	Volume (ml)	Total activity (μmol/min)	Total protein (mg)	Specific activity (μmol/mg min^{-1})	Yield (%)
Crude extract	12.8	1070	575	1.86	(100)
Dialyzed $(NH_4)_2SO_4$ fraction	2.0	570	34.8	16.4	53
HT fractions 49–70	61.1	436	23.0	18.9	40
HT fractions 42–48	32.9	77	1.98	38.9	7.2

[a] From packed cells equivalent to 760 mg, dry weight. Assays are for Reaction (-1).

gels are stained for protein. A single band, which coincides exactly with the major protein band, is observed on staining for reductase activity. Using the method of Hedrick and Smith[14] for evaluating molecular weight on polyacrylamide gels of varying porosity, *Pseudomonas* HMG-CoA reductase has a molecular weight of $(270 \pm 10) \times 10^3$.

Stoichiometry of HMG-CoA Reduction. Two moles of NADH are oxidized per mole of HMG-CoA consumed, and 1 mol each of mevalonate and of CoA are formed. The reaction thus has the stoichiometry shown in Reactions (1) and (−1).

Activities of Purified Enzyme. The purified enzyme catalyzes all four reactions, and during purification, parallel enrichment occurs for all four reactions. This strongly suggests that all are catalyzed by single protein.

Reversibility of HMG-CoA Reductase Reaction. While mammalian liver[15] and yeast[16] HMG-CoA reductases are reported to catalyze reductive deacylation of HMG-CoA to mevalonate essentially irreversibly, *Pseudomonas* and *Mycobacterium* HMG-CoA reductases readily catalyze Reaction (−1).[13] The purified enzyme is free of detectable HMG-CoA lyase or NADH oxidase activity, present in the crude extract and dialyzed $(NH_4)_2SO_4$ fraction. The ability of the purified enzyme to catalyze Reaction (−1) thus is not attributable to coupling to an energetically favored reaction. Synthesis of mevalonate from HMG-CoA by the purified enzyme was demonstrated by isolation of labeled mevalonate formed from [3-^{14}C]HMG-CoA and NADH.

Equilibrium Constant. Using the spectral assay, the time course for Reaction (1), unlike that for Reaction (−1), remains linear for protracted times. For

$$K = \frac{[\text{Mevalonate}][\text{NAD}^+]^2[\text{CoA}]}{[\text{HMG-CoA}][\text{NADH}]^2[\text{H}^+]^2}$$

K approximated 10^{13} to 10^{15} M^{-1}. Reaction (1) thus strongly favors mevalonate synthesis.

K_m Values. Table II lists K_m values for cosubstrates for all reactions catalyzed by the purified enzyme; HMG is not a substrate.

pH Optimum. Optimum activity for Reaction (−1) occurs at pH 9.2–9.6. An abrupt decline in activity at alkaline pH is in part attributable to loss of the proton of the SH group of CoA ($pK = 9.6$).[17] At higher CoA

[14] J. L. Hedrick and A. J. Smith, *Arch. Biochem. Biophys.* **126,** 155 (1968).

[15] F. Lynen, *Biosyn. Terpenes Sterols, Ciba Found. Symp. 1958,* p. 95 (1959).

[16] I. F. Durr and H. Rudney, *J. Biol. Chem.* **235,** 2572 (1960).

[17] L. Jaenicke and F. Lynen, *in* "The Enzymes" (P. D. Boyer, H. Lardy, and K. Myrbäck, eds.), 2nd ed, Vol. 3, p. 3. Academic Press, New York, 1959.

TABLE II
K_m VALUES[a]

	K_m (μM)					
	Reductant			Oxidant		
Reaction	Mevalonate	Mevaldate	NADH	HMG-CoA	NAD$^+$	CoA
1	—	—	32 ± 4	50 ± 7	—	—
−1	300 ± 30	—	—	—	270 ± 50	39 ± 9
$\frac{1}{2}$	—	9000 ± 3000[b]	360 ± 80	—	—	—
$-\frac{1}{2}$	—	610 ± 80	—	—	60 ± 10	46 ± 3

[a] Tabulated values are ± SEM.
[b] Line passes too close to origin to permit more accurate estimate of K_m.

concentrations, optimal activity occurs at pH 9.6, and considerable activity is present even at pH 10.5.

Inhibition by HMG. Reaction (−1) is 100-fold more sensitive to inhibition by HMG than is Reaction (1) (Table III). Although groups on the CoA moiety of HMG-CoA may be required for HMG-CoA binding, no enhancement of inhibition of reductive deacylation of HMG-CoA by HMG is observed when free CoA is added. Reactions ($\frac{1}{2}$) and ($-\frac{1}{2}$) both are strongly inhibited by HMG with $K_i : K_m$ ratios of the same order of magnitude as that for oxidative acylation of mevalonate.

For Reaction (−1), the carboxyl, the 3-methyl and the 3-hydroxy groups all are essential for substrate or inhibitor binding to the enzyme. Fimognari and Rodwell[13] suggested that these groups participated in binding of HMG-CoA and that 3-hydroxy-3-methylcarboxylic acids might also

TABLE III
COMPETITIVE INHIBITION BY HMG[a]

Reaction	K_i (μM)	K_m (μM)	K_i/K_m
1	3930	50	79
−1	220	300	0.73
$\frac{1}{2}$	2280	9000	0.25
$-\frac{1}{2}$	1320	610	2.2

[a] Velocities measured ± 2.86 mM sodium R,S-HMG under standard assay conditions except for concentrations of mevalonate (0.11 to 2.86 mM), mevaldate (0.57 to 5.7 mM), or HMG-CoA (57 to 86 mM). Inhibition is competitive with the above substrates in all cases. K_i values are calculated from double reciprocal plots of the data obtained. K_m data are from Table II.

competitively inhibit Reaction (1). While HMG is indeed a competitive inhibitor of Reaction (1), K_i (3.93 mM) is 18-fold above K_i for mevalonate oxidation (0.22 mM). More significantly, the $K_i : K_m$ ratio of 79 for Reaction (1) is over 100-fold higher than that for Reaction ($-$1), 0.73. Thus, HMG and similar acids, while effective inhibitors of Reaction ($-$1), would inhibit Reaction (1) only at low concentrations of HMG-CoA.

Mechanism. The stereo specificity of hydride transfer is not established. By analogy with yeast[18] and rat liver[19] HMG-CoA reductases, this may involve hydride transfer from the A side of NADH.

While the purified enzyme catalyzes Reactions ($\frac{1}{2}$) and ($-\frac{1}{2}$), it is not a general aldehyde or alcohol dehydrogenase (glyceraldehyde is neither oxidized nor reduced), and the failure to catalyze either oxidative acylation or reduction of glutarate semialdehyde (5-oxopentanoate) suggests a substrate specificity for mevaldate (3-hydroxy-3-methyl-5-oxopentanoate) similar to that for mevalonate.

While enzyme-bound mevaldate is a likely intermediate in all four reactions, formation of free mevaldate could not be detected during catalysis. Isotope from [^{14}C]mevalonate is not incorporated into a mevaldate trapping pool during Reaction ($-$1). Analogous observations have been made for Reaction (1) catalyzed by yeast HMG-CoA reductase.[16]

For discussion of probable enzyme-bound intermediates and comparative mechanistic aspects of HMG-CoA reductases, see Brown and Rodwell.[20]

[18] R. E. Dugan and J. W. Porter, *J. Biol. Chem.* **246**, 5361 (1971). See also this volume [29].
[19] A. S. Beedle, K. A. Munday, and D. C. Wilton, *Eur. J. Biochem.* **28**, 151 (1972).
[20] W. E. Brown and V. W. Rodwell, *in* "Dehydrogenases Requiring Nicotinamide Coenzymes" (J. Jeffery, ed.), pp. 232–272. Birkhauser, Basel, 1980.

[57] Assay of Enzymes That Modulate S-3-Hydroxy-3-Methylglutaryl-CoA Reductase by Reversible Phosphorylation

EC 1.1.1.34 Mevalonate : NADP$^+$ oxidoreductase (CoA-acylating)

By Thomas S. Ingebritsen and David M. Gibson

The catalytic efficiency of liver HMG-CoA reductase [hydroxymethylglutaryl-CoA reductase (NADPH), EC. 1.1.1.34] is modulated by reversible phosphorylation.[1-11] Reductase is converted from an active dephospho form (reductase a) to an inactive phospho form (reductase b) through the action of reductase kinase in the presence of Mg^{2+} and ATP. Reductase b is converted back to reductase a by incubation with reductase phosphatase. This latter enzyme appears to be identical to a broad specificity protein phosphatase[1-4] (originally identified as phosphorylase phosphatase), which also modulates the state of phosphorylation of key regulatory enzymes in glycogen metabolism (phosphorylase a, phosphorylase kinase a, and glycogen synthase b) and fatty acid synthesis (acetyl-CoA carboxylase).[12,13]

The activity of reductase kinase is also controlled by reversible phosphorylation.[1-5] This kinase is converted from an inactive dephospho form (reductase kinase b) to an active phospho form (reductase kinase a) by a second protein kinase (reductase kinase kinase) in the presence of MgATP. Like reductase, dephosphorylation of reductase kinase is catalyzed by the broad specificity protein phosphatase.[1-5] (This pattern of change has been confirmed by Beg et al.[11])

The bulk of reductase kinase and reductase kinase kinase activities is concentrated in the liver cytosol. Both kinases are distinct from cAMP-dependent protein kinase.[4]

Thus liver HMG-CoA reductase is controlled by the bicyclic system shown in Fig. 1. Methods for assaying the reductase modulating enzymes are described; a description of the enzyme preparations ordinarily employed follows.

[1] T. S. Ingebritsen, H.-S. Lee, R. A. Parker, and D. M. Gibson, *Biochem. Biophys. Res. Commun.* **81**, 1268 (1978).

[2] D. M. Gibson and T. S. Ingebritsen, *Life Sci.* **23**, 2649 (1978).

[3] T. S. Ingebritsen and D. M. Gibson, in "Molecular Aspects of Cellular Regulation" (P. Cohen, ed.), Vol. I, p. 63. Elsevier–North Holland Biomed. Press, Amsterdam, 1980.

[4] T. S. Ingebritsen, R. A. Parker, and D. M. Gibson, *J. Biol. Chem.* in press (1981).

[5] T. S. Ingebritsen, M. J. H. Geelen, R. A. Parker, K. J. Evenson, and D. M. Gibson, *J. Biol. Chem.* **254**, 9986 (1979).

[6] Z. H. Beg, D. W. Allmann, and D. M. Gibson, *Biochem. Biophys. Res. Commun.* **54**, 1362 (1973).

[7] J. L. Nordstrom, V. W. Rodwell, and J. J. Mitschelen, *J. Biol. Chem.* **252**, 8924 (1977).

[8] M. S. Brown, G. Y. Brunschede, and J. L. Goldstein, *J. Biol. Chem.* **250**, 2502 (1975).

[9] Z. H. Beg, J. A. Stonik, and H. B. Brewer, *Proc. Natl. Acad. Sci. U.S.A.* **75**, 3678 (1978).

[10] M. L. Keith, V. W. Rodwell, D. H. Rogers, and H. Rudney, *Biochem. Biophys. Res. Commun.* **90**, 969 (1979).

[11] Z. H. Beg, J. A. Stonik, and H. B. Brewer, *Proc. Natl. Acad. Sci. U.S.A.* **76**, 4375 (1979).

[12] E. Y. C. Lee, H. Brandt, Z. L. Capulong, and S. D. Killilea, *Adv. Enzyme Regul.* **14**, 467 (1976).

[13] P. Cohen, *Curr. Top. Cell. Regul.* **14**, 117 (1978).

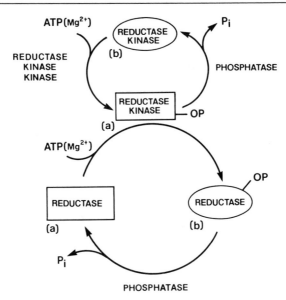

FIG. 1. Bicyclic system that controls liver hydroxymethylglutaryl-CoA reductase.

Assay Methods

HMG-CoA Reductase Assay

Principle. 3-Hydroxy-3-methylglutaryl-CoA [HMG-CoA] reductase is assayed by a modification of the method of Shapiro *et al.* [14] In this method, the enzyme is assayed by following the conversion of [3-^{14}C]HMG-CoA to mevalonic acid. The product is separated from [3-^{14}C]HMG-CoA by thin-layer chromatography.

Reagents

Buffer A: 1 mM EDTA, 250 mM NaCl, 5 mM dithiothreitol, and 50 mM orthophosphate, pH 7.4

Cofactor–substrate cocktail: 4.5 μmol of glucose 6-phosphate (potassium salt), 0.3 IU of glucose-6-phosphate dehydrogenase, 450 nmol of NADP$^+$, 50 nmol of DL-hydroxymethyl-[3-^{14}C]glutaryl-CoA (specific activity 1 dpm/pmol), 20,000 dpm DL-[2-^{3}H]mevalonic acid lactone (1639 dpm/pmol), and 4.4 μmol of EDTA dissolved in buffer A to give 50 μl

HCl, 6 N

Mevalonolactone, 0.5 M

[14] D. J. Shapiro, J. L. Nordstrom, J. J. Mitschelen, V. W. Rodwell, and R. T. Schimke, *Biochem. Biophys. Acta* **370**, 369 (1974).

Liquid scintillator: 40 ml of Permafluor (Packard) concentrated scintillator, 100 ml of methanol, 860 ml of toluene

Thin-layer chromatography plates: Whatman 20 × 20 cm, 19 channel, LK5D plates

Acetone–toluene, 1 : 1

Assay Procedure. Aliquots of the microsomal suspension (200–400 μg
of protein) are diluted to 100 μl with buffer A and preincubated in 1.5-ml
Eppendorf tubes for 5 min at 37°. The assay is started by addition of 50 μl
of the cofactor–substrate cocktail.

The incubation is continued for 15 min and then terminated by the
addition of 10 μl of 6 N HCl. Samples are incubated for an additional 30
min to lactonize the mevalonate. The denatured protein is then removed
by centrifugation, and 100 μl is transferred to a tube containing 10 μl of 0.5
M mevalonolactone. The entire volume of each sample is transferred to a
20 × 20 cm, 19-channel LK5D plate (Whatman) and developed in
acetone–toluene (1 : 1). The mevalonolactone bands are visualized in an
iodine chamber, destained, and scraped into scintillation vials. The [^{3}H]-
plus [^{14}C]mevalonolactone is counted in 10 ml of liquid scintillator. (The
value for [^{3}H]mevalonolactone is an index of its recovery.)

Units. One unit of reductase activity catalyzes the formation of 1 nmol
of mevalonic acid per minute at 37°.

HMG-CoA Reductase Kinase Assay

Principle. HMG-CoA reductase kinase activity is assayed by following the time-dependent inactivation of HMG-CoA reductase in reductase
kinase-deficient microsomes in the presence of MgATP and a suitable
source of reductase kinase.

Reagents

Buffer A: 1 mM EDTA, 250 mM NaCl, 5 mM dithiothreitol, and 50
mM orthophosphate, pH 7.4

Buffer B: 1 mM EDTA, 250 mM NaCl, 50 mM NaF, 5 mM dithiothreitol, and 50 mM orthophosphate, pH 7.4

Buffer C: 1.5 mM EDTA, 375 mM NaCl, 125 mM NaF, 7.5 mM
dithiothreitol, and 75 mM orthophosphate, pH 7.4

MgATP: 40 mM MgCl$_2$ and 20 mM ATP, pH 7.4

Reductase kinase and reductase-deficient microsomes: See later
sections.

Assay Procedure. Twenty microliters of reductase kinase-deficient
microsomes (200–400 mU of reductase activity) suspended in buffer A are
mixed with 20 μl of buffer C and 50 μl of reductase kinase appropriately
diluted in buffer B. The mixture is preincubated for 5 min at 37°, and the
reaction is then started by addition of 10 μl of MgATP. After incubation

for an appropriate period of time (up to 20 min), the reaction is stopped and the reductase assay is started by addition of 50 μl of the cofactor–substrate cocktail (see earlier) to which sufficient EDTA is added to bring the final concentration to 30 mM. The dose of added reductase kinase and the incubation time is adjusted so that 20–50% of the reductase is inactivated during the incubation. A blank assay should be run in the absence of added reductase kinase to correct for any inactivation due to reductase kinase endogenous to the microsomal preparation. In addition, when assaying reductase kinase in crude fractions it is also important to include a blank assay containing all the components except MgATP to correct for nonspecific inhibition of reductase by factors in these fractions.[15,16]

Units. One unit of reductase kinase catalyzes the inactivation of one unit of HMG-CoA reductase per minute at 37° under standard reductase kinase assay conditions.

HMG-CoA Reductase Kinase Kinase Assay

Principle. HMG-CoA reductase kinase kinase activity is assayed by following the time-dependent activation of HMG-CoA reductase kinase *b* in the presence of MgATP and a suitable source of reductase kinase kinase.

Reagents

Buffer A: 1 mM EDTA, 250 mM NaCl, 50 mM NaF, 5 mM dithiothreitol, and 50 mM imidazole, pH 7.4

NaF, 0.25 M

MgATP: 40 mM MgCl$_2$ and 20 mM ATP

Buffer B: 5 mM EDTA, 1.25 M NaCl, 25 mM dithiothreitol, and 250 mM orthophosphate, pH 7.4

Buffer C: 1 mM EDTA, 250 mM NaCl, 50 mM NaF, 5 mM dithiothreitol, and 50 mM orthophosphate, pH 7.4

Reductase kinase kinase and reductase kinase *b*: see later sections.

Assay Procedure. Twenty microliters of reductase kinase *b* (0.16–0.24 mg of protein) in buffer A is mixed with 20 μl of 0.25 M NaF and 50 μl of reductase kinase kinase and then preincubated for 5 min at 37°. The reaction is then started by addition of 10 μl of MgATP and the incubation is continued for 10 min at 37°. The incubation is terminated by addition of 20 μl of buffer B and the mixture is diluted with 380 μl of buffer C. A 50-μl aliquot is assayed for reductase kinase activity as described above. Because of the presence of endogenous reductase kinase kinase in the reduc-

[15] J. L. Nordstrom, Ph.D. thesis, Purdue Univ., West Lafayette, Indiana, 1976.
[16] G. C. Ness and M. H. Moffler, *Arch. Biochem. Biophys.* **189,** 221 (1978).

tase kinase *b* preparation, it is necessary to run a blank incubation without added reductase kinase kinase to correct for this activity. It is also important to run a blank incubation without reductase kinase *b*, but with reductase kinase kinase to correct for any reductase kinase in the kinase kinase preparation. The net reductase kinase activation due to added reductase kinase kinase equals total reductase kinase at the end of the incubation minus reductase kinase added with reductase kinase kinase minus reductase kinase in the blank incubation without added reductase kinase kinase. The total reductase kinase in the blank incubations usually represents less than 20% of the increase in reductase kinase due to added reductase kinase kinase. Since reductase kinase kinase is inhibited by high ionic strength,[1,4] it is important to add the enzyme to the incubation in a low ionic strength buffer. A suitable buffer is 5 mM Tris, containing 1 mM EDTA and 30 mM mercaptoethanol, pH 7.4.

Units. One unit of reductase kinase kinase catalyzes the conversion of 1 unit of reductase kinase *b* to reductase kinase *a* per minute at 37° under standard reductase kinase kinase assay conditions.

Phosphorylase Phosphatase Assay

Principle. Phosphorylase phosphatase activity is assayed by following the time-dependent inactivation of phosphorylase *a* by added protein phosphatase. The assay used is a modification of the procedure cited in Brandt *et al.*[17]

Reagents

Buffer A: 1 mM EDTA, 5 mM dithiothreitol, and 50 mM imidazole, pH 7.4

Buffer B: 5 mM EDTA, 0.5 mM dithiothreitol, 1 mg/ml bovine serum albumin (BSA), and 50 mM imidazole, pH 7.4

Buffer C: 7.67 mM EDTA, 12.5 mM dithiothreitol, 16.7 mM theophylline, and 50 mM imidazole, pH 7.4

Buffer D: 5 mM EDTA, 100 mM NaF, 1 mg/ml BSA, and 50 mM imidazole, pH 7.4

Substrate solution: 0.15 M glucose 1-phosphate, 2% glycogen, and 40 mM imidazole, pH 6.5

H_2SO_4, 0.072 M

1% Ammonium molybdate and 4% ferrous sulfate in 1 N H_2SO_4

Assay Procedure. To 30 μl of buffer C is added 50 μl of protein phosphatase in Buffer B. The mixture is preincubated at 37° for 5 min, and then

---

[17] H. Brandt, Z. L. Capulong, and E. Y. C. Lee, *J. Biol. Chem.* **250**, 8038 (1975).

the assay is started by addition of 20 μl of phosphorylase a (1 mg of protein per milliliter) in buffer A. The final composition of the assay mixture is 5 mM EDTA, 5 mM theophylline, 5 mM dithiothreitol, 0.5 mg/ml BSA, and 50 mM imidazole, pH 7.4. The reaction is stopped after 2–10 min by addition of 0.4 ml of buffer D. Phosphorylase a activity remaining in the assay mixture is then assayed by adding 50 μl of the mixture to 50 μl of substrate solution and incubating at 30° for 10 min. The phosphorylase a assay is terminated by addition of 2 ml of 0.072 M H_2SO_4, followed by 2 ml of 1% ammonium molybdate–4% ferrous sulfate in 1 N H_2SO_4. Inorganic phosphate released in the phosphorylase assay is determined by color development at 700 nm after 2 min. The time of incubation and the quantity of protein phosphatase in the phosphatase assay are adjusted to give 20–50% inactivation of phosphorylase referenced to a control incubation in which the phosphatase is omitted and 0.4 ml of buffer D is added at zero time.

Units. The definition of phosphorylase phosphatase units is based on the definition cited by Brandt *et al.*[17] One unit of phosphatase inactivates 0.2 mg of phosphorylase a (1 nmol of dimer) per minute at 37° under standard assay conditions.

HMG-CoA Reductase Phosphatase Assay

Principle. HMG-CoA reductase phosphatase is assayed by following the time-dependent activation of HMG-CoA reductase b in the presence of a suitable source of reductase phosphatase. The 34,000 MW broad-specificity phosphorylase phosphatase of Brandt *et al.*[12,17] (see later section) is used here for reductase activation.[1–5] The assay is a modification of that used for phosphorylase phosphatase.[17]

Reagents

Buffer A: 1 mM EDTA, 250 mM NaCl, 5 mM dithiothreitol, and 50 mM imidazole, pH 7.4
Buffer B: 5 mM EDTA, 0.5 mM dithiothreitol, 1 mg/ml BSA, and 50 mM imidazole, pH 7.4
Buffer C: 7.67 mM EDTA, 12.5 mM dithiothreitol, 16.7 mM theophylline, and 50 mM imidazole, pH 7.4
HMG-CoA reductase b: See later section.

Assay Procedure

Protein phosphatase (50 μl) in buffer B is mixed with 30 μl of buffer C and preincubated at 37° for 5 min. The assay is then started by addition of 20 μl of HMG-CoA reductase b (20 mg of protein per milliliter in buffer

A). The final composition of the assay buffer is 50 mM NaCl, 5 mM EDTA, 5 mM theophylline, 5 mM dithiothreitol, 0.5 mg/ml BSA, and 50 mM imidazole, pH 7.4. After an appropriate assay time (5–20 min) the reaction is stopped and the reductase assay is commenced by addition of 50 μl of the reductase assay cofactor–substrate cocktail in buffer A (supplemented with NaCl and EDTA to give final concentrations of 250 mM and 30 mM, respectively). Further reactivation of reductase is prevented by addition of the cocktail.[7,18] The dose of phosphatase and the incubation time are adjusted so that no more than 50% of the total reductase b is reactivated. When protein phosphatase preparations containing glycerol (as a stabilizer) are used, glycerol must be removed either by dialysis or by passing the phosphatase preparation successively over two columns of Sephadex G-25 (equilibrated with buffer B), since glycerol inhibits reductase phosphatase.[2] A blank incubation should be performed in the absence of added phosphatase to correct for the small amount of reactivation due to phosphatase endogenous to the reductase b preparation.

Units. One unit of reductase phosphatase catalyzes the formation of 1 unit of reductase a per minute at 37° under standard assay conditions.

Reductase Kinase Phosphatase Assay

Principle. Reductase kinase phosphatase is assayed by following the time-dependent inactivation of reductase kinase a in the presence of a suitable source of reductase kinase phosphatase. The assay is a modification of that used to assay phosphorylase phosphatase and reductase phosphatase.

Reagents

Buffer A: 1 mM EDTA, 5 mM dithiothreitol, and 50 mM imidazole, pH 7.4

Buffer B: 5 mM EDTA, 0.5 mM dithiothreitol, 1 mg/ml BSA, and 50 mM imidazole, pH 7.4

Buffer C: 7.67 mM EDTA, 12.5 mM dithiothreitol, and 50 mM imidazole, pH 7.4

Buffer D: 1.67 mM EDTA, 8.3 mM dithiothreitol, 83.3 mM NaF, 417 mM NaCl, and 83.3 mM orthophosphate, pH 7.4

Reductase kinase a: See later section.

Assay Procedure. Protein phosphatase (50 μl) in buffer B is mixed with 30 μl of buffer C and preincubated at 37° for 5 min. The assay is then started by adding 20 μl of fluoride-free reductase kinase a (50–100 μg, 2.11 units/mg) in buffer A. The final composition of the assay buffer is 5 mM EDTA, 5 mM dithiothreitol, 0.5 mg/ml BSA, and 50 mM imidazole, pH

[18] S. E. Saucier and A. A. Kandutsch, *Biochim. Biophys. Acta* **572**, 541 (1976).

7.4. Theophylline is omitted from the assay buffer because it inhibits reductase kinase activity. The incubation is terminated by addition of 150 μl of buffer D. The diluted mixture contains 3 mM EDTA, 7 mM dithiothreitol, 50 mM NaF, 250 mM NaCl, 10 mM imidazole and 50 mM orthophosphate, pH 7.4. A 20-μl aliquot of this mixture is then assayed for reductase kinase by the standard procedure. The dose of phosphatase and the time of incubation in the reductase kinase phosphatase are adjusted so that 20–50% of reductase kinase is inactivated during the assay. A blank incubation should be conducted in the absence of added phosphatase to correct for the small inactivation due to phosphatase endogenous to the reductase kinase preparation. Fluoride-free reductase kinase a is prepared by desalting reductase kinase a on a Sephadex G-25 column just prior to starting the assays. The reductase kinase a preparation used for these assays is soluble microsomal reductase kinase a.

Units. One unit of reductase kinase phosphatase catalyzes the inactivation of 1 unit of reductase kinase per minute at 37° under standard assay conditions.

Preparation of Enzyme Fractions

Materials

Buffer A: 1 mM EDTA, 250 mM NaCl, 5 mM dithiothreitol, and 50 mM orthophosphate, pH 7.4

Buffer B: 1 mM EDTA, 250 mM NaCl, 5 mM dithiothreitol, and 50 mM imidazole, pH 7.4

Buffer C: buffer B + 50 mM NaF

Buffer D: 4 mM EDTA, 30 mM mercaptoethanol, and 100 mM α-glycerophosphate, pH 8.2

Buffer E: 1 mM EDTA, 30 mM mercaptoethanol, and 5 mM Tris, pH 7.4

Buffer F: 5 mM EDTA, 0.5 mM dithiothreitol, and 50 mM imidazole, pH 7.4

Cycled Rats. Rats are maintained on a controlled lighting schedule in which the room is illuminated from 3:00 PM to 3:00 AM daily for at least 2 weeks prior to use. Reductase activity is maximal in the rats at 9:00 AM.

Other rat populations are maintained on a normal light cycle in which the room is illuminated from 6:00 AM to 6:00 PM (noncycled).

Microsomal HMG-CoA Reductase. Microsomes are prepared by the method of Nordstrom *et al.*[7] Briefly, cycled rats are sacrificed by decapitation at 9:00 AM (i.e., at the peak of the normal diurnal cycle 6 hr into the dark period) and the livers are immediately removed and placed in a beaker containing ice cold homogenization medium (300 mM sucrose and 10 mM mercaptoethanol). The livers are rinsed twice with the medium, minced with a pair of scissors, and homogenized with three strokes of a Potter–Elvehjem homogenizer. The homogenate is centrifuged twice at 12,000 g for 15 min saving the supernatant each time. The supernatant is then centrifuged at 100,000 g for 60 min. The pellet is resuspended in one-fifth the homogenate volume with buffer A or buffer B.

Reductase Kinase-Deficient Microsomal Reductase. Microsomes devoid of reductase kinase are prepared by a two-step procedure. Native microsomes suspended in buffer B are incubated at a concentration of 5–10 mg of protein per milliliter in buffer B at 37° for 2 hr to inactivate reductase kinase. Inactive reductase kinase is then extracted from the membranes by centrifuging the preparation at 100,000 g for 60 min, resuspending the microsomes in buffer B (5–10 mg of protein per milliliter) and reisolating the membranes by centrifugation at 100,000 g for 60 min. The microsomes are extracted three times and finally resuspended in buffer A to give a final concentration of 5–10 mg of protein per milliliter. The suspension is frozen in aliquots in liquid nitrogen and stored at $-70°$ until use.

Microsomal Reductase b. Reductase b is prepared by incubating microsomes suspended in buffer A (1–2 mg of protein per milliliter) supplemented with 4 mM MgCl$_2$ and 2 mM ATP at 37° for 20 min. The EDTA concentration is then brought to 10 mM, and the microsomal reductase is recovered from the incubation mixture by centrifugation at 100,000 g for 60 min and resuspended (5–10 mg of protein per milliliter) in buffer B. Control reductase a is prepared in a similar fashion except that ATP and MgCl$_2$ are omitted from the 20-min incubation. The reductase a and reductase b suspensions are frozen in liquid nitrogen and stored at $-70°$ until use.

Reductase Kinase a. Rat liver (obtained from noncycled rats) is homogenized in 3.5 volumes of a medium consisting of 300 mM sucrose, 10 mM mercaptoethanol, and 50 mM NaF using a Potter–Elvehjem homogenizer. The homogenate is centrifuged twice at 15,000 g for 20 min saving the supernatant each time. The final supernatant is then centrifuged at 100,000 g for 60 min. The resulting supernatant is saved as cytosolic reductase kinase a. The specific activity of reductase kinase in this fraction is 0.8 units/mg protein.

The microsomal pellet is resuspended in one-fifth the original homogenate volume of buffer C. The suspension is centrifuged at 100,000 g for 60 min, and the supernatant is saved as the first microsomal extract. The microsomal pellet is resuspended in buffer, and the extraction is repeated as above. The first and second extracts are combined and concentrated 10-fold by ultrafiltration using an Amicon PM-10 membrane. The resulting preparation of soluble microsomal reductase kinase *a* has a specific activity of 2 units/mg protein. This preparation has only a trace contamination of reductase kinase phosphatase activity.

Reductase Kinase b. Liver microsomes are prepared from noncycled rats as described for the preparation of microsomal reductase and then resuspended in one-fifth the original homogenate volume of buffer B (5–10 mg of protein per milliliter). The suspension is then incubated at 37° for 2 hr to inactivate reductase kinase (due to the action of endogenous phosphatase). At the end of the incubation, NaF is added to give a final concentration of 50 mM and the reductase kinase *b* is extracted from the membranes and concentrated as described for the preparation of soluble microsomal reductase kinase *a*.

Reductase Kinase Kinase. Reductase kinase kinase activity is concentrated primarily in rat liver cytosol. The specific activity of the enzyme in cytosol prepared as described for cytosolic reductase kinase a is 0.3 unit/mg protein.

Alternatively, reductase kinase kinase can be prepared by homogenizing liver from noncycled rats in 3.5 volumes of 300 mM sucrose and 10 mM mercaptoethanol. The homogenate is centrifuged twice at 15,000 g for 20 min, saving the supernatant each time. The pH of the final supernatant is adjusted to 6.1 with 1 N acetic acid. After 5 min, the fraction is centrifuged for 30 min at 12,000 g. The pellet is resuspended in buffer D and the pH is adjusted to 7.4 with 1 N acetic acid. The suspension is centrifuged at 100,000 g for 60 min. The resulting pellet is resuspended in buffer E (one-fifth the original homogenate volume) and then centrifuged at 100,000 g for 60 min. The pellet is resuspended in buffer E, and the extraction is repeated once more. The final pellet is resuspended in one-twentieth the original homogenate volume of buffer E. Although this preparation has a lower final specific activity (0.035 unit/mg), it is virtually free of reductase kinase activity.

Broad Specificity Protein Phosphatase. The phosphatase is purified from rat liver using the procedure of Brandt *et al.*[17] Briefly, fresh rat liver is homogenized in ice cold buffer F using a Waring blender. All subsequent steps are carried out at 4°. The pH of the homogenate is lowered to 5.8 by addition of 1 N acetic acid. Norite A (1 g/liter) is then added, and the mixture is immediately centrifuged at 10,000 g for 20 min. The supernatant is recovered, and the pH is adjusted to 7.2 by addition of Tris base. Solid $(NH_4)_2SO_4$ is then added to 70% saturation, the mixture is stirred for 30 min, then centrifuged at 10,000 g for 20 min. The precipitate is redissolved in a minimal volume of buffer F. The 0–70% $(NH_4)_2SO_4$ fraction is added to 5 volumes of room temperature ethanol with stirring. The mixture is immediately centrifuged for 5 min at 5000 g (at 4°) to collect the copious precipitate. The precipitate is resuspended in buffer F and then centrifuged at 16,000 g for 15 min. The supernatant is poured off and saved, and the ethanol precipitate is extracted a second time with buffer F. The two extracts are combined and dialyzed overnight against two changes of buffer F (10 volumes). The dialyzed extracts are then brought to 40% saturation with solid $(NH_4)_2SO_4$. After stirring for 30 min, the mixture is centrifuged for 20 min at 10,000 g and the pellet is discarded. The supernatant is brought to 75% saturation with solid $(NH_4)_2SO_4$. After sitrring for 30 min, this fraction is centrifuged for 20 min at 10,000 g. The pellet is dissolved in a minimal volume of buffer F plus 0.18 M NaCl and then dialyzed overnight against two changes of 20 volumes of the same buffer. The 40–75% $(NH_4)_2SO_4$ fraction is applied to a 1.5 × 28 cm column of DEAE–Sephadex A-50 preequilibrated with buffer F + 0.18 M NaCl. The column is washed with the same buffer until the optical density at 280 nm approaches baseline. The phosphatase is then eluted with buffer F + 0.24 M NaCl. The peak fractions are combined and diluted with buffer F to reduce the NaCl concentration to 0.1 M. The combined fractions are adsorbed to a small volume of DEAE–Sephadex A-50 in a batchwise manner as described by Brandt et al.[17] and applied to the top of a 1.5 × 22 cm column of DEAE–Sephadex A-50 equilibrated with buffer F + 0.1 M NaCl. The column is washed with one column volume of the same buffer, and the phosphatase is eluted with buffer F + 0.24 M NaCl. The peak fractions are combined, diluted with buffer F and again adsorbed to DEAE–Sephadex. The gel is packed into a small column and eluted with 0.4 M NaCl in buffer F. The phosphatase preparation is stored in 60% glycerol at $-28°$. The phosphatase at this stage is purified 1500-fold and has the following specific activities: phosphorylase a as substrate, 350 units/mg; reductase b as substrate, 32 units/mg; and reductase kinase a as substrate, 320 units/mg.

Acknowledgments

The authors wish to acknowledge the expert technical assistance of Rex A. Parker in the development of these procedures. This study was supported by Grants AM19199 and AM21278 from the National Institutes of Health, and grants from the American Heart Association, Indiana Affiliate, and the Grace M. Showalter Foundation.

[58] Assay of S-3-Hydroxy-3-Methylglutaryl-CoA Reductase

EC 1.1.1.34 Mevalonate : NADP$^+$ oxidoreductase (CoA-acylating)

By NANCY L. YOUNG and BRADLEY BERGER

S-3-Hydroxy-3-methylglutaryl-CoA + 2 NADPH + 2H$^+$ $\rightarrow$
R-3-mevalonic acid + 2 NADP$^+$ + CoA

The reaction catalyzed by HMG-CoA reductase is a regulated step in the biosynthetic pathway to cholesterol.

HMG-CoA reductase activity is commonly measured by a radiochemical assay[1] in which [^{14}C]HMG-CoA is converted to [^{14}C]mevalonic acid, and excess NADPH is generated in the assay from NADP and glucose 6-phosphate by glucose-6-phosphate dehydrogenase. [^{14}C]Mevalonic acid is converted to [^{14}C]mevalonolactone at the conclusion of the assay by addition of acid, and [^{14}C]mevalonolactone is separated from [^{14}C]HMG-CoA by thin-layer chromatography (TLC). [^{3}H]Mevalonic acid is included from the start of the assay as a recovery standard, and [^{3}H]- and [^{14}C]mevalonolactone are measured by scintillation counting.

General Guidelines

Reductase activity of a tissue can be measured in homogenate or in a subcellular fraction obtained by differential centrifugation. The preparation of choice varies with the tissue and depends on the sedimentation characteristics of the reductase, HMG-CoA cleavage enzymes, and reductase phosphatase, and on whether one wishes to maximize specific activity of reductase or total recovery.

The activity of HMG-CoA cleavage enzymes in a preparation should be considered because they can seriously interfere with the reductase assay by depleting substrate, by forming products that inhibit reductase, and by forming products that are isolated with mevalonolactone. In our experience with rat liver and small intestine and with human mononuclear cells, most cleavage activity sediments at low forces ($\sim$10,000 g) with mitochondria, although some remains suspended and a small amount sediments at high forces ($\sim$100,000 g) with microsomes. In some tissues, such as rat liver, the presence of high cleavage activity in homogenates, low-speed pellets, and even low-speed supernatant fractions makes the assay

[1] D. J. Shapiro, J. L. Nordstrom, J. J. Mitschelin, V. W. Rodwell, and R. T. Schimke. *Biochim. Biophys. Acta* **370**, 369 (1974).

of reductase activity in those fractions unreliable. Techniques for monitoring and for minimizing the effects of cleavage activity on the reductase assay are illustrated below in the application of the assay to mononuclear cells freshly isolated from human blood. These techniques are generally applicable to the assay of reductase activity in any tissue.

To determine whether cleavage activity or other factors diminish the production of mevalonate in the assay, microsomes from 2 mg of rat liver, and homogenate or subcellular fraction of the tissue of interest are assayed separately and together. If the amount of mevalonate produced by the mixture is less than the sum of the amounts produced by each alone, then inhibitors are present.

At the time of homogenization of a tissue, reductase is largely in an inactive phosphorylated form and is activated by endogenous phosphatase during subsequent handling of the homogenate.[2] This has been observed in the liver of a wide variety of vertebrates,[3,4] in mouse brain and fibroblasts,[4] and in rat small intestine.[5] In some cases, activation of reductase by endogenous phosphatase can proceed rapidly even in the cold. For example, activation is nearly complete during the time required for the isolation of microsomes from rat liver homogenate at 4°.[6] If phosphatase is primarily in the cytosol, as in rat liver, then reductase sedimenting during the first (low speed) centrifugation will be separated from phosphatase earlier and may be activated less completely than reductase sedimenting with microsomes during the second (high speed) centrifugation. Clearly, for reliable studies of the subcellular distribution of reductase, phosphatase activation must be controlled. This can be done either by adding a phosphatase inhibitor such as fluoride[6,7] or pyrophosphate[3] at the time of homogenization and measuring the inactivated level of reductase, or by adding exogenous phosphatase after fractionation and measuring the fully activated level of reductase.[8,9] NaF at 50 mM will inhibit phosphatase in rat liver homogenate at 4° but not at 25°.[10] Higher inhibitor concentrations are required at the higher temperature.[11] For technical details on control-

[2] D. M. Gibson and T. S. Ingebritsen, *Life Sci.* **23**, 2649 (1978).
[3] C. F. Hunter and V. W. Rodwell, *J. Lipid Res.* **20**, 1046 (1979).
[4] S. E. Saucier and A. A. Kandutsch, *Biochim. Biophys. Acta* **572**, 541 (1979).
[5] N. L. Young, unpublished observation, 1978.
[6] J. L. Nordstrom, V. W. Rodwell, and J. J. Mitschelen, *J. Biol. Chem.* **252**, 8924 (1977).
[7] J. Berndt, F. G. Hegardt, J. Bové, R. Gaumert, J. Still, and M. T. Cardó, *Hoppe-Seyler's Z. Physiol. Chem.* **357**, 1277 (1976).
[8] B. W. Philipp and D. J. Shapiro, *J. Lipid Res.* **20**, 588 (1979).
[9] M. S. Brown, J. L. Goldstein, and J. M. Dietschy, *J. Biol. Chem.* **254**, 5144 (1979).
[10] J. Berndt, R. Henneberg, and M. Löwel, *J. Lipid Res.* **20**, 1045 (1979).
[11] V. W. Rodwell, unpublished observations, 1979.

ling phosphatase activity, the reader is referred to the reports cited and to this volume [57].

The specific activity and total recovery of reductase are maximized in the same subcellular fraction from some but not all tissues. In rat liver, both are highest in microsomes because cleavage activity sediments at low speeds and reductase at high speed.[12] In human mononuclear cells, most cleavage activity sediments at low speeds; however, very little protein and essentially no reductase activity sediment at high speeds. Consequently, the high-speed centrifugation can be omitted, and the low-speed supernatant fraction provides the highest specific activity and total recovery.[12] In contrast, in rat small intestine, microsomes have the highest specific activity but homogenate has the highest total recovery, in part because reductase tends to sediment at low speeds along with a large portion of cell protein.[12,13]

Homogenates or subcellular fractions may be stored in liquid nitrogen before assaying for reductase activity. Freeze–thawing results in a decrease in activity that is reversed with a preincubation at 37° for 15–30 min. However, storage in an ordinary freezer or repeated freeze-thawing may lower activity irreversibly.[5]

The optimum time and temperature for preincubation and the linear ranges for response of activity to protein concentration and time of assay should be determined for the preparation that is to be assayed routinely.

Finally, reductase activity in some tissues has a marked diurnal rhythm linked to food intake. In liver and small intestine from *ad libitum* fed rats there is a maximum in reductase activity at mid-dark, and a broad minimum during the light period. Thus the time of day when the rat is killed may be an important variable in need of control. Housing rats in a windowless room with a reversed lighting schedule facilitates the study of rats during their dark period.

Application to Mononuclear Cells Freshly Isolated from Human Blood

Methods

Homogenization Medium (HM). HM contains 100 mM K$_2$HPO$_4$, 30 mM Na$_2$EDTA, NaOH to give pH 7.5, NaCl to give 220 mM Na$^+$, 10 mM dithiothreitol, and 0.2% (w/v) soybean trypsin inhibitor (type 1-S from Sigma Chemical Co.). HM × 5, in which the concentration of all compo-

---

[12] N. L. Young and S. L. Zuckerbrod, unpublished observation, 1978.
[13] S. R. Panini, G. Lehrer, D. H. Rogers, and H. Rudney, *J. Lipid Res.* **20**, 879 (1979).

nents is 5 times that given above, is used in preparing the substrate–cofactor solution (see below). HM is used immediately or is stored frozen for less than 1 month to avoid oxidation of dithiothreitol.

Isolation and Treatment of Cells. Mononuclear cells are isolated at 0 to 4° from 20 ml of EDTA-treated blood by the method of Boyum[14] using a commercial preparation such as Ficoll–Paque (Pharmacia Fine Chemicals). The cell pellet in a Nunc tube (T. W. Smith, New York, New York) is weighed (usually about 80 mg) and stored in liquid nitrogen. Subsequent steps in the treatment of cells are at 20 to 25° and cold is avoided. Cells are homogenized in 300 μl of HM in a No. 18 Potter–Elvehjem tissue grinder with a Teflon pestle (Kontes) driven by a motor at 600 rpm for 1 min. The homogenization is repeated after 90 min. The homogenate is centrifuged at 1000 g for 5 min; the supernatant is saved, and the pellet is discarded.

Protein Assay. Protein in the supernatant fraction is assayed by the fluorescamine method described by Young and Rodwell.[15] The usual yield is about 100 mg of protein per gram of cells.

[^{14}C]HMG-CoA. (3-R,S)[3-^{14}C]HMG-CoA is prepared by a modified version of the method described by Goldfarb and Pitot[16] from [3-^{14}C]HMG possessing a specific activity of at least 55 Ci/mol (Amersham Corp., Arlington Heights, Illinois). Recrystallization of HMG-anhydride is omitted, and the [^{14}C]HMG-CoA is purified by paper chromatography with butanol : acetic acid : water (5 : 3 : 2) to a radiochemical purity of at least 95% determined with TLC system I (see below). The final product is stored in 1 mM HCl at $-20°$.

Substrate–Cofactor Solution. Each assay contains 10 μl of HM $\times$ 5; 0.3 unit of glucose-6-phosphate dehydrogenase; 4.5 μmol of glucose 6-phosphate, 450 nmol of NADP$^+$ (all from Sigma Chemical Co., St. Louis, Missouri); 9 nCi (3-R,S)[5-^{3}H]mevalonic acid (having specific activity of 5.2 Ci/mmol, from New England Nuclear, Boston, Massachusetts); 3 nmol of [^{14}C]HMG-CoA; NaOH to give pH 7.5; and water to give a total volume of 50 μl. To avoid hydrolysis of HMG-CoA when adding NaOH, the pH should not exceed 8. The solution is stored in liquid nitrogen.

HMG-CoA reductase Assay. In a 0.5-ml polyethylene centrifuge tube, 100 μl of the supernatant fraction of mononuclear cell homogenate containing about 2.5 mg of protein is vortexed with 50 μl of substrate–cofactor solution and is then incubated at 37° for 15 min. An assay tube with supernatant plus substrate–cofactor solution lacking NADP$^+$ is used as a control for the formation of ^{14}C-labeled products independent of NADPH which are isolated with the reductase product. (Omitting or in-

[14] A. Boyum, *Nature (London)* **204,** 793 (1964).
[15] N. L. Young and V. W. Rodwell, *J. Lipid Res.* **18,** 572 (1977).
[16] S. Goldfarb and H. C. Pitot, *J. Lipid Res.* **12,** 512 (1971).

cluding NADP$^+$ is equivalent to omitting or including NADPH, since a regeneration system is present.) Assay tubes with HM in place of cell supernatant are used as controls for nonenzymatic cleavage of substrate. The reaction is stopped by adding 20 μl of 6 N HCl with vortexing. The tubes are kept at $-20°$ or on ice unless otherwise indicated to minimize cleavage of the substrate by the acid. After centrifugation (10,000 g, 25°, 1 min), [^{14}C]HMG-CoA in 10 μl of the supernatant solution is isolated by TLC system I (see below). The assay tubes are incubated at 37° for 15 min to convert mevalonic acid to mevalonolactone, and recentrifuged if necessary. [^{14}C]- and [^{3}H]mevalonolactone in the remaining supernatant solution are isolated by TLC system II. When a more highly purified reductase product is desired, mevalonolactone is converted to mevalonic acid on the TLC sheet and rechromatographed in system III.

Thin-Layer Chromatography. Silica gel (No. 13179) and cellulose (No. 13255) TLC sheets with plastic backs (Eastman Organic Chemicals) are scored in 4 and 16 vertical channels, respectively. Silica gel sheets are dried at 100° for 15 min just before use. Samples are applied 2.5 cm from the bottom, dried with a stream of cool air, and chromatographed to 2 cm from the top of the sheet in sandwich chambers (Sargent Welch, Springfield, New Jersey) placed out of drafts in a fume hood. Some HMG-CoA cleavage products, e.g., acetoacetate, are converted to volatile products by heat. Therefore, the developed cellulose sheets are kept cool and scraped for scintillation counting, as described below, as soon as possible.

SYSTEM I. A 10-μl sample is applied to one channel of a cellulose sheet and chromatographed immediately with butanol : acetic acid : water (7 : 2 : 3). The solvent is freshly prepared to avoid ester formation. The R_f of HMG-CoA is 0.3, and that of cleavage products is >0.5.

SYSTEM II. A 150-μl sample is applied to one channel of a silica gel sheet in successive portions of 50, 40, 30, 20, and 10 μl with complete drying between each portion to avoid buildup of residue at the perimeter of the spot, which decreases mevalonolactone recovery, and chromatographed with acetone : toluene (1 : 1). (Benzene was used originally but is toxic, so it was replaced by toluene with no change in the R_f of mevalonolactone.) The R_f of HMG-CoA and most of its cleavage products is $\leq$ 0.5, and that of mevalonolactone is 0.7.

SYSTEM III. After chromatography in system II, the sheet is sprayed evenly with 0.1 N NaOH until it glistens but does not run, placed in a prewarmed sandwich chamber horizontally in an oven at 60° for 10 min to convert mevalonolactone to mevalonic acid, thoroughly dried with a stream of cool air, and chromatographed in the *reverse* direction with ethanol : acetone (1 : 4). The R_f of mevalonate with the reverse chromatography is <0.03.

Scintillation Counting. For scintillation counting using the external standard channel ratio method for quench correction, it is essential that the radiolabeled compound be soluble in the scintillation cocktail. HMG-CoA is most soluble in a cocktail containing a basic solubilizer such as NCS, perhaps because hydrolysis to the soluble HMG is promoted. In contrast, in the presence of silica gel, mevalonate is most soluble in an acidified polar cocktail, perhaps because conversion to the more soluble lactone is promoted. After chromatography, mevalonolactone spontaneously hydrolyzes to mevalonic acid on the sheet (50% in 36 hrs).

For the construction of quench curves relating external standard channel ratio to counting efficiency, commercial sealed standards may be unsuitable. It is preferable to determine quench curves with two series of quenched standards prepared with known amounts of [^{3}H]- and [^{14}C]mevalonolactone and the size vial, the amount of silica gel, and the amount and type of scintillation cocktail used routinely in reductase assays.

The R_f and purity of radiolabeled compounds are verified by measuring radioactivity in 0.5-cm sections of the TLC sheets. Once reproducible mobilities and adequate purities are assured, wider sections are counted for routine assays as described below.

SYSTEM I. The bottom half of the TLC channel to R_f 0.5, which contains HMG-CoA, is scraped into one 20-ml scintillation vial; the top half above R_f 0.5, which contains all cleavage products (Fig. 1), is scraped into another vial; 10 ml of a scintillation cocktail composed of

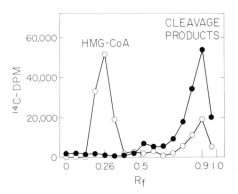

FIG. 1. Thin-layer chromatographic separation of [^{14}C]HMG-CoA from its cleavage products. A supernatant fraction of 6 mg of cells prepared in ethanolamine buffer according to the method of Fogelman *et al.*[18] was incubated with substrate–cofactor solution containing 30 μM HMG-CoA in a total volume of 50 μl at 37° for 1 hr. After acidification and centrifugation, the supernatant liquid was chromatographed in system I. The ^{14}C radioactivity (disintegrations per minute) in 1-cm sections of the TLC is shown as (●) for cell supernatant and as (○) for a control lacking cell supernatant.

OCS : NCS : H$_2$O (100 : 10 : 1) is added to each. (OCS, a nonpolar scintillation fluid, and NCS, a basic solubilizer, are from Amersham.)

SYSTEMS II AND III. The sheet is lightly sprayed with a fine mist of water, and the 3-cm region between R_f 0.6 and 0.8 is scraped into an 8-ml scintillation vial to which 20 μl of 6 N HCl and 4 ml of Scintiverse (Fisher Scientific Co.) are added.

Calculations. The HMG-CoA cleavage activity is calculated from the fraction of [^{14}C]HMG-CoA cleaved by subtracting the fraction cleaved in the controls lacking enzyme, then multiplying by the amount of HMG-CoA present initially in the assay and dividing by the assay time and the milligrams of protein in the assay or the grams of cells from which the sample was derived.

The HMG-CoA reductase activity is calculated as follows: After correcting for ^{3}H recovery, ^{14}C from the control lacking NADPH is subtracted from ^{14}C in the assay with the complete substrate–cofactor solution to give the NADPH-dependent ^{14}C, which is then divided by the specific activity of [^{14}C]HMG-CoA in the assay (at least 120 dpm/pmol), the assay time, and the milligrams of protein or the grams of tissue from which the sample was derived.

Discussion

Under conditions favorable for HMG-CoA lyase, i.e., with Tris buffer at pH 8.5 and Triton X-100, homogenate of freshly isolated mononuclear cells from human blood has lyase activity of about 20 nmol per milligram of protein per minute.[17] If lyase activity is not discouraged in some way, mevalonate formation in the reductase assay is reduced because of substrate depletion. In some tissues, most HMG-CoA cleavage activity can be separated from reductase by differential centrifugation. Unfortunately, this is not so easily accomplished with human mononuclear cells. Reductase activity in these cells remains in the supernatant fraction after high speed centrifugation, and the rather drastic treatment of freeze–thawing and homogenization required to break the cells results in the suspension of considerable cleavage activity in this fraction.[12] The nonionic detergent Kyro-EOB has been used to lyse human[18] and rat[15] mononuclear cells prior to reductase assays. Although it improves the recovery of reductase in microsomal pellets from rat cells,[15] it does not improve the recovery of reductase in the low-speed supernatant fraction of human cells.[12] However, advantage may be taken of the instability of cleavage enzyme activ-

[17] S. J. Wysocki and R. Hähnel, *Clin. Chim. Acta* **73**, 373 (1976).
[18] A. M. Fogelman, J. Edmond, J. Seager, and G. Popják, *J. Biol. Chem.* **250**, 2045 (1975).

ity in phosphate buffer in which reductase activity is stable. Thus cleavage activity in the low-speed supernatant fraction of mononuclear cells prepared in phosphate buffer pH 7.5 in the absence of detergent is reduced from about 100 to about 20 pmol per milligram of protein per minute after a 90-min preincubation at room temperature.[12] Cleavage activity is stable in Tris and in ethanolamine buffers,[12] so, although the latter buffer was used previously in assays of reductase activity in lymphocytes,[18,19] we feel that it should be avoided.

Even at the reduced level, cleavage activity is about 40 times reductase activity, so it is advisable to monitor substrate depletion in the reductase assay. A convenient TLC method for isolating HMG-CoA from its cleavage products (Fig. 1) has been developed for this purpose. With this method, total HMG-CoA cleavage is measured rather than the formation of a specific product, such as acetoacetate. This is of advantage with freshly isolated human mononuclear cells, which form several products including acetoacetate and hydroxymethylglutarate.[20]

Some unidentified products of [^{14}C]HMG-CoA cleavage migrate in the mevalonolactone region of TLC system II (Fig. 2). These are a small fraction of all cleavage products, but under some conditions, as when ethanolamine buffer is used according to a previously published method,[18] they exceed the small amount of mevalonolactone isolated (Fig. 2A and B). The amount of contaminating cleavage products is much smaller when phosphate buffer is used according to the present method (Fig. 2C and D), and is further reduced by an additional chromatographic step (Fig. 2E and F). In this procedure, mevalonolactone is chromatographed as usual, then hydrolyzed on the sheet to mevalonic acid and rechromatographed in the reverse direction with a more polar solvent system in which the cleavage products, but not mevalonic acid, migrate. The remaining contaminants are corrected for by subtracting the ^{14}C radioactivity recovered from controls in which NADP^{+} was omitted. Incidentally, ether extraction of mevalonolactone from the assay mixture before TLC is used in some laboratories, but in our hands results in a far greater contamination of the isolated mevalonolactone and a poorer recovery of [^{3}H]mevalonolactone.

The effect of excessive substrate depletion is illustrated in Fig. 3, which gives data for three experiments with the substrate dose response of reductase in mononuclear cells. In the first experiment, the assay time was 60 min and there was excessive substrate depletion when initial substrate concentration was less than 100 μM. At the lowest concentration, 2 μM, when 3% of the substrate was converted to mevalonate, 80% of the substrate was cleaved. This resulted in reduced average velocities of

[19] M. J. P. Higgins and D. J. Galton, Eur. J. Clin. Invest. 7, 309 (1977).
[20] K. Y. Shum and N. L. Young, unpublished observation, 1978.

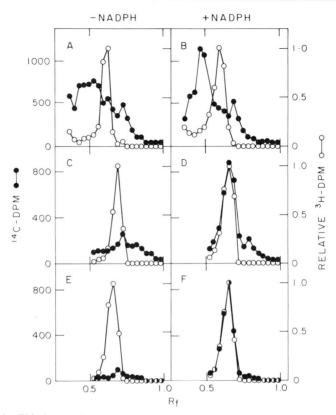

Fig. 2. Thin-layer chromatography of [³H]mevalonolactone and metabolites of [¹⁴C]HMG-CoA. In the first experiment, the supernatant fraction from 38 mg of cells prepared as described in Fig. 1 was incubated with substrate–cofactor solution containing [³H]mevalonate and [¹⁴C]HMG-CoA at 300 μM in a total volume of 150 μl for 60 min. After lactonization of mevalonic acid and centrifugation, the supernatant solution was chromatographed in TLC system II (panels A and B). In the second experiment, the supernatant fraction of 15 mg of cells was prepared in phosphate buffer as described in Methods, incubated with substrate–cofactor solution, and chromatographed as in the first experiment (panels C and D). Replicate assays of the second experiment were chromatographed in system III (panels E and F). Radioactivity (disintegrations per minute) in 0.5 cm sections of each TLC is shown as (●) for ¹⁴C radioactivity and as (○) for relative ³H radioactivity. Relative ³H radioactivity is the ratio of ³H radioactivity in a section to ³H radioactivity in the section containing the peak. The TLC runs for assay tubes with NADPH are shown on the right; those without NADPH are shown on the left. Note that including or omitting NADP⁺ is equivalent to including or omitting NADPH since a regenerating system was present.

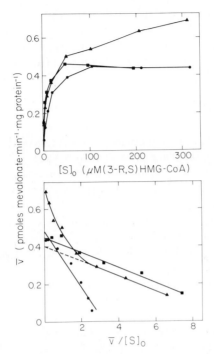

FIG. 3. Substrate dose response of HMG-CoA reductase in freshly isolated human mononuclear cells. In each of three experiments cells were isolated from 500 ml of blood from one of three individuals. The supernatant fraction of cell homogenates was prepared and assayed for HMG-CoA reductase activity as described in methods except that the (3-R,S)HMG-CoA concentration was varied from 2 to 300 μM, assays at each concentration were in triplicate, and each assay tube contained the supernatant fraction from 25 mg of cells. The assay time and the thin-layer chromatography system used to isolate the reductase product were 60 min and system III in experiment 1 (●), 15 min and system II in experiment 2 (▲), and 15 min and system III in experiment 3 (■). The data are plotted in the bottom panel according to Eadie–Hofstee. The K_m was estimated to be 7.75 ± 0.23, 2.12 ± 0.02, and 2.15 ± 0.15 μM (3-S)HMG-CoA from the slope of linear regressions using all eight data points in experiment 1, three data points for the lowest substrate concentrations in experiment 2, and all seven data points in experiment 3, respectively.

mevalonate formation, as can be seen by comparison with data from the second and third experiments (Fig. 3), in which the assay time was 15 min and at most 20% of the substrate was cleaved. Excessive substrate depletion raised the apparent K_m from 2.1 ± 0.1 to 7.8 ± 0.2 μM (3-S)HMG-CoA as can be seen by the increase in the slope of the data in the Eadie–Hofstee plot (Fig. 3, lower panel). Thus, monitoring the utilization of substrate by reductase to avoid spuriously high estimates of K_m as was suggested by Langdon and Counsell[21] is not always sufficient. With these cells, it is necessary to monitor utilization by all enzymes metabolizing HMG-CoA in the assay. Somewhat higher estimates for K_m of 3.5 and 5.3 μM (3-S)HMG-CoA for reductase in cultured human lymphocytes were obtained by LaPorte et al.[22] under assay conditions in which substrate conversion by reductase exceeded 20% when initial substrate concentration was $\leq 5 \mu M$. Cleavage activity was not monitored in this case but may be much lower in cultured than in freshly isolated cells (see below).

Figure 3 also illustrates the effect of contamination of the isolated reductase product by cleavage products. In the second experiment, where TLC system II was used to isolate mevalonolactone, more NADPH-dependent ^{14}C-labeled product was formed as the substrate concentration was raised from 50 to 300 μM. This translated to an upswing in the reciprocal plot (lower panel) near the origin. By contrast, in the first and third experiments, where TLC system III was used to obtain a more purified product, NADPH-dependent ^{14}C-labeled product did not increase above 100 μM HMG-CoA.

The optimal substrate concentration for the routine assay of reductase activity in these cells is, in our opinion, 20 μM (3-R,S)HMG-CoA. At this concentration and with a 15-min assay the amount of [^{14}C]mevalonate recovered is adequate for reliable scintillation counting, the velocity of mevalonate formation is 83% of maximal, about 8% of the substrate is cleaved, and 81% of the product isolated by TLC system II is [^{14}C]-mevalonolactone. That is, 87% of this product is NADPH-dependent, and 93% of the NADPH-dependent product remains after the second chromatography in TLC system III. The K_m values of the cleavage enzymes are higher than the K_m of reductase,[23] so as the substrate concentration is increased, mevalonate becomes a smaller proportion of all products formed. Thus at 100 μM (3-R,S)HMG-CoA only 40–50% of the ^{14}C-labeled product isolated by TLC system II is mevalonolactone; i.e., 60–70% is NADPH-dependent, and 72% of the NADPH-dependent product remains after the second chromatography.

In the three substrate-dose response experiments (Fig. 3), the mean coefficient of variation for assays in triplicate was 10% (range 1 to 27%). In view of the occasional high variation, duplicate assays which require a total of 40 ml of blood are advisable.

With the assay method described here, freshly isolated mononuclear cells from 4 normal men had reductase activities averaging 0.45 ± 0.02

[21] R. E. Langdon and R. E. Counsell, J. Biol. Chem. 251, 5820 (1976).
[22] M. Laporte, M. Astruc, C. Tabacik, B. Descomps, and A. Crastes DePaulet, FEBS Lett. 86, 225 (1978).
[23] N. L. Young, unpublished observation, 1976.

pmol of mevalonate per milligram of protein per minute. Higgins and Galton[19] using ethanolamine rather than phosphate buffer, observed a mean activity of 0.098 ± 0.005 in cells from 55 normal individuals.

Phosphatase activation of reductase in these cells has not been examined.

The assay of reductase in cultured human lymphocytes is considerably easier than in freshly isolated mononuclear cells. Expressed reductase activity is much higher in cultured cells,[24,25] and cleavage activity may be lower. When freshly isolated cells were incubated for 2 hr in a medium containing lecithin to promote efflux of cholesterol and hence an increase in reductase activity, cleavage activity decreased dramatically.[23] Since cleavage activity can interfere in the reductase assay, change in cleavage activity under various culture conditions could produce artifactual change in measured reductase activity. This effect was observed even in rat liver microsomes where the activity of an HMG-CoA utilizing enzyme lowered measured reductase activity and, what is worse, exhibited a physiological variation that was complementary to the variation in reductase activity.[26] To control for this possibility in the assay of reductase activity, substrate depletion and the purity of the isolated mevalonolactone can be monitored with the TLC procedures described here.

[24] H. J. Kayden, L. Hatam, and N. G. Beratis, *Biochemistry* **15**, 521 (1976).
[25] M. Astruc, M. Laporte, C. Tabacik, and A. Crastes DePaulet, *Biochem. Biophys. Res. Commun.* **85**, 691 (1978).
[26] G. C. Ness and M. H. Moffler, *Biochim. Biophys. Acta* **572**, 333 (1979).

Enzymes of Phospholipid Synthesis and Related Enzymes

[59] Lysophospholipase–Transacylase from Rat Lung

EC 3.1.1.5 Lysolecithin acylhydrolase

By H. van den Bosch, G. M. Vianen, and G. P. H. van Heusden

The 100,000 g supernatant of rat lung is capable of converting 1-acyl-sn-glycero-3-phosphocholine (lysophosphatidylcholine; lysoPC) into sn-glycero-3-phosphocholine, fatty acid, and sn-3-phosphatidylcholine. Recent evidence[1-3] has shown that a single enzyme is responsible for the formation of these products according to the reactions

$$\text{lysoPC} + \text{H}_2\text{O} \rightarrow \text{glycerophosphocholine} + \text{fatty acid} \qquad (1)$$
$$\text{lysoPC} + \text{lysoPC} \rightarrow \text{glycerophosphocholine} + \text{phosphatidylcholine} \qquad (2)$$

To explain these results a reaction mechanism has been proposed in which the enzyme first reacts with lysoPC to give glycerophosphocholine and an acyl-enzyme intermediate. The latter complex transfers the acyl group not only to water to yield free fatty acid, but also to another molecule of lysoPC to form phosphatidylcholine. In view of these activities the enzyme has been termed lysophospholipase–transacylase. This report will describe the purification and some properties of the enzyme from rat lung.

Assay Methods

Reagents

1-[1-¹⁴C]Palmitoyl-sn-glycero-3-phosphocholine, 2 mM, sonicate in water[4]
Potassium phosphate buffer, 0.5 M, pH 6.5
Chloroform
Chloroform/methanol (1:2, v/v)
Dole's extraction medium[5] (isopropanol, n-heptane, N/1 H₂SO₄, 40:10:1)
n-Heptane
Silica gel (Merck, Kieselgel 60, reinst)

[1] M. Abe, K. Ohno, and R. Sato, *Biochim. Biophys. Acta* **369**, 361 (1974).
[2] G. Brumley and H. van den Bosch, *J. Lipid Res.* **18**, 532 (1977).
[3] G. M. Vianen and H. van den Bosch, *Arch. Biochem. Biophys.* **190**, 373 (1978).
[4] 1-[1-¹⁴C]Palmitoyl-sn-glycero-3-phosphocholine is available from New England Nuclear, Boston, Massachusetts, with a specific radioactivity of 0.11 mCi/mg. It is diluted to the required specific radioactivity with unlabeled palmitoyllysophosphatidylcholine.
[5] V. P. Dole, *J. Clin. Invest.* **35**, 150 (1956).

METHODS IN ENZYMOLOGY, VOL. 71

Assay by Thin-Layer Chromatography

This assay is used when both the lysophospholipase [Reaction (1)] and transacylase [Reaction (2)] activity of the enzyme are to be measured. A chloroform (CHCl$_3$) solution containing 200 nmol of 1-[1-^{14}C]palmitoyl-lysophosphatidylcholine (120 dpm/nmol) is transferred to a test tube and evaporated to dryness with a stream of nitrogen or by rotating evaporation. After addition of 0.1 ml of 0.5 M potassium phosphate buffer, pH 6.5, the lysoPC substrate is solubilized by vortexing the tubes for 30 sec. The tubes are placed in ice and an appropriate amount of water and enzyme (0.1–0.4 mg of protein for the 100,000 g supernatant) is added to give a final volume of 0.5 ml. After rapid mixing the tubes are incubated at 37° for 10 min. The reaction is stopped by addition of 1.5 ml of CHCl$_3$/CH$_3$OH (1:2, v/v) from a dispenser. The lipids are extracted according to Bligh and Dyer[6] with two successive washes of the water/methanol layer with CHCl$_3$. The combined CHCl$_3$ extracts are transferred by means of a Pasteur pipette to a conical flask; 5 μl of a CHCl$_3$ solution containing 10 mg/ml of each lysoPC, phosphatidylcholine, and fatty acid (oleic acid)[7] as references is added, and the mixture is evaporated to dryness. The lipids are dissolved in a few drops of CHCl$_3$/CH$_3$OH (1:2, v/v) and transferred to 0.5 mm silica gel H thin-layer plates with the aid of a glass capillary. To achieve as quantitative a transfer as possible the conical flask is rinsed once with a few drops of the CHCl$_3$/CH$_3$OH mixture. The thin-layer plates are developed with CHCl$_3$/CH$_3$OH/CH$_3$COOH/H$_2$O (100:50:16:8, v/v),[8] dried, and stained with iodine vapor. The lysoPC, phosphatidylcholine, and fatty acid spots are scraped and transferred to scintillation vials containing 10 ml of a scintillation mixture consisting of 2 parts 0.5% (w/v) PPO and 0.03% (w/v) POPOP in toluene, 1 part Triton X-100 and 0.2 part water. Total recovery of radioactivity ranged from 80 to 100%. Radioactivity values are corrected to 100% recovery, assuming equal losses for all three components during the procedure. Enzymatic activities are expressed in milliunits, i.e., in nanomoles of product formed per minute. Radioactivity in phosphatidylcholine is divided by twice the specific activity of the lysoPC substrate to obtain the number of nanomoles formed. This assay procedure resulted in a molar ratio for the formation of free fatty acid and phosphatidylcholine of about 2.2.

During purification of the enzyme lysophospholipase–transacylase activity is monitored by measuring lysophospholipase activity [Reaction (1)]; when the ratio of lysophospholipase to transacylase activity is re-

[6] E. G. Bligh and W. J. Dyer, *Can. J. Biochem. Physiol.* **37**, 911 (1959).
[7] The use of oleic acid instead of palmitic acid facilitates detection by iodine staining.
[8] V. Skipski, R. Peterson, J. Sanders, and M. Barclay, *J. Lipid Res.* **4**, 227 (1963).

quired, we use the more tedious combined assay [Reactions (1) and (2)] described above.

Assay by Dole Extraction

When it is sufficient to determine only the lysophospholipase activity, e.g., for the distribution of enzymatic activity in column effluents, a modified Dole extraction procedure is used.[9] For convenience, a 2 mM stock solution of 1-[1-^{14}C]palmitoyl-lysoPC in H$_2$O is prepared by sonication for about 30 sec.[10] With the aid of an Eppendorf pipette, 0.1 ml of the substrate sonicate is transferred to a test tube. Otherwise, the assay conditions are as described above. The reaction is stopped by addition of 2.5 ml of Dole's extraction mixture[5] and approximately 100 mg of silica gel is added by means of a glass spatula. The mixture is vortexed for 15 sec. The silica gel binds and sediments the remaining lysoPC, which would otherwise partition to a small extent into the heptane phase. Then, 1.5 ml of heptane and 1.5 ml of water are added from dispensers. The mixture is vortexed for 15 sec, and the phases are allowed to separate. An aliquot of 1 ml of the upper heptane phase is transferred with an Eppendorf pipette to a scintillation vial containing 10 ml of toluene scintillation fluid. A blank incubation without enzyme is included in each series to account for nonenzymatic hydrolysis and traces free fatty acid present in the substrate sonicate. With this assay system other lysophospholipases, not possessing transacylase activity,[2] are detected as well. In working with crude subcellular fractions care must be taken to test column peaks showing lysophospholipase activity for transacylase activity by the thin-layer chromatography assay described above.

Spectrophotometric Assay

In principle, lysophospholipase activity can also be determined spectrophotometrically by recording the release of free thiol groups from thioester substrate analogs in the presence of thiol reagents.[11,12] This procedure, although in principle the most convenient assay method, is not used routinely for the lysophospholipase–transacylase purification. The purification procedures require the presence of β-mercaptoethanol as stabilizer. For the spectrophotometric assay, β-mercaptoethanol has to be

[9] S. A. Ibrahim, *Biochim. Biophys. Acta* **137**, 413 (1967).
[10] This sonicate can be stored indefinitely at $-16°$. After thawing at room temperature the solution is resonicated for about 10 sec, after which it is ready for use.
[11] A. J. Aarsman, L. L. M van Deenen, and H. van den Bosch, *Bioorg. Chem.* **5**, 241 (1976).
[12] A. J. Aarsman and H. van den Bosch, *FEBS Lett.* **79**, 317 (1977).

removed by dialysis. This results in severe loss of activity; so that about 30–40% of the original activity remains after 6 hr of dialysis against buffer A (see below) without β-mercaptoethanol.

Purification Procedure

Starting Material. The purification procedure previously described[2] can be carried out with 50 lungs. Freshly prepared[13] lungs from 50 Wistar rats are rinsed three times with homogenization buffer (0.25 M sucrose, 50 mM potassium phosphate, pH 6.8, 1 mM EDTA and 10% glycerol). The tissue is cut into small pieces with scissors, rinsed twice with buffer and homogenized in a Potter tube to give a 20–25% (w/v) homogenate. It has been shown that 68% of the total lysophospholipase activity and 89% of the total transacylase activity of rat lung is present in the 100,000 g supernatant.[2] This fraction is prepared by centrifugation of the homogenate for 20 min at 10,000 g in a Sorvall GSA rotor and subsequently for 60 min at 100,000 g in a Beckman L 5-65 ultracentrifuge using a type 30 rotor. The enzyme is purified from the resulting clear red supernatant in four steps; ammonium sulfate precipitation, chromatography on hydroxyapatite and DEAE-cellulose columns, and filtration on an AcA-44 column.

Ammonium Sulfate Precipitation. The supernatant is brought to 65% saturation with solid ammonium sulfate, stirred for 40 min and centrifuged for 20 min at 20,000 g. The precipitate is dissolved in about 80 ml of buffer A (50 mM potassium phosphate, pH 6.8, containing 10 mM β-mercaptoethanol and 10% glycerol). The resulting solution is dialyzed twice against 10 liters of buffer A. The precipitate that forms during the dialysis has less than 10% of the total lysophospholipase–transacylase activity and is removed by centrifugation.

Hydroxyapatite Chromatography. The dialyzate (100 ml) is pumped onto a hydroxyapatite column (4 × 10 cm) at a flow rate of 60 ml/hr. The column is washed with buffer until the eluate is free of protein (absorbance at 280 nm). This is followed by elution with a solution of 120 mM potassium phosphate, pH 6.8, containing 10 mM β-mercaptoethanol and 10% glycerol. Elution (flow rate 60 ml/hr) is continued until the absorbance at 280 nm is nearly zero.[14] The eluate is dialyzed overnight against 8 liters of buffer A.

---

[13] Fresh tissue has to be used. When rat lungs were frozen and collected over a period of months, upon thawing and work-up as described here the enzymatic activity was eluted from the columns in very broad peaks and no appreciable purification was obtained.

[14] Further elution with 400 mM potassium phosphate, pH 6.8, containing 10 mM β-mercaptoethanol and 10% glycerol yields a red fraction that has about 10% of the activity applied to the column, but with a considerably lower specific activity.

PURIFICATION OF LYSOPHOSPHOLIPASE–TRANSACYLASE FROM RAT LUNG

Fraction	Total protein[a] (mg)	Total activity[b] (nmol FA/min)	Specific activity (nmol/mg min^{-1})	Recovery of activity (%)	Purification fold	Ratio FA:PC[c]
100,000 g supernatant	2367	26198	11.0	100	—	3.8
Ammonium sulfate precipitate	1029	21945	21.3	84	1.9	3.1
Hydroxyapatite	223	10727	48.2	41	4.4	2.3
DEAE-cellulose	25	4841	193.6	18.5	17.5	2.2
AcA-44	1.2	2050	1673	7.8	151.1	2.4

[a] Protein was measured using a modified Bradford procedure as described previously by Vianen and van den Bosch.[3]

[b] The enzymatic activity was measured with the Dole extraction procedure.

[c] FA, fatty acid. PC, phosphatidylcholine.

DEAE–Cellulose Chromatography. The dialyzed solution (100 ml) is pumped onto a DEAE 52-cellulose column (2 × 9 cm). This column is washed with buffer until the eluate is free of protein. Absorbed proteins are eluted using a 220-ml linear NaCl gradient (0 to 400 mM) in buffer A. The column is operated at a flow rate of 30 ml/hr. The active fractions elute at approximately 150 mM NaCl. The pooled fractions (37.5 ml) are concentrated to 2 ml using an ultrafiltration device (Amicon, UM-10 filter).

AcA-44 Chromatography. The concentrated protein solution is applied to an AcA-44 column (1.3 × 98 cm), previously equilibrated with buffer A containing 0.5 M NaCl. The column is eluted with the equilibration buffer at a flow rate of 9 ml/hr. Fractions of 1 ml are collected. The active fractions are stored at −45°.

The results of a typical purification are given in the table.

Properties

A Single Enzyme with Dual Activities. As can be seen in the table, lysophospholipase and transacylase activity copurify with a constant ratio of fatty acid and phosphatidylcholine production in the last three purification steps. The idea of a single enzyme being responsible for the formation of both products is supported by acrylamide gel electrophoresis. Both in the absence and in the presence of sodium dodecyl sulfate (SDS) a single band is visible.[2,3] Figure 1 shows that the single protein band obtained after electrophoresis without SDS has both the lysophospholipase and transacylase activity in the ratio predicted by the table. Heat treatment and preincubation with increasing concentrations of diisopropyl-fluorophosphate of the purified enzyme resulted in identical inhibition of lysophospholipase and transacylase activity.[3]

Physical Properties. AcA-44 exclusion chromatography indicates a molecular weight of approximately 58,000, a value somewhat larger than previously reported.[2] However, SDS–disc gel electrophoresis consistently gives an estimated molecular weight of 78,000. The reason for this discrepancy is not known at present.

pH Optimum. Both the lysophospholipase and the transacylase activity are optimally expressed at pH 6–7 and follow a similar pH vs activity profile, except beyond pH 8.5, where the lysophospholipase reaction decreases more rapidly than the transacylase activity.

Inhibitors. The following inhibitors at the indicated concentrations inactivated the enzyme to various extents: diisopropylfluorophosphate (1 mM, 90%), iodoacetate (3 mM, 25%; 15 mM, 31%), N-ethylmaleimide (10 mM, 47%; 40 mM, 62%), NaCN (10 mM, 16%; 50 mM, 86%). In the

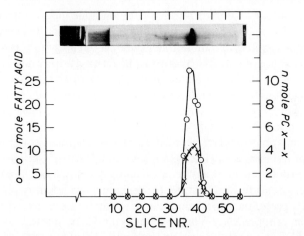

FIG. 1. Disc electrophoresis of lysophospholipase–transacylase. Experimental conditions: 30 μg of purified enzyme were applied to 7.5% polyacrylamide disc gels at pH 8.3 as described by Davis [B. J. Davis, *Ann. N. Y. Acad. Sci.* **121**, 404 (1964)]. One gel was stained [A. H. Reisner, *Anal. Biochem.* **64**, 509 (1975)] and the other sliced into 1-mm slices. These were soaked overnight in 0.1 ml of 0.5 M potassium phosphate buffer, pH 6.5, containing 10% (v/v) glycerol and 10 mM β-mercaptoethanol and assayed for lysophospholipase (O——O) and transacylase (X——X) activity.

presence of these inhibitors the ratio of fatty acid and phosphatidylcholine production remained the same as for the uninhibited enzyme. This ratio is subject to variation with substrate concentration, in that phosphatidylcholine formation still increases after fatty acid release has reached a plateau.[2] Excess free fatty acid inhibits both activities, but affects lysophospholipase activity to a greater extent. With 0.4 mM substrate the ratio fatty acid : phosphatidylcholine decreases from 2.1 in the absence, to 0.2 in the presence, of 2 mM fatty acid.[3]

Stability. The enzyme is unstable in the absence of glycerol and β-mercaptoethanol. During the purification, 10% glycerol (v/v) and 10 mM β-mercaptoethanol are included in the buffers. Under these conditions ammonium sulfate precipitation and DEAE–cellulose chromatography give reasonable recoveries of enzymatic activity, in contrast to a previous report[1] in which the same techniques used in the absence of glycerol were unsatisfactory. The purified enzyme, as obtained from the AcA-44 column, can be stored at $-45°$ with less than 10% loss of activity during several months. Repeated freezing and thawing of the enzyme, however, caused significant loss of activity.

Substrate Specificity. The enzyme is not specific for lysoPC. Hydrolysis and acyl transfer takes place also with lysophosphatidylethanolamine and lysophosphatidylglycerol. Monoglyceride and palmitoyl-CoA are not hydrolyzed. The enzyme does not require the presence of a free hydroxyl group in the acyl donor; deoxylysoPC and deoxylysophosphatidylethanolamine (with a CH_2 group replacing the HCOH group) are effectively hydrolyzed.[3] DeoxylysoPC has been used as an acyl donor to study the acceptor specificity of the enzyme. LysoPC, lysophosphatidylethanolamine, lysophosphatidylglycerol, and to a lesser extent lysophosphatidic acid serve as acyl acceptors. Monoglyceride, diglycerides, cholesterol, and the water-soluble compounds glycerol, sn-glycero-3-phosphate and sn-glycero-3-phosphocholine do not function as acyl acceptors in the acyl transfer reaction. With a series of lysoPCs having different saturated acyl chains there is a slight preference for palmitate over stearate and laurate to be incorporated into phosphatidylcholine. No preference for palmitate over myristate is found.

Reaction Mechanism. When the enzyme is incubated with 1-[9,10-$^3H_2$]stearoyl-sn-glycero-3-phospho-[^{14}C]methylcholine the $^3H : ^{14}C$ ratio in phosphatidylcholine is twice that of the starting lysoPC, even when the reaction is carried out in the presence of excess fatty acid. Reaction with ^3H-labeled lysoPC in the presence of [^{14}C]palmitate results in formation of phosphatidylcholine having only the ^3H label. These results demonstrate that both acyl chains in phosphatidylcholine originate from lysoPC. If the reaction is carried out in $H_2^{18}O$, the released palmitate shows the same atom percent excess $H_2^{18}O$ as the water, indicating an O-acyl cleavage mechanism in the hydrolysis of palmitoyllysoPC. The absence of ^{18}O in the transacylation product phosphatidylcholine provides strong evidence for a covalent acylenzyme intermediate in the acyltransfer reaction.[15]

Similar Enzymes

Enzymes catalyzing the formation of phosphatidylcholine from two molecules of lysoPC have been reported for rat liver,[16] yeast,[17] erythrocytes,[18] leukocytes,[19] rabbit[20] and mouse lung,[21] human plasma,[22] intestinal[23] and gastric mucosa.[24] With the exception of the gastric mu-

[15] G. P. H. van Heusden and H. van den Bosch, *Biochem. Biophys. Res. Commun.* **90**, 1000 (1979).
[16] J. F. Erbland and G. V. Marinetti, *Biochim. Biophys. Acta* **106**, 128 (1965).
[17] H. van den Bosch, H. A. Bonte, and L. L. M. van Deenen, *Biochim. Biophys. Acta* **98**, 648 (1965).
[18] E. Mulder, J. W. O. van den Berg, and L. L. M. van Deenen, *Biochim. Biophys. Acta* **106**, 118 (1965).
[19] P. Elsbach, *J. Lipid Res.* **8**, 359 (1967).
[20] F. H. C. Tsao and R. D. Zachman, *Pediat. Res.* **11**, 849 (1977).
[21] V. Oldenborg and L. M. G van Golde, *Biochim. Biophys. Acta* **441**, 433 (1976).
[22] P. V. Subbaiah and J. D. Bagdade, *Life Sci.* **22**, 1971 (1978).
[23] P. V. Subbaiah and J. Ganguly, *Ind. J. Biochem. Biophys.* **8**, 197 (1971).

cosa enzyme, none has been purified. The partially purified enzyme from gastric mucosa[24] shows both lysophospholipase and transacylase activities and exhibits several properties resembling the enzyme from rat lung described here.

[24] Y. N. Lin, M. K. Wassef, and M. I. Horowitz, *Arch. Biochem. Biophys.* **193**, 213 (1979).

[60] UDPgalactose: Ceramide Galactosyltransferase from Rat Brain

EC 2.4.1.45 UDPgalactose : (2-hydroxyacyl)sphingosine galactosyltransferase

By NENAD M. NESKOVIC, PAUL MANDEL, and SHIMON GATT

UDPgalactose + ceramide → cerebroside + UDP

Assay Method

Principle. The method is based on measuring the incorporation of a radioactive galactosyl residue of UDPgalactose into galactosyl ceramide (cerebroside). The lipid substrate, ceramide, is dispersed in an aqueous solution of the nonionic detergent Triton X-100. The product of the reaction (cerebroside) is separated from the unreacted UDPgalactose by partitioning between a chloroform-rich phase and an aqueous-methanol phase,[1] and its radioactivity is measured in a liquid scintillation spectrometer.

Reagents

Tris-HCl, 0.5 M, pH 8.0
$MgCl_2$, 0.1 M
Triton X-100, 1% (w/v), in chloroform–methanol, 2:1
Ceramide (containing α-hydroxy fatty acids), 2 mM, in chloroform–methanol, 2:1. This ceramide can be prepared from brain cerebrosides.[2]
Brain phosphatidylcholine,[3] 2 mg/ml, in chloroform–methanol, 2:1
UDP-[U-^{14}C]galactose purchased from New England Nuclear (Boston, Massachusetts) is diluted to approximately 5 μCi/μmol with

[1] J. Folch, M. Lees, and G. H. Sloane-Stanley, *J. Biol. Chem.* **226**, 497 (1957).
[2] N. S. Radin, this series, Vol. 28 [65].
[3] N. M. Neskovic, L. L. Sarlière, and P. Mandel, *Biochim. Biophys. Acta* **334**, 309 (1974).

unlabeled UDPgalactose (Sigma Chemical Co., Saint Louis, Missouri) and the concentration is adjusted to 0.5 mM.

Choroform–methanol, 2 : 1

"Pure solvents upper phase" (methanol–water–chloroform, 48 : 47 : 3)[1]

Procedure. The solutions of ceramide (0.05 ml), phosphatidylcholine (0.03 ml), and Triton X-100 (0.05 ml) are pipetted into a 12 × 75 mm tube, and the solvents are evaporated *in vacuo.* MgCl$_2$ (0.01 ml), buffer (0.02 ml), and water (0.1 ml) are added, and the tube contents is stirred on a Vortex cyclomixer. The tube is heated in a boiling water bath for 2 min, cooled to room temperature, and placed in an ice-water bath. Water (to a final volume of 0.25 ml), enzyme, and UDPgalactose (0.04 ml) are added, and the tube is incubated for 15 min at 27° with shaking. The reaction is terminated by the addition of 2.5 ml of chloroform–methanol, 2 : 1, followed by 0.25 ml of water. The tube is shaken vigorously on a Vortex cyclomixer and centrifuged at about 2500 rpm for 5 min; the upper phase is removed by aspiration. The lower phase is washed twice with 0.5 ml each of "pure solvents upper phase"; an aliquot is transferred to a counting vial and evaporated in an oven at 100°; then 10 ml of toluene-based scintillator (0.4% Omnifluor, Koch-Light Laboratories, Colnbrook, England) are added to the still warm vials and the radioactivity is determined in a liquid scintillation spectrometer.

Unit. One unit is defined as 1 nmol of galactose transferred per hour under the conditions described above.

Comments on the Assay Method

This procedure yields reproducible results provided an appropriate ratio of Triton X-100 to enzyme is maintained. In the absence of added phospholipid, the curve of velocity vs enzyme concentration is not a straight line, but a nonsymmetric sigmoid,[4] whose precise shape depends on the concentration of Triton X-100. The deviation from linearity in the initial portion of the curve (i.e., at high ratios of Triton X-100 to enzyme) suggests inhibition by excess detergent. The sigmoidal curve can be converted to a straight line by adding a heat-inactivated enzyme preparation (e.g., of microsomes or purified enzyme from step 5; see the table) to the reaction mixtures. Linearity is improved if a constant sum of active plus heat-inactivated enzymatic protein is maintained in all tubes. Linear curves are also obtained by replacing the heat-inactivated enzyme with lipids extracted from the enzymatic preparation.[4] The procedure de-

[4] N. M. Neskovic, P. Mandel, and S. Gatt, *in* "Enzymes of Lipid Metabolism" (S. Gatt, L. Freysz, and P. Mandel, eds.), p. 613. Plenum, New York, 1978.

scribed here, which is routinely used in our laboratory, is simplified by using a fixed quantity of phosphatidylcholine. This method provides a satisfactory linearity with respect to the quantity of enzyme, using up to 100 μg of microsomal protein of a 20-day-old rat, or 20 μg of protein of the purified enzyme from step 5 (see Purification Procedure).

The reaction rate is constant to about 25 min in the presence of 60 μg of microsomal protein or 3 μg of the purified enzyme.

Alternative Assay Methods

Several methods of UDPgalactose : ceramide galactosyltransferase assay have been described in which a ceramide dispersion was prepared without detergent. Morell and Radin[5] adsorbed ceramide onto solid particles (Celite). Brenkert and Radin[6] prepared the ceramide dispersion by mixing a ceramide solution in benzene with lyophilized enzyme source (brain homogenate or microsomes) and removing the solvent by evaporation. These methods were described in a previous article in this series.[2] Cestelli *et al.*[7] used a dispersion of a sonically irradiated mixture of ceramide, phosphatidylethanolamine, and phosphatidylcholine.

Purification Procedure

The procedure is based on the results of previous studies.[3,8] Several modifications were introduced in the final steps of purification, and glycerol-containing buffers were used. The initial stages of purification (see steps 1–3, of the table) are most conveniently carried out with 100 brains; in further stages two to three batches of freeze-dried powder from step 3 can be pooled. The best yields of ceramide galactosyltransferase during the extraction from microsomes are obtained with Cemulsol NP 12[9]; in steps 5–7 Cemulsol NP 12 can be replaced by Triton X-100. Unless otherwise stated, all operations are carried out at 4°.

Preparation of the Microsomal Fraction and Extraction. Brains of 100 rats 15–20 days old are washed with ice cold 0.9% NaCl, minced with scissors, and homogenized with nine volumes of 0.32 M sucrose containing 1 mM EDTA in a Potter–Elvehjem homogenizer. The homogenate is

[5] P. Morell and N. S. Radin, *Biochemistry* **8,** 506 (1969).

[6] A. Brenkert and N. S. Radin, *Brain Res.* **36,** 183 (1972).

[7] A. Cestelli, F. V. White, and E. Costantino-Ceccarini, *Biochim. Biophys. Acta* **572,** 283 (1979).

[8] N. M. Neskovic, L. L. Sarlieve, and P. Mandel, *Biochim. Biophys. Acta* **429,** 342 (1976).

[9] Nonionic detergent purchased from Rhône-Poulenc Chimie Fine, Persan, France; Cemulsol NP 12 is a nonyl phenol polyoxyethylene with the number of ethylene oxide residues ranging from 3 to 20.

PURIFICATION OF UDPGALACTOSE : CERAMIDE GALACTOSYLTRANSFERASE FROM 100 RAT
BRAINS

Step	Total activity (units)[a]	Total protein (mg)[b]	Specific activity (unit/mg)	Yield (%)
1. Extract	1587	992	1.6	100
2. DEAE-cellulose	1389	479	2.9	87
3. Solvent extraction	1240	400	3.1	78
4. Pronase treatment	1106	140	7.9	70
5. $(NH_4)_2SO_4$ precipitation	1016	62	16.4	64
6. DEAE-Sephadex	504	6.3	80	32
7. Ultrafiltration	196	1.2	164	12

[a] A unit of activity is defined as the incorporation of 1 nmol of galactose per hour.
[b] Determined by a modification of the Folin–Lowry method (Neskovic et al.[8]).

centrifuged in a Sorvall GSA rotor at 3000 rpm ($1500\,g_{max}$) for 10 min, the supernatant is taken off, and the pellet is resuspended in two–three volumes of 0.32 M sucrose–1 mM EDTA and centrifuged as above. The combined supernatants are centrifuged in a Sorvall GSA rotor at 9000 rpm ($13{,}000\,g_{max}$) for 20 min. The supernatant is poured off and centrifuged in a Beckman No. 30 rotor at 30,000 rpm ($106{,}000\,g_{max}$) for 60 min. The supernatant is discarded, and the microsomal pellet is suspended by homogenization in 0.1 M potassium phosphate buffer, pH 7.6, containing 0.2% (v/v) 2-mercaptoethanol and 2 mM EDTA, to make a final volume of 90 ml and a protein concentration of about 20 mg/ml. A 5% (v/v) solution of Cemulsol NP 12 in above buffer was added slowly to the mechanically stirred microsomal suspension, until 1% (v/v) final concentration of the detergent is obtained. The suspension is stirred for 30 min, diluted two-fold with water, and centrifuged in a Beckman No. 30 rotor at 30,000 rpm ($106{,}000\,g_{max}$) for 1 hr. The supernatant is carefully aspirated, and the pellet is discarded.

Negative Adsorption with DEAE-Cellulose. The supernatant is transferred into two 250-ml centrifuge bottles containing DE-52 slurry (about 80 ml per bottle of resin settled by gravity), previously equilibrated with 50 mM potassium phosphate buffer, pH 7.6, containing 0.25% (v/v) Cemulsol NP 12, 0.1% (v/v) 2-mercaptoethanol, and 1 mM EDTA. The mixture is stirred mechanically for 15 min and centrifuged at 1500 g for 5 min to sediment the resin. The supernatant is decanted through a loose plug of glass wool; the residues are washed with 20 ml each of the buffer, and the washings are combined with the first supernatant. The opalescent liquid is dialyzed against 2 × 5 liters of redistilled water for 16 hr.

Solvent Extraction. Redistilled solvents are used throughout this stage. The dialyzed DE-52 extract is freeze-dried, the residue is suspended in 60 ml of dry acetone precooled at $-20°$ and stirred for 2 min. After the centrifugation at $2500 g$ for 5 min at $-10°$, the acetone is decanted and the residue is treated once more with 60 ml of acetone. The residue is finally washed with 60 ml of benzene and freeze-dried to remove this solvent. The white powder is stored in a desiccator at $-20°$.

Pronase Treatment. The lipid-depleted powder is suspended with a Potter–Elvehjem homogenizer in 50 volumes of 0.1 M potassium phosphate buffer, pH 7.6, containing 0.1% (v/v) 2-mercaptoethanol and 1 mM EDTA. Aliquots (15 ml) of the suspension are irradiated sonically for 2 min at $0°$ with an MSE sonic oscillator using the 9.5 mm tip at 75% output. To this solution (about 6 mg of protein per milliliter) a 1% solution of Pronase (Koch-Light Laboratories, Colnbrook, England) is added to a final concentration of 0.2 mg/ml. The solution is slowly stirred with a magnetic stirrer for 16 hr and clarified in a Sorvall SS 34 rotor at 15,000 rpm (27,000 g_{max}) for 15 min.

Ammonium Sulfate Precipitation. The solution from step 4 is stirred with a magnetic stirrer and brought to 40% saturation by gradually adding solid ammonium sulfate. After stirring for 15 min, the liquid is centrifuged in a Sorvall SS 34 rotor at 15,000 rpm (27,000 g_{max}) for 15 min. The precipitated material forms a layer on the surface of the liquid. The underlying liquid is carefully aspirated, and the precipitate that adheres to the wall of the tube is dissolved with the aid of a Potter–Elvehjem homogenizer in about 15 ml of 50 mM Tris-HCl, pH 8.0, containing 1 mM EDTA, 0.1% (v/v) 2-mercaptoethanol, 20% (w/v) glycerol (abbreviated TEMG), and 0.25% (v/v) Cemulsol NP 12. The slightly opalescent, yellowish solution is dialyzed against two changes (500 ml each) of TEMG containing 0.1% (w/v) Cemulsol NP 12 overnight.

DEAE-Sephadex Chromatography. The ammonium sulfate fraction is diluted with TEMG containing 0.1% (w/v) Cemulsol NP 12 to give a protein concentration of about 1.2 mg/ml and applied to a column (6 × 20 cm) of DEAE-Sephadex A-25 previously equilibrated with the same buffer. Elution is carried out with the starting buffer and monitored spectrophotometrically at 280 nm. Eluate is collected fractionally and assayed for the enzyme activity. The fractions exhibiting the peak activity, which precedes slightly the major peak of light absorption at 280 nm, are pooled.

Ultrafiltration. The DEAE-Sephadex fraction is concentrated to 2–3 ml by ultrafiltration through a Diaflo PM-30 membrane.

Properties

Stability. The lyophilized preparation (step 3) could be stored in a desiccator at $-20°$ for several months without a noticeable loss of activity; the solubilized preparations showed considerable loss within several days at 0°. The presence of glycerol greatly improved the stability of both the membrane-bound and the solubilized enzyme preparations.[8] Thus, the fraction of step 5 (4–7 mg of protein per milliliter) could be stored at −20° in a buffer containing 50% (w/v) glycerol for a period of 4 months with a loss of activity less than 30%. Using the same conditions, more purified enzyme preparations (steps 6 and 7), usually containing less than 1 mg/ml of protein, lost about 30% of the original activity in 1 month. In the presence of glycerol the enzyme could be frozen and rethawed several times without appreciable effect on the stability of the enzyme.

Purity. Disc gel electrophoresis on polyacrylamide gels in the presence of Triton X-100[10] revealed two major protein bands; the enzyme activity migrated with the slow-moving band.[11]

Phospholipid Content. Phospholipids are present in the various enzyme preparations. The fraction of step 5 contained 2.1–4.0 mg of phospholipids per milligram of protein; the phospholipid content of step 7 ranged from 2.6 to 3.7 mg per milligram of protein. The phospholipids can be partially separated from the enzymatic protein by filtration through Sepharose 6B in the presence of the nonionic detergent Cemulsol NP 12.[8] About 50% of the total phospholipids in the enzyme preparation of step 5 accompanied the peak of enzyme activity, suggesting that lipoprotein complexes with molecular weights of 400,000–500,000 might be present.

Substrate Specificity. The membrane-bound as well as the partially purified enzyme shows a marked preference for the ceramide containing α-hydroxy fatty acids; in the enzyme assay described above, this substrate cannot be replaced by ceramide containing nonhydroxy fatty acids (see also Basu *et al.*[12]). Other lipids that might serve as galactose acceptors were incubated with the purified enzyme of step 7: no galactosyltransferase activity was observed with glucosylceramide, lactosylceramide, or asialo G_{M2} ganglioside.[4] Sphingosine, which is readily transformed to psychosine by brain homogenates and microsomes, yielded very low galactosyltransferase activity with the purified ceramide galactosyltransferase. In microsomes the ratio (activity with ceramide) : (activity with sphingosine) was about 3; it increased to about 80 in the purified enzyme of step 5, which was depleted of the nonionic detergent with SM-2 Bio-Beads.[13]

Warren *et al.*[14] tested the ceramide galactosyltransferase activity of rat brain with several synthetic ceramides containing α-hydroxy fatty acids of

[10] B. Dewald, J. H. Dulaney, and O. Touster, this series, Vol. 32 [8].
[11] N. M. Neskovic, unpublished results.
[12] S. Basu, A. M. Schultz, M. Basu, and S. Roseman, *J. Biol. Chem.* **246,** 4272 (1971).
[13] N. M. Neskovic, unpublished results.
[14] K. R. Warren, R. S. Misra, R. C. Arora, and N. S. Radin, *J. Neurochem.* **26,** 1063 (1976).

varying chain length. They found that the ceramide containing D-C_{7h} fatty acid was the best substrate and that increasing the chain length up to C_{18} decreased the reaction rate; with C_{22h} and C_{24h} ceramide the reaction rate increased again. Again, ceramides with nonhydroxy fatty acids yielded very low reaction rate.

Activation by Phospholipids. The addition of several phospholipids to the incubation mixture containing Triton X-100 increased the rates of the enzymatic reaction.[3,4] A similar stimulation was also observed using the membrane-bound microsomal enzyme in the assay system devoid of detergent.[15] Phosphatidylcholine and phosphatidylethanolamine produced the greatest stimulation.

It should be noted that specific requirement for a given phospholipid was not demonstrated in the above-mentioned studies. It is possible that exogenous phospholipids increase the reaction rate by a mechanism involving phospholipid–ceramide or phospholipid–detergent interactions.[4] Evidence for the phospholipid requirement of ceramide galactosyltransferase was provided by the inactivation of the enzyme after the phospholipase A or C action.[8]

Inhibitors. Bile salts, such as sodium deoxycholate, sodium taurocholate, or sodium taurodeoxycholate, strongly inhibit the ceramide galactosyltransferase reaction.[4] Among natural products, sphingosine has a marked, though yet unexplained, inhibitory effect.[12] Warren *et al.*[14] tested several synthetic analogs of ceramide as inhibitors of ceramide galactosyltransferase. Derivatives of 1-phenyl-2-amino-1,3-propanediol had a moderate inhibitory effect, the best inhibitor being octanoyl-D-*threo-p*-nitro-1-phenyl-2-amino-1,3-propanediol; some of these compounds may act as competitive inhibitors.

A protein inhibitor of ceramide galactosyltransferase was found in brain, kidney, spleen, and liver of rat and other animals[16]; it was extracted from microsomes and partially characterized.

Kinetic Parameters and pH Optimum. The K_m value for UDPgalactose with the purified enzyme ranged from 2.7 to 3.4 × 10^{-5} *M*. The K_m values for the lipid substrate, ceramide, are influenced by the dispersion state of this lipid. In the presence of 2 m*M* Triton S-100, biphasic kinetics was observed with increasing ceramide concentration, yielding two apparent K_m values of 1.8 × 10^{-5} and 1.5 × 10^{-4} *M*.[4]

The purified enzyme shows peak of activity between pH 7.8 and 8.2.

[15] E. Costantino-Ceccarini and K. Suzuki, *Arch. Biochem. Biophys.* **167**, 646 (1975).
[16] E. Costantino-Ceccarini and K. Suzuki, *J. Biol. Chem.* **253**, 340 (1978).

[61] Glycerolipid Acyltransferases from Rat Liver: 1-Acylglycerophosphate Acyltransferase, 1-Acylglycerophosphorylcholine Acyltransferase, and Diacylglycerol Acyltransferase

EC 2.3.1.51 Acyl-CoA: 1-acyl-*sn*-glycerol-3-phosphate *O*-acyltransferase
EC 2.3.1.23 Acyl-CoA: 1-acyl-*sn*-glycero-3-phosphocholine
                    O-acyltransferase
EC 2.3.1.20 Acyl-CoA: 1,2-diacyl-*sn*-glycerol *O*-acyltransferase

By Satoshi Yamashita, Kohei Hosaka, Yoshinobu Miki, and
Shosaku Numa

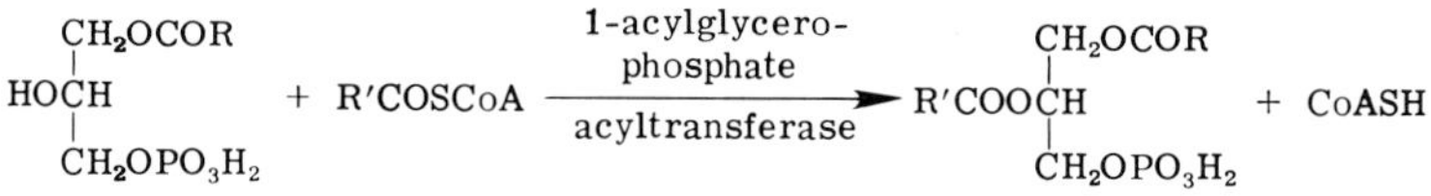

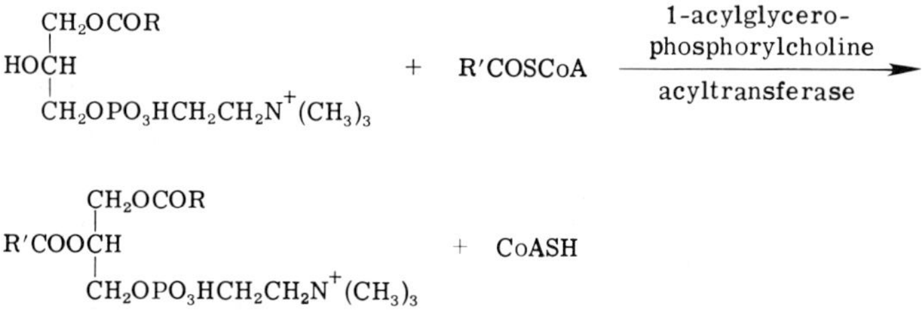

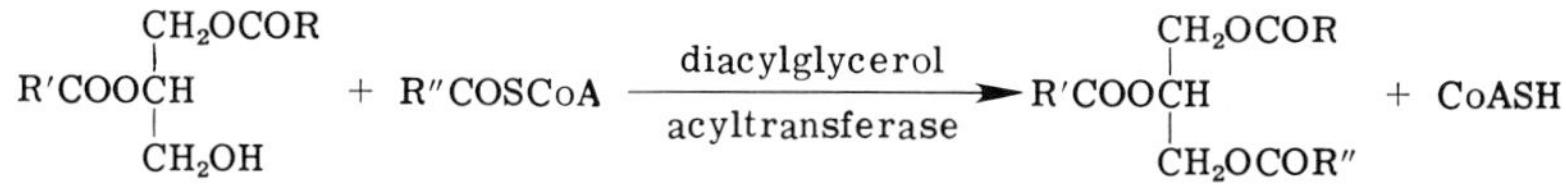

$$
\begin{array}{c}
CH_2OCOR \\
HOCH \\
CH_2OPO_3H_2
\end{array}
+ R'COSCoA
\xrightarrow[\text{acyltransferase}]{\substack{\text{1-acylglycero-}\\\text{phosphate}}}
\begin{array}{c}
CH_2OCOR \\
R'COOCH \\
CH_2OPO_3H_2
\end{array}
+ CoASH
$$

$$
\begin{array}{c}
CH_2OCOR \\
HOCH \\
CH_2OPO_3HCH_2CH_2N^+(CH_3)_3
\end{array}
+ R'COSCoA
\xrightarrow[\text{acyltransferase}]{\substack{\text{1-acylglycero-}\\\text{phosphorylcholine}}}
$$

$$
\begin{array}{c}
CH_2OCOR \\
R'COOCH \\
CH_2OPO_3HCH_2CH_2N^+(CH_3)_3
\end{array}
+ CoASH
$$

$$
\begin{array}{c}
CH_2OCOR \\
R'COOCH \\
CH_2OH
\end{array}
+ R''COSCoA
\xrightarrow[\text{acyltransferase}]{\text{diacylglycerol}}
\begin{array}{c}
CH_2OCOR \\
R'COOCH \\
CH_2OCOR''
\end{array}
+ CoASH
$$

The assay method, separation, and properties of 1-acylglycerophosphate acyltransferase, 1-acylglycerophosphorylcholine acyltransferase, and diacylglycerol acyltransferase are described herein. These acyltransferases play crucial roles in the formation of specific fatty acid distribution in naturally occurring glycerolipids. By sucrose density gradient centrifugation of microsomes resolved with Triton X-100, a nonionic detergent, these enzymes are separated from glycerophosphate acyltransferase,[1] and furthermore 1-acylglycerophosphate acyltransferase is separated from 1-acylglycerophosphorylcholine acyltransferase and diacylglycerol acyltransferase.

[1] S. Yamashita and S. Numa, this volume [64].

METHODS IN ENZYMOLOGY, VOL. 71

Assay Methods

1-Acylglycerophosphate Acyltransferase[2]

Principle.[3] The enzyme activity can be assayed by measuring the reaction of released CoA with 5,5'-dithiobis(2-nitrobenzoic acid) spectrophotometrically.

Reagents

Tris-HCl buffer, 0.5 M, pH 7.4
5,5'-Dithiobis(2-nitrobenzoic acid), 10 mM, neutralized with NaOH
Oleoyl-CoA, 0.5 mM
1-Acyl-*sn*-glycerol 3-phosphate, 1 mM, in 50 mM Tris-HCl buffer, pH 7.4, containing 0.25 M sucrose and 1 mM EDTA. This is prepared from phosphatidic acid, derived from egg phosphatidylcholine, by the action of *Trimeresurus flavoviridis* phospholipase A_2.

Procedure. The reaction mixture contains 100 μmol of Tris-HCl buffer, pH 7.4, 1 μmol of 5,5'-dithiobis(2-nitrobenzoic acid), 20 nmol of oleoyl-CoA, 50 nmol of 1-acyl-*sn*-glycerol 3-phosphate, and enzyme (up to 28 mU) in a total volume of 1 ml. The reaction is initiated by the addition of oleoyl-CoA after preincubation of the enzyme with all other components for 2–4 min. The increase in absorbance at 405 nm (or at 412 nm) is followed at 37° with a recording spectrophotometer; a molar absorbance of 13,600 M^{-1} cm^{-1} is used to calculate the activity. Control incubations are always conducted in the absence of 1-acyl-*sn*-glycerol 3-phosphate, and the observed rates are subtracted from the rates measured in the presence of the acyl acceptor.

1-Acylglycerophosphorylcholine Acyltransferase[2]

The enzyme activity can be assayed by the same principle as described for 1-acylglycerophosphate acyltransferase. The assay procedure is identical to that for 1-acylglycerophosphate acyltransferase, except that 1-acyl-*sn*-glycero-3-phosphorylcholine, instead of 1-acyl-*sn*-glycerol 3-phosphate, is used as acyl acceptor and that arachidonoyl-CoA, instead of oleoyl-CoA, is used as acyl donor; up to 12 mU of enzyme are added.

Diacylglycerol Acyltransferase[4]

Principle. The enzyme can be assayed by measuring the incorporation of radioactive palmitoyl residue into triglyceride, which is isolated by

[2] S. Yamashita, K. Hosaka, and S. Numa, *Eur. J. Biochem.* **38,** 25 (1973).
[3] W. E. M. Lands and P. Hart, *J. Biol. Chem.* **240,** 1905 (1965).
[4] K. Hosaka, U. Schiele, and S. Numa, *Eur. J. Biochem.* **76,** 113 (1977).

solvent extraction followed by Na_2CO_3-impregnated silica gel column chromatography.

Reagents

Tris-HCl buffer, 0.5 M, pH 7.4

$MgCl_2$, 0.1 M

Dithiothreitol, 0.5 M

Bovine serum albumin (fatty acid-free; Miles Laboratories), 10 mg/ml

Sucrose, 1 M

[1-^{14}C]Palmitoyl-CoA (770–1190 cpm/nmol), 2 mM

1,2-Diacyl-*sn*-glycerol, 20 mM. This is prepared from egg phosphatidylcholine by the action of *Clostridium welchii* phospholipase C and emulsified in 2 mg of Tween 20 per milliliter containing 1 mM dithiothreitol

n-Heptane/isopropanol/0.5 M H_2SO_4 mixture (10 : 40 : 1, v/v/v)

n-Heptane

$KHCO_3$, 0.1 M

NaCl, 1 M

Na_2CO_3-impregnated silica gel 60. This is prepared by drying a slurry containing 55 g of silica gel 60 (Merck) and 130 ml of 10 mM Na_2CO_3 at 110°.

n-Heptane/diethylether (3 : 1, v/v)

Bray's scintillator solution[5]

Procedure. The reaction mixture contains 10 μmol of Tris-HCl buffer, pH 7.4, 1.5 μmol of $MgCl_2$, 1 μmol of dithiothreitol, 0.2 mg of bovine serum albumin, 30 μmol of sucrose, 50 nmol of [1-^{14}C]palmitoyl-CoA, 2 μmol of 1,2-diacyl-*sn*-glycerol, 0.2 mg of Tween 20, and enzyme (up to 0.44 mU) in a total volume of 0.25 ml. After the addition of the 1,2-diacyl-*sn*-glycerol emulsion, the mixture is prewarmed at 37° for 10 min, and then the reaction is initiated by the addition of enzyme. After incubation at 37° for 5 min with shaking, the reaction is terminated by adding 1.5 ml of *n*-heptane/isopropanol/0.5 M H_2SO_4 (10 : 40 : 1, v/v/v). Thirty to sixty minutes later, 1 ml of 0.1 M $KHCO_3$ and 1 ml of *n*-heptane are added, and the mixture is shaken vigorously with a Vortex mixer. After the phases have separated, the upper phase is collected. The lower phase is extracted again with 1 ml of *n*-heptane in the same manner. The two *n*-heptane extracts are combined and washed with 1.5 ml of 1 M NaCl. The washed extract is applied to a Na_2CO_3-impregnated silica gel 60 column (0.6 × 5.3 cm) which is equilibrated with *n*-heptane, and the column is eluted with 4 ml of *n*-heptane/diethyl ether (3 : 1, v/v). The eluate is taken to dryness in

[5] G. A. Bray, *Anal. Biochem.* **1**, 279 (1960).

a counting vial, and the radioactivity is determined by scintillation counting in 5 ml of Bray's solution.

Units. One unit of glycerolipid acyltransferase activity is defined as the amount that catalyzes the formation of 1 μmol of the product per minute under the assay conditions described. Specific activity is expressed as units per milligram of protein. Protein is determined by the method of Lowry *et al.*[6] with bovine serum albumin as standard; particulate protein is solubilized with 48 mM sodium deoxycholate prior to the determination.

Separation Procedure[2,4,7]

The glycerolipid acyltransferases, like glycerophosphate acyltransferase,[1] are partially purified from microsomes resolved with Triton X-100. The detergent concentration required for the dissociation of the acyltransferases from the microsomal membrane varies with the individual enzymes. Glycerophosphate acyltransferase is most readily solubilized from the membrane with Triton X-100. An increase in the detergent concentration results in the detachment of 1-acylglycerophosphorylcholine acyltransferase and diacylglycerol acyltransferase from the membrane. 1-Acylglycerophosphate acyltransferase is hardly solubilized with Triton X-100. This effect can be exploited for the separation of glycerolipid acyltransferases from one another. For stabilization of the enzymes, 20% (v/v) ethylene glycol is used throughout the procedure. All operations are carried out at 0–4°.

Resolution of Rat Liver Microsomes. Rat liver microsomes are prepared as described for glycerophosphate acyltransferase[1] and resolved with Triton X-100 in the presence of 20 mM glycine-NaOH buffer, pH 8.6, and 20% (v/v) ethylene glycol at a protein concentration of 10 mg/ml; in case of preparing diacylglycerol acyltransferase, the resolution medium contains in addition 2 mM dithiothreitol. The adequate concentration of Triton X-100 varies somewhat with different batches of microsomes, ranging from 6 mM to 8 mM.[8]

Sepharose 2B Column Chromatography. The resolved microsomes (6 ml) are applied to a Sepharose 2B column (2.6 × 16 cm) equilibrated with 20 mM glycine–NaOH buffer, pH 8.6, containing 0.25 mM Triton X-100 and 20% (v/v) ethylene glycol; in case of preparing diacylglycerol acyl-

[6] O. H. Lowry, N. J. Rosebrough, A. L. Farr, and R. J. Randall, *J. Biol. Chem.* **193,** 265 (1951).
[7] S. Yamashita, N. Nakaya, Y. Miki, and S. Numa, *Proc. Natl. Acad. Sci. U.S.A.* **72,** 600 (1975).
[8] The average molecular weight of Triton X-100 is taken to be 628.

transferase, the buffer contains in addition 0.5 mM dithiothreitol. The column is eluted with the same solution, and 2-ml fractions are collected. The most turbid three fractions that appear near the void volume contain the majority of the activities of 1-acylglycerophosphate acyltransferase, 1-acylglycerophosphorylcholine acyltransferase, and diacylglycerol acyltransferase. These fractions are combined and subjected to sucrose density gradient centrifugation.

Sucrose Density Gradient Centrifugation. A two-layered sucrose density gradient is constructed; the lower phase is 0.3 ml of 2 M sucrose, and the upper phase is 3.4 ml of a linear gradient (0.5 to 1.1 M) containing 20 mM glycine–NaOH buffer, pH 8.6, and 20% (v/v) ethylene glycol; in case of preparing diacylglycerol acyltransferase, the upper phase contains in addition 0.5 mM dithiothreitol. Onto this gradient is applied 1.5 ml of the enzyme solution eluted from the Sepharose 2B column. The tube is centrifuged in a Beckman SW65L rotor at 65,000 rpm for 7 hr. 1-Acylglycerophosphorylcholine acyltransferase and diacylglycerol acyltransferase are found near the bottom of the gradient, whereas 1-acylglycerophosphate acyltransferase remains near the top of the gradient. Glycerophosphate acyltransferase is located in the middle of the gradient.[1] The results of a typical experiment are represented in Fig. 1. The specific activities of the partially purified glycerolipid acyltransferases are shown in the table. Although 1-acylglycerophosphorylcholine acyltransferase and diacylglycerol acyltransferase are not

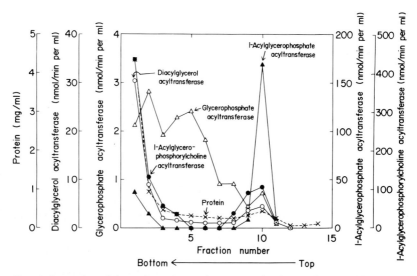

FIG. 1. Separation of the acyltransferases by sucrose density gradient centrifugation. For experimental details, see the text. From Hosaka *et al.*[4]

SPECIFIC ACTIVITIES OF PARTIALLY PURIFIED GLYCEROLIPID ACYLTRANSFERASES

Fraction	1-Acylglycerophosphate acyltransferase (mU/mg)	1-Acylglycerophosphorylcholine acyltransferase (mU/mg)	Diacylglycerol acyltransferase (mU/mg)
Microsomes	58	75	4.8
Partially purified enzyme	440	680	8.9

separated from each other by the procedure described, these two enzymes are distinguishable with respect to heat stability and sensitivity to sulfhydryl-binding reagents.[4]

Properties

Acyl-Donor Specificities.[2,4,7,9] 1-Acylglycerophosphate acyltransferase exhibits a significant specificity for monoenoic and dienoic acyl-CoA thioesters. The effectiveness of various acyl donors is in the order oleoyl-CoA > linoleoyl-CoA ≈ palmitoleoyl-CoA > palmitoyl-CoA ≈ myristoyl-CoA. Arachidonoyl-CoA, stearoyl-CoA, and lauroyl-CoA are poor substrates. 1-Acylglycerophosphorylcholine acyltransferase utilizes arachidonoyl-CoA most efficiently. Linoleoyl-CoA and oleoyl-CoA are also fairly effective substrates. Palmitoleoyl-CoA, lauroyl-CoA, myristoyl-CoA palmitoyl-CoA, and stearoyl-CoA are almost ineffective. On the other hand, diacylglycerol acyltransferase possesses a broad acyl-donor specificity, utilizing saturated, monoenoic, dienoic, and tetraenoic acyl-CoA thioesters efficiently.

Acyl-Acceptor Specificities.[9] The activities of 1-acylglycerophosphate acyltransferase and 1-acylglycerophosphorylcholine acyltransferase are essentially not affected by the fatty acid constituent of the respective acyl acceptors, except that the 1-stearoyl and 1-arachidonoyl derivatives are somewhat less effective acyl acceptors for both enzymes. 1-Acylglycerophosphorylcholine acyltransferase utilizes several acyl acceptors differing in the polar head group with the following order of effectiveness: 1-acyl-*sn*-glycero-3-phosphorylcholine > 1-acyl-*sn*-glycero-3-phosphoryldimethylethanolamine > 1-acyl-*sn*-glycero-3-phosphorylmonomethylethanolamine ≈ 1-acyl-*sn*-glycero-3-phosphorylethanolamine; 1-acyl-*sn*-glycero-3-phosphorylethanol and 1-acyl-*sn*-glycerol 3-phosphate are ineffective. In con-

[9] Y. Miki, K. Hosaka, S. Yamashita, H. Handa, and S. Numa, *Eur. J. Biochem.* **81**, 433 (1977).

trast, 1-acylglycerophosphate acyltransferase is highly specific for 1-acyl-*sn*-glycerol 3-phosphate. The fatty acid constituent of the acyl acceptors exerts essentially no effect on the acyl-donor specificity of 1-acylglycero-phosphate acyltransferase and 1-acylglycerophosphorylcholine acyltrans-ferase except in case of 1-arachidonoyl-*sn*-glycero-3-phosphorylcholine. Furthermore, the acyl-donor specificity of 1-acylglycerophosphoryl-choline acyltransferase is virtually not affected by the polar head group of the acyl acceptor.

Enzymatic Basis for Asymmetric Fatty Acid Distribution in Glycerolipids[2,4,7,9-12]

The acyl constituents of naturally occurring glycerolipids are generally distributed in a nonrandom manner, saturated fatty acids lying at the C-1 position of the glycerol moiety and unsaturated fatty acids at the C-2 position.[13,14] In this context, the above-mentioned substrate specificities of the glycerolipid acyltransferases, together with that of glycerophos-phate acyltransferase,[1] are schematically summarized in Fig. 2. The C-1 position of *sn*-glycerol 3-phosphate is first acylated by glycerophosphate acyltransferase with saturated acyl-CoA, such as palmitoyl-CoA or stearoyl-CoA. The C-2 position of the 1-acyl-*sn*-glycerol 3-phosphate formed is then acylated mainly with monoenoic or dienoic acyl-CoA, such as oleoyl-CoA or linoleoyl-CoA, to produce phosphatidic acid. 1-Acyl-glycerophosphate acyltransferase, which catalyzes this reaction, strictly selects 1-acyl-*sn*-glycerol 3-phosphate as acyl acceptor and does not intro-duce arachidonoyl-CoA at this step. Since this enzyme can utilize any 1-acyl-*sn*-glycerol 3-phosphate derivative regardless of which fatty acid is bound to the C-1 position, the 1-acyl-*sn*-glycerol 3-phosphate containing saturated fatty acid, which is formed by the action of glycerophosphate acyltransferase, is simply further acylated at the C-2 position. The phos-phatidic acid thus formed should therefore contain saturated fatty acid at the C-1 position and monoenoic or dienoic fatty acid at the C-2 position. Since diacylglycerol acyltransferase exhibits a broad acyl-donor spe-cificity, the C-3 position of 1,2-diacyl-*sn*-glycerol can be acylated by any acyl-CoA thioester present in the pool. Thus, it is evident that the non-

[10] S. Yamashita and S. Numa, *Eur. J. Biochem.* **31**, 565 (1972).

[11] S. Numa and S. Yamashita, *Curr. Top. Cell. Regul.* **8**, 197 (1974).

[12] S. Numa, T. Kamiryo, M. Mishina, S. Tashiro, S. Nakanishi, S. Horikawa, T. Tanabe, S. Yamashita, K. Hosaka, U. Schiele, and Y. Miki, *in* "Biochemical Aspects of Nutrition" (K. Yagi, ed.), p. 91. Japan Sci. Press, Tokyo, 1979.

[13] N. H. Tattrie, *J. Lipid Res.* **1**, 60 (1959).

[14] D. J. Hanahan, H. Brockerhoff, and E. J. Barron, *J. Biol. Chem.* **235**, 1917 (1960).

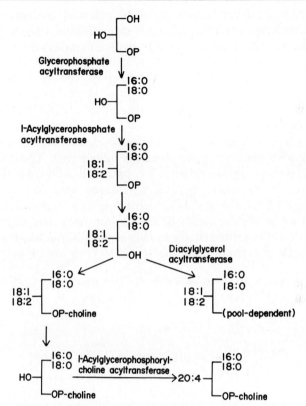

FIG. 2. Specific introduction of fatty acids into glycerolipids based on the substrate specificities of the acyltransferases. From Numa et al.[12]

random fatty acid distribution in monoenoic and dienoic species of glycerolipids is accomplished at the step of de novo synthesis of phosphatidic acid. On the other hand, glycerophospholipids in animal tissues are known to contain large amounts of arachidonic acid.[15] In this connection, the substrate specificity of 1-acylglycerophosphorylcholine acyltransferase appears to be of great importance. This enzyme strictly recognizes the polar basic group at the C-3 position of the acyl acceptor and utilizes arachidonyl-CoA as the best acyl donor. This finding strongly supports the view that a portion of the newly synthesized glycerophospholipids is once deacylated at the 2-position and then reacylated with arachidonyl-CoA to form tetraenoic species of glycerophospholipids via the deacylation–reacylation cycle.[16] Thus, it is concluded that the substrate spe-

[15] L. L. M. Van Deenen, in "Regulatory Functions of Biological Membranes" (J. Järnefelt, ed.), p. 72. Elsevier, Amsterdam, 1968.
[16] E. E. Hill and W. E. M. Lands, Biochim. Biophys. Acta 152, 645 (1968).

cificities of the acyltransferases make an essential contribution to the specific introduction of fatty acids into glycerolipids, effecting the non-random fatty acid distribution in naturally occurring glycerolipids.

Notes

2-Acylglycerophosphate Acyltransferase (EC 2.3.1.52; Acyl-CoA :2-Acyl-sn-glycerol-3-phosphate O-Acyltransferase). This enzyme is separated from glycerophosphate acyltransferase by sucrose density gradient centrifugation of rat liver microsomes resolved with Triton X-100[2]; it exhibits a predominant selectivity for saturated acyl-CoA thioesters.[2] Stearoyl-CoA is the most effective acyl donor, and palmitoyl-CoA is utilized fairly efficiently. 2-Acylglycerophosphate acyltransferase may also contribute to the formation of nonrandom fatty acid distribution in glycerolipids, although its activity in microsomes is much lower than that of 1-acylglycerophosphate acyltransferase.[17]

[17] H. Okuyama, H. Eibl, and W. E. M. Lands, *Biochim. Biophys. Acta* **248,** 263 (1971).

[62] 1,2-Diacylglycerol : CDPcholine Cholinephosphotransferase

(EC 2.7.8.2) and 1,2-Diacylglycerol : CDPethanolamine Ethanolaminephosphotransferase (EC 2.7.8.1) from Rat Liver

By HIDEO KANOH and KIMIYOSHI OHNO

Cholinephosphotransferase:

1,2-Diacylglycerol + CDPcholine ↔ Phosphatidylcholine + CMP

Ethanolaminephosphotransferase:

1,2-Diacylglycerol + CDPethanolamine ↔

phosphatidylethanolamine + CMP

The assay method and basic properties of cholinephosphotransferase have been previously described by Kennedy.[1] Since then, the studies of the two transferases have been hampered by the two main difficulties: the use of water-insoluble substrate, diacylglycerol, and the microsome-

[1] E. P. Kennedy, this series, Vol. 5 [65d].

bound nature of the enzymes. The two transferases, however, have been recently shown to possess different substrate specificity with regard to the acyl chain lengths[2-5] and unsaturation[3,5] of diacylglycerol. There are also accumulating data[6] showing that the activities of choline- and ethanolaminephosphotransferases reside in two different enzymes.

Assay Methods

Assay Method Using Diacylglycerol Suspension[4]

Reagents

Tris-HCl, 0.50 M, pH 8.0
Sodium deoxycholate, 20 mM (stored at room temperature)
Tween-20, 1.0%
MgCl$_2$, 0.20 M
CDP[Me-^{14}C]choline 5.0 mM; specific radioactivity, 0.2–0.4 μCi/ μmol
CDP[2-^{14}C]ethanolamine 5.0 mM; specific radioactivity, 0.2–0.4 μCi/μmol
Enzyme: microsomal fraction from rat liver, 5 mg of protein per milliliter
Diacylglycerol prepared from egg phosphatidylcholine by phospholipase C (*Clostridium welchii*) treatment.[1] Diacylglycerol, after being completely dried under N$_2$, is dissolved in *n*-hexane (10 μmol/ml). The formation of 1,3-isomer is negligible for 2–3 weeks when stored at $-20°$. Diacylglycerol suspension is prepared immediately before the experiments. For this purpose, an appropriate amount of the hexane solution is dried in a small conical tube under N$_2$. Per 1 μmol of the dried diacylglycerol are added 0.4 ml of Tris-HCl and 0.2 ml each of deoxycholate and Tween-20. The tube is then sonicated by a Branson sonifier at $0°$ three times each for 30 sec at maximum intensity using a microprobe.[7]

[2] R. Coleman and R. M. Bell, *J. Biol. Chem.* **252**, 3050 (1977).
[3] B. J. Holub, *J. Biol. Chem.* **253**, 691 (1978).
[4] K. Morimoto and H. Kanoh, *J. Biol. Chem.* **253**, 5056 (1978).
[5] H. Kanoh and K. Ohno, *Biochim. Biophys. Acta* **380**, 199 (1975).
[6] H. Kanoh and K. Ohno, *Eur. J. Biochem.* **66**, 201 (1976).
[7] The accurate concentration of diacylglycerol incubated can be determined by extracting the sonicated mixture with organic solvents as described in the text. The sonication at 25° or higher temperature is done to disperse fully saturated or chemically synthesized diacylglycerols containing long-chain fatty acids.[4] Sonication without deoxycholate[1] or the use of ethanol-dispersed diacylglycerol[2] has also been described.

Procedure.[4] Aliquots of the following reagents are added to a small centrifuge tube (1.4 × 10 cm) placed in an ice bath: 0.1 ml of the diacylglycerol suspension; 0.025 ml of $MgCl_2$; 0.025 ml of either CDP[Me-^{14}C]choline (for cholinephosphotransferase) or CDP[2-^{14}C]ethanolamine (for ethanolaminephosphotransferase); water and finally less than 0.02 ml of enzyme to give a final volume of 0.25 ml. The solution is incubated with vigorous shaking for 5 min at 30°. The reaction is started by adding the labeled substances or enzyme, but more conveniently by placing the tube containing the whole mixture in a shaking water-bath. The reaction is stopped by adding 2.5 ml of chloroform/methanol (1 : 2, v/v). After mixing thoroughly, 1 ml of chloroform is added to each tube. The tube is vortexed, and then 2 ml of 0.9% NaCl are added. After a vigorous shaking, the two phases are obtained by centrifuging for 5 min at 2000 rpm. The upper phase is sucked off carefully with a capillary pipette and discarded. The lower phase is washed twice with 3 ml of methanol/water (1 : 1, v/v) saturated with chloroform. Protein that precipitates at the interphase can be aspirated together with the upper phase or may be left untouched. The washed chloroform layer is dried completely under N_2 and redissolved in 1.0 ml of toluene/methanol (9 : 1, v/v). An aliquot, usually 0.5 ml, is taken for radioactivity determination in toluene scintillator. For both transferases, more than 95% of radioactivity in the washed chloroform phase is associated with phosphatidylcholine or phosphatidylethanolamine. Control incubations with all reaction components except for diacylglycerol should always be included. The incorporations of labeled substances without addition of diacylglycerol are subtracted from those obtained in the standard incubation. The enzyme activities are calculated from the specific radioactivities of labeled substances and the radioactivities in the washed chloroform phase. They are usually expressed as nanomoles of phospholipids formed per minute per milligram of protein or as milliunits per milligram of protein.

Comments. There is always some formation of phospholipids without added diacylglycerol. This is due to the microsomal endogenous diacylglycerol (about 10 nmol per milligram of protein).[8] The incorporation of labeled substrates independent of exogenous diacylglycerol is 8–10% for cholinephosphotransferase, and 10–15% for ethanolaminephosphotransferase, of the activities observed in the standard condition. In the case of ethanolaminephosphotransferase, Mn^{2+} is more effective than Mg^{2+}, although the two enzymes have been studied in most cases with Mg^{2+}. Diacylglycerol suspension prepared as described gives reasonably reproducible enzyme activities in repeated experiments. However,

[8] H. Kanoh and K. Ohno, *Biochim. Biophys. Acta* **326**, 17 (1973).

different microsomal preparations show considerably variable enzyme activities. When measured in the standard reaction conditions, specific enzyme activities of about 5 mU/mg and 3 mU/mg of microsomal protein are most often found for cholinephosphotransferase and ethanolaminephosphotransferase, respectively. Liver microsomes contain several enzyme activities that degrade the cytidine nucleotides[9] and lipids. These coexisting enzymes do not seriously affect the transferase activities measured in the standard condition. For both transferases, from 0.25 to 1.0 mM of diacylglycerol is effective almost to the same extent. Substrate inhibition of the two enzyme activities by the added diacylglycerol at 1.0 mM or higher concentrations is very often noticed.[4]

Assay Method Using Membrane-Bound Substrates

For assaying the two transferases, several attempts have been made to circumvent the solubility problem of diacylglycerol, and also to mimic the physiological state of the lipid substrate. In these studies, microsomes are pretreated in several ways to accumulate endogenous diacylglycerol. The treated microsomes are reincubated with labeled CDPcholine or CDPethanolamine to assess the transferase activities. Microsomes serve, therefore, as the source of enzymes as well as substrates. With the membrane-bound substrates, the transferases can be assayed without using detergents.

Four general principles are involved in these assays.

1. Microsomes are first incubated in the conditions for phosphatidate synthesis (in the presence of acyl-CoA and glycerophosphate). The accumulated phosphatidate is subsequently hydrolyzed by phosphatidate phosphatase from the liver cytosol to yield microsomes containing diacylglycerol.[10,11]
2. Microsomes are treated with phospholipase C.[12] The membrane-bound diacylglycerol derived from microsomal phospholipids is utilized by cholinephosphotransferase.
3. Before isolation of microsomes, rats are injected with radioactive precursors such as [Me-^{14}C]choline and [2-^{14}C]ethanolamine.[9] The prelabeled microsomes are incubated with CMP to study the release of CDPcholine or CDPethanolamine. This method uses the reversibility[1] of the action of both transferases, and the lipid sub-

[9] H. Kanoh and K. Ohno, *Biochim. Biophys. Acta* **306**, 203 (1973).
[10] M. P. Mitchel, D. N. Brindley, and G. Hübscher, *Eur. J. Biochem.* **18**, 214 (1971).
[11] H. J. Fallon, J. Barwick, R. G. Lamb, and H. van den Bosch, *J. Lipid Res.* **16**, 107 (1975).
[12] W. C. McMurray, *Can. J. Biochem.* **53**, 784 (1975).

strates are phosphatidylcholine or phosphatidylethanolamine in microsomal membranes.

4. The back-reaction of cholinephosphotransferase degrading microsomal phosphatidylcholine is utilized for generating membrane-bound diacylglycerol.[5,8] It is necessary to preincubate microsomes with diisopropylfluorophosphate to suppress the microsomal lipolytic activity, which actively degrades endogenous diacylglycerol. The microsomes recovered after incubation with CMP contain, per milligram of protein, about 90 nmol of diacylglycerol which has molecular species similar to those of original microsomal phosphatidylcholine.[8] The generated diacylglycerol is actively utilized by the two transferases as well as by diacylglycerol: acyl-CoA acyltransferease. Treatment of microsomes with diisopropylfluorophosphate does not affect the two transferase activities.

Comments. The amounts of membrane-bound diacylglycerols generated by these procedures are usually less than 100 nmol per milligram of protein. The activities of choline- and ethanolaminephosphotransferases using membrane-bound substrates are much higher than those estimated with diacylglycerol suspension.[5,8] Therefore, a linear incorporation of the labeled nucleotides into phospholipids is observed only for a very short incubation time, usually less than 5 min. In the treatment of microsomes with CMP,[8] or with phospholipase C,[13] more than half of the generated diacylglycerol is hydrolyzed by microsomal lipase, resulting in the simultaneous accumulation of free fatty acids. Interestingly, diacylglycerol species containing arachidonic acid is degraded more rapidly than the other species.[8,13]

Preparation of Enzymes

Preparation of Microsomes. Male or female rats, 200–250 g body weight, are used after an overnight fast. Livers are thoroughly perfused with ice-cold 0.9% NaCl, and then homogenized with 4 ml per gram wet weight of 0.25 M sucrose/1 mM EDTA (pH adjusted to 7.4 with 1 M Tris). The homogenates are first centrifuged at 10,000 g for 20 min at 0–4°. Microsomal fractions are then sedimented by centrifuging at 105,000 g for 1 hr. The microsomes are washed at least twice with either 0.1 M potassium phosphate (pH 7.4) or 1.15% KCl or 0.1 M Tris-HCl (pH 7.4).[9,14] The

[13] H. Kanoh and B. Åkesson, *Eur. J. Biochem.* **85,** 225 (1978).

[14] The washed microsomes contain, per milligram of protein, 0.63 μmol of total phospholipids, 0.38 μmol of phosphatidylcholine, 0.16 μmol of phosphatidylethanolamine, and 0.01 μmol of diacylglycerol. These lipid contents are much higher than those found in microsomes unwashed or washed with the sucrose solution only.

washed microsomes are finally suspended in the original sucrose solution or in 0.1 M Tris-HCl (pH 7.4)/1 mM EDTA (5 mg of protein per milliliter). The suspension is divided in 1–3 ml portions into small test tubes placed in an ice bath. The tubes are thoroughly flushed with N_2 and are quickly rubber-stoppered. When stored at below $-20°$, the activities of both transferases are stable for more than 1 month.

Solubilization of Transferases[6]

Materials

Microsomes suspended in 0.1 M Tris-HCl (pH 7.4)/1 mM EDTA (5 mg of protein per milliliter)

Total microsomal phospholipids suspended by sonication in 10 mM Tris-HCl (pH 7.4) at a concentration of 5 mM on the basis of lipid–phosphorus

Sodium deoxycholate, 16 mM and 20 mM, freshly prepared daily

80% Glycerol/2 mM dithiothreitol

Buffer A: 0.1 M Tris-HCl (pH 8.5)/1 mM EDTA/0.5 mM dithiothreitol

Buffer B: 50 mM Tris-HCl (pH 8.5)/20% glycerol/0.5 mM EDTA/0.5 mM dithiothreitol

All procedures are carried out at 0–4°. The enzyme activities are assayed with diacylglycerol suspension as described except that the assay mixture for cholinephosphotransferase contains 0.5 mM total microsomal phospholipids. The amount of deoxycholate in the diacylglycerol suspension is adjusted to give a final concentration of 2 mM in the assay mixture.

Step 1. Sonication at pH 7.4. To 4.0 ml of the microsomal suspension are added 2.0 ml each of 16 mM deoxycholate and 80% glycerol/2 mM dithiothreitol. The mixture is gently stirred and is sonicated for 3 min at maximum intensity using a microtip. The sonicated mixture is then centrifuged at 225,000 g for 40 min. The pellet obtained is washed once with buffer A, and suspended in the same buffer at a concentration of 5 mg of protein per milliliter. Both transferase activities are recovered in the pellet without loss of activities. The preparation can be stored at $-20°$ for at least a week.

Step 2. Sonication at pH 8.5. Two milliliters of step 1 preparation is mixed with 1.0 ml each of 80% glycerol/2 mM dithiothreitol and 20 mM deoxycholate. After standing in an ice bath for 30 min or longer, the mixture is sonicated and centrifuged as described in step 1. The saved supernatant contains about 70% of the recovered enzyme activities. There is always some inactivation of the enzymes at this step, which is to some extent variable depending on different microsomal preparations. The solubilized enzymes are extremely heat-sensitive, and 15–30% of both activities is lost upon storage overnight at 0–4°.

Step 3. Dialysis. Step 2 preparation is dialyzed against 100 volumes of buffer B for more than 20 hr. After being dialyzed, both enzymes are sedimented by centrifugation at 225,000 g, and the pelleted enzymes are again solubilized by sonication in buffer B containing 5 mM deoxycholate (3 mg of protein per milliliter).

Step 4. Sucrose Density Gradient Centrifugation. Step 3 preparation (1–2.0 ml) is layered over 29 ml of a linear 5 to 20% sucrose gradient prepared in buffer B containing 5 mM deoxycholate. The tubes are centrifuged at 60,000 g for 15 hr in a swinging bucket rotor, and 1-ml fractions are collected from the bottom of the tube. Both transferases form a single peak usually at the seventeenth fraction from the bottom. Considerable losses in activity of cholinephosphotransferase occur during this procedure (20–40% recovery), and over 70% of ethanolaminephosphotransferase activity is recovered. The results of the purification procedures are summarized in the table. The partially purified enzymes obtained at step 3 are, although apparently solubilized, still largely aggregates of proteins and phospholipids. Attempts to purify the enzyme further by gel filtration, ammonium sulfate fractionation, and DEAE-cellulose chromatography all failed.

Properties

Substrate Specificity. The two transferases are affected differently by the acyl compositions of diacylglycerol. Ethanolaminephosphotransferase shows a marked preference for sonicated[3] and, in particular, membrane-bound[5] 2-docosahexaenoyl containing diacylglycerol in contrast to cholinephosphotransferase, which shows no specificity for diacylglycerol species with different degrees of unsaturation. Acyl chain lengths of saturated fatty acids located at the C-1 position of the diacylglycerol also affect the two enzymes differently. As seen in Fig. 1, a series of 1-acyl, 2-oleoylglycerols, ranging from 1-lauroyl to 1-heptadecanoyl species, are equally well utilized by cholinephosphotransferase, whereas in the case of ethanolaminephosphotransferase, 1-heptadecanoyl and 1-stearoyl species are used most actively, and the introduction of saturated fatty acids shorter than heptadecanoic acid markedly reduces the enzyme activity. As described earlier,[1] fully saturated diacylglycerols like dipalmitoylglycerol are poorly used by both transferases. In the lung, dipalmitoylglycerol generated within the microsomal membranes is also poorly used by cholinephosphotransferase.[15] This is not due to the presence of

[15] M. G. Sarzala and L. M. G. Van Golde, *Biochim. Biophys. Acta* **441**, 423 (1976).

PURIFICATION OF CHOLINEPHOSPHOTRANSFERASE AND ETHANOLAMINEPHOSPHOTRANSFERASE FROM RAT LIVER MICROSOMES[a]

Fraction or step	Protein (mg)	Phospholipids [μmol(μmol/mg)]	Cholinephosphotransferase[b]		Ethanolaminephosphotransferase[b]	
			Specific activity [mU/mg(purification)]	Total activity [mU (recovery)]	Specific activity [mU/mg (purification)]	Total activity [mU (recovery)]
Microsomes	238	150 (0.63)	5.0 (1)	1190 (100)	2.6 (1)	619 (100)
Step 1. Sonication at pH 7.4 (particulate)	66.6	84.6 (1.3)	13.2 (2.6)	879 (74)	7.2 (2.8)	480 (78)
Step 2. Sonication at pH 8.5 (solubilized)	50.0	63.0 (1.3)	12.1 (2.4)	605 (51)	6.7 (2.6)	335 (54)
Step 3. Dialysis and sonication	25.0	40.8 (1.6)	20.9 (4.2)	523 (44)	12.1 (4.7)	303 (49)
Step 4. Sucrose density gradient contrifugation (peak fraction)	2.6	4.9 (1.9)	15.8 (3.2)	41 (3.2)	22.1 (8.5)	57 (9,2)

[a] Reproduced in part from H. Kanoh and K. Ohno.[6]

[b] Both enzyme activities were measured as described in the assay procedure except that the pH of the buffer was 8.5 and the reaction mixture for cholinephosphotransferase (but not ethanolaminephosphotransferase) contained 0.5 mM total microsomal phospholipids.

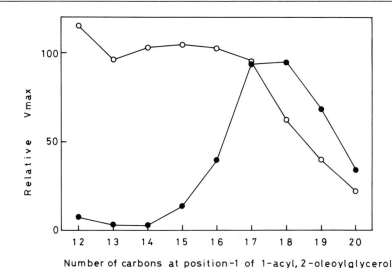

Number of carbons at position-1 of 1-acyl, 2-oleoylglycerol

FIG. 1. Effect of different 1-saturated, 2-oleoyl-*sn*-glycerols on cholinephospho-transferase and ethanolaminephosphotransferase. A series of 1-acyl, 2-oleoylglycerols containing odd- as well as even-numbered saturated fatty acids, ranging from lauric to arachidic acids, were tested as substrates for the solubilized cholinephosphotransferase (○) and ethanolaminephosphotransferase (●). The V_{max} values for dioleoylglycerols are taken as 100, and the relative values for the diacylglycerol species are presented. Adapted from Morimoto and Kanoh.[4]

saturated fatty acids at the C-2 instead of C-1 position, since several 1-oleoyl, 2-saturated types of diacylglycerol are effectively used by the two transferases.[4] Interestingly the acyl chain length specificity of the enzymes as presented in Fig. 1 becomes obscured when 1-unsaturated, 2-saturated types of substrates are used.[4]

Reversibility. When incubated with CMP, both transferases degrade phospholipids in microsomal membranes through their back-reactions, liberating diacylglycerol, and CDPcholine or CDPethanolamine.[9] The activity of the back-reaction of cholinephosphotransferase is much higher than that of ethanolaminephosphotransferase. Both transferases show similar K_m values for CMP (0.19 mM for cholinephosphotransferase and 0.14 mM for ethanolaminephosphotransferase). The product inhibition by CDPethanolamine (K_i 0.05 mM) is much more potent than that by CDPcholine (K_i 1.0 mM). In the back-reaction microsomal cholinephosphotransferase uses different phosphatidylcholine species equally well.[9]

Kinetic Parameters. With the solubilized enzymes (step 3 preparation), the K_m values for CDPcholine and CDPethanolamine are 36.4 μM and 22.0 μM, respectively. The K_m values for suspended diacylglycerol are 81 μM for cholinephosphotransferase and 63 μM for ethanolaminephosphotransferase. There is not much difference in the K_m values when 1-acyl, 2-oleoyl, or 1-oleoyl, 2-acylglycerols of different acyl chain lengths are tested.[4] CDPcholine and CDPethanolamine inhibit ethanolaminephosphotransferase and cholinephosphotransferase competitively, with a K_i value of 350 μM for both.

Activators and Inhibitors. Most of the properties are obtained from the solubilized enzymes (step 3 preparation).[6] In general the observations are reproduced with membrane-bound enzymes. Cholinephosphotransferase requires Mg^{2+} for its activity, and Mn^{2+} can partially replace Mg^{2+}. On the other hand, Mn^{2+} is much more effective than Mg^{2+} for ethanolaminephosphotransferase. Both enzymes are stimulated by deoxycholate. At a concentration higher than 2 mM deoxycholate inhibits cholinephosphotransferase but not ethanolaminephosphotransferase. A stimulatory effect of taurocholate on cholinephosphotransferase has also been noted.[16] Ethanolaminephosphotransferase is stable when treated with 0.5% Triton X-100 or when assayed in the reaction mixture containing up to 0.5% of the detergent. Cholinephosphotransferase is more sensitive to Triton X-100 in comparison to ethanolaminephosphotransferase. The inactivating or inhibitory effect of the detergent depends on the cations used for assaying cholinephosphotransferase. The enzyme activity assayed with Mg^{2+} is more labile to Triton X-100 in comparison to that assayed with Mn^{2+}. Cholinephosphotransferase is stimulated by total microsomal phospholipids or phosphatidylcholine, whereas ethanolaminephosphotransferase, solubilized or membrane-bound, is inhibited by the phospholipids. An observation concerning the lipid requirements is obtained by treating membrane-bound enzymes with phospholipase A_2.[17] Ethanolaminephosphotransferase activity is not significantly affected even when over 90% of microsomal phosphatidylcholine and phosphatidylethanolamine is degraded. In contrast, the activity of cholinephosphotransferase appears to be dependent on the membrane phosphatidylcholine.

Optimum pH. Both transferases give the highest activity at pH 8.0 to 8.5.[6]

Separation of the Two Transferases. The properties of the two enzymes described above indicate that two different enzymes are involved in the *de novo* syntheses of phosphatidylcholine and phosphatidylethanolamine. The two transferases are partially separated from each other by sucrose density gradient centrifugation of the solubilized enzymes (step 3 preparation) in the presence of Triton X-100.[6] Furthermore, cholinephospho-

[16] D. E. Vance, P. C. Choy, S. B. Farren, P. H. Lim, and W. J. Schneider, *Nature (London)* **270**, 268 (1977).

[17] K. Morimoto and H. Kanoh, *Biochim. Biophys. Acta* **531**, 16 (1978).

transferase itself appears to consist of Mg^{2+}-requiring and Mn^{2+}-requiring components. Since these data are obtained with a considerably inactivated enzyme preparation, further work is necessary to achieve a clear separation of the two enzymes. In this context, a solubilized preparation of ethanolaminephosphotransferase, which is free of cholinephosphotransferase activity, has been obtained by treating microsomes with octyl glucoside.[18]

Enzyme Localization in the Microsomal Membranes. Both choline- and ethanolaminephosphotransferases have been shown to be located on the cytoplasmic surface of the microsomal vesicles.[16,19]

[18] A. Radominska-Pyrek, M. Pilarska, and P. Zimniak, *Biochem. Biophys. Res. Commun.* **85**, 1074 (1978).

[19] R. Coleman and R. M. Bell, *J. Cell Biol.* **76**, 245 (1978).

[63] *sn*-Glycero-3-phosphate Acyltransferase from *Escherichia coli*

By MARTIN D. SNIDER

$$sn\text{-Glycero-3-P} + \begin{array}{l} \text{acyl-CoA} \\ \text{acyl-acyl carrier protein} \end{array}$$

$$\rightarrow 1\text{-acyl-}sn\text{-glycero-3-P} + \begin{array}{l} \text{CoASH} \\ \text{acyl carrier protein} \end{array}$$

Assay

Principle. Glycerophosphate acyltransferase activity is determined by following the conversion of [³H]glycero-3-P to lipid. The method described below is based on the chloroform solubility of the products, lysophosphatidic and phosphatidic acid. A filter paper disk assay, which depends on the trichloroacetic acid insolubility of the lipid products, has also been used.[1]

Reagents

    Assay buffer: 0.2 *M* Tris-HCl, pH 8.4, 0.8 *M* NaCl, 10 m*M* MgCl₂, 40% glycerol

    Bovine serum albumin, 20 mg/ml

    2-Mercaptoethanol, 100 m*M*

[1] H. van den Bosch and P. R. Vagelos, *Biochim. Biophys. Acta* **218**, 233 (1970).

Palmityl-CoA, 0.86 mM

sn-[2-³H]glycero-3-P, 12.5 mM, prepared by the enzymatic phosphorylation of [2-³H]glycerol[2]

Chloroform–methanol (2 : 1, v/v) containing 0.01 N HCl

KCl, 2 N

Triton X-100, 20%

Procedure. Membrane-bound enzyme may be assayed directly. However, detergent-solubilized preparations must be reconsititued with phospholipid prior to assay (see below).

Each incubation contains 0.1 M Tris HCl, pH 8.4, 0.4 M NaCl, 5 mM MgCl$_2$, 20% glycerol, 1 mg/ml bovine serum albumin, 5 mM 2-mercaptoethanol, 43 μM palmityl-CoA, 0.2 unit or less of enzyme, and 1.25 mM sn-[2-³H]glycero-3-P (added last) in a total volume of 0.2 ml. After 10 min of incubation at 30°, reactions are stopped by adding 3 ml of chloroform–methanol (2 : 1, v/v) containing 0.01 N HCl. The acid must be added as the Tris salt of the lysophosphatidic acid product is water soluble. Then 7 ml of 2 N KCl are added, the tubes vortexed and the phases are separated by centrifugation. The upper phases are aspirated, and the chloroform phases are washed with 7 ml of 2 N KCl. Aliquots (1 ml) of each lower phase are transferred to scintillation vials, and 10 μl of 20% Triton X-100 are added. After evaporation of the chloroform on a boiling water bath, incorporated radioactivity is determined by liquid scintillation counting. Glycerophosphate acyltransferase activity corresponds to incorporation of glycero-3-P into lipid dependent on palmityl-CoA. Specific incorporation is usually greater than 95% of the total.

Units. One unit of glycerophosphate acyltransferase is the amount of enzyme required to convert 1 nmol of glycero-3-P to lipid per minute at 30°. Specific activity is defined as the number of units per milligram of protein.

Reconstitution of Glycerophosphate Acyltransferase. As the enzyme is inactive in detergent extracts, reconstitution with unilamellar vesicles of *E. coli* phospholipid must be carried out prior to assay. Phospholipids extracted[3] from freshly grown cultures and stored in benzene at −20° are used. Lipids from stored or commercially frozen cells are unsuitable because they contain substantial amounts of deacylation products. The lipid is lyophilized from benzene solution, resuspended in water (20 mg/ml), and sonicated for 10 min at 4° under nitrogen, in a water-jacketed cell.[4] Aggregated lipid is removed by centrifugation at 10,000 g for 10 min, and the vesicles are stored at 4° for up to 2 weeks. To reconstitute glycero-

[2] Y.-Y. Chang and E. P. Kennedy, *J. Lipid Res.* **8**, 447 (1967).

[3] E. G. Bligh and W. J. Dyer, *Can. J. Biochem. Physiol.* **37**, 911 (1959).

[4] C.-S. Huang, *Biochemistry* **8**, 344 (1969).

phosphate acyltransferase activity, detergent-solubilized enzyme is mixed with phospholipid vesicles at a phospholipid-detergent ratio of 20 (w/w). Assays may be carried out immediately.

Purification

Particulate Preparation. Freshly grown cultures of *E. coli* should be used as enzyme source, since phospholipid degradation products present in stored and commercially available frozen cells cause low yields of glycerophosphate acyltransferase. Cells are harvested by centrifugation at 5000 g for 15 min, washed in 50 mM Tris-HCl, pH 8.4, 5 mM MgCl$_2$, and 5 mM 2-mercaptoethanol, and resuspended in the same buffer in one-fiftieth of the original culture volume. The cells are then broken by passage through a French pressure cell or by sonication at 4° for 5 min in a water-jacketed cell, and the cell debris is removed by centrifugation at 5000 g for 10 min. The membranes are then harvested by centrifugation at 100,000 g for 1 hr. The pellets are resuspended in buffer A (0.1 M Tris-HCl, pH 8.4, 0.4 M NaCl, 5 mM MgCl$_2$, 20% glycerol, and 5 mM 2-mercaptoethanol) and stored at −20°.

Differential Extraction. Purification in these steps results from the selective extraction of the acyltransferase from the membrane. A membrane suspension is centrifuged at 100,000 g for 1 hr. The pellet is resuspended with a Potter–Elvehjem homogenizer at 40 mg of protein per milliliter in 25 mM Tris-HCl, pH 8.4, 5 mM MgCl$_2$, 5 mM 2-mercaptoethanol (buffer B), containing either 0.5 mg or 1 mg or Triton X-100 per milliliter. The suspension is centrifuged at 100,000 g for 1 hr, and the supernatant is decanted. The pellet is then homogenized in twice the volume used for the first extraction in buffer B containing 1 or 2 mg of Triton X-100 per milliliter. After centrifugation at 100,000 g for 1 hr, the supernatant is saved and the pellet is discarded. This second Triton extract contains most of the glycerophosphate acyltransferase activity. Optimal detergent concentrations for these two extractions vary with bacterial strain and growth conditions and must be determined for each strain.

Fractionation with Polyethylene Glycol. One part of 50% polyethylene glycol 6000 is mixed with 4 parts of the second Triton extract, and the mixture is held on ice for 15 min. After centrifugation at 8000 g for 10 min, the supernatant is decanted and the pellet is discarded. The polyethylene glycol supernatant is frozen on Dry Ice and stored at −70°.

The purification is summarized in the table. Purification of 20- to 40-fold is achieved with good yield. In addition, this fractionation resolves glycerophosphate acyltransferase from the other acyltransferase involved in phospholipid synthesis; lysophosphatidate acyltransferase activity can-

PURIFICATION OF GLYCEROPHOSPHATE ACYLTRANSFERASE[a]

Fraction	Total volume (ml)	Total protein (mg)	Glycerophosphate acyltransferase		Yield (%)
			Total activity (units)	Specific activity (units/mg)	
Membranes	3.78	44.4	339	7.6	1.00
Second Triton extract[b]	3.78	3.2	271	83	0.80
Polyethylene glycol supernatant[b]	4.73	2.3	248	105	0.73

[a] Results from a purification with 2.5 g of *Escherichia coli* K12 as starting material.

[b] Fractions containing Triton X-100 were reconstituted with *E. coli* phospholipid immediately prior to assay.

not be detected in the polyethylene glycol supernatant fraction.[5] Ishinaga *et al.* have also reported a procedure for the partial purification of glycerophosphate acyltransferase.[6]

Recombinant plasmids carrying the gene for glycerophosphate acyltransferase, *plsB*, have recently been identified (R. M. Bell, personal communication).[7] Plasmid-containing strains having acyltransferase levels that are higher than wild type by five fold or more are useful as enzyme sources and yield partially purified fractions with correspondingly higher specific activities.

Properties

Phospholipid Requirement. Glycerophosphate acyltransferase appears to have an absolute requirement for phospholipid for activity. Enzyme can be extracted from the membrane in reconstitutable form with a variety of detergents, although the extracts contain no detectable activity. Best reconstitutions are obtained with vesicles of *E. coli* lipid or mixtures of phosphatidylethanolamine and phosphatidylglycerol or cardiolipin.[5] Active enzyme has been shown to be tightly associated with phospholipid.[5]

Stability. All the enzyme preparations described are stable frozen for at least 6 months. Reconstitutable activity in detergent extracts is extremely unstable (half-time of 5 min at 37°). Membrane-bound or reconstituted activity is stabilized by $MgCl_2$, glycerol, 2-mercaptoethanol, and NaCl or KCl; it is stable for at least 4 hr at 37° in buffer A.[5]

[5] M. D. Snider and E. P. Kennedy, *J. Bacteriol.* **130**, 1072 (1977).

[6] M. Ishinaga, M. Nishihara, and M. Kito, *Biochim. Biophys. Acta* **450**, 269 (1976).

[7] M. D. Snider, *J. Biol. Chem.* **254**, 7197 (1979).

pH. Glycerophosphate acyltransferase has a pH optimum of 8.5.[1,8]

Products. The sole product of glycerophosphate acyltransferase is 1-acyl-*sn*-glycero-3-P. The *sn*-2 isomer of lysophosphatidic acid is not found.[1,5,6,8]

Substrate Specificity. Both acyl-CoA and acyl-acyl carrier protein thioesters are active as acyl donors with reported K_m values of 50 and 70 μM respectively.[9,10] In accord with the predominance of saturated moieties in the *sn*-1 position of *E. coli* phospholipids, saturated acyl donors are 4- to 10-fold better substrates than unsaturated acyl donors. The preference for saturates is dependent on glycero-3-P concentration, highest activity ratios (saturates : unsaturates) being found at low glycero-3-P concentrations.[5,11]

K_m values for glycero-3-P from 35 to 560 μM have been reported.[1,5,9,12] This wide range is probably due to a dependence of the K_m on assay conditions.[11] Unlike enzymes from mammalian sources, dihydroxyacetone-P is not a substrate for glycerophosphate acyltransferase from *E. coli*.[1]

[8] H. Okuyama and S. J. Wakil, *J. Biol. Chem.* **248**, 5197 (1973).
[9] T. K. Ray and J. E. Cronan, Jr. *J. Biol. Chem.* **250**, 8422 (1975).
[10] D. R. Lueking and H. Goldfine, *J. Biol. Chem.* **250**, 4911 (1975).
[11] H. Okuyama, K. Yamada, H. Ikezawa, and S. J. Wakil, *J. Biol. Chem.* **251**, 2487 (1976).
[12] M. Kito, M. Lubin and L. I. Pizer, *Biochem. Biophys. Res. Commun.* **34**, 454 (1969).

[64] Glycerophosphate Acyltransferase from Rat Liver

EC 2.3.1.15 Acyl-CoA : *sn*-glycerol-3-phosphate *O*-acyltransferase

By SATOSHI YAMASHITA and SHOSAKU NUMA

$$
\begin{array}{c}
CH_2OH \\
| \\
HOCH \\
| \\
CH_2OPO_3H_2
\end{array}
\quad + RCOSCoA \rightarrow
\begin{array}{c}
CH_2OCOR \\
| \\
HOCH \\
| \\
CH_2OPO_3H_2
\end{array}
\quad + CoASH
$$

Phosphatidic acid, which is an essential intermediate in phospholipid biosynthesis,[1] is formed by sequential acylation of *sn*-glycerol 3-phosphate with intermediate formation of 1-acyl-*sn*-glycerol 3-phosphate.[2,3] This process is catalyzed by two distinct enzymes, glycerophosphate acyltransferase and 1-acylglycerophosphate acyl-

[1] A. Kornberg and W. E. Pricer, Jr., *J. Biol. Chem.* **204**, 345 (1953).
[2] S. Yamashita and S. Numa, *Eur. J. Biochem.* **31**, 565 (1972).

transferase. These two acyltransferases are separated from each other by sucrose density gradient centrifugation of microsomes resolved with Triton X-100, a nonionic detergent.[3] Described here are the assay method, partial purification, and properties of glycerophosphate acyltransferase as well as the separation of this enzyme from 1-acylglycerophosphate acyltransferase. The latter enzyme, together with other glycerolipid acyltransferases, is dealt with elsewhere in this volume.[4]

Assay Method[2]

Principle.[5] The enzyme activity can be assayed by measuring the rate of conversion of radioactive sn-glycerol 3-phosphate into chloroform/methanol-soluble material.

Reagents

Tris-HCl buffer, 0.5 M, pH 7.6
sn-[1(3)-^{14}C]Glycerol 3-phosphate (1000 cpm/nmol), 10 mM
Palmitoyl-CoA, 0.5 mM
CaCl$_2$, 0.1 M
Phosphatidylcholine, 20 mM: soybean phosphatidylcholine dispersed in 0.25 M sucrose containing 10 mM Tris-HCl buffer, pH 7.6, and 1 mM EDTA by sonic oscillation
Methanol saturated with sodium sn-glycerol 3-phosphate
Chloroform
HCl, 0.2 M
Methanol/0.2 M HCl (1 : 1, v/v)
Bray's scintillator solution[6]

Procedure. The reaction mixture contains 20 μmol of Tris-HCl buffer, pH 7.6, 0.25 μmol of sn-[1(3)-^{14}C]glycerol 3-phosphate (1000 cpm/nmol), 10 nmol of palmitoyl-CoA, 1.75 μmol of CaCl$_2$, 0.2 μmol of soybean phosphatidylcholine and enzyme (up to 0.18 mU) in a total volume of 0.35 ml. When microsomes are assayed, CaCl$_2$ and phosphatidylcholine are omitted. The reaction is initiated by the addition of enzyme. After incubation for 5 min at 20°, the reaction is terminated by adding 2 ml of methanol saturated with unlabeled sodium sn-glycerol 3-phosphate. Four milliliters of chloroform and 1 ml of 0.2 M HCl are added, and the tube is shaken vigorously. After the phases have separated, the upper phase is removed by suction, and the lower phase is washed three times with 4 ml of

[3] S. Yamashita, K. Hosaka, and S. Numa, *Proc. Natl. Acad. Sci. U.S.A.* **69**, 3490 (1972).
[4] S. Yamashita, K. Hosaka, Y. Miki, and S. Numa, this volume [61].
[5] E. E. Hill, D. R. Husbands, and W. E. M. Lands, *J. Biol. Chem.* **243**, 4440 (1968).
[6] G. A. Bray, *Anal. Biochem.* **1**, 279 (1960).

methanol/0.2 M HCl (1 : 1, v/v). The washed organic phase is taken to dryness and assayed for radioactivity by scintillation counting in 5 ml of Bray's solution.

Units. One unit of glycerophosphate acyltransferase activity is defined as that amount which catalyzes the incorporation of 1 μmol of *sn*-glycerol 3-phosphate into lipid per minute under the assay conditions described. Specific activity is expressed as units per milligram of protein. Protein is determined by the method of Lowry *et al.*[7] with bovine serum albumin as standard; particulate protein is solubilized with 48 mM sodium deoxycholate prior to the determination.

Purification Procedure[2]

All operations are carried out at 0–5°.

Preparation of Rat Liver Microsomes. Livers from Wistar-strain rats are rinsed with 0.25 M sucrose and homogenized in three volumes of 0.25 M sucrose with a Teflon–glass homogenizer. The homogenate is centrifuged at 10,000 g for 20 min. The supernatant is further centrifuged at 105,000 g for 60 min. The pellet is suspended in 0.25 M sucrose and resedimented by centrifugation at 105,000 g for 60 min. For a large-scale preparation of microsomes, the postmitochondrial supernatant is centrifuged at 44,500 g for 90 min. The pellet is suspended in 0.25 M sucrose and resedimented by centrifugation at 78,500 g for 90 min. The microsomes thus obtained are suspended in 0.25 M sucrose to yield a protein concentration of 30–40 mg/ml and, after being divided into small aliquots, are stored at $-20°$.

Resolution of Microsomes. A suspension of microsomes is added to a solution containing Triton X-100, glycine–NaOH buffer, pH 8.6, and dithiothreitol, the final concentrations of which are 6 mM,[8] 20 mM, and 2 mM, respectively, to give a protein concentration of 10 mg/ml (total volume, 13 ml). The tube is capped, inverted twice, and then allowed to stand for 2 hr.

Sepharose 2B Column Chromatography. The resolved microsomes are passed through a Sepharose 2B column (4.6 × 26 cm) equilibrated with 20 mM glycine–NaOH buffer, pH 8.6, containing 0.25 mM Triton X-100 and 0.5 mM dithiothreitol. The column is eluted with the same solution. The enzyme appears in the first fractions of the eluate that contain protein (30 ml).

Sucrose Density Gradient Centrifugation. Two-layered sucrose gra-

---

[7] O. H. Lowry, N. J. Rosebrough, A. L. Farr, and R. J. Randall, *J. Biol. Chem.* **193**, 265 (1951).

[8] The adequate concentration of the detergent varies slightly with different batches of microsomes. The average molecular weight of Triton X-100 is taken to be 628.

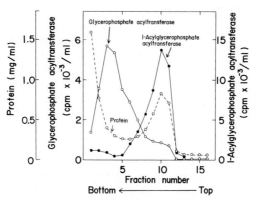

FIG. 1. Resolution of the phosphatidate-synthesizing system by sucrose density gradient centrifugation. For experimental details, see the text. Centrifugation was carried out at 65,000 rpm for 210 min in a Beckman SW65L rotor. In this experiment, 1-acylglycerophosphate acyltransferase activity was assayed at 20° by measuring the conversion of 1-palmitoyl-sn-[1(3)-^{14}C]glycerol 3-phosphate to [^{14}C]phosphatidate. From Yamashita et al.[3]

dients are constructed; the lower layer is 2 ml of 2 M sucrose, and the upper layer is 20 ml of a linear sucrose gradient (0.5 to 1.5 M) containing 20 mM glycine–NaOH buffer, pH 8.6, 0.25 mM Triton X-100 and 0.5 mM dithiothreitol. The enzyme solution is divided into three 10-ml portions, and each portion is layered onto each of triplicate two-layered gradients. The tubes are centrifuged at 24,500 rpm for 14 hr in a Beckman SW25.1 rotor. Alternatively,[3] 1.5 ml of the enzyme preparation is applied to a two-layered sucrose gradient that consists of 0.3 ml of 2 M sucrose and 3.4 ml of a linear sucrose gradient (0.5 to 1.1 M) containing 20 mM glycine–NaOH buffer, pH 8.6, and 0.5 mM dithiothreitol. The tube is centrifuged at 65,000 rpm for 210 min in a Beckman SW65L rotor. The latter procedure facilitates rapid isolation of the enzyme, thus preventing the enzyme from inactivation. As shown in Fig. 1, glycerophosphate acyltransferase is located between the two major protein peaks. The glycerophosphate acyltransferase preparation thus obtained is free from the activities of 1-acylglycerophosphate acyltransferase, 2-acylglycerophosphate acyltransferase, 1-acylglycerophosphorylcholine acyltransferase, and diacylglycerol acyltransferase as well as from the bulk of phospholipids and RNA. The separation of the glycerolipid acyltransferases from one another is described in a separate article of this volume.[4]

The results of a typical purification of glycerophosphate acyltransferase from rat liver microsomes are summarized in the table. Although glycerophosphate acyltransferase is mainly localized in micro-

[9] H. Eibl, E. E. Hill, and W. E. M. Lands, Eur. J. Biochem. 9, 250 (1969).

PURIFICATION OF GLYCEROPHOSPHATE ACYLTRANSFERASE FROM RAT LIVER[a]

Fraction	Protein (mg)	Specific activity (mU/mg)	Total activity (mU)
Microsomes	130	0.61	79
Triton-treated microsomes	130	0.53	69
Sepharose 2B eluate	32	1.2	38
Sucrose density gradient centrifugation	3.6	2.2	7.9

[a] From Yamashita and Numa.[2]

somes,[9] this enzyme is also present in mitochondria. Monroy et al.[10] have achieved a partial purification of the enzyme from rat liver mitochondria by cholate extraction and ammonium sulfate fractionation.

Properties[2]

pH Optimum. The enzyme has a broad pH optimum ranging from 6.6 to 9.0.

Activators and Inhibitors. The enzyme is activated by divalent cations. Ca^{2+} exhibits the strongest stimulatory effect. Mg^{2+} is slightly less effective. Mn^{2+} and Co^{2+} activate the enzyme less efficiently than Mg^{2+}. The enzyme is also activated by soybean phosphatidylcholine. Microsomal phosphatidylcholine and phosphatidylethanolamine are also effective, but microsomal phosphatidylserine and phosphatidylinositol are inhibitory. The enzyme is inhibited by sulfhydryl reagents like N-ethylmaleimide, p-chloromercuribenzoate, and 5,5'-dithiobis(2-nitrobenzoic acid).[11]

Kinetic Properties. The apparent Michaelis constant for sn-glycerol 3-phosphate is 0.2 mM under the standard assay conditions.[2] Because of a strong substrate inhibition, however, it is difficult to measure a Michaelis constant for acyl-CoA in the usual manner.[12]

Acyl-Donor Specificity and Reaction Product. The best acyl donor is palmitoyl-CoA, and stearoyl-CoA is also utilized fairly efficiently. Oleoyl-CoA, linoleoyl-CoA, and arachidonoyl-CoA are poor substrates. The reaction product formed from palmitoyl-CoA and sn-glycerol 3-phosphate is identified exclusively as 1-acyl-sn-glycerol 3-phosphate. This enzyme plays an essential role in the formation of the asymmetric distribution of fatty acids in naturally occurring glycerolipids as described elsewhere in this volume.[4]

[10] G. Monroy, H. C. Kelker, and M. E. Pullman, J. Biol. Chem. 248, 2845 (1979).
[11] W. E. M. Lands and P. Hart, J. Biol. Chem. 240, 1905 (1965).
[12] H. M. Abou-Issa and W. W. Cleland, Biochim. Biophys. Acta 176, 692 (1969).

[65] Phosphatidylglycerophosphate Synthase from *Escherichia coli*[1]

EC 2.7.8.5 CDPdiacylglycerol:*sn*-glycero-3-phosphate phosphatidyltransferase

By WILLIAM DOWHAN and TAKASHI HIRABAYASHI

Phosphatidylglycerophosphate synthase, which catalyzes reaction (1), is a membrane-associated enzyme found in a wide range of eukaryotic[2] and prokaryotic[3] organisms.

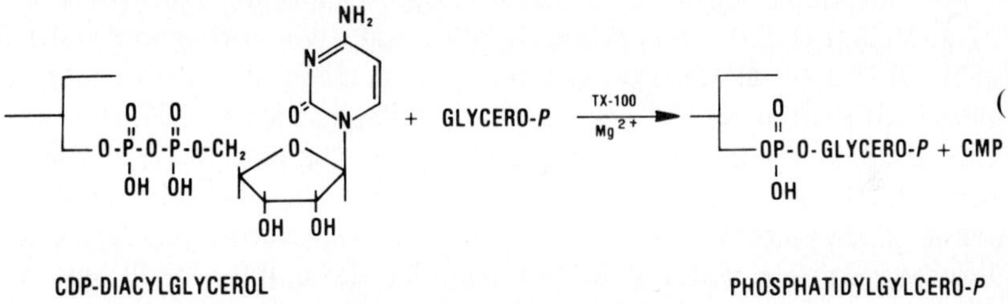

(1)

Chang and Kennedy[4] first solubilized and partially purified the enzyme from *Escherichia coli* membranes using detergent. The enzyme has been purified to near homogeneity from *E. coli* using substrate affinity chromatography.[5] The enzyme appears to be very similar in *Bacillus licheniformis,* from which it has been partially purified by a similar affinity procedure.[6]

Mutants of *E. coli*[7] have been isolated that have very low levels of phosphatidylglycerophosphate synthase, but appear to be phenotypically normal. These mutants have made possible the cloning of the *pgsA* gene locus, the structural gene for the enzyme, using multicopy number plasmids. The enzyme is overproduced in such plasmid-carrying strains by 15- to 20-fold over wild-type levels.[8] Although such strains will most likely be

[1] Supported by Grant 20478 from the National Institute of General Medical Sciences and by Grant AU-599 from the Robert A. Welch Foundation.

[2] H. van den Bosch, L. M. G. van Golde, and L. L. M. van Deenen, *Rev. Physiol.* **66,** 13 (1972).

[3] A. Dutt and W. Dowhan, *J. Bacteriol.* **132,** 159 (1977).

[4] Y.-Y. Chang and E. P. Kennedy, *J. Lip. Res.* **8,** 456 (1967).

[5] T. Hirabayashi, T. J. Larson, and W. Dowhan, *Biochemistry* **15,** 5205 (1976).

[6] T. J. Larson, T. Hirabayashi, and W. Dowhan, *Biochemistry* **15,** 974 (1976).

[7] M. Nishijima and C. R. H. Raetz, *J. Biol. Chem.* **254,** 7837 (1979).

[8] A. Ohta and W. Dowhan, unpublished results, 1979.

ISBN 0-12-181971-X

the best source of the enzyme, the enzyme from such strains has not yet been isolated. The enzyme has been extensively purified only from *E. coli* B cells although K12 strains should give similar results.

Assay Method

The enzyme catalyzes [reaction (1)] the displacement of CMP from CDPdiacylglycerol by *sn*-glycero-3-phosphate. dCDPdiacylglycerol is an equally good substrate for the enzyme. The reaction is dependent on magnesium ion and nonionic detergent such as Triton X-100.[4,5] The assay method most appropriate for crude extracts and for routine work is the measurement of chloroform-soluble radioactivity derived from labeled glycerophosphate.[4,5] The procedure is essentially identical to that outlined for the phosphatidylserine synthase.[9] The assay mixture consists of 0.25 M Tris-HCl (pH 8.0), 1% Triton X-100 (made w/v throughout), 0.1 M MgCl$_2$, 0.2 mM CDPdiacylglycerol, 0.5 mM *sn*-glycero-3-phosphate (labeled with either ^{3}H, ^{14}C, or ^{32}P at a specific activity of 200–1000 cpm/ nmol), and enzyme in a final volume of 0.1 ml. The reaction is allowed to proceed for 10 min at 37°. One unit of enzymatic activity is defined as the amount of enzyme that converts 1 μmol of *sn*-glycero-3-phosphate into chloroform-soluble material in 1 min under the above conditions. The assay is linear up to at least 50% conversion of the limiting substrate and in purified preparations goes to near 100% completion. The enzyme is usually diluted prior to assay using Triton X-100 (0.1%) containing buffers.

Purification of Enzyme

The major purification step relies on affinity chromatography on CDP diacylglycerol-Sepharose.[5,6] Preparation of sufficient amounts of CDPdiacylglycerol for this purpose can be accomplished as outlined previously.[9]

Preparation of Affinity Resins. Two types of affinity resins are employed in this purification—one is an oxidized form, and the other a reduced form, of the resin. The oxidized form of the resin consists of adipic acid dihydazide (available from Sigma) covalently attached to Sepharose 4B via CNBr coupling and to NaIO$_4$ oxidized CDPdiacylglycerol via an hydrazone[10] (see Fig. 1). The reduced form of the resin is prepared by NaBH$_4$ treatment of the oxidized resin. The exact chemical changes brought about by this treatment are not known, but the chromatographic properties of the resin are changed.

---

[9] W. Dowhan and T. Larson, this volume [66].
[10] R. Lamed, Y. Levin, and M. Wilchek, *Biochim. Biophys. Acta* **304**, 231 (1973).

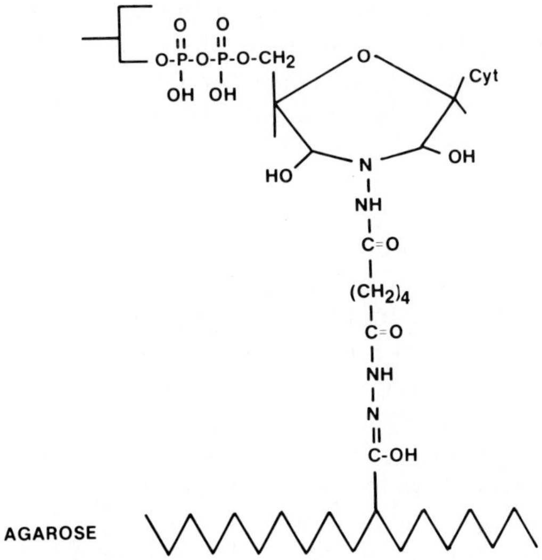

FIG. 1. Structure of CDPdiacylglycerol–Sepharose affinity column.

To prepare the resins[5,6] 100 ml of Sepharose 4B, prewashed with 1 M NaCl and then with water, are activated by mixing with 20 g of CNBr in 100 ml of water at 0° as described by Cuatrecasas.[11] The pH is immediately raised and maintained at 11 with 8 M NaOH. The temperature is maintained below 20° using an ice bath. After about 20 min the reaction is over, as indicated by cessation of proton release. The mixture is quickly brought to 0° by the addition of ice and rapidly filtered and washed with 1.2 liters of cold 0.1 M Na_2CO_3 (pH 9.5) on a sintered-glass funnel. The activated resin is suspended in 100 ml of 0.1 M Na_2CO_3 (pH 9.5) containing 9 g of adipic acid dihydrazide and mixed (17 hr at 4°) by gently inverting the container to prevent mechanical damage to the resin. The resin is washed extensively with water on a sintered-glass funnel and tested for the presence of covalently bound hydrazide. The resin should turn a dark red in 0.5 ml of saturated sodium borate containing several drops of 3% 2,4,6-trinitrobenzene sulfonate.

The ribosyl hydroxyls of CDPdiacylglycerol (0.36 mmol) are oxidized to aldehydes in 70 ml of 0.1 M sodium acetate (pH 4.5) and 0.5% Triton X-100 by the addition of $NaIO_4$ (0.47 mmol). The mixture is incubated in the dark first for 1 hr at room temperature and then overnight at 4°. Excess periodate is removed by the addition of 0.1 ml of glycerol. Conversion to the dialdehyde is usually 90% as judged by an increase in the mobility of the liponucleotide on silica gel thin-layer plates in chloroform–methanol–acetic acid–water (50 : 28 : 4 : 8).

[11] P. Cuatrecasas, *J. Biol. Chem.* **245**, 3059 (1970).

The oxidized liponucleotide solution and the Sepharose 4B–adipic acid dihydrazide are combined in 180 ml of 0.1 M sodium acetate (pH 5.0) containing 0.5% Triton X-100 and mixed at 4° overnight. The resin is then washed extensively with the reaction buffer containing 0.5 M KCl and finally with water. Analysis for total phosphate content[12] indicates between 1 and 2 μmol of covalently bound liponucleotide per milliliter of resin.

The reduced resin is prepared at 4° by three successive additions of NaBH$_4$ (0.4 g, each addition) at 1-hr intervals to 50 ml of the oxidized resin in two resin volumes of 0.5 M Tris-HCl (pH 8.0). The reduced resin is extensively washed with 1 M KCl before use.

Preparation of Cell Paste. Commercial preparations of *E. coli* B and K12 (Grain Processing, Muscatine, Iowa) have been found to be acceptable as a source for the enzyme. Cell pastes can be prepared by growth of any wild-type strain on rich medium supplemented with 0.5% glucose. Cell paste is stable at $-20°$ for years.

Preparation of Membrane Extract. All procedures are carried out at 5°. *Escherichia coli* cells (450 g wet weight) are suspended with the aid of a Waring blender in 950 ml of 20 mM potassium phosphate, pH 7.0, 10 mM MgCl$_2$, and 10 mM 2-mercaptoethanol. The cells are broken using a French pressure cell or by sonication. The cell membrane fraction is collected by a 2-hr centrifugation at 100,000 g. The pellet is suspended in 750 ml of the above buffer containing 4% Triton X-100 using a Waring blender. To prevent excessive foaming of the preparation the blender jar is filled to the top and capped to exclude air. After standing for 2 hr the insoluble material is removed by centrifugation at 100,000 g for 1 hr.

Affinity Chromatography. The phosphatidylglycerophosphate synthase activity is adsorbed to the oxidized form of the affinity resin by a batch procedure. The resin (50 ml settled volume) is mixed in succession for 5 hr each with two 250-ml batches of the Triton X-100 extract. The supernatant after treatment with the resin contains between 5 and 10% of the original enzymatic activity and virtually all of the CDPdiacylglycerol hydrolase and phosphatidylserine decarboxylase activities. The affinity resin is then packed into a 2 × 16 cm column. The remaining Triton extract is passed through the column until only 70% of the applied activity binds. The column is then washed (50 ml/hr) with 250 ml of buffer A (20 mM potassium phosphate, pH 7.0, 10 mM 2-mercaptoethanol and 10 mM MgCl$_2$) containing 0.1% Triton X-100 and 0.5 M KCl followed by 400 ml of buffer A containing 0.1% Triton X-100 and 1.0 M KCl. The enzyme is then eluted at 10 ml/hr with 300 ml of buffer A containing 0.1% Triton X-100, 0.5 M KCl, and 0.8 M hydroxylamine-HCl (adjusted to pH 7 with KOH); the

[12] G. R. Bartlette, *J. Biol. Chem.* **234**, 466 (1959).

hydroxylamine displaces the liponucleotide ligand from the column, thus releasing the enzyme.

The pooled synthase activity is dialyzed extensively against several changes of buffer B (20 mM potassium phosphate, pH 7.0, 10 mM MgCl$_2$, 0.5 mM dithiothreitol, and 0.1% Triton X-100). The enzyme is then applied (50 ml/hr) to a 1.4 × 16 cm column of reduced affinity resin in buffer B; about 5% of the enzyme is not retained by the column. The resin is washed (50 ml/hr) with 300-ml portions of buffer B containing first 0.1 M KCl, then 0.4 mM CDPdiacylglycerol containing 0.1 M KCl, and finally 0.1 M KCl. The synthase elutes in the second and third washes.

Concentration of Enzyme. The synthase does not bind to conventional ion exchange resins, so concentration by step elution from such resins is not possible. Ultrafiltration of large volumes of eluent also results in concentration of the detergent. In order to concentrate the enzyme, the pooled sample is diluted with two volumes of buffer B and adsorbed at 50 ml/hr to a 1.4 × 9.5 cm column of oxidized affinity resin. The column is washed with 150 ml of buffer B containing 1 M KCl, and then the enzyme is eluted with 150 ml of buffer B containing 0.8 M hydroxylamine and 0.5 M KCl as described earlier. The enzyme pool is exhaustively dialyzed against buffer B, concentrated by ultrafiltration (PM-10 filter), and passed through a 2 × 16 cm DEAE-Sephadex (A-50) column equilibrated and eluted with buffer B. The enzyme emerges at the void volume with most of the lipid retained by the column. The final enzyme pool is concentrated by ultrafiltration as above.

Enzyme Purity. The purification scheme summarized in the table indicates an overall purification of 6000-fold from crude extracts. The preparation appears to be at least 85% pure based on sodium dodecyl sulfate–polyacrylamide gel electrophoresis. The major protein species has an apparent molecular weight of 24,000. The purification method selects for enzymes that bind CDPdiacylglycerol. There are no other known

PURIFICATION OF PHOSPHATIDYLGLYCEROPHOSPHATE SYNTHASE FROM *Escherichia coli*

Step	Total volume (ml)	Total protein (mg)	Specific activity (units/mg)	Yield (%)
1. Broken cells	1300	89,000	0.003	100
2. Cell membrane	840	40,600	0.005	74
3. Triton X-100 extract	690	13,800	0.014	67
4. Oxidized affinity resin	200	—	—	47
5. Reduced affinity resin	470	—	—	30
6. DEAE-Sephadex	4.7	3	18.6	20

CDPdiacylglyerol-dependent enzymes in samples prepared as outlined above.

Properties

Kinetic Properties. The phosphatidylglycerophosphate synthase has a strong dependence on divalent metal ion (magnesium being preferred) and a nonionic detergent such as Triton X-100 for activity. The apparent K_m for magnesium ion in 1% Triton X-100 is 50 mM, concentrations above 150 mM being inhibitory. The dependence on Triton X-100[5] is unlike many other enzymes which are stimulated by nonionic detergents.[9] Maximal activity is attained at about 0.2% Triton X-100, but concentrations up to 6% are not inhibitory. Therefore the enzyme appears to be insensitive to the ratio of detergent to lipid substrate in the assay and displays classical kinetics at all detergent and substrate concentrations which result in the formation of uniform mixed micelles.

The enzyme appears to catalyze a sequential Bi Bi reaction.[5] CDPdiacylglycerol apparently binds first and also acts as an uncompetitive inhibitor at high concentration. The K_m for *sn*-glycero-3-phosphate is 0.32 mM, and for CDPdiacylglycerol it is 46 μM (34 μM for the deoxy derivative). Evidence for this mechanism comes largely from kinetic studies, isotope exchange reactions, and the lack of hydrolase activity toward either the lipid substrate or the lipid product. The enzyme catalyzes an isotope exchange reaction between phosphatidylglycerophosphate and glycerophosphate only in the presence of CMP.

Detergent Binding. To prevent aggregation the enzyme requires the presence of nonionic detergent, which apparently binds to the enzyme.[5] The Stokes' radius of the enzyme in the presence of detergent is very large relative to its sedimentation coefficient as determined by sucrose gradient centrifugation. The mass to charge ratio of the enzyme in the presence of Triton X-100 appears to be high, since the enzyme does not electrophoretically enter a 5% polyacrylamide gel. The addition of the anionic liponucleotide substrate to the electrophoretic sample allows the enzyme to enter the gel. The mobility of the enzyme is roughly proportional to the amount of substrate added. The liponucleotide presumably adds ionic character to the enzyme–detergent complex formed under these conditions.

Synthetic and Analytical Uses. By using the properly labeled liponucleotide or *sn*-glycero-3-phosphate, various radiolabeled derivatives of phosphatidylglycerophosphate can be made. Since the reaction proceeds to near 100% completion and there are no hydrolytic activities present, no purification of the chloroform-soluble material is required if excess glycerophosphate is used. Removal of Triton X-100 can be accomplished by silica gel thin-layer chromatography in chloroform–methanol–acetic acid (65 : 25 : 8). Similarly the enzyme can be used to quantitate the levels of cytidine liponucleotides in various preparations by using excess labeled glycerophosphate in the assay.

[66] Phosphatidylserine Synthase from *Escherichia coli*[1]

EC 2.7.8.8 CDPdiacylglycerol : L-serine *O*-phosphatidyltransferase

By WILLIAM DOWHAN and TIMOTHY LARSON

The phosphatidylserine synthase of several gram-negative organisms[2] including *Escherichia coli*[3] catalyzes the formation of phosphatidylserine by reaction (1).

The enzyme is unique among phospholipid biosynthetic enzymes in these organisms in that it is found tightly associated with the ribosomal fraction rather than the membrane fraction[2-4] in crude broken-cell preparations. The enzyme appears to be markedly different in gram-positive organisms such as the bacilli in that it is membrane associated.[5,6] The enzyme has not been found in higher eukaryotes[7] but appears to be present in yeast.[8]

The enzyme has been most extensively studied in *E. coli*. Mutants, temperature-sensitive in the enzyme,[9] have been isolated that fail to make

[1] Supported by Grant GM 20478 from the National Institute of General Medical Sciences and by Grant AU-599 from the Robert A. Welch Foundation.
[2] A. Dutt and W. Dowhan, *J. Bacteriol.* **132**, 159 (1977).
[3] J. Kanfer and E. P. Kennedy, *J. Biol. Chem.* **239**, 1720 (1964).
[4] C. R. H. Raetz and E. P. Kennedy, *J. Biol. Chem.* **247**, 2008 (1972).
[5] P. H. Patterson and W. J. Lennarz, *J. Biol. Chem.* **246**, 1062 (1971).
[6] A. Dutt and W. Dowhan, unpublished results, 1978.
[7] H. van den Bosch, L. M. G., van Golde, and L. L. M. van Deenen, *Rev. Physiol.* **66**, 13 (1972).
[8] M. R. Steiner and R. L. Lester, *Biochim. Biophys. Acta* **260**, 222 (1972).
[9] C. R. H. Raetz, *J. Biol. Chem.* **251**, 3242 (1976).

phosphatidylserine at the restrictive temperature and eventually filament and die as their phosphatidylethanolamine pool drops relative to the polyglycerophosphatides. The availability of such mutants has made possible the isolation of hybrid plasmids carrying the *pss* gene locus,[10] the structural gene for the phosphatidylserine synthase. These plasmids correct the temperature-sensitive phenotype of mutant strains and restore enzymatic activity to levels that are 10- to 20-fold above that of wild-type strains.

The enzyme has been purified to apparent homogeneity, as judged by sodium dodecyl sulfate–gel electrophoresis, from *E. coli* B,[11] *E. coli* K12,[10] and K12[10] strains carrying multicopy-number hybrid plasmids that contain the *pss* gene locus. The enzyme appears to be very similar from all these sources as judged by specific activity of purified preparations, ribosomal affinity, and minimum subunit molecular weight. Purification of the enzyme is accomplished by polyethylene glycol–dextran phase partition[12] to separate the enzyme from nucleic acid followed by specific CDPdiacylglycerol elution from phosphocellulose to obtain the homogeneous enzyme.[11]

Assay Method

Phosphatidylserine synthase catalyzes the displacement of CMP from CDPdiacylglycerol (or dCDPdiacylglycerol) by serine, as shown in reaction (1). The enzyme can be assayed by the incorporation of radioactive serine into chloroform-soluble material (the method discussed below[4,11]). This method is most appropriate for routine assay of both crude and purified preparations. The enzyme can also be assayed by a continuous spectrophotometric method,[13] which couples the release of CMP to NADH oxidation via CMP kinase, pyruvate kinase, and lactate dehydrogenase. This assay is most appropriate for purified preparations owing to interfering activities in crude extracts.

The final assay mixture consists of 0.67 mM CDPdiacylglycerol (fatty acid composition appears to be irrelevant), 0.5 mM L-serine (either 3-[3]H- or 3-^{14}C-labeled at a specific activity of 200–1000 cpm/nmol), 1 mg/ml bovine serum albumin, 0.1% Triton X-100 (made w/v throughout), 0.1 M potassium phosphate, pH 7.4, and enzyme in a final volume of 0.06 ml. To assure uniform dispersement of lipid substrate, a stock solution of Triton

---

[10] C. R. H. Raetz, T. J. Larson, and W. Dowhan, *Proc. Natl. Acad. Sci. U.S.A.* **74**, 1412 (1977).

[11] T. J. Larson and W. Dowhan, *Biochemistry* **15**, 5212 (1976).

[12] C. R. H. Raetz and E. P. Kennedy, *J. Biol. Chem.* **249**, 5038 (1974).

[13] G. M. Carman and W. Dowhan, *J. Lipid Res.* **19**, 519 (1978).

X-100 and lipid is briefly sonicated prior to dilution into the assay mix. The reaction is usually initiated by the addition of enzyme and carried out at 30° in a 12-ml polypropylene tube. The reaction is terminated after 10 min by the addition of 0.5 ml of methanol (0.1 N in HCl) followed by the addition of 1.5 ml of chloroform and 3 ml of 1 M MgCl$_2$. After mixing and separation of the phases by centrifugation, 1.0 ml of the chloroform phase is removed, evaporated to dryness at 65° in a scintillation vial, and counted for radioactivity. One unit of enzymatic activity is defined as the amount of enzyme that converts 1 μmol of serine to chloroform-soluble material in 1 min under the above conditions.

L-[U-^{14}C]Serine can be used in the assay once the phosphatidylserine decarboxylase has been removed from the preparation (after the streptomycin precipitation step). The enzyme, especially when near homogeneity, is sensitive to dilution in low ionic strength buffers (below 0.1 M KCl) in the absence of Triton X-100 and tends to adhere to, or be inactivated by, glass surfaces. Therefore, dilution of the enzyme for assay is usually done in the presence of 0.1 M potassium phosphate, pH 7.4, 0.1% Triton X-100 containing 1 mg/ml bovine serum albumin in a plastic tube. The reaction under the above conditions is linear up to 50% conversion of the serine to product when pure enzyme is employed. The reaction will proceed to near 95% completion based on the limiting substrate when the purified enzyme is used; complete conversion is probably not observed owing to the inherent hydrolase activity[11,12] (about 1% of the synthetic rate) of the synthase toward both its lipid substrate and product. Crude extracts of *E. coli* contain a membrane-associated CDPdiacylglycerol hydrolase activity[14,15] that can affect the linearity of the assay with time owing to the consumption of lipid substrate; the hydrolase can be inhibited by AMP (1 mM).

Purification Procedure

The phosphatidylserine synthase can be purified to near homogeneity using the method of Raetz and Kennedy[12] to separate the enzyme from ribosomes coupled with specific elution from phosphocellulose using CDPdiacylglycerol as developed by Larson and Dowhan[11]; therefore, gram quantities of the lipid substrate are required for purification of the enzyme. The source of the enzyme routinely used is *E. coli* K12 strain RA324,[10] which carries the ColE1-*E. coli* DNA hybrid plasmid pLC34-44

[14] C. R. H. Raetz, C. B. Hirschberg, W. Dowhan, W. T. Wickner, and E. P. Kennedy, *J. Biol. Chem.* **247**, 2245 (1972).
[15] C. R. H. Raetz, W. Dowhan, and E. P. Kennedy, *J. Bacteriol.* **125**, 855 (1976).

(carrying the *pss* gene locus) isolated from the Clarke and Carbon[16] collection of such plasmids. This strain has a specific activity about 15-fold higher than wild-type strains. Recently, hybrid plasmids under lambda replication control have been constructed that produce enzyme at 150-fold higher levels than wild-type strains.[17]

Growth and Storage of Organism. Strain RA324 is stored at $-70°$ in 40% glycerol. Selection and growth of the hybrid ColE1-containing strain requires colicin E1 toxin to select against loss of the plasmid. Colicin E1 toxin[18] can be prepared from any *E. coli* strain, such as W3110 or JC411, carrying ColE1 plasmid by growth in 2 liters of 0.5% yeast extract and 1% tryptone to a cell density of 5×10^8 cells/ml at $37°$. Mitomycin C is added to 1 μg/ml, and the culture is vigorously aerated in the dark by shaking at $37°$ for 8–12 hr. The following procedures are carried out at $5°$. The cells are collected by centrifugation at 10,000 g for 15 min and suspended in 50 ml of 0.1 M potassium phosphate (pH 7.0) containing 1.0 M NaCl. The cells are disrupted by sonication, and the cell debris is removed by centrifugation at 10,000 g for 20 min. The supernatant is brought to 40% saturation with ammonium sulfate (22.6 g/100 ml), and the precipitate is removed by centrifugation (10,000 g, 20 min). The supernatant is then brought to 60% saturation with ammonium sulfate (12 g/100 ml) and then again centrifuged. The pellet is dissolved in 5 ml of 0.1 M potassium phosphate (pH 7.0) and dialyzed against the same buffer. After treatment with chloroform, the aqueous phase can be stored in 50% glycerol at $-20°$. The units of killing activity per milliliter are defined by the fold dilution required to prevent formation of a clear plaque by one drop of diluent on a lawn of colicin E1-sensitive cells.[19]

To prepare RA324 cell paste, single colonies that are resistant on LB agar to 50 units of colicin E1 toxin per milliliter are grown overnight at $37°$ in liquid medium (500 ml) containing 1% yeast extract, 1% glucose, 0.2 M potassium phosphate, pH 7.3, and 50–100 units of colicin E1 toxin per milliliter. An overnight culture with a specific activity of at least 0.07 unit/mg is used to inoculate a 12-liter culture of the above medium lacking colicin E1 in a New Brunswick fermentor. Cells are grown with high aeration and maintenance of pH above 7.0 at $37°$ to near stationary phase. The yield of cells is about 8–10 g wet weight per liter of growth medium. Fermentation can be scaled up appropriately.

Preparation of CDPdiacylglycerol. CDPdiacylglycerol can be prepared in large quantities at a reasonable cost starting with either commercially

[16] L. Clarke and J. Carbon, *Cell* **9**, 91 (1976).
[17] A. Ohta, K. Louie, and W. Dowhan, unpublished results, 1979.
[18] S. A. Schwartz and D. R. Helinski, *J. Biol. Chem.* **246**, 6318 (1971).
[19] H. R. Herschman and D. R. Helinski, *J. Biol. Chem.* **242**, 5360 (1967).

prepared phosphatidylcholine or that isolated directly from eggs.[20] The lipid is most reproducibly converted to phosphatidic acid using freshly prepared cabbage phospholipase D, although commercial sources can be used. Phosphatidic acid (commerical sources can be used at a much higher cost) is then converted to CDPdiacylglycerol by reaction with CMPmorpholidate by a modification[21] of a procedure of Agranoff and Suomi.[22] A crude preparation of phospholipase D from cabbage is obtained by homogenizing 100 g of fresh cabbage leaves in 200 ml of ice-cold 0.1 M sodium acetate (pH 5.7) in a blender. The supernatant containing the lipase is collected by centrifugation at 10,000 g for 30 min and stored at $-20°$.

Four grams of phosphatidylcholine (from egg, Sigma, type IC-E) are dissolved in 450 ml of diethyl ether. Subsequently, 400 ml of water, 50 ml of 1 M sodium acetate (pH 5.7), 60 ml of 1 M CaCl$_2$, and 100 ml of the crude phospholipase D (or Calbiochem B grade, 80 units) are added, and the mixture is stirred at room temperature. The hydrolysis can be monitored by silica gel thin-layer chromatography (chloroform–pyridine–formic acid, 50 : 20 : 7, v/v/v; phosphatidylcholine, R_f = 0.25; phosphatidic acid, R_f = 0.7). Complete hydrolysis is usually obtained after one overnight incubation. The ether phase is collected and evaporated to dryness. The residue is dissolved in 100 ml of chloroform[23] and washed twice with 50 ml of methanol–water (1 : 1). The chloroform phase is placed in two 500-ml plastic centrifuge bottles to each of which 450 ml of precooled ($-20°$) methanol is added. The calcium salt of phosphatidic acid is allowed to precipitate overnight at $-20°$ and is collected by centrifugation at 10,000 g for 30 min. The yield is routinely 3.0–3.5 g of the calcium salt of phosphatidic acid.

The calcium salt of phosphatidic acid (1 g) is converted to the free acid by dissolving in 15 ml of chloroform to which is added 15 ml of methanol and 10 ml of water. The pH of the equilibrated upper phase is adjusted to between 2 and 3 with 1 N HCl. The mixture is centrifuged to separate the phases, and the chloroform phase collected. The chloroform phase is washed with 16 ml of methanol–water (1 : 3) and evaporated to dryness in a 250-ml round-bottom flask. CMPmorpholidate[24] (400 mg, Sigma) is

---

[20] M. A. Wells and D. J. Hanahan, this series, Vol. 14, p. 178.

[21] W. Dowhan and A. Radominska-Pyrek, unpublished results, 1979.

[22] B. W. Agranoff and W. D. Suomi, *Biochem. Prep.* **10**, 46 (1963).

[23] Chloroform contains variable amounts of polar materials that particularly affect the silica gel chromatography; therefore, chloroform is equilibrated prior to use three times with one volume of water and then dried over magnesium sulfate.

[24] The quality of commercial CMPmorpholidate is variable. The formation of a precipitate or a gel during the reaction is due to unidentified impurities in the CMPmorpholidate and results in a lower yield. Precipitate should be removed before purification of the product.

added to the flask, followed by 10 ml of benzene. After mixing, the solvent is evaporated on a rotary evaporator leaving a thin film. The film is dissolved in 20 ml of anhydrous pyridine in a glass-stoppered flask by stirring at 37°. The stirring is continued for 2 days, during which time two portions of 200 mg each of CMPmorpholidate are added at approximately 15-hr intervals. The progress of the reaction is checked by silica gel thin-layer chromatography (chloroform–methanol–acetic acid, 65 : 25 : 8; CDPdiacylglycerol $R_f = 0.04$, phosphatidic acid $R_f = 0.58$). When approximately 90% conversion is obtained, the reaction mixture is evaporated to dryness and the residue is dissolved in 100 ml of chloroform and washed twice with 50 ml of methanol–water (1 : 1). After separation of the phases by centrifugation, the lower phase is taken to dryness. The yield of the crude CDPdiacylglycerol is about 1.7 g.

The crude product in 5 ml of chloroform is applied to a 3×30 cm activated silica gel column (Silica Woelm 63-200, Nutritional Biochemicals) preequilibrated with the same solvent. The column is eluted with chloroform collecting 25-ml fractions which are checked by silica gel thin-layer chromatography in the system described above. All the CDPdiacylglycerol is usually found in fractions 2–6, whereas the phosphatidic acid remains on the column. The yellowish, CDPdiacylglycerol-containing fractions are pooled and evaporated to dryness. The resulting pyridinium salt is converted to the free acid by dissolving in 15 ml of chloroform and 15 ml of methanol and adjusting the pH (checked by pH paper) to 2, using $2 N$ HCl. Water (10 ml) is added; after shaking and separating the phases (upper phase should be near pH 2), the upper phase is removed and the lower phase is washed once with methanol : water (3 : 2). The free acid is converted to the calcium salt by vigorously stirring the chloroform phase with 0.5 ml of $1 M$ CaCl$_2$. The chloroform phase is collected, washed with 4 ml of methanol plus 10 ml of water and evaporated to dryness. The residue is dissolved in 50 ml of chloroform and added to 450 ml of methanol precooled to $-20°$. A white precipitate is allowed to form overnight at $-20°$ which is collected by centrifugation, redissolved in chloroform and reprecipitated. The yield of pure CDPdiacylglycerol as the calcium salt is 0.8 g.

Since the calcium salt is not easily dispersed in water, it is converted to the ammonium salt for use. The product is first converted to the free acid as described immediately above, except that all the aqueous additions are made 10 mM in EDTA. Complete conversion to the free acid is particularly important at this stage because residual calcium makes dispersion of the salt of CDPdiacylglycerol in water difficult. The collected chloroform phase is adjusted to pH 7–8 (checked by pH paper) by dropwise addition of $1 N$ NH$_4$OH in methanol. Solvent is removed by evaporation and finally by drying under vacuum for 30 min. The diammonium salt of

CDPdiacylglycerol (97% yield) is dispersed in water by sonication[24a] at a concentration of up to 50 mM and stored at $-20°$. This salt is also readily soluble in chloroform–methanol (1 : 1).

CDPdiacylglycerol concentration is determined by its optical density at 280 nm (molar extinction of 12,800 M^{-1} cm^{-1}) in water at pH 2. The 280 : 260 absorption ratio should be 2.1 under these conditions, owing to the presence of cytidine. Concentration determined by optical density is generally within 5% of the value determined by complete enzymatic conversion to phosphatidylserine using excess radiolabeled serine in the assay. The molar ratios of cytidine : ester : phosphate : enzyme substrate are typically close to the theoretical values of 1 : 2 : 2 : 1. Phosphate and ester determinations are routinely performed by the methods of Bartlett[25] and Snyder and Stephens,[26] respectively.

Preparation of Ribosome-Bound Enzyme. All the following procedures, except where noted, are carried out at 5°. Cells from *E. coli* strain RA324 (150 g) are suspended with the aid of a Waring blender in a final volume of 1 liter of 10 mM potassium phosphate, pH 7.4, 10 mM 2-mercaptoethanol, 2 mM Na$_2$EDTA. The cells are broken using a French pressure cell (sonication can also be used), and the cell supernatant is obtained after centrifugation at 13,500 g for 2 hr. The supernatant is thoroughly mixed with 325 ml of 5% streptomycin sulfate–25% Triton X-100. After standing for 1 hr, the precipitate is collected by a 20-min centrifugation at 10,000 g. The pellet is suspended by homogenization in 450 ml of 20 mM potassium phosphate, pH 7.4, containing 10 mM 2-mercaptoethanol and 5 M NaCl.

Polymer Partitioning. To separate the enzyme from nucleic acid, partitioning between aqueous phases of polyethylene glycol and dextran is employed. To the above solution is added 100 ml of 30% (w/w) polyethylene glycol (Carbowax, PEG 6000) and 50 ml of 20% (w/w) dextran (Sigma, average molecular weight 500,000). The mixture is stirred for 30 min, and the phases are separated by centrifugation at 10,000 g for 10 min. The upper polyethylene glycol-rich phase is saved, and the lower dextran-rich phase and any material at the interface are discarded.

The enzyme is precipitated by slowly adding with stirring 19 g of ammonium sulfate per 100 ml of the above polyethylene glycol solution. The temperature is maintained between 10° and 15° to allow solubilization of the salt. The solution is then centrifuged (15°) at 10,000 g for 20 min. The solution separates into a lower aqueous phase and an upper polyethylene glycol phase; the precipitated enzyme is found at the inter-

---

[24a] If the solution does not clear, residual calcium is present. Add an excess of Na$_2$EDTA (pH 8.0) to clear the solution and remove the CaEDTA by dialysis against water adjusted to pH 8.0 with NH$_4$OH.

[25] G. R. Bartlett, *J. Biol. Chem.* **234**, 466 (1959).

[26] F. Snyder and N. Stephens, *Biochim. Biophys. Acta* **34**, 244 (1959).

face between these phases. The upper and lower phases are carefully removed by suction while allowing the precipitate to collect on the walls of the centrifuge tube. The precipitate is suspended in 50 ml of 0.2 M potassium phosphate, pH 7.4, containing 10 mM 2-mercaptoethanol and 10% glycerol. The suspension is cleared by centrifugation (10,000 g for 20 min), and the pellet is extracted with 25 ml of additional buffer. The supernatants are pooled.

Phosphocellulose Column. These procedures are carried out at room temperature. Phosphocellulose (Whatman P-11) is washed with base and acid prior to use,[27] adjusted to pH 7.4, and equilibrated with 0.1 M potassium phosphate, pH 7.4. A 5 × 5 cm phosphocellulose column is poured and equilibrated with 4 column volumes of 0.05 M potassium phosphate, pH 7.4, containing 10% glycerol, 0.1% Triton X-100, and 1 mg/ml bovine serum albumin; pretreatment of the column with albumin increases yield by twofold. The enzyme solution from the preceding step is diluted with two volumes of 0.1 M NaCl, 0.5% Triton X-100, 10% glycerol and then applied (400 ml/hr) to the column. The column is washed (400 ml/hr) first with 300 ml of 0.1 M potassium phosphate, pH 7.4, 1% Triton X-100, 0.5 mM dithiothreitol, and 10% glycerol containing 0.65 M NaCl, followed by 150 ml of the same buffer but with only 0.1% Triton X-100 and 0.5 M NaCl. The enzyme is finally eluted (150 ml/hr) with 200 ml of 0.1 M potassium phosphate, pH 7.4, containing 0.5 mM dithiothreitol, 10% glycerol, 0.1% Triton X-100, 0.5 M NaCl, and 0.4 mM CDPdiacylglycerol. The enzyme is essentially homogeneous but dilute at this point. Concentration can be accomplished by ultrafiltration using an Amicon XM-100 filter, but detergent and lipid are also concentrated. Alternatively, step elution from DEAE-cellulose concentrates the enzyme and removes the bulk of the detergent and lipid.

DEAE-Cellulose Column. The phosphocellulose pool of enzymatic activity is dialyzed against eight volumes of 20 mM potassium phosphate, pH 7.0, containing 0.2% Triton X-100, 0.5 mM dithiothreitol and 10% glycerol for at least 10 hr. The dialyzate is applied (70 ml/hr) to a 2.5 × 12 cm DEAE-cellulose column equilibrated with the dialysis buffer. The column is washed with 150 ml of the equilibration buffer containing 0.1 M NaCl and finally with the same buffer (25 ml/hr) containing 1.2 M NaCl to elute the enzyme. The pooled fractions of enzymatic activity are stable for at least a year in 25% glycerol at $-20°$.

Degree of Purity. The overall purification as summarized in the table is about 325-fold for plasmid-containing strains to a final specific activity of about 34–40 units/mg. The yield of protein is about 6 mg per 100 g wet weight of cell paste. The enzyme can be purified from wild-type cells

[27] C. G. Kurland, S. J. S. Hardy, and G. Mora, this series, Vol. 20, p. 381.

PURIFICATION OF PHOSPHATIDYLSERINE SYNTHASE FROM *Escherichia coli*

Step	Total volume (ml)	Total protein (mg)	Specific activity (units/mg)	Yield (%)
1. Broken cells	1000	22,400	0.12	100
2. Cell supernatant	945	18,100	0.14	97
3. Streptomycin sulfate	450	5,400	0.4	83
4. Polymer partitioning	505	4,100	0.32	50
5. Ammonium sulfate	232	1,300	0.82	42
6. Phosphocellulose	124	—	—	20
7. DEAE-Sephadex	5	8.5	39	13

using the above procedure to the same degree of purity, but the yield of protein will be much less (ca. 0.4 mg per 100 g of cell paste).

The enzyme preparation appears to be better than 95% homogeneous, as judged by sodium dodecyl sulfate (SDS)–gel electrophoresis, exhibiting a minimum molecular weight of 54,000. It is important to fully denature the enzyme by incubation at 100° for 5 min in 8 M urea, 1% SDS, and 1 mM 2-mercaptoethanol prior to electrophoresis. The preparation eluted from DEAE-cellulose does contain varying amounts of phospholipid, primarily phosphatidic acid and traces of phosphatidylglycerol (see below), which are generated enzymatically from CDPdiacylglycerol.

Properties of Phosphatidylserine Synthase

Enzymatic Activity. The enzyme from *E. coli* has a broad pH optimum ranging from 7.0 to 8.5 and is dependent on ionic strength[28] (0.3 or higher being optimal) and on a nonionic detergent such as Triton X-100[11,28] for activity. The dependence on Triton X-100 and lipid substrate is not directly proportional to the molar concentration of either component. Rather, the activity increases in a saturable manner with increasing lipid substrate provided the molar ratio of detergent to substrate is held constant; the molar ratio of detergent to liponucleotide should be at least 8 : 1, and the components must be dispersed in a uniform mixed micelle form by sonication in order to obtain linear relationships. The enzymatic activity is proportional to the mole fraction of liponucleotide in the detergent mixed micelle, since the activity decreases as the amount of detergent is increased at constant liponucleotide concentration above the point where uniform mixed micelles are formed. The dependence of activity on detergent concentration, lipid substrate concentration, and ionic strength paral-

[28] G. M. Carman and W. Dowhan, *J. Biol. Chem.* **254**, 8391 (1979).

lels very closely the requirements for the physical binding of the enzyme to detergent-substrate mixed micelles.[28]

Physical Properties. The enzyme has an apparent minimum subunit molecular weight of 54,000[11] as judged by SDS–gel electrophoresis. The undenatured enzyme appears to have a molecular weight near 500,000 in the absence of detergent based on glycerol gradient centrifugation.[28] The enzyme binds Triton X-100, and the apparent binding constant of this interaction increases with the ionic strength of the medium and with increasing liponucleotide concentration relative to detergent.[28]

Other Reactions. The phosphatidylserine synthase appears to catalyze a Ping-Pong reaction that may proceed through a common enzyme–phosphatidyl (Enz-Ptd) intermediate as shown in reaction (2).

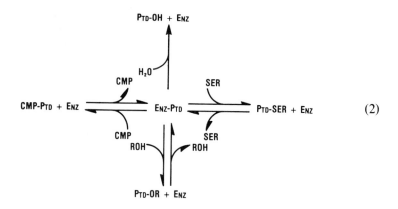

$$\text{(2)}$$

Therefore, the enzyme can catalyze various partial reactions,[11,12] such as the exchange of CMP with CDPdiacylglycerol or serine with phosphatidylserine. In addition, at about 1% of the synthetic rate the enzyme can catalyze the hydrolysis of CDPdiacylglycerol or phosphatidylserine to phosphatidic acid and either CMP or serine, respectively. Finally, at even slower rates, the enzyme can transfer the phosphatidyl moiety to other alcohols, such as glycerol and glycerophosphate.

These partial reactions can be monitored[12] under the standard assay conditions by using the appropriately labeled compound. The CMP exchange can be followed in the absence of serine by quantitating the amount of chloroform-soluble label (as CDPdiacylglycerol) formed in the presence of 1 mM labeled CMP. Similarly, by substituting 1 mM phosphatidylserine for the liponucleotide, serine exchange can be monitored. Finally, the hydrolase reactions can be quantitated by the amount of water–soluble counts released from either cytidine-labeled CDPdiacyl-glycerol or serine-labeled phosphatidylserine, respectively, in the absence of other substrates.

Preparative Uses. The phosphatidylserine synthase can be used to prepare radiolabeled substrates useful in assaying other enzymes. By using DL-[1-^{14}C]serine (usually only available as the racemic mixture, but the enzyme is specific for the L-isomer) as a substrate, phosphatidylserine specifically labeled in the serine carboxyl group[3] can be obtained for assay of phosphatidylserine decarboxylase.[29] The CMP–CDPdiacylglycerol exchange reaction discussed above can be used to synthesize cytidine-labeled CDPdiacylglycerol, which can be used to assay the membrane-bound CDPdiacylglycerol-specific hydrolase[14,15] of *E. coli.*

[29] W. Dowhan, W. T. Wickner, and E. P. Kennedy, *J. Biol. Chem.* **249**, 3079 (1974).

[67] Phosphatidylserine Decarboxylase from *Escherichia coli*

EC 4.1.1.65 Phosphatidylserine carboxy-lyase

By EDWARD HAWROT

Phosphatidylserine → phosphatidylethanolamine + CO_2

Assay Method

Principle. The activity of phosphatidylserine decarboxylase can be readily determined by measuring the release of radioactive CO_2 from phosphatidyl[1-^{14}C]serine.[1,2]

Reagents

Phosphatidyl[1-^{14}C]serine (20,000–100,000 cpm/μmol), 1 mM
Triton X-100, 1%
Potassium phosphate buffer, 0.5 M, pH 7.0
Potassium hydroxide, 2 N
Sulfuric acid, 0.5 N

Procedure. The substrate, phosphatidyl[1-^{14}C]serine, prepared enzymatically[1] from CDPdipalmitin[3] and [1-^{14}C]serine, is purified by DEAE-cellulose chromatography as described by Tyhach *et al.*[4] The

[1] J. Kanfer and E. P. Kennedy, *J. Biol. Chem.* **239**, 1720 (1964).
[2] W. Dowhan, W. T. Wickner, and E. P. Kennedy, *J. Biol. Chem.* **249**, 3079 (1974).
[3] B. W. Agranoff and W. D. Suomi, *Biochem. Prep.* **10**, 47 (1963).
[4] R. J. Tyhach, R. Engel, and B. E. Tropp, *J. Biol. Chem.* **251**, 6717 (1976).

specific activity of the labeled dipalmitoylphosphatidylserine can be adjusted by mixing with unlabeled phosphatidylserine from bovine brain (Sigma Chemical Co.). The assay incubation is carried out at 37° in a 25-ml Erlenmeyer flask stoppered with a rubber serum cap and equipped with a center well containing a folded piece of filter paper (Whatman No. 1, 2 × 3 cm) wet with 0.05 ml of 2 N KOH. The assay mixture contains 0.05 ml of Triton X-100, 0.1 ml of potassium phosphate buffer (pH 7.0), 0.1 ml of phosphatidyl[1-^{14}C]serine, and an appropriate amount of enzyme in a final volume of 0.5 ml. After 10 min the reaction is stopped by the addition of 0.5 ml of 0.5 N H_2SO_4 from a syringe through the rubber cap. After another 30–60 min the paper is removed and placed in a mixture of 10 ml of Patterson–Greene scintillation fluid[5] and 1 ml of water; the radioactivity is determined by scintillation counting. Controls for nonenzymatic decarboxylation are performed by omitting the enzyme addition.

Phosphatidylserine decarboxylase can, alternatively, be assayed by detecting the released CO_2 in a gas chromatographic assay.[6] Radioactively labeled phosphatidylserine can also be obtained from temperature-sensitive *E. coli psd* mutants grown in [1-^{14}C]serine under conditions of phosphatidylserine accumulation.[7]

Units. One unit of decarboxylase activity is the amount of enzyme which can decarboxylate 1 μmol of phosphatidylserine per minute at 37°.

Purification Procedure[2]

Step 1. Cell-Free Extract. Frozen *E. coli* B, full log cells are obtained from Grain Processing Inc., Muscatine, Iowa. The cell paste (1.1 kg) is suspended in 1.6 liters of 0.1 M potassium phosphate buffer (pH 6.8) containing 5 mM $MgSO_4$ and 10 mM mercaptoethanol, and the cells are ruptured by passing the suspension four times through a Manton–Gaulin press.

Step 2. Membrane Extract. Phosphatidylserine decarboxylase is a membrane-bound enzyme. In order to isolate the membranes, the cell-free extract is centrifuged at 45,000 g for 5 hr at 4°. The pellet, containing the cell envelope fraction, is suspended at 4° in 970 ml of 0.1 M potassium phosphate buffer (pH 7.15), containing 5% Triton X-100 and 10 mM mercaptoethanol, and extracted for up to 12 hr. Unextracted material is removed by centrifugation for 90 min at 45,000 g and 4°.

Step 3. Acetone Precipitation. Glycerol and 0.5 N acetic acid are added to the membrane extract to bring the pH to between 5.1 and 5.4 and the

[5] M. S. Patterson and R. C. Greene, *Anal. Chem.* **37**, 854 (1965).
[6] T. G. Warner and E. A. Dennis, *J. Lipid Res.* **14**, 595 (1973).
[7] E. Hawrot and E. P. Kennedy, *J. Biol. Chem.* **253**, 8213 (1978).

glycerol content to 20%. Cold acetone is added with stirring to a final concentration of 70%, and the precipitate is collected by centrifugation for 10 min at 14,000 *g* and 4°. This pellet is gently suspended with a Waring blender in 3 liters of 10 m*M* sodium acetate buffer, pH 5.0, containing 1% Triton X-100, 15% glycerol, and 10 m*M* mercaptoethanol, and stirred for 15 min. Insoluble material is removed by centrifugation (20 min at 14,000 *g*), and the pH of the extract is adjusted to between 7.0 and 7.4 with a saturated solution of Tris-free base.

Step 4. Chromatography on DEAE-Cellulose. The extract is applied at a flow rate of 1.0–1.5 liters per hour to a DEAE-cellulose column (9 × 28 cm, Whatman DE-52, at room temperature) equilibrated with 10 m*M* potassium phosphate buffer, pH 7.4, containing 1% Triton X-100, 10% glycerol, and 10 m*M* mercaptoethanol. After an additional wash with 500 ml of the initial equilibrating buffer, the enzyme is eluted by a linear gradient of NaCl (0 to 0.6 *M* in 12 liters) in this same buffer delivered at a rate of 1.0 liter/hr. The activity is recovered in approximately 1 liter at about 0.17 *M* NaCl. The peak fractions are diluted with 2.5 volumes of distilled water and applied to a similarly equilibrated DEAE-cellulose column (2.4 × 4 cm) at a flow rate of 800 ml/hr. The enzyme is concentrated by eluting this column with 10 m*M* potassium phosphate buffer, pH 7.4, containing 0.1% Triton X-100, 10% glycerol, 10 m*M* mercaptoethanol, and 0.6 *M* NaCl. The activity is recovered in a volume of 10–20 ml.

Step 5. Gel Filtration on Sephadex G-150. The enzyme solution is applied to a Sephadex G-150 column (3 × 90 cm, room temperature) equilibrated with 10 m*M* potassium phosphate buffer, pH 7.4, containing 0.1% Triton X-100, 10% glycerol, 10 m*M* mercaptoethanol, and 50 m*M* NaCl. The column is developed by upward flow at a rate of 30 ml/hr, and the activity elutes in the void volume.

Step 6. Chromatography on QAE-Sephadex. The partially purified enzyme from approximately 3.9 kg of initial cell paste is pooled and adjusted to pH 6.0 with 0.5 *N* acetic acid. This solution is applied to a QAE-Sephadex column (2.4 × 40 cm, Pharmacia A-25, at room temperature) equilibrated with 20 m*M* potassium phosphate buffer, pH 6.0, containing 0.5% Triton X-100, 10% glycerol, 10 m*M* mercaptoethanol, and 50 m*M* NaCl. The column is washed with 60 ml of this buffer, and the enzyme is eluted with a linear gradient of NaCl (50 m*M* to 0.5 *M* in 1.0 liter) in the same buffer applied at 50 ml/hr. The active fractions are concentrated as in step 4 by first diluting with 2.5 volumes of distilled water and then adsorbing the enzyme onto a DEAE-column (1.2 × 2.7 cm). The enzyme is eluted with 10 m*M* potassium phosphate buffer, pH 7.4, containing 0.1% Triton X-100, 10 m*M* mercaptoethanol, 1% glycerol, and 0.6 *M* NaCl.

PURIFICATION OF PHOSPHATIDYLSERINE DECARBOXYLASE FROM *Escherichia coli*[a]

Step	Total protein	Specific activity (units/mg)	Yield (%)
1. Cell-free extract	1150 g	0.014	100
2. Triton extract	147 g	0.063	60
3. Acetone precipitation	44 g	0.139	39
4. DEAE-cellulose	4.8 g	0.695	21
5. Sephadex G-150	392 mg	6.75	17
6. QAE-Sephadex	55 mg	27.0	10
7. Sucrose gradient centrifugation	20 mg	48.0	6
8. Agarose	14.3 mg	49.0	5

[a] Cumulative results taken from Dowhan *et al.*[2]

Step 7. Sucrose Gradient Centrifugation. The enzyme concentrate (0.45 ml per gradient) is layered on top of a 5 to 20% sucrose gradient (in 10 mM potassium phosphate buffer, pH 7.4, 0.1% Triton X-100, 10 mM mercaptoethanol) prepared in 12-ml polyallomer centrifugation tubes (9.5 cm long). The samples are centrifuged for 28 hr at 200,000 g at 20°. The decarboxylase activity migrates about three-quarters of the distance down the gradient. The purity of the fractions is determined by acrylamide gel electrophoresis in sodium dodecyl sulfate (SDS).

Step 8. Agarose Gel Filtration. The pooled fractions are filtered through a 4% agarose column (1.8 × 40 cm) in 10 mM potassium phosphate buffer, pH 7.4, containing 10% glycerol, 1% Triton X-100, and 10 mM mercaptoethanol, and 0.6 M NaCl at a flow rate of 8 ml/hr. The decarboxylase elutes with a K_{av} of 0.46. A summary of the purification steps is provided in the table.

Purity. Based on polyacrylamide gel electrophoresis in Triton X-100 and in SDS, the decarboxylase is nearly homogeneous (90–95% pure) after purification through step 8. The final yield is about 5%, with a specific activity of 49 units per milligram of protein under the standard assay conditions (a 3600-fold purification). The subunit molecular weight is about 35,000.[2]

Overproducing Strains. Genetic studies on mutants in the structural gene for phosphatidylserine decarboxylase have led to the identification of *E. coli* strains in which the levels of decarboxylase may be 40–50 times higher than wild-type levels.[8] The elevated decarboxylase activity in these strains is due to a gene dosage effect resulting from the insertion of the

[8] R. J. Tyhach, E. Hawrot, M. Satre, and E. P. Kennedy, *J. Biol. Chem.* **254**, 627 (1979).

bacterial *psd* gene into an independently replicating ColE1 plasmid.[9] For purification purposes, it is an advantage to start with a plasmid-bearing strain such as A324/pLC8-47[8] since only a few hundredfold purification is required to obtain near homogeneous decarboxylase and the last two steps of the purification can usually be eliminated. Decarboxylase purified from the two sources has comparable specific activities, $K_{m(app)}$ values for phosphatidylserine, and mobility during SDS–polyacrylamide gel electrophoresis. There is some evidence though, that a portion of the decarboxylase in overproducing strains has an altered mode of attachment to the membrane.[8]

Properties

Stability. Phosphatidylserine decarboxylase is stabilized by low temperature, neutral pH, and glycerol. In the pH range of 6.0 to 7.8, and in the presence of 10% glycerol and 0.1% Triton X-100, the decarboxylase is stable for 1.5 years at $-25°$, for several months at $0°$, and for at least 1 week at room temperature.[2] The membrane-bound enzyme is much more resistant to heat inactivation than the Triton-extracted enzyme.[2,7]

Specificity. Under the standard assay conditions, the half-maximal rate of decarboxylation occurs at a concentration of 23 μM phosphatidylserine. The decarboxylase appears to have little specificity with regard to the fatty acids present in the substrate and there is no detectable affinity for DL-serine, O-phospho-L-serine, or L-α-glycerophosphorylserine.[2]

Effect of pH. Maximum activity is obtained between pH 6.5 and 7.5.

Effect of Detergents. Besides being required to keep the decarboxylase soluble, a nonionic detergent such as Triton X-100 is also needed in the assay system. No activity is observed in the absence of Triton, maximal activity occurring at a molar ratio of Triton to phosphatidylserine of about 6.1.[10] It appears that phosphatidylserine incorporated into mixed micelles with detergent is accessible to the soluble decarboxylase whereas phosphatidylserine in lipid multibilayers is not. It has been shown that the phosphatidylserine in erythrocyte ghost membranes is not decarboxylated when erythrocyte ghosts are incubated with purified decarboxylase in the absence of added surfactant.[10] Ionic detergents inhibit decarboxylase activity.

Inhibitors. The purified decarboxylase does not have the absorption spectrum characteristic of pyridoxal-containing enzymes. However, the decarboxylase is irreversibly inactivated by treatment with a number of reagents that attack carbonyl groups. These include hydroxylamine, NSD

[9] L. Clarke and J. Carbon, *Cell* **9**, 91 (1976).
[10] T. G. Warner and E. A. Dennis, *J. Biol. Chem.* **250**, 8004 (1975).

1055, sodium borohydride, and cyanoborohydride. It appears that decarboxylase activity is dependent on a covalently bound residue of pyruvic acid.[11]

Mutant Variants of Decarboxylase. Mutations in the structural gene for phosphatidylserine decarboxylase provide a number of structurally altered forms of the enzyme.[12,13] The available mutant forms of the decarboxylase are thermolabile, with glycerol providing some protection against heat inactivation.[7,13] Thermal inactivation occurs more rapidly with the enzyme extracted from the membrane.

[11] M. Satre and E. P. Kennedy, *J. Biol. Chem.* **253**, 479 (1978).
[12] E. Hawrot and E. P. Kennedy, *Proc. Natl. Acad. Sci. U.S.A.* **72**, 1112 (1975).
[13] E. Hawrot and E. P. Kennedy, *Mol. Gen. Genet.* **148**, 271 (1976).

[68] CTP: Phosphocholine Cytidylyltransferase from Rat Liver

EC 2.7.7.15 CTP: Phosphocholine cytidylyltransferase

By DENNIS E. VANCE, STEVEN D. PELECH, and PATRICK C. CHOY

$$CTP + (CH_3)_3\overset{+}{N}—CH_2CH_2OPO_3^{2-} \rightleftharpoons CDPcholine + PP_i$$

The assay and partial purification of this enzyme was discussed in this series in 1969.[1] Since that time the enzyme has been highly purified from rat liver and the kinetic properties have been determined.[2] The lipid requirements for activation of the enzyme have been studied.[3,4] In addition, the activity of the enzyme appears to play a central role in the regulation of phosphatidylcholine biosynthesis.[5-7]

Assay Method

Principle. Many different approaches to the assay of this enzyme have been published. We have tried these and other methods. However, the assay described below remains the method of choice in our laboratories. It

[1] G. B. Ansell and T. Chojnacki, this series, Vol. 14 [21].
[2] P. C. Choy, P. H. Lim, and D. E. Vance, *J. Biol. Chem.* **252**, 7673 (1977).
[3] P. C. Choy and D. E. Vance, *J. Biol. Chem.* **253**, 5163 (1978).
[4] P. C. Choy, S. B. Farren, and D. E. Vance, *Can. J. Biochem.* **57**, 605 (1979).
[5] D. E. Vance and P. C. Choy, *Trends Biochem. Sci.* **4**, 145 (1979).
[6] D. E. Vance, E. M. Trip, and H. B. Paddon, *J. Biol. Chem.* **255**, 1064 (1980).
[7] P. C. Choy, H. B. Paddon, and D. E. Vance, *J. Biol. Chem.* **255**, 1070 (1980).

is based on the conversion of [Me-^{3}H]phosphocholine to CDP[Me-^{3}H]choline, separation of radioactive substrate and product by thin-layer chromatography, and determination of the radioactivity in CDPcholine.

Reagents

Tris buffer, 1 M, adjusted to pH 6.5 with succinic acid

CTP, 40 mM dissolved in H$_2$O

Magnesium acetate, 120 mM

[Me-^{3}H]Phosphocholine, 10 mM (20 μCi/μmol). The tritium derivative of this substrate is not commercially available, although [Me-^{14}C]phosphocholine can be purchased. We prepare the tritium derivative from [Me-^{3}H]choline chloride.[8] Two millicuries of the labeled choline are added to a test tube (13 × 100 mm), and the solvent is evaporated under a stream of N$_2$. Thirty microliters of MgCl$_2$ (100 mM), 30 μl of ATP (100 mM), 10 μl of 1 M Tris-HCl, pH 8.0, 10 μl of H$_2$O, and 200 μl of choline kinase were added to the tube. The kinase is a preparation from baker's yeast (Sigma grade II). One unit of enzyme is dissolved in 3 ml of H$_2$O, concentrated to 1 ml by ultrafiltration (Amicon Centriflo membrane cones, type CF 50A), and stored at $-20°$. The reaction mixture is incubated at 37° for 1 hr, boiled for 2 min, and centrifuged at low speed for 5 min. The supernatant is applied as two 4 cm streaks to silica gel G thin-layer plates and developed in the solvent system CH$_3$OH : 0.6% NaCl : NH$_4$OH (50 : 50 : 5, v/v/v). The phosphocholine has an R_f of about 0.23 and migrates ahead of the choline. It is identified by scraping 0.5 cm lengths of silica gel into different scintillation vials and adding 2 ml of H$_2$O. Aliquots (2 ml) are removed, and the radioactivity is determined. The major radioactive samples are combined and centrifuged. The supernatant is lyophilized to about 4 ml, and phosphocholine is added to give the correct specific radioactivity. The yield is greater than 80%.

ACS liquid scintillation fluid from Amersham

Glacial acetic acid

CDPcholine (5 mg/ml distilled H$_2$O)

SilicAR^r TLC-7GF thin-layer plates (20 × 20 cm) from Mallinckrodt

Phospholipid. The residue from a chloroform–methanol extract of 10 g of rat liver is washed twice with 25 ml of ice-cold acetone at 0°. The acetone is aspirated, and the precipitate is dried under reduced pressure. The phospholipid is dissolved in CHCl$_3$/CH$_3$OH (2 : 1) to give a solution containing 20 mg/ml.

Procedure. When lipid is required, it is added to a small test tube

[8] H. B. Paddon and D. E. Vance, *Biochim. Biophys. Acta* **488**, 181 (1977).

(10 × 75 mm), and the organic solvent is evaporated under a stream of N_2. Tris-succinate buffer (7.5 μl, 7.5 μmol), magnesium acetate (5 μl, 0.6 μmol) CTP (5 μl, 0.05 μmol) 0–72 μl of enzyme, and 0–72 μl of distilled H_2O are added to the tube. The tube is warmed to 37° and gently agitated on a Vortex mixer for 15–20 sec (necessary only when lipid has been added) and put into a shaking water bath at 37°. After 2 min, 10 μl of [Me-^{3}H]phosphocholine (0.1 μmol) are added (final volume 0.1 ml). The reaction is stopped after 15 min by immersion of the tubes in boiling water for 2 min and the tubes are centrifuged for 10 min at 4500 g. The thin-layer plates are spotted with 20 μl of CDPcholine as carrier. Sixty microliters of the assay mixture are spotted as a 2.5 cm streak on the silica gel plates. Each plate will accommodate seven assays. No special pretreatment of the thin-layer plate is required. The plate is developed for about 2 hr in the solvent system $CH_3OH : 0.6\%$ $NaCl : NH_4OH$ (50 : 50 : 5, v/v/v). After the solvent has evaporated from the plate, the CDPcholine is visualized under shortwave ultraviolet light. The silica is scraped into a scintillation vial, and 2 ml of H_2O are added. Acetic acid (100 μl) is added to prevent chemiluminescence, and 10 ml of scintillation fluid are added. The radioactivity is determined in a liquid scintillation counter. One unit of enzyme activity is defined as 1 μmol of product formed per minute.

Preparation of Enzyme

The purification takes advantage of the aggregation property of the cytidylyltransferase in cytosol.[2,4] Fresh cytosol contains mostly a low molecular weight (L) form of the enzyme. Upon storage at room temperature the enzyme aggregates to a high molecular weight (H) form. Chromatography of the H form on Sepharose 6B eliminates lower molecular weight proteins. The H form is dissociated into the L form and rechromatographed on the Sepharose 6B column; it yields a highly purified enzyme.[2]

Livers from two rats (200–250 g) are removed, rinsed in isotonic saline, cut into pieces, and homogenized in four volumes (ml/g liver) of isotonic saline with 10 strokes of a Potter–Elvehjem homogenizer. The homogenate is centrifuged at 100,000 g (38,000 rpm in a Beckman Ti 70 rotor) for 60 min. Phenylmethylsulfonyl flouride (17 mg per 100 ml of supernatant) is added to the cytosol, which is stored at 21° for 9 hr and subsequently stored at 4° overnight. The cytosol is adjusted to 20% $(NH_4)_2SO_4$ saturation (at 4°) by dropwise addition over 15 min of a solution of 100% $(NH_4)_2SO_4$. After 30 min the precipitate is removed by centrifugation at 10,000 g for 10 min and further $(NH_4)_2SO_4$ is added to bring the supernatant to 30% saturation; it is stored for 1 hr in an ice bath. After centrifugation at 10,000 g for 10 min, the precipitate is dissolved in 4 ml of

20 mM Tris-HCl, 0.1 M NaCl, and 1 mM dithiothreitol, pH 7.0 (buffer A). The sample is applied to a Sepharose 6B column (2.5 × 30 cm) equilibrated with the same buffer. The enzyme is eluted from the column in 4 ml fractions with a flow rate of about 50 ml/hr. The fractions are assayed for cytidylyltransferase activity without the addition of lipid. The enzyme elutes in the void volume which is slightly milky. Two or three fractions with the major amount of activity are combined and stored in an ice bath. A 1% solution of sodium dodecyl sulfate (SDS) is added dropwise to 4 ml of the pooled fractions at 0° to a final concentration of 0.05%. The solution is stored in an ice bath for 2 hr. The enzyme is applied to a Sepharose 6B column (2.5 × 30 cm) equilibrated with buffer A that also contained 0.005% SDS, and the column is eluted at a flow rate of 50 ml/hr. The fractions (6 ml) are assayed in the presence of 0.6 mg per tube of phospholipid. Undissociated cytidylyltransferase elutes in the void volume, and a second peak of activity elutes after another 6 to 7 fractions. The second fraction of enzyme activity is pooled and concentrated by ultrafiltration over an Amicon XM-100 membrane. The dissociation and chromatography step are repeated. A summary of a recent purification is shown in the table. The most difficult part of this preparation is the dissociation of the H form to the L form. Although we have tried many different procedures to achieve dissociation, the procedure described has consistently given the highest yield. The addition of higher concentrations of SDS completely inactivates the enzyme. Although the yield of enzyme is low, the procedure has been very reproducible.

Properties

Lipid Requirements. The most striking feature of the cytidylyltransferase is its requirement of lipid for significant activity.[2] There is lipid present in cytosol, so the stimulation by exogenous lipid is only 6- to 10-fold, and in aged cytosol only 1- to 2-fold. The purified enzyme is stimulated 20- to 30-fold by certain lipids.[3] Lysophosphatidylethanolamine is the best lipid activator for the rat liver enzyme. Phosphatidylserine, phosphatidylinositol, and phosphatidylglycerol are also good activators. Lysophosphatidylcholine, and to a lesser extent phosphatidylcholine, are potent inhibitors of the enzyme.[3]

Molecular Weight. The L form of the enzyme has a molecular weight of 2.0×10^5 as estimated by Sepharose 4B chromatography.[2] It aggregates in the cytosol to form high molecular weight species with a median value of 1.2×10^6. The lipid responsible for this aggregation appears to be diacylglycerol.[4] The functional significance, if any, of the two different forms of the cytidylyltransferase is not clear.

Optimum pH. The cytosolic enzyme has a broad pH range with an

PURIFICATION OF CTP: PHOSPHOCHOLINE CYTIDYLYLTRANSFERASE FROM RAT LIVER

Fraction	Volume (ml)	Protein (mg)	Total activity (units)	Specific activity (units/mg) × 10³	Recovery (%)	Purification (fold)	Total activity[b] (units)	Specific activity[b] (units/mg) × 10³
Cytosol	152	2100	0.420	0.20	100	1	3.27	1.56
Aged cytosol	152	2040	3.47	1.70	826	8.5	3.20	1.57
(NH₄)₂SO₄ 20–30%	12	281	1.46	5.20	348	26	2.67	9.50
First Sepharose 6B	22.8	57	0.85	14.9	202	74.5	1.36	23.9
Second Sepharose 6B	36	0.67	0.07[a]	105[a]	17[a]	525	0.07	105

[a] The purified enzyme has no activity unless assayed in the presence of phospholipid.
[b] All fractions were assayed in the presence of phospholipid.

$(NH_4)_2SO_4$ 20–30%

optimum of 6.0. The partially purified H form and the purified L form also have a broad pH range with the optimum at 7.0.

Kinetic Properties.[2] The true Michaelis constants for the L form of the enzyme (assayed in the presence of 0.2 mg of phospholipid per 0.1 ml) were 0.21 mM for CTP, 0.17 mM for phosphocholine, 0.21 mM for CDPcholine, and 0.004 mM for inorganic pyrophosphate. The partially purified H form had similar kinetic properties except that the value for CDPcholine was 0.64 mM. The equilibrium constant for the reaction is 0.80 at pH 6.4.

Kinetic analysis has been made of the activation of the cytidylyltransferase by oleoyl-lysophosphatidylethanolamine (O-LPE) and its inhibition by oleoyl-lysophosphatidylcholine (O-LPC).[9] Half-maximal activation by O-LPE occurred at 0.3 mM. The O-LPE lowered the K_m for CTP from 2 mM at 0.1 mM O-LPE to 0.5 mM at 0.4 mM O-LPE. The O-LPE had no effect on the K_m for phosphocholine. O-LPC inhibition was not competitive with phosphocholine or O-LPE and increased the K_m for CTP 2-fold. The activation by O-LPE may result largely from an effect on the affinity of the enzyme for CTP.

[9] S. B. Farren, P. C. Choy, and D. E. Vance, unpublished experiments.

[69] Conversion of Phosphatidylethanolamine to Phosphatidylcholine

EC 2.1.1.17 S-Adenosyl-L-methionine : phosphatidylethanolamine
N-methyltransferase

By Dennis E. Vance and Wolfgang J. Schneider

Phosphatidylethanolamine + 3 S-adenosyl-L-methionine →
     phosphatidylcholine + 3 S-adenosyl-L-homocysteine

Phosphatidylethanolamine methyltransferase was covered in this series in 1969.[1] The enzyme is tightly bound to the endoplasmic reticulum of liver, and progress with its purification has been hindered by the absence of a suitable solubilization procedure. Two methods have been published for release of the enzyme from microsomes with 0.2% Triton X-100.[2,3] This development has allowed for a characterization of the partially purified enzyme(s) from rat liver[2] and, it is hoped, will pave the way for a homogeneous preparation of the enzyme(s).

[1] J. Bremer, this series, Vol. 14 [22].
[2] W. J. Schneider and D. E. Vance, *J. Biol. Chem.* **254**, 3886 (1979).
[3] Y. Tanaka, O. Doi, and Y. Akamatsu, *Biochem. Biophys. Res. Commun.* **87**, 1109 (1979).

Assay Method

Principle. The assay measures the transfer of [^{3}H]methyl groups from *S*-adenosylmethionine to phosphatidylethanolamine, phosphatidyl-*N*-monomethylethanolamine, or phosphatidyl-*N*,*N*-dimethylethanolamine. The lipid products are extracted with butanol and separated from each other by thin-layer chromatography.

Reagents

Tris-HCl, 1.25 *M*, pH 9.2
Cysteine, 10 m*M*
[Me-^{3}H]-*S*-adenosylmethionine, 0.68 m*M* (100–200 μCi/μmol)
Lipid (4.5–6 mg) is suspended in 3 ml of 0.06% Triton X-100 in 5 m*M* Tris-HCl, pH 9.2, by sonication at 40°.[2] The suspension is stored at 2–4° and discarded after 4 days.
Enzyme: subcellular fractions of rat liver

Procedure. The final volume of the assay is varied between 0.1 and 1 ml so as to give a final concentration of 0.04% Triton X-100. The directions provided here are for a final volume of 0.5 ml in a small test tube (13 × 100 mm). Tris-HCl buffer (50 μl), cysteine (50 μl), *S*-adenosylmethionine (50 μl), and lipid suspensions (325 μl) are added to the tube with 0–25 μl of 0.9% saline. After warming the mixture to 37°, enzyme (0–25 μl) is added and incubated at 37° for 5–40 min. Longer incubation times are used so long as formation of product is linear with incubation time. The incubations are terminated by the addition of five volumes of 0.1 *N* HCl, and the lipids are extracted immediately by the addition of 2 ml *n*-butanol. The two phases are mixed for 30 sec and then centrifuged at 1000 *g* for 5 min. The butanol is removed and evaporated under a stream of nitrogen, and 10 μg of each carrier lipid are added to each sample. Phosphatidylethanolamine and its *N*-methylated derivatives are separated by thin-layer chromatography on silica gel G thin-layer plates with a solvent system of CHCl$_3$: CH$_3$OH : CH$_3$COOH (70 : 30 : 4, v/v/v). The lipids are visualized with iodine vapor, which is subsequently allowed to evaporate. The silica gel is scraped into liquid scintillation vials and 1 ml of water is added to deactivate the silica. After 30 min 10 ml of an aqueous scintillation fluid is added and, after the silica has settled, radioactivity in the sample is determined. One unit of enzyme activity is defined as 1 μmol of methyl group transferred per minute.

The identity of the product can be confirmed by two-dimensional thin-layer chromatography.[4] Alternatively, the lipid can be hydrolyzed with 6

[4] S. L. Katyal and B. Lombardi, *Lipids* **11**, 513 (1976).

N HCl at 100° for 3 hr, and the N-methylated bases can be separated by thin-layer chromatography.[2]

Investigators should be aware of an active thiol methyltransferase present in rat liver microsomes. An assay for this activity has been described.[5,6] Our solubilized enzyme(s) contained significant amounts of thiol methyltransferase (2.1 nmol mg^{-1} min^{-1}).[2]

Preparation of Enzyme

Rat liver is homogenized in 0.9% NaCl (4 ml/g liver) using 10 strokes of a Potter–Elvehjem homogenizer. The homogenate is centrifuged at 10,000 g for 20 min. The supernatant is centrifuged at 100,000 g for 60 min. The resulting microsomal pellet is resuspended in 50 mM Tris-HCl, pH 8.65, 5 mM cysteine.

If a solubilized preparation is required, microsomes are fractionated on a stepped gradient consisting of a 3-ml layer of 1.5 M sucrose and a 12-ml layer of 0.32 M sucrose. The gradient is centrifuged at 94,000 g (27,000 rpm with a Beckman SW27.1 rotor) for 20 min. The band above the interface of the two sucrose solutions is removed with a Pasteur pipette and diluted with an equal volume of 50 mM Tris-HCl, pH 8.65, 5 mM cysteine; the microsomes are sedimented at 100,000 g for 60 min. The microsomal pellet is resuspended in a mixture containing 50 mM Tris-HCl, pH 8.65, 5 mM cysteine, and 0.2% Triton X-100 at a protein concentration of about 7 mg/ml. This suspension is sonicated at 4° for 15 min with a probe sonicator and centrifuged at 100,000 g for 2 hr. The supernatant (7.5 ml at a time) is applied to a column of octyl-Sepharose CL-4B (1.7 × 6.6 cm) equilibrated with 50 mM Tris-HCl, pH 8.65, 5 mM cysteine, and 1 M NaCl. After 40 ml have eluted, the buffer is changed to 50 mM Tris-HCl, pH 8.65, 5 mM cysteine, and 0.2% Triton X-100. The column fractions are assayed with the substrate phosphatidyl-N-monomethylethanolamine; the most active fractions are combined and concentrated to 4 ml with ultrafiltration using an Amicon XM-100 filter membrane. Buffer without Triton X-100 is added periodically so that the concentration of this detergent is always below 0.6%. The soluble enzyme is subsequently applied to a Sepharose 6B column (2.6 × 40 cm) equilibrated in 50 mM Tris-HCl, pH 8.65, 5 mM cysteine, 0.1% Triton X-100. All three N-methyltransferase activities elute as a single peak that corresponds to a molecular weight of 2.0 × 10^5. These fractions are combined and assayed immediately. The purification is summarized in Table I. The preparation is free of thiol methyltransferase activity.

[5] R. T. Borchardt and Chao Fu Cheng, *Biochim. Biophys. Acta* **522,** 340 (1978).
[6] R. A. Weisiger and W. B. Jakoby, *Arch. Biochem. Biophys.* **196,** 631 (1979).

TABLE I

PARTIAL PURIFICATION OF PHOSPHATIDYLETHANOLAMINE N-METHYLTRANSFERASE(S)[a]

Fraction	Volume (ml)	Protein (mg)	Phospholipid (mg)	Total activity[b] (units) $\times 10^3$	Specific activity[b] (units/mg) $\times 10^3$	Recovery (%)	Purification (fold)
1. Homogenate	25	972	—	19.4	0.02	100	1
2. Microsomes	12.5	110	—	14.3	0.13	74	6.5
3. Microsomes from sucrose gradient	10.0	72	33	11.5	0.16	59	8.0
4. Triton X-100 extract	15.0	45	27	5.0	0.11	26	5.5
5. Octyl-Sepharose	24.0	12	3.84	3.5	0.29	18	14.5
6. Sepharose 6B	10.5	3.2	0.84	2.0	0.63	10	31.5

[a] Data were taken from Table I of Schneider and Vance[2] with permission of the *Journal of Biological Chemistry*.

[b] Activity for this table is calculated with phosphatidyl-N-monomethylethanolamine as substrate minus the activity without the addition of exogenous lipid. The activities with no addition of lipid and addition of exogenous phosphatidylethanolamine or phosphatidyl-N,N-dimethylethanolamine are reported by Schneider and Vance[2].

Properties of Partially Purified Enzyme(s) from Rat Liver

Stability. The activity of the enzyme(s) is not stable, and 30% of the activity is lost after 12 hr at 2°.[2] The solubilized enzyme is inactivated after freezing at −70°. The stability is dependent on a sulfhydryl reducing agent. The instability of the solubilized enzyme remains a major problem.

Optimum pH. The activities increase in nearly a linear fashion between pH 6.0 and 9.5.[2] The enzyme activities are erratic at higher pH values; this is probably a function of the alkaline hydrolysis of S-adenosylmethionine.

Cofactor Requirements. None have been detected.

Kinetic Properties.[2] These analyses are complicated because the product of one methylation reaction can be subsequently methylated. Thus, methylation of phosphatidylethanolamine results in the formation of the N-monomethyl derivative, which can itself be methylated to the N,N-dimethyl derivative. Despite these problems estimates of kinetic parameters have been reported[2] and are summarized in Table II.

Conversion of Phosphatidylethanolamine to Phosphatidylcholine by Other Mammalian Systems

The N-methyltransferase(s) has been solubilized from mouse liver microsomes.[3] In this procedure, the microsomes were first treated with 0.3% sodium deoxycholate and the enzyme(s) subsequently was solubilized with 0.2% Triton X-100.

Two N-methyltransferases have been observed associated with microsomes from bovine adrenal medulla.[7] The first acts on phosphatidylethanolamine to form phosphatidyl-N-monomethylethanolamine; it has a relatively low specific activity (3.9×10^{-6} unit/mg) with a pH optimum of 6.5 and requires Mg^{2+}. The second converts the phosphatidyl-N-monomethylethanolamine to phosphatidyl-N,N-dimethylethanolamine and phosphatidylcholine and has a pH optimum of 10 but no requirement for Mg^{2+}. The specific activity of the latter enzyme was not reported.

A similar system has been described for membranes from rat erythrocytes.[8] The phosphatidylethanolamine-N-methyltransferase had considerably lower activity than observed in the adrenal microsomes (between 1.6×10^{-7} and 7.6×10^{-8} unit/mg). The second transferase had similarly low activity (3.2×10^{-7} unit/mg). Evidence was presented that the initial methylation of phosphatidylethanolamine occurs on the inside of the membrane and the subsequent methylations occur on the outside of the

[7] F. Hirata, O. H. Viveros, E. J. Diliberto, and J. Axelrod, *Proc. Natl. Acad. Sci. U.S.A.* **75**, 1718 (1978).

[8] F. Hirata and J. Axelrod, *Proc. Natl. Acad. Sci. U.S.A.* **75**, 2348 (1978).

TABLE II

KINETIC PROPERTIES OF PARTIALLY PURIFIED PHOSPHATIDYLETHANOLAMINE N-METHYLTRANSFERASE[a]

Lipid substrate	Product	K_m for lipid (mM)	K_m for S-adenosylmethionine (μM)	K_i for S-adenosylhomocysteine (μM)	Initial velocity or V_{max} (unit/mg) $\times 10^3$
Phosphatidyl-ethanolamine	Phosphatidyl-N-monomethyl-ethanolamine	—	—	—	0.08
Phosphatidyl-N-monomethyl-ethanolamine	Phosphatidyl-N,N-dimethyl-ethanolamine	0.08[b]	22[b]	4.9[b]	0.69
Phosphatidyl-N,N-dimethyl ethanolamine	Phosphatidyl-choline	0.45[b]	16	6.7	0.88

[a] Taken from Schneider and Vance[2] with permission of the *Journal of Biological Chemistry*.
[b] Apparent K_m or K_i values.

membrane. Furthermore, it was concluded that this methylation of phosphatidylethanolamine increased the fluidity of the erythrocyte membrane.[9] Recent calculations have shown that in the experiments where the fluidity was decreased from 1.62 to 1.09 poise[9] only 0.0015% of the phosphatidylethanolamine was converted to the N-monomethyl derivative.[10] Furthermore, of the phosphatidylcholine molecules present in the red cell membrane, only 0.00044% were formed via methylation of phosphatidylethanolamine in the 1-hr incubation.[10] How such minuscule changes in phospholipid content can alter the fluidity of the red cell is not clear.

A number of other physiological effects have been attributed to methylation of phosphatidylethanolamine.[11-16] These reports also show extremely low activities (pmol/mg) for the methylation reactions.

More recently, the conversion of phosphatidylethanolamine to phosphatidylcholine has been studied in brain homogenates.[17] Methylation of phosphatidylethanolamine was observed at pH 6.8, but was twice as active at pH 8.0. Thus, it would appear that, in brain, an N-methylating enzyme with a low pH optimum does not occur.

Control of N-Methylation of Phosphatidylethanolamine

Very little information is available on the control of phospholipid methylation. We have reported an increase in this activity in livers from choline-deficient rats.[18] The reason for this increased activity is not clear. In another study with primary cultures of rat hepatocytes, the concentration of phosphatidylethanolamine can be varied by the amount of ethanolamine in the culture medium.[19] The incorporation of [Me-14C]methionine into phospholipid varied directly with the concentration of phosphatidylethanolamine.[19] The implication is that the concentration of phosphatidylethanolamine in the hepatocyte influences its own methylation.

[9] F. Hirata and J. Axelrod, Nature (London) 275, 219 (1978).
[10] D. E. Vance and B. de Kruijff, Nature (London) 288, 277 (1980).
[11] W. J. Strittmatter, F. Hirata, and J. Axelrod, Science 204, 1205 (1979).
[12] F. Hirata, W. J. Strittmatter, and J. Axelrod, Proc. Natl. Acad. Sci. U.S.A. 76, 368 (1979).
[13] F. Hirata, B. A. Corcoran, K. Venkatasubramanian, E. Schiffmann, and J. Axelrod, Proc. Natl. Acad. Sci. U.S.A. 76, 2640 (1979).
[14] F. Hirata, J. Axelrod, and F. T. Crews, Proc. Natl. Acad. Sci. U.S.A. 76, 4813 (1979).
[15] A. Bhattacharya and B. K. Vonderhaar, Proc. Natl. Acad. Sci. U.S.A. 76, 4489 (1979).
[16] M. C. Pike, N. M. Kredich, and R. Snyderman, Proc. Natl. Acad. Sci. U.S.A. 76, 2922 (1979).
[17] R. Mozzi and G. Porcellati, FEBS Lett. 100, 363 (1979).
[18] W. J. Schneider and D. E. Vance, Eur. J. Biochem. 85, 181 (1978).
[19] B. Åkesson, FEBS Lett. 92, 177 (1978).

Our understanding of the enzymology and control of the methylation of phosphatidylethanolamine is still in its infancy. The major difficulty for research in this area is the small amount of this membrane-bound activity that is present in liver and other tissues. Since two procedures for solubilization of the enzyme(s) have been developed recently,[2,3] we can expect more rapid developments in the future.

[70] Separation of Base Exchange Enzymes from Brain with Special Reference to L-Serine Exchange

By T. Miura, T. Taki, and J. N. Kanfer

The only available mechanism for the formation of phosphatidylserine by animal tissues is by "base exchange" reactions catalyzed by enzymes stimulated by Ca^{2+} and possessing somewhat alkaline pH optima. The reactions do not result in the net synthesis of lipid and are most readily monitored by incorporation of radioactive precursors into the corresponding phospholipid.

L-Serine incorporation into lipid by rat liver mitochondria was originally reported in 1959, and it was suggested that this may have occurred by a reversal of phospholipase activity.[1] Subsequently a similar reaction was observed for choline incorporation, which was stimulated by Ca^{2+}, and the possibility that it was due to a reversal of phospholipase D action or an "exchange reaction" was suggested.[2] The "base exchange" enzymes have been detected in many animal, plant, and insect tissues. Evidence has accumulated suggesting that the incorporation of the three bases commonly employed for such studies, L-serine, ethanolamine, and choline, are catalyzed by separate enzymes.[3]

Solubilization and Separation of the Base Exchange Enzymes

The starting material for these preparations is rat brain microsomes.

Reagents

Solution A: .8% Miranol H2M (available from the Miranol Chemical

---

[1] G. Hübscher, R. P. Dils, and W. F. R. Pover, *Biochim. Biophys. Acta* **36**, 518 (1959).

[2] R. P. Dils and G. Hübscher, *Biochim. Biophys. Acta* **46**, 505 (1961).

[3] T. Taki, T. Miura, and J. N. Kanfer, "Enzymes of Lipid Metabolism" (S. Gatt, L. Freysz, and P. Mandel, eds.), p. 301. Plenum, New York, 1978.

Co., Irvington, New Jersey), 0.5% sodium cholate, 20% glycerol in
5 mM HEPES pH 7.23

Solution B: 0.4% Miranol H2M and 0.2% sodium cholate, 20%
glycerol, 1 mM 2 mercaptoethanol in 5 mM HEPES, pH 7.23

Solution C: Asolectin microdispersion (6 µg of P per milliliter), 20%
glycerol, 1 mM 2 mercaptoethanol in 5 mM HEPES pH 7.23

Solution D: 20% glycerol, 1 mM 2 mercaptoethanol in 5 mM HEPES,
pH 7.23

Solution E: 20% glycerol, 1 mM β-mercaptoethanol, Asolectin microdispersion (6 µg P/ml), 0.5% Brig 58 in 5 mM HEPES, pH 7.23

DEAE-cellulose 32

Affi-gel 102 (Bio-Rad Labs, Richmond, California)

L-[U-14C]Serine and nonradioactive L-serine

[14C]Ethanolamine and nonradioactive ethanolamine

[14C]Choline and nonradioactive choline

Asolectin (Associated Concentrates, New York City), microdisper-
sions prepared according to this series, Vol. 10 [70]

Ammonium sulfate

Nitrocellulose membrane filter HA, 0.45 µm, 2.5 cm (Millipore Co.,
Bedford, Massachusetts)

Trichloroacetic acid, 10%

Trichloroacetic acid, 5%

CaCl₂

HEPES, pH 7.23

EDTA

Bovine serum albumin

2-Mercaptoethanol

Procedure

Brains were obtained from rats 22–29 days of age and homogenized in
nine volumes of 0.32 M sucrose containing 2 mM HEPES and 1 mM
EDTA, pH 8.0. Homogenization was performed with a Polytron instrument (Kinematica, CMBH, Lucerne, Switzerland) at maximum speed for
20 sec. The homogenate was centrifuged at 12,000 g for 10 min, and the
supernatant was centrifuged at 56,000 g for 60 min. The pellet thus obtained was suspended in the homogenization medium with a Potter–
Elvehjem homogenizer and designated the microsomal fraction.

Solubilization of the Enzyme. The microsomal fraction was suspended
in solution A at about 2.5 mg of protein per milliliter. The suspension was
allowed to stand for 15 min in an ice bath and then centrifuged at 165,000 g
for 60 min. In order to measure the enzyme activity present, the sol-

ubilized supernatant was precipitated with 60% saturation of $(NH_4)_2SO_4$ and dissolved in a small volume of solution C. The solution was dialyzed overnight against 100 volumes of solution B (AS-supernatant). The dialyzed solution was used to assay for base-exchange activities.

Ammonium Sulfate Fractionation of the Solubilized Supernatant. The supernatant obtained after solubilizing with solution A was fractionated with $(NH_4)_2SO_4$ in the presence of Asolectin dispersion (6 μg of P/ml). The ammonium sulfate fractionation was performed in an ice bath, adjusting the pH to 7.2–7.4 with concentrated NH_4OH. Precipitated material was recovered 20 min later by centrifugation and dissolved in solution C. The solution was dialyzed overnight against 100 volumes of solution D. The dialyzed solutions were used to assay base-exchange activities.

Separation of the Individual Base-Exchange Enzymes

1. Sepharose 4B Column. The fraction that precipitated between 35 and 55% $(NH_4)_2SO_4$ saturation was dissolved in solution B (approximately 15 mg of protein/ml). A 5- to 6-ml aliquot was applied to a Sepharose 4B column (2.6 × 40 cm) equilibrated with a solution of 0.25% Miranol H2M in solution C and eluted with the same solvent at a flow rate of 18 ml/hr. Enzyme-containing fractions were precipitated with 60% saturation of $(NH_4)_2SO_4$ in the presence of Asolectin dispersion (6 μg of P/ml). The precipitate was dissolved in solution B (approximately 10 mg of protein/ml). This solution was dialyzed against 30 volumes of solution B. Sepharose 4B column removed large membrane fragments that also were found to contain the bulk of the choline incorporation activity. The turbid front fractions contained more than 60% of the charged proteins. This is followed by a broad peak (110–190 ml) containing almost all the ethanolamine and serine incorporation activities. The choline incorporation activity showed a completely different distribution. More than two-thirds of this activity was in the turbid front fractions and the remainder was in the broad peak containing the enzymes for ethanolamine and serine incorporation. The presence of Asolectin dispersion in the elution solution was required to protect the enzymes. The eluate contained almost all the activities toward ethanolamine and serine, and these activities were precipitated with $(NH_4)_2SO_4$ in the presence of Asolectin dispersion.

Affi-Gel 102 Column. The dialysand was applied 1 hr later to a column (1.8 × 20 cm) of Affi-Gel 102, which had been equilibrated with a solution of 0.5% Brij 58 in solution C. The base-exchange activities were eluted both with this solution and a solution of 0.25% Miranol H2M in solution C; a typical elution pattern is shown in Fig. 1. With a Brij-58 solution as the eluent, the bulk of charged proteins is not retained by the column but

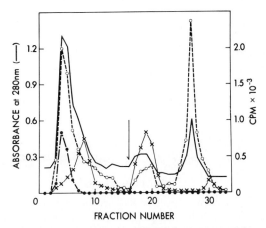

FIG. 1. Chromatography of base-exchange enzymes on an Affi-Gel 102 column. The eluted fractions from a Sepharose 4B column, which contained almost all activities for ethanolamine and serine incorporation, were concentrated with 60% saturation of $(NH_4)_2SO_4$ and dissolved in solution B. This solution was dialyzed for 1 hr against 30 volumes of 0.25% Miranol H2M in solution D and applied to an Affi-Gel 102 column. The elution was conducted at a flow rate of 40 ml/hr with 0.5% Brij 58 in solution C until the point indicated by the arrow and then continued with 0.25% Miranol H2M in solution C. Further details of the procedure are described in the text. Fractions of 5 ml were collected, and aliquots of 20 μl of each fraction were used to assay incorporation activities for ethanolamine (O - - - O), serine (● – – ●), and choline (×——×). Activities are shown as counts per minute of ^{14}C-labeled base incorporated per 20 μl of each fraction.

emerges as a slightly turbid solution. Some activity for ethanolamine and serine is found in these fractions (fractions 3–6) with a 26 and 18% yield, respectively. Elution with Miranol H2M resulted in removal of the adsorbed proteins and a small peak of choline and ethanolamine activities. This was followed by a major peak of ethanolamine incorporation activity (fractions 23–30) that was essentially free of activity toward the other two bases.

DEAE-Cellulose Column. Fractions 4–6 of the Affi-Gel 102 column were combined, and L-serine was added to make the final concentration 0.1 mM; after 15 min, the sample was applied onto a column of DEAE-cellulose (DE-32), which had been equilibrated with a solution of 0.5% Brij-58 in solution C and 0.1 mM L-serine. The column was washed with a small amount of the same medium. Elution was conducted in stepwise manner with the same eluent containing 0.15 M NaCl followed by 0.75 M NaCl. Serine incorporation activity was eluted with 0.15 M sodium chloride, completely overlapping the elution of the bulk of proteins when elution was conducted by a Brij-58 solution in the absence of L-serine. When the column chromatographic procedures were conducted in the

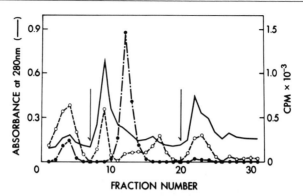

FRACTION NUMBER

FIG. 2. Chromatography on a DEAE-cellulose column of the eluent from an Affi-Gel 102 column (fractions 4–6). Fractions 4–6 of an Affi-Gel 102 column were applied to a DEAE-cellulose column. After washing the column with 0.5% Brij 58 in solution C containing 0.1 mM L-serine, the enzymes were eluted at a flow rate of 50 ml/hr in a stepwise manner with the same eluent containing 0.15 M NaCl followed by 0.75 M NaCl as indicated by the first arrow (fraction No. 8). The increase in the concentration of sodium chloride is indicated by the second arrow (fraction No. 20). Fractions of 3 ml were collected, and aliquots of 20 μl of each fraction were used to assay incorporation activities for ethanolamine (O---O) and serine (●---●). The activity is shown as described in the legend of Fig. 1.

presence of 0.1 mM L-serine, one-fourth of charged proteins was eluted with 0.15 M sodium chloride (Fig. 2). Serine incorporation activity occurred as a peak well separated from the bulk of proteins with a recovery of 97% of applied activity (fractions 11–14). There was only negligible ethanolamine incorporation activity associated with this peak.

Purification of the Serine Base Exchange Enzyme

The isolation of the rat brain particulates and their solubilization is accomplished as described in the preceding section. Solution E was composed of 20% glycerol, 5 mM HEPES, pH 7.2, 1 mM 2-mercaptoethanol, 1 mM EDTA.

Sepharose 4B Column Chromatography. The pellet obtained as the 35–60% saturation $(NH_4)_2SO_4$ fraction was dissolved in 10 ml of 0.4% Miranol H2M/0.4% sodium cholate in solution E. A Sepharose 4B column (2.6 × 40 cm) was equilibrated prior to use with 300 ml of the elution medium [0.25% Miranol H2M/0.125% sodium cholate/asolectin microdispersion (200 μg of phospholipid-phosphorus/100 ml elution medium)] in solution E. A 10–11.5 ml enzyme sample was applied to the Sepharose 4B column, and fractions of approximately 6 ml were collected; a typical elution profile is shown in Fig. 3. Approximately 65–70% of the serine and ethanolamine incorporating activities was eluted just after the first protein

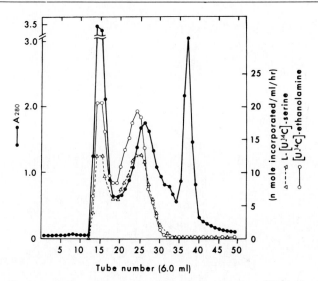

FIG. 3. Chromatography of the base exchange enzyme on a Sepharose 4B column. Fractions of 6 ml were collected, and aliquots of 15 μl of each fraction were used to assay the incorporation activity for ethanolamine (O——O) and for L-serine (△--- △). The elution of protein (●——●) is based on absorbance at 280 nm.

peak, which was quite turbid and contained about 30–35% of applied activity. The amount of enzyme activity present in the first protein peak was increased if asolectin microdispersion was added at the step of $(NH_4)_2SO_4$ precipitation. The second peak of enzyme activity present in tubes 18–27 is pooled and applied to a DEAE-cellulose column.

The processing of brain tissue through the Sephadex 4B column step was accomplished in one day.

DEAE-Cellulose Column I. The enzyme-rich fractions from the Sephadex column were pooled (about 60 ml) and applied to a DEAE-cellulose column (1.5 × 10 cm), which had been previously equilibrated with a solution containing 0.25% Miranol H2M/0.125% cholate/asolectin/ 0.1 M NaCl, in solution E. The column was eluted in a stepwise manner by increasing the concentration of NaCl, and a typical elution profile is shown in Fig. 4. The major portion of the serine and ethanolamine enzyme activity was eluted together with 0.2 M NaCl. The ratio between serine and ethanolamine activities was changed from the Sepharose 4B column to the DEAE-cellulose column, suggesting that the serine enzyme is more stable than the ethanolamine enzyme. The contents of tubes 10–14 were pooled; L-serine was added and applied to a second DEAE-cellulose column.

DEAE-Cellulose Column II. The fractions containing the base exchange

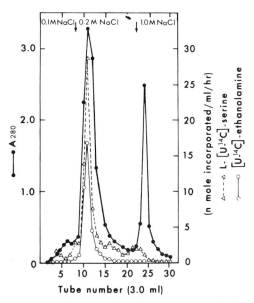

FIG. 4. Chromatography of the base exchange enzymes on first DEAE-cellulose column. A 60-ml sample from Sepharose 4B column was applied. Elution was conducted by the stepwise increase of the NaCl concentration. NaCl, 0.2 M, was added at fraction No. 8 and a 1.0 M NaCl solution at fraction No. 21. Symbols are the same as for Fig. 3.

enzymes were combined, and sufficient L-serine was added to bring the final concentration to 1 mM. The sample was applied to a second DEAE-cellulose column (1.2 × 8 cm), which had been equilibrated with a solution containing 0.25% Miranol H2M/0.125% sodium cholate/0.14 M NaCl/1 mM serine, in solution E. Elution was performed with stepwise increasing of the NaCl concentration; a typical profile is presented in Fig. 5. Approximately 32% of serine base exchange enzyme activity applied to this column was eluted with 0.2 M NaCl. Ethanolamine incorporation enzyme activity was undetectable in these fractions (40–45). The separation and yield of enzyme activities are summarized in the table. The serine enzyme from the DEAE-cellulose column II is very unstable; therefore, all experiments on its properties using pooled fractions 41, 42, and 43 were executed within 24 hr.

Assay Procedure

The basic constituents of the reaction mixture were 10 μmol of HEPES, pH 7.23, Asolectin dispersion (25 μg of P), either 7.2 nmol (0.24 μCi) of ethanolamine, 20.2 nmol (0.71 μCi) of serine, or 64.8 nmol (0.67 μCi) of choline, 2 μmol of CaCl$_2$ for ethanolamine and choline incorpora-

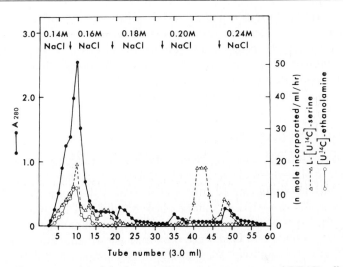

FIG. 5. Chromatography of the base exchange enzymes on second DEAE-cellulose column. The eluted enzyme from the first DEAE-cellulose column (fractions 10–14) was collected. The NaCl concentration of the sample was adjusted to contain 0.14 M NaCl and 1 mM L-serine. Symbols are the same as for Fig. 3.

tions or 6 μmol of CaCl$_2$ for serine incorporation, 50 μg of bovine serum albumin, and enzyme protein in a total volume of 0.24 ml. The reaction tubes were incubated at 37° for 15 min with shaking and terminated by the addition of 1 ml of ice-cold 10% trichloroacetic acid. The contents of the tube were filtered onto a nitrocellulose membrane filter prewashed once with 3 ml of 5% trichloroacetic acid (HA 0.45 μm, 2.5 cm; Millipore Co.,

PURIFICATION OF L-SERINE EXCHANGE ENZYME FROM RAT BRAIN

	Serine exchange	
Enzyme fraction	Specific activity (nmol/mg protein per hour)	Yield (%)
Microsome	9.45	100
Soluble enzyme	7.0	74
Soluble enzyme (150,000 g supernatant)	9.05	57
35–60% (NH$_4$)$_2$SO$_4$	2.0	8.1
Sepharose 4B	10.2	35
DEAE-cellulose I	13.7	14
DEAE-cellulose II	370[a]	2–5.6

[a] Protein content was calculated from absorption at 280 nm.

Bedford, Massachusetts) and washed with 20 ml of ice-cold 5% trichloroacetic acid. The prewashing of the filter membrane was important both for lowering the nonspecific radioactivity trapped on the membrane and for diminishing the variations of the radioactivity measured. The filter was then dissolved in Aquasol (New England Nuclear, Boston, Massachusetts), and the radioactivity was determined.

Properties of the Serine Base Exchange Enzyme

The optimum pH was approximately 8.0. The incorporation rate of L-serine into phospholipid was linear up to 20 min incubation time, and the activity was maximum at 10 mM CaCl$_2$. The calculated K_m value for L-serine was 0.4 mM. Ethanolamine phospholipid was the most effective acceptor for L-serine incorporation, particularly ethanolamine plasmalogen. The K_m values obtained were 0.25 mM for ethanolamine plasmalogen, 0.25 mM for pig liver phosphatidylethanolamine, and 0.66 mM for egg yolk phosphatidylethanolamine. Neither ethanolamine nor choline inhibited the L-serine exchange activity. There was no detectable conversion of phosphatidylcholine or phosphatidylethanolamine to phosphatidic acid by the partially purified enzyme.

[71] Phospholipid-Synthesizing Enzymes from Castor Bean Endosperm

By T. S. MOORE, JR.

The tissues principally utilized for investigations of phospholipid biosynthesis in plants have been castor bean endosperm and spinach leaves, although cauliflower florets and onion roots also have been useful. Spinach was employed for much of the earlier work[1] and castor bean has proved to be useful more recently.[2-9] Castor bean endosperm is an excel-

[1] M. Kates and M. O. Marshall, *in* "Recent Advances in the Chemistry and Biochemistry of Plant Lipids" (T. Galliard and E. I. Mercer, eds.), p. 115. Academic Press, New York, 1975.
[2] J. M. Lord, T. Kagawa, T. S. Moore, and H. Beevers, *J. Cell Biol.* **57**, 659 (1973).
[3] T. S. Moore, J. M. Lord, T. Kagawa, and H. Beevers, *Plant Physiol.* **52**, 50 (1973).
[4] T. S. Moore, *Plant Physiol.* **54**, 164 (1974).
[5] T. S. Moore, *Plant Physiol.* **56**, 177 (1975).
[6] T. S. Moore, *Plant Physiol.* **57**, 382 (1976).
[7] T. S. Moore, *Plant Physiol.* **60**, 754 (1977).
[8] J. C. Sexton and T. S. Moore, *Plant Physiol.* **62**, 978 (1978).
[9] S. A. Sparace and T. S. Moore, *Plant Physiol.* **63**, 963 (1979).

lent source for organelle purifications, and it is capable of rapid organelle biogenesis over a period of 4–5 days following germination.[10] Relatively high specific activities for phospholipid biosynthesis are obtained without further purification, and cellular compartmentation of biosynthesis is readily determined. No extensive purifications beyond organelle and organelle membrane isolation have been reported from plant tissues.

Organelle Purification from Castor Bean Endosperm

The procedure is based on methods described for the isolation of glyoxysomes.[11] It involves gently disrupting the tissue by chopping with a single razor blade, removing whole cells and cell debris by filtration and low-speed centrifugation, and purification of the organelles on sucrose density gradients.[2,8]

Purification Procedure

Reagents

Homogenization medium: 150 mM Tricine buffer, pH 7.5, 10 mM KCl, 1 mM MgCl$_2$, 1 mM EDTA, pH 7.5, 16% sucrose (w/w)
Sucrose, 60% (w/w) containing 3 mM EDTA, pH 7.5
Sucrose, 32% (w/w) containing 3 mM EDTA, pH 7.5
Sucrose, 20% (w/w) containing 3 mM EDTA, pH 7.5

Procedure

Step 1. Gradient Preparation. Prepare the gradients prior to starting the tissue homogenization. Introduce 2.0 ml of 60% sucrose solution into a 37.5 ml centrifuge tube. Add 10 ml of 60% sucrose to the mixing chamber of the gradient maker and 10 ml of 32% sucrose to the other chamber. Allow the 60% sucrose to drop into the centrifuge tube (against the side) at about 1 drop per second; once this has started, open the stopcock between the chambers. The solutions are mixed by bubbling nitrogen or air through the mixing chamber. After the gradient has been formed, it is stored in a refrigerator until just prior to its use.

Step 2. Tissue Preparation. Select 15 castor bean seedlings, 4–4.5 days old, which have been grown in moist vermiculite in the dark at 30° and 75–80% relative humidity. Remove the roots and rinse the endosperm in

[10] H. Beevers, *in* "Recent Advances in the Chemistry and Biochemistry of Plant Lipids" (T. Galliard and E. I. Mercer, eds.), p. 287. Academic Press, New York, 1975.
[11] H. Beevers and R. W. Breidenbach, this series, Vol. 31, p. 565.

water to remove all the adhering vermiculite. Blot the tissue and split the endosperm halves; remove the cotyledons and remainder of the embryo and discard them. Place the endosperm halves into a petri dish that is kept cool on ice.

Step 3. Homogenization. Add 11 ml of homogenization medium to the endosperm in the petri dish. Homogenize the tissue by chopping with a razor blade at about 4–5 strokes/sec for 15 min. The petri dish should be maintained at a slight angle on the ice, so that the medium flows to one side and the tissue can be concentrated over a smaller area in the medium. This leads to more efficient homogenization.

Step 4. Removal of Cells and Cell Debris. Filter the homogenate through two layers of nylon or Dacron cloth (200–600 pores/cm^2) and then centrifuge the suspension at 250–300 g for 10 min. The centrifugation may be omitted for improved recovery of all organelles, especially the plastids, but then the bottom of the gradient tube will contain considerable cell debris.

Step 5. Gradient Separation. Carefully layer 10 ml of 20% sucrose onto the tops of the gradients prepared in step 1. Avoid excessive mixing, as the microsomal fraction is concentrated at the step between the 20% and 32% sucrose. Remove the lipid from the top of the centrifuged homogenate by repeatedly touching the top edge of a small test tube to the surface. A portion of the lipid adheres to the edges of the tube and is held across the opening by surface tension. Wipe off the lipid after each portion is removed. It is difficult to remove all of the lipid, but most can be eliminated in this manner. Decant the homogenate into a tube and then carefully layer 5 ml onto the surface of the 20% sucrose.

Step 6. Ultracentrifugation. Centrifuge the sample in a swinging-bucket rotor at 53,000 g (R_{av}) for 4 hr. A shorter time period (2–3 hr) may be sufficient, but the user must determine if true density equilibrium of the organelles has been attained.

Step 7. Fractionation. Following centrifugation, carefully remove the lipid layer from the top as in step 4. The gradient may now be fractionated with a standard commercial fractionator, by collecting drops after puncturing the bottom of the tube, or by drawing individual bands up into a syringe equipped with a long, hooked, 18-gauge needle. In the latter case, the needle is inserted through the band until the tip is at its bottom, and then the syringe is filled. This requires considerable practice, but can be quite accurate.

Notes

The purity of the microsomal fraction and mitochondrial fractions obtained as described above is quite high (Fig. 1). Measurements of

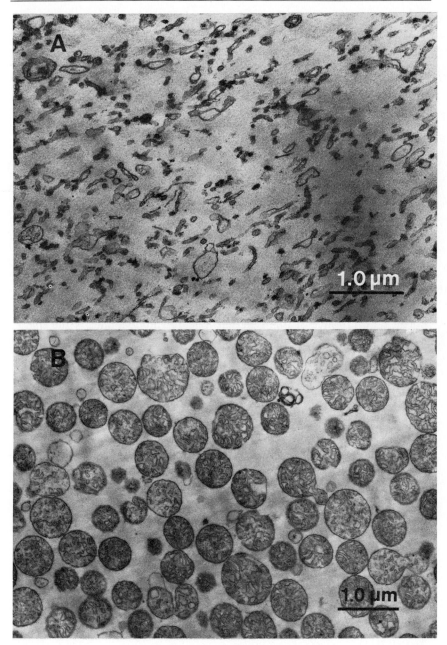

FIG. 1. Electron micrographs of the microsomal (A) and mitochondrial (B) fractions isolated from castor bean endosperm 4 days after germination. A, ×20,450; B, ×16,500.

marker enzymes indicate the occurrence of the endoplasmic reticulum at an average density of 1.12 g/cm³, mitochondria at 1.18 g/cm³, plastids at 1.22–1.23 g/cm³, and glyoxysomes at 1.24 g/cm³ (Fig. 2).[10]

The purpose of the step between 20 and 32% sucrose is to concentrate the microsomal fraction. When gradients are made utilizing 15 ml each of 16 and 60% (w/w) sucrose containing 150 mM Tricine buffer, pH 7.5, and 1 mM EDTA, pH 7.5,[2] the microsomal fraction ranges in density from about 1.11 to 1.14 g/cm³. Such spread out distributions allow for investigation of subfractions of the endoplasmic reticulum.

The microsomal fraction prepared as described above consists of membrane vesicles with no attached ribosomes. Maintenance of some ribosomal attachment is possible by introducing 3 mM MgCl₂ into the Tricine-containing gradients described above.[2] Under these conditions, the average density of the microsomal fraction shifts to 1.16 g/cm³, reflecting the increased density due to the ribosomal attachment.

CDPcholine : 1,2-Diacylglycerol Cholinephosphotransferase (EC 2.7.8.2)

sn-1,2-Diglyceride + CDPcholine → phosphatidylcholine + CMP

The lipid product of this reaction, phosphatidylcholine, is the major phospholipid of most plant cell membranes.[12,13]

Assay Method

Principle. Incorporation of radioactivity from CDP[1,2-¹⁴C]choline into phospholipids is determined by extracting the chloroform-soluble phospholipid products by a modification of the method of Bligh and Dyer.[6,14]

Reagents

Tris-HCl buffer, 25 mM, pH 7.5
Dithiothreitol, 5 mM
MgCl₂, 10 mM
CDP[1,2-¹⁴C]choline (ICN, Irvine, California), 1 mM, 5 mCi/mmol
HCl, concentrated
Chloroform–methanol–H₂O (1 : 2 : 0.3, by volume)
Chloroform
KCl, 1.0 M

Assay Procedure. Add 0.1 ml each of buffer, dithiothreitol, MgCl₂, and

[12] C. Hitchcock and B. W. Nichols, "Plant Lipid Biochemistry," Academic Press, New York, 1971.
[13] M. Kates, *Adv. Lipid Res.* **8**, 225 (1970).
[14] E. G. Bligh and W. J. Dyer, *Can. J. Biochem. Physiol.* **37**, 911 (1959).

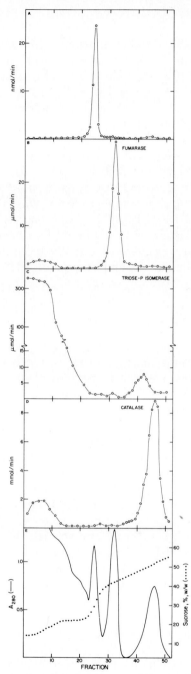

Fig. 2. Marker enzyme profiles as measured in gradients prepared as described in the text. Marker enzymes used were as follows: (A) CDPcholine:diglyceride phosphorylcholinetransferase, endoplasmic reticulum; (B) fumarase, mitochondria; (C) triose-P isomerase, cytosol and plastids[2,4]; (D) catalase, glyoxysomes; Data of G. Troyer, S. A. Sparace, and J. C. Sexton.

CDPcholine solutions, and sufficient water to give a final volume of 0.5 ml after further additions, to a test tube for each sample to be assayed. Incubate at 37° for 5 min. Start the reaction by adding enzyme. Incubate with shaking at 37° for 30 min and stop the reaction by adding 0.05 ml of HCl followed by 3.3 ml of the chloroform–methanol–H_2O mixture. Incubate at room temperature for 1 hr.

Extraction Procedure. Centrifuge the above mixture at about 2500 rpm in a tabletop or similar centrifuge and decant the solution into a 10 ml, conical, pennyhead-stoppered centrifuge tube. Add 1.0 ml of chloroform and 2.0 ml of H_2O. Stopper and shake vigorously. Recentrifuge and aspirate off and discard the upper, aqueous layer. Wash the lower chloroform layer twice with 3.0 ml of 1.0 M KCl and once with H_2O. Break any emulsion by centrifugation and aspirate off the aqueous layers.

Radioactivity Determination. Quantitatively transfer the chloroform fraction into a scintillation vial and evaporate. Add scintillation fluid and measure the radioactivity.

Properties

Linearity of Reaction. The activity of the microsomal fraction enzyme from castor bean is constant for up to 30 min and is proportional to the protein concentration up to 0.45 mg/0.5 ml.[6] With the spinach enzyme, phosphatidylcholine synthesis is linear for 15 min and is proportional to the protein concentration up to 4.0 mg/1.6 ml.[15]

Substrate Affinities. The apparent K_m of the castor bean enzyme for CDPcholine is 9.7 μM.[6] The estimated K_m of the spinach leaf enzyme is 10 μM.[15] No K_m has been determined for diglyceride, since endogenous levels in the membranes appear to support a high reaction rate. Several diglycerides stimulate the activity of the spinach leaf enzyme, but dipalmitin and distearin do not.[15] Dipalmitin and diglycerides from pig and egg stimulate the activity of the castor bean endosperm enzyme.[6]

pH Optimum. The castor bean[6] and spinach leaf[15] enzymes have pH optima of 7.5 in Tris-HCl buffer containing Mg^{2+}. With the spinach enzyme the pH optimum is 8.0 in the presence of Mn^{2+}.

Metal Ions. The activity of the castor bean endosperm enzyme[6] is nine times greater with Mg^{2+} than with Mn^{2+}. The maximum activity of the spinach enzyme is similar with both cations.[15]

Inhibitors. Several sulfhydryl reagents inhibit the spinach enzyme, but of several such reagents tested, only p-chloromercuribenzoate inhibits the castor bean enzyme. Elimination of dithiothreitol from the assay results in

---

[15] K. A. Devor and J. B. Mudd, *J. Lipid Res.* **12**, 403 (1971).

considerable loss of activity of the spinach enzyme but has little effect on the castor bean enzyme.[6,15]

Stability. The castor bean enzyme loses 40% of its activity at 37° after 5 hr.[16] Spinach enzyme activity[15] at 30° is stabilized by dithiothreitol and Mn^{2+}.

Mitochondrial Enzyme. The mitochondrial enzyme from castor bean has properties similar to those of the enzyme from microsomal fraction.[16] The apparent K_m for CDPcholine is about 8 μM, and the optimum concentration for Mg^{2+} is 10 mM. The mitochondrial enzyme is less stable at 37° than the microsomal fraction enzyme, decaying 95% in 5 hr.

Intracellular Compartmentation. CDPcholine : 1,2-diacylglycerol cholinephosphotransferase has been reported to occur in microsomal,[15,17] endoplasmic reticulum,[6] Golgi,[18,19] and mitochondrial[9] fractions isolated from plant cells. In the castor bean, the microsomal fraction contains 98% of the total activity, the mitochondria the other 2%.[16] Golgi represent at most a very small fraction of the organelles of this tissue. On the other hand, Golgi activity is 25% of the endoplasmic reticulum activity of *Pisum sativum* stem tissue.[19]

CDPethanolamine : 1,2-Diacylglycerol Ethanolaminephosphotransferase (EC 2.7.8.1)

sn-1,2-Diglyceride + CDPethanolamine →

phosphatidylethanolamine + CMP

CDPethanolamine : 1,2-diacylglycerol ethanolaminephosphotransferase has been demonstrated to occur in spinach leaf[17,20] and in microsomal fraction from castor bean endosperm.[21]

Assay Method

Principle. The phospholipid product of CDP[1,2-^{14}C]ethanolamine incorporation is extracted into chloroform by a modification of the method of Bligh and Dyer.[14,21]

Reagents

MES buffer, 50 mM, pH 6.5
$MgCl_2$, 15 mM
Dithiothreitol, 5 mM

[16] S. A. Sparace and T. S. Moore, *Plant Physiol.,* in press.
[17] M. O. Marshall and M. Kates, *Can. J. Biochem.* **52,** 469 (1974).
[18] D. J. Morre, S. Nyquist, and E. Rivera, *Plant Physiol.* **45,** 800 (1970).
[19] M. J. Montague and P. M. Ray, *Plant Physiol.* **59,** 225 (1977).
[20] B. A. Macher and J. B. Mudd, *Plant Physiol.* **53,** 171 (1974).
[21] S. A. Sparace, L. K. Wagner, and T. S. Moore, *Plant Physiol.,* in press.

CDP[1,2-^{14}C]ethanolamine (ICN, Irvine, California), 75 μM, 3.0 mCi/mmol

Chloroform–methanol–H_2O (1 : 2 : 0.3, by volume)

Chloroform

KCl, 1.0 M

Assay Procedure. Pipette 0.1 ml each of buffer, $MgCl_2$, dithiothreitol, and CDPethanolamine into each reaction tube. Add water to provide a final volume of 0.5 ml. Incubate the tubes at 37° for 5 min; the reaction is then started by adding the enzyme. Incubate the tubes while shaking at 37° for 30 min, then stop the reaction by adding the chloroform–methanol–H_2O mixture.

Extraction and Radioactive Assay. The extraction and radioactivity measurements are as described for the CDPcholine : 1,2-diacylglycerol cholinephosphotransferase assay.

Properties

Linearity of Reaction. The rate of the reaction is maintained for up to 40 min, after which it decreases rapidly. The rate increase is linear with increasing protein concentration up to 0.15 mg/0.5 ml, the highest protein concentration tested.[21] The spinach enzyme activity falls off after 5–10 min; it is proportional to the protein concentration up to 0.2–0.4 mg of protein/0.5 ml.[20]

Substrate Affinities. The apparent K_m of the castor bean enzyme for CDPethanolamine is 6.0 μM,[21] and that for spinach is 20 μM.[20] The enzyme from castor bean endosperm reaches its maximum velocity with about 50–60 μM CDP ethanolamine, whereas the spinach enzyme attains V_{max} at 100 μM.[20,21]

pH Optimum. The pH optimum for the castor bean enzyme[21] is 6.5 in the presence of either MES or HEPES buffer and Mn^{2+}. The spinach enzyme[17,20] pH optimum is 8.0 with Tris buffer and Mn^{2+} and 7.0–7.5 with Mg^{2+} substituted for Mn^{2+}.

Metal Ions. Mg^{2+} is the preferred cation for the castor bean endosperm enzyme, maximum activity being achieved above 3 mM.[21] Several other cations are ineffective. The spinach leaf enzyme[20] is most active between 0.6 mM[20] and 2 mM[17] Mn^{2+}. Mg^{2+} stimulates most between 8 mM[17] and 25 mM.[20]

Inhibitor Effects. Dithiothreitol has little or no effect on maintaining the activity of the castor bean and spinach enzymes.[20,21] CDPcholine is a competitive inhibitor of the castor bean enzyme; it also strongly inhibits the spinach enzyme activity.[21]

Intracellular Compartmentation. Ninety-eight percent of the transferase activity of castor bean is bound to the microsomal fraction, the remainder being mitochondrial.[21] The enzyme is largely microsomal in spinach.[17,20]

CDPdiacylglycerol : *myo*-Inositol Phosphatidyltransferase (EC 2.7.8.11)

CDP-*sn*-1,2-diacylglycerol + *myo*-inositol →
$$\text{1-phosphatidylinositol} + \text{CMP}$$

This enzyme has been described from castor bean, cauliflower, and spinach.[1,8,22]

Assay Method

Principle. Incorporation of *myo*-[2-^{3}H]inositol into the chloroform-soluble product, phosphatidylinositol, is measured.[8]

Reagents

Tris-HCl buffer, 250 mM, pH 8.5
MnCl$_2$, 7.5 mM
CDPdipalmitoylglyceride, 5.0 mM
myo-[2-^{3}H]Inositol (New England Nuclear), 6.0 mM, 1.79 mCi/mmol
Chloroform–methanol–water (1 : 2 : 0.3 by volume)
Chloroform
KCl, 1.0 M

Assay Procedure. Add 0.1 ml each of buffer, MnCl$_2$, CDPdiglyceride, and *myo*-inositol to each reaction tube, with adequate water for a final volume of 0.5 ml after further additions. Mix, then incubate the tube at 37° for 5 min. Add the enzyme and incubate with shaking at 37° for 30 min. Stop the reaction by adding 3.3 ml of the chloroform–methanol–water mixture.

Extraction and Radioactivity Determination. Extract the product and measure the radioactivity as described for CDPcholine : 1,2-diacylglycerol cholinephosphotransferase.

Properties

Linearity of Reaction. The rate of the reaction is constant for about 40 min, after which it decreases rapidly. The rate is proportional to the protein concentration up to 0.075 mg/0.5 ml.

Substrate Affinities. The apparent K_m for *myo*-inositol is 0.30 mM and for CDPdipalmitoylglyceride is 1.35 mM.[8] CDPdiglyceride concentrations

---

[22] S. Sumida and J. B. Mudd, *Plant Physiol.* **45**, 712 (1970).

greater than 1.0 mM inhibit the reaction. The K_m values for the enzyme from cauliflower florets are 0.045 mM and 0.27 μM for CDPdiglyceride and inositol, respectively.[22]

pH Optimum. The pH optimum for the castor bean enzyme is 8.5[8]; for cauliflower it is 9.0.[22]

Metal Ions. A divalent cation is absolutely required,[8] the preferred cation for the castor bean enzyme being Mn^{2+}. Mn^{2+} is most effective at 1.5 mM, Mg^{2+} only partially replaces Mn^{2+}, and Ca^{2+}, Fe^{2+}, and Co^{2+} are completely ineffective.[8] The cauliflower enzyme has similar requirements.[22] The spinach enzyme[1] requires Mn^{2+}; the optimal concentration is 5 mM, and it cannot be replaced by Mg^{2+} or Ca^{2+}.

Detergent Effects. Triton X-100 inhibits the castor bean activity at concentrations as low as 0.01% (w/w), while deoxycholate slightly stimulates the activity at 0.01–0.1% (w/w) and inhibits at higher concentrations.[8] The spinach enzyme is stimulated about fourfold by Triton X-100.[1]

Thermal Stability. The cauliflower enzyme loses 50% of its activity when incubated 30 min at 30°.[22]

Intracellular Compartmentation. The castor bean enzyme has been detected only in the microsomal fraction.[8]

Phosphatidylinositol : *myo*-Inositol Phosphatidyltransferase

Phosphatidylinositol + *myo*-Inositol* $\leftrightarrows$

phosphatidylinositol* + *myo*-inositol

In plants, the enzyme, that exchanges inositol for inositol, has been demonstrated only in castor bean endosperm.[22]

Assay Method

Principle. The incorporation of *myo*-[2-³H]inositol into chloroform-soluble phosphatidylinositol is measured.[23]

Reagents

HEPES buffer, 250 mM, pH 8.0
MnCl$_2$, 125 mM
myo-[2-³H]Inositol (New England Nuclear), 2.5 mM, 1.79 mCi/mmol
Chloroform–methanol–H$_2$O (1 : 2 : 0.3, by volume)

---

[23] J. C. Sexton and T. S. Moore, *Plant Physiol.,* in press.

Chloroform
KCl, 1.0 M

Assay Procedure. Add 0.1 ml each of buffer, $MnCl_2$, and substrate to each assay tube. Add water so that the final volume after further additions will be 0.5 ml. Incubate the tubes at 37° for 5 min, start the reaction by adding the enzyme, incubate with shaking at 37° for 1 hr, and stop the reaction by pipetting 3.3 ml of the chloroform–methanol–H_2O solution into each tube.

Extraction and Radioactivity Measurement. Extract the product and measure its radioactivity as described for CDPcholine : 1,2-diacylglycerol cholinephosphotransferase.

Properties

Linearity of Reaction. The reaction is linear with time for 90 min.[23] Increasing the protein concentration produces an exponential increase in the rate of the reaction except when cytidine 5'-monophosphate is present, when the rate is proportional to the protein concentration.

Substrate Affinity. The apparent K_m for *myo*-inositol is 26 μM.[23] *Scyllo*-inositol, choline, and ethanolamine do not compete with *myo*-inositol.[23]

Phospholipid Effects. Phosphatidylinositol purified from soybean does not affect the reaction, but 0.6 μM phosphatidylinositol from yeast and 0.6 μM phosphatidylinositol isolated from castor bean double the reaction rate.[23] Phosphatidic acid and phosphatidylethanolamine have no effect on the reaction, but phosphatidylserine, phosphatidylglycerol, and phosphatidylcholine stimulate up to 50%.[23]

pH Optimum. The optimum pH is 8.0 with either HEPES or Tris-MES buffers.[23]

Metal Ions. A metal cation is an absolute requirement for activity; Mn^{2+} promotes maximum activity at 15 mM; Mg^{2+} is only about 10% as effective. Ca^{2+} is ineffective, but counters the Mn^{2+} promotion.[23]

Detergent Effects. Triton X-100 stimulates the rate of the reaction up to about 0.025% (w/w). Deoxycholate inhibits at concentrations as low as 0.01% (w/w).[23]

Nucleotide effects. CMP, CDP, and CTP stimulate the reaction about 15-fold at 40 μM. CDPcholine, CDPethanolamine, and CDPdiglyceride stimulate to a lesser extent at the same concentration. Cytidine, 3',5'-cCMP, other nucleotides, orthophosphate, and pyrophosphate have no effect.[23]

Intracellular Compartmentation. This exchange activity has been obtained only in the microsomal fraction of castor bean endosperm.[23]

S-Adenosyl-L-methionine : Phosphatidylethanolamine N-Methyltransferase
(EC 2.1.1.17)

Phosphatidylethanolamine + 3 S-adenosylmethionine →
phosphatidylcholine + 3 S-adenosylhomocysteine

The reaction has been demonstrated in castor bean endosperm,[6] spinach leaf,[17] and potato tuber.[24]

Assay Method

Principle. The radioactivity of the chloroform-soluble product, phosphatidylcholine, is determined following reaction with S-adenosyl-L-[Me-^{14}C]methionine.

Reagents

Tris-HCl buffer, 250 mM, pH 9.0 (or glycine-NaOH, 500 mM, pH 9.0)

Phosphatidylethanolamine, 1.0 mM

S-Adenosyl-L-[Me-^{14}C]methionine (New England Nuclear), 1.0 mM, 1 mCi/mmol

HCl, concentrated

Chloroform–methanol–H$_2$O (1 : 2 : 0.3, by volume)

Chloroform

KCl, 1.0 M

Assay Procedure. Add 0.1 ml each of buffer, phosphatidylethanolamine, and S-adenosylmethionine, plus water for a final volume after further additions of 0.5 ml, to each assay tube. Incubate at 37° for 5 min and then start the reaction by adding enzyme. Incubate with shaking at 37° for 30 min. Stop the reaction by adding 50 µl of HCl followed by 3.3 ml of the chloroform–methanol–H$_2$O mixture. Incubate this mixture at room temperature for 1 hr.

Extraction and Radioactivity Determination. Extract the product and determine its radioactivity as described for CDPcholine : 1,2-diacylglycerol cholinephosphotransferase.

Properties

Linearity of Reaction. With the castor bean endosperm enzyme, the reaction rate is linear with time for up to 30 min. The rate of the reaction is proportional to the protein concentration up to 0.4 mg/0.5 ml.[6]

Substrate Affinities. The castor bean enzyme exhibits an apparent K_m

---

[24] W. Tang and P. A. Castelfranco, *Plant Physiol.* **43**, 1232 (1968).

for S-adenosylmethionine of 31 μm.[6] Maximum stimulation of the reaction rate by added phosphatidylethanolamine is 33% at 0.2 mM, but phosphatidylmonomethylethanolamine (PMME) and phosphatidyldimethylethanolamine (PDME) increase the rate of the reaction by 148% and 161%, respectively, at the same concentration.[6] Phosphatidylethanolamine makes up 26.6% of the total phospholipid of these membranes and may thus be abundant in the vicinity of the enzyme.[25] The spinach leaf enzyme[17] is not stimulated by phosphatidylethanolamine, but the synthesis of PDME and phosphatidylcholine is increased in the presence of PMME. Formation of phosphatidylcholine is increased in the presence of PDME.

Products. The products of the methylation reaction with the castor bean endosperm enzyme are phosphatidylcholine (52%) and PDME (48%).[6] With the spinach enzyme, the bulk of the radioactivity is incorporated into phosphatidylcholine, with a small amount appearing in the PMME fraction;[17] no activity is found in PDME unless PMME is added as a precursor.[17] PDME is found in lesser amounts than the other products in aged potato slices.[24]

pH Optimum. Maximum activity of the castor bean enzyme occurs at pH 9.0 in both Tris-HCl and glycine-NaOH buffers.[6] The spinach enzyme has an optimum pH of 8.0 with Tris-HCl buffer.[17]

Metal Ions. Mg^{2+} does not affect the castor bean enzyme activity, but Ca^{2+} and Mn^{2+} both inhibit slightly. The spinach enzyme[17] is not stimulated by Na^+, K^+, Mg^{2+}, or Mn^{2+}.

Inhibitors. The castor bean enzyme activity is inhibited by addition of S-adenosylhomocysteine.[6] N-Ethylmaleimide and iodoacetamide have no effect at concentrations up to 1 mM, but 1 mM p-chloromercuribenzoate inhibits 60%.[6]

Detergent Effects. Triton X-100 and deoxycholate inhibit the castor bean enzyme at concentrations as low as 0.01% and 0.04% (w/w), respectively.[6]

Intracellular Compartmentation. Approximately 90% of the castor bean enzyme activity is in the microsomal fraction, the other 10% is mitochondrial.[6] Most of the activity in spinach leaves is microsomal.[17]

Phosphatidylethanolamine : L-Serine Phosphatidyltransferase

Phosphatidylethanolamine + L-serine →

phosphatidylserine + ethanolamine

---

[25] R. P. Donaldson and H. Beevers, *Plant Physiol.* **59**, 259 (1977).

The synthesis of phosphatidylserine occurs in plants both by the Ca^{2+}-stimulated exchange reaction catalyzed by phosphatidylethanolamine : L-serine phosphatidyltransferase and by CDPdiglyceride : L-serine phosphatidyltransferase. The former is found in castor bean endosperm[5] and pea seedlings,[26] the latter in spinach leaves.[17]

Assay Procedure

Principle. The Ca^{2+}-stimulated exchange reaction is assayed by observing the incorporation of L-[3-^{14}C]serine into the chloroform-soluble lipid fraction.[5]

Reagents

HEPES buffer, 200 mM, pH 7.8
$CaCl_2$, 10 mM
L-[3-^{14}C]Serine (ICN, Amersham/Searle Corp.), 2.0 mM, 2.5 mCi/ mmol (a chloroform-soluble impurity occurred in the L-[3-^{14}C]serine so aqueous solutions of the substrate were routinely washed 4–5 times with chloroform, evaporated with dry nitrogen gas, and then resuspended in water prior to use)
Chloroform–methanol–H_2O (1 : 2 : 0.3, by volume)
Chloroform
KCl, 1.0 M

Assay Procedure. Add 0.1 ml each of buffer, $CaCl_2$, and L-serine plus an appropriate amount of water to give a final volume of 0.5 ml after further additions to each assay tube. Incubate the tubes at 30° for 5 min, then start the reaction by adding enzyme. Incubate with shaking at 30° for 60 min. Stop the reaction by adding 3.3 ml of the chloroform–methanol–H_2O mixture. Incubate on ice for 30 min.

Extraction and Radioactivity Determination. Extract the chloroform-soluble product and measure its radioactivity as described for CDPcholine : 1,2-diacylglycerol cholinephosphotransferase.

Properties

Linearity of Reaction. The castor bean enzyme activity remains constant for 2 hr; it is proportional to the protein concentration up to 200 μg/0.5 ml.[5]

Substrate Affinities. The K_m for L-serine is about 20 μM.[5] Ethanolamine competes with L-serine, but D-serine and choline do not.[5] Phosphatidylethanolamine and CDPdiglyceride increase the reaction rate by 50%.[5]

[26] S. L. Vandor and K. E. Richardson, *Can. J. Biochem.* **46**, 1309 (1968).

Products. Two products have been observed, phosphatidylserine (48%) and phosphatidylethanolamine (52%),[5] so a phosphatidylserine decarboxylase appears to be present.

pH Optimum. The optimum pH is 7.8 in the presence of 2 mM $CaCl_2$.[5] At higher concentrations of Ca^{2+}, the optimum is shifted to values in the range of 7–7.5.

Metal Ions. The optimum concentration for Ca^{2+} is 2 mM.[5] Mg^{2+}, Mn^{2+}, Cu^{2+}, Co^{2+}, and Zn^{2+} do not substitute for calcium.

Intracellular Compartmentation. The exchange reaction is localized in the microsomal fraction of castor bean endosperm[5] and occurs in a microsomal fraction from etiolated pea seedlings.[26]

CDPdiacylglycerol : Glycerophosphate Phosphatidyltransferase (EC 2.7.8.5) *plus* Phosphatidylglycerophosphate Phosphohydrolase (EC 3.1.3.27)

(a) *sn*-glycerol-3-phosphate + CDP-diglyceride →
    3-*sn*-phosphatidyl-1'-*sn*-glycerol-3'-phosphate + CMP
(b) 3-*sn*-phosphatidyl-1'-*sn*-glycerol-3'-phosphate →
    3-*sn*-phosphatidyl-1'-*sn*-glycerol + P_i

The first *in vitro* demonstration of phosphatidylglyerol synthesis in plants utilized a crude homogenate from spinach leaves;[27] since then the reactions have been demonstrated in organelles from cauliflower florets,[28] spinach leaves,[1,29] and castor bean endosperm.[4]

Assay Method

Principle. The reaction is measured by determining the amount of [U-^{14}C]-*sn*-glycerol-P incorporated into chloroform-soluble phosphatidylglycerol.[4]

Reagents

Tris-HCl, 2.0 M, pH 7.3
$MnCl_2$, 40 mM
Triton X-100, 0.6% (w/w)
CDPdiglyceride, 0.8 mM
[U-^{14}C]-*sn*-glycerolphosphate (ICN, New England Nuclear), 4.0 mM,
    1.9 mCi/mmol
Chloroform
Methanol
KCl, 1.0 M

[27] P. S. Sastry and M. Kates, *Can. J. Biochem.* **44,** 459 (1966).
[28] R. Douce and J. Dupont, *C. R. Acad. Sci. Ser. D* **268,** 1657 (1969).

Assay Procedure. Add 25 μl each of buffer, $MnCl_2$, detergent, and substrates to a tube for each sample to be assayed. Add sufficient water so that the final volume after further additions will be 0.2 ml. Incubate the reaction mixture at 30° for 5 min and then start the reaction by the addition of enzyme. Incubate with shaking at 30° for 1.5 hr, and stop the reaction by adding 2.0 ml of methanol. Incubate on ice for 1 hr.

Extraction Procedure. Add 2 ml of chloroform to the assay tubes and mix. Centrifuge at 2500 rpm in a tabletop centrifuge and then transfer the supernatant to a 10 ml, pennyhead stopper conical centrifuge tube. Wash twice with 5-ml aliquots of 1.0 M KCl and twice with distilled water. Break the emulsion after each wash by centrifugation in a tabletop centrifuge.

Radioactivity Determination. Transfer the chloroform fraction into a scintillation vial, evaporate, add the scintillation cocktail, then measure the radioactivity.

Properties

Linearity of Reaction. Activities of the enzymes in both the microsomal fraction and mitochondria of castor bean endosperm are constant for up to 2 hr. Their activities are proportional to the amount of protein up to 10 μg/0.2 ml.[4] The activity of the enzyme from spinach microsomes is proportional to the protein concentration up to about 0.4 mg of protein per milliliter and substrate incorporation is proportional to time for at least 1 hr.[1,29]

Substrate Affinities. The apparent K_m of the castor bean endosperm enzymes in both organelles for glycerol-P is about 50 μM.[4] The apparent K_m for CDPdiglyceride is 2–3 μM.[4]

The spinach enzymes have an apparent K_m for glycerol-P of 0.25 mM.[1,29] Phosphatidylglycerol-P synthesis reaches a steady state above 0.53 mM glycerol-P.[1,29] Maximum enzyme activity occurs at 0.04 mM CDPdiglyceride; at higher concentrations, inhibition occurs.

Products. Phosphatidylglycerol is the only product observed with the castor bean enzymes.[4] Both phosphatidylglycerol and phosphatidylglycerol-P were products of the spinach enzymes, being 90–98% and 2–10% of the total, respectively.[1,29]

pH Optimum. The overall reactions of both the castor bean and spinach enzymes have an optimum pH of 7.3[1,4,29]

Metal Ions. Mn^{2+} stimulates both the microsomal fraction and mitochondrial enzymes of castor bean more than other cations; maximum activity occurs at 5 mM.[4] $MgCl_2$ is effective at a lower concentration (1–2 mM), but the maximum activities are less than with 5 mM Mn^{2+}.

[29] M. O. Marshall and M. Kates, *Biochim. Biophys. Acta* **260**, 558 (1972).

Detergent Effects. Triton X-100 stimulates the castor bean enzymes from both organelles at concentrations up to 0.075% (w/w), above which the activity decreases until at 0.15% (w/w) the activity is equal to that in the absence of the detergent.[4]

Inhibitor Effects. HgCl$_2$, iodoacetate, and NaF all strongly inhibit the castor bean enzyme activities, but omission of dithiothreitol from the homogenization and reaction media has little effect.[4] Similar results were obtained with the cauliflower enzymes,[28] but the spinach enzyme activities apparently are not inhibited by these reagents.[1,29]

Intracellular Compartmentation. The enzyme activities in castor bean endosperm are 57% in the mitochondrial and 43% in the microsomal fractions.[4] In spinach the activities are both microsomal and mitochondrial, and in cauliflower the activities are mitochondrial.[1,28,29]

Section V

Hydrolases

Vol. XXVIII [117]. Thermal Fractionation of Serum Hexosaminidases: Applications to Heterozygote Detection and Diagnosis of Tay-Sach's Disease. M. M. Kaback.

Vol. XXVIII [118]. Ganglioside G_{M1} β-Galactosidase. H. R. Sloan.

Vol. XXVIII [119]. Spingomyelinase from Human Liver (Sphingomyelin Cholinephosphohydrolase). H. R. Sloan.

Vol. XXVIII [20]. Cerebroside Sulfate Sulfatase (Arylsulfatase A) from Human Urine. J. L. Breslow and H. R. Sloan.

Vol. XXXV [23]. Hormone-Sensitive Triglyceride Lipase from Rat Adipose Tissue. J. C. Khoo and D. Steinberg.

Vol. XXXV [24]. Carboxylesterases from Pig and Ox Liver. N. P. B. Dudman and B. Zerner.

Vol. XXXV [25]. Carboxylesterases from Chicken, Sheep, and Horse Liver. K. Scott and B. Zerner.

Vol. XXXV [26]. Appendix: An Accurate Gravimetric Determination of the Concentration of Pure Proteins and the Specific Activity of Enzymes. R. L. Blakeley and B. Zerner.

Vol. XXXV [27]. Phospholipase D from Peanut Seeds. M. Heller, N. Mozes, and E. Maes.

Vol. L [47]. Enzymic Diagnosis of Sphingolipidoses. Kunihiko Suzuki.

Vol. L [56]. Glucocerebrosidase from Human Placenta. F. S. Furbish, H. E. Blair, J. Shiloach, P. G. Pentchev, and R. O. Brady.

Vol. L [57]. Ceramide Trihexosidase from Human Placenta. J. W. Kusiak, J. M. Quirk, and R. O. Brady.

[72] Triglyceride Lipase from Porcine Pancreas[1]

EC 3.1.1.3   Triacylglycerol acylhydrolase

By HOWARD L. BROCKMAN

Pancreatic lipase (EC 3.1.1.3) acts in the duodenum to convert water-insoluble glycerides into species that can be readily transported through the intestinal wall. Specifically, it catalyzes the sequential reactions

$$\text{Triglyceride} + H_2O \rightleftharpoons 1,2\text{-diglyceride} + \text{fatty acid}$$
$$1,2\text{-Diglyceride} + H_2O \rightleftharpoons 2\text{-monoglyceride} + \text{fatty acid}$$

Although other glycerides with primary acyl groups can serve as substrates, triglycerides and 1,2-diglycerides are probably the predominant substrates encountered *in vivo*.

The pancreas contains other enzymes, such as monoglyceride lipase, lysophopholipase, carboxylesterase, and cholesterol esterase, that also catalyze the hydrolysis of di- and triglycerides, but are distinct from triglyceride lipase. They are generally less specific with respect to primary versus secondary esters, have lower turnover numbers, readily utilize water-soluble substrates and substrate analogs, and are sensitive to the serine esterase inhibitor diisopropylfluorophosphate.[2]

Triglyceride lipase has been purified to homogeneity from pancreas or pancreatic juice of a number of species. The first purification of lipase free of endogenous lipid was achieved by starting with a solvent-extracted preparation of porcine pancreas,[3] and over the years modifications of the purification have been described.[4,5] The porcine enzyme is the most thoroughly studied and has been used extensively as a model for studying the action of soluble enzymes on water-insoluble substrates. From this work came detailed knowledge of how bile salts and colipase regulate the adsorption of the enzyme to the lipid–water interface.[6] In turn, this

[1] This work was supported by Grant HL 08214 from the NIH, U. S. Public Health Service and by The Hormel Foundation. The manuscript was prepared during the tenure of an Established Investigatorship of the American Heart Association. The assistance of W. E. Momsen and Robert A. Sikkink is gratefully acknowledged.
[2] H. Brockerhoff and R. G. Jensen, "Lipolytic Enzymes." Academic Press, New York, 1974.
[3] R. Verger, G. H. de Haas, L. Sarda, and P. Desnuelle, *Biochim. Biophys. Acta* **188**, 272 (1969).
[4] C. W. Garner, Jr. and L. C. Smith, *J. Biol. Chem.* **247**, 561 (1972).
[5] M. Rovery, M. Boudouard, and J. Bianchetta, *Biochim. Biophys. Acta* **525**, 373 (1978).
[6] W. E. Momsen and H. L. Brockman, *J. Biol. Chem.* **251**, 384 (1976).

knowledge has recently been exploited to develop a purification scheme which considerably reduces the time and effort required to isolate pancreatic triglyceride lipase in milligram amounts.[7]

Assay Method

Principle. A number of methods have been used to assay lipases.[2] In general triglyceride lipases exhibit very poor activity toward chromogenic substrates. For the pancreatic enzymes titrametric assays are normally employed, but higher sensitivity can be obtained using radioactively labeled substrates. From extensive studies of the pancreatic enzymes it is known that optimum activity and assay linearity are obtained if both bile salts and colipase, a protein cofactor from pancreas, are included in the assay. A convenient, reliable assay is that described by Erlanson and Borgström,[8] as modified by Momsen and Brockman.[7] With this procedure the rate of hydrolysis of emulsified tributyrin is measured titrametrically in the presence of taurodeoxycholate and an excess of colipase. Because this procedure utilizes a partially soluble substrate that contains secondary as well as primary ester groups, it does not discriminate against other lipolytic and esterolytic enzymes and cannot, therefore, be used reliably to measure triglyceride lipase in all tissue preparations. For pancreatic extracts the assay is valid because treatment of the extract with diisopropylfluorophosphate inhibits other enzymes that hydrolyze tributyrin.

Reagents

Assay buffer: Tris-HCl, 2 mM, pH 6.5, 6.0 mM sodium taurodeoxycholate, 0.15 M NaCl, 1 mM CaCl$_2$, and 0.3 mM NaN$_3$
NaOH, approximately 5 mM, standardized
Tributyrin, purified as described below if necessary
Colipase, purified as described below

Preparation of Colipase for Assays. A partially purified preparation of colipase, free of lipase activity, can be obtained from a small amount of the solvent-extracted porcine pancreas used as a source of lipase (see below). The procedure is adapted from that described by Canioni et al.[9] for the preparation of homogeneous colipase. A 2 g aliquot of the delipidated powder of pancreas, which is described below, is stirred in 20 ml of 0.1 M H$_2$SO$_4$ for 1 hr at 24° and centrifuged and the supernatant is adjusted to pH 7.0 with 5 M NaOH. This and subsequent centrifugations were for

[7] W. Momsen and H. L. Bockman, *J. Lipid Res.* **19**, 1032 (1978).
[8] C. Erlanson and B. Borgström, *Scand. J. Gastroenterol.* **5**, 293 (1970).
[9] P. Canioni, R. Julien, J. Rathelot, H. Rochat, and L. Sarda, *Biochimie* **59**, 919 (1977).

15 min at 27,000 g. After recentrifugation, 3.3 g of $(NH_4)_2SO_4$ per milliliter are slowly added to the supernatant, the solution is stirred for 30 min at 4°, and the pellet obtained after centrifugation is dissolved in the 2 mM Tris-HCl buffer, pH 6.5, to give a volume of 10 ml. To this solution 10 ml of ethanol are added slowly, with stirring, at 4°; stirring is continued for 1 hr. After centrifugation, 80 ml of cold ethanol are added to the supernatant, and the mixture is again stirred for 1 hr at 4° and centrifuged. The resulting pellet is dissolved in the assay buffer described above to give a volume of 5 ml. This preparation is stable for at least several months at $-20°$ and contains no detectable lipase activity.

Purification of Tributyrin. Commercial tributyrin may contain free butyric acid in amounts sufficient to lower the pH of the weakly buffered assay mixture by more than 0.5 pH unit. If it is present, butyric acid can be easily removed with Florisil (Fisher, F-100, 60–100 mesh). The amount of standard NaOH necessary to bring 0.5 ml of tributyrin suspended in 3 ml of water to pH 7 is determined titrametrically, and from this the mole percent of butyric acid in the tributyrin is calculated. The stock tributyrin is mixed with 8 g of Florisil per milliliter of tributyrin per mole percent of butyric acid (typically 0.3 g/ml) and centrifuged briefly in glass or polypropylene tubes to remove the Florisil. After being decanted from the pellet the tributyrin is stable for over a year at $-20°$.

Assay Procedure. Into a 20-ml screw-cap culture tube are placed 0.375 ml of tributyrin (1.28 mmol) and 10 ml of assay buffer; the mixture is shaken vigorously or placed for a few seconds on a vortex mixer. A 2.0-ml aliquot is placed in a glass reaction vessel containing a magnetic stirring bar, and 0.01 ml of a 1 : 100 dilution of the colipase solution in assay buffer is added. The reaction vessel is placed in a recording pH stat with stirring at 1000 rpm; the temperature is controlled at 25°, and the pH is adjusted to 6.5 if necessary. After the reaction is initiated with an aliquot of lipase, the pH stat recording of volume of standardized base consumed versus time will be linear for up to 20 min and the rate of tributyrin hydrolysis will be directly proportional to added lipase up to 0.375 μmol of base consumed per minute.

Because the reaction is interfacial and an excess of surface and colipase are present, the rate of hydrolysis of tributyrin is independent of the volume of assay mixture used. The basic procedure described above is, therefore, readily adapted to any type of pH stat. It may be necessary, however, to increase the amount of colipase and the base concentration used if the volume is more than tripled. This assay can also be used to monitor colipase activity if excess colipase-free lipase replaces the colipase used above.

Purification Procedure for Porcine Pancreatic Lipase

Step 1. Preparation of Delipidated Pancreas.[3] Fresh porcine pancreas is obtained as soon as possible after slaughter and frozen with Dry Ice. Without the tissue being thawed, the whitish fat is removed and 600 g of defatted pancreas are homogenized in an explosion-proof blender for 1 min with, successively, three times 1800 ml of chloroform–*n*-butanol (9:1, v/v) at 25°, two times 1200 ml of chloroform–*n*-butanol (4:1, v/v) at 4°, two times 1200 ml of acetone at 4°, and two times 1200 ml of diethyl ether at 25°. After each homogenization with chloroform–butanol the solvents are decanted from the gummy tissue residue as completely as possible. By the second acetone treatment the consistency of the residue permits more complete solvent removal by using vacuum filtration in a Büchner funnel 18.5 cm in diameter with Whatman No. 1 filter paper. After the second homogenization with ether, the "cake" remaining on the filter is broken up and dried overnight under vacuum (water aspirator) at room temperature. If kept dry, the resulting powdery material is stable for several months at $-20°$. The yield is typically 80 g.

Step 2. Extraction of Lipase and Affinity Purification Using Hydrophobic Glass Beads. Prior to or during extraction of delipidated pancreas, 4 kg of waterproof glass beads (class III, size MS:XL, $-270 + 1000$ mesh, Cataphote Division, Ferro Corp., Jackson, Mississippi) are equilibrated with occasional stirring for 1 hr in 3.0 liter of 10 mM potassium phosphate buffer, pH 6.5, 0.05 mM taurodeoxycholate at 24°. Just before use they are transferred to a 30.5 cm tabletop Büchner funnel (Scientific Products) fitted with rapid-flow filter paper, washed with another 3.0 liter of buffer, followed by a final equilibration with 6.0 liters of 10 mM phosphate buffer, pH 6.5, containing 0.2 mM taurodeoxycholate, 1 mM EDTA, and 10 mM mercaptoethanol at 24°. For this and subsequent operations, the flow rate is controlled at 200 ml/min with a peristaltic pump. Lipase is solubilized by mixing 20 g of delipidated pancreas with 200 ml of 1 mM potassium phosphate buffer, pH 7.5, 0.1 M NaCl at 4°, after which 0.2 ml of 1.0 M diisopropylfluorophosphate in isopropanol is added with stirring. After 5 min the pH is adjusted to 8.5–9.0, with 1.0 M NaOH; the solution is stirred for 1 hr at 4° and centrifuged for 15 min at 27,000 g. The pH of the supernatant is adjusted to 6.5 with 1.0 M HCl and recentrifuged to remove the precipitate that forms. The extract is added to 8.0 liters of the final equilibration buffer at 24° and passed through the bed of beads. This is followed immediately by another 10 liters of the same buffer and by 3.0 liters of 10 mM potassium phosphate, pH 7.5, 0.1 M NaCl, 1 mM EDTA, 10 mM mercaptoethanol, 3.5 mM taurodeoxycholate. The last, 3 liter, fraction is collected and concentrated by ultrafiltration at 4° to approximately 100 ml (Amicon H1P10 hollow fiber, flow rate 2 liter/hr) and 0.05 ml of 1 M diisopropylfluorophosphate in isopropanol is added. After a 10-min incubation at 4°, the concentrate is diluted 10- to 20-fold with 5 mM Tris-HCl, pH 8.0; the concentration is repeated to remove the bile salt and phosphate buffer. This process is repeated twice, the final dilution also containing 3.3 mM CaCl$_2$. This solution should be stored at $-20°$ if it is not chromatographed on DEAE-cellulose immediately.

Step 3. Chromatography on DEAE-Cellulose. The concentrated solution from step 2 may be applied directly to a small DEAE-cellulose column[7] or, as described below, three batches of such concentrated solutions may be combined and chromatographed together. The three batches are applied to a 2.5 × 22 cm column packed with DEAE-cellulose (Whatman DE-52) and equilibrated with 5 mM Tris-HCl 3.3 mM CaCl$_2$, pH 8. The column is washed with 25 ml of this buffer and the enzyme is then eluted by a 2400-ml linear gradient from 0 to 0.1 M NaCl in the same buffer. The flow rate is 240 ml/hr, and 19-ml fractions are collected. The enzyme activity elutes in a double peak with maxima at 43 ± 3 and 57 ± 4 mM NaCl. The tubes are normally combined to yield fractions I, II, and III, representing most of the first peak, the intermediate region of overlap, and the remainder of the second peak of activity, respectively.

A typical purification summary is shown in the table. For steps 1 and 2, the results for the three separate preparations are given as averages ± standard deviation to indicate the reproducibility of the method. Fraction I from the DEAE chromatography is essentially homogeneous pancreatic lipase B containing less than 2 mol % colipase; typically it represents 40% of the lipase recovered.[7] The fraction is dialyzed against deionized water, lyophilized, and stored at $-20°$. Fraction III contains lipase A and B in approximately a 1 : 1 ratio with colipase. Fraction II also contains significant colipase. These fractions may be lyophilized and stored like fraction I or colipase may be removed as described in step 4.

Step 4. Removal of Colipase (Optional). Colipase can be removed from fractions 2 and 3 by chromatography on concanavalin A–Sepharose,[4] gel filtration in the presence of guanidine hydrochloride and mercaptoethanol,[10] gel filtration of 5-thio-2-nitrobenzoyl lipase,[11] or chromatography on hydroxyapatite.[12] The use of concanavalin A–Sepharose is convenient, but the reported recovery of lipase activity is only 30%. However, in our hands this can be raised to 50–80% by slowing the flow rate during the elution step. Up to 2 × 10^5 units of lipase–colipase mixture

[10] J. Donnér, *Acta Chem. Scand. Ser. B* **30**, 430 (1976).
[11] M. F. Maylie, M. Charles, C. Gache, and P. Desnuelle, *Biochim. Acta* **229**, 286 (1976).
[12] J. Rietsch, F. Pattus, P. Desnuelle, and R. Verger, *J. Biol. Chem.* **252**, 4313 (1977).

PURIFICATION OF TRIGLYCERIDE LIPASE FROM PORCINE PANCREAS

Procedure	Volume (ml)	Total activity[a] (10^{-4} × units)	Total protein (mg)	Specific activity (units/mg)	Yield (%)	Colipase (mol %)
Extraction of delipidated pancreas[b]	190 ± 7	96 ± 13	1.1 ± 0.1 × 10^4 [c]	87	100	—
Concentration of 3.5 mM taurodeoxycholate eluate from beads[b]	107 ± 16	35 ± 14	220 ± 90[c]	1590	36	—
DEAE chromatography						
Fraction I	189	21.1	33.0[d]	6360	7.3	≤1.2
Fraction II	170	9.7	17.1[d]	5610	3.3	12.3
Fraction III	224	12.4	24.0[d]	5120	4.3	62.0

[a] One unit = 1 μmol of butyric acid released per minute.
[b] Volume, activity, and protein are expressed as average ±SD for the results from three preparations subsequently combined for chromatography on DEAE-cellulose.
[c] Protein was determined from A_{280}/A_{260} [O. Warburg and W. Christian, *Biochem Z.* **310**, 384 (1942)].
[d] Protein was determined using an extinction coefficient of 1.33 OD per milligram of lipase per milliliter at 280 nm.[14]

from the DEAE column are applied to a 2.5 × 4 cm column of concanavalin A–Sepharose equilibrated with 5 mM Tris-HCl, pH 7.5, 1.0 mM MgCl$_2$, at a flow rate of 30–50 ml/hr. The column is washed with 100 ml of the same buffer, and lipase is eluted at 5 ml/hr with 5% methyl-D-mannoside in 5 mM Tris-HCl, pH 7.5. The elution volume is 70–100 ml. The colipase content is typically less than 2 mol %, and the specific activity is comparable to that shown in the table for fraction I. Neither the adsorption of the enzyme nor its elution profile is affected by the presence of taurodeoxycholate above or below its critical micelle concentration. After this step the lipase is dialyzed and lyophilized as described above.

Comments on Purification

1. Although it would be convenient to use commercially available preparations of solvent-extracted pancreas, they are unsuitable[3,13] except for that supplied by Choay-Chimie, Paris, France.[5]

2. The solvent extraction procedure represents an explosion hazard and should be performed in a hood using an explosion-proof blender.

3. In contrast to more involved preparations,[4] this procedure can be performed using phenylmethylsulfonyl fluoride in place of the more highly toxic diisopropylfluorophosphate with no apparent change in the properties of the lipase obtained.

4. The preparation can be readily scaled up or down provided that the bed height of the hydrophobic beads is about 4 cm and the solution flow rates are 0.3–0.5 ml/cm^2 of funnel surface per minute.

5. Colipase contamination is characteristic of lipase purified by most procedures. Unless absolutely pure lipase is necessary for particular experiments, it is desirable to have colipase present. It helps to stabilize the enzyme and allows it to function in the presence of detergents and hydrophobic proteins.[14]

Properties of the Enzyme

Purity. Fraction I lipase from DEAE is homogeneous as judged by polyacrylamide gel electrophoresis and comigrates with lipase B prepared by other methods. Fractions II and III contain predominantly lipase B with lesser amounts of lipase A and colipase.[7] Enzyme purified by this procedure contains no lipid-soluble phosphate, but 0.4–10 mol of phosphate are present even if phosphate buffer and diisopropylfluorophosphate are omitted from the purification procedure. Fatty acid was not found at the detection limit of our procedure (10 mol % or 0.05% by weight) but

[13] C. W. Garner and L. C. Smith, *Arch. Biochem. Biophys.* **140**, 503 (1970).
[14] B. Borgström and C. Erlanson, *Gastroenterology* **75**, 382 (1978).

small amounts of bile salt (13 mol %) may be present. The bile salt determination represents an upper limit because the protein itself contributes to the bile salt measurement.

Physical Properties. The molecular weight of pancreatic lipase determined by gel filtration, SDS–polyacrylamide electrophoresis, and ultracentrifugation is normally given as 50,000 although published values range from 45,000 to 52,000.[3,5,10] $E_{280}^{1\%} = 13.3$,[15] $\bar{v} = 0.72$–0.73 cm^3/g,[5,10] $D_{20,w} = 6.7 \times 10^{-7}$ cm^2/sec, $s_{20,w} = 4.0 \times 10^{-13}$ sec, $f/f_0 = 1.23$, the Stokes radius $= 3.03$ nm, and the isoelectric point $= 5.18$.[10]

In solution lipase binds to colipase with a dissociation constant of $5 \times 10^{-7} M$, and 4 mM taurodeoxycholate increases this to approximately $10^{-5} M$.[10] At a hydrophobic surface–water interface the value is less than $2.8 \times 10^{-9} M$.[6] Binding of lipase alone to a hydrophobic surface occurs with a dissociation constant of $1.8 \times 10^{-8} M$.[16] Taurodeoxycholate binds cooperatively to lipase with a Hill coefficient of 4 and a dissociation constant of $1.4 \times 10^{-15} M^4$ (ref. 6), although long-term equilibrium dialysis has shown that as many as 12 molecules of bile salt can be bound per molecule of enzyme.[17]

Chemical Properties. Pancreatic lipase is a single polypeptide of about 470 residues; the sequence of the first 307 have been published.[18] The previously noted carbohydrate residues[4] were attached to Asn-166, and the serine residue that is essential for interfacial adsorption of the enzyme, and probably for catalysis as well, is part of the sequence Ile-Gly-His-Ser-Leu-Gly-Ser-His-Ile (residues 149–157). In light of the affinity of the enzyme for apolar surfaces, the strongly apolar sequence Ala-Ala-Pro-Ile-Ile-Pro-Asn-Leu-Gly-Phe-Gly-Met (residues 206–217) is of interest. The only apparent difference between the lipase A and B isozymes is the occurrence of sialic acid in the former.[19]

Catalytic Properties. By definition lipolytic enzymes act at lipid–water interfaces, although they may also utilize soluble substrates. Studies of lipase and related enzymes have established that catalysis proceeds in two steps, interaction of the substrate with the interface followed by catalysis within the interfacial plane. It follows, therefore, that the regulation of catalysis is a function not only of the chemical structure of the substrate, but also of the physical properties of the interface at which catalysis occurs. It should be noted that the interaction of lipase with the interface

---

[15] R. Verger, L. Sarda, and P. Desnuelle, *Biochim. Biophys. Acta* **242**, 580 (1971).

[16] H. Brockman, F. J. Kézdy, and J. H. Law, *J. Biol. Chem.* **248**, 4965 (1973).

[17] B. Borgström and J. Donnér, *Biochim. Biophys. Acta* **450**, 352 (1976).

[18] J. D. Bianchetta, J. Bidaud, A. A. Guidoni, J. J. Bonicel, and M. Rovery, *Eur. J. Biochim.* **97**, 395 (1979); also A. Guidoni, J. Bonicel, J. Bianchetta, and M. Rovery, *Biochimie* **61**, 641 (1979).

[19] T. H. Plummer and L. Sarda, *J. Biol. Chem.* **248**, 7865 (1973).

may differ from the classical notion of enzyme-substrate binding in that the binding of the enzyme to the interface may well involve one or more substrate molecules or other molecules not serving as substrates for lipolysis.

The chemical specificity of lipase is for esters of primary alcohols, with an absence of stereospecificity with regard to the sn-1 and sn-3 positions of glycerol.[2] Because catalysis of soluble substrates occurs at relatively low rates, more recent work has focused on the role of interfacial structure in catalysis and its regulation. For example, in the presence of bile salt, lipase loses its affinity for the substrate and bile salt–water interface.[6] Likewise, hydrophobic proteins can interfere with lipase adsorption.[14] In each case lack of adsorption implies lack of catalysis, because the substrate is located at the interface. The binding of colipase is unaffected by these agents, and at interfaces it has a high affinity for lipase; hence, colipase allows catalysis to proceed in the presence of bile salts and other proteins.

Studies with substrates in monolayers at the air–water interface have shown that both the binding of lipase to the interface and subsequent interfacial catalysis exhibited an atypical dependence on the two-dimensional substrate concentrations. More recently it has been shown that binding and catalysis are also modified by the presence of phospholipid at the substrate–water interface.[20] These findings are consistent with a proposal based on physical and kinetic data, which states that the glycerol moieties of substrates may exist in a finite number of discrete conformational states[21] and that the enzyme could exhibit binding and/or catalytic preference for substrate in a particular conformation. The distribution of molecules among these states would be a function of both the composition and packing density of the molecules in the monolayer.

[20] G. Pieroni and R. Verger, J. Biol. Chem. 254, 10090 (1979).
[21] W. Momsen, J. M. Smaby, and H. L. Brockman, J. Biol. Chem. 254, 8855 (1979).

[73] Hormone-Sensitive Lipase from Chicken Adipose Tissue Including the Separation and Purification of Monoglyceride Lipase

By John C. Khoo and Daniel Steinberg

The term hormone-sensitive lipase (HSL) is used to refer to the triglyceride lipase in adipose tissue whose activity is rate-limiting with regard to mobilization of fatty acids from depot triglycerides. Partial purification of the enzyme from rat adipose tissue was described in an earlier volume.[1] The enzyme was characterized as a large, lipid-rich complex that hydrolyzed triglycerides, diglycerides, monoglycerides, and cholesterol esters. Indirect evidence indicated that more than one enzyme protein might be present in the complex. Activation and phosphorylation of the enzyme by cAMP-dependent protein kinase was demonstrated and characterized.[2] The degree of activation obtainable with the rat enzyme was only about 50%, making it difficult to characterize the activation process in more detail or to study the deactivation that was presumed to occur by dephosphorylation. Purification of the enzyme from rat adipose tissue is described in this volume [74].

The analogous enzyme from chicken adipose tissue is activated to a much greater extent by protein kinase (up to 50-fold), which makes the chicken enzyme more useful in studies of the activation–deactivation processes.[3,4] Like the enzyme from rat adipose tissue, chicken HSL in crude preparations exists in the form of a large, lipid-rich complex. Treatment with Triton X-100 dissociates the complex, allowing the application of conventional purification techniques.[5] In the following sections we describe first the characteristics of chicken HSL with respect to activation by protein kinase and deactivation by phosphoprotein phosphatase; second, the purification of the triglyceride and diglyceride hydrolase activities, which appear to represent a single enzyme protein; third, the resolution and purification of a separate monoglyceride hydrolase activity.

Assay Methods

Principle. Free [^{14}C]oleic acid produced by the hydrolysis of tri[^{14}C]oleoylglycerol, di[^{14}C]oleoyglycerol, mono[^{14}C]oleoglycerol, or cholesteryl[1-^{14}C]oleate is assayed by measuring the rate of appearance of free [^{14}C]oleic acid. The [^{14}C]oleic acid is separated from unhydrolyzed esters using liquid–liquid partition and is assayed for radioactivity.

Reagents

Triolein, diolein, monoolein (containing [1-^{14}C]oleic acid distributed randomly among the acylated positions), and cholesteryl[1-^{14}C]oleate are diluted with nonradioactive triolein, diolein, mono-

[1] J. C. Khoo and D. Steinberg, this series, Vol. 35 [23].
[2] J. K. Huttunen and D. Steinberg, *Biochim. Biophys. Acta* **239,** 411 (1971).
[3] J. C. Khoo and D. Steinberg, *J. Lipid Res.* **15,** 602 (1974).
[4] J. C. Khoo, P. J. Sperry, G. N. Gill, and D. Steinberg, *Proc. Natl. Acad. Sci. U.S.A.* **74,** 4843 (1977).
[5] L. Berglund, J. C. Khoo, D. Jensen, and D. Steinberg, *J. Biol. Chem.* **255,** 5420 (1980).

olein, and cholesteryl oleate and used at a final concentration of 0.1 mM in the assay mixture.

Dehydrated Amberlie (Mallinckrodt, IRA-400 C.P.) in RN(CH$_3$)$_3$$^+Cl^-$ form is prepared as described previously[1] and used to remove trace amounts of free [^{14}C]oleic acid when the blank values in the labeled ester mixtures are unacceptably high.

Fatty acid extraction mixture: chloroform–methanol–benzene, (1 : 2.4 : 2, v/v).

Preparation of Emulsions of Tri[^{14}C]oleoyglycerol, Di[^{14}C]oleoyglycerol, Mono[^{14}C]oleoglycerol, and Cholesteryl[^{14}C]oleate. Stock solutions of 12 μmol of unlabeled substrate per milliliter containing approximately 1 × 10^7 dpm of ^{14}C-labeled substrate per milliliter are prepared and stored at 4° in absolute ethanol. To prepare substrate mixture for 10 assays, 4.9 ml of H$_2$O and 2 ml of 0.2 M sodium phosphate, pH 7.0, containing 10% bovine serum albumin (for triglyceride and cholesterol hydrolase assays) or 2% bovine serum albumin (for diglyceride and monoglyceride hydrolase assays) are added to a 50-ml disposable plastic tube. Then, 0.1 ml of stock substrate is added while the tube is swirled by hand. The final concentration of ethanol is 2%.[6] In studies involving activation of lipase by cAMP-dependent protein kinase, 5 mM phosphate buffer is used instead of 50 mM.[7]

Assay of Acylhydrolases. The substrate mixture (0.7 ml) is added to 0.1 ml of enzyme mixture in a 13 × 120 mm disposable test tube and incubated for 30 min at 30°. The reaction is stopped by the addition of 3 ml of the fatty acid extraction mixture (see above), followed by the addition of 0.1 ml of 1 N NaOH (final pH, 11.5). The mixture is vortexed vigorously for at least 15 sec and then centrifuged at 3000 rpm at room temperature for 10 min. An aliquot of 1.6 ml from the upper aqueous phase is assayed for radioactivity.

Enzyme Characterization and Purification

Before describing the purification scheme in detail, the characterization of the enzyme in a crude isoelectric precipitate fraction will be described, including some aspects of the protein kinase-catalyzed activation.

Preparation of the pH 5.2 Precipitate Fraction (5.2 P Fraction). It is important that fresh adipose tissue be used in order to obtain a preparation that can be activated readily by protein kinase. Chicken fat obtained from

---

[6] J. C. Khoo, D. Steinberg, J. J. Huang, and P. R. Vagelos, *J. Biol. Chem.* **251**, 2882 (1976).

[7] J. C. Khoo, D. Steinberg, and E. Y. C. Lee, *Biochem. Biophys. Res. Commun.* **80**, 418 (1978).

local slaughterhouses has not been suitable in our hands either for studies of activation-deactivation or for purification. Fat, laying hens (White Leghorn) are decapitated and the adipose tissue is quickly removed from the abdominal region and from around the gizzard. The adipose tissue is immediately homogenized in a Waring blender for 30 sec at 15° in two volumes of a buffer solution containing 25 mM Tris-HCl, 1 mM EDTA, and 0.25 M sucrose at pH 7.4. The homogenate is centrifuged at 1000 g for 10 min, and most of the fat cake is removed. The infranatant fraction is centrifuged at 100,000 g for 1 hr. The 100,000 g supernatant fraction (S_{100}) is brought to pH 5.2 by dropwise addition of acetic acid (0.2 M) with constant stirring at 4°. After 10–15 min the resulting precipitate is collected by centrifugation at 1000 g for 10 min at 4°. The precipitate is redissolved in 25 mM Tris, 1 mM EDTA, and 20% glycerol, pH 7.4 (buffer A) to one-thirtieth of the original volume of the S_{100} fraction. This 5.2 P fraction is stable for many months at −70°.

Activation of Acylhydrolases with cAMP-Dependent Protein Kinase

Unlike phosphorylase kinase and glycogen synthase, which have been shown to be phosphorylated both by Ca^{2+}-dependent protein kinases and cAMP-independent protein kinases,[8–12] the activation–phosphorylation of HSL is highly specific for cAMP-dependent protein kinase. Although highly purified cGMP-dependent protein kinase has been shown to activate HSL, the concentration of kinase required is at least ten times greater than that of cAMP-dependent protein kinase.[4,13] (Since the level of cGMP and cGMP binding proteins is extremely low in adipose tissue,[14] the activation of HSL by cGMP-dependent protein kinase does not appear to be of physiological importance.) The HSL of chicken adipose tissue is unique in the magnitude of its activation by cAMP-dependent protein kinase—as high as 50-fold compared to 30–80% for the HSL from other mammalian species.[2,13,15] However, as described below, HSL loses its capacity to be activated by protein kinase during later steps in purification.

[8] Y. Takai, A. Kishimoto, Y. Iwasa, Y. Kawahara, T. Mori, and Y. Nishizuka, J. Biol. Chem. 254, 3692 (1979).
[9] A. K. Srivastava, D. M. Waisman, C. O. Brostrom, and T. R. Sodering, J. Biol. Chem. 254, 583 (1979).
[10] A. A. DePaoli-Roach, P. J. Roach, and J. Larner, J. Biol. Chem. 254, 12062 (1979).
[11] H. G. Nimmo and P. Cohen, FEBS Lett. 47, 162 (1974).
[12] J. H. Brown, B. Thompson, and S. E. Mayer, Biochemistry 16, 5501 (1977).
[13] J. C. Khoo and G. N. Gill, Biochim. Biophys. Acta 584, 21 (1979).
[14] T. M. Lincoln, C. L. Hall, C. R. Park, and J. D. Corbin, Proc. Natl. Acad. Sci. U.S.A. 73, 2559 (1976).
[15] J. C. Khoo, A. A. Aquino, and D. Steinberg, J. Clin. Invest. 53, 1124 (1974).

The activation mixture of 0.1 ml contains 5 mM Mg^{2+}, 0.5 mM ATP, 10 μM cAMP, and a sample of 5.2 P enzyme in buffer A. The protein per assay is 50–100 μg for triglyceride and cholesterol ester hydrolases and 5–10 μg for diglyceride and monoglyceride hydrolases. Exogenous protein kinase is not ordinarily added because there is sufficient endogenous protein kinase to effect maximal activation in 2–5 min at 30°. In control tubes, both ATP and cAMP are omitted. The activation reaction is terminated by the addition of 0.7 ml of the acylhydrolase assay mixture containing 2 mM EDTA, and the mixture is incubated for an additional 30 min at 30°. The degree of activation can be greatly enhanced by heating the 5.2 P fraction at 50° in the presence of 20% glycerol for 20 min prior to incubation with protein kinase.[16] The ionic strength in the assay mixtures is also critically important. The activation of diglyceride lipase activity with protein kinase can vary from as much as 10-fold when the assay is carried out in 5 mM phosphate buffer to as little as 10% when assayed in 50 mM phosphate buffer. The difference is due to the enhancement by salt of the activity of the basal form of the enzyme; the activated form is not nearly so sensitive to salt concentrations. Similar considerations apply to triglyceride and cholesterol ester hydrolases.

Unlike the other three acylhydrolase activities, the monoglyceride lipase can be activated only 20–80%. Also, the degree of activation is enhanced neither by prior heat treatment of the 5.2 P fraction nor by assaying at low ionic strength.[7]

Studies of Reversible Deactivation. The fully activated, phosphorylated triglyceride, diglyceride, and cholesterol ester hydrolases in the 5.2 P fraction can be deactivated either by endogenous "lipase phosphatase"[3,6] or by phosphoprotein phosphatases (highly purified phosphoprotein phosphatases from rabbit liver or bovine heart or phosphoprotein phosphatase partially purified from rat epididymal fat pad[4,7,16]). Prior to studies using exogenous phosphatases, the endogenous phosphatase activity can be first removed by ATP-Sepharose affinity chromatography or by gel filtration on 4% agarose. The HSL complex emerges in the void volume while the phosphatase activity is retained. Heat treatment of the chicken 5.2 P at 50° for 20 min (in the presence of 20% glycerol) also partially inactivates endogenous lipase phosphatase.[16]

The HSL in the 5.2 P fraction is fully activated by incubation with ATP–Mg^{2+}, cAMP, and endogenous protein kinase for 5 min at 30°. The fully activated HSL is freed from ATP–Mg^{2+} and cAMP immediately by gel filtration on a Sephadex G-25 column. The activated enzyme emerging in the void volume is collected. Deactivation is followed at 30° in the

[16] D. L. Severson, J. C. Khoo, and D. Steinberg, *J. Biol. Chem.* **252**, 1484 (1977).

presence of 5 mM Mg^{2+} removing 0.1 ml samples at intervals. Control tubes without any additions are included to monitor irreversible denaturation of HSL activity. Reversibility of the deactivation is tested by incubating the deactivated HSL with cAMP, ATP–Mg^{2+}, and added exogenous protein kinase for 5 min at 30°. The HSL activity should approach the level of activity in the incubated control tube.

Purification of Chicken Hormone-Sensitive Lipase (HSL)[5]

Lipoprotein lipase (LPL) in the 5.2 P fraction (see above) is removed by affinity chromatography on heparin-Sepharose as described by Khoo *et al.*[6] The LPL-free 5.2 P fraction is incubated with buffer A containing 0.6 mM Triton X-100 and 50 mM NaCl (buffer B) for 24 hr at 4°. The detergent-treated 5.2 P fraction is vortexed and centrifuged at 10,000 g. Protein is determined by the Coomassie blue dye-binding method of Bradford using thyroglobulin as standard.[17] It is not possible to use A_{280} to monitor protein concentration during the purification procedure because of the phenoxy group in Triton. The presence of 20% glycerol and Tris-HCl also interferes with protein determination by the conventional method of Lowry *et al.*[18]

Step 1. Chromatography on 4% Agarose. Eighteen milliliters of the detergent-treated 5.2 P fraction is chromatographed on a column of 4% agarose (2.5 × 50 cm) equilibrated and eluted with buffer B. The acylhydrolases are eluted at a flow rate of 9 ml per hour and 5.5 ml per fraction. Triglyceride, diglyceride, monoglyceride, and cholesterol ester hydrolase activities cofractionate and elute at about $V_e = 2.2 \times V_0$.

Step 2. Chromatography on Hydroxyapatite. Pooled fractions containing acylhydrolase activities are loaded on a hydroxyapatite column (1.6 × 7.0 cm) and washed with 2 volumes of buffer B. As shown in Fig. 1, most of the monoglyceride lipase activity is eluted stepwise with buffer B containing 0.04 M phosphate. (This distinct monoglyceride lipase can be further purified as described in the section on the purification of monoglyceride lipase.) The hydroxyapatite column is eluted using a linear gradient of phosphate from 0.04 to 0.14 M in buffer B (60 ml in each gradient chamber) at a flow rate of 40 ml/hr. Fractions of 6 ml are collected. Diglyceride, triglyceride, and cholesterol ester hydrolase activities cofractionate and elute at about 0.08 M phosphate.

Step 3. Chromatography on DE-52. Enzyme-containing fractions from step 2 are dialyzed against buffer B and then loaded on a DE-52 column

[17] M. M. Bradford, *Anal. Biochem.* **72**, 248 (1976).
[18] O. H. Lowry, N. J. Rosebrough, A. L. Farr, and R. J. Randall, *J. Biol. Chem.* **193**, 265 (1951).

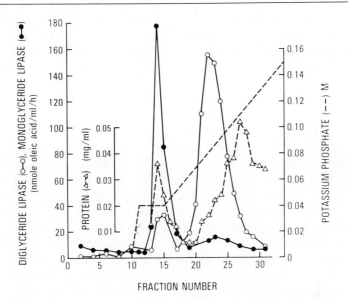

FIG. 1. Resolution of hormone-sensitive triglyceride–diglyceride lipase from mono-glyceride lipase by chromatography on hydroxyapatite. The pooled enzyme fractions from 4% agarose chromatography of the pH 5.2 precipitate fraction were chromatographed on a 1.6 × 7.0 cm column, beginning with stepwise elution of monoglyceride lipase activity with buffer B containing 0.04 M phosphate and followed by a linear gradient (60 ml of buffer B containing 0.04 M phosphate and 60 ml of buffer B containing 0.15 M phosphate).

(1.6 × 4.9 cm). The column is washed with two volumes of buffer B before eluting the acylhydrolase activities with a linear gradient of NaCl (0.05 M to 0.45 M, 40 ml in gradient chamber) at a flow rate of 1.5 ml/min. Fractions of 4 ml are collected. The peak of diglyceride lipase activity is eluted at 0.3 M NaCl. Triglyceride lipase activity, although low, cofrac-tionates with diglyceride lipase activity. The residual monoglyceride lipase activity (not separated in step 2) is eluted at 0.15 M NaCl. The cholesterol esterase activity does not cochromatograph exactly with either diglyceride lipase or monoglyceride lipase but is eluted at about 0.22 M NaCl. However, the activity is low and it is not certain whether or not it is a distinct enzyme.

Step 4. Chromatography on TEAE-Cellulose. Fractions containing diglyceride–triglyceride lipase activity from the previous DE-52 column are pooled and diluted with an equal volume of buffer B without NaCl (buffer C) to yield a final NaCl concentration of 0.15 M. The enzyme is loaded on a TEAE-cellulose column (1.6 × 2.3 cm). The column is washed with two volumes of buffer C containing 0.15 M NaCl and is then eluted with a linear gradient of NaCl from 0.15 to 0.6 M (20 ml in each gradient chamber) at a flow rate of 1.5 ml/min, and fractions of 2 ml are collected. The peak of diglyceride lipase activity is eluted at 0.4 M NaCl. Triglyceride lipase activity, although very low, copurifies with diglyceride lipase activity. Fractions containing enzyme activities are pooled and dialyzed.

The overall recovery in a typical purification was 17% with approximately 127-fold purification in the pooled fraction. The specific activity in the peak tube represented a 325-fold purification. However, since the diglyceride lipase activity is monitored by the release of [¹⁴C]oleic acid, and since the purified enzyme fraction is almost free of monoglyceride lipase, there is considerable accumulation of [¹⁴C]monoolein during the assay. After correction for this, the true degree of purification in the peak tube was about 600-fold. When the purified enzyme was subjected to electrophoresis in sodium dodecyl sulfate (SDS)–acrylamide gel (a gradient gel of 7.5 to 15% acrylamide),[19] one dominant protein band of molecular weight 42,000–45,000 was observed.[5] Triglyceride and diglyceride hydrolase activities copurify without change in relative activities (1 : 20). The relative monoglyceride lipase activity in the purified triglyceride-diglyceride lipase fraction is less than 2% that in the original starting material. It is concluded that triglyceride lipase activity and diglyceride lipase activity probably are referable to a single enzyme protein, i.e., hormone-sensitive lipase, while the monoglyceride activity initially associated with the complex is loosely bound and represents a different enzyme protein.

Purification of Monoglyceride Lipase

As described in the preceding section, a distinct monoglyceride lipase activity can be resolved from hormone-sensitive lipase at the hydroxyapatite chromatography step. In this section, we describe the further purification of this enzyme. Similar purification methods are used, including the sizes of columns, the flow rates, and the volumes collected per fraction at each chromatography step. The major difference is that monoglyceride lipase is eluted at different NaCl concentrations during ion-exchange chromatography. Briefly, the monoglyceride lipase activity is eluted at 0.04 M phosphate from the hydroxyapatite chromatography; at 0.23 M from the first DE-52 column, using a linear gradient of 0 to 0.27 M NaCl; at 0.2 M from the TEAE-cellulose chromatography, using a linear gradient of 0.1 to 0.32 M NaCl; and at 0.2 M from the second DE-52 chromatography, using a linear gradient of 0.075 to 0.3 M NaCl. At the final step of purification there is a low level of diglyceride lipase activity that cofrac-

---

[19] U. K. Laemmli, *Nature (London)* **227**, 680 (1970).

tionates with the monoglyceride lipase activity at a constant ratio of 1 : 3.5. The purified monoglyceride lipase has no detectable activity against triolein. The specific activity of monoglyceride lipase is constant over the entire peak. The degree of purification is about 92-fold with 14% recovery.

The purified monoglyceride lipase shows a single dominant protein band of molecular weight 45,000 in SDS–acrylamide gel electrophoresis.[5] This protein band corresponds to one of the major protein bands in the crude 5.2 P fraction. It has a molecular weight of 45,000 and is associated with monoglyceride lipase activity in two-dimensional polyacrylamide gel electrophoresis using isoelectrofocusing in 4% acrylamide tube gel (first dimension) followed by SDS–acrylamide gel electrophoresis in the second dimension according to the procedure of O'Farrell.[20] However, urea is omitted in the first dimension in order to avoid inhibition of enzyme activity.

Both the 42,000–45,000 MW protein of hormone-sensitive triglyceride–diglyceride lipase and the 45,000 MW protein of monoglyceride lipase can be phosphorylated by cAMP-dependent protein kinase in the presence of $[\gamma-^{32}P]ATP$.[5] However, this phosphorylation is not accompanied by enzyme activation in the more purified fractions. The stimulation of basal HSL activity by high ionic strength is also lost in the purified preparations. One possible explanation for the loss of activation by protein kinase and salt is that it reflects changes induced by treatment with detergent. A second possibility is that during purification some components that are crucially involved in activation are resolved.

HSL in Other Species. Belfrage et al.[21] have purified HSL from rat adipose tissue. They reported that the molecular weight of rat HSL is 86,000, about twice that of chicken HSL. Using methods similar to those used to purify chicken HSL, we have confirmed that the rat HSL is associated with a protein of 84,000 MW in two-dimensional polyacrylamide gel electrophoresis.[22] In the case both of chicken and rat adipose tissue a distinct monoglyceride lipase has been resolved from HSL. The purified chicken monoglyceride lipase has a molecular weight of 45,000 whereas that of the purified rat enzyme is 32,900.[23] The chicken monoglyceride lipase is activated by protein kinase in crude preparation, whereas the rat enzyme is not activated either in crude preparations or in more purified preparations.[24]

[20] P. H. O'Farrell, *J. Biol. Chem.* **250**, 4007 (1975).
[21] P. Belfrage, B. Jergil, P. Strålfors, and H. Tornqvist, *FEBS Lett.* **75**, 259 (1977).
[22] J. C. Khoo, L. Berglund, D. Jensen, and D. Steinberg, *Biochim. Biophys. Acta* **619**, 440 (1980).
[23] H. Tornqvist and P. Belfrage, *J. Biol. Chem.* **251**, 813 (1976).
[24] R. A. Heller and D. Steinberg, *Biochim. Biophys. Acta* **270**, 65 (1972).

The HSL from adipose tissue of human, orangutan, rhesus monkey, cat, rat, hamster, mouse, ground squirrel, pigeon, and turkey are found to be activated by cAMP-dependent protein kinase. The magnitude of activation is higher for avian species.[25] We also found that HSL from rat brown adipose tissue and from adipocytes derived from cultured 3T3-L1 cells is activated about twofold by protein kinase.[26]

Acknowledgment

This work was supported by NIH grant HL 22053 awarded by the National Heart, Lung and Blood Institute. The authors are indebted to Dr. Lars Berglund for his work on purification and to Mr. Dennis Jensen and Mr. Ed Wancewicz for their superb technical assistance.

[25] J. C. Khoo and D. Steinberg, unpublished results, 1980.

[26] M. Kawamura, D. F. Jensen, E. V. Wancewicz, L. L. Joy, J. C. Khoo, and D. Steinberg, *Proc. Natl. Acad. Sci. U.S.A.*, in press (1980).

[74] Hormone-Sensitive Lipase from Adipose Tissue of Rat

By Gudrun Fredrikson, Peter Strålfors, Nils Östen Nilsson, and Per Belfrage

Triacylglycerol → diacylglycerol and fatty acid
Diacylglycerol → monoacylglycerol and fatty acid
Monoacylglycerol → glycerol and fatty acid

Hormone-sensitive lipase (HSL)[1] catalyzes the step that, under normal conditions, is rate-limiting in the breakdown of the stored lipids of adipose tissue: the hydrolysis of the first ester bond of the triacylglycerol. It also catalyzes the hydrolysis of the diacylglycerol produced and at least part of the monoacylglycerol.[2,3] The activity of the enzyme, and thus the overall lipolysis rate in the adipocyte, changes in response to a variety of hormones.[4] This regulation has, for a number of years, been thought to be associated with a cyclic AMP and protein kinase-dependent activation mechanism.[5] However, only recently has it been possible to demonstrate

[1] M. Vaughan, J. E. Berger, and D. Steinberg, *J. Biol. Chem.* **239**, 401 (1964).

[2] P. Belfrage, B. Jergil, P. Strålfors, and H. Tornqvist. *FEBS Lett.* **75**, 259 (1977).

[3] H. Tornqvist, P. Nilsson-Ehle, and P. Belfrage. *Biochim. Biophys. Acta* **530**, 474 (1978).

[4] M. Vaughan and D. Steinberg, in "Handbook of Physiology," Section 5, "Adipose Tissue" (A. E. Renold and G. F. Cahill, Jr., eds.), p. 239. Williams & Wilkins, Baltimore, Maryland, 1965.

[5] J. K. Huttunen, D. Steinberg, and S. E. Mayer, *Proc. Natl. Acad. Sci. U.S.A.* **67**, 290 (1970).

directly that the enzyme protein is phosphorylated in cell-free systems[2,6] and also in intact adipocytes.[7] The physiological relevance of this phosphorylation is verified by the finding that the extent of phosphorylation of the enzyme in intact adipocytes is altered by norepinephrine and insulin in close association with changes of its activity.[8] Purification of the enzyme from adipose tissue of chicken is described in this volume [73].

Assay Method

Principle. The hydrolysis of the ester bond of emulsified [1(3)-^{3}H]oleoyl-2-O-oleylglycerol, a monoether analog of dioleoylglycerol, is catalyzed by HSL.[8,9] The substrate emulsion is stabilized with phospholipids, and bovine serum albumin is used as fatty acid acceptor. The [^{3}H]oleic acid released is separated from remaining substrate as potassium [^{3}H]oleate into the upper phase of a liquid–liquid partition system and determined by liquid scintillation. The rate of release of [^{3}H]oleic acid is a measure of the activity of the enzyme.

Several problems are encountered when measuring HSL activity in crude adipose tissue extracts. Release of ^{3}H-labeled fatty acid from labeled acylglycerol substrate will not be a correct measure of enzyme activity when endogenous substrate lipids are present in significant amounts. Release of fatty acids from these lipids must be assayed instead.[1] Crude tissue preparations contain lipoprotein lipase[3] and monoacylglycerol lipase.[10] Since monoacylglycerols are formed during the hydrolysis of tri- and diacylglycerols by HSL, the action of highly active monoacylglycerol lipase will add to the measured fatty acid release. Lipoprotein lipase is active against the same lipid substrates as HSL, and specificity for the latter enzyme has generally had to rely on specific incubation conditions.

Most of these difficulties are overcome using the assay method described below: substrate for monoacylglycerol lipase is not formed from monoacylmonoalkylglycerol and, under conditions used in the assay, pH 7.1 and no CII-activator, little lipoprotein lipase activity is measured.[11] In

[6] P. Belfrage, G. Fredrikson, N. Ö. Nilsson, and P. Strålfors, *in* "Les Colloques de l'INSERM: Obesity—Cellular and Molecular Aspects" (G. Ailhaud, ed.) INSERM, Vol. 87, p. 161, Paris, 1979.
[7] P. Belfrage, G. Fredrikson, N. Ö. Nilsson, and P. Strålfors, *FEBS Lett.* **111**, 120 (1980).
[8] N. Ö. Nilsson, P. Strålfors, G. Fredrikson and P. Belfrage, *FEBS Lett.* **111**, 125 (1980).
[9] H. Tornqvist, P. Björgell, L. Krabisch, and P. Belfrage, *J. Lipid Res.* **19**, 654 (1978).
[10] H. Tornqvist and P. Belfrage, *J. Biol. Chem.* **251**, 813 (1976).
[11] P. Belfrage and G. Fredrikson, unpublished data, 1979.

comparison with other ways to prepare lipid substrates,[12,13] in our hands, phospholipid emulsions provide more stable and uniform substrates, rendering a higher reproducibility to the enzyme assay. Detergent inhibition is less with monoacylmonoalkyl- or diacylglycerol than with the corresponding monoacyldialkyl- or triacylglycerol compounds.[14] Thus the former substrates are preferable when the enzyme preparation contains detergent.

Reagents

[³H]Oleoyl-2-*O*-oleylglycerol (5.0 mg/ml) in redistilled benzene stored at 4°

Oleoyl-2-*O*-oleylglycerol (50 mg/ml) in heptane, −20°

Egg yolk phospholipid (20 mg/ml) in chloroform, −20°

Tris-HCl, 20 mM, pH 7.4, 0.15 M NaCl (Tris-HCl buffer)

Bovine serum albumin, 20% (w/v) in 20 mM Tris-HCl, pH 7.4, 0.15 M NaCl

Tris-HCl, 20 mM, pH 7.4, 1 mM EDTA, 1 mM dithioerythritol, 0.02% (w/v) bovine serum albumin (Tris-HCl with 0.02% albumin)

Methanol–chloroform–heptane, 1.41 : 1.25 : 1 (v/v/v), with 3 g of oleic acid per liter as fatty acid carrier

Potassium–carbonate buffer, 0.1 M, pH 10.5

Liquid scintillator: Instagel-toluene, 1 : 1 (v/v)

Preparation of Reagents. Oleoyl-2-*O*-oleylglycerol (labeled and unlabeled) is synthesized, then purified with silicic acid chromatography to more than 99.5% purity.[9,15] The purity is checked regularly by thin-layer chromatography, as blanks (free [³H]oleic acid) must be kept low. Bovine serum albumin is defatted by treatment with activated charcoal[16] and dialyzed. Egg yolk phospholipid is purified by elution with methanol from silicic acid.

Preparation of Substrate Emulsion. [³H]Oleoyl-2-*O*-oleylglycerol (approximately 20 × 10⁶ cpm) and oleoyl-2-*O*-oleylglycerol in a combined amount of 12.1 mg (20 μmol) are mixed with 0.6 mg of phospholipid in a small glass vial and solvents are carefully evaporated with dry N$_2$. Two milliliters of Tris-HCl buffer (room temperature) are added, and the mixture is sonicated. Since the sonication procedure is critical for obtaining

---

[12] J. C. Khoo and D. Steinberg, this series, Vol. 35 [23].

[13] J. C. Khoo, D. Steinberg, J. J. Huang, and P. R. Vagelos. *J. Biol. Chem.* **251**, 2882 (1976).

[14] G. Fredrikson, P. Strålfors, N. Ö. Nilsson, and P. Belfrage. Manuscript submitted.

[15] [1(3)-³H]Oleoyl-2-*O*-oleylglycerol and 1(3)-oleoyl-2-*O*-oleylglycerol are synthesized from 1,3-benzylidene glycerol, oleylmethane sulfonate, and [9,10-³H]oleoylchloride (oleoylchloride) as described.[9] 2-*O*-Oleylglycerol for the above synthesis can be purchased from Supelco Inc., Bellefonte, Pennsylvania.

[16] R. Chen, *J. Biol. Chem.* **242**, 173 (1967).

reproducible emulsions, the following points must be observed. It is essential that the sonifier energy output be the same each time an emulsion is prepared. The energy output varies from one sonifier microtip to another, and the tips are also rapidly degraded by cavitation. Therefore, with each tip, the setting of the sonifier must be calibrated against sonic pressure under the same conditions as those used for emulsion preparation. The calibration has to be repeated occasionally to correct for the degradation of the tips.

The substrate is sonified with a setting corresponding to 2.5–3 g of sonic pressure for 2 × 1 min. With a Branson model B 12, equipped with an exponential microtip, this generally corresponds to a setting of 2–3. After cooling to room temperature, 1.6 ml of Tris-HCl buffer and 0.4 ml of 20% albumin in Tris-HCl buffer are added and sonication is continued for 4 × 30 sec at a setting corresponding to approximately 3 g of sonic pressure (setting 3–4), now with the vial immersed in an ice bath. The substrate emulsion is used within 3–4 hr of preparation.

Procedure. Substrate emulsion, 100 μl, is added to enzyme sample and Tris-HCl with 0.02% albumin, to a final incubation volume of 200 μl. The mixture is incubated at 37°. We have found it convenient to use disposable glass tubes, 14 × 100 mm; shaking (see below) can then be performed without sealing the tubes. Detergent present in the enzyme sample will inhibit HSL activity; the enzyme must therefore be appropriately diluted to a noninhibitory detergent concentration. $C_{13}E_{12}$[17] at a final concentration of less than 0.001% and C_8E_6[18] at less than 0.008% do not cause significant inhibition. Amount of enzyme and incubation time are selected to keep substrate hydrolysis below 10%.

The incubation is terminated by addition of 3.25 ml of methanol–chloroform–heptane followed by 1.05 ml of potassium carbonate buffer. The tubes are shaken vigorously in a vortex mixer for 15 sec, and the two phases are separated by centrifugation (2000 rpm for 20 min). During these procedures the temperature should always be kept the same, since the partition coefficient of potassium oleate is temperature dependent.[19] A sample (1 ml) of the upper phase (2.5 ml) is taken for scintillation counting in 10 ml of Instagel-toluene. During sampling the walls of the tubes must not be touched with the pipette tip, since they are covered with a surface

---

[17] $C_{13}E_{12}$ is a polydisperse preparation of alkyl polyoxyethylenes with the indicated average composition; C = alkyl carbons, E = oxyethylene units. The detergent is available from AB Berolkemi, Stenungsund, Sweden.

[18] C_8E_6, a mixture of C_8E_5 (<25%), C_8E_6 (>50%), and C_8E_7 (<25%), has been synthesized essentially as described in J. M. Corkill, J. F. Goodman, and R. H. Ollewill, *Trans. Faraday Soc.* **57**, 1627 (1961).

[19] P. Belfrage and M. Vaughan, *J. Lipid Res.* **10**, 341 (1969).

film of substrate. Enzyme activity can be calculated after determining the partition coefficient of potassium oleate.[19]

Units. One unit of enzyme activity is equivalent to 1 μmol of fatty acid produced per minute at 37°. Specific activity is expressed as units per milligram. Protein was determined by a scaled-down version of the method of Lowry *et al.*[20] after precipitation with trichloroacetic acid.[21]

Other Substrates. A dioleoylglycerol emulsion can be prepared and used exactly as described above.[9] Trioleoylglycerol and its mono- and dialkylanalogs are prepared in a similar manner.[22] Monooleoylglycerol in an emulsified form can be prepared as described elsewhere[3] and can be used for HSL assay of tissue preparations not containing monoacylglycerol lipase.

Purification of Enzyme[14]

The purification procedure described below results in an enzyme of more than 50% protein purity after approximately 2000-fold purification. The table shows a summary of the procedure. Except when noted all steps were carried out at 4°. The pH was measured at room temperature unless otherwise indicated.

Step 1. Preparation of Adipose Tissue Homogenate. Epididymal fat pads from 25 rats (male Sprague–Dawley, 200–220 g) were homogenized in two volumes of 0.25 M sucrose, 1 mM EDTA, pH 7.4, in a Potter–Elvehjem homogenizer with a Teflon pestle rotating at 200 rpm (20 strokes) followed by homogenization in a ground-glass tissue grinder (10 strokes). The homogenate was centrifuged in a swing-out rotor at 110,000 g for 45 min, the flotated fat cake is removed by slicing the tubes, and the supernatant is filtered through glass wool in order to remove as much as possible of remaining fat.

Step 2. Precipitation at pH 5.2. The pH of the supernatant was decreased to 5.2 by dropwise addition of 0.2 M acetic acid; after 20 min the precipitate formed was collected by centrifugation (12,000 rpm for 30 min). The precipitate was suspended in 20 mM Tris-HCl, pH 7.4, 1 mM EDTA, 1 mM dithioerythritol, 0.25 ml/g adipose tissue, with a ground-glass homogenizer. The suspension, the pH 5.2 precipitate fraction, was stored at −70° until used. Under these conditions the enzyme activity was stable for months.

[20] O. H. Lowry, N. J. Rosebrough, A. L. Farr, and R. J. Randall, *J. Biol. Chem.* **193**, 265 (1951).
[21] H. Tornqvist and P. Belfrage, *J. Lipid Res.* **17**, 542 (1976).
[22] H. Tornqvist, L. Krabisch, and P. Belfrage, *J. Lipid Res.* **13**, 424 (1972).

SUMMARY OF PURIFICATION OF HORMONE-SENSITIVE LIPASE FROM RAT ADIPOSE TISSUE[a]

Fraction	Volume (ml)	Total protein (mg)	Total enzyme (units)[b]	Specific activity (units/mg protein)	Yield (%)	Purification (fold)
I. 110,000 g supernatant	562	1200	77.5	0.065	100	1.0
II. pH 5.2 precipitate	50	355	72.1	0.203	93	3.0
III. Detergent-solubilized pH 5.2 precipitate	130	336	66.7	0.199	86	3.0
IV. First QAE-Sephadex chromatography	436	17.9	40.1	2.24	52	34
Pressure ultrafiltration	45	16.8	32.8	1.95	42	30
V. Second QAE-Sephadex chromatography	218	2.9	20.5	7.07	26	110
VI. Hydroxyapatite chromatography with detergent exchange	24	2.2	14.4	6.55	19	100
VII. Affinity chromatography on Ultrogel AcA 34	180	0.04[c]	5.1	127	6.5	1950

[a] Epididymal fat pads (160 g) from 200 rats were used as starting material.
[b] Micromoles of fatty acids released per minute at 37°.
[c] Protein estimated by scanning densitometry of sodium dodecyl sulfate–polyacrylamide gel electrophoresis.

Step 3. Detergent Solubilization. The pH 5.2 precipitate fraction was solubilized in the following way: Sucrose, 16.9 g, NaCl, 494 mg, dithioerythritol, 20 mg, and EDTA, 48 mg, were dissolved in 50 ml of pH 5.2 precipitate suspension, obtained from 200 rats. This mixture was added in portions to 69 ml of $C_{13}E_{12}$, 12% (w/v), in 30 mM Tris-HCl, pH 7.97 (10°). The solution was sonicated briefly, at a setting giving a sonic pressure of approximately 4 g, between each addition, the temperature being kept between 10° and 15°. Tris-HCl buffer was added to a final volume of 130 ml, and sonication was continued for 6 × 30 sec with temperature controlled as above by immersion in an ice bath. The solubilization resulted in a visually almost clear solution.

Step 4. First QAE-Sephadex Chromatography. Gradient sievorptive chromatography as described by Kierkegaard[23] was employed in this purification step and the next. In both steps the chromatography was performed at 10° with all buffers at a pH of 7.97 (10°). QAE-Sephadex, in 0.2 M NaCl, was packed in a column, 10 × 40 cm, and equilibrated successively with 10 volumes of 0.1 M Tris-HCl, 10 volumes of 40 mM Tris-HCl with 15 mM NaCl, and 2 volumes of 40 mM Tris-HCl with 15 mM NaCl, 0.2 mM EDTA, 1 mM dithioerythritol, 20% glycerol (w/v), 0.2% $C_{13}E_{12}$ (w/v). The solubilized pH 5.2 precipitate fraction from step 3 was applied, and the column was eluted with the same buffer, but containing 65 mM NaCl and 40% glycerol (w/v), at a flow rate of 300 ml/hr. Glycerol with 30 mM KH_2PO_4 and 1 mM dithioerythritol was added continuously to the eluate through a mixing chamber, to a glycerol concentration of 50% (w/v) and a pH of approximately 7 in order to stabilize the enzyme activity. The HSL was eluted at approximately 40 mM NaCl. Fractions containing more than 40 mU of enzyme activity per milliliter were pooled. NaCl was added to approximately 60 mM final concentration, and the pooled enzyme was concentrated approximately 10-fold by pressure ultrafiltration through a PM-10 Diaflo membrane at 4°.

Step 5. Second QAE-Sephadex Chromatography. The concentrated enzyme from step 4 was dialyzed against 3 liters of 40 mM Tris-HCl, 50 mM NaCl, 0.2 mM EDTA, 1 mM dithioerythritol, 20% glycerol (w/v), 0.2% $C_{13}E_{12}$ (w/v) for 3 hr, with two changes of buffer. The sample volume increased to 63 ml during the dialysis. QAE-Sephadex in a 5 × 86 cm column was equilibrated successively with 10 volumes of 0.1 M Tris-HCl, 10 volumes of 40 mM Tris-HCl, 20 mM NaCl, and 2 volumes of 40 mM Tris-HCl, 20 mM NaCl, 0.2 mM EDTA, 1 mM dithioerythritol, 20% glycerol (w/v), 0.2% $C_{13}E_{12}$ (w/v). A 130-ml linear gradient, 20 to 50 ml of NaCl, in the last equilibration buffer was run into the column, followed by

[23] L. H. Kierkegaard, *in* "Methods of Protein Separation" (N. Catsimpoolas, ed.), p. 279. Plenum, New York, 1976.

the dialyzed sample and 50 mM NaCl in the same buffer as above, at a flow rate of 100 ml/hr. The density of the first 100 ml of the elution buffer was increased by addition to 5% sucrose (w/v) in order to avoid mixing with the sample. Glycerol with KH_2PO_4 and dithioerythritol was added to the eluate continuously as described above. The HSL eluted at approximately 40 mM NaCl. Fractions with more than 50 mU of enzyme activity per milliliter were pooled.

Step 6. Hydroxyapatite Chromatography with Detergent Exchange. Enzyme from step 5 was dialyzed overnight against 50% glycerol (w/v), 1 mM dithioerythritol. The dialyzed enzyme was applied to a hydroxyapatite (spheroidal, BDH) column, 1.6 × 12 cm, equilibrated with 10 volumes of 10 mM potassium phosphate buffer, pH 7.0, 1 mM dithioerythritol, 30% glycerol (w/v), 0.8% C_8E_6,[18] at a flow rate of 50 ml/hr. The column was washed with 4 volumes of the equilibration buffer and eluted with the same buffer made 0.5 M with potassium phosphate. Fractions containing more than 40 mU of enzyme activity per milliliter were pooled. By this step the enzyme was sixfold concentrated, and the detergent $C_{13}E_{12}$ was exchanged for C_8E_6. The pooled enzyme was then immediately subjected to the following fractionation step.

Step 7. Affinity Chromatography on Triacylglycerol-Containing Ultrogel AcA 34. Ultrogel AcA 34[24] was packed in a column, 5 × 89 cm, and equilibrated with 2 volumes of 5 mM potassium phosphate buffer, pH 7.0, 0.15 M NaCl, 1 mM dithioerythritol, 30% glycerol (w/v), 0.2% $C_{13}E_{12}$ (w/v) followed by 2 volumes of the same buffer but with 0.8% C_8E_6 instead of $C_{13}E_{12}$. Enzyme from step 6, 24 ml, was applied to the column and eluted with the buffer containing C_8E_6. The enzyme eluted immediately before the total column volume. Fractions containing more than 20 mU/ml were pooled. The enzyme was stored at −70° after addition of glycerol and $C_{13}E_{12}$ to final concentrations of 50% (w/v) and 0.2% (w/v), respectively.

Concentration of Purified Enzyme. Pooled enzyme from step 7, with

---

[24] Ultrogel AcA 34, batch No. 9592, was manufactured by l'Industrie Biologique Française (IBF), Gennevilliers, France, and obtained from LKB, Stockholm, Sweden. This batch, and other batches prepared up to 1976, contains approximately 0.3 mg of soybean oil triacylglycerol per milliliter of wet gel. The triacylglycerol has been included in the gel during the manufacturing process. This process was changed by IBF during 1977, and batches produced after this date do not contain triacylglycerols and cannot be used in step 7 without further treatment. Triacylglycerols can, however, be introduced into such a gel in the following way: The gel is washed in water followed by 75% ethanol. Soybean oil (30 g per 100 ml of gel) suspended in ethanol is added, and the mixture is stirred carefully overnight. The gel is then quickly poured into 10 volumes of 1 M NaCl and washed extensively with this salt solution to remove excess lipid. It is further equilibrated as described in step 7.

added glycerol and $C_{13}E_{12}$, was dialyzed and subjected to hydroxyapatite chromatography as in step 6. Buffers contained 50% (w/v) glycerol and 0.2% $C_{13}E_{12}$ instead of C_8E_6 to increase the stability of the enzyme. The enzyme was approximately 10-fold concentrated with a yield of 50%. This enzyme preparation was used in studies of substrate specificity and in activation experiments. It was stored at $-70°$.

Comments on the Purification Procedure. Hormone-sensitive lipase has been difficult to purify, mainly owing to its amphiphilic character and to its low tissue concentration (1–2 μg per gram of rat adipose tissue). In the purification procedure described, some points require special attention. The solubilization of the pH 5.2 precipitate with detergent (step 3) must be complete, as indicated by a clear or almost clear solution. Protein–lipid aggregates otherwise contaminate the enzyme during the first QAE-Sephadex chromatography (step 4) and cause irreversible aggregation during its subsequent concentration by pressure ultrafiltration. It may therefore be necessary to add more detergent, dilute the sample, and/or remove contaminating lipids by ultracentrifugation in step 3. The gradient sievorptive chromatography employed in steps 4 and 5 is critically dependent on the pH and, owing to the large temperature coefficient of Tris-HCl buffers, on the temperature. $C_{13}E_{12}$ in the AcA 34 column chromatography will prevent the adsorption of HSL in step 7. The enzyme will then elute at a K_{av} of 0.45 with much less purification. The capacity of the triacylglycerol-containing Ultrogel AcA 34 gel is rather low and decreases with the use of the gel. If it is exceeded, part of the enzyme will also appear at this K_{av}. Therefore it is necessary to test each batch of gel[24] before it is used.

Monoacylglycerol lipase[25] will elute with the unretained protein peak during the first QAE-Sephadex chromatography (step 4). This fraction can be used as the starting material for the purification of this enzyme, as described elsewhere.[25]

Properties

Stability. The concentrated enzyme from step 7 lost less than 20% of its activity in 1 month if stored at $-70°$, in 50% (w/v) glycerol and 0.2% $C_{13}E_{12}$, at neutral pH. In 30% glycerol and 0.8% C_8E_6 (i.e., the conditions of the step 7 enzyme preparation), 60% of the enzyme activity was lost in 1 week at $-70°$. The enzyme is stable only in the neutral pH range; at pH 8 its activity is halved in 24 hr at 10°.

Physical Properties.[14] The HSL from step 7 has an apparent minimum molecular weight of 84,000 by sodium dodecyl sulfate (SDS)–polyacryla-

[25] H. Tornqvist and P. Belfrage, this volume [75].

mide gel electrophoresis. The enzyme of the step 7 preparation constitutes more than 50% of stainable protein; some minor bands possibly
represent proteolytic fragments of HSL. Detergent binding to a monomer
of the enzyme is suggested by its gel chromatographic behavior. Depending on the type of detergent and the exact conditions of the chromatography, it is eluted in the molecular weight range from 70,000 to 150,000
(determined with globular reference proteins). The pI of a partially purified enzyme has been found to be 6.7.[2]

Substrate Specificity. The purified lipase hydrolyzed emulsified tri-, di-,
and monooleoylglycerol at the approximate relative rates of 1 : 10 : 4. The
monoalkyl analogs of the tri- and dioleoylglycerol were hydrolyzed 10 and
25% more rapidly, respectively, than the corresponding acylglycerols.

Inhibitors. The lipase activity is reversibly inactivated by detergents
(as stated above). NaF, 10 mM, inhibits the activity by 50% after 30 min
of preincubation at 37°; 20 μM diisopropylfluorophosphate, 10 μM HgCl$_2$,
and 20 μM p-chloromercuribenzoic acid also inhibit enzymatic activity by
50%. The enzyme protein from step 7 is readily labeled by incubation with
[^{3}H]diisopropylfluorophosphate.[14]

Phosphorylation of Hormone-Sensitive Lipase. The enzyme protein is
phosphorylated as follows: The catalytic subunit of cyclic AMP-
dependent protein kinase from rat adipose tissue[26] (3 μg/ml) is preincubated[27] in 0.3 M Tris-HCl, pH 7.3, containing 30 mM MgCl$_2$, 3 mM
dithioerythritol, and 0.3 mM ATP, for 15 min at 21°. [γ-^{32}P]ATP and
enzyme (0.5–5 μg) are then added to final concentrations of 0.1 M Tris-
HCl, 10 mM MgCl$_2$, 1 mM dithioerythritol, and 0.1 mM [γ-^{32}P]ATP (approximately 10^3 cpm/pmol); the incubation is continued for 30 min at 37°.
The reaction is terminated by precipitation in 10% trichloroacetic acid.
Distribution of radioactivity after polyacrylamide gel electrophoresis in
SDS is determined by autoradiography of the dried gels using Agfa Osray
M 3 film. By this procedure, approximately 0.5 mol of phosphate per mole
of HSL 84,000 MW polypeptide is incorporated. With enzyme samples
from steps 5 and 6, HSL is the major [^{32}P]phosphopeptide; after step 7
only HSL is phosphorylated.

Activation of Hormone-Sensitive Lipase. The activity of HSL against
trioleoylglycerol is increased 50–70% by the following procedure: The
enzyme sample (2–10 mU) is incubated for 15 min at 37° in 0.1 M Tris-
HCl, pH 7.0, 37°, with 0.5 mM ATP, 10 mM MgCl$_2$, 1 mM dithioerythritol, and the catalytic subunit of cyclic AMP-dependent protein kinase
(100 ng) in a final volume of 30 μl. Lipase activity is then determined by

---

[26] P. Strålfors, G. Fredrikson, N. Ö. Nilsson, and P. Belfrage. Manuscript submitted.
[27] The preincubation is done in order to saturate phosphorylatable sites on the protein kinase
with unlabeled phosphate.

incubation for 30 min at 37° with 0.25 mM emulsified tri[³H]oleoylglycerol as substrate in a final volume of 200 μl (as described above).

Acknowledgments

This work has been supported by grants from Påhlssons Foundation, Segerfalks Foundation, Novo Insulin Foundation, the Swedish Diabetes Foundation, the Medical Faculty of the University of Lund, and the Swedish Medical Research Council (Grant No. 3362).

[75] Monoacylglycerol Lipase from Rat Adipose Tissue

EC 3.1.1.23 Glycerol-monoester acylhydrolase

By HANS TORNQVIST and PER BELFRAGE

Monoacylglycerol [1(3)- or 2-isomer] + H_2O → glycerol + fatty acid

Rat adipose tissue contains very active long-chain monoacylglycerol hydrolase activity. Most of this is due to monoacylglycerol lipase[1] an enzyme first described by Vaughan et al.[2] The enzyme has been extensively purified and partially characterized.[3] The physiological role of the enzyme is not clear; its activity is not influenced by the nutritional state of the animal or by hormones.[1] It has been suggested that monoacylglycerol lipase catalyzes the hydrolysis of monoacylglycerols derived both from the hormone-sensitive lipase-catalyzed intracellular lipolysis and from the lipoprotein lipase-catalyzed hydrolysis of chylomicron and very low-density lipoprotein lipids.[1] Thus, the enzyme may account for the observation that monoacylglycerols do not accumulate in adipose tissue.

Assay Method

Principle. Free [³H]glycerol produced by hydrolysis of 1(3)-mono-oleoyl[³H]glycerol in mixed micelles with nonionic detergent is isolated with a simple liquid–liquid partition system, and the radioactivity is determined by liquid scintillation counting.[4] Hormone-sensitive lipase[1,5] and

[1] H. Tornqvist, P. Nilsson-Ehle, and P. Belfrage. *Biochim. Biophys. Acta* **530**, 474 (1978).
[2] M. Vaughan, J. E. Berger, and D. Steinberg. *J. Biol. Chem.* **239**, 401 (1964).
[3] H. Tornqvist and P. Belfrage. *J. Biol. Chem.* **251**, 813 (1976).
[4] H. Tornqvist, L. Krabisch, and P. Belfrage. *J. Lipid Res.* **15**, 291 (1974).
[5] G. Fredrikson, P. Strålfors, N. Ö. Nilsson, and P. Belfrage, this volume [74].

lipoprotein lipase[1] are also active against long-chain monoacylglycerols, but contribute relatively little to this enzyme activity in crude adipose tissue extracts. The activity of these enzymes is strongly inhibited by the concentration of nonionic detergent used in the assay, rendering it highly specific for monoacylglycerol lipase. We prefer to use the assay described here because of its high sensitivity compared to assays with unlabeled substrate and because the reaction products can be rapidly determined in a simple way. Monooleoylglycerol is preferred as substrate over other monoacylglycerols for several technical reasons.[4]

Synthesis of Substrate. Monooleoyl[^{3}H]glycerol is synthesized as follows[4]: Sodium mono[^{3}H]glycerate is obtained by allowing [1(3)-^{3}H]-glycerol (10 mCi, 1 mmol) to react with NaOH at 150°. The glycerol is dissolved in 1 ml of methanol, 1 ml of 1 M NaOH is added, and the solvents are evaporated with dry nitrogen at 150°; the reaction is allowed to continue for another 30 min at 150°. The sodium mono[^{3}H]glycerate is dried overnight in a desiccator. Ethyl acetate (20 ml), which has been dried with molecular sieve type 4A to remove water and trace amounts of alcohols, is thoroughly mixed with the compound in a closed vial, and the glycerate is subsequently acylated at room temperature for 2.5 hr with 400 μl of oleoyl chloride. The reaction is interrupted by addition of water, and the lipids are extracted with diethyl ether, which is evaporated after drying over Na_2SO_4. Unlabeled monooleoylglycerol (0.7 g) is added and the monooleoyl[^{3}H]glycerol is purified by silicic acid column chromatography. The compound is applied to the column (30 mg of lipid per gram of silicic acid) in hexane. Tri- and dioleoylglycerol and oleic acid are eluted with up to 25% diethyl ether in hexane and the monooleoylglycerol is subsequently eluted with 50% diethyl ether in hexane. Fractions containing more than 99.5% pure monooleoylglycerol are pooled, the solvent is evaporated, and the lipid is dissolved in dry redistilled benzene at 4° to minimize hydrolysis during storage. Purity is regularly checked by thin-layer chromatography on silicic acid. The monooleoyl[^{3}H]glycerol so obtained should have a specific activity of about 2 μCi/μmol. It is sufficient for at least 5000 enzyme determinations.

Reagents

Monooleoyl[^{3}H]glycerol, 5 mg/ml in benzene, stored at 4°
Monooleoylglycerol, 20 mg/ml in heptane, $-20°$
Tris-HCl, 0.2 M, pH 8.0, with $C_{13}E_{12}$ nonionic detergent,[6] 0.8% (w/v), 4°

---

[6] $C_{13}E_{12}$ is a polydisperse preparation of alkyl polyoxyethylenes with the indicated average composition; C = alkyl carbons, E = oxyethylene units. Detergents of this type are available from AB Berolkemi, Stenungsund, Sweden. Other commercially available detergents

Methanol–chloroform–heptane, 1.41 : 1.25 : 1 (v/v/v)
NaCl, 2%, w/v
Instagel[7]–toluene 1 : 1 (v/v)

Preparation of Monooleoyl[³H]glycerol Substrate. Monooleoyl-[³H]glycerol (approximately 4×10^6 cpm) is mixed with monooleoyl-glycerol to a combined amount of 14.2 mg (40 μmol) in a small glass vial, and the solvents are evaporated with dry N_2. The substrate is then dispersed by sonication on ice in 4.0 ml of the Tris-HCl buffer with detergent, until a water-clear solution is obtained. Conditions for sonication are not critical. If not used the same day, the substrate solution is frozen at $-20°$ and resonicated before use.

Procedure. Enzyme and buffer, in a total volume of 100 μl, is incubated with 100 μl of the above substrate solution in a disposable test tube (10 $\times$ 140 mm) for 5–60 min at 21° or 37°. The amount of enzyme and incubation time are chosen such that total hydrolysis does not exceed 10%. The reaction is terminated by adding 3.25 ml of the methanol–chloroform–heptane solution and 1.05 ml of the sodium chloride solution. The tubes are shaken vigorously in a vortex mixer, and two phases are separated by centrifugation. Without delay, a 0.5 ml sample of the upper methanol–water phase (2.5 ml) containing all the [³H]glycerol produced is transferred to a counting vial with 5 ml of Instagel-toluene, and the radioactivity is determined.

Units. One unit of enzyme activity corresponds to the release of 1 μmol of glycerol per minute at 21° unless otherwise indicated.

Purification Procedure[3]

An overall purification of the enzyme of about 2500-fold is obtained by the purification procedure described below.[8] The table shows a summary of a purification. Unless otherwise indicated all steps were carried out at

---

of similar composition, e.g., those of the Lubrol or Brij series, or alkylphenoxy derivatives, e.g. Triton X-100, can also be used in the assay below. Optimal concentration may be different. Detergents containing aromatic groups, however, have strong UV absorbance and are thus not suitable for use during protein fractionation.

[7] Scintillator solution is commercially available from Packard Instruments Inc.

[8] Monoacylglycerol lipase can also be prepared starting from a fraction obtained during the purification of hormone-sensitive lipase from rat adipose tissue,[5] using a slightly different procedure. During gradient sievorptive chromatography on QAE-Sephadex of the latter enzyme, monoacylglycerol lipase is eluted with the unretained proteins. After dialysis and concentration by pressure ultrafiltration, the pooled enzyme is subjected to isoelectric focusing at pH 6–8 (step 6), but in the LKB 440 ml electrofocusing column. The enzyme is then, after concentration by pressure ultrafiltration, taken through the Sephadex G-150 gel filtration step (step 5) and finally again electrofocused at pH 8–6 (step 7).

PURIFICATION OF MONOACYLGLYCEROL LIPASE FROM RAT ADIPOSE TISSUE[a]

Steps	Volume (ml)	Total protein (mg)	Specific activity (units/mg protein)	Yield (%)	Purification (fold)
1. 110,000 g supernatant	822	720	0.14	100	1
2. pH 5.2 precipitate	85	220	0.43	93	3.1
3. Solubilized pH 5.2 precipitate	87	220	0.42	91	3.0
4. TEAE chromatography	338	47	1.5	70	11
5. Ultrafiltration, PM-30 gel filtration (Sephadex G-150)	72	4.7	11	52	79
6. First isoelectric focusing, pH 6–8	6.5	0.11	270	29	1900
7. Second isoelectric focusing, pH 8–6	6.8	—	—	16	—
Gel filtration of peak fractions (Sephadex G-50)	4.0	0.04	350	13	2500

[a] Epididymal fat pads, 120 g, from 100 rats were used as enzyme source. Activity units are micromoles per minute at 21°. Data are from Tornqvist and Belfrage.[3]

4°. Reagents, buffers, and water of highest available purity were used because the enzyme contains functionally important free SH groups. All solutions contained 1 mM EDTA and 1 mM dithiothreitol and, from step 3 on, the nonionic detergent $C_{13}E_{12}{}^6$ (0.2% w/v), unless otherwise stated. All pH values were measured at 21°.

Steps 1 and 2. Crude Fractionation of Adipose Tissue Homogenate. Epididymal fat pads from 10 Sprague–Dawley rats (200–250 g) were homogenized in 10 ml of 0.25 M sucrose, pH 7.4, per gram of tissue, with a glass homogenizer with a rotating Teflon pestle for at least 10 min and further with a ground-glass grinder for another 10 min. The homogenate was then centrifuged in a swing-out rotor at 110,000 g for 45 min. The floating fat cake was removed by cutting the tubes, and the clear supernatant solution containing more than 80% of total enzymatic activity was decanted. (A more concentrated homogenate, e.g., 1 g of adipose tissue per two volumes of sucrose solution, can be prepared, but a slightly lower yield of enzyme in the 110,000 g supernatant is obtained.)

The combined supernatants were precipitated at pH 5.2 by addition of 0.2 M acetic acid. After 30 min, the precipitate formed was collected by centrifugation and suspended in 20 mM Tris-HCl, pH 7.4, 1.0 ml/g adipose tissue. Suspensions of pH 5.2 precipitate from 100 rats were prepared and stored at −70° until further use.

Step 3. Detergent Solubilization. $C_{13}E_{12}$ nonionic detergent was added to the pH 5.2 precipitate suspension to a concentration of 0.2% (w/v); the mixture was sonicated on ice with a Branson Model B12 sonifier at setting 3 for 2 min (in 30-sec intervals to avoid excessive heating) until the suspension became visually clear.

Step 4. TEAE Column Chromatography. TEAE-cellulose was prepared for chromatography according to the manufacturer's instructions. It was then washed with distilled water several times, and fines were removed by sedimentation. The TEAE-cellulose was packed in a column (4 × 16 cm) and equilibrated by running 10 volumes of 20 mM Tris-HCl, pH 7.4, through the column. The sample, approximately 90 ml of step 3 material was adjusted to pH 7.4 and applied to the column. The enzyme was eluted (180 ml/hr) with about 800 ml of starting buffer, and the main part of the contaminating proteins were retained on the column. Fractions containing more than 0.1 unit of enzyme activity per milliliter were pooled.

Step 5. Sephadex G-150 Gel Filtration. Dry Sephadex G-150 superfine gel beads were fractionated to get more uniform size allowing better separation at higher elution rates.[9] A column (5.0 × 70 cm) was packed and equilibrated in 20 mM Tris-HCl, pH 7.0.

The pooled enzyme peak fractions from the TEAE-cellulose column were concentrated by pressure ultrafiltration through a PM-30 Diaflo membrane (cutoff limit: MW ≃ 30,000) to 20 ml and applied to the gel filtration column. This was eluted at 10–15 ml/hr, and 10-ml fractions were collected. The enzyme was obtained at an elution volume of approximately 550 ml. Fractions containing more than 0.2 unit of enzyme activity per milliliter were pooled.

Step 6. First Isoelectric Focusing. An LKB 110 ml electrofocusing column was filled as follows: (*a*) anode solution (lower end): sucrose, 57% (w/v), phosphoric acid, 1% (v/v); (*b*) cathode solution: ethanolamine, 1% (v/v); (*c*) 100 ml of a linear sucrose gradient made up from 50 ml of sucrose, 46% (w/v), containing Ampholine,[10] pH 6–8, 1.6% (w/v), and 50 ml Ampholine solution, 0.4% (w/v). The pooled enzyme from step 5 was included in the sucrose gradient, if necessary after concentration by pressure ultrafiltration, as described above. The isoelectric focusing was performed at a constant effect of 5 W with a maximum of 1400 V or 10 mA for 80 hr. The column was emptied from the bottom with a peristaltic pump (50 ml/hr) and 1-ml fractions were collected. Fractions containing approximately 80% of total enzyme activity were combined.

Step 7. Second Isoelectric Focusing. A second isoelectric focusing col-

[9] R. Ekman, B. G. Johansson, and U. Ravnskov, *Anal. Biochem.* **70**, 628 (1976).
[10] Ampholine ampholytes, LKB, Stockholm, Sweden.

umn was prepared exactly as described above, except that electrodes were reversed [anode solution: phosphoric acid, 1% (v/v); cathode solution: ethanolamine, 1% (v/v) and sucrose, 57% (v/v)]. Fractions containing approximately 80% of total enzyme activity were pooled. Most of the contaminating proteins had slightly higher pI values than monoacylglycerol lipase; care should be taken not to include these with the pooled enzyme.

The purified enzyme was then run through a Sephadex G-50 Superfine column (1.6 × 40 cm) in 20 mM phosphate buffer, pH 7.0, to remove sucrose and Ampholine, and the pooled enzyme was stored at −70°.

Properties[3]

Purity. The enzyme prepared by the above procedure accounts for more than 85% of the stainable protein after SDS–polyacrylamide gel electrophoresis.

Molecular Properties. Owing to the small amount of enzyme protein obtained, no extensive molecular characterization has been performed. The SDS-polyacrylamide gel electrophoresis indicated a minimum molecular weight of 32,900. The purified enzyme has a pI of 7.2 (4°) and a Stokes' radius of 39 Å. It should be observed that the latter value presumably is a measure of an enzyme protein–detergent complex.

Stability. The enzyme is stable for several months at −70° in 20 mM Tris-HCl or phosphate buffer, pH 7.0, containing 0.2% $C_{13}E_{12}$, 1 mM dithiothreitol, and 1 mM EDTA. Under these conditions half-lives were approximately 9 hr at 37°, 36 hr at +21°, and 9 days at +4°. Removal of the detergent,[11] or dithiothreitol, drastically reduces enzyme stability. The enzyme rapidly loses activity if heated above +44°.

Inhibitors. The enzyme is strongly inhibited by SH reagents such as mercuric chloride and *p*-chloromercuribenzoate. Diisopropylfluorophosphate (10 μM) totally inhibits its activity, indicating a functional serine residue.

Substrate Specificity. The enzyme catalyzes the hydrolysis of 1(3)- and 2-monooleoylglycerol at equal rates (apparent K_m at 21° = 0.2 mM). It has no activity against long-chain di- and triacylglycerols, cholesterol ester, or lysophosphatidylcholine. The enzyme catalyzes the hydrolysis of a variety of medium- and long-chain monoacylglycerols.[4] Using aqueous dispersions of monodecanoyl- and monododecanoylglycerol it has been found that the enzyme requires an organized substrate structure, a minimal substrate interphase, to be brought into proper alignment for optimal

[11] P. Strålfors, H. Tornqvist, B. Jergil, and P. Belfrage, *Biochim. Biophys. Acta* **533**, 90 (1978).

catalytic action.[12,13] Because of this characteristic "activation" by lipid–water interphases, monoacylglycerol lipase should be classified as a "true" lipase, like phospholipase A_2 and pancreatic lipase.[14]

Acknowledgments

This work was supported by grants from A. Påhlssons Foundation, Malmö; P. Håkanssons Foundation, Eslöv; Segerfalks Foundation, Helsingborg; The Medical Faculty, University of Lund, Sweden and the Swedish Medical Research Council (project No. 3362).

[12] H. Tornqvist and P. Belfrage, unpublished observation.
[13] H. Tornqvist. Thesis, University of Lund, Sweden, 1975.
[14] R. Verger and G. H. de Haas, *Annu. Rev. Biophys. Bioeng.* **5**, 77 (1976).

[76] Cutinases from Fungi and Pollen

By P. E. KOLATTUKUDY, R. E. PURDY, and I. B. MAITI

Cutinase is an enzyme that catalyzes hydrolysis of an insoluble biopolyester, cutin, the structural component of plant cuticle. This polyester is composed of the following fatty acids: ω-hydroxyfatty acids, dihydroxypalmitic acid, saturated and Δ^{12}-monounsaturated 18-hydroxy-9,10-epoxy C_{18} acids, and saturated and Δ^{12}-monounsaturated 9,10,18-trihydroxy C_{18} acids.[1] The actual composition of cutin is dependent on the species; but, in general, fast growing plants appear to contain predominantly the C_{16} acids, particularly dihydroxypalmitate, whereas in slower growing plants a mixture of C_{16} and C_{18} acids is found (Fig. 1). Cutinase hydrolytically releases all types of monomers from the polymer.

Preparation of the Polymer

Mature Golden delicious apple fruits are cut into quarters, and as much internal tissue as possible is removed from the slices. The peel is boiled for a few hours 'in water to remove as much internal tissue as possible and collected by filtration through cheesecloth. To remove the remaining internal tissue from the cuticular layer the peel is boiled for 4 hr in an aqueous solution containing 4 g of oxalic acid and 16 g of ammonium oxalate per liter. The cuticular layers collected by filtration through cheesecloth are thoroughly washed with water and mixed with an excess of a 2 : 1 mixture of chloroform and methanol (20 ml/g wet weight). The solid material collected by filtration is reextracted twice with the

[1] P. E. Kolattukudy, *Recent Adv. Phytochem.* **11**, 185 (1972).

METHODS IN ENZYMOLOGY, VOL. 71

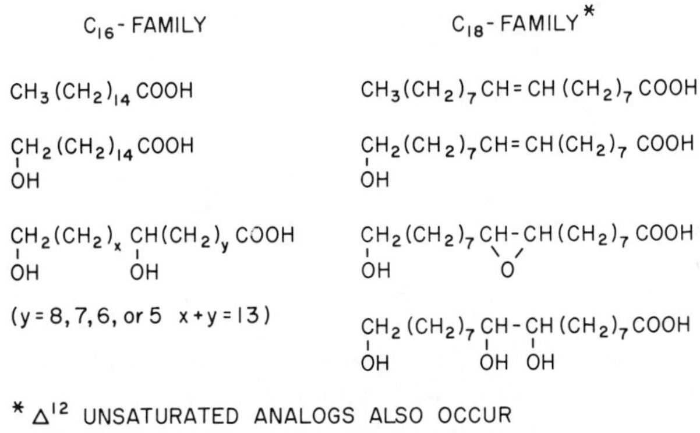

C$_{16}$- FAMILY

C$_{18}$- FAMILY[*]

CH$_3$(CH$_2$)$_{14}$COOH

CH$_3$(CH$_2$)$_7$CH=CH(CH$_2$)$_7$COOH

CH$_2$(CH$_2$)$_{14}$COOH
|
OH

CH$_2$(CH$_2$)$_7$CH=CH(CH$_2$)$_7$COOH
|
OH

CH$_2$(CH$_2$)$_x$ CH(CH$_2$)$_y$ COOH
| |
OH   OH

CH$_2$(CH$_2$)$_7$ CH-CH(CH$_2$)$_7$ COOH
| \ /
OH   O

(y = 8, 7, 6, or 5   x+y = 13)

CH$_2$(CH$_2$)$_7$ CH-CH(CH$_2$)$_7$COOH
| | |
OH   OH OH

[*] Δ^{12} UNSATURATED ANALOGS ALSO OCCUR

Fig. 1. Cutin acids.

chloroform : methanol mixture and then subjected to Soxhlet extraction with chloroform for 48 hr to remove cuticular waxes. After air drying, the cuticular material is treated with a mixture of cellulase (*Aspergillus niger*, Sigma Chem. Co., 5 g/liter) and pectinase (Sigma Chem Co., 1 g/liter) in 0.05 M sodium acetate buffer, pH 4.0, at 30° for 16 hr with shaking. The solids recovered from the reaction mixture are subjected to the solvent extractions, the enzyme treatment followed by the solvent extraction as above. The final residue is air dried and powdered with a Wiley mill; the powder, which passes through a 60-mesh filter, is collected. This powder is subjected to a final extraction with chloroform in a Soxhlet extractor.

Fungal Cutinases

Assay Method

Method 1

Principle. Release of soluble radioactive materials from labeled insoluble polymer is used as a measure of hydrolysis.

Reagents

Suspension of labeled cutin (see below)

Glycine-NaOH, 50 mM, pH 10.0

Enzyme solution

Preparation of Radioactive Cutin. Three procedures can be used to prepare labeled cutin.

1. Exposure of powdered cutin to ^{3}H$_2$. About 0.5 g of powdered (60 mesh) Golden delicious apple cutin prepared as described above is exposed to ^{3}H$_2$ for 6 hr at room temperature, and the labile ^{3}H is removed by

repeatedly washing the material with water. [This procedure is best carried out by commercial firms, such as New England Nuclear, because of the high levels of $^3H_2$ involved.] Since cutin, tritiated in this manner, releases ^3H into a variety of solvents it is extremely important to remove exchangeable ^3H and soluble tritiated compounds before using this material for enzyme assays. For this purpose the labeled cutin is subjected to Soxhlet extraction with water, dioxane, and chloroform as solvents (1–3 days each). Unless this cycle of extraction with solvents is repeated several times, ^3H will leach out into solvents and thus give too high a background in enzyme assays. The thoroughly extracted cutin powder is mixed with 60-fold excess of unlabeled cutin, and the mixture is homogenized in a Ten Broeck homogenizer with water and centrifuged. (Since the exposure to ^3H would give batch to batch variation in specific activity, appropriate amounts of unlabeled cutin should be added to give the desired final specific radioactivity.) The residue is thoroughly washed with acetone and subjected to Soxhlet extraction with acetone for 3 days. The final powder is dried and stored at $-20°$. Prior to enzyme assays the desired amount of the labeled cutin powder is homogenized in water with a Ten Broeck homogenizer; the mixture is placed in a 15-ml graduated centrifuge tube and centrifuged. The pellet is washed three times (by suspension and centrifugation) with water followed by three washes with acetone. The solid recovered is soaked in acetone overnight, followed by the same washing procedure as above. The resulting residue is soaked in 50 mM glycine–NaOH buffer, pH 10.0, for several hours and subsequently washed several times with the buffer. At this stage the specific radioactivity can be adjusted to the desired level by homogenizing unlabeled cutin with the radioactive cutin in a Ten Broeck homogenizer. After such a procedure the finely powdered material is washed several times with water or the buffer, and then it is suspended in a convenient volume of buffer.

2. Preparation of biosynthetically labeled cutin. Young rapidly expanding apple fruits (3–4 cm in diameter) are picked and washed thoroughly with water. With a cork borer, cores (11 mm in diameter) are punched out into distilled water; from each core, skin disks (<1 mm thick) are excised and placed in water. After thoroughly washing the disks with water, they are gently blotted dry. In a 125-ml Erlenmeyer flask 25 disks are placed in 0.75 ml of water containing 125 μCi (2.1 μmol) of sodium [1-^{14}C]acetate. Each disk is individually bathed in the solution. The disks are spread at the bottom of the flask and incubated for 6 hr at 30° with gentle shaking. The disks are thoroughly washed with water, and 400-disk batches are placed in 1.5 liters of 50 mM acetate buffer, pH 4.0, containing 7.5 g of cellulase (*Aspergillus niger*, Sigma Chem. Co.) and 1.5 g of pec-

tinase (fungal, Sigma Chem. Co.). After incubation with shaking overnight, the cuticle disks, which separate from the tissue, are collected and any adhering tissue is carefully scraped away from each disk. Then the cuticular disks are soaked in an excess of a 2 : 1 mixture of chloroform and CH_3OH, thoroughly washed with the same solvent, and Soxhlet-extracted with chloroform overnight to remove soluble waxes. The enzyme treatment and solvent extraction are repeated. The resulting radioactive cutin disks are ground in acetone with a Ten Broeck homogenizer and mixed with similarly homogenized unlabeled cutin so that the specific radioactivity is about 50,000 cpm/3 mg. (To determine the specific activity the finely powdered cutin is subjected to hydrogenolysis with $LiAlH_4$,[2] and the amount of ^{14}C in the lipid products is measured. If the ^{14}C in the aqueous layer is over 5% of the total ^{14}C further treatment with cellulase–pectinase is needed.) The final powdered material is washed twice with acetone, three times with deionized water, and finally with the buffer used for enzyme assay.

3. Preparation of [1-^{14}C]palmitoyl-cutin. [1-^{14}C]Palmitic acid (50 μCi, 59 Ci/mol) is refluxed with 3 ml of thionyl chloride for 2 hr; the excess thionyl chloride is removed with a rotary evaporator under reduced pressure. The resulting [1-^{14}C]palmitoyl chloride is dissolved in 7 ml of dry (over sodium) benzene; 2 g of apple cutin powder (60 mesh, dried over P_2O_5) and 2 ml of dry pyridine are added, and the mixture is kept at room temperature with occasional shaking for 24 hr in a desiccator containing P_2O_5. Cutin powder, recovered from the reaction mixture by filtration, is thoroughly washed with a 2 : 1 mixture of chloroform and CH_3OH. The residue is Soxhlet extracted with absolute methanol and chloroform (48 hr each) and subsequently washed three times each with acetone and water and finally with the buffer used for enzyme assay until no more ^{14}C is released.

Assay Procedure. To a small test tube, 0.4 ml of the suspension of labeled cutin (4 mg) in 50 mM glycine–NaOH buffer, pH 10.0, and 0.1 ml of enzyme ($\sim$1 μg of protein) are added; the test tube is incubated at 30° for 10 min with shaking in a gyrating water bath shaker. The reaction is terminated by the addition of 0.2 ml of 1 M HCl, and the mixture is filtered through a glass wool plug placed in a Pasteur pipette. Two 0.5-ml portions of acetone are passed through to remove any soluble material noncovalently bound to the cutin powder. Aliquots (0.5 ml) of the combined filtrate are mixed with 10 ml of ScintiVerse and assayed for radioactivity in a scintillation counter. A control is run with boiled enzyme or without enzyme, and the value obtained is subtracted from the experimental value. When tritiated cutin, obtained by exposure to $^3H_2$, is used a significant

[2] T. J. Walton and P. E. Kolattukudy, *Biochemistry* **11**, 1885 (1972).

background value is obtained even after using extensive cleanup procedures. Alternatively the reaction mixture is acidified and extracted with chloroform three times, and the pooled organic phase is evaporated to dryness. The residue is dissolved in 10 ml of ScintiVerse and assayed for radioactivity in a scintillation counter. This latter method minimizes problems created by release of labeled impurities, such as carbohydrates, present in the labeled cutin substrate.

Method 2

Principle. Cutinase can also be measured spectrophotometrically by following the absorbance change at 405 nm indicating the release of *p*-nitrophenol from *p*-nitrophenyl esters of short-chain fatty acids used as model substrates.[3]

Reagents

Substrate solution (see below)
Enzyme solution
Triton X-100, 2 g/500 ml of water
Sodium phosphate buffer, 0.1 M, pH 8.0

Preparation of Substrate Solution. Into a 100-ml beaker, 21 mg of *p*-nitrophenyl butyrate and 160 mg of Triton X-100 are transferred with ethyl ether as solvent. After the solvent is evaporated off with a stream of N_2, 50 ml of water are added and the beaker is heated on a steam bath until the substrate forms a liquid layer. The mixture is subjected to ultrasonic treatment with the large probe of Biosonic III for 1 min at 40% of the maximum power. Upon cooling a clear solution is obtained, but with longer-chain esters slight turbidity is observed.

Assay Procedure. Into a cuvette the following reagents are transferred: phosphate buffer, 1.6 ml; Triton X-100 solution, 0.2 ml; enzyme solution 0.2 ml; substrate solution 1.0 ml. After the contents are mixed well, absorbance at 405 nm is measured in a recording spectrophotometer. Initial linear rates are used for calculating the rate of reaction.

Preparation of Crude Enzyme. Fusarium solani pisi (ATCC 38136) is grown on potato dextrose agar. Before autoclaving, two pea leaves are placed in each petri dish so that the fungus will maintain its virulence. After about 10 days of growth a spore suspension is prepared in distilled water and 10^7 to 10^9 spores are introduced to Roux culture bottles each containing 0.5 g of apple cutin powder suspended in a mineral medium.[4]

[3] R. E. Purdy and P. E. Kolattukudy, *Arch. Biochem. Biophys.* **159**, 61 (1973).
[4] L. Hankin and P. E. Kolattukudy, *J. Gen. Microbiol.* **51**, 457 (1968).

The bottles (60–70) are incubated at 23–25° for about 10–12 days, and the extracellular fluid is collected by filtration under suction with a Büchner funnel and filter paper. The filtrate (5–6 liters) is frozen in stainless steel pans and lyophilized. [Direct ammonium sulfate precipitation can also be used, but at the low protein concentrations ($<$400 μg/ml) found in the extracellular fluid the recovery of the enzyme is higher if the lyophilization technique is used.] The residue is dissolved in a minimal volume of water and dialyzed against 10 mM sodium phosphate buffer, pH 7.0, to remove the excess salt. To this solution powdered ammonium sulfate is added with stirring until 50% saturation is achieved. The precipitate is collected by centrifugation at 12,000 g for 10 min, dissolved in a minimum volume of 50 mM sodium phosphate buffer, pH 7.0, and dialyzed overnight against the same buffer. This protein solution is concentrated to 6 ml by ultrafiltration with an Amicon UM-10 membrane. Alternatively, the ammonium sulfate precipitation step can be omitted and the protein solution can be concentrated by ultrafiltration, but, owing to the large volume, a much longer concentration period is needed.

Gel Filtration with Sephadex G-100. The protein solution is passed through a Sephadex G-100 column (4.0 × 49 cm) in 50 mM phosphate buffer, pH 8.0, and fractions (10 ml) are assayed for cutinase using both the spectrophotometric method with p-nitrophenyl butyrate and p-nitrophenyl palmitate as substrates and with labeled cutin. With p-nitrophenyl butyrate two peaks of enzymic activity are found, of which only the one representing the larger protein catalyzes hydrolysis of p-nitrophenyl palmitate whereas only the fraction representing the smaller protein catalyzes hydrolysis of cutin. Because of the presence of phenolics associated with the proteins, the two enzymatic activities might not be completely resolved in all cases. The fractions containing cutin hydrolase activity are pooled, concentrated by ultrafiltration with a UM-10 membrane, and dialyzed against 100 mM Tris-HCl buffer, pH 8.3.

QAE-Sephadex Chromatography. The protein solution is passed through a QAE-Sephadex A-25 column (3 × 21.5 cm) equilibrated with 100 mM Tris-HCl buffer, pH 8.3. This step should remove all the visible color from the solution. To obtain optimum yields, the minimum amount of QAE-Sephadex required to remove the color should be used. Fractions containing cutinase activity are pooled and concentrated as before and dialyzed overnight against 5 mM citrate–10 mM phosphate buffer, pH 5.0.

SP-Sephadex Chromatography. The protein solution is applied to an SP-Sephadex column (1.8 × 15 cm) equilibrated with the citrate–phosphate buffer, pH 5.0. After washing the column with two bed volumes of buffer, a linear gradient of 0 to 0.3 M NaCl in a total volume of 400 ml of the citrate–phosphate buffer is applied. The fractions containing enzymatic activity are pooled, concentrated by ultrafiltration, and dialyzed against 0.1 M Tris-HCl buffer pH 8.0. In some cases SP-Sephadex chromatography may not be needed to get apparently homogeneous enzyme. In the case of *Fusarium solani pisi* SP-Sephadex separates two isozymes.[5,6] It is advisable to follow the purification using electrophoresis in each new case so that this step can be used if needed.

Using the procedure described here, cutinase has been purified from *Fusarium solani pisi*,[5] *Fusarium roseum culmorum*,[7] *Fusarium roseum sambucinum*,[8] *Ulocladium consortial*,[8] *Helminthosporum sativum*,[8] and *Streptomyces scabies*[8] (Table I). In general, 6- to 16-fold purification is accomplished with 16–40% recovery of the enzymatic activity, and the product is electrophoretically homogeneous.

Properties of the Fungal Enzyme

General Molecular Properties. In all cases the molecular weight is in the neighborhood of 25,000 with a Stokes' radius of about 20 Å. Although the enzyme is a single peptide, the presence of a proteolytic nick has been detected by sodium dodecyl sulfate (SDS)-polyacrylamide electrophoresis in cutinase II from *F. solani pisi*, cutinase from *F. roseum culmorum, F. roseum sambucinum*, and *U. consortial*. One disulfide linkage, but no free SH, is present in the enzyme from *F. solani pisi*, and reduction of this disulfide results in inactivation of the enzyme.[9]

Amino Acid and Carbohydrate Composition of the Enzyme. Fungal cutinases thus far examined are quite similar in their amino acid composition and contain one residue each of methionine and tryptophan. They are all glycoproteins containing 4–6% carbohydrates. The carbohydrates are attached by *o*-glycosidic linkages via serine, threonine, β-hydroxyphenylalanine, and β-hydroxytyrosine,[10,11] but all of these amino acids are not involved in all cutinases.[8] The N-terminal glycine has D-glucuronic acid attached via an amide linkage.[12]

Catalytic Properties. Optimum pH of cutinase is 10.0 for the hydrolysis of cutin. Hydrolysis of radioactive cutin (release of soluble labeled materials) is linear up to about 1 μg/ml of enzyme and 15 min of incubation.

[5] R. E. Purdy and P. E. Kolattukudy, *Biochemistry* **14**, 2824 (1975).
[6] R. E. Purdy and P. E. Kolattukudy, *Biochemistry* **14**, 2832 (1975).
[7] C. L. Soliday and P. E. Kolattukudy, *Arch. Biochem. Biophys* **176**, 334 (1976).
[8] T. S. Lin and P. E. Kolattukudy, *Physiol. Plant Pathol.* **17**, 1 (1980).
[9] T. S. Lin, C. L. Soliday, and P. E. Kolattukudy, unpublished results, 1978.
[10] T. S. Lin and P. E. Kolattukudy, *Biochem. Biophys. Res. Commun.* **72**, 243 (1976).
[11] T. S. Lin and P. E. Kolattukudy, *Arch. Biochem. Biophys.* **196**, 255 (1979).
[12] T. S. Lin and P. E. Kolattukudy, *Biochem. Biophys. Res. Commun.* **75**, 87 (1977).

TABLE I
PURIFICATION OF EXTRACELLULAR FUNGAL CUTINASES

Step	Fusarium roseum sambucinum		Ulocladium consortial		Helminthosporum sativum		Streptomyces scabies		Fusarium solani pisi	
	Total protein (mg)	Specific activity[a] (10^3 dpm/µg)	Total protein (mg)	Specific activity[a] (10^3 dpm/µg)	Total protein (mg)	Specific activity[a] (10^3 dpm/µg)	Total protein (mg)	Specific activity[a] (10^3 dpm/µg)	Total protein (mg)	Specific activity[a] (10^4 dpm/µg)
Extracellular fluid	660	5.0	410	0.11	210	0.11	310	0.10	1250	1.2
0–50% $(NH_4)_2SO_4$ precipitate	449	6.0	148	0.18	69	0.13	102	0.13	502	2.0
Sephadex G-100	233	7.0	77	0.30	46	0.40	60	0.35	223	4.1
Sephadex QAE-A25	25	28	10	1.7	4	1.00	6	1.20	71	7.6
Sephadex SP-C25	20	31	7	1.8	3	1.20	4	1.20	58,[b] 25[c]	7.8

[a] The specific activity is expressed as the amount of ^{3}H released per minute per microgram of protein from labeled cutin.
[b] Cutinase I.
[c] Cutinase II.

With the insoluble polymeric substrate, the usual kinetic parameters such as K_m cannot be determined. For operational purposes 6–8 mg of cutin per milliliter gives maximal rates of hydrolysis of cutin.

With the model substrate, p-nitrophenyl butyrate, assays are done at pH 8.0 because nonenzymatic hydrolysis is too rapid above this pH. With this substrate, linear rates of hydrolysis are obtained up to 3 min and 15 ng of enzyme per milliliter.

Substrate Specificity. The enzyme catalyzes hydrolysis of the polyester into oligomers and hydrolysis of the oligomers into monomers.[6] With model synthetic substrates, it has been determined that cutinases have a high degree of specificity for primary alcohol esters, although cyclohexyl-hexadecanoate is readily hydrolyzed but cholesteryl hexadecanoate is not hydrolyzed.[6] With p-nitrophenyl esters of fatty acids, cutinase shows a high degree of preference for short chains; as the acyl chain reaches C_{12}, V decreases to <1% of that observed with C_2, whereas K_m shows very little difference (Table II).

Inhibitors. Organic phosphates such as diisopropylfluorophosphate and paraoxon (at micromolar concentrations) are potent inhibitors of the enzyme because the enzyme contains one "active" serine per mole, which is involved in catalysis.[6] Thiol-directed reagents and chelators have no effect, as there is no free thiol group in the enzyme and no metal ion is involved.[8]

Immunological Comparison. Rabbit antibodies prepared against cutinase I from *F. solani pisi* cross-react with cutinase II isolated from the same organism, but Ouchterlony double-diffusion analysis shows spurs indicating that the two enzymes are immunologically not identical.[7] With the anticutinase I, immunoprecipitation is not detected with the cutinases from any of the other organisms. However, cutinase from *F. roseum culmorum* is just as sensitive to inhibition by rabbit anticutinase I as is cutinase I itself. Cutinase from *F. roseum sambucinum* is also partially inhibited by anticutinase I, but the enzymatic activity of cutinases from the other organisms are not affected by the antibodies prepared against cutinase I from *F. solani pisi.*

Pollen Cutinase

The pollen tube has to penetrate a cuticular barrier, which encloses the mature stigma, before it can grow through the pistil on its way to the ovary. In some plant species the incompatibility reaction involves inability of the pollen tube to penetrate the cuticle.[13] Therefore it has been

---

[13] J. Heslop-Harrison, *Annu. Rev. Plant Physiol.* **26**, 403 (1975).

TABLE II
CHAIN-LENGTH SPECIFICITY OF CUTINASES FOR HYDROLYSIS OF FATTY ACID ESTERS OF
p-NITROPHENOL[a]

Chain length of acyl moiety	Fungal cutinase				Pollen cutinase	
	Cutinase I		Cutinase II		K_m $(10^{-5}\ M)$	V (mol/g, min^{-1})
	$K_m\ (10^{-4}\ M)$	V	$K_m\ (10^{-4}\ M)$	V		
C_2	68.0	1.5	97.0	2.3	3.33	25.0
C_4	3.5	1.0	7.5	1.2	2.90	20.0
C_6	8.9	0.46	8.6	0.2	2.20	11.0
C_8	8.8	0.17	5.9	0.06	2.50	28.6
C_{10}	4.8	0.038	3.6	0.011	2.00	20.0
C_{12}	5.6	0.008	4.5	0.002	2.10	15.4
C_{14}	—	—	—	—	2.10	9.1
C_{16}	—	—	—	—	1.92	15.4
C_{18}	—	—	—	—	1.82	7.4

[a] These results are from Purdy and Kolattukudy[6] and Maiti et al.[17]

suggested that cutinase is involved in the penetration process and in the incompatibility reaction. Indirect evidence for the existence of such an enzyme has been presented,[14,15] and recently it was shown that mature pollen of nasturtium (*Tropaeolum majus*) readily released cutinase into an aqueous medium without involving protein synthesis.[16] This cutinase has been purified to apparent homogeneity.[17]

Enzyme Assay

The procedures described under fungal cutinases are used for routine assays; the spectrophotometric assay with the model substrate, p-nitrophenyl butyrate, is the most convenient one. However, 0.1 M sodium phosphate buffer, pH 6.5, containing 1 mM dithioerythritol (DTE) should be used for enzyme assay with radioactive cutin as the substrate. When a crude enzyme preparation from a different plant is examined, it is advisable to check the results obtained from the spectrophotometric method with a tracer assay using radioactive cutin, because enzymes, other than cutinase, contained in such preparations might catalyze the hydrolysis of the model substrate but not of cutin.

[14] W. Heinen and H. F. Linskens, *Nature* (*London*) **191,** 1416 (1961).
[15] H. F. Linskens and W. Heinen, *Z. Bot.* **50,** 338 (1962).
[16] M. Shaykh, P. E. Kolattukudy, and R. W. Davis, *Plant Physiol.* **60,** 907 (1977).
[17] I. B. Maiti, P. E. Kolattukudy, and M. Shaykh, *Arch. Biochem. Biophys.* **196,** 412 (1979).

Purification

Pollen. Nasturtium plants are grown from seeds (Chas H. Lilly Co.) in a 1:1:1 mixture of sand:perlite:mica peat under Grow-lux lights (~1200 ftc) supplemented with incandescent lights with a 16-hr light period per day. The plants produce flowers within 2 months after germination. Pollen, collected from mature (dehisced) flowers with an aspirator and a pollen trap, are kept frozen at $-20°$.

Leaching of the Enzyme from Pollen. Pollen grains (15 g) are incubated in 225 ml of a medium containing 0.29 M sorbitol, 1 mM DTE, and 0.02 M sodium phosphate buffer, pH 6.5, for 5 hr at 26° with shaking in a gyrating water bath. (The pollen grains do not germinate under these conditions.) The mixture is chilled in an ice bath and all subsequent operations are performed at 4°. The pollen grains are removed by centrifugation at 12,000 g for 10 min, and the supernatant is filtered through glass wool to remove any floating material. After dialysis against 0.02 M sodium phosphate buffer, pH 6.5, containing 1 mM DTE, the protein solution is concentrated to 15 ml by ultrafiltration through a Diaflo PM-10 membrane.

Gel Filtration with Sephadex G-100. The protein solution is filtered through a Sephadex G-100 column (3.5 cm × 110 cm) equilibrated with 0.02 M sodium phosphate buffer, pH 6.5, containing 1 mM DTE. The fractions containing enzyme activity are pooled and concentrated by ultrafiltration as above and dialyzed against 0.01 Tris-HCl buffer, pH 7.0, containing 1 mM DTE.

QAE-Sephadex. The protein solution is passed through a QAE-Sephadex (Q-25-120) column (10 × 1.5 cm) equilibrated with the same buffer. Under these conditions most of the pigments are retained, but cutinase is not. The fractions containing the enzymatic activity are pooled and dialyzed against 1% glycine containing 1 mM DTE.

Isoelectric Focusing. The protein solution is subjected to isoelectric focusing at 3° using a 110 ml column (LKB 8100 Stockholm) and 1% carrier ampholytes pH 4–6 as described in the LKB instruction manual. The sample is applied with a 0.15 to 1.5 M sucrose gradient containing 1 mM DTE. After applying 900 V for 46 hr, 1.5-ml fractions are collected. The fractions containing enzymatic activity are pooled, concentrated by ultrafiltration as before, and dialyzed against 0.02 M sodium phosphate buffer, pH 6.5, containing 1 mM DTE. Alternatively, the concentrate can be passed through a 30 × 1.5 cm Sephadex G-50 column to remove the Ampholine.

The procedure described here results in over 40-fold purification with a 14% recovery (Table III). Instability of the enzyme is a significant problem during purification; cation and anion exchange chromatography with SP-

TABLE III
PURIFICATION OF CUTINASE FROM NASTURTIUM POLLEN

Step	Protein (mg)	Enzyme activity (Δ_{405}/min)	Specific activity (Δ_{405}/min/mg)	Recovery (%)	Purification (fold)
Crude	304	3600	11.8	100	1.0
Sephadex G-100	23	2040	87.2	57	7.4
QAE-Sephadex	4.5	850	193.2	24	16.3
Isoelectric focusing	1.0	500	500.0	14	42.2

Sephadex, CM-Sephadex, CM-Cellulose, phosphocellulose, and DEAE-cellulose are not suitable because even in the presence of glycerol, which stabilizes the enzyme against inactivation, only a few percent of the enzymatic activity can be recovered after each step.

Purity. The enzyme is apparently homogeneous, as polyacrylamide gel electrophoresis with and without SDS shows a single protein band.

Properties of Pollen Cutinase

General Properties. This enzyme is a single 40,000 MW peptide with a pI of 5.45. It contains about 7% carbohydrates attached via alkali-stable linkage(s), presumably via asparagine. The proportion of acidic amino acid residues is much higher in this protein than that found in fungal cutinases.

With cutin as the substrate the enzyme shows a rather sharp pH optimum at 6.8. Linear rates are obtained up to 40 μg of protein per milliliter and 5 hr of incubation time. Rate of hydrolysis increases linearly up to about 6 mg of cutin per milliliter, and subsequent increase in substrate gives no increase in the rate.

With p-nitrophenyl butyrate as the substrate, the pH optimum is at 8.0 and linear rates are obtained up to 1.2 μg of enzyme per milliliter and 6 min incubation time.

Substrate Specificity. Pollen cutinase shows specificity for esters of primary alcohols, although esters of alkane-2-ol are also hydrolyzed at slower rates. Fatty acid esters of p-nitrophenol are hydrolyzed, and, unlike the fungal cutinases, the pollen enzyme shows comparable K_m and V for C_2 to C_{18} esters (Table II).

Inhibitors. Pollen cutinase is very sensitive to inhibition by thiol-directed reagents such as p-hydroxymercuribenzoate and N-ethylmaleimide, but it is completely insensitive to organic phosphates, such as diisopropylfluorophosphate, which reacts with "active" serine. In this regard the pollen enzyme is quite unlike that from fungi.

Phenylglyoxal inhibits the enzyme whereas metal ions and chelators have no effect.

Acknowledgments

This work was supported in part by Grant PCM 77-00927 from the National Science Foundation and a grant from the Washington Tree Fruit Commission. Drs. C. L. Soliday, T. S. Lin, and Mashouf Shaykh made valuable contributions to the methods described.

[77] Sterol Ester Hydrolase from Rat Pancreas

EC 3.1.1.13 Sterol-ester acylhydrolase

By LINDA L. GALLO

$$\text{Sterol + fatty acid} \xrightleftharpoons{\text{cholic acid}} \text{sterol ester}$$

Sterol ester hydrolase, commonly called cholesterol esterase, is secreted in the pancreatic juice by the exocrine pancreas. The enzyme catalyzes the synthesis and hydrolysis of sterol esters. In the intestinal lumen, hydrolysis predominates; in the intestinal mucosa, which takes up the esterase from the lumen, the equilibrium lies in favor of esterification.

Assay Methods

The activity of this enzyme is conveniently measured by radioactive or colorimetric assay. In both, the rate of appearance (cholesterol ester → cholesterol + fatty acid) or disappearance (cholesterol + fatty acid → cholesterol ester) of free cholesterol is determined.

Method I. Radiochemical Assay

Principle. Radioactive assay of this enzyme has been described previously in this series.[1] Briefly, either the amount of radioactive cholesterol ester formed by enzymatic esterification of [^{14}C]cholesterol with fatty acid in the presence of cholic acid or the amount of free [^{14}C]cholesterol formed by enzymatic hydrolysis of radioactive cholesterol ester in the presence of cholic acid is determined. During the incubation, aliquots of the enzyme–substrate mixture are removed and extracted in acetone : ethanol, 1 : 1 v/v, [^{14}C]cholesterol separated from [^{14}C]cholesterol

---

[1] G. V. Vahouny and C. R. Treadwell, this series, Vol. 15 [23].

ester by thin-layer chromatography on silicic acid, and the two lipid bands corresponding to authentic standards are counted for radioactivity in a liquid scintillation counter.

Method II. Colorimetric Assay

Principle. In the colorimetric assay, cholesterol is incubated with fatty acid and cholesterol esterase in the presence of the cofactor, taurocholate. During the incubation, aliquots of the enzyme substrate mixture are removed and added to isopropanol. The free cholesterol concentration in the isopropanol extracts is determined enzymatically using a single aqueous reagent containing cholesterol oxidase and peroxidase.

$$\text{Cholesterol} + O_2 \xrightarrow[\text{oxidase}]{\text{cholesterol}} \Delta^4\text{-cholestenone} + H_2O_2$$

$$H_2O_2 + \text{phenol} + \text{4-aminoantipyrine} \xrightarrow[\text{peroxidase}]{\text{horseradish}} \text{rose-colored product}$$

The cholesterol oxidase converts cholesterol into Δ^4-cholestenone and H_2O_2. Peroxidase couples the H_2O_2 with phenol and 4-aminoantipyrine to yield a stable rose-colored product that absorbs at 500 nm.

Reagents for Cholesterol Ester Synthesis

Lipid mixture
  Cholesterol, 15.5 mM
  Oleic acid, 93.0 mM in diethyl ether (anhydrous, peroxide-free)
Assay buffer mixture
  potassium phosphate, 0.154 mM, pH 6.2
  sodium taurocholate, 20 mM
  NH$_4$Cl, 66 mM
  Bovine serum albumin, fraction V, fatty acid poor, 4 mg/ml
Enzyme: pancreatic juice, 10 mg protein/ml; pancreatic homogenate, 10% in potassium phosphate buffer, 0.154 M, pH 6.2; or cholesterol esterase in potassium phosphate buffer, pH 6.2, obtained from any step of enzyme purification

Reagent for Cholesterol Determination[2,3]

Complete cholesterol reagent:
  Potassium phosphate, 0.1 M, pH 7.0
  Sodium azide, 2.0 mM
  Phenol, 14.0 mM

[2] W. Richmond, C. Allain, L. S. Poon, and S. G. Chan, *Clin. Chem.* **20,** 470 (1974).
[3] P. N. Tarbutton and C. R. Gunter, *Clin. Chem.* **20,** 724 (1974).

4-Aminoantipyrine, 0.82 mM
Triton X-100, 0.5 ml/liter
Cholesterol oxidase, 90 units/liter (Boehringer-Mannheim)
Horseradish peroxidase, 33,000 units/liter (Boehringer-Mannheim)
The complete cholesterol reagent is stable for at least 4 weeks when stored at 4°.

Preparation of Substrate.[4] For each enzyme assay to be performed, 1.5 ml of the lipid mixture are transferred to a Potter–Elvehjem homogenization tube. The ether is evaporated to dryness under a stream of nitrogen, 3 ml of assay buffer mixture are added, and the lipid–buffer components are homogenized to provide a homogeneous emulsion. It is most convenient to prepare 6–36 ml of substrate emulsion.

Procedure. Three milliliters of this emulsion containing 23.3 μmol of cholesterol, 139.5 μmol of oleic acid, 60 μmol of sodium taurocholate, 198.0 μmol of NH$_4$Cl, and 12.0 mg of bovine serum albumin are transferred to an Erlenmeyer flask (25 ml) and preincubated in a shaker bath at 37° for 15 min. The enzyme in 0.154 M potassium phosphate buffer, pH 6.2, is added to give a final incubation volume of 6.0 ml. Under these conditions, the rate of esterification is linear with incubation time within the limits of 5 and 50%. Enzyme activity must be diluted to fall within this range. One-milliliter samples are withdrawn initially and at three intervals during the total incubation time, which may vary from 15 to 90 min, depending on enzyme activity. Each aliquot withdrawn is added to approximately 3 ml of isopropanol contained in a 5-ml volumetric flask, the mixture is brought to a rolling boil (boileezers will prevent bumping), allowed to cool, and brought to volume with isopropanol. The extract is transferred to a centrifuge tube and centrifuged at 14,000 g for 10 min at 4° to precipitate denatured protein. Eighty microliters of the room-temperature supernatant are added to 3.0 ml of the complete cholesterol reagent in a colorimeter tube. The contents are mixed and allowed to stand for at least 10 min, but no longer than 90 min (color slowly fades after 90 min). Then the absorbance of the rose-colored complex is measured at a wavelength of 500 nm. The reagent blank is prepared to contain 80 μl of isopropanol. A standard curve is constructed by addition of 80 μl of each of several standard solutions of cholesterol (10–100 μg) in isopropanol to 3.0 ml of the complete cholesterol reagent.

Units. One unit of cholesterol esterase activity is defined as the amount of activity that converts or releases 1 μmol of cholesterol per hour.

Preparation of Enzyme.[5] Cholesterol esterase is conveniently prepared from fresh or frozen pancreas (rat, guinea pig, rabbit, cow, dog, human).

[4] L. L. Gallo and R. Atasoy, *J. Lipid Res.* **19**, 913 (1978).
[5] K. B. Calame, L. Gallo, E. Cheriathundam, G. V. Vahouny, and C. R. Treadwell, *Arch. Biochem. Biophys.* **168**, 57 (1975).

Eighty grams of pancreas trimmed of connective tissue are a convenient amount to process. A 20% homogenate of pancreas in 0.154 M NaCl containing 0.5% digitonin (add 5 N KOH to solubilize; then readjust pH to 6.2) is incubated with shaking for 30 min at 37°. After centrifugation at 100,000 g for 30 min, 80% of the total activity of the whole homogenate is recovered in the supernatant.

Purification Procedure

Method I.[5] Standard Chromatographic Purification

Step 1. Ammonium Sulfate Fractionation. This step and all subsequent steps are carried out at 4° except where noted. Solid ammonium sulfate is added to the digitonin extract of cholesterol esterase to give 30% saturation. After 20 min of stirring, the precipitated protein is removed by centrifugation at 15,000 g for 15 min. The supernatant from the centrifugation is adjusted with solid ammonium sulfate to give 45% saturation, and the stirring and centrifugation sequence is repeated. Approximately 60% of the enzymatic activity is recovered with a twofold purification in the precipitated fraction between 30 and 45% saturation. The precipitated protein is dissolved in a minimal amount (approximately 30 ml for 80 g of original pancreas) of 10 mM sodium phosphate buffer, pH 6.2, containing 0.05% soybean trypsin inhibitor. The protein fraction is desalted on a Sephadex G-25 column equilibrated with 10 mM sodium phosphate buffer, pH 6.2. This and all subsequent buffers, unless otherwise indicated, are 1 mM in dithiothreitol to protect enzyme protein sulfhydryl groups, and 0.02% in sodium azide to prevent bacterial growth. The dissolved protein from ammonium sulfate precipitation, prior to desalting, may be stored frozen for months without loss of enzymatic activity.

Step 2. Hydroxyapatite Chromatography. BioGel HT (Bio-Rad) is equilibrated with 10 mM sodium phosphate buffer, pH 6.2, and packed in a 4 × 50 cm column, bed volume, 400 ml. A constant flow rate of 30–35 ml/hr is maintained with a peristaltic pump. The desalted fraction (60–100 ml from 80 g of original pancreas) is applied to the column and 1.5 liter of 3 M KCl in 10 mM sodium phosphate buffer, pH 6.2, passed through to elute basic proteins. Then the enzyme is eluted by a linear sodium–potassium phosphate buffer (pH 6.2) gradient from 10 to 400 mM contained in a total volume of 2500 ml. The eluent is collected in fractions (10 ml/tube) and assayed for enzyme activity and protein.[6] The enzyme elutes near the end of the gradient. This step represents a 10- to 15-fold purification of the activity from the ammonium sulfate fraction with 40–60% recovery in the

[6] M. M. Bradford, *Anal. Biochem.* **72**, 248 (1976).

high specific activity fractions. The high specific activity fractions are pooled (250–350 ml) and concentrated to a volume of approximately 5 ml by Diaflo filtration through a PM-10 membrane.

Step 3. Sephadex G-200 Chromatography. The property of the enzymatically inactive cholesterol esterase subunits, molecular weight 70,000, to aggregate in the presence of taurocholate yielding the enzymatically active polymer, molecular weight 412,000, which disaggregates upon removal of the taurocholate, is exploited. Sephadex G-200 is equilibrated with 50 m*M* sodium phosphate buffer, pH 6.2, and packed in a 2.5 × 100 cm column, bed volume, 500 ml. The column is equilibrated and eluted with the same buffer. A flow rate of 6–8 ml/hr is maintained with a hydrostatic pressure of 10 cm. Enzyme subunits from the preceding step contained in approximately 5 ml of buffer are incubated in 5 m*M* sodium taurocholate (2.75 mg of solid sodium taurocholate added per milliliter of solution) for 5 min at 37° to produce the enzyme aggregate that is applied to the column. The column eluent is collected in fractions (5 ml/tube), which are assayed for activity and protein.[6] The enzyme elutes near the void volume, which is determined by measuring the elution volume of a 0.2% (w/v) solution of Blue-Dextran 2000. The enzymatically active fractions are pooled (50–100 ml), and the taurocholate is removed from the enzyme by treatment with the bile salt binding resin, cholestyramine (1 mg/ml pooled sample), for 30 min at 20°. The resin is removed by filtration, and the cholestyramine treatment and filtration are repeated.

The re-formed enzyme subunits in the filtrate are precipitated by addition of solid ammonium sulfate to give 90% saturation. The conditions for isolation of the precipitated protein are as described in step 1.

The protein is dissolved in 50 m*M* sodium phosphate buffer (dithiothreitol omitted), pH 6.2, and rechromatographed on Sephadex G-200 as described above.

The change in molecular size between subunit and active enzyme permits a 16-fold purification of subunit in the two successive gel filtration steps. The results of a typical purification are summarized in Table I and show a 500-fold increase in specific activity with a 4.7% recovery of enzymatic activity from the crude extract.

Method II. Affinity Chromatographic Purification

Preparation of Anti-Cholesterol Esterase IgG (Immune IgG).[7] New Zealand white rabbits are immunized by standard procedures[8] with choles-

[7] L. Gallo, E. Cheriathundam, G. V. Vahouny, and C. R. Treadwell, *Arch. Biochem. Biophys.* **191**, 42 (1978).
[8] J. S. Garvey, N. E. Cremer, and D. H. Sussdorf, "Methods in Immunology" Benjamin, New York, 1977.

TABLE I
SUMMARY OF THE PURIFICATION OF SUBUNITS OF STEROL ESTER HYDROLASE[a]

Step	Total enzyme units[b]	Enzyme specific activity[c]	Overall yield (%)
1. Digitonin-solubilized supernatant	45,024	5.0	100
2. $(NH_4)_2SO_4$ fraction, 30–45% saturation	25,810	9.6	57
3. Hydroxyapatite chromatography	12,600	163	25
4. Concentration on Diaflo PM-10	11,680	—	23
5. Sephadex G-200 chromatography	8,200	269	16
6. Removal of taurocholate with cholestyramine	4,880	—	9.7
7. $(NH_4)SO_4$ precipitation	3,810	—	7.6
8. Sephadex G-200 chromatography	2,380	2612	4.7

[a] Data from Calame et al.[5] Data are for 150 g of rat pancreas (wet weight).

[b] Micromoles of cholesterol esterified per hour.

[c] Specific activity = enzyme units per milligram of protein.

terol esterase purified by Method I. For each rabbit immunized, 200 μg of purified, active enzyme aggregate (MW 412,000) in 0.5 ml of 0.05 M sodium phosphate buffer, pH 6.2, are emulsified with an equal volume of Freund's complete adjuvant. Each rabbit is initially injected in each thigh muscle with 0.5 ml of emulsion. Booster immunizations follow at 3, 4, and 5 weeks after the primary immunization. Each booster injection contains 200 μg of cholesterol esterase in 0.5 ml sodium phosphate buffer, pH 6.2 and is administered via the ear vein. When double immunodiffusion assay[9] of the rabbit serum against the enzyme reveals antibody production (6 weeks after the primary immunization), large (40–50 ml of blood per rabbit) blood collections are made by ear arterial bleeding with a butterfly needle. Serum is separated from the cellular components of the blood by standard methodology,[8] and sodium azide (1 mg/ml) is added to arrest bacterial growth.

The anti-cholesterol esterase IgG is purified from serum by a two-step procedure. The first step employs batchwise ion-exchange chromatography on DEAE cellulose,[10,11] which removes 85% of the total serum protein as follows.

Preswollen and precycled anion-exchange cellulose DE-52 (Whatman) (3.2 g per milliliter of undiluted serum) is stirred into 0.02 M sodium phosphate buffer, pH 6.5 (5 ml per gram of resin), containing 0.02%

[9] O. Ouchterlony and L. A. Nilsson, in "Handbook of Experimental Immunology" (D. M. Weir, ed.), p. 19.1. Blackwell, Oxford, 1973.

[10] A. E. Reif, Immunochemistry 6, 723 (1969).

[11] J. Gergely, D. R. Stanworth, R. Jefferis, D. F. Normansell, C. S. Henney, and G. I. Pardow, Immunochemistry 4, 101 (1967).

sodium azide. The pH of the mixture is adjusted and kept at 6.5 with 1.0 N HCl. The mixture is allowed to settle for 30 min, and the supernatant and fines are decanted. The cycle of resuspension, settling, and fine decantation is repeated two additional times. The gel is then filtered on a Büchner funnel that contains two layers of Whatman No. 1 filter paper.

Immune serum is diluted with double-distilled water (three volumes of water per volume of serum) to reduce its ionic strength and stirred into the wet gel (4.4 g per milliliter of undiluted serum). The mixture is allowed to equilibrate for 1 hr at 4°, with manual stirring every 10 min. After equilibration, the mixture is filtered on a Büchner funnel containing two layers of Whatman No. 1 filter paper. The resin cake is rapidly washed with 0.02 M sodium phosphate buffer, pH 6.5, containing 0.02% sodium azide (the volume of buffer is equal to that of the original, undiluted serum) and filtered. This wash and filtration sequence is repeated seven times. The effluent from this step (approximately 500 ml per 50 ml of original serum) is concentrated 25-fold by ultrafiltration on a Diaflo PM-10 membrane.

The second step of the purification is gel filtration of the concentrated effluent on several columns packed with Sephadex G-200.[12] The gel for each column is prepared as follows: Eighteen grams of Sephadex G-200 are swollen (at room temperature for 72 hr) in 1.0 liter of double-distilled water containing 0.02% sodium azide. After equilibration, as described by the manufacturer (Pharmacia), with 0.15 M sodium chloride containing 0.02% sodium azide, the gel is packed in a 2.5 × 100 cm column, bed volume, 500 ml. The gel bed is stabilized by washing with two bed volumes of 0.15 M NaCl containing 0.02% sodium azide, at a hydrostatic pressure of 10 cm. The void volume of each column is determined by measuring the elution volume of a 0.2% (w/v) solution of Blue-Dextran 2000.

A 5-ml aliquot of the partially purified IgG concentrate from DEAE-cellulose chromatography is applied to each column and eluted with 0.15 M NaCl containing 0.02% sodium azide. The IgG peak is identified by double immunodiffusion assay,[9] with goat IgG fraction against rabbit IgG. The IgG containing fractions from this step are pooled and concentrated by ultrafiltration on a PM-10 membrane to half of the original serum volume. The purified IgG preparation contains 10 mg of protein per milliliter of solution. Seven percent of the original serum protein is recovered as IgG, and 5.0 mg of IgG are recovered per milliliter of original serum.

Coupling of Anti-Cholesterol Esterase IgG to Sepharose Matrix. CNBr-activated Sepharose (Pharmacia) is reswollen and washed according to the manufacturer with 0.001 M HCl (200 ml per gram of gel). Each gram of dry gel swells to 3.5 ml of matrix that is ready for coupling.

[12] K. F. Watson, R. C. Nowinski, A. Yaniv, and S. Spiegleman, *J. Virol.* **10**, 951 (1972).

Purified IgG in 0.15 M NaCl with 0.02% sodium azide is transferred into coupling buffer (0.1 M sodium phosphate buffer in 0.5 M NaCl with 0.02% sodium azide, pH 6.5) by passing it through Sephadex G-25 in a 1.6 × 30 cm column, bed volume 30 ml, equilibrated with the coupling buffer. The IgG solution in coupling buffer is then mixed with the reactivated Sepharose matrix (10 mg of IgG per milliliter of reswollen, reactivated matrix), in approximately equal volumes. The mixture is rotated end-over-end in a container attached to a slowly revolving wheel at 4° for 24 hr.

The IgG coupled to the matrix is washed on a sintered-glass filter apparatus with several solutions as follows: coupling buffer (500 ml), ethanolamine at pH 9.0 (a volume equal to twice the volume of each individual IgG-matrix mixture), coupling buffer (500 ml), 0.1 M sodium acetate buffer in 0.5 M NaCl with 0.02% sodium azide, pH 4.0 (500 ml), and a final wash with coupling buffer (500 ml).

The wash effluents are concentrated by ultrafiltration on a Diaflo PM-10 membrane, and the protein content is assayed in order to estimate the amount of IgG that does not bind covalently to the gel.

Step 1. Ammonium Sulfate Fractionation. This step and subsequent steps are carried out at 4°. A 30–45% ammonium sulfate fraction is prepared from the digitonin extract of cholesterol esterase from pancreas, desalted by Sephadex G-25 chromatography, and concentrated by ultrafiltration as described in steps 1 and 2, Method I.

Step 2. Affinity Chromatography. [13] Sepharose matrix (21.5 g dry; 75 ml swollen) covalently bound to immune IgG (1025 mg), i.e., anti-cholesterol esterase IgG, is packed in a 2.6 × 40 cm column, bed volume, 75 ml. A constant flow rate of 12 ml/hr is maintained with a peristaltic pump. The column is equilibrated with two bed volumes of starting buffer (0.1 M sodium phosphate buffer in 0.5 M sodium chloride with 0.02% sodium azide, pH 6.2). The desalted ammonium sulfate concentrate containing cholesterol esterase (approximately 500 units) in starting buffer (1–5 ml) is applied.

After the sample enters the column bed, the flow is stopped for 1 hr to allow time for immune IgG-enzyme binding. Then the column is washed with two bed volumes of starting buffer to remove protein (enzymatic and nonenzymatic) which has not bound. The enzyme is eluted with two bed volumes of eluting buffer (0.35 M potassium phosphate buffer in 0.5 M sodium chloride with 0.02% sodium azide, pH 6.2). The column eluent is collected in fractions (5 ml/tube) which are assayed for activity and protein.[6] The results of a typical purification are summarized in Table II and show that the single affinity step provides a 38-fold purification of the

[13] P. W. Jacobson and L. Gallo, unpublished results, 1979.

TABLE II
SUMMARY OF AFFINITY PURIFICATION PROCEDURE[a]

Step	Total activity (units)[b]	Total protein (mg)	Specific activity[c] (units/mg)	Yield (%)
1. (NH₄)₂SO₄ fraction, 30–45%	500	34.3	14.5	100
2. Sepharose-immune IgG	483.5	0.85	569	96.7

[a] Data for an aliquot of an $(NH_4)_2SO_4$ fraction equivalent to 4.0 g of rat pancreas (wet weight).
[b] Micromoles of cholesterol esterified per hour.
[c] Specific activity = enzyme units per milligram of protein.

activity in the ammonium sulfate fraction with a recovery of 96.7% of the applied activity. The overall yield from the digitonin-solubilized enzyme is 55%, the only loss of activity occurring during ammonium sulfate fractionation.

The column is regenerated by the passage of two bed volumes of starting buffer and can be reused repeatedly without deterioration of the enzyme binding capacity of the immobilized IgG.

NOTE: The amount of IgG required to bind 500 units of enzyme activity will depend upon the avidity of the particular IgG used.

Properties

Purity of the Enzyme. The purity of the cholesterol esterase preparation is tested by SDS–polyacrylamide gel electrophoresis (pH 7.1). The product from either purification scheme[5,13] moves as a single protein band upon staining when 64 μg of protein are applied to the gel. The difference in specific activity calculated for the enzyme purified by Method I (specific activity = 2612) and Method II (specific activity = 569) represents, in the former case, a determination based upon amino acid composition of the purified enzyme and, in the latter case, upon protein content as determined by protein assay. The specific activities of the purified products from either procedure are comparable when a regular protein assay is employed to determine protein content.

Stability. Rat pancreas cholesterol esterase is more stable in the aggregated than in the subunit form.[5] The aggregate is protected from inactivation due to pH changes,[14] trypsin or chymotrypsin proteolysis,[15] extremes

[14] S. K. Murthy and J. Ganguly, *Biochem. J.* **83**, 460 (1962).
[15] G. V. Vahouny, S. Weersing, and C. R. Treadwell, *Biochim. Biophys. Acta* **98**, 607 (1965).

TABLE III
AMINO ACID COMPOSITION OF THE SUBUNITS OF STEROL ESTER HYDROLASE FROM RAT
PANCREAS AND RAT PANCREATIC JUICE

	Sterol ester hydrolase subunits		
	Pancreatic tissue[5]		
Amino acid	Residues/mole	Next integers	Pancreatic juice[a16]
Alanine	56.8	57	71
Arginine	32.7	33	31
Aspartic	76.1	76	71
Cysteine	11.3	11	6
Glutamic	67.0	67	63
Glycine	73.0	73	67
Histidine	14.2	14	12
Isoleucine	29.7	30	30
Leucine	45.2	45	49
Lysine	35.0	35	42
Methionine	2	2	10
Phenylalanine	25.3	25	26
Proline	42.7	43	39
Serine	65.9	66	56
Threonine	42.1	42	36
Tyrosine	0	0	0–2[b]
Valine	43.9	44	49

[a] Using an average molecular weight of 69,000.
[b] Values of 0 and 2 were obtained in two separate analyses.

of temperature (50°),[5] and guanidine hydrochloride treatment,[5] whereas the subunit undergoes rapid inactivation. The purified enzyme aggregate has been stored for 10 months at −20° without loss of activity (in buffer, no dithiothreitol).

Physical Properties. Cholesterol esterase is secreted in the pancreatic juice as an enzymatically inactive subunit with an apparent molecular weight of 70,000. In the presence of physiological concentrations (5 mM) of cholic acid or its conjugates (other bile salts not effective), the subunits aggregate to form an enzymatically active hexamer with an apparent molecular weight of 412,000.[5,16] Investigation of the kinetics of subunit aggregation and activation reveals a half-time in the range of 15–20 sec. Studies of cholic acid binding to the subunit suggest the occurrence of a cooperative process when the moles of cholic acid bound per mole of subunit is unity. This is evidenced by a large increase in cholic acid binding over a narrow cholic acid concentration range.[5] Thus, upon binding of

[16] J. Hyun, C. R. Treadwell, and G. V. Vahouny, *Arch. Biochem. Biophys.* **152**, 233 (1972).

cholic acid to available sites on the subunit protein, a conformational change in the subunit protein is suggested that makes additional cholic acid binding sites available.

Chemical Properties. The amino acid composition of enzyme purified from pancreas and pancreatic juice has been determined and is shown in Table III. Tyrosine is notably absent.

Inhibition studies[16] reveal that phenylmethylsulfonyl fluoride and *p*-chloromercuribenzoate inhibit cholesterol esterase activity, suggesting that serine and cysteine residues are essential. Additional activators and inhibitors of enzymatic activity have been discussed previously in this series.[1]

Immunological Comparison of Cholesterol Esterases.[7] Rat pancreas cholesterol esterase has been immunologically compared with cholesterol esterase solubilized from rat intestine, aorta, adrenal, and liver and with cholesterol esterase from the pancreas of rabbit, dog, cow, and guinea pig. Anti-rat pancreas cholesterol esterase serum cross-reacts with the cholesterol esterase of intestine with a pattern of complete immunological identity in the immunodiffusion assay.[9] However, cholesterol esterases, isolated from the other rat tissues and from the pancreas of the other species, give no evidence of cross-reactivity.

Optimum pH. The optimum pH for esterification is 6.2, and for hydrolysis 6.6, in the assay methods described.

Acknowledgment

This work was supported by NIH Grants HL 02033 and AM 17269 from the U.S. Public Health Service.

[78] Phospholipases A₁ from Lysosomes and Plasma Membranes of Rat Liver

EC 3.1.1.32 Phosphatidate 1-acylhydrolase

By Moseley Waite, R. Hanumantha Rao, Harry Griffin, Richard Franson, Craig Miller, Patricia Sisson, and Jeanie Frye

Phospholipases A_1 are hydrolytic enzymes that catalyze the removal of the acyl group from position 1 of phosphoglycerides. The two phospholipases A_1 described in this chapter have absolute positional specificity but will degrade more than one lipid. Within the phospholipid class, phosphatidylethanolamine is the substrate most readily attacked. These two phospholipases A_1 vary considerably in their mode of action and undoubtedly in their physiological function. The plasmalemma enzyme readily acts on neutral glycerides, in particular the monoacylglycerols. Under certain conditions it will act on triacylglycerols and probably is the same as the enzyme isolated from plasma termed the hepatic triacylglycerol lipase.[1,2]

Assay Methods

The substrates for the assay of these enzymes are prepared in the laboratory, although some commercially prepared substrates are available (Applied Science). Studies on the specificity of the enzymes for acyl groups indicate that phosphatidylethanolamine containing unsaturated acyl chains (at least at position 2) is more useful than phosphatidylethanolamine containing saturated fatty acids. The same is true when monoglyceride is used to assay the phospholipase A_1 of the plasma membrane.

The Preparation of $[1\text{-}^3H]\text{-}$ or $[1\text{-}^{14}C]Palmitoyl\text{-}2\text{-}acyl\text{-}sn\text{-}3\text{-}glycerophosphorylethanolamine$ ($[1\text{-}^3H]Phosphatidylethanolamine$)

The 2-acyl lysophosphatidylethanolamine is prepared by I_2 cleavage of plasmalogen phosphatidylethanolamine (Serdary) based on the procedure originally described by Lands and Merkl.[3] Ten milligrams of plasmalogen phosphatidylethanolamine are dissolved in 1.5 ml of 33% methanol in chloroform and then added to 1 ml of 20 mM phosphate buffer, pH 6.6, that contains 90 μmol of iodine (this does not completely dissolve). After 5 min the reaction is terminated by the addition of 3 ml of ethanol and sufficient $Na_2S_2O_3$ to decolorize the solution. The products are immediately extracted for reacylation with 3 ml of chloroform in order to minimize migration of the acyl group. To this 1 mCi of [9,10-^{3}H]palmitic acid or [1-^{14}C]palmitic acid [specific activity, 12 mCi/μmol or 52 μCi/μmol, respectively] is added and the solvents are blown off with N_2. Ten milliliters of a solution that contains 100 mg of ATP, 10 mg of CoA, 10 μmol of $MgCl_2$, and 1 mmol of Tris-HCl, pH 7.4, is added to the dried lipid and the mixture is sonicated under N_2 until the lipid is well dispersed (usually 5 min). A 2 ml suspension of freshly prepared rat liver microsomes (15 mg/ml) resuspended in 0.1 M Tris-HCl, pH 7.4, is added to the sonicated suspension and the mixture is incubated under N_2 for 30 min at 37° with

[1] H. Jansen and W. C. Hulsmann, *Biochim. Biophys. Acta* **369**, 387 (1974).
[2] C. Ehnholm, W. Shaw, H. Greten, and W. V. Brown, *J. Biol. Chem.* **250**, 6756 (1974).
[3] W. E. Lands and I. Merkl, *J. Biol. Chem.* **238**, 898 (1963).

shaking. The reaction is terminated by the addition of 24 ml of methanol, 12 ml of chloroform is added, and the mixture is gently shaken. This mixture is then broken into two phases by the addition of 12 ml of water and 12 ml of chloroform, the chloroform layer removed, the methanol–water phase washed two times with chloroform, and the combined chloroform layers dried over Na_2SO_4. The lipids are then dried using a rotary evaporator and redissolved in 2–3 ml of chloroform for spotting on two 5 mm silica gel H plates. The plates are then developed in chloroform–methanol–water–acetic acid, 60:40:2:4 (by volumes) (system I), dried under N_2 and the lipids visualized with an ultraviolet lamp. No fluorescent indicator is necessary owing to the high amount of lipid present. Usually phosphatidylethanolamine has an R_f of 0.60; this will vary, however, with atmospheric temperature and humidity. The silicic acid that contains phosphatidylethanolamine is scraped from the plate and extracted with 6 ml of a mixture of chloroform–methanol–water, 1:2:1 (v/v/v). When well mixed, 1 volume of water and 1 volume of chloroform are added, and the lipid containing chloroform is removed. The silicic acid and methanol–water layer is washed twice with chloroform; the combined chloroform layer is filtered through a paper filter and dried over Na_2SO_4. This can be stored at $-20°$ for 3–4 months without appreciable degradation; for longer storage, liquid N_2 or a $-70°$ freezer is recommended. The purity of the preparation should be checked by chromatography in the original system as well as a system that better separates phosphatidylethanolamine from phosphatidylserine [chloroform–methanol–7 M NH$_4$OH, 65:35:5 (v/v/v)]. Any contaminating phosphatidylserine can be removed by DEAE column chromatography.[4] The preparation is usually better than 95% pure with 90% of the [^{3}H]palmitic acid at position 1, as determined by the degradation of the phosphatidylethanolamine by snake venom phospholipase A_2.[5] The [^{3}H]phosphatidylethanolamine is diluted with nonradioactive egg yolk or rat liver phosphatidylethanolamine to a final specific radioactivity of 2×10^5 cpm/μmol prior to use. Phosphatidylethanolamine that contains [1-^{14}C]linoleic acid at position 2 can be prepared in exactly the same manner except that 5 μmol of 1-acyl-sn-3-glycerophosphorylethanolamine is used as the substrate. This is commercially available from several companies or can be prepared by hydrolysis of phosphatidylethanolamine by the phospholipase A_2 in snake venom.[5]

The Preparation of Monoacylglycerols from [2-^{3}H]Glycerol

A simple assay for the plasma membrane phospholipase A_1 is based on the degradation of 1-oleoyl-[2-^{3}H]glycerol. Since no commercial prepara-

[4] G. Rouser, G. Kritchevsky, A. Yakamoto, G. Simon, C. Galli, and A. J. Bauman, this series, Vol. 14, p. 272.
[5] M. Waite and L. L. M. van Deenen, *Biochim. Biophys. Acta* **137**, 498 (1967).

tion of this is available at this time, it too is prepared in the laboratory. The simplicity of the assay more than justifies the commitment to the synthesis of the substrate.

To prevent polyacylation, the glycerol was first protected as its acetone ketal, then acylated, and finally deprotected. Overall yields of monoacylglycerol via this sequence are such that it may be equally efficient simply to acylate the unprotected glycerol and chromatographically separate the polyacyl products from the monoacyl product. This latter approach was not investigated.

[2-³H]Glycerol Acetone Ketal (2,2-Dimethyl-4-tritio-4-hydroxymethyl-1,3-dioxolane). The procedure reported here is an adaptation of one used by Baer and Fischer[6] to protect mannitol. A 5 mCi sample of [2-³H]-glycerol (Amersham) was diluted with 0.20 mol (18.4 g) of unlabeled glycerol and combined with a previously filtered solution of 30 g (0.22 mol) of zinc chloride (anhydrous) in 90 ml of dry (distilled from anhydrous calcium chloride) acetone. The resulting solution was stirred for 12 hr at ambient temperature and treated with a solution of 45 g of potassium carbonate in 60 ml of water. This produced a lumpy, white precipitate, which was broken up with a glass rod and overlaid with 300 ml of ether. The mixture was stirred intermittently for 3 hr and suction-filtered. The filtrate and a 400-ml ether–acetone, 1 : 1 (v/v) wash of the filter cake were combined and dried over anhydrous potassium carbonate. Rotary evaporation of the volatile materials and fractional distillation of the resulting residue at 12 mm pressure produced 11.9 g of glycerol acetone ketal (bp 84.5–86°) spectroscopically and chromatographically identical to a known sample. The activity of the product was 2.6×10^4 cpm/μmol.

Monoacyl [2-³H]Glycerols. A 3.0 mmol sample of the [2-³H]glycerol acetone ketal was dissolved in 6 ml of dry (distilled from potassium hydroxide) pyridine and cooled to 0° under a nitrogen atmosphere. As this solution was magnetically stirred, it was treated dropwise (by syringe) with a solution of 3.3 mmol of the acyl chloride (Nu Chek Prep) in 6 ml of dry [stored over Na Pb alloy, "dri Na" (Fisher)] benzene. The reaction mixture was warmed and stirred at ambient temperature for 3 hr. It was then diluted with 60 ml of benzene and washed successively with two 50-ml portions of 1 N hydrochloric acid and one 50-ml portion of 3% aqueous sodium bicarbonate. The benzene extract was dried over anhydrous sodium sulfate and rotary evaporated.

In order to remove the acetone ketal the residue from evaporation was dissolved in 25 ml of acetone–methanol, 2 : 1 (v/v) and combined with 5 ml of 1 N hydrochloric acid. This mixture was refluxed under nitrogen for 1.5 hr, cooled, and poured into 10 ml of saturated aqueous sodium bicarbonate. Two 60-ml chloroform extracts were taken, dried (anhydrous sodium sulfate) and evaporated. Column chromatographic purification of

[6] E. Baer and H. O. L. Fischer, *J. Biol. Chem.* **128,** 463 (1939).

the residue on silica gel with an increasing gradient of ether in hexane produced a pure monoacyl [2-³H]glycerol product. Monomyristin, monostearin, monoolein, monolinolein, and monolinoelaidin were prepared from [2-³H]glycerol acetone ketal in this manner, and yields averaged 50%.

Reagents for the Assays

Sodium acetate-acetic acid (Fisher)
Tris-HCl (Fisher)
CaCl₂ (Fisher)
EDTA (Fisher)
Chloroform and methanol for extraction and chromatography are distilled before use

Lysosomal Phospholipase A₁

Assay of the Lysosomal Phospholipase A₁

A substrate cocktail for the assay is prepared as follows. For each incubation, 100 nmol of [³H]phosphatidylethanolamine in the chloroform solution is added to a thin-wall round-bottom test tube. The chloroform is removed under a stream of N₂, and 0.2 ml of 50 mM sodium acetate buffer, pH 4.0, containing 5.0 mM sodium EDTA is added for each incubation. This is then sonicated for 2 min in a bath sonicator to suspend the substrate. Usually 2–4 ml of the cocktail is prepared for a series of assays. A 0.2-ml aliquot of the cocktail is then removed and added to a conical 12-ml test tube, then the enzyme sample and water are added to bring the total volume to 1.0 ml. The tubes are incubated at 37° in a shaking water bath for a time period ranging up to 30 min. The amount of protein added ranges from about 1 μg (most highly purified fractions) to 25 μg (crude lysosomes). After the incubation, 3 ml of a mixture of chloroform–methanol, 1 : 2 (v/v) is added and the lipids are extracted into chloroform by the addition of 1.0 ml of water and 1.0 ml of chloroform. The chloroform solution is evaporated with N₂, and the lipid is redissolved in 0.2 ml of chloroform and spotted on a 2 cm strip of a 20 × 20 silica gel G plate along with standard lysophosphatidylethanolamine, phosphatidylethanolamine, and free fatty acids, and the plate is developed in system I. After the plates are dried, the lipids are visualized by I₂ vapor, and the silicic acid containing lysophosphatidylethanolamine, phosphatidylethanolamine, and free fatty acid are scraped into scintillation vials and counted in toluene–Triton X-100–water, 2 : 1 : 0.2 (v/v/v) that contains 5 g of Omnifluor (New England Nuclear) per liter. The phospholipase A₁ activity is taken as the percentage of free fatty acid in the total of the three compounds recovered. In some situations it is important to use a mixed labeled substrate ([1-³H]palmitoyl-[2-¹⁴C]linoleoyl-*sn*-3-glycerophosphorylethanolamine) as the substrate. This quantitates phospholipase A₁ (products: ³H-labeled fatty acid and [^{14}C]lysophosphatidylethanolamine), phospholipase A₂ (products: ^{14}C-labeled fatty acid and [^{3}H]lysophosphatidylethanolamine), and lysophospholipase (estimated as the difference between the total amount of free fatty acid and lysophosphatidylethanolamine recovered). All three activities are present in lysosomes. The reaction is fairly linear until about 25% of the substrate is hydrolyzed; beyond that point, product inhibition becomes a major problem.

Purification

Reagents

Triton WR 1339 (Rutgers Chemical Corp.)
Amicon concentrator with YM-10 filter
Spectra grade n-butanol (Fisher)
Fraction collector and ultraviolet monitor
Ultrogel AcA 34 (LKB Instruments)
Ampholine column, 110 ml, and ampholytes (LKB Instruments)
DEAE BioGel A (Bio-Rad)

Isolation of Lysosomes. Initial attempts to isolate the lysosomal phospholipase A₁ from whole rat liver homogenates failed because it is very difficult to find detectable activity in the homogenates. This is due to the inhibitory effect of soluble proteins on the phospholipase A₁.[7] (It is necessary, therefore, to isolate the enzyme from purified lysosomes.) For this, the use of Triton WR-1339 to alter the density of the lysosomes is invaluable.[8] Twenty to forty male Sprague–Dawley rats (180–200 g) are injected in the tail vein with 0.9 ml of a 20% (w/v) Triton WR 1339 in a saline solution. After 4 days the rats are killed, and the livers are removed, minced, and homogenized in 80 ml per rat of ice-cold 0.25 M sucrose containing 20 mM Tris-HCl, pH 7.4, using a Potter–Elvehjem homogenizer. Four passes are made with the pestle turning at 200 rpm. All subsequent procedures are carried out at 0–4°. The homogenate is then centrifuged at 200 g for 10 min, and the pellet containing cellular debris and nuclei is washed twice with a total of 80 ml of 0.25 M sucrose per rat. The combined supernatant fluids are then centrifuged at 16,000 g for 30 min to obtain the crude lysosomal pellet. This pellet is resuspended in a total volume of about 5 ml per rat liver with 45% sucrose (w/v, final density), which provides the bottom layer of a discontinuous sucrose gradient. Eighteen milliliters of the resuspended lysosomal pellet is pipetted into a tube for the Beckman SW27 rotor. A gradient is formed over the resuspended lysosomes that consists of 9 ml of 34.5% (w/w) and 9 ml of 14.3% (w/w) sucrose. The preparation is then centrifuged at 27,000 rpm for 2 hr, and the lysosomes filled with Triton WR 1339 float to the 14.3%/34.5% sucrose interface. The

[7] M. Waite, H. D. Griffin, and R. Franson, *in* "Lysosomes in Biology and Pathology" (J. T. Dingle and R. T. Dean, eds.), Vol. 5, p. 257. North-Holland Publ., Amsterdam, 1976.
[8] A. Trouet, this series, Vol. 31, p. 323.

lysosomes are collected with a Pasteur pipette, diluted with an equal volume of 0.25 M sucrose, and centrifuged at 100,000 g for 20 min. The resulting pellet is resuspended in 20 mM Tris-HCl and dialyzed overnight against 10 mM Tris-HCl, pH 7.4, that contains 1 mM EDTA to remove the sucrose and rupture the lysosomes. Based on marker enzyme analysis, the resuspended lysosomes are purified about 40-fold over the homogenate. The dialysis procedure is crucial to the subsequent steps of purification, since it apparently allows some degradation of phospholipase A_1 that results in a reasonably homogeneous form of the enzyme as determined by isoelectric focusing (to be discussed later).

Lipid Extraction. The lysosomal preparation is removed from the dialysis tubing, and the membranes are sedimented by centrifugation at 100,000 g for 1 hr. The soluble fraction that contains all the detectable phospholipase A_1 also contains lipids and Triton WR-1339 that must be removed if detailed studies are to be done on the interaction of the enzyme with lipid substrates. If this type of study is not a goal, this step may be omitted since the presence of trace amounts of lipid does not influence the purification, as far as we have been able to determine.

To remove the lipid, the soluble fraction is diluted to a protein concentration of 1 mg/ml with 10 mM Tris-HCl, pH 7.4. One-half volume of n-butanol at $-20°$ is added dropwise with continuous stirring as the solution is maintained at $-2°$. After 1 hr of stirring, the phases are separated by centrifugation and the aqueous phase is reextracted for 1 hr with n-butanol saturated with water. The aqueous layer is then concentrated severalfold using the Amicon concentrator with a YM 10 filter and dialyzed overnight against two changes of 50 mM Tris-HCl, pH 7.4.

DEAE-BioGel Chromatography. After dialysis the preparation is applied to a 1 × 20 cm column of DEAE-BioGel A that has been equilibrated with 50 mM Tris-HCl, pH 7.4. The column is initially washed with the Tris-HCl buffer at a flow rate of 8–10 ml/hr until no ultraviolet absorbing material is detected at 280 nm using an LKB ultraviolet monitor. This usually is 30–35 ml. The column is then washed with a linear 0 to 100 mM NaCl gradient in the Tris-HCl buffer made from two 80-ml reservoirs. The phospholipase A_1 elutes with about 40 mM NaCl, slightly ahead of the phospholipase A_2. The phospholipase A_1 in preparations of lysosomes that are not exhaustively dialyzed in the initial step does not bind tightly to the column and is eluted prior to the initiation of the NaCl gradient. The fractions containing the phospholipase A_1 are pooled and concentrated to 5–6 ml for the gel filtration step.

Gel Filtration on BioGel AcA 34. Five milliliters of the concentrated phospholipase A_1 preparation is applied to a 2.5 × 90 cm column of BioGel AcA 34 that is equilibrated with 50 mM Tris-HCl, pH 7.4. The gel filtration is carried out at a flow rate of about 10 ml per hour, and a convenient fraction size is 3 ml. The major peak of phospholipase A_1

elutes slightly after the major protein peak but before any remaining phospholipase A_2. The phospholipase A_1 is somewhat diffuse as judged by comparison with standard protein or by comparison with phospholipase A_2. Consequently, the investigator will have a choice to make; it is possible to have either a near quantitative recovery with only a two- to threefold purification or a recovery of 40–50% with a five- to six-fold purification. If the latter is chosen, it is possible essentially to free the preparation of contaminating phospholipase A_2 and lysophospholipase at this point. The pooled fractions are then concentrated to 5 ml and used for the isoelectric focusing.

Isoelectric Focusing. The density gradient for the isoelectric focusing is 5% to 50% sucrose (w/w) that contains 2.77% (v/v) of pH 4–6 ampholine and 0.27% (v/v), pH 6–8 ampholine. The anode solution is 1% H_2SO_4 in 60% sucrose (w/w). The anode buffer layer is 1.25% (v/v) of pH 2.5–4 ampholine in 60% sucrose (w/w) and the cathode solution consists of 2.5% (v/v) of pH 6–8 ampholine. The focusing column is filled as follows: Five milliliters of the anode solution is introduced into the anode chamber and is released into the focusing chamber. The anode chamber is then closed, and 10 ml of the anode solution is used to fill the anode chamber. Next, 5 ml of the anode buffer is carefully pipetted onto the anode solution through the focusing chamber, and the chamber is filled with the sucrose gradient that contains the pH 4–6 ampholine. The gradient is formed using a two-chambered linear gradient former that is pumped with a peristaltic pump. Approximately 10 ml of the cathode buffer is then used to cover the cathode wire. The anode chamber is then opened, and the column is prefocused at 15 W (up to 1600 V) for 1 hr. To apply the sample, the power is cut, the chamber is opened, a 5.0-ml aliquot of the solution is removed from the column one-third the distance from the top; the sample in a mixture of 1% of pH 4–6 and 1% of pH 6–8 ampholine in sucrose is added through a syringe at the point where the gradient sample was removed. The amount of sucrose added to the solution is sufficient to equal the concentration of the solution removed, as determined by its refractive index. The column is then focused for 14–18 hr at 15 W with the voltage ranging up to 1600 V. The power is then cut, and 2.0-ml fractions are collected from the bottom of the column. The major band of activity is pH 4.8, but a considerable amount is spread toward the anode. This spread is dependent on the initial preparation and dialysis of the lysosomal preparation as described later.

It is somewhat easier to mix the enzyme with the initial sucrose gradient solution. However, the recovery of phospholipase A_1 activity is somewhat lower and the peak of activity is more diffuse. It is possible to shorten the focusing time to 8 hr without influencing the results; however, an overnight run is convenient.

Results of the Purification Procedure and Enzyme Stability. The results

TABLE I
PURIFICATION OF PHOSPHOLIPASE A_1 FROM RAT LIVER

Step	Protein (mg)	Total activity (nmol/min)	Specific activity (nmol/mg/min)	Purification (fold)	Recovery (%)
Crude lysosome	216	13,824	64		100
Soluble fraction	123	16,515	134	2.1	120
Butanol extract	72	6,658	91	1.4	48
DEAE	6.0	2,331	387	6.1	17
Ultrogel AcA 34	1.36	1,574	1157	18.0	11
Isoelectric focusing	0.063	403	6383	100.0[a]	3

[a] With respect to homogenate, 4000-fold.

of a typical purification of the phospholipase A_1 from rats is given in Table I. In general, the enzyme is purified about 100-fold from the lysosomal fraction. Considering that the lysosomes are about 40-fold purified from the crude homogenate, the phospholipase A_1 is roughly 4000-fold purified over the homogenate.

The yield of activity and quantity of protein recovered is rather low, ranging from 2 to 6% of the original activity. There are two major reasons for this. First, although the enzyme is stable for periods of 2–3 months in ice or when frozen during the initial stages, after isoelectric focusing the activity is lost in a matter of days. In pilot studies, we have found that some component of the ampholines appears to cause this inactivation, and we are currently working on methods to rapidly remove all traces of the ampholines. The second reason for low recovery of the enzyme from ion exchange and gel filtration is the selection of the fraction for the subsequent steps. When the activity from the column is totaled, we usually recover 70–90% of the original activity. However, it has been our experience that it is not possible to increase the yields significantly without a sacrifice in purity that is unacceptable for our purposes. Sodium dodecyl sulfate (SDS)–polyacrylamide gel electrophoresis of the final product suggests that the enzyme is better than 75% pure, and in some preparations only a single band on these is found. In those cases the final specific enzymatic activity is about 6–7 μmol/mg per minute.

Properties of the Enzyme

Isoelectric Point and Molecular Size. The phospholipase A_1 is heterogeneous as shown by isoelectric focusing, gel filtration, and electrophoresis on SDS gels. The pI range from about 2.8 to 4.8, depending on the treatment of the lysosomes. If the lysosomes are frozen and thawed once, the membranes removed by centrifugation, and the soluble fraction immediately isoelectrically focused, very little of the form with a pI = 4.8 is recovered; the most is spread between 2.8 and 4.2. Although we have not yet been able to define the exact nature of the change to the enzyme molecule, it appears most likely that either saccharidases or proteases in the lysosomes are partially degrading the phospholipase A₁ (interestingly, the character of the phospholipase A₂ is unchanged). This does not alter the recovery of activity, however. Likewise, the size of the enzyme is altered somewhat; the apparent molecular weight of the enzyme ranges from about 68,000 to a minimum of 40,000 as determined on Ultrogel. However, these values must be regarded with great caution since SDS-gel electrophoresis gives a molecular weight as low as 32,000. This difference in molecular size, we believe, is the result of the presence of a component, such as carbohydrate, that causes the molecule to be rather asymmetric. This component apparently is lost or modified during the autolysis, accounting for the change in the pI and the size of the enzyme.

Characteristics of the Enzyme Preparation. The pH optimum of the reaction is 4.0; only one-half of the maximal activity is found at pH 3.0 or pH 5.0, and essentially no activity is found above or below those values. Metal ions inhibit the reaction, and for that reason EDTA usually is added to the reaction mixture. For example, 1.0 mM FeCl₃ completely inhibits, 10 mM CaCl₂ gives 50–60% inhibition, and 100 mM NaCl causes 20–30% inhibition. This inhibitory effect is related to the surface charge on the substrate liposome, as determined by microelectrophoresis.

The preferred substrate is phosphatidylethanolamine; phosphatidylcholine is degraded at one-third the rate of phosphatidylethanolamine. The acidic phospholipids, such as phosphatidylserine or phosphatidylinositol, are not degraded to any significant extent. The addition of charge amphipaths, such as dicetyl phosphate and cetyl trimethylammonium bromide, regulates the activity of the enzyme by changing the charge on the liposome, as determined by microelectrophoresis.[9] This does not account for the observed substrate specificity, however, since the adjustment of the surface charge on phosphatidylcholine liposomes with dicetyl phosphate to equal that of phosphatidylethanolamine does not allow equal hydrolytic rate by the phospholipase A₁.

Both products of the reaction, lysophosphatidylethanolamine and free fatty acid, inhibit the reaction. This appears to be more than a surface dilution effect since 30 nmol of lysophosphatidylethanolamine or 40 nmol of free fatty acid (compared with 100 nmol of substrate) causes about 75% inhibition. This probably accounts for nonlinear kinetics above 20–25%

[9] R. Franson and M. Waite, *Biochemistry* **17**, 4029 (1978).

hydrolysis. Likewise some soluble proteins significantly inhibit the reaction. For example, 10 μg of bovine serum albumin or 50 μg cytochrome c per milliliter causes a 50% inhibition. Likewise, the proteins in a 100,000 g supernatant fraction from rat liver are inhibitory, probably accounting for our inability to find lysosomal phospholipase A_1 activity in crude homogenates. This is not true for all proteins, however; ovalbumin at a concentration of 50 μg/ml does not have any effect.

Plasmalemma Phospholipase A_1 (MGAT)

This phospholipase A_1 is more active on monoacylglycerol than phospholipid. Also, it actively catalyzes a transacylation if a lipid acyl acceptor is present. For these reasons, we term the enzyme monoacylglycerolacyltransferase (MGAT) and generally use monoacylglycerol as the substrate. Hereafter the enzyme is referred to as MGAT.

Assay of MGAT

Hydrolysis

$$[1\text{-}^3H]\text{Phosphatidylethanolamine} + H_2O \rightarrow$$
$$[^3H]\text{fatty acid} + \text{lysophosphatidylethanolamine} \quad (1)$$
$$[1\text{-}^{14}C]\text{Oleoylglycerol} + H_2O \rightarrow [^{14}C]\text{fatty acid} + \text{glycerol} \quad (2)$$

Transacylation

$$2\ [1\text{-}^{14}C]\text{Oleoylglycerol} \rightarrow$$
$$[1\text{-}^{14}C]\text{oleoyl-[2- or 3-}^{14}C]\text{oleoylglycerol} + \text{glycerol} \quad (3)$$
$$[1\text{-}^{14}C]\text{Oleoylglycerol} + [1\text{-}^3H]\text{phosphatidylethanolamine} \rightarrow$$
$$[1\text{-}^{14}C]\text{oleoyl-[2- or 3-}^3H]\text{acylglycerol}$$
$$+ \text{lysophosphatidylethanolamine} \quad (4)$$

The diglyceride produced via transacylation is also a substrate for hydrolysis. As a consequence, diacylglycerol is a transitory product and eventually free fatty acid and glycerol are the only products. Generally, the activity is measured using [1-^{14}C]oleoylglycerol (commercially available from Applied Science) or 1-oleoyl-[2-^{3}H]glycerol as substrate rather than [1-^{3}H]phosphatidylethanolamine. During the purification procedures, it is convenient to use the combination of [1-^{14}C]oleoylglycerol and [1-^{3}H]phosphatidylethanolamine to demonstrate that the two activities copurify. The description given is for the mixed substrates in the assay; either can be used singly under the same conditions.

The substrate cocktail is prepared by drying the chloroform solutions of [1-^{14}C]oleoylglycerol (500 nmol; 15,000 cpm per assay) and [1-^{3}H]phosphatidylethanolamine (100 nmol; 45,000 cpm per assay) in round-bottom test tubes with a stream of N_2. For each assay 0.2 ml of a solution of 0.5 M Tris-HCl, pH 8.1, and 25 mM CaCl$_2$ is added to the dried lipids, and the mixture is sonicated under N_2 for 2 min in a sonic bath. The substrate suspension is then added to a conical centrifuge tube, along with 1–50 μg of enzyme protein (depending on source and degree of purity) and water to bring the total volume to 1.0 ml. The reaction mixture is then incubated for 5–20° min in a shaking water bath at 37°. The lipids are extracted, as described for the lysosomal phospholipase A_1. The chromatography system used is ether–petroleum ether–formic acid, 40 : 60 : 1.5 (v/v/v); monoglyceride, 1,2- and 1,3-diglycerides, and free fatty acid are used as standards. The silicic acid that contains the phospholipid fraction at the origin, monoglyceride, the two diglycerides, and free fatty acid are scraped and counted using settings on the scintillation counter optimized for simultaneous counting of ^{3}H and ^{14}C. The activity on [1-^{3}H]phosphatidylethanolamine is taken as the percentage of the total ^{3}H recovered, which is found in ^{3}H-labeled free fatty acid (hydrolysis) and [^{3}H]diglyceride (transacylation). The activity on [1-^{14}C]oleoylglycerol is caluclated in the same manner using the ^{14}C-labeled fractions. The reaction is usually linear up to 40–50% hydrolysis of either substrate.

The assay using 1-oleoyl-[2-^{3}H]glycerol as substrate is identical to that described above except that only monoacylglycerol is used as substrate and the amount of [^{3}H]glycerol liberated is determined by counting an aliquot of the methanol–water layer. This assay has the obvious advantage of eliminating the chromatography and scraping of silicic acid. However, it is not possible to determine the relative amounts of hydrolysis and transacylation that occur.

Purification

Materials for Preparation:
Heparin–Sepharose 4B affinity column
Ultrogel AcA 40
Amicon concentrator with YM 10 filter
Fraction collector and ultraviolet monitor

The preparation of the heparin–Sepharose 4B is described by Olivecrona et al.[10] and modified as follows (this must be carried out in a well ventilated hood): Two hundred milligrams of heparin (Wilson Laboratory) are dissolved in 50 ml (0.1 M) of NaHCO$_3$. Seventy-five milliliters of solid Sepharose 4B that has been thoroughly washed with water are suspended in a total volume of 150 ml with water. Five grams of CNBr are dissolved

[10] T. Olivecrona, T. Egelrud, P.-H. Iverius, and A. Lindhal, *Biochem. Biophys. Res. Commun.* **3**, 524 (1971).

in water, the gel suspension is added with continuous stirring, and the pH is maintained between 10.5 and 11.5 with 2 M NaOH. The pH is monitored with a pH meter, and the mixture is kept below 20° by cooling the beaker in a cold water bath. After 5 min (the pH should be relatively constant) the whole suspension is added to 1.5 liters of ice-cold water, and the suspension is filtered on a sintered-glass funnel. After washing with an additional 2 liters of water, the gel is washed once with 1 liter of cold 0.1 M NaHCO$_3$. (Do not allow the activated gel to dry.) The activated gel is then stirred with the heparin solution for 16 hr at 0–4° in a total volume of 150 ml of 0.1 M NaHCO$_3$. Next, 7.5 ml of ethanolamine are added, the mixture is stirred for 4 hr, and finally is washed with 1 liter of water, 1 liter of 0.5 M NaCl, and 3 liters of water.

Source of Enzyme. The enzyme can be purified from isolated plasma membranes or from a heparin-perfusate from the liver. The latter is more suitable for enzyme purification, since the starting material has a specific enzymatic activity roughly 10 times that obtained from plasma membranes and the yield per liver is severalfold higher than that of the plasma membrane. After the initial step, the procedure for the two is identical.

Isolation of plasma membranes is carried out as described in Volume 31 of this series except that the final washes are omitted.[11] Generally 10 rats are used at one time. The membranes from the 1.16/1.18 interface are diluted with an equal volume of 10 mM Tris, pH 7.4, and pelleted at 100,000 g for 1 hr to remove the sucrose. This pellet is then resuspended in 10 mM Tris, pH 7.4, that contains 0.2 mg of heparin (Sigma) per milliliter. After 5 min the membranes are pelleted again and the supernatant solution is used as the starting point for purification. The specific activity of the enzyme on monoacylglycerol is about 30 nmol/mg per minute, and the activity on phosphatidylethanolamine is about 10 nmol/mg per minute.

Liver Perfusion. The procedure for the liver perfusion is described in detail by Seglen.[12] Basically, a rat is injected with pentobarbital [0.1 ml/100 g of a preparation of 1 g/ml (Barber Veterinary Supply, Richmond, Virginia)]. Ether can be used as well, but when several rats are done at one time the effect of ether on the experimenter is questionable. The rat is then tied to a surgery board, and the abdominal wall of the rat is opened. The portal vein is cannulated with a needle connected to a 250-ml reservoir filled with Hanks' buffer. The superior vena cava is cannulated in the same way except the tubing is connected to a collecting reservoir but is initially clamped off. The inferior vena cava is then cut, and the blood is washed from the liver by pumping the Hanks' buffer in through the portal

[11] P. Emmelot, C. J. Bos, R. P. van Hoeven, and W. J. van Blitterswijk, this series, Vol. 31, p. 75.
[12] P. O. Seglen, *Methods in Cell Biology* **13**, 29 (1976).

vein at a rate of 30 ml/min. Once the buffer is clear, the inferior vena cave is clamped and the superior vena cava is opened, and the perfusion is continued for another 2–3 min. A solution of 50 mg of heparin per milliliter is then added to the perfusion reservoir (0.2 ml per 250 ml of Hanks' buffer) and a new collection reservoir is used. In 2 min about 90% of the heparin-releasable MGAT is recovered and serves as the starting point for purification of the enzyme. Usually the perfusate from 20–40 rats is used. This usually takes 2–3 days to prepare, so the perfusates are frozen at −20° until all is collected.

Heparin Affinity Chromatography. The combined perfusates are thawed and concentrated 10-fold on an Amicon YM-10 filter. The concentrate is then mixed with an equal volume of the heparin–Sepharose in 10 mM sodium Veronal buffer, pH 7.4, that contains 0.16 M NaCl and is stirred in an ice bath for 10 min. (The concentration of the heparin–Sephadex has not been measured. Routinely the Sephadex is allowed to settle completely and the Veronal buffer is decanted.) The mixture is then poured into a column and washed with a column volume of the Veronal buffer with 0.16 M NaCl. The enzyme is then eluted from the column with a 1000-ml linear salt gradient made from 0.16 M to 1.5 M NaCl in Veronal buffer. The enzyme elutes at about 0.65 M NaCl. The active fractions are then combined and concentrated using the Amicon YM-10 filter to about 6–7 ml.

Gel Filtration. The concentrate is then mixed with Triton X-100 to a final concentration of 0.2% and applied to a column of Ultrogel AcA 44 equilibrated in 50 mM Tris, pH 7.4, that contains 0.2% Triton X-100. The enzyme is eluted with the same buffer and comes off the column at the void volume. This procedure separates the enzyme from possible lipid contaminants and from the major protein contaminant, which has a molecular weight of about 65,000, probably albumin.

Heparin–Sepharose Chromatography. The preparation is immediately applied to a second heparin affinity column that is 1 × 10 cm. The conditions for eluting the enzyme are exactly as for the first column except that the volumes used are reduced 10-fold. The enzyme elutes in two peaks, one with the initial wash and the second with 0.7–1.0 M NaCl; the second peak is broader than that obtained with the initial heparin affinity chromatography. It appears that the lack of binding of some of the enzyme to the column is not the result of column overload. This is based on the following observations: first, reapplication of the first peak to the column gives the same result; second, at this stage of purification, there should be none of the heparin of the original perfusion remaining; and third, the ratio of protein to column volume is lower than that of the first heparin affinity binding.

TABLE II

PURIFICATION OF MONOACYLGLYCEROLACYLTRANSFERASE (MGAT)

Step	Protein (mg)	Total activity (μmol/min)	Specific activity (μmol/min/mg)	Purification (fold)	Recovery (%)
Concentrated perfusate	131	85.1	0.65		
Heparin affinity column	9.0	25.2	2.80	4.3	29.6
Ultrogel Heparin affinity column	3.6	36.3	10.1	15.5	42.7
Peak I	0.73	13.1	17.9	27.5	15.4
Peak II	0.18	3.7	20.6	31.7	4.3
Total (I + II)	0.91	16.8	—	—	19.7

Results of the Enzyme Purification and Stability. Table II gives the results of a purification of MGAT from the perfusate of 40 rats. The enzymatic activity was measured using [1-^{14}C]oleoylglycerol; however, the activity on [1-^{3}H]phosphatidylethanolamine has been demonstrated to copurify.[13] The loss at the initial step in part is the result of the spread of the enzyme in a rather large number of tubes. However, even when this is taken into consideration, over half of the enzyme is either inactivated or bound very tightly to the column. The apparent increase in enzyme recovered following the Ultrogel step is probably the result of the presence of small amounts of Triton X-100 in the assay, which slightly stimulates the activity on monoacylglycerol and phosphatidylethanolamine.[13] At the initial steps the enzyme is stable in frozen form or in ice for periods of several months. After the treatment with Triton X-100, the enzyme is much more labile and activity is lost within a period of 1–2 weeks. The second peak from the second affinity column is particularly labile, losing all activity within 2–3 days.

Other procedures have been used including affinity chromatography on lipophilic columns of phenylalanine Sepharose and concanavalin A. Although the enzyme binds to these columns and can be recovered in good yield, we have not yet achieved a significant increase in purity and therefore are not included.

Properties of the Enzymes

The SDS-PAGE gel electrophoresis of the peak I in 8% gels shows a major band with a molecular weight slightly greater than albumin (about

[13] M. Waite, P. Sisson, and R. El-Maghrabi, *Biochim. Biophys. Acta* **530**, 592 (1978).

75,000) with a minor band with roughly twice the mobility of the major band. It is not possible at this point to determine whether this is a monomer–dimer interaction, but we have demonstrated that the enzyme does tend to aggregate. The protein from peak II does not enter the gel and remains at the top even when 5% gels are used; this we believe is the result of aggregation. Electrofocusing of the preparation from either the perfusate or the plasma membrane shows a heterogeneous pattern with pI values ranging from 3.8 to 5.6. This could be the result of carbohydrate constituents.

The enzyme from the perfusate is most active on monoacylglycerol; the activity on phosphatidylethanolamine is roughly one-half to one-third of that. The enzyme is also active on diacyl- and triacylglycerols, but at a rather low rate.[13] The ratio activities on the various substrates will vary among preparations for reasons not yet known. Indeed, we have not yet shown any activity of the enzyme on the plasma membrane on triacylglycerol. This is true using hepatocyte suspensions,[14] isolated plasma membranes, or the enzyme solubilized from the isolated plasma membranes by heparin.[15,16] It is interesting, however, that these preparations, when treated with Triton and purified through the second heparin affinity column, will act on triacylglycerol.[13]

The enzyme purified from the perfusate or from the plasma membrane is active on substrates in a variety of physical forms. Substrate preparations studied include chylomicra remnant particles, low density lipoprotein, high density lipoprotein, albumin, and liposome.[14,17] Of the various phospholipids tested, phosphatidylethanolamine and lysophosphatidylethanolamine are preferred; phosphatidylcholine is degraded at about 50% of the maximal rate. The strongly acidic lipids, such as phosphatidylserine, phosphatidylinositol, and phosphatidylglycerol, are very poor substrates.[18]

The enzyme is active over a wide pH range, from pH 6.5 to 9.5. Optimal activity is between 8.0 and 9.0. $CaCl_2$ stimulates the enzyme severalfold when it is associated with the plasma membrane. However, there is only a slight stimulation by $CaCl_2$ on the solubilized enzyme, suggesting that Ca^{2+} is not involved in the catalytic event.

[14] R. El-Maghrabi, M. Waite, L. L. Rudel, and V. L. King, *Biochim. Biophys. Acta* **572,** 52 (1979).
[15] M. Waite and P. Sisson, *J. Biol. Chem.* **248,** 7985 (1973).
[16] M. Waite and P. Sisson, *J. Biol. Chem.* **249,** 6401 (1974).
[17] R. El-Maghrabi, M. Waite, and L. L. Rudel, *Biochem. Biophys. Res. Commun.* **81,** 82 (1978).
[18] J. Newkirk and M. Waite, *Biochim. Biophys. Acta* **298,** 562 (1973).

[79] Isolation of Phospholipase A₂ from Red Cell Membranes of Sheep

EC 3.1.1.4 Phosphatide 2-acylhydrolase

By Peter Zahler and Ruth Kramer

Red cell membranes from sheep contain a phospholipase A_2 (EC 3.1.1.4) with the following characteristic properties: marked preference for phosphatidylcholine, specificity for the fatty acid at the 2-position, requirement for Ca^{2+}, alkaline pH optimum, activation by various detergents, and stability against denaturating agents.[1] The enzyme is present in ruminant erythrocytes and is thought to play a role in maintaining the very low phosphatidylcholine content of the ruminant red cell membrane.[2]

Soluble phospholipase A from different sources has been purified by conventional methods.[3,4] These procedures required a number of steps, including ammonium sulfate fractionations and a variety of column chromatographic methods. A new purification technique was developed by Rock and Snyder,[5] who devised an affinity adsorbent to isolate soluble phospholipase A_2 from snake venom, using an alkyl ether analog of lecithin as ligand. Based on the fact that this enzyme required Ca^{2+} for substrate binding,[6] adsorption to the column was obtained in the presence of Ca^{2+} and subsequent elution was performed with buffer containing EDTA.

Less information is available concerning the isolation of membrane-bound phospholipases, either of the A_1 or A_2 type.[4] Whereas in recent years membrane-bound phospholipase A_1 has been purified to near homogeneity from microorganisms,[7,8] membrane-bound phospholipase A_2, to our knowledge, has been only partially purified from rat liver mitochondria.[9]

The erythrocyte membrane contains a variety of protein components.[10] Most of them have been characterized in terms of their arrange-

[1] R. Kramer, B. Jungi, and P. Zahler, *Biochim. Biophys. Acta* **373**, 404 (1974).

[2] R. F. A. Zwaal, R. Flückiger, S. Moser, and P. Zahler, *Biochim. Biophys. Acta* **373**, 416 (1974).

[3] S. Gatt and Y. Barenholz, *Annu. Rev. Biochem.* **42**, 61 (1973).

[4] H. van den Bosch, *Annu. Rev. Biochem.* **43**, 243 (1974).

[5] C. O. Rock and F. Snyder, *J. Biol. Chem.* **250**, 6564 (1975).

[6] M. A. Wells, *Biochemistry* **11**, 1030 (1972).

[7] C. J. Scandella and A. Kornberg, *Biochemistry* **10**, 4447 (1971).

[8] M. Nishijma, Y. Akamatsu, and S. Nojima, *J. Biol. Chem.* **249** (1974).

[9] M. Waite and P. Sisson, *Biochemistry* **12**, 2377 (1971).

[10] R. L. Juliano, *Biochim. Biophys. Acta* **300**, 341 (1973).

ment in the membrane.[11] The major polypeptides of the human erythrocyte membrane have been purified and biochemically characterized.[12]
Many minor proteins, however, have not yet been investigated in this
respect. Phospholipase A$_2$ from sheep red cell membranes is considered
as a minor component.[1,2] The major obstacles for the isolation of this
enzyme have been (a) small quantity of enzyme present; (b) difficulties in
finding suitable methods for a stepwise purification of this membrane
protein; (c) the need for continued presence of detergents to maintain the
enzyme in a soluble state. The current investigation presents a procedure
for separation of phospholipase A$_2$ from the bulk of membrane proteins
and describes the further purification of this enzyme to near homogeneity
by affinity chromatography according to the method of Rock and Snyder.[5]

Materials and Methods

Materials. Sheep blood was provided by the municipal slaughterhouse,
usually within 1 hr of the death of the animal. Sephadex G-75 and AH-
Sepharose 4B were obtained from Pharmacia, Uppsala. Molecular weight
markers for dodecyl sulfate gel calibration were bovine serum albumin
(Calbiochem), pepsin (Fluka), trypsin (Sigma), myoglobin (Sigma), and
hemoglobin (Fluka). Triton X-100, deoxycholate, cholate, and dodecyl
sulfate were purchased from Merck. L-α-Phosphatidylcholine from egg
yolk was obtained from Koch-Light (grade I). *rac*-1-(11-Carboxy)undecyl-
2-hexadecylglycero-3-phosphocholine was obtained from R. Berchtold,
Biochemical Laboratory, Berne. Phosphatidyl[^{14}C]choline (1.8 Ci/mmol,
uniformly labeled, from *Chlorella pyrenoidosa*) and dodecyl [^{35}S]sulfate
(10 Ci/mol) were products of New England Nuclear Co. Inorganic salts
and organic solvents were reagent grade from Fluka, Merck, and Sigma.

Assay of Phospholipase A$_2$ Activity. Phospholipase activity was determined by incubating 0.3 μmol of phosphatidylcholine (made up to a
specific radioactivity of 1.873×10^5 cpm/μmol with phosphatidyl-
[^{14}C]choline) with appropriate amounts of enzyme in the presence of 10
mM glycylglycine, pH 8, 4 mM CaCl$_2$, and 0.5% cholate (w/v) in a total
volume of 0.47 ml. After incubation at 37° for 60 min with shaking, the
reaction was stopped by adding 30 μl of 0.2 M EDTA and 2.15 ml of
chloroform–methanol (5:8, v/v). Lipids were then extracted by the
method of Renkonen *et al.*[13] The chloroform layer was sampled for total
radioactivity. Aliquots were chromatographed on silica gel HR plates

[11] T. L. Steck, *J. Cell Biol.* **62**, 1 (1974).
[12] V. T. Marchesi, H. Furthmayr, and M. Tomita, *Annu. Rev. Biochem.* **45**, 667 (1976).
[13] O. Renkonen, T. U. Kosunen, and O. V. Renkonen, *Ann. Med. Exp. Biol. Fenn.* **41**, 375
(1963).

(Merck) with chloroform–methanol–water–acetic acid (14 : 6 : 1 : 0.5, v/v) as developing solvent to separate phosphatidylcholine, lysophosphatidylcholine, and free fatty acids. The chromatograms were stained with iodine vapor, and individual lipid spots were scraped off into counting vials. After addition of 5 ml of methanol and 10 ml of a scintillation fluid [toluene containing 7 g of 2-(4'-*tert*-butylphenyl)-5-(4''-biphenylyl-1,3,4)oxadiazole (Ciba-Geigy) per liter] the radioactivity was measured in a Packard Tri-Carb scintillation counter (Model 2450). The percentage of phosphatidyl[^{14}C]choline degraded to lysophosphatidyl[^{14}C]choline was calculated, and the phospholipase activity was expressed as micromoles or nanomoles of phosphatidylcholine hydrolyzed per hour.

Analytical Methods. Protein was estimated by the method of Lowry *et al.*[14] using bovine serum albumin as standard or by absorption at 280 nm. Corrections for the effect of detergents and glycylglycine on the Lowry method were carried out.[15] For protein determination in samples containing Triton X-100, dodecyl sulfate was included in the alkaline copper reagent for the Folin reaction, as indicated by Dulley and Grieve.[16] Lipids were extracted with chloroform–methanol (2 : 1, v/v) by the method of Folch *et al.*[17] The phospholipids were separated by two-dimensional thin-layer chromatography using the procedure of Broekhuyse,[18] and the phospholipid phosphorus was determined according to Bartlett.[19] Electrophoresis in dodecyl sulfate polyacrylamide gels was done according to Fairbanks *et al.*[20] For preparative dodecyl sulfate gel electrophoresis the enzyme was solubilized in the absence of dithiothreitol, since sulfhydryl reagents were shown to destroy the lecithinase activity.[1] Solubilized lecithinase (10–20 μg of protein) was layered on each of two gels. One gel was stained and destained to localize the protein band; the other gel was sliced and eluted with 0.1% dodecyl sulfate solution as described by Weber and Osborn.[21] The eluents and aliquots of the solubilized enzyme were dialyzed against 0.5% cholate buffer and then assayed for enzymatic activity.

[14] O. H. Lowry, N. J. Rosebrough, N. J. Farr, and R. J. Randall, *J. Biol. Chem.* **193**, 265 (1951).
[15] T. H. Ji, *Anal. Biochem.* **52**, 517 (1973).
[16] J. R. Dulley and P. A. Grieve, *Anal. Biochem.* **64**, 136 (1975).
[17] J. Folch, M. Lees, and G. H. Sloane-Stanley, *J. Biol. Chem.* **226**, 497 (1975).
[18] R. M. Broekhuyse, *Clin. Chim. Acta* **23**, 457 (1969).
[19] G. R. Bartlett, *J. Biol. Chem.* **234**, 466 (1959).
[20] G. Fairbanks, T. L. Steck, and D. F. H. Wallach, *Biochemistry* **10**, 2606 (1971).
[21] K. Weber and M. Osborn, *J. Biol. Chem.* **244**, 4406 (1969).

Preparative Procedures

Scheme 1 shows a summary of the isolation procedure. In step 1 a large portion of the extrinsic proteins, such as spectrin, is removed from the membranes.

In step 2 the membranes are solubilized by SDS. Hereby the enzyme is inactivated; if, however, SDS is replaced by cholate, the enzymatic activity can be fully restored (see step 3). Cholate is not suited for solubilization because most of the enzyme remains in the pellet. Although Triton X-100 solubilizes most of the enzyme in an active form, it is not suited for affinity chromatography.

In step 3 SDS is exchanged for cholate by gel chromatography. A large portion of protein aggregates and elutes in the excluded volume of the eluate, whereas the phospholipase remains monomeric and is retarded. The SDS elutes in the included volume at the end.

In step 4 the phospholipase peak of the Sephadex G-75 eluate in cholate is further purified at the affinity column.

Membrane Preparation. Fresh sheep blood, anticoagulated with acid citrate–dextrose, was centrifuged in a Sorvall GSA rotor at 600 g for 10 min. The red cells were washed three times with two volumes of 310 mOsM phosphate buffer (pH 8) at 600 g and 0–4° for 10 min. After each wash the leukocytes, forming the "buffy coat," were removed as thoroughly as possible by aspiration and one-fifth of the original volume of the erythrocytes was sacrificed. One flask (450 ml) of blood yielded 80–90 ml of washed erythrocytes, which were lysed in 20–30 volumes of 10 mOsM phosphate buffer, pH 8, while standing at 0–4° for 1 hr. The hemolyzate was centrifuged at 18,000 g and 0–4° for 30 min. The ghost pellets were washed five times with 10 mOsM phosphate, pH 8, in a Sorvall SS-34 rotor at 40 000 g and 0–4° for 15 min to yield washed ghosts that were cream-pink colored. The sticky pellets beneath the loose ghost pellets, which contain protease activity,[20] were carefully removed after each wash of the ghosts.

Step 1. Extractions of Membranes. Sheep erythrocyte membranes were extracted with (A) 0.1 mM EDTA, pH 8; (B) 1 mM glycylglycine, pH 8; (C) 0.5 M NaCl in 5 mM phosphate, pH 8; (D) 0.75 M NaI, pH 8; and (E) 0.01 M NaOH by the procedures of Fairbanks *et al.*[20] (A,B), Kahlenberg and Walker[22] (C,D), and Steck and Yu[23] (E). For this purpose packed ghosts equivalent to 7.5 mg of membrane protein were suspended in 7 ml of the appropriate solution. After incubation as indicated in the references

[22] A. Kahlenberg and C. Walker, *J. Biol. Chem.* **251**, 1582 (1976).
[23] T. L. Steck and J. Yu, *J. Supramol. Struct.* **1**, 220 (1973).

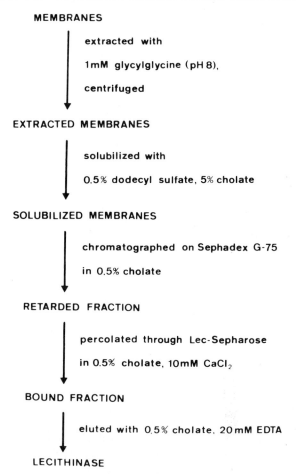

MEMBRANES

    extracted with

    1mM glycylglycine (pH 8),

    centrifuged

EXTRACTED MEMBRANES

    solubilized with

    0.5% dodecyl sulfate, 5% cholate

SOLUBILIZED MEMBRANES

    chromatographed on Sephadex G-75

    in 0.5% cholate

RETARDED FRACTION

    percolated through Lec-Sepharose

    in 0.5% cholate, 10mM CaCl$_2$

BOUND FRACTION

    eluted with 0.5% cholate, 20mM EDTA

LECITHINASE

SCHEME 1. Summary of isolation procedure.

(procedures A and B at 37° for 15 min, procedures C and D on ice for 30 min, and procedure E, none), the membrane suspension were centrifuged (60 min, 150,000 g, at 0–4°) in a Beckman L-2. The sedimented membrane material of procedures A, C, D, and E was washed with 7.5 mM phosphate (pH 7.5) and suspended by homogenization in the same buffer. The supernatant and pellet fractions and untreated membranes (control membranes) were dialyzed against 10 mM glycylglycine (pH 8) overnight, and aliquots were assayed for protein content and phospholipase activity.

Extractions with solutions A and B resulted in highest increase of specific activity with very little loss of total activity.[24] In order to avoid EDTA we used glycylglycine buffer B for the final purification procedure.

Step 2. Solubilization of Membranes. Various detergents have been tested for maximal solubilization of the enzyme. Triton X-100, 0.2%, at pH 8 resulted in 60% release of phospholipase activity from membranes (2.5 mg of protein per milliliter) into the supernatant obtained by centrifuging at 150,000 g, and 0–4° for 60 min, but no binding of the solubilized enzyme to the affinity column could be obtained. Cholate poorly solubilized the enzyme; however, when the solubilized enzyme was passed through the affinity column, all of the phospholipase bound to the ligand. Dodecyl sulfate inactivated the enzyme and solubilized all of the membrane; when the SDS was replaced by cholate the enzyme regained full activity.[24]

Extracted membranes were solubilized in 10 mM glycylglycine (pH 8) containing 1% dodecyl sulfate (5 mg of protein per milliliter). An equal volume of the same buffer containing 10% cholate and 0.4 M KCl was added to give a final concentration of 2.5 mg of protein per milliliter, 0.5% dodecyl sulfate, 5% cholate, and 0.2 M KCl. The solution was stirred at 20° for 1 hr. Upon termination of the incubation, solubilization of the membranes was complete, leaving no material in a sedimentable form (60 min, 150,000 g at 0–4°).

Step 3. Gel Filtration on Sephadex G-75. The solubilized membranes (not more than 60 ml) were applied to a Sephadex G-75 column (1250 cm × 7 cm), which had been equilibrated with 10 mM glycylglycine, pH 8, containing 0.5% cholate, 0.2 M KCl, and 0.05% NaN$_3$ (buffer A). Elution was performed with 5000 ml of buffer A at 4°. Fractions of 30 ml were collected at a flow rate of 250 ml/hr. The column chromatography was routinely followed by determination of the absorbance at 280 nm and assay of phospholipase activity (Fig. 1).

Step 4. Affinity Chromatography

i. Preparation of the affinity adsorbent. The alkyl ether analog of lecithin [rac-1-(11-carboxy)undecyl-2-hexadecylglycero-3-phosphocholine] was coupled to AH-Sepharose 4B through the carboxyl group using a modified carbodiimide coupling procedure of Rock and Snyder.[5] Freeze-dried powder of AH-Sepharose 4B (5 g) was swollen and washed as recommended by the supplier to give about 20 ml of gel volume. The ligand (800 μmol) was dissolved in 10 ml of tetrahydrofuran–water, pH 7 (1 : 1, v/v). The carbodiimide

[24] R. M. Kramer, C. Wüthrich, C. Bollier, P. R. Allegrini, and P. Zahler, Biochim. Biophys. Acta 507, 381 (1978).

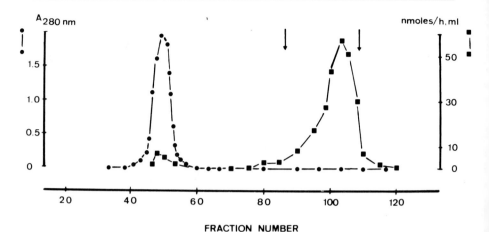

FIG. 1. Sephadex G-75 chromatography of dodecyl sulfate/solubilized sheep red cell membranes in the presence of cholate. The membrane solution containing 0.5% sodium dodecyl sulfate was applied to a Sephadex G-75 column and eluted with buffer containing 0.5% cholate and 0.2 M KCl. Phospholipase activity (■——■) was determined using the standard assay with 0.5 ml of the fractions as indicated. ●——●, Protein.

[1-cyclohexyl-3-(2-morpholinoethyl)carbodiimide metho-p-toluene sulfonate] (6 mmol) was dissolved in 30 ml of tetrahydrofuran–water, pH 4.5 (1 : 1, v/v). The combined ligand and carbodiimide solution was added to 20 ml of packed gel, and the pH was adjusted to 5.0. The reaction was allowed to proceed at room temperature for 30 hr with gentle stirring. The pH of the reaction mixture decreased to 4.5 during the reaction. The gel was washed with 200 ml of tetrahydrofuran–water (1 : 1, v/v) followed by 200 ml of methanol, 400 ml of 1 M NaCl, 800 ml of distilled water, and 200 ml of 10 mM glycylglycine, pH 8. It was then equilibrated and packed in the first column buffer (10 mM glycylglycine, pH 8, containing 0.5% cholate, 10 mM CaCl$_2$, 0.2 M KCl, and 0.05% NaN$_3$). Coupling of the ligand was examined by digesting aliquots (50 μl) of the washed gel with H$_2$SO$_4$ and determining inorganic phosphorus according to Bartlett.[19] The amount of coupled ligand was 2–3 μmol per milliliter of swollen gel.

ii. *Chromatography.* The enzymatically active fractions from the Sephadex G-75 eluates were combined and CaCl$_2$ was aeded to a final concentration of 10 mM. The resulting solution was passed through a dialkylphosphatidylcholine-Sepharose column (15 cm × 0.8 cm) at a flow rate of 20 ml/hr at 4°. The column was washed extensively with buffer A containing 10 mM CaCl$_2$. Elution of the bound phospholipase was then carried out with 20 mM EDTA in buffer A. Fractions of 10 ml were collected at a flow rate of 10 ml/hr. The desired fractions were dialyzed

PURIFICATION OF SHEEP ERYTHROCYTE MEMBRANE PHOSPHOLIPASE A₂[a]

Fraction	Protein (mg)	Total activity (μmol/hr)	Specific activity (μmol/mg hr^{-1})	Overall yield (%)	Purification (fold)
I. Membranes	300	36.0	0.12	100	—
II. Extracted membranes	200	34.2	0.17	95	1.4
III. Sephadex G-75 filtrate	4	27.9	6.98	78	58
IV. Affinity column eluate	0.085	24.1	283.5	67	2835

[a] Protein content and enzymatic activity of the fractions were determined as described in Materials and Methods.

overnight against buffer A to remove the EDTA. The enzymatic activity was determined with aliquots of the fractions (0.2 ml) using the standard assay.

Column fractions containing the enzymatic activity were pooled and concentrated using an Amicon ultrafiltration cell equipped with a PM-10 membrane. The concentrated enzyme solution (2–3 ml) was assayed for enzymatic activity and then dialyzed against water (adjusted to pH 9 with NaOH) to remove the cholate. Aliquots of the resulting slightly turbid enzyme preparation were lyophilized. The lyophilized samples were resuspended either in distilled water for protein determination, phospholipid analysis, and dodecyl sulfate gel electrophoresis or in buffer A for measuring enzymatic activity. A similar procedure was used for activity, protein, and phospholipid measurements on the enzymatically active, pooled fractions from the Sephadex G-75 chromatography.

The yield and the degree of purification of phospholipase A₂ by the above procedure are given in the table.

Properties of the Enzyme

Enzyme Purity. The purity of the isolated phospholipase was investigated by polyacrylamide gel electrophoresis in the presence of dodecyl sulfate. The protein was present as a single band indicating that the polypeptide was electrophoretically pure. Moreover, upon preparative gel electrophoresis the phospholipase activity was recovered in the region corresponding to the protein band of the stained gel.

Molecular Weight. Sephadex G-75 chromatography indicates that the phospholipase has a molecular weight of approximately 12,000. Based on migration relative to marker proteins, an apparent molecular weight of 18,500 was estimated by dodecyl sulfate electrophoresis.

Specificity. The enzyme shows a pronounced preference for phosphatidylcholine. Reaction velocities (nmol/μg min^{-1}) were as follows: PC, 8.0; PG, 1.3; PE, 1.0; PS, 0.0. It also shows specificity for long-chain unsaturated fatty acids in the *Sn-2* position, namely, for PC $C_{18:2}$, 4.2 $\times$ 10^{-2}; $C_{20:4}$, 6.5 $\times$ 10^{-2}; $C_{22:4}$, 42 $\times$ 10^{-2}).[25]

Cofactors. The enzyme requires Ca^{2+} for activity. It is inactive at a concentration below 10 μM and fully active above 500 μM Ca^{2+}. Mg^{2+} may partially substitute for Ca^{2+}.[26]

Localization. The enzyme is localized on the external side of the red cell membrane of all ruminants; however, it deeply penetrates or even spans the membrane.[26]

Function. The main function of the enzyme is still unclear; however, it is responsible for the extremely low content of PC in red cell membranes of all ruminants.

[25] J. Jimeno-Abendano and P. Zahler, *Biochim. Biophys. Acta* **573**, 266 (1979).
[26] E. Frei and P. Zahler, *Biochim. Biophys. Acta* **550**, 450 (1979).

[80] Phospholipase A₂ from Bee Venom

EC 3.1.1.4 Phosphatide acylhydrolase

By R. C. COTTRELL

Diacylglycerophosphatide + H_2O →
$\qquad$ 1-acylglycerophosphatide + free fatty acid

Assay Method

Principle. Since the main phospholipase activity in bee venom has been shown to be of the A₂ type,[1] assay methods that do not distinguish between A₁ and A₂ activities are suitable for most purposes. A number of electrophoretically distinct species, however, have been shown to contain phospholipase activity[1] that may not all be of the A₂ type. Clearly this consideration would become of importance in any study involving the isolation of minor components with phospholipase activity.

The most convenient assay methods rely on the use of commercially available materials and equipment[1,2] but will probably be superseded by

[1] R. A. Shipolini, G. L. Callewaert, R. C. Cottrell, S. Doonan, and C. A. Vernon, *Eur. J. Biochem.* **20**, 459 (1971).
[2] G. H. de Haas, N. M. Postema, W. Nieuwenhuizen, and L. L. M. van Deenen, *Biochim. Biophys. Acta* **159**, 103 (1968).

the spectrophotometric procedure of Aarsman et al.[3] if the acyl thioester substrate needed becomes more readily available. Unfortunately, for studies of certain aspects of the enzyme's behavior the acyl thioester substrates are not suitable. An accurate and practicable conductimetric assay has been developed for work with the more hydrophobic substrates.[4]

The method outlined here is straightforward and does not require prior synthesis of the substrate. The solutions used in the assay appear to be homogeneous, and the reproducibility is good. None of the problems of maintaining the pH electrode in a satisfactory working condition, experienced with the heterogeneous assay of de Haas et al.,[2] occur. Commercially available pH stat equipment (e.g., Radiometer, Copenhagen) capable of handling nonaqueous solutions is used. The pH stat is connected to a suitable recorder and set to register the volume of titrant added against time.

Reagents

Ovolecithin: A variety of suitable materials are available commercially. The level of purity required will depend on the application envisaged but should generally be at least 60% phosphatidylcholine. Synthetic diacyl-L-α-phosphatidylcholines are also available and of high purity. Solvent is removed in a stream of nitrogen, and the waxy solid is dissolved in n-propanol to give a final concentration of 10 mg/ml.

Electrode vessel solution: The substrate solution in n-propanol is mixed with a solution of 500 μM CaCl$_2$ and 50 μM EDTA in the ratio 3:1 v/v, and 2 ml are placed in the electrode vessel. This solution is maintained at 25° and stirred magnetically.

Titrant: A solution of NaOH (0.01 N) in n-propanol:water (3:1, v/v) is used as titrant.

Enzyme: An amount equivalent to about 2 μg of pure enzyme is used.

Procedure. The pH of the electrode vessel is brought to pH 8.0 automatically with the titrant solution. On addition of phospholipase the fatty acid released causes a drop in pH that is automatically corrected by the addition of the alkaline titrant. The initial slope of the plot of alkali volume added vs time is proportional to enzyme activity.

Purification Procedure

CAUTION: Bee venom and particularly phospholipase A are highly allergenic in the lyophilized form.

[3] A. J. Aarsman, L. L. M. van Deenen, and H. van den Bosch, *Bioorg. Chem.* **5**, 241 (1976).
[4] D. Drainas and A. J. Lawrence, *Eur. J. Biochem.* **91**, 131 (1978).

A number of procedures have been reported[1,5,6] for isolating phospholipase A from bee venom. The method given here produces a high degree of purification of the major phospholipase A component from other protein and peptide constituents of the venom. The complete resolution of the major phospholipase A component from other phospholipase active material is somewhat tedious[1] and will not be described here. It is not clear whether a number of distinct phospholipases are present in commercial lyophilized bee venom (possibly arising from the pooling of venom from more than one genetically distinct insect population) or whether the various "isoenzymes" observed represent structural modifications of a single parent molecule arising either during the collection and storage of the venom or as a result of posttranslational modifications of the polypeptide chain.

Step 1. Lyophilized venom (700 mg) is dissolved in water (7 ml) and force dialyzed at 4° until the volume remaining in the dialysis tubing is reduced to approximately 0.7 ml. This procedure removes many of the small peptides present in the venom, but "mellitin" (which constitutes 50% of the dry weight of the venom[7]) is not completely removed because of aggregation in solution to the tetramer of molecular weight approximately 11,000.

Step 2. The material from step 1 is diluted to 3 ml with ammonium formate buffer (0.1 M, pH 4.6) and applied to a column (1.6 × 100 cm) of Sephadex G-50 (fine) equilibrated in the same buffer. The absorbance at 254 nm of the eluent is monitored, and fractious of 1.0 ml are collected at a flow rate of 11 ml/hr. Phospholipase A is eluted at about 50% of the included volume of the column and constitutes the second largest ultraviolet peak. Fractions containing phospholipase A activity are pooled and lyophilized, yielding about 80 mg.

Step 3. The material from step 2 is dissolved in 0.8 ml of Tris-citrate buffer (0.05 M, pH 8.25), and the pH is readjusted to pH 8.25 using a solution of 10% w/v Tris base. The solution is centrifuged to remove insoluble material and applied to a column (0.9 × 15 cm) of SE-Sephadex G-25 equilibrated in the same buffer. The buffer is passed through the column until the absorbance at 254 nm drops to below 0.1. A salt gradient is then applied by passing 0.3 M Tris-citrate buffer into a closed 2-ml mixing vessel containing starting buffer. Fractions of 0.4 ml are collected at a flow rate of 2.0 ml/hr. Phosphilipase A is eluted as the second largest peak at about two column volumes. Fractions containing phospholipase

[5] E. Haberman and K. G. Reiz, *Biochem. Z* **341**, 451 (1965).
[6] D. Munjal and W. B. Elliott, *Toxicon* **9**, 403 (1971).
[7] E. Haberman, *Science* **177**, 314 (1972).

activity are pooled and lyphilized to yield about 25 mg of a nearly color-less powder that is a mixture of the phosphalipase A isoenzymes. Further purification to resolve the isoenzymes has been reported.[1]

Properties

Specificity. The major phospholipase A activity in bee venom has been shown to be of the A$_2$ type by examining the fatty acid released during the hydrolysis of 1-oleoyl-2-isolauroyl phosphatidylethanolamine.[1] The only free fatty acid released was identified as isolauric acid by re-versed phase thin-layer chromatography. The enzyme is active against long fatty acid chain phosphatidylcholines and ethanolamines but also acts on the 1,2-dihexanoylphosphatidylcholine.[1] In contrast to the enzyme from porcine pancreas,[8] the bee venom enzyme will attack molecularly dis-persed 1,2-dihexanoylphosphatidylcholine more readily than the micellar solution.[1] The bee venom enzyme is also active against liposomes of dilauroyl-, dimyristoyl-, and dipalmitoylphosphatidylcholine at tempera-tures other than the transition temperature from the liquid crystalline to the gel phase, whereas the pancreatic enzyme acts only near this tempera-ture.[9] The enzymes from these two sources also differ markedly in their ability to penetrate monolayers of short-chain phospholipids.[10,11] The min-imum structural requirements for a substrate of the bee venom enzyme are likely to be similar to other phospholipases of the A$_2$ class.[12]

Stability. The lyophilized bee venom enzyme is extremely stable at room temperature. In aqueous solution the enzyme is extremely stable to heating.[13]

Effect of pH. The apparent optimum pH for the hydrolysis of 1,2-dihexanoylphosphatidylcholine in 75% propanol, 25% water is pH 8.0.[1]

Activation and Inhibition. Purified bee venom phospholipase A appears to retain some calcium since activity is observed without the addition of calcium ions to the assay medium but the addition of EDTA inhibits the enzyme.[1] On addition of calcium a small increase in activity is observed. A considerably larger increase in activity follows the inclusion of small

[8] G. H. de Haas, P. P. M. Bonsen, W. A. Pieterson, and L. L. M. van Deenen, *Biochim. Biophys. Acta* **239**, 252 (1971).
[9] J. A. F. op den Kamp, J. de Gier, and L. L. M. van Deenen, *Biochim. Biophys. Acta* **345**, 253 (1974).
[10] R. Verger, M. C. E. Mieras, and G. H. de Haas, *J. Biol. Chem.* **248**, 4023 (1973).
[11] W. A. Pieterson, J. C. Vidal, J. J. Volwerk, and G. H. de Haas, *Biochemistry* **13**, 1455 (1974).
[12] L. L. M. van Deenen and G. H. de Haas, *Biochim. Biophys. Acta* **70**, 538 (1963).
[13] C. Nair, J. Hermans, D. Munjal, and W. B. Elliott, *Toxicon* **14**, 35 (1976).

quantities of EDTA (of the order of one-tenth the calcium concentration) in addition to calcium.[1] This effect presumably arises from the chelation of inhibiting metal ions that bind both to EDTA and to the enzyme more tightly than calcium. Magnesium, strontium, and barium can replace calcium as activators, but with decreased efficiency.[1] Curiously, the activation observed toward some substrates after treatment of the enzyme with acylating agents[14] appears to be related to the effect of calcium.[15] The activation by, for example, fatty acid imidazolides involves the irreversible addition of a single fatty acid acyl residue to the protein molecule[15] and may operate by affecting the ability of the enzyme to penetrate the interface between the aqueous phase and an insoluble substrate phase.[15] It has also been suggested that the fatty acid stabilizes the dimeric form of the enzyme and that it is the dimer which is active.[16] The enzyme has, indeed, been shown to exist as a dimer in concentrated aqueous solution.[1]

Neither O-isopropylfluoromethylphosphonate nor diisopropylphosphofluoridate inhibit the enzyme.[1] Guanidination with O-methylisourea in alkaline solution and treatment with iodoacetate are also without effect on activity.[17] Maleic anhydride treatment, however, is inhibitory.[17]

Primary Structure. The amino acid sequence of the major phospholipase A component has been reported[18] together with the position of attachment and composition of the carbohydrate side chain[18] and the position of the disulfide bridges.[19] Although no extensive sequence homologies are apparent between the bee venom enzyme and the enzymes from porcine, equine, and elapidae sources,[20] it is of interest to note that the sequence Thr-His-Asp at residues 47–49 of the porcine enzyme is reproduced at residues 33–35 of the bee venom enzyme. The residues His-Asp are thought to play an essential role in the catalytic activity of the porcine enzyme.[21] As yet the three-dimensional structure of the bee venom enzyme is not available, so the functional significance of residues 34–35 cannot be assessed.

[14] D. Drainas, G. R. Moores, and A. J. Lawrence, *FEBS Lett.* **86,** 49 (1978).
[15] D. Drainas and A. J. Lawrence, *Eur. J. Biochem.* **91,** 131 (1978).
[16] M. F. Roberts, R. A. Deems, and E. A. Dennis, *Proc. Natl. Acad. Sci. U.S.A.* **74,** 1950 (1977).
[17] R. C. Cottrell, Ph.D. thesis, University of London (Biological Chemistry) 1971, pp. 51–54.
[18] R. A. Shipolini, G. L. Callewaert, R. C. Cottrell, and C. A. Vernon, *Eur. J. Biochem.* **48,** 465 (1974).
[19] R. A. Shipolini, S. Doonan, and C. A. Vernon, *Eur. J. Biochem.* **48,** 477 (1974).
[20] W. C. Puijk, H. M. Verheij, and G. H. de Haas, *Biochim. Biophys. Acta* **492,** 254 (1977).
[21] J. Drenth, C. M. Enzing, K. H. Kalk, and J. C. A. Vessies, *Nature (London)* **264,** 373 (1976).

[81] Phospholipase A$_2$ from Cobra Venom (*Naja naja naja*)

EC 3.1.1.4 Phosphatide 2-Acylhydrolase

By RAYMOND A. DEEMS and EDWARD A. DENNIS

1,2-Diacyl-*sn*-glycero-3-phosphorylcholine + H$_2$O
$$\underset{\rightleftharpoons}{\overset{Ca^{2+}}{}} \text{1-acyl-}sn\text{-glycero-3-phosphorylcholine + fatty acid}$$

Cobra venom contains 8–14 forms of phospholipase A$_2$,[1-3] some of which are easily separated from the other venom components.[4] However, the isolation of individual forms of the enzyme is much more difficult, and the procedures that have been employed are inefficient and laborious.[2,3] Although some of these forms may represent experimental artifacts or catalytically insignificant differences in the proteins,[5,6] there are several whose physical and catalytic properties are dramatically different.[2,3,7] Therefore, detailed studies of the characteristics and catalytic mechanism of phospholipase A$_2$ require the use of a pure single form of the enzyme. Because these forms may also differ in how they act on membranes,[7] a pure single form is needed also when the enzyme is to be used as a probe of membrane structure. Since the cobra venom phospholipase A$_2$ has been extensively employed in both types of studies, it is important to have an easily obtainable, well characterized, pure form of this enzyme. The following relatively simple procedure[6] produces a high yield of a major form of phospholipase A$_2$ whose characteristics have been studied in some detail.

Assay Method

Principle. Phospholipase A$_2$ catalyzes the hydrolysis of the fatty acid ester in the 2-position of phospholipids. The substrate used in the standard assay is egg yolk phosphatidylcholine that has been solubilized in micelles of the nonionic detergent Triton X-100. The course of the enzymatic reaction is followed by titrating the fatty acid that is released.

[1] B. M. Braganca and Y. M. Sambray, *Nature (London)* **216**, 1210 (1967).
[2] J. I. Salach, P. Turini, R. Seng, J. Hauber, and T. P. Singer, *J. Biol. Chem.* **246**, 331 (1971).
[3] J. Shiloah, C. Klibansky, and A. de Vries, *Toxicon* **11**, 481 (1973).
[4] T. Cremona and E. B. Kearney, *J. Biol. Chem.* **239**, 2328 (1964).
[5] B. T. Currie, D. E. Oakley, and C. A. Broomfield, *Nature (London)* **220**, 371 (1968).
[6] R. A. Deems and E. A. Dennis, *J. Biol. Chem.* **250**, 9008 (1975).
[7] J. I. Salach, R. Seng, H. Tisdale, and T. P. Singer, *J. Biol. Chem.* **246**, 340 (1971).

Reagents

Egg phosphatidylcholine (purified from fresh egg yolks by the method of Singleton *et al.*[8]), 200 mM in $CHCl_3$
Triton X-100, 200 mM
$CaCl_2$, 200 mM
KOH, 5 mM

Procedure.[9] The free fatty acid that is produced in the enzymatic reaction is titrated with 5 mM KOH using a Radiometer pH stat apparatus consisting of a TTT60 titrator, an ABU 13 autoburette (0.25 ml burette), a pHm 62 pH meter with a combined glass–calomel electrode, and an REA recorder with an REA 300 bridge. A Lauda water bath is used to maintain the temperature at 40°. The total volume of the assay is 2.0 ml. The reaction is carried out at pH 8.0 with continuous stirring. A stream of nitrogen is passed over the assay mixture to prevent CO_2 absorption.

The pK_a of free fatty acids is 5–6 and, although the pK_a may be higher in aggregates, at pH 8.0 the fatty acids should be almost completely ionized. Therefore, the rate of the enzymatic reaction, d[fatty acid]/dt, is equal to the number of equivalents of KOH required to maintain the solution at pH 8.0 divided by the time interval. The reaction is generally run for 2.0 min. The KOH titrant is standardized with potassium acid phthalate.

The standard assay mixture contains 5 mM egg phosphatidylcholine, 20 mM Triton X-100, and 10 mM $CaCl_2$. The solution of egg phosphatidylcholine is placed in a homogenizing vessel, and the chloroform is removed under a stream of nitrogen. The sample is then placed under vacuum for 10 min to remove the last traces of chloroform and any stabilizer, usually ethanol, that is present in the chloroform. The Triton X-100 and $CaCl_2$ solutions are added, and the volume is brought to the appropriate level with deionized H_2O. The phospholipid is dispersed by either vortexing or homogenization. The solubilization of the phospholipid can be accelerated by heating the mixture to 40–50°. An aliquot of the assay mixture is placed in a polyethylene titration vessel and after thermal equilibrium is reached, usually within 5 min, the solution is brought to pH 8.0 with the titrator. The reaction is initiated by adding an appropriate amount of enzyme solution to the reaction mixture.

The base delivery tip and electrode are rinsed in sodium hypochlorite (Clorox), 0.1 N HCl and then deionized H_2O between each assay. This procedure is essential for reproducibility and presumably cleans the surfaces of any residual phospholipid, Triton X-100, or enzyme.

[8] W. S. Singleton, M. S. Gray, M. L. Brown, and J. L. White, *J. Am. Oil Chem. Soc.* **42**, 53 (1965).
[9] E. A. Dennis, *J. Lipid Res.* **14**, 152 (1973).

One unit of activity is the amount of enzyme required to hydrolyze 1 μmol of phospholipid per minute. The protein concentration is determined by the method of Lowry et al.[10] It should be noted that the Lowry procedure is used routinely because of its sensitivity and simplicity. However, it overestimates[11] the phospholipase A_2 present by 50% when bovine serum albumin is used as the standard and when compared to protein determinations using amino acid analysis, dry weights, and refractive index.[11] The biuret overestimates the protein by 20%.[11] Since the correction factor is valid only for the purified phospholipase A_2, it cannot be used during the purification procedure. Thus, only Lowry protein values are given in this chapter, and any physical or kinetic parameter that depends upon the protein concentration must have the protein term divided by 1.5 to get the parameter in terms of the absolute protein concentration.

The titration apparatus can measure rates between 20 and 300 nmol min^{-1} to an accuracy of $\pm 5\%$. The upper limit can be raised by appropriate changes in the burette size and KOH concentration. Decreasing the lower limit is more difficult because of limitations in the equipment and difficulties with CO_2 absorption that occur when very low concentrations of KOH are used.

While the reproducibility of the entire assay procedure is usually $\pm 5\%$, several factors can dramatically increase this value. In particular, it is important to ensure that the correct amount of phospholipid is in solution and that it is in the same physical state from one assay to the next. Thus, each time an assay mixture is prepared the same procedure should be followed exactly. Care should be taken to ensure that all the phospholipid has been solubilized and that the assay mixture is clear and devoid of any large masses of phospholipid. Also, since the assay mixture is not saturated in substrate (see the kinetics section), small changes in the concentration of detergent or phospholipid can affect the observed rates.

Purification Procedure

The basic purification scheme was first utilized by Braganca et al.[12] It was modified by Deems and Dennis,[6] and the final step was added by Roberts et al.[13] The table summarizes the results of a typical purification.

CAUTION. The lyophilized venom is usually a fine, light powder that is

[10] O. H. Lowry, N. J. Rosebrough, A. L. Farr, and R. J. Randall, J. Biol. Chem. 193, 265 (1951).

[11] P. L. Darke, A. A. Jarvis, R. A. Deems, and E. A. Dennis, Biochim. Biophys. Acta 626, 154 (1980).

[12] B. M. Braganca, Y. M. Sambray, and R. C. Ghadially, Toxicon 7, 151 (1969).

[13] M. F. Roberts, R. A. Deems, and E. A. Dennis, J. Biol. Chem. 252, 6011 (1977).

PURIFICATION OF PHOSPHOLIPASE A_2 FROM COBRA VENOM (*Naja naja naja*)

Step	Volume (ml)	Activity (units)	Protein (mg)	Specific activity (units mg^{-1})	Recovery (%)	Purification (fold)
1. Supernatant crude venom	550	757,350	9779	77	100	1
2. Neutralized $HClO_4$ precipitate	1360	618,800	3631	170	82	2.2
3. Supernatant after centrifugation	1360	463,760	1360	341	61	4.4
4. CM-cellulose	149	460,680	846	544	61	7.1
5. Sephadex G-100	220	403,046	661	610	53	7.9
6. DEAE-cellulose	440	360,800	319	1129	47	15

easily dispersed in the air. Extreme caution should be used to prevent this from occurring. Gloves and a mask are recommended for anyone handling the powdered venom. The LD_{50}, measured by subcutaneous injection of mice, rises from 0.7 mg kg^{-1} of body weight for the dissolved venom to 6.2 at the end of step 2.[12] The LD_{50} of the main peak from the Sephadex G-100 column is greater than 20 mg kg^{-1} of body weight.[12] The LD_{50} has not been determined for the final enzyme preparation, nor for venom that has been inhaled or ingested.

Step 1. The lyophilized crude venom from the cobra *Naja naja naja* (Pakistan), obtained from the Miami Serpentarium, is dissolved in glass-distilled H_2O at a protein concentration of 20 mg ml^{-1}. Undissolved material is removed by centrifugation (8000 g, 4°, 10 min).

Step 2. Four volumes of 6% perchloric acid are added to the clear supernatant at 4° with stirring. After the solution has stood for several hours the precipitate is collected by centrifugation (8000 g, 4°, 20 min). The precipitate is dissolved in the minimum amount of H_2O required to yield a clear solution. The final protein concentration is usually between 1 and 3 mg ml^{-1}. The pH of the solution is brought to 8.0 with 1 N NaOH.

Step 3. Since a white precipitate exists between pH 5 and 9, the solution is centrifuged (8000 g, 4°, 10 min), and the precipitate is discarded. The loss of activity in this step varies from 5 to 30%, whereas the protein loss is 30 to 50%. There is always an increase in specific activity.

Step 4. The clear supernatant is eluted from a CM-cellulose[14] (Whatman CM 11) column (4 × 50 cm) that has been equilibrated with 5 mM sodium phosphate buffer, pH 7.5. The major active peak comes off close to the breakthrough volume of the column and only requires the use of 5

[14] All the column materials were prepared according to the manufacturer's directions.

mM phosphate for elution. Two other peaks can be eluted from the CM-cellulose column with 100 and 500 mM phosphate buffers at pH 7.5. These peaks contain between 10 and 20% of the original activity. The major CM-cellulose peak is lyophilized and then dissolved in 100–200 ml of deionized H$_2$O.

Step 5. This protein is then eluted in several 50-ml batches from a Sephadex G-100 column (2.5 × 90 cm) with 50 mM sodium phosphate buffer, pH 7.5. Two peaks are routinely obtained on this column. The first comes in the void volume and contains less than 10% of the protein and only a trace of activity. The second peak contains most of the activity and protein. A smaller inactive peak, eluting after the major peak, is found in some preparations. The second peaks are combined, lyophilized, and dissolved in H$_2$O, 100–200 ml. This solution is dialyzed for 24 hr against deionized H$_2$O.

Step 6. The enzyme is then passed through a DEAE-cellulose (Whatman DE-11) column (4 × 40 cm)[13] that has been preequilibrated in 5 mM phosphate buffer at pH 7.0. A step gradient (0.1 M NaCl, 0.3 M NaCl) is used to elute two protein peaks. The first peak elutes in 5 mM phosphate, pH 7.0, containing 0.10 M NaCl. This peak represents about 20% of the protein and has a specific activity of about 100 units mg^{-1} and is discarded. The second peak elutes with 0.3 M NaCl, contains about 60% of the protein and has a specific activity of between 600 and 1100 units mg^{-1}.

Comment. We have recently found that desalting and a small additional purification of the enzyme can be achieved by elution of the DEAE peak on Affi-Gel Blue chromatography.[11]

Properties

Purity. This preparation of phospholipase A$_2$ gives single, coincidental activity and protein peaks on narrow-range isoelectric focusing and Sephadex G-100 gel filtration.[13] A single protein band is also found with both native and sodium dodecyl sulfate disc gel electrophoresis.[6,15]

Physical Properties. The isoelectric point of the enzyme is 5.1.[13] The $E_{280}^{1\%}$ is 14.5, at pH 8.0,[13] based upon the Lowry protein determination and 22.0 based upon the corrected protein concentration (see note in Assay Method section).[11] The enzyme has a broad plateau of activity between pH 7 and 9 and a pK_a between 5 and 6.[13]

The enzyme undergoes a concentration-dependent aggregation at pH 7.5.[6] Below 0.05 mg ml^{-1} phospholipase A$_2$ exists as a monomer, between

[15] A. A. Jarvis, P. L. Darke, C. R. Kensil, R. A. Deems, and E. A. Dennis, unpublished experiments.

about 0.1 and 2 mg ml⁻¹, it exists predominantly as a dimer, and above about 5 mg ml⁻¹ very large aggregates of the enzyme appear.

Estimates of the enzyme's monomer molecular weight were obtained from Sephadex G-100 gel filtration, sodium dodecyl sulfate disc gel electrophoresis, and analytical ultracentrifugation.[6] Numerous experiments using all three techniques and sequence analogy with enzymes from related snake venoms[11] indicate that the molecular weight is 13,000 ± 2000 g mol⁻¹.

Stability and Storage. The cobra venom phospholipase A_2, like many other phospholipases, has 6–7 disulfide bonds and is remarkably stable. The following are a few of the more notable examples.[15] Heating an enzyme solution at 100° for 10 min at pH 3–4 produces only a 5% loss in activity. The enzyme is soluble in many organic solvents (for example, chloroform, acetone, benzene, or ether) yet regains almost full activity when added to the standard assay mixture. Its elution profile and apparent Stokes' radius on Sephadex G-100 gel filtration does not change significantly, whether it is run in 10 mM Tris buffer or 6 M urea. The enzyme retains 60% of its activity when the normal assay solution contains 6 M guanidine-HCl. The complete alkylation of the disulfide bonds is very difficult even under very severe conditions. In light of these observations, caution should be used in planning any experiments that depend upon the denaturation of the protein.

Concentrated solutions of the enzyme stored at room temperature or at 4° develop a precipitate and lose activity with time. Similar results are obtained when the protein is lyophilized. Solutions of lyophilized enzyme are often opalescent but can be cleared by Millipore filtration, gel filtration, or centrifugation. The loss of protein ranges from 5 to 20%. In light of the enzymes's stability, these precipitates are probably due to its concentration-dependent aggregation rather than to its denaturation. To circumvent these problems the enzyme is routinely stored frozen in solutions of less than 2 mg ml⁻¹. Preparations stored in this fashion have shown no loss of activity nor any signs of precipitation for up to 12 months.[6]

Specificity. The enzyme catalyzes the hydrolysis of the fatty acid ester of L-phospholipids at the 2-position (denoted as *sn*-2 specificity).[6] Enzymatic activity is much greater toward substrates that are present as part of an interface than toward the same substrates as monomers.[16] The exact nature of the interface also exerts a strong influence on the observed rates of hydrolysis. In the micellar system employed here the effects of the phos-

[16] M. F. Roberts, A.-B. Otnaess, C. A. Kensil, and E. A. Dennis, *J. Biol. Chem.* **253**, 1252 (1978).

pholipid on the interface can be minimized. The chain length and degree of unsaturation of the fatty acid groups of the phospholipid also affect the enzymatic rates. This effect could be due to an enzymatic specificity. It might also represent the influence of changes in the interface caused by the different fatty acid compositions of the phospholipids.

When a single kind of phospholipid is present in the assay mixture, the specificity of the enzyme is[16] phosphatidylcholine > phosphatidyl-ethanolamine > phosphatidylserine. The introduction of various "activator" lipids into the assay causes the following change in specificity[17]: phosphatidylethanolamine > phosphatidylcholine. These activators need not be incorporated into the substrate micelle; their presence in the bulk phase of the solution is sufficient to cause this effect. The best activator compounds for phosphatidylethanolamine found so far are phosphatidyl-choline and sphingomyelin.[18]

Kinetic Parameters. The kinetic analysis of phospholipase activity toward interfacial substrates is very complex.[19,20] The enzymatic rate depends not only on the phospholipid's bulk concentration, but also on its concentration in the interface. The surface concentration in the assay system presented here is a function of the mole ratio of Triton X-100 to phospholipid. If this mole ratio is held constant at 2 : 1, an apparent K_m of 2–5 mM and an apparent V_{max} of 2000 units mg^{-1} are obtained.[9] The relationship of these values to classical enzyme kinetic constants is not clear. However, it appears that in order to saturate the enzyme the phospholipid concentration should be 10 mM or greater, and the mole ratio of Triton X-100 to phospholipid should be less than 2 : 1. The physical characteristics of this micellar system prevents these conditions from being employed in the standard assay. Since the standard assay system is not saturating in substrate, the assay has an increased sensitivity to small differences in both phospholipid and Triton X-100 concentrations, as well as to subtle changes in the interface.

Inhibitors. This phospholipase A$_2$ exhibits an absolute requirement for Ca^{2+} with a K_D of 0.15 mM; Ba^{2+} and Sr^{2+} are competitive inhibitors of this cofactor. They both have a K_D of about 0.6 mM.[13] p-Bromophenacyl bromide and its analogs are potent inhibitors that covalently modify a histidine residue. The apparent rate constant of inhibition by p-bromophenacyl bromide is 0.13 min^{-1}.[21]

[17] M. F. Roberts, M. Adamich, R. J. Robson, and E. A. Dennis, *Biochemistry* **18,** 3301 (1979).

[18] M. Adamich, M. F. Roberts, and E. A. Dennis, *Biochemistry* **18,** 3308 (1979).

[19] E. A. Dennis, *Arch. Biochem. Biophys.* **158,** 485 (1973).

[20] R. A. Deems, B. R. Eaton, and E. A. Dennis, *J. Biol. Chem.* **250,** 9013 (1975).

[21] M. F. Roberts, R. A. Deems, T. C. Mincey, and E. A. Dennis, *J. Biol. Chem.* **252,** 2405 (1977).

The effects of fatty acids on the enzymatic activity are complex. Not only do they interact directly with the enzyme, but also they can influence enzymatic activity through their effects on the interface. Under various conditions they can be either activators or inhibitors.[15]

Comment. Cobra venom phospholipase A_2 has been used extensively in studies of the structure and function of membranes. The variability of enzymatic rates and specificities in response to the lipid composition and nature of the interface complicate the interpretation of these experiments unless stringent controls are performed.

Acknowledgments

Financial support was provided by NIH Grant GM 20,501 and NSF Grant PCM 76-21552.

[82] Phospholipase C from *Clostridium perfringens*

EC 3.1.4.3 Phosphatidylcholine cholinephosphohydrolase

By T. Takahashi, T. Sugahara, and A. Ohsaka

1,2-Diacyl-*sn*-glycero-3-phosphatide + H_2O →
$\qquad\qquad$ 1,2-diacylglycerol + phosphorylalcohol

Phospholipase C (α-toxin) of *Clostridium perfringens* is one of the most active bacterial phospholipases C that degrade phospholipids in biomembranes. This phospholipase C is strongly lethal, necrotizing, and hemolytic,[1] in contrast to most bacterial enzymes including *Bacillus cereus* phospholipase C,[2-5] which are entirely devoid of such activities.

[1] M. G. Macfarlane and B. C. J. G. Knight, *Biochem. J.* **35**, 884 (1941).
[2] R. F. A. Zwaal, B. Roelofsen, P. Comfurius, and L. L. M. van Deenen, *Biochim. Biophys. Acta* **233**, 474 (1971).
[3] D. M. Molnar, *J. Bacteriol.* **84**, 147 (1962).
[4] C. Little, B. Aurebekk, and A.-B. Otnaess, *FEBS Lett.* **52**, 175 (1975).
[5] H. Ikezawa and R. Taguchi, this volume [84].

Purification

Various attempts have been made to purify *C. perfringens* phospholipase C,[6-14] but only a few have succeeded in isolating the enzyme with a high specific activity in a reasonable yield.[6,7] We describe here two alternative purification procedures—one based on affinity chromatography on agarose-linked egg yolk lipoprotein[7] and the other including ion-exchange and molecular sieve chromatography.[6]

Method 1. Purification by Affinity Chromatography[7]

This procedure is a unique and elegant method for the purification of the enzyme and is applicable to phospholipases C from other sources.[4]

All the purification steps described below are carried out at 4°, unless otherwise indicated.

Step 1. Preparation of Crude Phospholipase C. *Clostridium perfringens* strain PB6K (ATCC No. 10543) is grown in a peptone medium containing meat particles,[15] which adsorb the θ-toxin produced by the organism and stimulate phospholipase C production. The medium contains the following constituents per liter (a modification of the medium of Murata *et al.*[16]): 45 g of proteose peptone (Difco), 14.32 g of $Na_2HPO_4 \cdot 12 H_2O$, 1.36 g of KH_2PO_4, 0.2 g of $MgSO_4 \cdot 7 H_2O$, 0.5 g of yeast extract (Difco), 5.75 mg of $ZnSO_4 \cdot 7 H_2O$, 3.94 mg of $MnCl_2 \cdot 4 H_2O$, 10 g of fructose, 7 mg of $FeSO_4 \cdot 7 H_2O$, 1 g of $KHCO_3$, 0.01 ml of thioglycolic acid, and 5 g of meat particles (Difco). The medium is adjusted to pH 7.6–7.8 with 10 N NaOH and sterilized. After cultivation for 5 hr at 37°, the organisms are removed by centrifugation. The culture filtrate, containing antitoxin-binding power[17] of usually 9–11 Lv/ml, is precipitated with solid $(NH_4)_2SO_4$ between 20% and 80%[18] saturation. The precipitate is dis-

[6] Y. Yamakawa and A. Ohsaka, *J. Biochem. (Tokyo)* **81**, 115 (1977).
[7] T. Takahashi, T. Sugahara, and A. Ohsaka, *Biochim. Biophys. Acta* **351**, 155 (1974).
[8] R. Möllby and T. Wadström, *Biochim. Biophys. Acta* **321**, 569 (1973).
[9] K. Mitsui, N. Mitsui, and J. Hase, *Jpn. J. Exp. Med.* **43**, 65 (1973).
[10] W. L. Stahl, *Arch. Biochem. Biophys.* **154**, 47 (1973).
[11] A. Casu, V. Pala, R. Monacelli, and G. Nanni, *Ital. J. Biochem.* **20**, 166 (1971).
[12] M. V. Ispolatovskaya, *Vopr. Med. Khim.* **17**, 137 (1971).
[13] B. A. Diner, *Biochim. Biophys. Acta* **198**, 514 (1970).
[14] G. F. Shemanova, E. V. Vlasova, V. S. Tsvetkov, A. I. Loginov, and F. B. Levin, *Biokhimiya* **33**, 130 (1968).
[15] S. Kameyama and K. Akama, *Jpn. J. Med. Sci. Biol.* **24**, 9 (1971).
[16] R. Murata, A. Yamamoto, S. Soda, and A. Ito, *Jpn. J. Med. Sci. Biol.* **18**, 189 (1965).
[17] See Assay Methods: Turbidimetric Method. Lv stands for lecithovitellin reaction.
[18] Fifty percent saturation of ammonium sulfate is sufficient to precipitate the enzyme quantitatively.[6]

solved in a minimum volume of 0.05 M Tris-HCl buffer (pH 7.5) and dialyzed exhaustively against the buffer. The dialysis residue is centrifuged to remove the precipitate formed and the supernatant fluid is concentrated through a Diaflo membrane UM-2 (Amicon Corp.). The crude enzyme thus obtained, containing per milliliter about 70 mg of protein and 300 Lv, is stable for at least a year when stored at $-20°$.

Step 2. Affinity Chromatography on Agarose-Linked Egg Yolk Lipoprotein

PREPARATION OF AGAROSE-LINKED EGG YOLK LIPOPROTEIN. The egg yolk lipoprotein is coupled to Sepharose 4B activated with BrCN[19] according to the method of Cuatrecasas *et al.*[20] with slight modifications. Sepharose 4B (settled volume 130 ml) is suspended in an equal volume of distilled water, and solid BrCN (30 g) is added to the suspension. The mixture is immediately adjusted to pH 11, and this pH is maintained for 20 min with 5 N NaOH. The gel is collected and washed with 1.5 liters of ice-cold 0.1 M sodium borate–NaOH buffer (pH 9.6) on a Büchner funnel. The washed gel is added to 130 ml of an egg yolk lipoprotein solution, which has been prepared by homogenizing one egg yolk and 5 g of Super Cel (Wako Pure Chemicals) in 200 ml of the same buffer and then by centrifuging the mixture at 8000 rpm for 15 min to discard the precipitate. The coupling reaction is allowed to proceed overnight under gentle stirring in a cold room. The conjugated gel collected is washed successively with a few liters each of distilled water, 0.5 M NaCl–0.05 M Tris-HCl buffer (pH 7.5), 0.02% sodium deoxycholate (SDC)–0.5 M NaCl–0.05 M Tris-HCl buffer (pH 7.5), and finally 30% (v/v) glycerol–0.05 M Tris-HCl buffer (pH 7.5). The conjugated gel thus prepared is stable for 1 month when stored in 0.05 M Tris-HCl buffer (pH 7.5) at 4° and is stable for more than 1 month when stored in the 30% (v/v) glycerol-supplemented buffer at $-20°$. The conjugated gel can be used repeatedly (at least four times) without losing any capacity when regenerated by washing with 30% (v/v) glycerol-supplemented Tris-HCl buffer, pH 7.5.

AFFINITY CHROMATOGRAPHY OF CRUDE PHOSPHOLIPASE C. A 10-ml portion of crude phospholipase C is mixed with 30% (v/v) glycerol, then applied to a column (1.5 × 8 cm) of Sepharose-linked egg yolk lipoprotein previously equilibrated with 30% (v/v) glycerol–0.05 M Tris-HCl buffer, pH 7.5. As shown in Fig. 1, not only phospholipase C activity but also lethal and hemolytic activities are strongly adsorbed onto the column, and

---

[19] BrCN-activated Sepharose is commercially available from Pharmacia Fine Chemicals, Uppsala, Sweden.

[20] P. Cuatrecasas, M. Wilchek, and C. B. Anfinsen, *Proc. Natl. Acad. Sci. U.S.A.* **61**, 636 (1968).

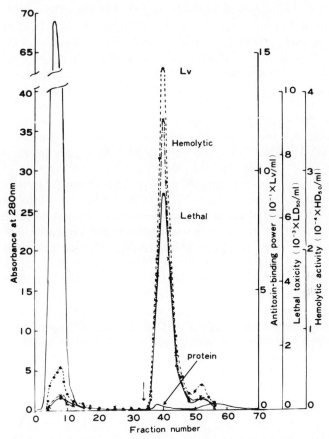

FIG. 1. Affinity chromatography of crude phospholipase C (α-toxin) on Sepharose-linked egg yolk lipoprotein. A mixture consisting of 10 ml of crude phospholipase C (70.5 mg of protein per milliliter) and 4 ml of glycerol was applied to a column (1.5 × 8 cm) of Sepharose-linked egg yolk lipoprotein, previously equilibrated with 30% glycerol-containing 0.05 M Tris-HCl buffer (pH 7.5). After the bulk of proteins had been percolated with the same buffer, elution was initiated with the buffer supplemented with 0.5 M NaCl and 0.02% sodium deoxycholate. The vertical arrow indicates the place where the elution was started. Four-milliliter fractions were collected at a flow rate of 50–60 ml/hr. ⸻, $A_{280\,nm}$; ●---●, antitoxin-binding power; ○——○, lethal toxicity; X--·X, hemolytic activity. Reprinted from Takahashi *et al.*[7]

the bulk of the applied protein passes through the column. Phospholipase C activity, together with lethal and hemolytic activities, is quantitatively eluted with an eluent system of 0.05 M Tris-HCl (pH 7.5)–30% (v/v) glycerol, containing 0.5 M NaCl and 0.02% SDC. Omission of any of the three constituents of the eluent, NaCl, SDC, or glycerol, results in a lower recovery of enzyme activity.

Step 3. Gel Filtration on Sephadex G-100. The phospholipase C fraction pooled from several runs of affinity chromatography is concentrated through a Diaflo membrane UM-2, and this is dialyzed exhaustively against 0.05 M Tris-HCl buffer, pH 7.5. To remove the remaining SDC and glycerol as well as contaminating proteins, the dialysis residue (54 mg of protein in 2 ml) is submitted to gel filtration in a column (2 × 100 cm) of Sephadex G-100 equilibrated with 0.05 M Tris-HCl buffer, pH 7.5. The fractions containing phospholipase C are combined, concentrated, and dialyzed against the buffer. The purified enzyme thus prepared is stable for at least 3 months when stored at −20° and at concentrations higher than 1–2 mg of protein per milliliter.

As shown in Table I, approximately 200-fold purification is attained from the crude phospholipase C (2500-fold purification from the culture filtrate) with a yield of about 60% with respect to phospholipase C activity in terms of lecithin-hydrolyzing and antitoxin-binding capacities and also with respect to lethal and hemolytic activities. The increase in potency of enzymatic, lethal, and hemolytic activities run parallel throughout the purification, indicating that the three activities reside in a single entity. The purified phospholipase C, being free from θ-hemolytic activity, is homogeneous as judged by polyacrylamide gel electrophoresis with or without sodium dodecyl sulfate (SDS), immunodiffusion, and also ultracentrifugation. The preparation has the highest specific activity among all the preparations so far reported.

Method 2. Purification by a Conventional Method[6]

The purification method described below includes successive chromatography on CM-Sephadex, DEAE-Sephadex, and Sephadex G-100. The method is suitable for large-scale preparations.

Step 1. Preparation of Crude Enzyme. Culture filtrate (10 liters) obtained as described in Method 1 is concentrated to about 1 liter by ultrafiltration through Hollow Fiber (Amicon HIP 10; MW cutoff 10,000), resulting in about 10-fold increase in specific activity. To the concentrated solution, 1.0 M calcium acetate is added to a final concentration of 0.1 M, the pH of the mixture being maintained at 7.0 by the addition of 5 N NaOH. After the precipitate formed has been removed by centrifugation, solid $(NH_4)_2SO_4$ is added to the supernatant fluid to 50% saturation. The resulting precipitate is collected by centrifugation, and this is dissolved in a small volume (30–50 ml) of 0.05 M Tris-HCl buffer (pH 7.5) containing 0.15 M NaCl. The crude enzyme thus obtained, which contained 5 g of protein and more than 80% of the enzyme activity of the original culture filtrate, was stable for more than 6 months when stored at −20°. The increase in specific activity was about 45-fold.

Step 2. CM-Sephadex Column Chromatography. The crude enzyme is dialyzed exhaustively against 0.05 M acetate buffer (pH 5.0) containing 5% (v/v) ethanol. The dialysis residue (4.56 g of protein), after centrifugation to remove the precipitate, is applied to a column (2 × 25 cm) of CM-Sephadex C-50 equilibrated with the same buffer. Elution is carried out stepwise, first with the buffer (600 ml) and then with the buffers containing 0.05 M (500 ml), 0.1 M (400 ml), and 1.0 M NaCl (500 ml). The activities recovered in eluates with buffers 1–4 were 2.3, 17.5, 29.2, and 2.2%, respectively. The enzyme emerging at 0.05 M NaCl was found to be in a polymer form(s), while that emerging at 0.1 M NaCl was proved to be in a monomer form(s). Only the enzyme in the monomer form(s) is further purified in the subsequent steps.

Step 3. DEAE-Sephadex Column Chromatography. The material from the previous step (0.1 M NaCl eluate) is collected by precipitation with ammonium sulfate at 60% saturation, and this is dissolved in a minimum volume of 0.05 M Tris-HCl buffer (pH 7.5) containing 0.05 M NaCl and 5% (v/v) ethanol. The solution containing 135 mg of protein, after exhaustive dialysis against the same buffer, is applied to a column (1.5 × 20 cm) of DEAE-Sephadex A-50 equilibrated with the buffer. The column is eluted with about 400 ml of the buffer, followed by stepwise elution with 0.125 M and 1.0 M NaCl in 0.05 M Tris-HCl buffer (pH 7.5) containing 5% (v/v) ethanol. Elution with the 0.125 M NaCl-containing buffer is continued until the enzymatic activity in the eluate falls below 5 Lv/ml. At this NaCl concentration, about 70% of the input activity is recovered.

Step 4. Gel Filtration on Sephadex G-100. The material from the preceding step (0.125 M NaCl eluate), after concentration by ultrafiltration and precipitation with ammonium sulfate, is applied to a column (2.6 × 90 cm) of Sephadex G-100, superfine, equilibrated with the same buffer. The column is eluted with the buffer. Phospholipase C was eluted as a single peak with some preceding protein and activity. Fractions that proved to be homogeneous in polyacrylamide gel electrophoresis were pooled and precipitated with ammonium sulfate at 60% saturation; the resulting suspension was stored at 4° until use. The purified enzyme is homogeneous as judged by polyacrylamide gel electrophoresis, whether or not denaturing reagents are present, and also in immunodiffusion.

The purification and yield at each step are summarized in Table II. The specific lecithin-hydrolyzing activity of the purified preparation is comparable to that of the preparation obtained by affinity chromatography (Method 1).[7]

Properties

Multiple Molecular Forms. Heterogeneity of *C. perfringens* phospholipase C in the purified as well as crude preparations has been shown

TABLE I

SUMMARY OF PURIFICATION OF PHOSPHOLIPASE C (α-TOXIN) BY METHOD 1[a]

Step	Protein (mg)	Lecithin-hydrolyzing activity[b]			Antitoxin-binding power			Lethal toxicity			Hemolytic activity		
		Total activity (units $\times 10^{-4}$)	Specific activity (units/mg)	Yield (%)	Total activity (Lv $\times 10^{-4}$)	Specific activity (Lv/mg)	Yield (%)	Total activity (LD$_{50}$ $\times 10^{-5}$)	Specific activity (LD$_{50}$ $\times 10^{-2}$/mg)	Yield (%)	Total activity (HD$_{50}$ $\times 10^{-6}$)	Specific activity (HD$_{50}$ $\times 10^{-3}$/mg)	Yield (%)
Starting material, (NH$_4$)$_2$SO$_4$ precipitate	7050	6.3	8.9	100	3.0	4.3	100	11.0	1.56	100	14	2	100
Affinity chromatography	54	4.1	759	65	2.4	444	80	8.6	159	78	12	222	86
Gel filtration on Sephadex G-100	21	3.4	1620	54	2.1	1000	70	6.3	300	57	8.6	409	61

[a] Reprinted from Takahashi et al.[7]

[b] Lecithin-hydrolyzing activity was determined by the phosphatase method (Ohsaka and Sugahara[37]).

TABLE II

SUMMARY OF PURIFICATION OF PHOSPHOLIPASE C (α-TOXIN) BY METHOD 2^a

| Step | Protein (A_{280nm}) | Antitoxin-binding power | | | Lecithin-hydrolyzing activity specific activity (mole P/$A_{280\,nm}$ min^{-1}) |
		Total activity (Lv $\times$ 10^{-3})	Recovery (%)	Specific activity (Lv/$A_{280\,nm}$)	
Crude toxin	4560	81.3	100	17.8	—
CM-Sephadex NaCl, 0.1 M (monomer fraction)	135.1	24.3	29.9	179.9	—
NaCl, 0.05 M (polymer fraction)	480.2	14.2	17.5	29.6	—
DEAE-Sephadex	45.6	17.0	20.9	372.8	—
Sephadex G-100	23.0	12.3	15.1	534.8	2317.1^b

[a] Reprinted from Yamakawa and Ohsaka.[6]
[b] The value corresponds to 1575 μmol of P per minute per milligram (BSA eq.) of enzyme.

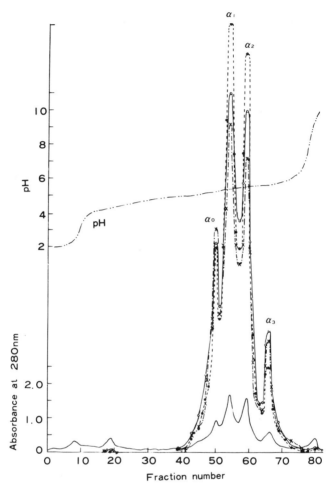

Fig. 2. Isoelectric focusing of the purified phospholipase C (α-toxin). The purified enzyme (20 mg of protein) prepared by Method 1 was subjected to isoelectric focusing with carrier ampholytes (pH 4–6), which was conducted in a gradient of glycerol (0% to 50%, v/v) for 20.7 hr[30] at 4° with a final potential drop of 840 V. Among the multiple monomer forms of enzyme separated, three major ones and one minor one were designated as α_0-, α_1-, α_2-, and α_3-toxins (phospholipases C) as indicated. From the control run, a small peak of $A_{280\,nm}$ focused on pH 2.4 was proved to have originated from the ampholytes themselves. ————, A_{280nm}; ●----●, antitoxin-binding power; ○——○, lethal toxicity; X–·–X, hemolytic activity; —·—, pH. Reprinted from Takahashi et al.[7]

by electrophoresis[12,21] and column chromatography[22,23] and more clearly by isoelectric focusing.[7,8,24-29] Some part of the observed multiplicity of enzyme may be ascribed to the formation of polymer(s), because under some experimental conditions a part of the enzyme is found to exist in active polymerized forms besides the monomer form.[6] Furthermore, the monomer form of enzyme consists of multiple molecular forms[7]; upon isoelectric focusing with carrier ampholytes (pH 4–6) in a gradient of glycerol (0 to 50%, v/v), the phospholipase C monomer purified by affinity chromatography is resolved into three major molecular forms and a few minor ones (Fig. 2).[7,30] Distribution of protein exactly coincides with that of enzymatic, lethal, and hemolytic activities, in support of the supposition that the three activities reside in a single entity. The four molecular forms are designated as α_0-, α_1-, α_2- and α_3-toxins (phospholipases C), their isoelectric points being 5.2, 5.3, 5.5, and 5.6, respectively.[7] The isolated multiple forms are distinguishable niether from one another nor from the parent enzyme (α-toxin), not only by polyacrylamide gel electrophoresis with or without SDS and immunodiffusion but also by their susceptibility to heat and pH. The structural difference of these multiple forms of enzyme has not yet been elucidated.

Molecular Weight. Various values for the molecular weight of *C. perfringens* phospholipase C have been reported.[6,31,32] The values range from 30,000 to 106,000, depending on different preparations and also different methods of estimation used. Much of this discrepancy may be accounted for by the presence of active polymers of enzyme.[6,31] The molecular

[21] M. V. Ispolatovskaya, *Biokhimiya* **29**, 869 (1964).

[22] V. Macchia and I. Pastan, *J. Biol. Chem.* **242**, 1864 (1967).

[23] M. V. Ispolatovskaya and G. A. Levdikova, *Biokhimiya* **27**, 631 (1962).

[24] C. J. Smyth and T. Wadström, *Anal. Biochem.* **65**, 137 (1975).

[25] C. J. Smyth and J. P. Arbuthnott, *J. Med. Microbiol.* **7**, 41 (1974).

[26] R. A. Bird, M. G. Low, and J. Stephen, *FEBS Lett.* **44**, 279 (1974).

[27] R. Mölby, C.-E. Nord, and T. Wadström, *Toxicon* **11**, 139 (1973).

[28] T. Sugahara and A. Ohsaka, *Jpn. J. Med. Sci. Biol.* **23**, 61 (1970).

[29] A. W. Bernheimer, P. Grushoff, and L. S. Avigad, *J. Bacteriol.* **95**, 2439 (1968).

[30] Electric focusing was run to the time point where the current became constant (duration about 20.5 hr). To obtain a quantitative recovery of the activity, great care was taken to avoid any prolonged run over this period. It was noted that the enzyme, when resolved into multiple forms, was readily inactivated unless an equal volume of 0.5 *M* Tris-HCl buffer (pH 7.5) containing glycerol (50%, v/v), gelatin (2 mg/ml), and Co^{2+} (2×10^{-5} *M*) was added; with these additions the time required for 50% loss of activity was prolonged from nearly a week to a month when stored at $-20°$.

[31] Y. Yamakawa, T. Takahashi, T. Sugahara, and A. Ohsaka, *in* "Animal, Plant, and Microbial Toxins" (A. Ohsaka, K. Hayashi, and Y. Sawai, eds.), Vol. 1, p. 409. Plenum, New York, 1976.

[32] R. Möllby, *in* "Bacterial Toxins and Cell Membranes" (J. Jeljaszewicz and T. Wadström, eds.), p. 367. Academic Press, New York, 1978.

weight for the extensively purified enzyme (monomer) was estimated to be 43,000 by SDS–polyacrylamide gel electrophoresis,[6,7] although the same preparation gave an apparent molecular weight of 31,000 by gel filtration.[6] The enzyme may interact in some way with dextran gel, resulting in underestimation of the molecular weight in gel filtration experiments.[6,8,33,34] The sedimentation coefficient of the enzyme was estimated to be 3.9 S.[7]

Stability. When purified enzyme was incubated for 1 hr at 37° in 0.01 M buffer (I = 0.1) of various pH values, it retained full activity between pH 5 and pH 10.[7] Upon heating for 5 min in 0.1 M sodium borate-HCl buffer (pH 7.0), purified enzyme exhibited an anomalous response to heat treatment, as did crude enzyme.[7] The enzyme is extensively inactivated at 60–70°, but is reactivated at higher temperatures with almost complete restoration of activity at 100°.[7,35] On dilution the enzyme is easily inactivated through surface denaturation.[1,21,35,36]

pH Optimum. When phosphatidylcholine is used as a substrate, the pH optimum lies between 7.0 and 8.0.[1,11,36,37]

Substrate Specificity. *Clostridium perfringens* phospholipase C shows a broad specificity toward various phospholipids. The rate of hydrolysis by the enzyme is in the following decreasing order[10]: lecithin > sphingomyelin > phosphatidylserine. The enzyme can also degrade phosphatidylethanolamine,[10,38–43] lysophosphatidylcholine,[10,40] choline plasmalogen,[44–47] ceramide phosphorylethanolamine[38,48] and ceramide

[33] A. W. Bernheimer and P. Grushoff, *J. Gen. Microbiol.* **46**, 143 (1967).

[34] G. Teodorescu, J. Bittner, and A. Ceacăreanu, *Arch. Roum. Pathol. Exp. Microbiol.* **29**, 541 (1970).

[35] L. D. S. Smith and M. V. Gardner, *Arch. Biochem.* **25**, 54 (1950).

[36] P. C. Zamecnik, L. E. Brewster, and F. Lipmann, *J. Exp. Med.* **85**, 381 (1947).

[37] A. Ohsaka and T. Sugahara, *J. Biochem.* (*Tokyo*) **64**, 335 (1968).

[38] K. Saito and K. Mukoyama, *Biochim. Biophys. Acta* **164**, 596 (1968).

[39] A. D. Bangham and R. M. C. Dawson, *Biochim. Biophys. Acta* **59**, 103 (1962).

[40] T. Takahashi and H. H. O. Schmid, *Chem. Phys. Lipid* **2**, 220 (1968).

[41] M. Matsumoto, *J. Biochem.* **49**, 23 (1961).

[42] J. de Gier, G. H. de Haas, and L. L. M. van Deenen, *Biochem. J.* **81**, 33p (1961).

[43] The hydrolysis of phosphatidylethanolamine has been the point of discussion. Although some workers failed to demonstrate the hydrolysis of this phospholipid without addition of lecithin[10,42] or lysolecithin,[40] some others detected the hydrolysis without these additions.[38,39]

[44] G. B. Ansell and S. Spanner, *Biochem. J.* **97**, 375 (1965).

[45] E. L. Gottfried and M. M. Rapport, *J. Biol. Chem.* **237**, 329 (1962).

[46] J. Y. Kiyasu and E. P. Kennedy, *J. Biol. Chem.* **235**, 2590 (1960).

[47] G. M. Gray and M. G. Macfarlane, *Biochem. J.* **70**, 409 (1958).

[48] T. Hori, I. Arakawa, M. Sugita, and O. Itasaka, *J. Biochem.* (*Tokyo*) **64**, 533 (1968).

aminoethylphosphonate.[38,48] Synthetic lecithins with different lengths of aliphatic side chain are hydrolyzed.[49] Various K_m values for the enzyme have been reported with egg lecithin as substrate: 1.6×10^{-5} (Casu *et al.*[11]), 2.1×10^{-4} (Jain and Cordes[50]), and 1.2×10^{-3} M (Ohsaka and Sugahara[37]). The marked difference among these values may be ascribed to the different assay systems used. A chromogenic synthetic substrate, *p*-nitrophenylphosphorylcholine, is hydrolyzed with the K_m value of 0.2 M.[51] Pastan *et al.*[52] claimed that *C. perfringens* produced, besides the hitherto known phospholipase C with a substrate preference for lecithin, another phospholipase C with a preference for sphingomyelin. However, the existence of a sphingomyelin-specific enzyme has been questioned.[7,31,32]

Requirement of Detergent and Calcium Ions for Maximal Enzymatic Activity. Since purified phospholipids as substrates are sparingly soluble in water, attempts have been made to make the enzymatic reaction proceed in an ethereal medium[53] or in an aqueous medium containing a detergent.[37,39] The enzyme activity in an aqueous medium is greatly influenced by physical states of phospholipid micelles, namely, the charge and size of micelles.[39,50,54,55] For the assay of phospholipase C activity, detergents such as SDC have been frequently used in the presence of Ca^{2+} ions to disperse phospholipids.[37,56] The rate of hydrolysis of lecithin by the enzyme is greatly enhanced by SDC and, to a lesser extent, by sodium cholate and sodium dehydrocholate.[6] Not only the ratio of detergent to lecithin but also that of detergent to Ca^{2+} ions is of importance for the maximum hydrolysis, an optimal molar ratio of SDC to substrate being about 0.5 or higher for dipalmitoyllecithin and 1.0 or higher for egg lecithin, in the presence of 5–10 mM Ca^{2+} ions.[6] The conflicting results reported on the effect of detergents on enzymatic hydrolysis of phospholipids may be due to the fact that the molar ratio of detergent to phospholipid used for the assay was not always optimal.

Activators and Inhibitors. Clostridium perfringens phospholipase C is activated greatly by Ca^{2+} and Mg^{2+} ions[1,36,37,39] and, to a lesser extent, by

[49] L. L. M. van Deenen, G. H. de Haas, C. H. T. Heemskerk, and J. Meduski, *Biochem. Biophys. Res. Commun.* **4**, 183 (1961).

[50] M. K. Jain and E. H. Cordes, *J. Membrane Biol.* **14**, 119 (1973).

[51] S. Kurioka and M. Matsuda, *Anal. Biochem.* **75**, 281 (1976).

[52] I. Pastan, V. Macchia, and R. Katzen, *J. Biol. Chem.* **243**, 3750 (1968).

[53] D. J. Hanahan and R. Vercamer, *J. Am. Chem. Soc.* **76**, 1804 (1954).

[54] R. M. C. Dawson, *in* "Form and Function of Phospholipids" (G. B. Ansell, J. N. Hawthorne, and R. M. C. Dawson, eds.), B. B. A. Library, Vol. 3, p. 97. Elsevier, Amsterdam, 1973.

[55] R. M. C. Dawson, N. L. Hemington, N. G. A. Miller, and A. D. Bangham, *J. Membrane Biol.* **29**, 179 (1976).

[56] S. Kurioka and P. V. Liu, *Appl. Microbiol.* **15**, 551 (1967).

Co^{2+}, Mn^{2+} and Zn^{2+} ions.[6,36] The stimulating effect of Ca^{2+} ions is strongly suppressed by Cd^{2+}, Pb^{2+}, Ni^{2+}, and Hg^{2+} ions.[6] The enzyme activity is inhibited by some metal-chelating agents, such as EDTA and o-phenanthroline, and this inhibition is completely or partly restored by the addition of divalent cations such as Zn^{2+}, Co^{2+}, and Mn^{2+} ions, but not by Ca^{2+} ions.[57,58] Considering the fact that the organisms absolutely require Zn^{2+} ions for the production of phospholipase C,[59] it may be stated that the enzyme is a zinc metalloenzyme[51,60] which requires Ca^{2+} ions for manifestation of its activity.[60] The enzyme is inhibited by the following reagents: fluoride,[1,41] ferricyanite,[39] SDS[1,51] and reducing agents[61] such as glutathione, cysteine, and thioglycolic acid. The enzyme is also inhibited by polyamines, such as spermine, spermidine, and putrescine.[62] Phosphonate[63] and phosphinate[64] analogs of lecithin are competitive inhibitors of this enzyme.

Biological Activities. *Clostridium perfringens* phospholipase C is considered to play the most important role in the pathogenesis of gas gangrene.[60] Only a few attempts have been made to correlate the enzymatic activity with biological activities.[7,28,65,66] The highly purified phospholipase C as well as all of the multiple forms of enzyme isolated by isoelectric focusing had a similar potency with respect to any of the following biological activities: lethal toxicity,[7] hemolytic activity,[7] vascular permeability-increasing activity,[65] and *in vitro* platelet-aggregating activity.[66] It may be justifiable, therefore, to conclude that these biological activities reside in a single entity (phospholipase C). Many studies have dealt with the effects of the enzyme on isolated cells and biomembranes as well as the *in vivo* effects.[32,67,68] Much caution must be taken, however, in interpreting the results obtained with enzyme preparations of different purities.

[57] M. V. Ispolatovskaya, *Biokhimiya* **35**, 434 (1970).
[58] M. Moskowitz, M. W. Deverell, and R. McKinney, *Science* **123**, 1077 (1956).
[59] H. Sato and R. Murata, *Infect. Immun.* **8**, 360 (1973).
[60] M. V. Ispolatovskaya, in "Microbial Toxins" (S. Kadis, T. C. Montie, and S. J. Ajl, eds.), Vol. IIA, p. 109. Academic Press, New York, 1971.
[61] M. V. Ispolatovskaya and L. V. Klimachova, *Biokhimiya* **31**, 491 (1966).
[62] A. M. Sechi, L. Cabrini, L. Landi, P. Pasquall, and G. Lenaz, *Arch. Biochem. Biophys.* **186**, 248 (1978).
[63] A. F. Rosenthal and M. Pousada, *Biochim. Biophys. Acta* **164**, 226 (1968).
[64] A. F. Rosenthal, S. V. Chodsky, and S. C. H. Han, *Biochim. Biophys. Acta* **187**, 385 (1969).
[65] T. Sugahara, T. Takahashi, S. Yamaya, and A. Ohsaka, *Toxicon* **15**, 81 (1977).
[66] T. Sugahara, T. Takahashi, S. Yamaya, and A. Ohsaka, *Jpn. J. Med. Sci. Biol.* **29**, 255 (1976).
[67] A. Ohsaka, M. Tsuchiya, C. Oshio, M. Miyairi, K. Suzuki, and Y. Yamakawa, *Toxicon* **16**, 333 (1978).
[68] A. Ohsaka, T. Sugahara, T. Takahashi, and S. Yamaya, in "Toxins: Animal, Plant and Microbial" (P. Rosenberg, ed.), p. 1031. Pergamon, Oxford, 1978.

Assay Methods

Various methods for the assay of phospholipase C have been reported: (1) turbidimetric method,[16,69] (2) acid-soluble phosphorus method,[1,6,37] (3) manometric method,[36] (4) titrimetric method,[53] (5) phosphatase method,[37,56,70] (6) p-nitrophenylphosphorylcholine method,[51,71] (7) organic solvent extraction method,[13] and (8) bioluminescence method.[72] None of these methods is strictly specific for phospholipase C. To characterize an enzyme as a phospholipase C, products of enzymatic reaction must be identified.

Because of their practicability, we chose the assay methods 1 and 2 for further description. For routine assays of enzyme activity during purification, the turbidimetric method[16,69] has widely been used; it is based on determination of the turbidity developed from the enzymatic reaction with egg yolk. The method, although simple to perform, has serious faults in that the reaction mechanism is obscure and the preparation of egg yolk having a uniform quality is difficult to obtain. To determine phospholipase C activity more precisely, the acid-soluble phosphorus method[6] is preferred with lecithin as a substrate.

Turbidimetric Method[16]

Egg yolk substrate is prepared according to Murata *et al.*[16] with some modifications. A mixture of one egg yolk, 5 g of Super Cel (Wako Pure Chemicals), and 400 ml of 0.03 M sodium borate–boric acid buffer (pH 7.1) containing 0.047 M NaCl and 5 mM Ca^{2+} ions is homogenized, then centrifuged at 8000 rpm for 15 min to remove Super Cel. The opaque supernatant fluid is used as substrate. It can be used for a week if stored at 4°.

Procedure. Turbidimetric assay of phospholipase C is carried out by the method of Murata *et al.*[16] with modifications. The method is based on the measurement of the turbidity ("egg yolk reaction") produced by the enzyme. Two-tenths milliliters of the egg yolk substrate are added to an equal volume of an enzyme solution, and this mixture is incubated for 15 min at 37°. The reaction is terminated by the addition of 1.1 ml of gas gangrene antitoxin[73] [0.5 International Units (IU)/ml]. The turbidity developed is measured at 450 nm. Phospholipase C activity in egg units (EU) of the test solution is determined from the resulting turbidity by comparing it to the standard curve constructed with dilutions of a glycerinated refer-

[69] W. E. van Heyningen, *Biochem. J.* **35,** 1246 (1941).

[70] E. L. Krug, N. J. Truesdale, and C. Kent, *Anal. Biochem.* **97,** 43 (1979).

[71] S. Kurioka, *J. Biochem. (Tokyo)* **63,** 678 (1968).

[72] S. Ulitzur and M. Heller, *Anal. Biochem.* **91,** 421 (1978).

[73] Preparations of gas gangrene antitoxin are commercially available from Behring Werke AG, Marburg, West Germany and from Chiba Serum Institute, Ichikawa, Japan.

ence phospholipase C (α-toxin) containing known lethal toxicity in LD_{50}. One egg unit is defined as the amount of a given enzyme equivalent to 3 LD_{50} of the reference enzyme.

The egg yolk reaction is used also for determining the end point of the reaction between serial dilutions of the enzyme (α-toxin) and a given concentration of gas gangrene antitoxin. The enzyme activity can be expressed, therefore, in terms of antitoxin-binding power. To 0.5 ml each of serial dilutions of an enzyme solution is added 0.5 ml of antitoxin containing 0.2 IU/ml. The mixture is kept standing for 30 min at room temperature to permit the formation of antigen–antibody complex. To the mixture, 0.25 ml of the egg yolk substrate is added, and this is incubated for 2 hr at 37°. One Lv unit of phospholipase C is defined as the amount of enzyme neutralized by exactly 1 IU of antitoxin. The antitoxin-binding power of the test solution in Lv units is calculated by multiplying, by the factor of 0.2, the reciprocal of the highest dilution showing minimum egg yolk reaction.

Acid-Soluble Phosphorus Method[6]

Lecithin used as substrate is isolated from egg yolk by successive chromatography on alumina and silicic acid columns.[74] The preparation of lecithin is sonicated at 28 KHz for 15–30 min in 0.25 M Tris-HCl buffer, pH 7.5.

Procedure. A test tube containing 0.2 ml of 10 mM egg lecithin, 0.1 ml of 0.02 M SDC (0.01 M for dipalmitoyllecithin) and 0.1 ml of 0.03 M CaCl₂ is preincubated for 5 min at 37°. To this mixture is added 0.1 ml of enzyme in 0.05 M Tris-HCl buffer, pH 7.5, containing 0.1% bovine serum albumin to avoid denaturation of enzyme. After incubation for 5 min, the reaction is terminated by adding 0.1 ml of 50% trichloroacetic acid. The reaction mixture is subjected to vigorous shaking with 2.5 ml of chloroform–methanol (2 : 1, v/v), followed by centrifugation at 3000 rpm for 15 min. From the upper methanol–water phase (1.33 ml), which contains phosphorylcholine, a 0.2-ml portion is taken into a Pyrex tube. A half milliliter of 60% $HClO_4$ is added to the tube, which is then heated for 60 min at 160–170°. The inorganic phosphate in the hydrolyzate is determined by the method of Eibl and Lands[75] with some modifications: 3.0 ml of distilled water are added to the hydrolyzate, and the mixture in the test tube is heated for 10 min in a boiling water bath. After cooling with running tap water, 0.5 ml of a 2.5% aqueous solution of ammonium molybdate and 0.05 ml of 1% Triton X-100 are added to the mixture, and the final volume

[74] C. H. Lea, D. N. Rhodes, and R. D. Stoll, *Biochem. J.* **60**, 353 (1955).
[75] H. Eibl and W. E. M. Lands, *Anal. Biochem.* **30**, 51 (1969).

is adjusted to 5.0 ml with distilled water. After agitation, the mixture is left to stand for 30–45 min at room temperature, and the resulting turbidity is measured at 510 nm. KH_2PO_4 is used as a standard inorganic phosphate. One unit of phospholipase C is defined as the amount of enzyme liberating 1 μmol of phosphorylcholine per minute under the specified conditions. The present assay method is much more sensitive than that reported previously[37]; as little as 0.05 EU of enzyme can be determined. When sphingomyelin is used as a substrate, the concentration of SDC used is 0.012 M instead of 0.02 M.[76]

[76] Y. Yamakawa and A. Ohsaka, unpublished results, 1977.

[83] Phospholipase C from *Bacillus cereus*

EC 3.1.4.3 Phosphatidylcholine cholinephosphohydrolase

By CLIVE LITTLE

Phosphatidylcholine + H_2O → 1,2-diacylglycerol + phosphorylcholine

The enzyme normally referred to as phospholipase C from *B. cereus* should more properly be termed the phosphatidylcholine-hydrolyzing phospholipase C to distinguish it from the more recently discovered and characterized phosphatidylinositol-hydrolyzing phospholipase C of *B. cereus*.[1,2]

Assay Methods

Ottolenghi has outlined the assay procedures for this enzyme.[3] In our experience, methods based on the continuous titration of acid formed by the liberation of phosphorylcholine during the reaction are by far the most convenient. The substrate used may be either a pure phosphatidylcholine[4] or a very crude egg yolk lipoprotein preparation,[5,6] the latter being more suitable for routine assays.

[1] M. W. Slein and G. F. Logan, Jr., *J. Bacteriol.* **90**, 69 (1965).
[2] H. Ikezawa, M. Yamanegi, R. Taguchi, T. Miyashita, and T. Ohyabu, *Biochim. Biophys. Acta* **450**, 154 (1976).
[3] A. C. Ottolenghi, this series, Vol. 14 [35].
[4] C. Little, *Acta Chem. Scand. B* **31**, 267 (1977).
[5] R. F. A. Zwaal, B. Roelofsen, P. Comfurius, and L. L. M. van Deenen, *Biochim. Biophys. Acta* **233**, 474 (1971).
[6] R. F. A. Zwaal and B. Roelofsen, this series, Vol. 32, p. 154.

Reagents

Egg yolk lipoprotein, prepared essentially as described by Zwaal and Roelofsen[6]: 4 washed egg yolks together with 400 ml of 0.15 M NaCl, 10 mM CaCl$_2$, 0.1 mM ZnCl$_2$ are stirred rapidly with a glass rod to break open the yolks and then stirred with a magnetic stirrer at room temperature for 1 hr. The suspension is then centrifuged at 20,000 g for 30 min at 4°. The supernatant is poured off and may be kept for up to 1 week at 4°.

Sodium deoxycholate, 10% (w/v) in water

NaOH, 0.02 N

Procedure. Five milliliters of egg yolk supernatant and 0.1 ml of sodium deoxycholate are mixed in a reaction vessel thermostatted at 23°. The mixture is adjusted to pH 7.5 by addition of NaOH. Enzyme is then added, and the release of acid is followed by continuous titration using a suitable automatic titrator with the end point set at pH 7.5. The rate of addition of 0.02 N NaOH is constant during the course of addition of at least 2.5 ml of titrant. For assays taking no longer than 10 min, no correction for the absorption of atmospheric CO_2 is necessary even when assaying activities as low as those found in the culture filtrate of *B. cereus*. One unit of phospholipase C activity is defined as liberating 1 μmol of titratable H^+ per minute under the above conditions; at pH 7.5, this can be taken as representing the hydrolysis of 1 μmol of phospholipid per minute.[4]

If automatic titration equipment is not available, the substrate decrease assay described by Zwaal and Roelofsen is recommended.[6]

Purification Procedure

Unless otherwise stated, all operations are carried out at 4°.

Preparation of Affinity Gel. The procedure is that published by Takahashi *et al.*[7] with minor modifications.

BrCN-activated Sepharose 4B (180 g; obtained commercially or prepared as described by March *et al.*[8]) is allowed to settle out from suspension in 0.1 M sodium borate buffer (pH 9.6), and the supernatant is decanted off. Egg yolk lipoprotein emulsion, 200 ml, is prepared by homogenizing for 1 min in a Waring blender one egg yolk, 5 g of Hyflo Celite, and 200 ml of 0.1 M sodium borate buffer (pH 9.6) and then centrifuging (13,000 g, 15 min) and discarding the pellet. The egg yolk lipoprotein is added to the activated Sepharose and the mixture is gently stirred for about 18 hr at 4°. The mixture is filtered through two layers of filter

[7] Takahashi, T. Sugahara, and A. Ohsaka, *Biochim. Biophys. Acta* **351,** 155 (1974).
[8] S. C. March, I. Parikh, and P. Cuatrecasas, *Anal. Biochem.* **60,** 149 (1974).

paper in a large Büchner funnel, then washed with 1.5 liters of distilled water followed by 1.5 liters of 0.15 M NaCl buffered with 2.9 mM sodium 5,5'-diethylbarbiturate (pH 7.4) containing 0.1 mM Zn^{2+} at room temperature. This agarose-coupled lipoprotein can be stored at 4° for several months in the above buffered saline containing 30% (w/v) glycerol. This amount of gel is adequate for the isolation of enzyme from 20 liters of bacteria.

Step 1. Bacterial Supernatant. Bacillus cereus strain ATCC 10987 AB-1[9] is grown by rotation (200 rpm) in 2-liter cottonwool-plugged conical flasks containing 1 liter of a sterilized medium comprising 10 g of yeast extract, 10 g of bacteriological peptone, 5 g of NaCl, and 0.4 g of $Na_2HPO_4 \cdot 4$ H_2O.[5] The medium is adjusted to pH 7.4 before sterilization. Medium at about 24° may be inoculated directly with bacteria from an agar plate, and the bacteria may be grown for 17 hr at 27°. Alternatively, medium at 37° may be inoculated from an overnight preculture, and the bacteria may be grown at this temperature for 4–5 hr. The cultures are then centrifuged (11,000 g, 10 min), and the supernatants are collected.

Step 2. Affinity Chromatography I. Cooled supernatant, 10–12 liters, is allowed to percolate through a column (4.7 × 6 cm) of agarose-linked lipoprotein at a flow rate around 10 ml/min. The column is then washed with two column volumes of 0.15 M NaCl containing 2.9 mM sodium 5,5'-diethylbarbiturate and 0.1 mM Zn^{2+}. Enzyme is eluted from the column with a freshly prepared solution of 8 M urea in the above saline–barbiturate–Zn^{2+} buffer at a flow rate of 2–3 ml/min, and fractions of 10 ml are collected. Fractions containing enzyme activity are pooled and dialyzed twice against 4-liter portions of the saline–barbiturate–Zn^{2+} buffer and then concentrated to about 50 ml in Amicon ultrafiltration equipment using a UM-10 membrane.

Step 3. Heat Treatment. The enzyme solution is placed in a 250-ml conical flask containing a thermometer and, with continuous swirling, is dipped in a bath of boiling water until 70° is reached. The enzyme solution is maintained at 68–70° for 10 min and then cooled on an ice bath. The solution is filtered to remove the precipitate formed.

Step 4. Affinity Chromatography II. The filtrate is allowed to percolate through a column (2.5 × 6 cm) of agarose-linked lipoprotein at a flow rate of 1–2 ml/min. The column is washed with two column volumes of the saline–barbiturate–Zn^{2+} buffer. Enzyme activity is eluted with a freshly prepared solution of 8 M urea in the saline–barbiturate–Zn^{2+} buffer at a flow rate of 0.5–1 ml/min; 5-ml fractions are collected. Active fractions are pooled and dialyzed first against the saline–barbiturate–Zn^{2+} buffer for at

[9] A.-B. Otnaess, C. Little, and H. Prydz, *Acta Pathol. Microbiol. Scand. B* **82**, 354 (1974).

least 2 hr and then for 45–60 min against 0.01 M sodium phosphate buffer at pH 8.0.

Step 5. DEAE-Sephadex Treatment. Immediately after the above dialysis the enzyme solution is passed through a column (2.0 × 5 cm) of DEAE-Sephadex A-50 previously equilibrated against the above phosphate buffer. The column is washed with one column volume of the phosphate buffer. The entire eluent of the column is collected as one batch, dialyzed overnight against the saline–barbiturate–Zn^{2+} buffer, and finally stored at $-20°$ at a concentration around 1–2 mg/ml. Under these conditions the enzyme can be kept indefinitely.

Crystallization.[10] After step 5 the enzyme solution is adjusted to 60% saturation of ammonium sulfate. The precipitated enzyme is centrifuged (13,000 g, 10 min), and the pellet is dissolved in water to a concentration of 3–5 mg of protein per milliliter. Solid ammonium sulfate is added to a concentration of about 1.4 M, and crystallization is brought about by vapor phase diffusion at room temperature against a solution of 2.5 M ammonium sulfate. Tetragonal prism crystals are usually obtained in 1–2 weeks. Crystals suitable for X-ray diffraction analysis can be produced in this way.[10]

Comments. A summary of the results of a typical purification is given in the table. In contrast with the comments of Gerasimene *et al.*,[11] we find the above method extremely reliable and reproducible; the average yield of our last 12 purifications is 75% with a standard deviation of ± 18%. The affinity gel should not be used more than once. This purification is that published by Little *et al.*[12] with small modifications.

Properties

Purity. Enzyme samples after stage 5 are homogeneous either in analytical disc gel electrophoresis (nondenaturing) or in sodium dodecyl sulfate (SDS)–gel electrophoresis under both reducing and nonreducing conditions[12] and are sufficiently pure to be used in limited amino acid sequencing work.[13]

Stability. Despite possessing no disulfide cross links,[13] phospholipase C is an extremely stable enzyme as regards both thermal denaturation[13] and denaturation by chaotropic agents[14,15]; indeed, the enzyme is catalytically

[10] E. Hough, C. Little, and K. Jynge, *J. Mol. Biol.* **121**, 567 (1978).
[11] G. B. Gerasimene, A. A. Glemzha, V. V. Kulene, Y. Y. Kulis, and Y. P. Makaryunaite, *Biokhimiya* **42**, 919 (1977).
[12] C. Little, B. Aurebekk, and A.-B. Otnaess, *FEBS Lett.* **52**, 175 (1975).
[13] A.-B. Otnaess, C. Little, K. Sletten, R. Wallin, S. Johnsen, R. Flengsrud, and H. Prydz, *Eur. J. Biochem.* **79**, 459 (1977).
[14] C. Little, *Biochem. J.* **175**, 977 (1978).
[15] C. Little and S. Johansen, *Biochem. J* **179**, 509 (1979).

PURIFICATION OF PHOSPHOLIPASE C FROM *Bacillus cereus*

Step	Volume (ml)	Total activity (units)	Specific activity (units/ mg protein)	Recovery (%)
1. Supernatant	10,500	57,700	0.8	100
2. Affinity chromatography I	113	53,500	790	93
3. Heat treatment	45	52,700	1250	91
4. Affinity chromatography II	40	47,800	1500	83
5. DEAE-Sephadex treatment + concentration	25	43,800	1510	76

active in 8 M urea.[14] Thermally coagulated, totally inactive enzyme can be dissolved and unfolded in solutions of guanidinium chloride and then refolded back to the native, fully active conformation.[15] The enzyme is further stabilized by the presence of free Zn^{2+}.[13,15,16] In the absence of free Zn^{2+}, the enzyme is readily inactivated by exposure to low concentrations of certain divalent cations, especially Cu^{2+}, Cd^{2+}, and Ni^{2+}, that displace the zinc in the structure. The enzyme is also readily inactivated by exposure to low concentrations of semipolar organic solvents.

Structure and Molecular Weight. Phospholipase C is a simple monomeric protein[13,16] containing two tightly bound Zn atoms.[17] Ultracentrifugation and sodium dodecyl sulfate gel electrophoresis indicate a molecular weight of 23,000.[16] This value is consistent with the amino acid composition of the enzyme.[13] Circular dichroism studies suggest a secondary structure comprising 30–36% α-helix, 24–30% β structure, and the remainder aperiodic.[14] The active site seems to contain lysine,[18] carboxyl,[19] histidine,[20] and arginine[21] residues, the latter apparently necessary for substrate binding.

Substrate Specificity. With pure phospholipids in deoxycholate-mixed micelles, pure phospholipase C readily degrades phosphatidylcholine, phosphatidylethanolamine, and phosphatidylserine with little or no significant degradation of sphingomyelin, phosphatidylglycerol, cardiolipin,

[16] A.-B. Otnaess, H. Prydz, E. Bjørklid, and Å. Berre, *Eur. J. Biochem.* **27,** 238 (1972).
[17] C. Little and A.-B. Otnaess, *Biochim. Biophys. Acta* **391,** 326 (1975).
[18] B. Aurebekk and C. Little, *Biochem. J.* **161,** 159 (1977).
[19] C. Little and B. Aurebekk, *Acta Chem. Scand. B* **31,** 273 (1977).
[20] C. Little, *Biochem. J* **167,** 399 (1977).
[21] B. Aurebekk and C. Little, *Int. J. Biochem.* **8,** 757 (1977).

phosphatidylinositol, lysophosphatidylcholine, and lysophosphatidyletha-nolamine.[13,22] However, the same substrate specificity is not necessarily shown when the enzyme degrades phospholipids in real biological membranes. Shukla *et al.*[23,24] reported the degradation by the enzyme of sphingomyelin in the human red cell membrane. In addition we have found that electrophoretically pure recrystallized phospholipase C readily degrades sphingomyelin both in the myelin sheath[25] and in the membrane of human red cells in the spheroechinocyte I morphological form.[26] The same enzyme samples are about 2000-fold more active with pure phosphatidylcholine-deoxycholate micelles than with sphingomyelin-deoxycholate micelles. The degradation of cardiolipin in *B. subtilis*[27] and cardiolipin and also phosphatidyldiglyceride in the membranous lipids of *Escherichia coli*[28] by the highly purified enzyme have also been reported.

pH Optimum and Kinetics. During the hydrolysis of insoluble substrates such as tissue thromboplastin phospholipids or long-chain phosphatidylcholine, activity optima have been noted at pH 6.6 and pH 8.0.[16] With solutions of short-chain, water-soluble phosphatidylcholine, phospholipase C shows a single peak of activity at pH 7.5–8.0.[4] The enzyme shows a marked preference for hydrolyzing substrates at a lipid–water interface, and anomalous kinetic behavior is shown with short-chain phosphatidylcholine around the critical micelle concentration.[4]

Although it is inactivated by exposure to chelating agents,[16,17,29] the enzyme does not require the presence of free metal ions for catalytic activity and is active even in the presence of EDTA for short periods. *o*-Phenanthroline, however, very rapidly removes the metal atoms from the structure, thereby causing inactivation.[17]

[22] M. F. Roberts, A.-B. Otnaess, C. A. Kensil, and E. A. Dennis, *J. Biol. Chem.* **253**, 1252 (1978).
[23] S. D. Shukla, M. M. Billah, R. Coleman, J. B. Finean, and R. H. Michell, *Biochim. Biophys. Acta* **509**, 48 (1978).
[24] S. D. Shukla, R. Coleman, J. B. Finean, and R. H. Michell, *Biochim. Biophys. Acta* **512**, 341 (1978).
[25] K. Gwarsha, C. Little, and M. G. Rumsby, *Biochem. Soc. Trans.* **7**, 990 (1979).
[26] C. Little and M. G. Rumsby, *Biochem. J.* **188**, 39 (1980).
[27] J. A. F. Op den Kamp, M. T. Kauertz, and L. L. M. van Deenen, *J. Bacteriol.* **112**, 1090 (1972).
[28] R. D. Mavis, R. M. Bell, and P. R. Vegelos, *J. Biol. Chem.* **247**, 2835 (1972).
[29] A. C. Ottolenghi, *Biochim. Biophys. Acta* **106**, 510 (1965).

[84] Phosphatidylinositol-Specific Phospholipase C from *Bacillus cereus* and *Bacillus thuringiensis*

By Hiroh Ikezawa and Ryo Taguchi

Phosphatidylinositol + H_2O →

1,2-diacylglycerol + D-*myo*-inositol 1,2-cyclic phosphate

Phosphatidylinositol-specific phospholipase C was discovered in the culture broth of *Bacillus cereus*[1,2] and characterized as an entity different from so-called "phospholipase C," i.e., phosphatidylcholine-hydrolyzing phospholipase C. This enzyme releases ectoenzymes such as alkaline phosphatase from plasma membrane of eukaryotic cells, *in vitro* and *in vivo*.[1-3] Recently, Taguchi *et al.*[4] found that *B. thuringiensis* was the richest source among phosphatidylinositol-specific phospholipase C-producing bacteria tested, and they purified this enzyme from the culture broth of *B. thuringiensis* to a homogeneous state by polyacrylamide gel electrophoresis.

Assay Methods

Principles. Mixed micelles of phosphatidylinositol and a detergent, such as sodium deoxycholate (or Triton X-100), are used for the enzyme assay of bacterial phosphatidylinositol-specific phospholipase C. *myo*-Inositol 1,2-cyclic phosphate, one of the reaction products, is estimated after elimination of remaining phosphatidylinositol by extraction with chloroform–methanol–HCl (66:33:1). After decomposition of *myo*-inositol phosphate by the method of Fiske and SubbaRow,[5] phosphorus is determined by the method of Eibl and Lands.[6]

In addition to the above-mentioned method, the assay of ectoenzyme releasing activity is also available as a specific biological method.

[1] M. W. Slein and G. F. Logan, Jr., *J. Bacteriol.* **90**, 69 (1965).

[2] H. Ikezawa, M. Yamanegi, R. Taguchi, T. Miyashita, and T. Ohyabu, *Biochim. Biophys. Acta* **450**, 154 (1976).

[3] T. Ohyabu, R. Taguchi, and H. Ikezawa, *Arch. Biochem. Biophys.* **190**, 1 (1978).

[4] R. Taguchi, Y. Asahi, and H. Ikezawa, *Biochim. Biophys. Acta* **619**, 48 (1980).

[5] H. Fiske and Y. SubbaRow, *J. Biol. Chem.* **66**, 375 (1925).

[6] H. Eibl and W. E. M. Lands, *Anal. Biochem.* **30**, 51 (1969).

1. Phosphorus-Release Assay[2]

Reagents

Phosphatidylinositol emulsion, 10 mM, sonicated for 3 min before use
Sodium borate, 100 mM, pH 7.5
Sodium deoxycholate, 0.8%
Chloroform–methanol–HCl, 66:33:1 (v/v/v)
H_2SO_4, 6 N
H_2O_2, 60%
Triton X-100, 0.36%
Ammonium molybdate, 2.5%

Assay Procedure. Reaction mixture containing 0.1 ml of 10 mM phosphatidylinositol, 0.1 ml of 0.8% sodium deoxycholate, 0.2 ml of 100 mM sodium borate (pH 7.5) and 0.1 ml of the enzyme solution (approximately 5 mU of the enzyme diluted with 0.1% bovine serum albumin) are incubated at 37° for 20 min. The reaction is terminated by adding 2.5 ml of chloroform–methanol–HCl (66:33:1) and mixing. After centrifugation at 2000 g for 5 min, a 0.4-ml aliquot is transferred from the methanol–water layer into a test tube. Then 0.3 ml of 6 N H_2SO_4 is added to each sample and heated at 170° for 30 min. Next a few drops of 60% H_2O_2 are added and the mixture is heated at 170° for another 120 min. After cooling at room temperature, 2.5 ml of distilled water and 0.1 ml of 0.36% Triton X-100 are added and mixed vigorously. About 20 min after addition of 0.3 ml of ammonium molybdate, turbidity is measured at 660 nm.

Units. A unit of enzyme is defined as the amount of enzyme causing the hydrolysis of 1 μmol of substrate per minute. Specific activity is defined as units per milligram of protein.

Comments. Phosphatidylinositol is most conveniently prepared from bakers' yeast by Trevelyan's method.[7] Purified substrate usually contains di- and triphosphoinositides, but the bacterial enzymes mainly hydrolyze monophosphoinositide during a short incubation period under the conditions described above. Thus liberated water-soluble phosphorus is equivalent to hydrolyzed substrate.

2. Alkaline Phosphatase-Releasing Activity[2]

Rat kidney is excised and cut into small pieces with scissors. From these pieces, slices are cut with a Feather razor blade. These slices are washed several times with 0.25 M sucrose. To 4 ml of each slice suspension (containing 0.2 g slices, wet weight), 10–20 mU of the enzyme are added, and the preparation is incubated for 60 min at 37°. A 2-ml aliquot is

[7] W. E. Trevelyan, *J. Lipid Res.* **7**, 445 (1966).

withdrawn from the reaction mixture, cooled, and centrifuged immediately at 40,000 g for 20 min. Alkaline phosphatase activity of the resulting supernatant is assayed according to Engström's method.[8]

Alkaline Phosphatase Assay[8]

Reagents

p-Nitrophenylphosphate, 2 mM, and MgCl$_2$, 5 mM, in NaHCO$_3$-Na$_2$CO$_3$, 50 mM, pH 10.1
Trichloroacetic acid, 15%
NaOH, 1 N

Procedure. To 4.8 ml of 2 mM substrate solution is added 0.2 ml of enzyme solution; the mixture is incubated at 37° for 10 min. The reaction is terminated by adding 1 ml of 15% trichloroacetic acid. The mixture is filtered on Toyo No. 2 filter paper, and 3 ml of each filtrate is transferred into another test tube containing 1 ml of 1 N NaOH. Absorbance at 400 nm is measured spectrophotometrically.

3. Acetylcholinesterase-Releasing Activity

Fresh bovine blood samples are centrifuged at 2000 g for 10 min, and plasma and buffy coat are removed by aspiration. The collected erythrocytes are washed three or four times with an isotonic phosphate buffer (150 mM NaCl in 5 mM sodium phosphate buffer, pH 7.6). The washed erythrocytes are suspended in and diluted with the same buffer to a final concentration of 10% (v/v). To each 2-ml aliquot of 10% bovine erythrocyte suspension is added 0.1 ml of enzyme solution (about 100 mU), and the mixture is incubated for 30 min at 37°. The reaction is terminated by centrifugation at 2000 g for 3 min. The supernatant is further centrifuged at 40,000 g for 20 min. Acetylcholinesterase activity in the resulting supernatant is determined by the method of Ellman *et al.*[9]

Acetylcholinesterase Assay[9]

Reagents

Sodium phosphate, 100 mM, pH 7.5
5,5'-Dithiobis(2-nitrobenzoic acid) (DTNB), 10 mM in 100 mM sodium phosphate, pH 7.0

[8] L. Engström, *Biochim. Biophys. Acta* **92**, 71 (1964).
[9] G. L. Ellman, K. D. Coartney, A. Valentine, Jr., and R. M. Featherstone, *Biochem. Pharmacol.* **7**, 88 (1961).

Acetylthiocholine chloride, 12.5 mM

Procedure. To each 0.1-ml aliquot of enzyme solution are added 0.8 ml of sodium phosphate (100 mM, pH 7.5), 50 μl of DTNB, and 50 μl of acetylthiocholine. The reaction is followed spectrophotometrically at 412 nm at room temperature.

Comments

Bovine and sheep erythrocytes are more sensitive to the enzyme action than pig or horse erythrocytes. When erythrocyte ghosts are used for assay, much lower levels of the enzyme (20–30 mU) are required.

Although centrifugation at 105,000 g for 90 min is better in the separation of cell debris, 40,000 g for 20 min is sufficient for the assays of the release of these ectoenzymes.

Purification of Enzyme from *Bacillus cereus*

Bacillus cereus IAM 1208 is grown in flasks containing 10 liters of a sterilized medium, which consists of 100 g of polypeptone, 100 g of yeast extract, 50 g of NaCl, 4 g of Na_2HPO_4, and water up to 10 liters. The medium is adjusted to pH 7.0 with 1 M NaOH. Medium is inoculated with 1–1.5% (v/v) of the preculture of the organism and incubated at 37° for 6 hr (the late logarithmic phase of growth) with shaking on a rotary shaker (the culture of Procedure I).

In the culture using a jar fermentor, 40 liters of the same medium are inoculated with 2.5% (v/v) of the preculture of the organism, and incubated at 37° for 5 hr with aeration of 10 liters per minute (the culture for Procedure II).

The cells are removed by centrifugation, and ammonium sulfate is added to the supernatant at 4° to 80% saturation. The resulting precipitate is collected by continuous centrifugation at 25,000 g.

Procedure I.[2] The precipitate is dissolved in distilled water and dialyzed against 0.01 M Tris-HCl buffer, pH 8.5. The dialyzate is applied to a column of DEAE-cellulose (4 × 14 cm). The column is washed with 1 liter of 0.3 M Tris-HCl buffer, pH 8.5. The active fractions are obtained by elution with a linear gradient of 1.2 liters of Tris-HCl buffer (pH 8.5) from 0.3 to 1.2 M. The fractions are pooled and dialyzed against 0.01 M acetate buffer, pH 5.5.

The dialyzed DEAE-cellulose eluate is applied to a column of CM-Sephadex C-50 (4 × 15 cm). After washing the column with 0.01 M acetate buffer (pH 5.5), the active fractions are obtained by elution with 800 ml of a linear gradient from 0 to 0.4 M NaCl in the same buffer. The active

TABLE I

PURIFICATION OF PHOSPHATIDYLINOSITOL-SPECIFIC PHOSPHOLIPASE C FROM *Bacillus cereus* (PROCEDURE I)[2]

Step	Total protein (mg)	Total activity (units)	Specific activity (units/mg)	Recovery (%)	Purification (fold)
Supernatant	20000	620	0.031	100	1
$(NH_4)_2SO_4$, 80% saturated	4015	612	0.152	98.7	4.9
DEAE-cellulose	82	177	2.16	28.5	69.6
CM-Sephadex	3.8	52.2	13.74	8.4	443

fractions are pooled and concentrated. A typical purification is shown in Table I.

Procedure II.[3,5] The precipitate is dissolved in distilled water and dialyzed against 5 mM Tris-maleate buffer, pH 6.5. This material is then applied to a column of CM-Sephadex C-50 (4 × 19 cm). Both phosphatidylinositol-specific phospholipase C and sphingomyelinase activities are obtained in the breakthrough effluent; they are thereby separated from the major part of phosphatidylcholine-hydrolyzing phospholipase C. The breakthrough fractions are then concentrated and dialyzed against 0.02 M Tris-HCl buffer, pH 8.5. The dialyzed material is applied to a column of DEAE-cellulose (4 × 19 cm). The column is washed with the same Tris buffer, then eluted with a linear gradient from 0 to 0.3 M NaCl in the same buffer. Phosphatidylinositol-specific phospholipase C activity is completely separated from the activity of sphingomyelinase, the latter activity being eluted from the column much earlier than the former. The fractions containing phosphatidylinositol-specific phospholipase C are pooled and concentrated. A typical purification is shown in Table II.

Comments. Procedure I will give a highly purified preparation, if a good yield of enzyme is sacrificed in the chromatographic steps by removal of less active fractions. The preparation then shows a single protein band on analytical polyacrylamide gel electrophoresis and is free of phosphatidylcholine-hydrolyzing phospholipase C and sphingomyelinase. However, exposure to a low pH such as 5.5 partially inactivates phosphatidylinositol-specific phospholipase C during the chromatographic run on CM-Sephadex, resulting in low specific activity and poor yield.

On the other hand, Procedure II produces an enzyme preparation with higher activity in better yield. This preparation will sometimes show a minor activity of phosphatidylcholine-hydrolyzing phospholipase C,

TABLE II

PURIFICATION OF PHOSPHATIDYLINOSITOL-SPECIFIC PHOSPHOLIPASE C FROM *Bacillus cereus* (PROCEDURE II)[3]

Step	Total protein (mg)	Total activity (units)	Specific activity (units/mg)	Recovery (%)	Purification (fold)
Supernatant	162,000	8748	0.054	100	1
(NH₄)₂SO₄, 80% saturated	3,250	8560	2.63	97.9	49
CM-Sephadex	550	4150	7.55	47.4	140
DEAE-cellulose	88	2100	23.9	24.0	447

which is unrelated to the activity of phosphatidylinositol-specific phospholipase C.[3]

The introduction of gel filtration with Sephadex G-75 column is more convenient when carried out in the final step of purification; the conditions of gel filtration are the same as for enzyme purification from *B. thuringiensis*.

Properties of the Enzyme from *Bacillus cereus*

Isoelectric Point and Molecular Weight. Isoelectric focusing with ampholite (pH 5–8) indicates a p*I* value of 5.4.[10] The molecular weight is estimated to be 29,000 by gel filtration with Sephadex G-75.

Substrate Specificity. Pure phosphatidylinositol-specific phospholipase C of *B. cereus* specifically hydrolyzes phosphatidylinositol[2,11] and lysophosphatidylinositol.[11] The K_m value of phosphatidylinositol is 1.4 m*M* at pH 7.5 in the presence of 0.16% sodium deoxycholate.[2] The enzyme does not hydrolyze the following at a significant rate: phosphatidylcholine,[2,11] phosphatidylethanolamine,[2,11] phosphatidylglycerol,[11] phosphatidylserine,[11] phosphatidic acid,[11] and sphingomyelin[2,11] either in individual pure micelles or in lipid extracts from yeast autolyzate.

Optimum pH and the Influence of Effectors. The enzyme is maximally active at pH 7.2–7.5, with purified yeast phosphatidylinositol as substrate.[2] The stimulation by sodium deoxycholate is maximal at 0.16% with 2 m*M* substrate.

In contrast to phosphatidylcholine-hydrolyzing phospholipase C or sphingomyelinase, the enzyme does not require divalent metal ions for its

[10] H. Ikezawa, M. Mori, T. Ohyabu, and R. Taguchi, *Biochim. Biophys. Acta* **528**, 247 (1978).
[11] R. Sundler, A. W. Alberts, and P. R. Vagelos, *J. Biol. Chem.* **253**, 4175 (1978).

activity in that the enzyme activity is not influenced by 0.2 mM EDTA, nor by 0.08 mM o-phenanthroline.[2] Divalent metal ions such as Ca^{2+}, Mg^{2+}, Mn^{2+}, and Zn^{2+} are inhibitory at 2–10 mM.[11] The enzyme activity is significantly inhibited by high concentrations of NaCl.[11]

Monoiodoacetic acid (IAA), p-chloromercuribenzoate (PCMB), and glutathione are without effect on the enzyme activity at 0.08 mM,[2] although both are inhibitory at higher concentrations.

Stability. The enzyme is fairly unstable in the acidic medium. It is inactivated by heating.

Purification of Enzyme from *Bacillus thuringiensis*[4]

Step 1. Bacillus thuringiensis IAM 11607 or IAM 12077 is grown in a jar fermentor containing 20 liters of a sterilized medium composed of 200 g of polypeptone, 200 g of yeast extract, 100 g of NaCl, 8 g of K_2HPO_4, and water up to 20 liters. The medium is adjusted to pH 7.0 with 1 M NaOH. The medium is inoculated with 2.5% (v/v) of the preculture of the organism, and incubated at 37° for 7 hr (the late logarithmic phase of growth) with aeration at a rate of 10 liters per minute. The cells are removed by centrifugation.

Step 2. To the supernatant is added ammonium sulfate at 4° to 60% saturation. The precipitate is collected by continuous centrifugation at 28,000 g, dissolved in distilled water, and dialyzed against 0.01 M acetate buffer, pH 5.2.

Step 3. The dialyzed material is applied to a column of CM-Sephadex C-50 (4 × 45 cm), washed with 0.01 M acetate buffer (pH 5.2), and eluted sequentially with 1.3 liters of 0.1 M and 2 liters of 0.2 M NaCl in the same buffer. Phosphatidylinositol-specific phospholipase C activity is eluted with 0.2 M NaCl.

Step 4. The concentrated CM-Sephadex eluate is applied to a Sephadex G-75 column (2.6 × 100 cm) equilibrated with 0.01 M Tris-HCl (pH 7.5) and eluted with the same buffer. Phosphatidylinositol-specific phospholipase C activity will appear in a minor protein fraction between 250 and 350 ml of effluent.

Step 5. The concentrated Sephadex G-75 eluate is subjected to isoelectric focusing with carrier ampholite (pH 5–8) in a gradient of sucrose (0–50%, v/v) according to the method of Vesterberg and Svensson.[12] The sample is placed in the center of a 110-ml column (LKB-Produkter, Stockholm, Sweden), and electrolysis is carried out at 4° for 26 hr at 900 V. After the electrolysis, 2-ml fractions are collected and subjected to pH measurement at room temperature and enzyme assay. The activity of

[12] O. Vesterberg and H. Svensson, *Acta Chem. Scand.* **20,** 820 (1966).

TABLE III
PURIFICATION OF PHOSPHATIDYLINOSITOL-SPECIFIC PHOSPHOLIPASE C FROM
Bacillus thuringiensis[4]

Step	Total protein (mg)	Total activity (units)	Specific activity (units/mg)	Recovery (%)	Purification (fold)
1. Supernatant	12000	1690	0.14	100	1
2. (NH₄)₂SO₄, 60% saturation	1190	1530	1.29	90.5	9
3. CM-Sephadex	89.8	1050	11.7	62.1	84
4. Sephadex G-75	3.97	480	121	28.4	846
5. Isoelectric focusing	1.47	460	312	27.2	2230

phosphatidylinositol-specific phospholipase C appears as a single protein peak. The purified enzyme thus obtained is homogeneous on polyacrylamide gel electrophoresis. A typical purification is shown in Table III.

Comment. Actually, there is no appreciable protease activity in the culture broth of *B. thuringiensis* IAM 12077. However, the strain IAM 11607 produces significant amount of an alkaline protease (optimal pH 10.0) in the culture broth. Since the protease activity is completely inhibited by 0.1 m*M* EDTA without any substantial effect on phosphatidylinositol-specific phospholipase C activity, it is better to elute the columns of CM-Sephadex and Sephadex G-75 with the buffers containing 0.1 m*M* EDTA during the purification of the enzyme from the culture supernatant of *B. thuringiensis* IAM 11607. Gel filtration on Sephadex G-75 completely separates the activity of phosphatidylinositol-specific phospholipase C from protease activity, which is eluted in the fractions close to the void volume.

Properties of the Enzyme from *Bacillus thuringiensis*[4]

Isoelectric Point and Molecular Weight. Isoelectric focusing in the final step of purification indicates a p*I* value of 5.4. The molecular weight is estimated to be 23,000 by gel filtration with Sephadex G-75.

Substrate Specificity. The purified phosphatidylinositol-specific phospholipase C of *B. thuringiensis* hydrolyzes phosphatidylinositol and lysophosphatidylinositol. Phosphatidylcholine, phosphatidylethanolamine, phosphatidylserine, phosphatidylglycerol, and sphingomyelin are not hydrolyzed by the enzyme.

Optimum pH and Influence of Several Effectors. Phosphatidylinositol-hydrolyzing activity of the enzyme is optimum around pH 7.5; it is stimulated greatly by sodium deoxycholate or Triton X-100 and slightly by Tween 20. Cetyltrimethylammonium bromide is without effect. Maximum stimulation by sodium deoxycholate or Triton X-100 occurs at 0.16% or 0.1%, respectively, when the concentration of phosphatidylinositol is 2 mM; EDTA or o-phenanthroline does not show any inhibitory effect. The enzyme activity is inhibited by 50% in the presence of 5 mM IAA or 1 mM PCMB. Divalent metal ions such as Ca^{2+}, Mg^{2+}, and Zn^{2+} are inhibitory at 0.1–10 mM. Much higher concentrations (0.1–0.5 M) of NaCl and KCl are also inhibitory. Inhibition by KCl is more pronounced than that by NaCl. Inhibition by these salts is stronger in the presence of 0.1% Triton X-100 than 0.16% sodium deoxycholate.

Stability. The enzyme is relatively stable at pH 5.0–7.0, but loses its activity readily at pH lower than 4.8. The enzyme is also thermostable between pH 7.2 and 9.7, but is completely inactivated at pH 4.8 by heating at 50° for 10 min.

Ectoenzyme Releasing Activity from Plasma Membrane

Slein and Logan[1] first suggested that phosphatidylinositol-hydrolyzing activity of *B. cereus* was associated with a phosphatasemia factor, which

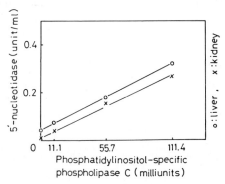

FIG. 1. 5'-Nucleotidase release from slices of rat kidney or liver by phosphatidylinositol-specific phospholipase C of *Bacillus cereus*. The reaction conditions are essentially the same as described in Assay Methods (Alkaline Phosphatase-Releasing Activity). The enzyme was added to slice suspensions containing 0.2-g slices of rat liver or kidney in 1.8 ml of 0.25 M sucrose, and the mixture was incubated for 90 min at 37°. The reaction mixtures were cooled in an ice bath and then centrifuged immediately at 40,000 g for 20 min. 5'-Nucleotidase activity released in the supernatant was determined as follows: To 0.8 ml of a solution containing 11 mM AMP, 5 mM $MgCl_2$, and 55 mM Tris-HCl buffer (pH 8.5), was added 0.2 ml of enzyme solution, and the mixture was incubated at 37° for 15 min. The reaction was terminated by adding 1 ml of 10% trichloroacetic acid. The liberated inorganic phosphate was determined by the method of Eibl and Lands.[6]

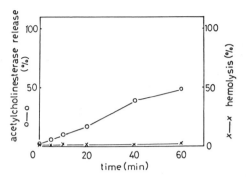

F<small>IG</small>. 2. Acetylcholinesterase release from bovine erythrocytes by phosphatidylinositol-specific phospholipase C of *Bacillus cereus*. The reaction conditions are the same as described in Assay Methods (Acetylcholinesterase-Releasing Activity). Acetylcholinesterase and hemoglobin release in supernatant are measured after centrifugation.

liberated alkaline phosphatase from animal tissues or cell homogenates. Ikezawa *et al.*[2] demonstrated that purified phosphatidylinositol-specific phospholipase C of *B. cereus* did really release alkaline phosphatase from rat kidney slices. The phosphatase release increased linearly with the incubation time and was proportional to the amount of phospholipase added. Furthermore, the releasing activity was not enhanced by addition of either phosphatidylcholine-hydrolyzing phospholipase C or sphingomyelinase of *B. cereus*. Ohyabu *et al.*[3] reported that an almost equal amount of purified anti-(phosphatidylinositol-specific phospholipase C)-IgG was needed to neutralize completely both phosphatidylinositol-hydrolyzing and phosphatase-releasing activities in the purified enzyme preparation, showing that these two activities can be ascribed to the same protein. They also demonstrated that this enzyme exhibited alkaline phosphatase-releasing activity *in vivo*. Taguchi and Ikezawa,[13] and Low and Finean[14,15] reported that the enzymes from *Clostridium novyi* and *Staphylococcus aureus* also liberate alkaline phosphatase as well as other ectoenzymes, such as 5'-nucleotidase and acetylcholinesterase from plasma membrane.

Phosphatidylinositol-specific phospholipase C of *B. cereus* also releases 5'-nucleotidase from kidney and liver slices, as shown in Fig. 1, and liberates acetylcholinesterase from bovine erythrocytes without causing hemolysis, as shown in Fig. 2. Bovine red cells treated with this phospholipase become osmotically fragile, as shown in Fig. 3. The pattern of changes in osmotic fragility is distinguished from those obtained by treat-

[13] R. Taguchi and H. Ikezawa, *Arch. Biochem. Biophys.* **186**, 196 (1978).
[14] M. G. Low and J. B. Finean, *Biochem. J.* **167**, 281 (1977).
[15] M. G. Low and J. B. Finean, *FEBS Lett.* **82**, 143 (1977).

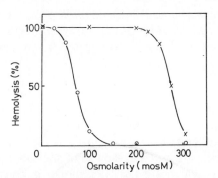

FIG. 3. Osmotic fragility change of bovine erythrocytes, by the action of phosphatidylinositol-specific phospholipase C of *Bacillus cereus*. As an index of osmotic fragility, the percentage of hemolysis of control (O——O) or of the enzyme-treated (×——×) (acetylcholinesterase release, 20%) bovine erythrocytes is measured at various osmolarities.

ments with other phospholipase C.[16] The increase in osmotic fragility of the red cells induced by phosphatidylinositol-specific phospholipase C may be a reflection of acetylcholinesterase release from erythrocyte membrane. Phosphatidylinositol-specific phospholipase C of *B. thuringiensis* induces the release of alkaline phosphatase,[4] 5'-nucleotidase, and acetylcholinesterase.[17] Thus the ability to release ectoenzymes is a common feature of phosphatidylinositol-specific phospholipase C from bacteria. Probably, these ectoenzymes exist in close contact with phosphatidylinositol on the plasma membrane, where this phospholipid may display specific, regulatory functions.

[16] R. Taguchi, M. Mizuno, M. Inoue, and H. Ikezawa, in preparation.
[17] R. Taguchi, M. Mizuno, M. Inoue, and H. Ikezawa, in preparation.

[85] Phosphatidylinositol-Specific Phospholipase C from *Staphylococcus aureus*

By MARTIN G. LOW

Phosphatidylinositol → diglyceride + inositol phosphate

Staphylococcus aureus produces two distinct phospholipase C activities—one specific for sphingomyelin, the other for phosphatidylinositol.[1] The sphingomyelinase has been purified by several investigators and identified as the staphylococcal β-hemolysin.[2] In contrast

[1] H. M. Doery, B. J. Magnusson, J. Gulasekharam, and J. E. Pearson, *J. Gen. Microbiol.* **40**, 283 (1965).
[2] G. M. Wiseman, *Bacteriol Rev.* **39**, 317 (1975).

the phosphatidylinositol-specific phospholipase C has had no cytotoxic activity attributed to it in spite of the observation that it was produced only by coagulase-positive (i.e., "pathogenic") strains of S. aureus.[1]

Assay Methods

At present there is no quantitative assay for this enzyme that is completely suitable. The enzyme has relatively low activity against pure phosphatidylinositol, and its activity is markedly inhibited by the addition of detergents. This has resulted in the use of sonicated phospholipid mixtures as substrates, which are not well suited to kinetic analysis. Two methods are described here: the first is recommended only for the comparison of specific activties during purification; the second is recommended for the rapid assay of multiple fractions, e.g., column eluates.

Principle. Phospholipase C hydrolyzes the phosphodiester linkage of phosphatidylinositol to release diglyceride and water-soluble inositol phosphate. The unhydrolyzed phospholipid is removed by extraction into chloroform whereas the inositol phosphate remains in the upper phase, where it is determined by measurement of organic phosphorus (Method 1) or radioactivity (Method 2).

Method 1[3]

Reagents. Soybean lipids[4] (20 mg/ml) are homogenized in water and then dialyzed extensively against water to remove water-soluble phosphorus compounds. The resulting lipid suspension (approximately 1.6 mM phosphatidylinositol) is stored frozen at $-20°$; small aliquots are sonicated for 10 min immediately before use.

Procedure. Soybean lipid suspension (0.5 ml), 100 mM HEPES–NaOH buffer, pH 7.0 (0.2 ml), and enzyme sample are incubated in a final volume of 1 ml for 2 hr at 37°. The reaction is stopped by cooling on ice followed by the immediate addition of 6 ml of chloroform–methanol–HCl (100 : 100 : 0.6, v/v/v) and 1.7 ml of 1 M HCl. The samples are mixed vigorously, and the phases are separated by centrifugation. One milliliter of the upper aqueous phase is assayed for organic phosphorus.[5] The substrate and sometimes the enzyme sample contain significant amounts of water-soluble phosphorus (equivalent to approximately 1 unit of activity), and therefore blanks should be included in the assay and the appropriate corrections made. This assay obviously cannot be used with samples of

[3] M. G. Low and J. B. Finean, *Biochem. J.* **154**, 203 (1976).
[4] Phosphatidylcholine Type II-S from Sigma Chemical Co., St. Louis, Missouri.
[5] G. R. Bartlett, *J. Biol. Chem.* **234**, 466 (1959).

enzyme containing large amounts of phosphate; culture supernatants should be dialyzed against phosphate-free buffer before assay. One unit of activity releases 0.1 μmol of water-soluble phosphorus in 2 hr at 37°.

Method 2

Reagents. [6] Rat liver is homogenized in two volumes of 0.25 M sucrose, 0.5 mM EDTA, 5 mM Tris-acetate, pH 7.4; centrifugation is performed twice at 10,000 g for 10 min, and the resulting supernatant is centrifuged at 100,000 g for 1 hr. The 100,000 g microsomal pellet was resuspended in sucrose–EDTA–Tris (1 ml/g liver) to give a phospholipid concentration of about 10 μmol/ml. Nine milliliters of microsomes, 0.1 ml of 100 mM MnCl$_2$, 1 ml of 100 mM HEPES–NaOH, pH 7.4, and 200 μCi of myo-[^{3}H] inositol (12.5 Ci/mmol) are incubated at 37° for 90 min, and the lipids are extracted with chloroform–methanol (2 : 1, v/v).[7] myo-Inositol, 200 μmol, is added at the partition and washing stages to facilitate removal of [^{3}H]inositol from the chloroform phase. The extract is dried down, dissolved in chloroform (approximately 2 μmol of lipid phosphorus per milliliter) and stored at $-20°$ under nitrogen. When prepared by this method, more than 98% of the radioactivity cochromatographs with phosphatidylinositol during thin-layer chromatography. When required, 0.5-ml aliquots of the ^{3}H-labeled microsomal lipid solution are dried down and sonicated in 1 ml of 100 mM HEPES–NaOH, pH 7.0.

Procedure. Sonicated ^{3}H-labeled microsomal lipids (0.05 ml), enzyme sample and water (to a total volume of 0.2 ml) are incubated at 37° for 10 min. The reaction is stopped by cooling on ice followed by the addition of 1 ml of chloroform–methanol–HCl (100 : 100 : 0.6) and 0.3 ml of 1 M HCl. The phases are separated by centrifugation, an aliquot (0.5 ml) of the upper phase is sampled into a scintillation vial, aqueous scintillation fluid is added, and radioactivity is determined. Methanol and traces of chloroform (which give a small amount of quenching) can be removed by evaporation before the addition of scintillation fluid; however, this lengthens the procedure considerably and does not improve reproducibility.

Enzyme Purification

The procedure described here is a modified version of a previously published method.[8]

[6] M. G. Low and D. B. Zilversmit, *Biochim. Biophys. Acta* **596,** 223 (1980).

[7] This series, Vol. 14 [44].

[8] M. G. Low and J. B. Finean, *Biochem. J.* **162,** 235 (1977).

Preparation of Culture Supernatant. The media used for growing *S. aureus* contained the diffusate from 88 g of Difco proteose peptone, KH_2PO_4 (4 g), K_2HPO_4 (4 g), $MgSO_4 \cdot 7 H_2O$ (1.6 g), lactic acid (16 g), and water to a total volume of 4 liters. The pH is adjusted to 7.4 with ammonium hydroxide and the media is autoclaved. Medium, 0.5 ml, in a culture tube is inoculated from a nutrient agar slope of *Staphylococcus aureus* (Newman) and incubated in a shaking water bath at 37° for 24 hr. This culture is then used to inoculate a fresh aliquot (0.5 ml) of medium; this process of subculturing is repeated 3–5 times. The final culture is used to inoculate 50 ml of medium in a 250-ml flask; after incubation at 37° for 24 hr, 10-ml aliquots are used to inoculate 1 liter of media in a 2-liter flask. The 2-liter flasks are incubated at 37° for 20–24 hr on an orbital shaker at approximately 200 rpm, and the bacteria are removed by centrifugation. All subsequent operations are carried out at 4°.

Chromatography on Amberlite CG-50. Sodium chloride (5 g/liter) is added to the culture supernatant (specific activity approximately 1 unit/mg), and the pH is adjusted to 5.5 with acetic acid. The culture supernatant is loaded onto a column (2.6 × 90 cm) of Amberlite CG-50 (100–200 mesh) equilibrated with 0.15 *M* NaCl, 0.05 *M* sodium acetate, pH 5.5 (buffer A), at a flow rate of approximately 150 ml/hr. The column is washed with 2–3 liters of buffer A and eluted with a linear gradient formed from 500 ml of buffer A and 500 ml of 1 *M* trisodium citrate at a flow rate of approximately 50 ml/hr; 10-ml fractions are collected. Phospholipase C activity is eluted as a broad peak (approximately 350–700 ml eluted), which is pooled and stored at −20°.

Chromatography on Sephadex G-75. The CG-50 eluates prepared from two batches of culture supernatant (see above) are thawed and dialyzed at 4° against 3 × 5 volumes of 50 m*M* Tris-acetate, pH 7.4 (at 4°), 0.02% sodium azide. The dialyzate is centrifuged at 10,000 *g* for 20 min, and the supernatant is concentrated in an Amicon ultrafiltration cell using a PM-10 membrane to a final volume of approximately 80 ml. The concentrate is loaded onto a column (5 × 200 cm) of Sephadex G-75, equilibrated in 50 m*M* Tris-acetate, pH 7.4, 0.02% sodium azide. The column is eluted at a flow rate of approximately 40 ml/hr, and 20-ml fractions are collected. The fractions are assayed for phospholipase C activity, and the peak fractions (V_e/V_o approximately 1.8) are pooled. The purified phospholipase C (specific activity approximately 6000 units/mg) can be stored at 4° for several months or lyophilized without substantial loss of activity. Polyacrylamide gel electrophoresis shows the phospholipase C to be 80–90% pure.

Properties[9]

The molecular weight of the phospholipase C is not certain, as values of 20,000 (by gel filtration) and 33,000 (by gel electrophoresis) have been obtained.[8] The phospholipase C is inhibited by $0.15 M$ salt solutions (KCl, NaCl, or NH_4Cl), by thiol reagents ($HgCl_2$, $CuSO_4$, or p-chloromercuribenzoate), or by detergents (Triton X-100, sodium deoxycholate, or cetyltrimethylammonium bromide).[3,8,10]

The phospholipase C activity is not stimulated by the addition of calcium ions and is not inhibited by chelators such as EDTA, EGTA, or 1,10-phenanthroline.[3,10] The other bacterial phosphatidylinositol-specific phospholipases[11] do not require metal ions either, suggesting a similar mechanism of action. Clearly they are quite different from the mammalian phosphatidylinositol-specific phospholipases C[12] and the other bacterial phospholipases C,[13] which require metal ions for maximal activity.

The phospholipase C is unable to hydrolyze detectable amounts of other phospholipids, such as phosphatidylcholine, phosphatidylethanolamine, phosphatidylserine, or sphingomyelin, in erythrocyte membranes or lipid extracts under conditions where extensive phosphatidylinositol hydrolysis occurs.[3] The activity against pure phosphatidylinositol is lower (approximately 10- to 20-fold) than that observed when lipid mixtures, such as soybean phospholipids or total rat liver lipid extract, which contain phosphatidylinositol, are used as substrates.[3]

Although the phospholipase C readily hydrolyzes phosphatidylinositol in erythrocyte ghosts, it is unable to hydrolyze this phospholipid in intact erythrocytes.[8] On the basis of these results it has been suggested that phosphatidylinositol is located at the inner surface of the erythrocyte membrane.[8] The phospholipase C is unable to hemolyze mammalian erythrocytes.[8]

This phospholipase C,[14-16] in common with the other phosphatidylinositol-specific phospholipases C of bacterial origin,[11] is able to release alkaline phosphatase, 5'-nucleotidase, and acetylcholines-

[9] Enzyme properties have been determined with both purified and partially purified phospholipase C; see original references for details.
[10] M. G. Low, unpublished work, 1977.
[11] See this volume [84].
[12] R. H. Michell, *Biochim. Biophys. Acta* **415**, 81 (1975).
[13] A. C. Ottolenghi, this series, Vol. 14 [35].
[14] M. G. Low and J. B. Finean, *Biochem. J.* **167**, 281 (1977).
[15] M. G. Low and J. B. Finean, *FEBS Lett* **82**, 143 (1977).
[16] M. G. Low and J. B. Finean, *Biochim. Biophys. Acta* **508**, 565 (1978).

terase from membranes. This effect has not been demonstrated with any other type of phospholipase, and for this reason it has been suggested that phosphatidylinositol is involved in the attachment of these enzymes to membranes. In view of these observations it it probable that this enzyme will be very useful for the study of membrane structure.

[86] Phospholipase D from Rat Brain

EC 3.1.4.4 Phosphatidylcholine phosphatidohydrolase

By T. TAKI and J. N. KANFER

Phospholipase D of plant tissues was described in a previous volume.[1] All phospholipase D preparations examined produce phosphatidic acid and the polar portion of the phospholipid substrate according to the following reaction:

Phosphatidylcholine (or phosphatidylethanolamine) + $H_2O \rightarrow$
phosphatidic acid + choline (or ethanolamine)

The properties of the plant enzyme were reviewed recently.[2] Phospholipase D of mammalian tissues is firmly bound to membranes and is virtually undetectable in homogenates but becomes demonstrable after detergent solubilization.[3,4]

Assay Method

Principle. The assay is based on the estimation either of phosphatidic acid formed from radioactive phosphatidylcholine or of choline or ethanolamine liberated from the corresponding base-labeled phospholipid.

Reagents

HEPES buffer, 1 *M*, pH 6.0, pH 8.0
$CaCl_2$ 0.1 *M*
Lyophilized rat brain powder
Solution A: 0.8% Miranol H2M, 0.5% cholate, 5 m*M* HEPES, pH 8.0
Solution B: 0.25% Miranol H2M, 0.125% cholate, 20% sucrose, 5

[1] M. Kates and P. S. Sastry, this series, Vol. XIV [36].
[2] M. Heller, *Adv. Lipid Res.* **16,** 267 (1978).
[3] M. Saito and J. N. Kanfer, *Biochem. Biophys. Res. Commun.* **53,** 391 (1973).
[4] M. Saito and J. N. Kanfer, *Arch. Biochem. Biophys.* **169,** 318 (1975).

METHODS IN ENZYMOLOGY, VOL. 71

ISBN 0-12-181971-X

mM HEPES (pH 7.0), and 1 mM mercaptoethanol
Solid ammonium sulfate
DEAE-52 cellulose powder
Sepharose 4B
Sephadex G-25
Silica gel G, 250 μm thin-layer chromatography plates
Chloroform, methanol, acetone, acetic acid, ammonia

Procedure. Phosphatidylcholine microdispersion is prepared with [U-^{14}C]lecithin, usually 10 μCi, and 4.2 mg of bovine lecithin suspended in 1 ml of *n*-butanol–20% sodium cholate (87 : 13, v/v) to give a final specific activity of 2.6×10^3 cpm/nmol. This solution is dialyzed against daily changes of 10 mM HEPES, pH 7.2, containing 1 mM EDTA for 1 week. Any insoluble material is removed by centrifugation as outlined elsewhere in this series.[5]

The incubation mixtures contain 10 μmol of HEPES buffer, pH 6.0 (10 μl), 1 μmol of $CaCl_2$ (10 μl), 350–400 nmol of [U-^{14}C]phosphatidylcholine (about 2000–4000 dpm/nmol) (80–100 μl), and up to 100 μg of enzyme protein in a final volume of 240 μl. The reaction is carried out at 37° for 30 min in air in a shaking water bath. It is stopped with 5 ml of chloroform–methanol (2 : 1, v/v) containing carrier phosphatidic acid. Material soluble in water is removed according to the procedure of Folch *et al.*[6] The final washed lower chloroform phase obtained after this treatment is transferred to a silica gel plate previously heat-activated at 100° for 1 hr. The lipids are separated two-dimensionally, chloroform, methanol, and ammonia (65 : 35 : 4) being employed in the first dimension and chloroform, acetone, methanol, acetic acid, and water (50 : 20 : 10 : 10 : 5) in the second. The radioactive products are detected by autoradiography, and the material on the plate corresponding to phosphatidic acid is removed and counted.

Preparation of Enzyme

Extraction. Lyophilized rat brain powder, 1.5 g, is homogenized with 200 ml of 5 mM HEPES buffer, pH 8.0, and the suspension is centrifuged at 100,000 g for 30 min. The supernatant is discarded, the pellet is homogenized with 160 ml of solution A, and the mixture is centrifuged at 100,000 g for 30 min. The supernatant is brought to 70% saturation of ammonium sulfate by addition of 69.76 g of the solid salt, and the mixture is allowed to remain at 4° for 30 min. The precipitate is harvested by centrifugation at 16,000 g for 20 min and suspended in 10 ml of solution B

[5] This series, Vol. 10 [70].
[6] Folch *et al.*, this series, Vol. 14 [44].

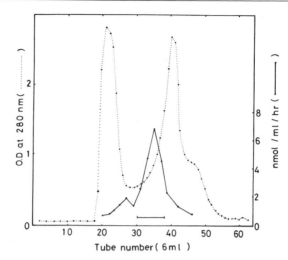

Fig. 1. Sepharose 4 B column chromatography of phospholipase D. A portion of each column fraction (40 μl) was assayed as described.

after removal of the supernatant, and any unsolubilized material is removed by centrifugation. At this point nearly all the phospholipase D activity is recovered in the supernatant with only negligible quantities of the base exchange activities. The enzyme has a tendency to aggregate, and therefore the presence of detergents is required at all stages of purification.

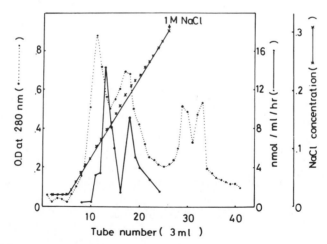

Fig. 2. DEAE-cellulose column chromatography I of phospholipase D.

Chromatographic Separations

A Sepharose 4B column (2.6 × 42 cm) preequilibrated with solution B containing 1 mM EDTA is loaded with 10 ml of the detergent-solubilized ammonium sulfate precipitate. Elution is continued with this solution and 6-ml fractions are collected at a rate of 15–18 ml/hr. A typical profile is shown in Fig. 1. Individual tubes are assayed for phospholipase D activity, and active fractions are pooled and brought to 70% of saturation with solid ammonium sulfate. Prior to further purification the precipitate is dissolved in solution B containing 1 mM EDTA and desalted on a Sephadex G-25 column preequilibrated with the same solution. The protein in the void volume of this column is applied to a DEAE-cellulose (DE-52) column (1.5 × 12 cm) preequilibrated with solution B. Elution is effected with a linear gradient starting with 40 ml of 0.05 N and 40 ml of 0.3 N NaCl in solution A. Fractions of 3 ml are collected at a rate of 30 ml/hr. A typical elution profile is presented in Fig. 2. The fractions obtained between 0.17 and 0.2 N NaCl that contain phospholipase D activity are pooled and concentrated by addition of solid ammonium sulfate to 70% saturation. The precipitate obtained is dissolved in solution B and desalted on a Sephadex G-25 column. The effluent is applied to a DEAE-cellulose (DE-52) column (1.5 × 8 cm) previously equilibrated with 0.1 N NaCl in solution B, and the enzyme is eluted with a linear gradient of NaCl starting with 25 ml of 0.1 N and 25 ml of 0.25 N NaCl in solution B. Fractions of 2 ml are collected at a rate of 35 ml/hr. The elution profile is shown in Fig. 3. The apparent overall enrichment is 240-fold (see the table).

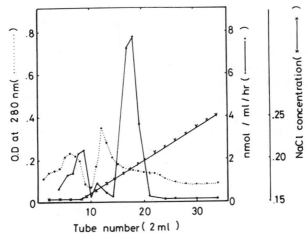

FIG. 3. DEAE-cellulose column chromatography II of phospholipase D.

PURIFICATION OF PHOSPHOLIPASE D FROM RAT BRAIN

Step	Phospholipase D	
	Activity (nmol/mg protein per hour)	Enrichment
Homogenate	0.4–0.5	1
70% $(NH_4)_2SO_4$ precipitate	3.0–5.0	7.5–10
Sepharose 4B column	22–24	55–60
DEAE-cellulose column I (second peak)	40–48	100
DEAE-cellulose column II	120	240

Properties

Stability. The 70% ammonium sulfate precipitate of the detergent extract tends to aggregate and therefore cannot be stored. The material from the Sepharose 4B column is stable for at least 1 month at 4°. The early peak of activity from the first DEAE-cellulose column is very unstable and resists further attempts at purification; the second peak is stable for about 1 week at 4°.

Effect of pH. The optimum pH is 6.0 with HEPES or β,β-dimethylglutarate buffers.

Specificity and Substrate Affinity. The most highly purified preparation hydrolyzes phosphatidylcholine and phosphatidylethanolamine to yield phosphatidic acid, but does not cleave lysophosphatidylcholine. There is no detectable base exchange activity associated with the purified preparation. The extraction procedure from lyophilized brain powder appears to eliminate these activities. K_m values are 0.75 mM for naturally occurring brain lecithin, 0.78 mM for dipalmitoyllecithin, and 0.91 mM for phosphatidylethanolamine.

Activators and Inhibitors. Added salts are not required for enzyme activity, but 5 mM $CaCl_2$ or $FeCl_2$ cause 65–80% stimulation. $FeCl_3$ and pCMB at 5 mM cause complete inhibition of activity, whereas EDTA at the same concentration has no effect. All detergents tested, including bile acids, sodium dodecyl sulfate, Tweens, and Tritons at 1 mg/ml, cause nearly complete loss of activity.

Section VI

Miscellaneous

[87] Lecithin-Cholesterol Acyltransferase from Human Plasma

EC 2.3.1.43 Lecithin : cholesterol acyltransferase

By YUKIO DOI and TOSHIRO NISHIDA

Lecithin–cholesterol acyltransferase (EC 2.3.1.43) is produced by the liver and secreted into the circulation.[1,2] The enzyme catalyzes primarily the transfer of an acyl group from the carbon-2 position of phosphatidylcholine to cholesterol. Furthermore, the cholesterol esters produced by this enzymatic reaction are necessary for the maintenance of the normal structure of plasma lipoproteins.[3,4] In view of the vital importance of lecithin–cholesterol acyltransferase in the metabolism and structure of plasma lipoproteins, many attempts have been made toward purification. Only recently have several groups obtained a highly purified enzyme yielding a single main band on both disc and sodium dodecyl sulfate (SDS)–polyacrylamide gel electrophoreses.[5–8] However, previously published purification methods were hampered by the complexity of the procedures employed or by the use of prolonged ultracentrifugation. We developed a convenient purification method better suited to routine use.[9] This method was modified as described in this chapter.

Assay Method

Principle. Single bilayer vesicles of egg phosphatidylcholine-[7α-^{3}H]cholesterol mixture (molar ratio, 6 : 1) are incubated with enzyme fractions in the presence of a cofactor peptide, apolipoprotein A-I (apo-A-I). To determine the amount of [7α-^{3}H]cholesterol ester produced, the lipids are extracted and separated by thin-layer chromatography. The radioactivity in the cholesterol and cholesterol ester fractions is determined with a liquid scintillation spectrometer.

[1] J. A. Glomset, *J. Lipid Res.* **9**, 155 (1968).
[2] J. A. Glomset, *in* "Blood Lipids and Lipoproteins" (G. Nelson, ed.), p. 745. Wiley, New York, 1972.
[3] J. A. Glomset and K. R. Norum, *Adv. Lipid Res.* **11**, 1 (1973).
[4] T. Forte, A. Nichols, J. A. Glomset, and K. R. Norum, *Scand. J. Clin. Lab. Invest.* **33**, Suppl. 137, 121 (1974).
[5] J. J. Albers, V. G. Cabana, and Y. Dee Barden Stahl, *Biochemistry* **15**, 1084 (1976).
[6] K. G. Varma and L. A. Soloff, *Biochem. J.* **155**, 583 (1976).
[7] L. Aron, S. Jones, and C. J. Fielding, *J. Biol. Chem.* **253**, 7220 (1978).
[8] J. Chung, D. A. Abano, G. M. Fless, and A. Scanu, *J. Biol. Chem.* **254**, 7456 (1979).
[9] K. Kitabatake, U. Piran, Y. Kamio, Y. Doi, and T. Nishida, *Biochim. Biophys. Acta* **573**, 145 (1979).

METHODS IN ENZYMOLOGY, VOL. 71

Reagents

Phosphate (KH_2PO_4–Na_2HPO_4) buffer (ionic strength 0.1, pH 7.4)
  containing 0.025% EDTA and 2 mM NaN_3
[7α-^{3}H]Cholesterol
Unlabeled cholesterol, 99+%
Egg phosphatidylcholine, prepared by the method of Singleton *et al.* [10]
  and purified by silicic acid column chromatography [11]
Acetic acid, reagent grade
Chloroform, reagent grade
Diethyl ether, reagent grade
Ethanol, reagent grade
Hexane, reagent grade
2-Mercaptoethanol, reagent grade
Methanol, reagent grade
Bovine serum albumin, crystallized and lyophilized
Apo-A-I, prepared from human high density lipoproteins [12] by delipi-
  dation with ethanol–diethyl ether [13] followed by gel filtration on
  Sephadex G-150 and DEAE-cellulose column chromatography at
  4° [14]
Liquid scintillator: 5 g of 2′,5-diphenyloxazole (PPO) and 100 mg of
  1,4-bis[2-(5-phenyloxazolyl)]benzene (POPOP) in 1 liter of toluene.

Preparation of Substrate Vesicles. Phosphatidylcholine–cholesterol vesicles are prepared essentially according to the method of Batzri and
Korn. [15] The vesicles prepared by their method appear to give more reproducible results in the enzyme assay than the sonicated dispersion. [6] An
ethanol solution, 0.5 ml, containing 9 μmol of egg phosphatidylcholine and
1.5 μmol of [7α-^{3}H]cholesterol is rapidly injected into 9.5 ml of phosphate
buffer through a Hamilton syringe. The injection is made below the liquid
surface with vigorous stirring using either a magnetic stirrer or Vortex
mixer under a stream of N_2. The phosphatidylcholine–cholesterol dispersion is dialyzed against phosphate buffer to remove ethanol and then
diluted to 10 ml with phosphate buffer. The removal of ethanol is neces-

[10] W. S. Singleton, M. S. Gray, M. L. Brown, and J. L. White, *J. Am. Oil Chem. Soc.* **42**, 53 (1965).
[11] A. D. Bangham, M. W. Hill, and N. G. A. Miller, *in* "Methods in Membrane Biology" (E. D. Korn, ed.), p. 1. Plenum, New York, 1974.
[12] R. A. Muesing and T. Nishida, *Biochemistry* **10**, 2952 (1971).
[13] V. Shore and B. Shore, *Biochemistry* **6**, 1962 (1967).
[14] R. L. Jackson, H. N. Baker, O. D. Taunton, L. C. Smith, C. W. Garner, and A. M. Gotto, *J. Biol. Chem.* **248**, 2639 (1973).
[15] S. Batzri and E. D. Korn, *Biochim. Biophys. Acta* **298**, 1015 (1973).

sary for the assay of highly purified enzyme. It is observed that the presence of 5% ethanol in the dispersion inhibits the activity of pure enzyme but not of crude enzyme preparations.[16]

Procedure. The assay mixture consists of 100 μl of vesicle solution (90 nmol of egg phosphatidylcholine and 15 nmol of $[7\alpha\text{-}^3\text{H}]$cholesterol, specific activity 4 mCi/mmol), 15 μg of apo-A-I, 4 mM 2-mercaptoethanol, 0.7 mM EDTA, 2.5 mg of bovine serum albumin, and the sample of enzyme fraction diluted to a final volume of 250 μl with phosphate buffer. The vesicle solution and apo-A-I are preincubated overnight at 4° or for 30 min at 37° under N_2 prior to addition to the assay mixtures. The mixtures, in 8-ml screw-cap tubes flushed with N_2 are incubated for 1 hr at 37° with mechanical shaking. After incubation 1 ml of methanol is added to each tube, and the mixtures are heated at 55° under N_2 for 15 min. This is followed by the addition of 3 ml of hexane–chloroform (4 : 1, v/v) with subsequent vigorous shaking for 15 min under N_2. To each tube is then added 1 ml of water; the mixtures are again shaken vigorously for 10 min under N_2. After centrifugation, the upper hexane–chloroform phase is removed and evaporated to dryness in a vacuum oven at 30°. The residue is dissolved in diethyl ether (50–100 μl) containing carrier cholesterol and cholesterol ester, 1 μg each per microliter, and applied on silicic acid-impregnated, glass fiber sheets (type ITLC-SA, Gelman Instrument Company, Ann Arbor, Michigan). The lipids are separated using hexane–diethyl ether–acetic acid (70 : 30 : 1, v/v/v) as the developing solvent, and the radioactivity in the cholesterol and cholesterol ester fractions is determined by cutting the lipid spots and counting in a liquid scintillation spectrometer. The number of nanomoles of cholesterol ester produced is calculated from the distribution of radioactivity in the lipid spots and the initial concentrations of labeled cholesterol. The results are corrected for values obtained with control assay mixtures treated in the same manner. The amount of enzyme fraction in an assay mixture is chosen to give the esterification of 0.4–2.0 nmol of cholesterol per hour. Under these assay conditions, the rate of cholesterol esterification is linear during a 1-hr incubation period.

Enzyme Units. One unit of enzyme designates the esterification of 1 nmol of unesterified cholesterol per hour under the standard assay conditions described above. Specific activities of various enzyme fractions are given by the numbers of enzyme units per milligram of protein. The protein contents are determined by the method of Lowry *et al.*[17]

---

[16] Y. Doi and T. Nishida, unpublished results, 1978.
[17] O. H. Lowry, N. J. Rosebrough, A. L. Farr, and R. J. Randall, *J. Biol. Chem.* **193**, 265 (1951).

Purification Procedure[18]

All operations are carried out at 4° or in an ice bath. Chromatographic runs are monitored at 280 nm with a double-beam recording spectrophotometer. All centrifugations are performed with a refrigerated high-speed centrifuge. Unless specified otherwise, phosphate (KH_2PO_4–Na_2HPO_4) buffer, pH 7.4, ionic strength 0.1, containing 0.025% EDTA and 2 mM sodium azide, is used. Human plasma containing 17 mM trisodium phosphate, 3 mM citric acid, 3 mM monobasic sodium phosphate, and 0.5% dextrose is obtained from a local hospital blood bank. Nonturbid plasma samples, 1–2 weeks old, are primarily used for the separation of lecithin–cholesterol acyltransferase. The following procedures are exemplified for 1 liter of plasma.

Step 1. Dialysis of Plasma. One liter of human plasma is centrifuged at 8000 g for 10 min to remove remaining cells and dialyzed against 0.025% EDTA (pH 7.4) containing 2 mM sodium azide.

Step 2. Dextran Sulfate Treatment. Dextran sulfate (Pharmacia Fine Chemicals) is dialyzed against glass-distilled water and lyophilized or concentrated prior to use. To the stirred dialyzed plasma, 14 g of dextran sulfate are added. This is followed by the addition of $CaCl_2$ to the final concentration of 0.1 M. The mixture is stirred gently for 15 min and then centrifuged at 16,000 g for 15 min to remove insoluble dextran sulfate–lipoprotein complex.

Step 3. Butanol–Ammonium Sulfate Treatment. To the stirred dextran sulfate supernatant fraction, solid sucrose is added to the concentration of 100 g per liter followed by the addition of solid $(NH_4)_2SO_4$ (extreme purity; Heico, Delaware Water Gap, Pennsylvania) to 40% saturation. One volume of precooled 1-butanol is added to two volumes of the solution, and this mixture is vigorously agitated with an Omni-mixer (DuPont Instruments, Newtown, Connecticut), four times for 30 sec each. The mixture is then centrifuged at 16,000 g for 15 min. The protein precipitated at the solvent interface is collected and suspended in 1 liter of precooled 40% saturated $(NH_4)_2SO_4$ solution containing 10% sucrose, followed by the addition of 500 ml of precooled 1-butanol. The mixture is agitated and centrifuged. The butanol–$(NH_4)_2SO_4$ treatment of the interfacial precipitate is repeated with one-half volumes of the solutions. The final interfacial precipitate obtained is suspended in 450 ml of phosphate buffer containing 10% sucrose and dissolved by stirring for 30 min with a magnetic stirrer under N_2. The turbid enzyme solution is dialyzed exhaustively against phosphate buffer to remove butanol, $(NH_4)_2SO_4$, and sucrose. The precipitate formed during dialysis is removed by centrifugation. The enzyme is present in the supernatant solution.

[18] Y. Doi, Y. Furukawa, and T. Nishida, *Fed. Proc., Fed. Am. Soc. Exp. Biol.* **38,** 334 (1979).

Step 4. DEAE-Sephadex Chromatography. The enzyme fraction is applied to a DEAE-Sephadex A-50 column (3 × 40 cm), which was previously equilibrated with phosphate buffer. The column is washed with 400 ml of phosphate buffer or until the eluate shows no significant absorbance at 280 nm. The enzyme fraction is eluted with phosphate buffer containing 0.16 M NaCl. The flow rate for the column is approximately 0.7 ml/min. The enzyme fraction eluted immediately after the main protein peak is collected in approximately 300 ml and dialyzed without delay against phosphate buffer.

Step 5. Second DEAE-Sephadex Chromatography.[19] A DEAE-Sephadex A-50 column (1.7 × 25 cm) is equilibrated with phosphate buffer. The enzyme solution (approximately 325 ml) is applied to the column followed by washing with five column volumes of phosphate buffer containing 0.07 M NaCl. The outlet of the column is connected to cellulose hollow fibers of a Beaker or Mini Beaker dialyzer (The Dow Chemical Company) and then to an ultraviolet monitor. The enzyme is eluted with phosphate buffer containing 0.12 M NaCl at a flow rate of 0.3 ml/min. During the enzyme elution, precooled deionized water is passed through the dialyzer chamber to reduce the ionic strength of the enzyme fraction passing through the fiber bundle. The enzyme fraction, which is eluted at the trailing end of the main protein peak, is collected in approximately 250 ml and dialyzed against 1 mM sodium phosphate buffer (pH 6.8) containing 4 mM 2-mercaptoethanol.

Step 6. Hydroxyapatite Chromatography. The dialyzed fraction (approximately 250 ml) is applied to a hydroxyapatite column (3 × 6 cm) equilibrated with 1 mM sodium phosphate buffer (pH 6.8). A protein load of 0.5–1.5 mg per milliliter of bed volume is used. The column is washed with four column volumes of 1 mM sodium phosphate buffer and then eluted with 4 mM sodium phosphate buffer (pH 6.8). The flow rate for the elution is 0.4 ml/min. Approximately 50% of the enzyme activity applied is recovered in a 60-ml eluate. Since the enzyme is stable only in a buffer of very low ionic strength,[20] the hydroxyapatite fraction is dialyzed against 0.4 mM sodium phosphate buffer containing 4 mM 2-mercaptoethanol or is transferred into a medium of very low ionic strength during the concentration through an ultrafiltration membrane.

The purification of lecithin–cholesterol acyltransferase from 1 liter of human plasma is summarized in Table I.

---

[19] This procedure can be omitted if the enzyme fraction obtained by small-scale hydroxyapatite chromatography, which is conducted in a similar manner as step 6 with 3 mg of protein on a 1.2 × 2.6 column, is not contaminated with a protein having a molecular weight of approximately 33,000.

[20] Y. Furukawa and T. Nishida, *J. Biol. Chem.* **254**, 7213 (1979).

TABLE I

PURIFICATION OF LECITHIN–CHOLESTEROL ACYLTRANSFERASE

Fraction	Total volume (ml)	Total protein (mg)	Total activity (units)[a]	Specific activity (units per mg protein)	Yield (%)	Purification (fold)
Plasma	1000	70,000	550,000	8[b]	100	1
Dextran sulfate supernatant	1200	55,000	500,000	9	90	1.1
Butanol–(NH$_4$)$_2$SO$_4$ precipitate	500	2,700	295,000	110	54	14
DEAE-Sephadex eluate	300	80	187,000	2,300	34	288
Second DEAE-Sephadex eluate	250	16	133,000	8,300	24	1,000
Hydroxyapatite eluate	60	0.41	66,000	161,000	12	20,000

[a] One unit of enzyme catalyzed the esterification of 1 nmol of unesterified cholesterol per hour at 37° with lecithin–cholesterol vesicles as a substrate.

[b] The enzyme activity of the plasma was assayed using 2 μl of plasma in 250 μl of incubation mixture. The assay mixture without the plasma was kept at 0° for 12 hr prior to the incubation at 37°. The rate of cholesterol esterification was calculated assuming the equilibration of vesicle cholesterol (15 nmol) and plasma unesterified cholesterol (3.2 nmol). Considerable difficulties exist in assessing the plasma enzyme level by the currently available methods as reviewed by Norum [K. R. Norum, *Scand. J. Clin. Lab. Invest.* **33** Suppl. 137, 7 (1974)], and summarized by Varma and Soloff.[6] The specific activities measured for the plasma by various workers showed considerable variation, mainly reflecting the differences in the assay methods, especially in the substrates used. We observed that when phosphatidylcholine–cholesterol vesicles were used as a substrate in the presence of 15 μg of apo-A-I, the rate of cholesterol esterification was linear for concentrations less than 2 μl of plasma in 250 μl of incubation mixture. Under such conditions, almost no enzyme activity was detected in the absence of added apo-A-I. It would appear that the plasma must be diluted sufficiently to eliminate or minimize the influence of cofactors and inhibitors present in the plasma, on the enzyme activity.

Comments about the Purification Procedure

Dialysis of Plasma and Dextran Sulfate Treatment. The enzyme appears to be associated with high-density lipoproteins (HDL) in human plasma.[2] The dialysis against 0.025% EDTA, however, results in the dissociation of the enzyme–lipoprotein complex due to a reduction in the ionic strength.[16] Subsequent dextran sulfate treatment allows a maximal precipitation of lipoproteins, leaving the enzyme in the supernatant fraction. It should be mentioned that this procedure may not be applicable to the plasma from other species, as the affinity of the enzyme for HDL may vary. The enzyme may not necessarily be dissociated from HDL by the dialysis of plasma from other species against 0.025% EDTA. The primary difference reflected here is probably due to the different distribution of apo-HDL and the surface organization asimilarities of the lipoproteins.

The precipitated lipoprotein–dextran sulfate fraction from dialyzed human plasma may be utilized for separation and purification of low-density lipoproteins (LDL) and HDL.[21] Instead of simultaneous precipitation of all lipoproteins by dextran sulfate, LDL- and HDL-rich fractions may be sequentially precipitated for subsequent purification. For this purpose, 0.18 g of dextran sulfate is added to the dialyzed plasma, and then $CaCl_2$ is added to a final concentration of 0.02 M. The precipitate consisting primarily of LDL–dextran sulfate complex is removed by centrifugation at 13,000 g for 15 min. To the supernatant fraction, 13 g of dextran sulfate are added, followed by $CaCl_2$ to a final concentration of 0.1 M. The precipitate containing HDL–dextran sulfate complex is removed by centrifugation at 20,000 g for 30 min.

Butanol–$(NH_4)_2SO_4$ Treatment. This treatment is similar to that previously employed to the central colorless fraction obtained by centrifugation of plasma at $d = 1.21$ g/cm³.[22] The use of 40% ammonium sulfate saturation, instead of 50% saturation, reduces the amount of contaminating protein in the interfacial precipitate without altering the enzyme yield. Although the crude enzyme fraction is relatively stable in the presence of $(NH_4)_2SO_4$ or 1-butanol alone, their combination in this procedure reduces enzyme stability. Therefore, the butanol–$(NH_4)_2SO_4$ treatment should be performed with a minimum of delay. Sucrose is included in the treatment to help minimize enzyme denaturation. When all procedures are carried out promptly, as much as 75% of the activity in the dextran sulfate fraction is recovered from the final interfacial precipitate.

DEAE-Sephadex Chromatography. The effect on the elution patterns of the enzyme and contaminants produced by varying the NaCl concentra-

[21] M. Burstein and H. R. Scholnick, *Adv. Lipid Res.* **11**, 67 (1973).
[22] A. K. Soutar, H. J. Pownall, A. S. Hu, and L. C. Smith, *Biochemistry* **13**, 2828 (1974).

tion in phosphate buffer is shown in Fig. 1. In the presence of 0.14 M NaCl
the enzyme is well separated from the main protein peak. Raising the
concentration to 0.16 M results in a narrower separation, however, with
the enzyme elution occurring in less volume. Further increasing the NaCl
concentration to 0.20 M results in considerable overlapping of the enzyme
and contaminant peaks. On the basis of these observations, the NaCl
concentration of 0.16 M was chosen for elution of the enzyme in the
standard procedure. This corresponds to an ionic strength of 0.26 for the
eluting solvent. A linear salt concentration gradient in buffer does not
separate the enzyme from the main contaminant protein as effectively as
the stepwise elution.

When the enzyme fraction at this step is applied directly to hydroxyapatite chromatography, the SDS–gel electrophoresis of the hydroxyapatite fraction often shows some contamination of the enzyme by a
protein having an apparent molecular weight of approximately 33,000. It
accounts for 2–10% of the total stainable material. This contaminant can

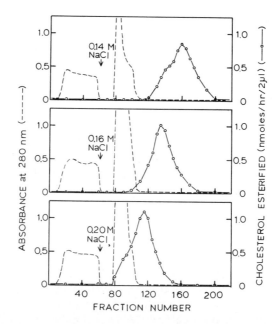

FIG. 1. DEAE-Sephadex chromatography of butanol–$(NH_4)_2SO_4$ fraction. The enzyme
fraction (300 ml, 5.3 mg of protein per milliliter) was applied to a DEAE-Sephadex A-50
column (3 × 40 cm) equilibrated with phosphate buffer. The column was washed with phosphate buffer and then eluted with phosphate buffer containing NaCl at the concentration of
(a) 0.14 M, (b) 0.16 M, and (c) 0.20 M. The flow rate was 0.7 ml/min, and 7.5-ml fractions
were collected.

be removed by carrying out a second DEAE-Sephadex chromatography prior to the hydroxyapatite chromatography.

Second DEAE-Sephadex Chromatography. After sample application, the column is washed with phosphate buffer containing 0.07 M NaCl to remove a major portion of the contaminant. A minor portion is removed by the subsequent elution with phosphate buffer containing 0.12 M NaCl. This salt concentration in the washing buffer is critical because an increase in the salt concentration causes a substantial elution of the enzyme in the washings. Excessive washing of the column also results in some loss of the enzyme. Thus the optimal degree of washing must be determined experimentally for a particular column (e.g., its bed volume, diameter, and height) and for the protein concentration as well as the volume of sample applied. The gel permeation chromatography of both the hydroxyapatite and the first DEAE-Sephadex fractions through Sephadex G-150 and Sephacryl S-200 columns is not as effective as the additional DEAE-Sephadex chromatography in removing the contaminant.

The hollow fiber dialyzer connected to the column is necessary to prevent the enzyme inactivation during the chromatography. Without the dialyzer often as much as 70% of the activity in the sample applied is lost. Although in the standard procedure for enzyme purification the second DEAE-Sephadex chromatography is performed between the first DEAE-Sephadex and hydroxyapatite chromatography, it may be carried out after the hydroxyapatite chromatography. In this case it is absolutely essential to reduce the ionic strength of the enzyme eluate by connecting the column outlet to a hollow fiber dialyzer or some other in-line dialysis device. The hydroxyapatite fraction is much more rapidly subject to inactivation than the first DEAE-Sephadex fraction upon exposure to salt medium, especially at air–water interface,[20] e.g., at the surface of the droplets formed at the column outlet and at the liquid surface produced in fraction collector tubes.

Hydroxyapatite Chromatography. Prior to hydroxyapatite chromatography, the second DEAE-Sephadex fraction is dialyzed against 1 mM sodium phosphate buffer (pH 6.8) containing 4 mM 2-mercaptoethanol, which is included to protect sulfhydryl groups essential for enzyme activity from oxidation.[20] In the subsequent hydroxyapatite chromatography, the omission or inclusion of 2-mercaptoethanol does not change the enzyme yield. Therefore, for convenience, 2-mercaptoethanol is not included in the elution buffer; its presence interferes with the detection of the enzyme at 280 nm with an absorbance monitor. When the dialyzed sample is applied to the hydroxyapatite column and washed with 1 mM sodium phosphate buffer, all proteins are absorbed onto the column. Although a broad peak with a strong absorbance at 280 nm appears immediately after the sample application, the peak is due to 2-mercaptoethanol present in the dialyzed sample. The fractions corresponding to the peak do not contain enzyme activity or show the presence of proteins on SDS–polyacrylamide gel electrophoresis. Upon elution of the enzyme with 4 mM sodium phosphate buffer, 50–60% of the enzyme activity applied is obtained as a sharp single peak. When the column is eluted with a gradient of phosphate buffers with the concentrations increasing linearly from 1 to 15 mM, no other peaks are detected. The elution of the enzyme with 4 mM phosphate buffer is preferred to the gradient elution because the latter causes a greater dilution of the enzyme fraction.

When the hydroxyapatite chromatography is conducted in the presence of 0.1 M or higher concentrations of NaCl as described by other investigators,[6,8,23] the enzyme elution requires phosphate buffer of higher concentrations. The purity of the enzyme obtained is comparable to that eluted with 4 mM phosphate buffer in the absence of NaCl. However, the presence of NaCl in a higher concentration of phosphate buffer facilitates the denaturation of the highly purified enzyme.[20] This inactivation may be minimized by reducing the ionic strength of the medium during enzyme elution by utilizing an in-line dialysis device. If the hydroxyapatite chromatography is used as the final step in the purification, a buffer of low ionic strength may be preferred.

Properties

Purity of the Enzyme. The enzyme prepared by the procedure described shows essentially a single band on polyacrylamide gel electrophoresis,[9] which is carried out according to the method of Davis[24] in a 7.5% slab gel using an apparatus described by O'Farrell.[25] Densitometric recording of the stained gel at 560 nm indicates that the enzyme protein accounts for 97–98% of the total stainable material. The superimposition of the enzyme activity with the protein band is confirmed by the enzyme assay for the extract obtained from sequential slices of a gel section, which is adjacent to the section subsequently stained.[9] The enzyme activity also comigrates with the protein band on electrophoresis in 5 and 10% polyacrylamide gels. The enzyme fraction incubated in 8 M urea containing 10 mM dithiothreitol for 30 min at room temperature also gives a single band on polyacrylamide gel electrophoresis in the presence of 8 M urea. Upon SDS–polyacrylamide gel electrophoresis, the enzyme fraction exhibits a single band located at a position slightly lower than that of bovine

[23] J. J. Albers, J. Lin, and G. P. Roberts, *Artery* **5**, 61 (1979).
[24] B. L. Davis, *Ann. N. Y. Acad. Sci.* **121**, 404 (1964).
[25] P. H. O'Farrell, *J. Biol. Chem.* **250**, 4007 (1975).

serum albumin.[9] Based on the specific activity, expressed in nanomoles of cholesterol esterified per hour per milligram protein, the final purification of the enzyme is 20,000-fold over the starting plasma with approximately 12% yield.

Stability. Crude enzyme preparations are stable in media differing widely in the ionic strength. However, the stability of purified enzyme is dependent upon the ionic strength of the medium.[20] In contrast to a rapid inactivation of the purified enzyme in buffers of ordinary ionic strength, the enzyme is remarkably stable in buffers of very low ionic strength. The use of very low ionic strength media is therefore important in the final stage of enzyme purification and in the storage of purified enzyme. The rapid inactivation of the enzyme in buffers of ordinary ionic strength, e.g., 39 mM phosphate buffer, is effectively prevented by apo-A-I. The substrate phosphatidylcholine–cholesterol vesicles also stabilizes the enzyme against inactivation, but to a lesser extent than apo-A-I. Although bovine serum albumin at low concentrations shows less stabilizing effect than the vesicles, the effect increases with increasing albumin concentration. It appears that the enzyme inactivation occurs primarily at the "air"–water interface and that media of higher ionic strength facilitate the extensive unfolding of the enzyme at the interface. Such an unfolding may be prevented by apo-A-I and, less effectively, by phosphatidylcholine–cholesterol vesicles and albumin.

The following precautions are necessary for enzyme purification and for all experiments with highly purified enzyme.[20]

1. The enzyme fractions after the first DEAE-Sephadex chromatography are susceptible to the interfacial denaturation in buffers of ionic strength commonly used. The enzyme inactivation and subsequent decrease in the specific activity can be minimized by following the procedure and cautions described in previous sections.
2. The purified enzyme should be stored, preferably at a high concentration in media of very low ionic strength, at 4° under N$_2$ with minimal "air"–water interface. Under such conditions the enzyme retains its original activity for at least 2 months. The purified enzyme cannot be stored in the frozen state because of its lability to freezing and thawing. Glycerol may be added to the purified enzyme in very low ionic strength media in order to allow storage at lower temperatures (e.g., −20°), and reduce the exposure of the enzyme at the interface by increasing the medium viscosity.
3. Since the enzyme activity is rapidly lost in media of ordinary ionic strengths, the experiments should be carried out in the presence of either apo-A-I or albumin.

4. If these "stabilizers" cannot be used, the exposure of the enzyme at the "air"–water interface must be minimized.

5. When the experiments involve chromatographic or centrifugal separation, the enzyme fractions obtained should be deposited directly into an apo-A-I or albumin solution or into the assay medium.

6. When the enzyme is added to solutions, especially those without stabilizers, it should be injected beneath the liquid surface to minimize inactivation at the interface. The enzyme should never be deposited on the walls of test tubes that are exposed to air, especially if glass tubes are used.

Physical Properties. The apparent molecular weight of purified lecithin–cholesterol acyltransferase determined by SDS–gel electrophoresis varies from 65,000 to 69,000.[5,7-9] A molecular weight of approximately 59,000 is obtained by sedimentation equilibrium analysis in the presence or absence of guanidine hydrochloride or mercaptoethanol.[8] In spite of the apparent homogeneity of the purified enzyme evidenced by both disc and SDS–polyacrylamide gel electrophoreses, isoelectric focus-

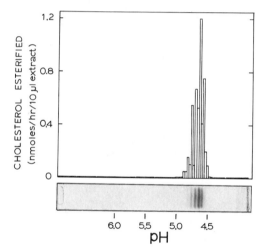

Fig. 2. Isoelectric focusing of lecithin–cholesterol acyltransferase. Two samples of a hydroxyapatite fraction (10 μg of protein each) were applied to adjacent wells of a 4.8% polyacrylamide gel slab (100 × 150 × 0.8 mm) containing 2% ampholine (pH 4–6) and 10% sucrose. After isoelectric focusing for 6 hr at 400 V, the pH 4.5–5.0 region of one of the vertical sections was sliced into 0.1 cm segments and the remainder into 1.0 cm segments. Each segment was extracted with 0.5 ml of phosphate buffer, and the enzyme activity was determined. The isoelectric focusing pattern was determined by protein staining of the adjacent vertical section of the gel slab. To determine the pH gradient, 5 mm sequential segments were obtained from both edges of the gel slab and extracted with 0.5 ml of 10 mM KCl.

ing in polyacrylamide gel gives four major bands and at least two minor bands as shown in Fig. 2. The isoelectric points of the bands determined at 4° range from 4.5 to 4.8. The comigration of the enzyme activity with the protein bands is confirmed. Similar multiple bands with the isoelectric points ranging from 5.1 to 5.5 are obtained.[23] The microheterogeneity or multiple forms of the enzyme may reflect a variation in the carbohydrate chains. However, the possibility that various ampholine molecules produce specific enzyme–ampholine complexes with different charge characteristics cannot be ruled out. The $A_{1cm}^{1\%}$ of the enzyme at 280 nm in phosphate buffer (ionic strength, 0.1, pH 7.4) is 20.0. This value is based on the protein content determined by the method of Lowry *et al.*[17] with crystalline bovine serum albumin dried over phosphorus pentoxide as a standard. The rather large coefficient may reflect high tryptophan content.

Chemical Properties. The amino acid composition of the purified enzyme is determined. The results are listed in Table II together with those reported by other investigators. Although the data are generally in agreement, some variations exist in the content of proline, glycine, valine, leucine, and tryptophan. Differences may reflect the enzyme purity, source of the plasma, purification method, and accuracy of the determination. The enzyme is a glycoprotein and reported to contain 24% carbohydrate by weight, consisting of 31 mol of mannose, 30 mol of galactose, 17 mol of glucosamine, and 13 mol of sialic acid per mole (59,000 g) of enzyme protein.[8]

Cofactor Peptides. Apo-A-I is known to exhibit a cofactor activity.[26,27] The enzyme fraction prepared by the present method shows an absolute requirement for apo-A-I with phosphatidylcholine–cholesterol vesicles as a substrate; no enzyme activity is detected in the absence of this protein cofactor. Although the enzymatic reaction requires the simultaneous presence of apo-A-I and the enzyme on the substrate vesicles, apo-A-I is not involved in the association of the enzyme with the substrate vesicles.[20] It is possible that apo-A-I may be involved as a cofactor by interacting reversibly with the enzyme at the water–vesicle interface. Such an interaction may lead to the activation of the enzyme and the optimal orientation of substrate molecules necessary for the enzymatic reaction. Apo-C-I also shows the cofactor activity but is less effective than apo-A-I with egg yolk or unsaturated phosphatidylcholine substrates.[23,27] On the other hand, apo-A-II, C-II, C-III, and D are shown to inhibit the activation of the enzyme.[23,27]

[26] C. J. Fielding and P. E. Fielding, *FEBS Lett.* **15,** 355 (1971).

[27] A. K. Soutar, C. W. Garner, H. N. Baker, J. T. Sparrow, R. L. Jackson, A. M. Gotto, and L. C. Smith, *Biochemistry* **14,** 3057 (1975).

TABLE II

AMINO ACID COMPOSITIONS OF LECITHIN–CHOLESTEROL ACYLTRANSFERASE

Amino acid	Moles/10^5 g protein			
	Present study[a]	Aron et al.[7]	Albers et al.[23][b]	Chung et al.[8]
Lysine	28	31	26	31
Histidine	27	26	22	26
Arginine	40	38	34	41
Aspartic acid	80	74	75	82
Threonine	48[c]	44	46	55
Serine	49[c]	50	48	56
Glutamic acid	83	92	80	93
Proline	84	76	64	78
Glycine	74	88	77	85
Alanine	48	51	49	60
Half-cystine	8[d]	ND[f]	6	8
Valine	61	46	51	61
Methionine	18	16	15	16
Isoleucine	37	28	32	39
Leucine	106	92	89	102
Tyrosine	40	38	32	29
Phenylalanine	39	37	35	32
Tryptophan	22[e]	ND	ND	15

[a] Values are the average of three independent determinations conducted on 24- and 72-hr hydrolyzates by the method of Spackman et al. [D. H. Spackman, W. H. Stein, and S. Moore, Anal. Chem. **30**, 1190 (1958)].

[b] Calculated from the data presented as moles of amino acids per 1000 mol of protein.

[c] Obtained by zero-time extrapolation.

[d] Average of two independent determinations carried out on the samples that were reduced and S-carboxymethylated essentially according to the method of Hirs (C. H. W. Hirs, this series, Vol. 11 [20]).

[e] Determined after alkaline hydrolysis according to the method of Hugli and Moore [T. E. Hugli and S. Moore, J. Biol. Chem. **247**, 2828 (1972)].

[f] ND, not determined.

Inhibitors. The enzyme is inhibited by a number of agents.[1] A reversible inhibition by a sulfhydryl blocking agent, dithiobis(2-nitrobenzoic acid), has been widely utilized for the enzyme assay in plasma.[28] Although low concentrations of 2-mercaptoethanol or dithiothreitol enhance the enzyme activity by preventing the oxidation of the sulfhydryl groups, high concentrations of the reducing agents cause a decline in the enzyme activity.[20] This decline could be attributed to the cleavage of intramolecular disulfide bridges. A serine agent, diisopropyl fluorophosphate is shown to

[28] K. T. Stokke and K. R. Norum, *Scand. J. Clin. Lab. Invest.* **27**, 21 (1971).

inactivate crude preparations[1,27] as well as homogeneous preparations of the enzyme.[7,20]

Substrate Specificity. With respect to the acyl donor, the enzyme shows a high degree of specificity for phospholipids containing a basic nitrogen atom.[29] The reaction rate increases by N-methylation, phosphatidyl-choline being the most efficient acyl donor. Mainly the fatty acid esterified at the carbon-2 position of the phospholipid can be transferred,[30] this fatty acid, however, can be of various chain lengths and degrees of saturation.[22,27,31] In contrast to the narrow substrate specificity toward the acyl donor, the enzyme exhibits a broad specificity toward the acyl acceptor. Not only sterols, but also long-chain primary alcohols show the acceptor activity.[9] In order for sterols to exhibit acyl acceptor activity, the hydroxyl group at carbon-3 must be in the β configuration, and the A and B rings must be in the planar, *trans* configuration.[32] These structural features may be important in providing the optimal orientation for the hydroxyl group in the transacylation at the active site of the enzyme, which is bound to substrate vesicles or lipoproteins. In contrast, the hydrophobic side chain of sterols at carbon-17 is not required for acyl acceptor activity. Sterols in which the side chain is more bulky and rigid owing to the presence of an additional methyl or ethyl group are less active than cholesterol. It appears that the presence of the side chain, especially when bulky or rigid, tends to interfere with the acceptor activity of sterols, possibly by reducing their mobility in substrate vesicles. On the other hand, the presence or the absence and the location of the double bond in the B ring are not important for effective acyl acceptor activity. Water may also be considered as an acyl acceptor in view of a phospholipase A_2-like activity that was demonstrated with a partially purified enzyme[33] and confirmed with pure enzyme preparations.[7,9] However, a relatively low phospholipase A_2-like activity suggests a limited accessibility of water molecules to the enzyme active site. The transfer of the acyl group to all effective acyl acceptors, as well as the phospholipase A_2 activity, requires the presence of the cofactor peptide, apo-A-I.[7,9,32]

Acknowledgments

This work was supported by funds from the National Institutes of Health, U.S. Public Health Service (HL-17597), Illinois Heart Association, and Illinois Agricultural Experiment Station. We are indebted to Dr. Y. Nakagawa (University of Chicago) for the amino acid analysis and to Mr. F. K. Robinson for valuable assistance in the preparation of the enzyme.

[29] C. J. Fielding, *Scand. J. Clin. Lab. Invest.* **33,** Suppl. 137, 15 (1974).
[30] J. A. Glomset, *Biochim. Biophys. Acta* **65,** 128 (1962).
[31] D. S. Sgoutas, *Biochemistry* **11,** 293 (1972).
[32] U. Piran and T. Nishida, *Lipids* **14,** 478 (1979).
[33] U. Piran and T. Nishida, *J. Biochem. (Tokyo)* **80,** 887 (1976).

[88] 1,2-Diacyl-sn-glycerol: Sterol Acyl Transferase from Spinach Leaves (*Spinacia olerecea* L.)

By R. E. GARCIA and J. B. MUDD

The synthesis of sterol esters in animals has been reported to follow three mechanisms: (*a*) phosphatidylcholine : cholesterol acyltransferase[1]; (*b*) fatty acyl-CoA : cholesterol acyltransferase[2]; and (*c*) reversal of cholesterol esterase.[3] In the plant kingdom Bartlett *et al.*[4] reported sterol ester biosynthesis in *Phycomyces blakesleeanus* to be catalyzed by a phosphatidylcholine : sterol acyltransferase. In yeast the synthesis appears to be by the acyl-CoA pathway.[5] In our studies,[6-10] we found that [^{14}C]acyl-labeled phosphatidylcholine gave rise to labeled sterol esters, but product analysis showed that sterol ester biosynthesis continued after phosphatidylcholine was depleted in the reaction mixture. A major labeled compound in this study was diacylglycerol, and we tested this as an acyl donor. It was by far the most efficient acyl donor of those tested.

Sterol + diacylglycerol → sterol ester + monoacylglycerol

Assay Method

Principles

The enzyme preparation is added to a mixture of sterol and 1,2 diacyl-sn-glycerol in a micellar system that includes Triton X-100 and phosphatidylcholine in addition to the substrates. Either the sterol or the acyl moieties of the 1,2 diacyl-sn-glycerol is radioactive. After the incubation period, the reaction mixture is partitioned into aqueous and chloroform phases. An aliquot of the chloroform phase is subjected to thin-layer chromatography (TLC). The sterol ester product is located by iodine

[1] J. A. Glomset, *J. Lipid Res.* **9**, 155 (1968).
[2] D. S. Goodman, D. Deykin, and T. Shiratori, *J. Biol. Chem.* **239**, 1335 (1964).
[3] J. Hyun, H. Kothari, E. Henn, J. Mortensen, C. R. Treadwell, and G. V. Vahouny, *J. Biol. Chem.* **244**, 1937 (1969).
[4] K. Bartlett, M. J. Keat, and E. I. Mercer, *Phytochemistry* **13**, 1107 (1974).
[5] S. Taketani, T. Nishino, and H. Katsuki, *Biochim. Biophys. Acta* **575**, 148 (1979).
[6] R. E. Garcia and J. B. Mudd, *Plant Physiol.* **61**, 354 (1978).
[7] R. E. Garcia and J. B. Mudd, *Plant Physiol.* **61**, 357 (1978).
[8] R. E. Garcia and J. B. Mudd, *Plant Physiol.* **62**, 348 (1978).
[9] R. E. Garcia and J. B. Mudd, *Arch. Biochem. Biophys.* **190**, 315 (1978).
[10] R. E. Garcia and J. B. Mudd, *Arch. Biochem. Biophys.* **191**, 487–493 (1978).

spray or by autoradiography. The sterol ester area is scraped off the plate, and the radioactivity is determined by scintillation counting either with or without elution of the sterol ester from silica gel.

Reagents

Radioactive sterols such as [4-^{14}C]sitosterol can be obtained from Amersham/Searle Corporation, Arlington Heights, Illinois; [^{14}C] Diacylglycerols can be obtained from Dhom Products, Los Angeles, California. Nonradioactive sterols can be purchased from Applied Science Laboratories, State College, Pennsylvania. Nonradioactive diacylglycerols were obtained from Nu-Chek-Prep, Elysian, Minnesota. All other chemicals were obtained from Sigma Chemical Company, St. Louis, Missouri.

Procedure

Preparation of Mixed Micelles. Micelles containing Triton X-100, phosphatidylcholine, diacylglycerol, and cholesterol were prepared by adding 1.2 mg of phosphatidylcholine, 1.2 mg of diacylglycerol, 1.2 mg of sterol, and 12 mg of Triton X-100 to a test tube. The organic solvents in which the compounds were dissolved were removed by evaporation under reduced pressure. Either the sterol or the diacylglycerol was radioactive. Distilled water was added dropwise to the detergent–lipid mixture with frequent mixing until the volume was 4 ml. The resulting mixed micelles appeared like a perfectly clear solution and remained so when stored at 5°.

Reaction Mixtures. The standard assay mixture contained 70 mM MES-NaOH, pH 7, 75 μM cholesterol (10^5 cpm), 52 μM diacylglycerol, 40 μM phosphatidylcholine, 480 μM Triton X-100, 10 mg of bovine serum albumin, and 2.5 mg of the enzyme preparation in a final volume of 1 ml. The concentrations of detergent and lipids correspond to 0.1 ml of the mixed micelle solution added to each 1 ml of reaction mixture. Reactions were started by addition of the enzyme and run in a Dubnoff metabolic shaker at 30° for 10 min. The reaction was stopped by addition of 3 ml of chloroform : methanol (1 : 2, v/v). Isolation of the lipid fraction followed the method of Bligh and Dyer.[11]

Separation of Products. (a) Separation of sterol esters from other lipids: The lipid sample in chloroform was applied to a thin-layer plate coated with silica gel G. The plates were developed in one dimension in benzene : chloroform (40 : 60, v/v). Sterol ester (R_f 0.75) was well separated

---

[11] E. G. Bligh and W. J. Dyer, *Can. J. Biochem. Physiol.* **37**, 911 (1959).

from sterol (R_f 0.1). The sterol ester spot was scraped off for assay by scintillation counting. (*b*) Separation of molecular species of sterol ester: Cholesterol esters with fatty acid substituents of varying degrees of saturation were separated by TLC on silica gel G plates impregnated with 2% (w/w) $AgNO_3$. The plates were developed in benzene : hexane (1 : 1, v/v) as described by Goodman *et al.*[2] Radioactive areas were detected by autoradiography and scraped off the plates, and the sterol esters were eluted with chloroform. The eluted lipid was assayed in a scintillation counter.

Preparation of Enzyme

Differential centrifugation of the leaf homogenate (specific activity, 0.2 nmol/mg protein per minute) showed the 20,000 g pellet to have the highest amount of enzyme of the fractions tested (34% of enzyme units, 1.4 nmol/mg protein per minute), but the highest specific activity was in the 88,000 g pellet (23% of enzyme units, 2.2 nmol/mg protein per minute). These two pellets were examined by electron microscopy and were devoid of chloroplasts, mitochondria, and fragments of these organelles. Endoplasmic reticulum, ribosomes, and other membrane vesicles were present in the active fractions. No further attempt has been made to determine the subcellar location of the enzyme.

Acetone Powder. Petioles were removed from leaves of spinach (*Spinacia oleracea* L.), the leaves were washed in distilled water and chilled at 5°. The leaves (200 g) were minced in a food chopper and then homogenized in a Waring blender in 300 ml of a solution containing 450 mM sucrose, 10 mM Tris-HCl, pH 7.5, and 1 mM EDTA at 0° with four bursts of 2 sec duration. The homogenate was filtered through three layers of cheesecloth. The fraction sedimenting between 3000 and 20,000 g was collected and suspended in 4 ml of 100 mM Tris-HCl, pH 7.5. The suspension was added slowly to 70 ml of rapidly stirred acetone at $-15°$ or 0°. Acetone at 0° tends to give better extraction of lipids. The precipitated protein was sedimented by centrifugation, washed by resuspension in cold acetone, and sedimented again by centrifugation. The pellet was then dried under reduced pressure at room temperature. This procedure gave approximately 400 mg of acetone powder from 200 g of leaves. The dried powder was stored at $-15°$. Preparation of the acetone powder is useful because it removes lipids that could participate in the enzyme-catalyzed reaction and because the preparation can be stored; it does not improve the specific activity of the enzyme.

Acetone–Ether Powder. The procedure described above was modified in that the pellet obtained in the first acetone precipitation was suspended in diethyl ether at 0° rather than in acetone. The suspension was then centrifuged, and the resulting pellet was dried under reduced pressure.

The acetone or acetone–ether powders were resuspended in buffer for addition to reaction mixtures: 50 mg of acetone or acetone–ether powder were suspended in 2 ml of 100 mM MES–NaOH buffer and mixed with a Potter–Elvehjem glass homogenizer.

Properties

Stability. The acetone and acetone–ether powders are stable for several weeks when stored in a desiccator at $-15°$.

Substrate Specificity. Cholesterol, sitosterol, and campesterol do not seem to be distinguished by the enzyme. Cholesterol and sitosterol were compared at several time points and several concentrations. All three sterols were tested in competition studies and appeared to be equivalent. These results are reasonably consistent with analyses of sterol esters isolated from plant tissue.[12]

The testing of diacylglycerols with only one type of fatty acid indicated the following preference for fatty acid transfer $16:1 > 18:3 > 18:1 > 18:2 > 16:0 > 18:0 \geq$ control. This order was found both in time course and diacylglycerol concentration studies. In all cases it could be readily demonstrated by $AgNO_3$ TLC of the sterol esters, that the fatty acid of the supplied diacylglycerol was transferred to the sterol. However, when mixtures of diacylglycerol were used in mixed micelles, e.g., 13 μM each of $16:0$, $18:1$, $18:2$, and $18:3$ diacylglycerol, the preference was $18:3 > 16:0 > 18:2 > 18:1$, the most striking feature being the greatly increased efficacy of $16:0$ diacylglycerol when present in the mixture of diacylglycerols.[9] It is not known whether there is preferential transfer of an acyl group from a particular position of the diacylglycerol. Neither mono- nor triacylglycerol is a suitable substrate for sterol ester biosynthesis.

Other Substrates. The enzyme extract shows several acyltransferase activities. Both di- and monoacylglycerol are hydrolyzed in the presence of the enzyme (transfer to water). In the presence of ethanol, ethyl esters of fatty acids are formed, particularly from monoacylglycerol (transfer to ethanol). In the presence of diacylglycerol, some triacylglycerol is formed, indicating acyl transfer from one diacylglycerol to the vacant hydroxyl of another.

Activators and Inhibitors. Tween 80 was not as satisfactory as Triton X-100 for the preparation of micelles. Phosphatidylcholine was indispens-

---

[12] J. V. Torres and F. Garcia-Olmedo, *Biochim. Biophys. Acta* **409**, 367 (1975).

able in the micelles for activity to be observed. Digitonin was an effective inhibitor of the reaction.

In earlier studies of sterol ester biosynthesis by enzyme preparations from spinach leaves, a pH optimum of 6.0 in MES–NaOH buffer was determined.[7] But diacylglycerol was not added to the reaction mixtures; it was generated from phospholipids by the action of phospholipase D and phosphatidic acid phosphatase. In fact it is difficult to measure a strong dependence on added DG at pH 6. Two methods can be employed to show this dependence. The simplest is to raise the pH to 7 to avoid the action of phospholipase D and phosphatidic acid phosphatase. The second is the judicious use of metal ions that will inhibit the phospholipid metabolizing enzymes while having a minimal effect on the 1,2-diacylglycerol:sterol acyltransferase. For this purpose Zn^{2+}, Cu^{2+}, and Fe^{2+} are particularly effective at 1 mM concentration.

[89] Aldehyde Dehydrogenases from Liver

EC 1.2.1.3 Aldehyde:NAD$^+$ oxidoreductase

By Regina Pietruszko and Takashi Yonetani

Aldehyde dehydrogenase (EC 1.2.1.3) catalyzes irreversible dehydrogenation of a large variety of aldehydes to corresponding carboxylic acids utilizing NAD as coenzyme. The enzyme occurs in a variety of organs, but its concentration is highest in the liver.[1,2] It was first demonstrated in beef liver by Racker (1949),[3] followed 20 years later by purification to homogeneity of a single enzyme from horse liver.[4] Subsequent purifications, however, have conclusively demonstrated that aldehyde dehydrogenase occurs in multiple molecular forms (isozymes) that differ in primary structure, catalytic properties, electrophoretic mobility, and subcellular distribution.[5-8] Two aldehyde dehydrogenases have been purified to homogeneity from horse[5] and sheep[6] liver and demonstrated to belong to

[1] R. A. Deitrich, *Biochem. Pharmacol.* **15**, 1911 (1966).
[2] F. Simpson and R. Lindahl, *J. Exp. Zool.* **207**, 383 (1979).
[3] E. Racker, *J. Biol. Chem.* **177**, 883 (1949).
[4] R. I. Feldman and H. Weiner, *J. Biol. Chem.* **247**, 260 (1972).
[5] J. Eckfeldt, L. Mope, K. Takio, and T. Yonetani, *J. Biol. Chem.* **251**, 236 (1976).
[6] K. E. Crow, T. M. Kitson, A. K. H. MacGibbon, and R. D. Batt, *Biochim. Biophys. Acta* **350**, 121 (1974).
[7] N. J. Greenfield and R. Pietruszko, *Biochim. Biophys. Acta* **483**, 35 (1977).
[8] J. D. Hempel and R. Pietruszko, Manuscript in preparation.

two different subcellular compartments: mitochondria and cytoplasm. For purification of the sheep liver enzymes subcellular fractionation was carried out prior to enzyme isolation,[6] and the horse enzymes were copurified and then demonstrated to belong to cytoplasm and mitochondria by a small-scale subcellular fractionation.[9] Aldehyde dehydrogenase has also been isolated from beef liver[10] and from rat liver microsomes.[11]

Catalytic and molecular properties of cytoplasmic and mitochondrial enzymes isolated from horse[5,9] and sheep[6,12] livers are listed in Table I. The two enzymes from each species differ slightly in molecular weights, electrophoretic mobility, and K_m values for NAD. Their differences are more pronounced when K_m values for short-chain aliphatic aldehydes are compared; however, the greatest difference is observed in susceptibility to disulfiram (tetraethylthiuram disulfide, a drug used to cause alcohol aversion in man). The cytoplasmic enzyme is highly susceptible to disulfiram inhibition, whereas the mitochondrial enzyme is relatively insensitive.[5,13] The first two attempts to isolate human aldehyde dehydrogenase also resulted in partial purification of a single enzyme.[14,15] However, relatively recently, two aldehyde dehydrogenases with properties analogous to those of horse and sheep enzymes have been purified to homogeneity from human postmortem livers.[7] During postmortem conditions, lysis of cell membranes occurs, and the enzymes are isolated from what appears to be a cytoplasmic fraction. However, comparison with horse enzymes indicates strongly that enzyme 1 (E_1, with properties almost identical with those of the horse cytoplasmic F_1 enzyme), is probably also cytoplasmic whereas enzyme 2 (E_2) is probably mitochondrial (compare data in Table I).

Aldehyde dehydrogenases are sulfhydryl enzymes that are sensitive to atmospheric oxygen. The cytoplasmic enzymes are particularly unstable when exposed to air, even when EDTA and excess reducing agent (mercaptoethanol or dithiothreitol) are present. The yield and specific activity of aldehyde dehydrogenase is greatly improved by carrying out purification under nitrogen. As long as oxygen is carefully excluded, the homoge-

[9] J. Eckfeldt and T. Yonetani, *Arch. Biochem. Biophys.* **175**, 717 (1976); J. Eckfeldt, K. Takio, L. Mope, and T. Yonetani, *in* "The Role of Acetaldehyde in the Actions of Ethanol" (K. O. Lindros and C. J. P. Eriksson, eds.), p. 19. Academic Press, New York, 1975.

[10] W. Leicht, F. Heinz, and B. Freimuller, *Eur. J. Biochem.* **83**, 189 (1978).

[11] H. Nakayasu, K. Mihara, and R. Sato, *Biochem. Biophys. Res. Commun.* **83**, 697 (1978).

[12] A. K. H. MacGibbon, R. L. Motion, K. E. Crow, P. D. Buckley, and L. F. Blackwell, *Eur. J. Biochem.* **96**, 585 (1979).

[13] T. M. Kitson, *Biochem. J* **151**, 407 (1975).

[14] R. J. Kraemer, and R. A. Deitrich, *J. Biol. Chem.* **243**, 6402 (1968).

[15] A. H. Blair and F. H. Bodley, *Can. J. Biochem.* **47**, 265 (1969).

TABLE I
CATALYTIC AND MOLECULAR PROPERTIES OF ALDEHYDE DEHYDROGENASES FROM HORSE, SHEEP, AND HUMAN LIVERS

Properties	Aldehyde dehydrogenases					
	Horse		Sheep		Human	
	F_1	F_2	Cytoplasmic	Mitochondrial	E_1	E_2
Molecular weight	230,000[a]	240,000[a]	212,000[b]	205,000[b]	245,000[c]	225,000[g]
Subunit weight	52,000[a]	53,000[a]	53,000[b]	53,000[b]	54,000[c]	54,000[c]
Extinction coefficient, 280 nm $A_{1cm}^{0.1\%}$	0.95[a]	1.05[a]	—	0.99[d]	0.96[c]	1.00[c]
K_m NAD (μM)	3.0[a]	30[a]	12.5[e]	77[e]	40[c]	70[c]
K_m Acetaldehyde (μM)	70[a]	0.2[a]	—	—	30[c]	3.0[c]
Disulfiram inhibition	Stoichiometric[a]	sl.[a]	Stoichiometric[f]	sl.[f]	Stoichiometric[c]	sl.[c]
Isoelectric point	6[g]	5[g,h]	5.25[b]	—	4.6[c,i]	4.8[c,i]
K_d NADH (μM)	10[g]	5[g]	1.2[b]	0.05[d]	—	130[j]
pH Optimum	—	—	—	—	8.6[c]	9.5[c]

[a] Eckfeldt et al.[5]

[b] MacGibbon et al.[12]

[c] Greenfield and Pietruszko.[7]

[d] G. J. Hart and F. M. Dickinson, Biochem. J. 163, 261 (1977).

[e] Crow et al.[6]

[f] Kitson.[13]

[g] K. Takio, Y. Sako, and T. Yonetani, in "Alcohol and Aldehyde Metabolizing Systems" (R. G. Thurman, T. Yonetani, J. R. Williamson, and B. Chance, eds.), p. 115. Academic Press, New York, 1974.

[h] Feldman and Weiner.[4]

[i] Isoionic point, the K_m values are at pH 7 except for sheep liver enzymes, which are at pH 9.3.

[j] R. S. Sidhu and A. H. Blair, J. Biol. Chem. 250, 7899 (1975).

neous enzymes are stable, losing less than 5% activity per month (for horse F_1 enzyme) and about 5% per year with human E_1 enzyme.

For purification of aldehyde dehydrogenases from animal tissues, classical procedures of salt fractionation followed by ion exchange chromatography on substituted celluloses and Sephadexes, combined with gel filtration, were employed. Two different approaches are possible with animal tissues:

1. Subcellular fractionation can be carried out before purification.
2. The enzymes can be copurified, since they are easily separable during these procedures.

Only the second approach has been possible with human aldehyde dehydrogenases purified from autopsy material, where membranes are already lysed. The key step in this purification is chromatography on 5'-AMP Sepharose 4B, which provides greater than 10-fold purification in a single step (the ion exchange steps are necessary to decrease concentration of proteins competing for 5'-AMP Sepharose binding sites). The affinity chromatography step also separates the E_1 and E_2 isozymes. From 600 g of human liver, approximately 360 mg of E_1 and 450 mg of E_2 are obtained within 1 week, representing 60% yield. The yield can be further improved by maintaining stricter anaerobic conditions.

Aldehyde Dehydrogenases from Horse Liver

Since 5'-AMP Sepharose 4B is expensive, we feel that a procedure employing strictly classical techniques, as well as the affinity chromatography procedure, should be described. A classical procedure was employed for purification of horse liver aldehyde dehydrogenases, which preceded chronologically the human enzyme purification and on which the human procedures were partly based.

Assays

Activity Assay. During purification the assay of Feldman and Weiner[4] is used; this contains, at final concentration: 0.1 M sodium pyrophosphate buffer, pH 9.0; 450 μM NAD; 68 μM propionaldehyde. One unit of enzyme activity is defined as the amount of enzyme producing 1.0 μmol of NADH per minute.

Protein Assay. Protein is assayed during purification by the method of Lowry *et al.*,[16] using crystalline lyophilized bovine serum albumin (Sigma)

[16] O. H. Lowry, N. J. Rosebrough, A. L. Farr, and R. J. Randall, *J. Biol. Chem* **193**, 265 (1951).

as standard. Protein concentration of the homogeneous isozymes can also be determined by absorption at 280 nm (see Table I).

Materials. Coenzymes are purchased from Sigma. Whatman CM-52 and DE-52 are obtained from Reeve Angel, and BioGel A-1.5 m is from Bio-Rad.

Polyacrylamide gel electrophoresis.[17] This procedure is used for horse isozyme identification. Thin-slab, rather than conventional, disc is employed; all staining is done with Coomassie brilliant blue.

Purification Procedure

Fresh horse livers are perfused with approximately 8 liters of 0.15 M NaCl to remove blood. After removal of the capsule and large blood vessels, the liver is divided into 800 g pieces and frozen for later use.

Extraction. Liver (800 g) is partially thawed and minced in a meat grinder. All subsequent procedures are performed below 5°, using nitrogen-saturated buffers and under nitrogen, where possible. To the minced liver, 1 liter of 0.25% 2-mercaptoethanol containing 2 mM EDTA is added, and the mixture is homogenized for 3 min, using a Torax Tissumizer. The insoluble debris is removed by centrifugation at 13,000 g for 60 min.

Ammonium Sulfate Fractionation. To the extract 2 ml of 2-mercaptoethanol and 209 g of solid ammonium sulfate per liter are added. After dissolution of ammonium sulfate, the slurry is allowed to stand for 60 min before centrifugation at 24,000 g for 60 min. To the supernatant, 129 g of solid ammonium sulfate per liter and 1 ml of 2-mercaptoethanol per liter are added. After dissolution of ammonium sulfate, the slurry is allowed to stand for 90 min, followed by centrifugation at 24,000 g for 30 min. The pellet is suspended in approximately 400 ml of 10 mM sodium phosphate buffer, pH 6.3, containing 0.25% 2-mercaptoethanol and 1 mM EDTA. This material is then dialyzed against several 10-liter changes of the same buffer.

CM-Cellulose Chromatography. The dialyzed material is recentrifuged at 24,000 g for 30 min to remove denatured protein. The supernatant is applied to a CM-cellulose column (4 × 15 cm) equilibrated with 10 mM sodium phosphate, pH 6.3, containing 0.25% 2-mercaptoethanol and 1 mM EDTA. The column is eluted with the same buffer. The eluate immediately following the void volume contains aldehyde dehydrogenase activity. The active eluate is dialyzed against several 10-liter changes of 5 mM

[17] B. J. Davis, *Ann. N. Y. Acad. Sci.* **121,** 404 (1964).

imidazole hydrochloride buffer, pH 7.2, containing 0.25% 2-mercapto-ethanol.

Separation of F_1 and F_2 Isozymes using DEAE-Cellulose Chromatography. After dialysis the enzyme is applied to a DEAE-cellulose column (5 × 20 cm) equilibrated with the same imidazole buffer. Aldehyde dehydrogenase is bound to DEAE column in these conditions. After removal of extraneous proteins by washing the column with this buffer (250 ml), aldehyde dehydrogenases are eluted using a gradient prepared from 1 liter of 5 mM imidazole hydrochloride, pH 7.2, containing 0.25% 2-mercapto-ethanol and 1 liter of imidazole hydrochloride, pH 7.2, containing 0.25% 2-mercaptoethanol and 200 mM sodium chloride. The gradient resolves the isozymes: F_1 is eluted before F_2.

Final Purification of F_1 Isozyme. The fractions containing F_1 isozyme (identified as F_1 by polyacrylamide gel electrophoresis) are concentrated on DEAE-cellulose using the above imidazole buffer. The sample is then applied onto a BioGel A-1.5m column (2.5 × 90 cm) equilibrated with 10 mM sodium phosphate buffer, pH 5.5, containing 0.25% 2-mercaptoethanol and 1 mM EDTA and eluted with the same buffer. The fractions containing F_1 are combined and stored under nitrogen at 4°.

Final Purification of F_2 Isozyme. Fractions from DEAE-cellulose column containing F_2 isozyme are combined and concentrated on another DEAE-cellulose column, or by ultrafiltration. The sample is then applied to the same BioGel A-1.5m column, this time equilibrated with 10 mM sodium phosphate, pH 5.8, containing 0.25% 2-mercaptoethanol and 1 mM EDTA, and eluted with the same buffer. The active fractions are then applied to DEAE-cellulose column (2.5 × 18 cm) equilibrated with the same pH 5.8 buffer. The bound F_2 isozyme is eluted by a linear gradient using 300 ml of 10 mM sodium phosphate, pH 5.8, and 300 ml of 200 mM sodium phosphate, pH 5.8, both containing 2.5% 2-mercaptoethanol and 1 mM EDTA. The fractions containing F_2 isozyme are combined and stored at 4° under nitrogen.

Results of a typical preparation are summarized in Table II. Approximately 100–200 mg of each isozyme can be obtained in a homogeneous state from 800 g of liver, representing approximately 10% yield.

Aldehyde Dehydrogenases from Human Liver

Materials. The affinity chromatography resin, 5'-AMP-Sepharose 4B, as well as CM-Sephadex C-50 and DEAE-Sephadex A-50 ion exchangers, are from Pharmacia Fine Chemicals, and DEAE BioGel from Bio-Rad; NAD is from Sigma. Propionaldehyde, from Eastman Organic Chemicals,

TABLE II

PURIFICATION OF TWO ALDEHYDE DEHYDROGENASE ISOZYMES FROM HORSE LIVER[a]

F_1/F_2 mixture	Total protein[b] (g)	Total activity[c] (μmol NADH min^{-1})	Specific activity (μmol NADH/g min^{-1})
Crude extract	61	1400	23
$(NH_4)_2SO_4$ 35%	38	1100	29
$(NH_4)_2SO_4$ 55%	24	800	33
CM-cellulose chromatography	12	500	42
F_1 isozyme			
DEAE-cellulose chromatography, imidazole buffer	0.61[d]	100	160
Gel filtration chromatography	0.20[d]	74	370
F_2 isozyme			
DEAE-cellulose chromatography, imidazole buffer	1.33[d]	200	150
Gel filtration chromatography	0.23[e]	131	570
DEAE-cellulose chromatography, phosphate buffer	0.10[e]	80	800

[a] From Eckfeldt et al.[5]
[b] By the Lowry method,[16] except as noted below.
[c] In the pH 9 sodium pyrophosphate buffer system with propionaldehyde substrate.
[d] Using $E_{280} = 0.95$ mg ml^{-1} cm^{-1} for the F_1 isoenzyme.
[e] Using $E_{280} = 1.05$ mg ml^{-1} cm^{-1} for the F_2 isozyme.

is redistilled before use; 3% v/v solutions in water are stable for at least 3 months when stored at 4°, as tested by gas chromatography and by enzymatic assay using human E_2 aldehyde dehydrogenase. All other chemicals are reagent grade.

Assays

The assay procedures employed for the human enzyme purification are the same as those employed for the horse liver enzymes and described in detail in the preceding sections.

Starch Gel Electrophoresis. Gels containing 11.5% w/v Otto Hiller Electrostarch in 5 mM phosphate buffer, pH 7.0 (the reservoirs contained 50 mM buffer), are electrophoresed overnight at a constant voltage of approximately 100 V. Enzymes are visualized in 0.1 M Tris-HCl buffer, pH 8.6, containing 0.3 mg of NAD per milliliter, 13.6 mM propionaldehyde Nitro Blue tetrazolium, and phenazine methosulfate. Nigrosin is employed for protein staining on gels. Aldehyde dehydrogenases migrate anodally; E_1 migrates somewhat slower than E_2.

Purification Procedure

Human livers obtained at autopsy (6–24 hr after death) can be used immediately or stored frozen at $-70°$ until use. The following buffers are employed during purification:

Buffer 1: 30 mM sodium phosphate, 1 mM EDTA, 1% v/v 2-mercaptoethanol, pH 6.0

Buffer 2: 30 mM sodium phosphate, 1 mM EDTA, 1% v/v 2-mercaptoethanol, pH 6.8

Buffer 3: 100 mM sodium phosphate, 1 mM EDTA, 1% v/v 2-mercaptoethanol, pH 8.0

Buffer 4 (optional): 5 mM sodium phosphate, 1 mM EDTA, 1% v/v 2-mercaptoethanol, pH 6.0

All buffers are deaerated by vacuum and saturated with nitrogen during cooling.

Extraction. Fresh or partially defrosted liver, 600 g, is homogenized in a Sears meat grinder, followed by Brinkmann Polytron with 600 ml of buffer 1. The resulting slurry is left to stand at $4°$ for 2 hr and then centrifuged at 13,000 g to remove insoluble debris. The supernatant is dialyzed against two changes of 10 times its volume of buffer 1 and centrifuged again to remove denatured proteins.

CM-Sephadex Chromatography. The centrifuged and dialyzed supernatant is applied to a 50 × 4 cm column of CM-Sephadex C-50 equilibrated with buffer 1. The enzymes E_1 and E_2 appear in the eluate immediately following the void volume. Fractions containing the enzyme activity are pooled.

There are two more aldehyde dehydrogenases in the human liver (E_3 and E_6) with high (mM) K_m values for acetaldehyde that are retained by CM-Sephadex. If these enzymes are desired, they can be eluted with a gradient consisting of 1.5 liters of buffer 1 in the mixing chamber and 1.5 liters of buffer 1 containing NaCl in the buffer reservoir. The enzymes are eluted in two peaks (detected by the assay as above, but with 13.6 mM propionaldehyde and 5 mM EDTA); the first peak eluted is E_6 and the second peak is E_3, both of which have so far been only partially purified.

DEAE-Sephadex Chromatography. The pH of the CM-Sephadex eluate containing E_1 and E_2 is raised to 6.8 by addition of small aliquots of saturated Tris base and applied to a 30 × 4 cm column of DEAE-Sephadex A-50 equilibrated with buffer 2. The column is washed with buffer 2 until all nonbound protein is eluted as monitored by ultraviolet absorbtion at 280 nm. The E_1 and E_2 isozymes are eluted by a salt gradient applied to the column with 1000 ml of buffer 2 in the mixing chamber and 1000 ml of buffer 2 containing 1 M NaCl in the buffer reservoir.

5'-AMP-Sepharose Chromatography. The pH of pooled active fractions from DEAE-Sephadex column is adjusted back to 6.0 with 1 M monobasic sodium phosphate. The sample is applied to 2 × 35 cm column prepared from 25 g of dry 5'-AMP-Sepharose 4B, previously swelled and equilibrated with buffer 1. The column is washed with buffer 1 until all extraneous proteins are removed, and E_2 is eluted with buffer 3. The elution with this buffer is continued until a sharp peak (visualized by a monitor at 280 nm) is completely eluted. E_1 isozyme, completely free from E_2, is then eluted with buffer 3 containing 0.5 mg of NADH per milliliter.

Fractions at the end of the E_2 peak may contain traces of E_1 (visualized by gel electrophoresis), while E_1 is completely free from E_2. The details of protein and activity are shown in Table III which represents an average of four purifications. The specific activity of E_1 is 0.58 μmol/mg per minute while that of E_2 has been found to vary between 1.5 to 2.5 nmol/mg per minute. The overall yield of 60%, shown in Table III, can be increased if more stringent precautions are taken to replace air by nitrogen. In one preparation where a large amount of nitrogen was used the yield was 100%.

The E_1 and E_2 isozymes obtained this way are homogeneous as visualized by starch gel electrophoresis and by examination of peptide maps and of cyanogen bromide cleavage products. Both enzymes have been stored for 2 years without any loss of activity in buffer 1 containing 25% v/v glycerol, under nitrogen at $-10°$.

TABLE III

PURIFICATION OF ALDEHYDE DEHYDROGENASES FROM HUMAN LIVER[a]

Procedure	Total protein (g)	Total activity (μmol NADH/min)	Specific activity (μmol NADH/mg per minute)	Yield (%)
Extraction and dialysis	24	1134	0.05	100
CM-Sephadex	12	931	0.08	82
DEAE-Sephadex	6	676	0.11	60
5'-AMP Sepharose 4B				
E_1	0.25	135	0.54	12
E_2	0.4	530	1.3	47

[a] Data represent average values from four different human livers; protein determined by the method of Lowry et al.[16] The procedure is an unpublished modification by D. Reed, J. D. Hempel, and R. Pietruszko of the procedure described by Greenfield and Pietruszko.[7]

If ultrapure E_1 and E_2 (free from the last 5% impurities) are required, they can be obtained as follows:

Ultrapure E_1. Dialyze E_1 in totally anaerobic conditions in buffer 4 and apply on DE-BioGel column (2.5 × 40 cm) equilibrated with the same buffer. Elute by gradient containing 500 ml of the same buffer in the mixing chamber and 500 ml of 100 mM sodium phosphate containing 1 mM EDTA and 1% v/v 2-mercaptoethanol, pH 6.0 (in anaerobic conditions).

Ultrapure E_2. Dialyze E_2 in totally anaerobic buffer 1 and apply onto 2 × 40 cm column of DE-BioGel equilibrated with buffer 1. Elute with gradient consisting of 500 ml of buffer 1 in the mixing chamber and 500 ml of 200 mM sodium phosphate containing 1 mM EDTA and 1% v/v 2-mercaptoethanol, pH 6.0. Extreme precautions have to be taken to exclude air during these purification stages, or multiple peaks with lower specific activities will be eluted.

[90] 3-Methylcrotonyl-CoA Carboxylase from *Achromobacter*

EC 6.4.1.4 3-Methylcrotonyl-CoA : carbon-dioxide ligase (ADP-forming)

By ULRICH SCHIELE and FEODOR LYNEN*

The oxidative degradation pathway of isovaleric acid (3-methylbutyric acid) takes the normal route until it reaches 3-methylcrotonyl-CoA. Since β-oxidation is not possible with this compound, the following reaction takes place, which is catalyzed by the 3-methylcrotonyl-CoA carboxylase:

$$
\begin{array}{c}
H_3C \\
\quad\quad C{=}CH{-}\overset{\displaystyle O}{\overset{\|}{C}}{-}SCoA + HCO_3^- + ATP \rightleftharpoons \\
H_3C
\end{array}
\quad
\begin{array}{c}
H_3C \\
\quad\quad C{=}CH{-}\overset{\displaystyle O}{\overset{\|}{C}}{-}SCoA + ADP + P_i \\
H_2C \\
\quad ^-OOC
\end{array}
$$

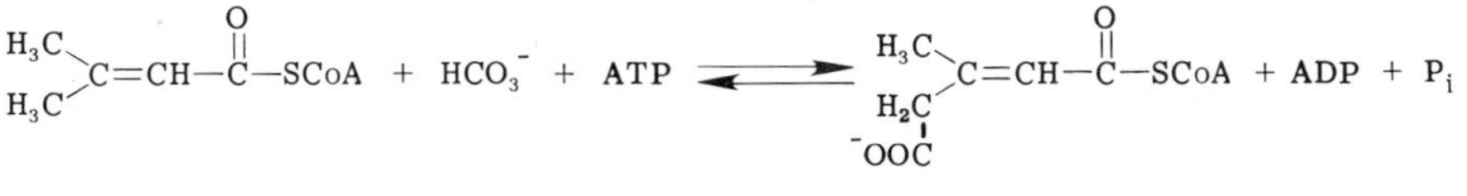

The reaction sequence continues to yield hydroxymethylglutaryl-CoA, which is split to acetoacetate and acetyl-CoA.

3-Methylcrotonyl CoA carboxylase from *Achromobacter* has a molecular weight of 700,000–760,000 and contains four molecules of biotin per complex.[1,2] The enzyme has the ability to carboxylate free biotin. This

---

* The manuscript was written after the death of Feodor Lynen.

[1] R. H. Himes, D. L. Young, E. Ringelmann, and F. Lynen, *Biochem. Z.* **337,** 48 (1963).

[2] R. Apitz-Castro, K. Rehn, and F. Lynen, *Eur. J. Biochem.* **16,** 71 (1970).

property led Lynen and co-workers to the elucidation of the reaction mechanism of biotin-containing carboxylases.[3,4] Carboxylations of this type proceed in two steps via carboxybiotin:

$$\text{ATP} + \text{HCO}_3^- + \text{biotin-enzyme} \overset{\text{Mg}^{2+}}{\rightleftharpoons} \text{CO}_2^- \text{-biotin-enzyme} + \text{ADP} + \text{P}_i$$
$$\text{CO}_2^- \text{-biotin-enzyme} + \text{substrate} \rightleftharpoons \text{substrate-CO}_2^- + \text{biotin-enzyme}$$

The intermediate carboxybiotin has the following structure:

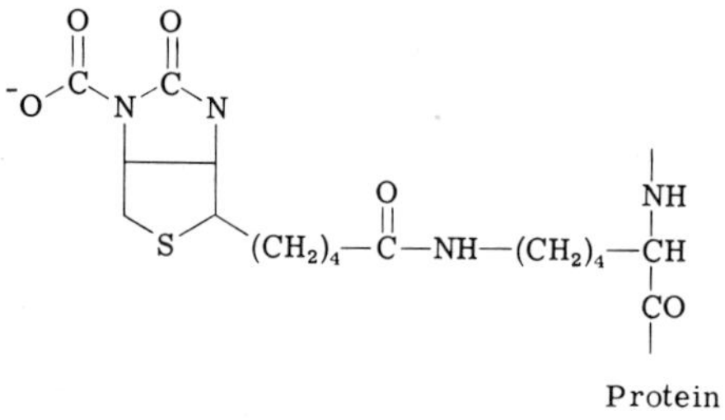

Assay Method

Principle. The enzymatic activity is measured spectrophotometrically by following the oxidation of NADH in the presence of pyruvate kinase, phosphoenolpyruvate, and lactate dehydrogenase.[3] The test system is suitable for the crude extract.

The same assay system is used to measure the first partial reaction. In this case biotin is added instead of 3-methylcrotonyl-CoA. The affinity of the 3-methylcrotonyl-CoA carboxylase for biotin is quite low. The enzyme is not saturated at 30 mM biotin, the concentration used in the experiment described below.

Reagents

Tris-HCl buffer, 1 M, pH 8.0
MgCl$_2$, 0.1 M
KHCO$_3$, 0.1 M
ATP, 0.01 M
Phosphoenolpyruvate, 0.01 M
NADH, 0.01 M
Bovine serum albumin, 10 mg/ml
Lactate dehydrogenase, 5 mg/ml
Pyruvate kinase, 2 mg/ml
3-Methylcrotonyl-CoA, 0.01 M
Biotin, 0.3 M

[3] F. Lynen, J. Knappe, E. Lorch, G. Jütting, E. Ringelmann, and J. P. Lachance, *Biochem. Z.* **335**, 123 (1961).
[4] J. Knappe, E. Ringelmann, and F. Lynen, *Biochem. Z* **335**, 168 (1961).

3-Methylcrotonyl-CoA was synthesized according to Knappe *et al.*[5] by the reaction of 3-methylcrotonic anhydride with CoA.

Procedure. The cuvette (diameter = 1 cm) contains in a final volume of 2 ml : 100 mM Tris-HCl pH 8.0, 4 mM MgCl$_2$, 5 mM KHCO$_3$, 0.5 mM ATP, 0.75 mM phosphoenolpyruvate, 0.15 mM NADH, 1 mg of bovine serum albumin, 50 μg of lactate dehydrogenase, 20 μg of pyruvate kinase, and 3-methylcrotonyl-CoA carboxylase or its subunits. A blank value is obtained. The reaction is then started by adding 20 μl of 0.01 M 3-methylcrotonyl-CoA.[1] The test of the first partial reaction is started by adding 200 μl of 0.3 M biotin in place of 3-methylcrotonyl-CoA.

One unit of 3-methylcrotonyl-CoA carboxylase is defined as the amount of enzyme that carboxylates 1 μmol of 3-methylcrotonyl-CoA per minute at 25°.

Purification

Cultivation of Bacteria.[2,3,6] *Achromobacter* IV S was isolated by Schlegel from sheep rumen. The bacteria can grow in a minimal medium with isovaleric acid as sole carbon source. Since the carboxylation of 3-methylcrotonyl-CoA is a crucial step in the catabolism of isovaleric acid, the bacteria must synthesize large amounts of the enzyme.

Composition of the Meat Extract Agar.[7] One liter contains 10 g of meat extract (E. Merck, Germany), 10 g of Pepton (Difco), 9 g of NaCl, 6 g of Na$_2$HPO$_4$ · 12H$_2$O, and 20 g of agar. The pH is adjusted to 7.4 with KOH.

Composition of the Isovaleric Acid Medium. One liter of the medium contains 5 ml of isovaleric acid, 5 g of NaCl, 1 g of (NH$_4$)$_2$SO$_4$, 1.8 g of KH$_2$PO$_4$, 100 mg of MgSO$_4$ · 6 H$_2$O, 50 mg of CaCl$_2$ · 2 H$_2$O, 10 mg of FeSO$_4$ · 7 H$_2$O, 7.5 mg of EDTA, 5 mg of MnCl$_2$ · 4 H$_2$O, and 1 mg of (NH$_4$)$_2$MoO$_4$ · H$_2$O. The solution is adjusted to pH 7.4 with KOH.

The medium is diluted to half its concentration for making isovaleric acid medium agar.

Cultivation Procedure. The bacteria can be stored at $-16°$. They are first grown on meat extract agar, then transferred to isovaleric acid medium agar. The cultivation is continued in shake culture. Finally, the bacteria are grown at 31° in a 10 liter aerated culture flask and later in five 25 liter aerated culture flasks. At this stage 5 × 10^7 cpm (1.65 μmol) of [^{14}C]-biotin is added. The bacteria begin to incorporate biotin 12 hr after the addition. Growth ceases when aeration is too strong. For this reason the bacteria are first aerated very weakly. As growth increases, aeration is

[5] J. Knappe, H. G. Schlegel, and F. Lynen, *Biochem. Z* **335**, 101 (1961).
[6] U. Schiele, R. Niedermeier, M. Stürzer, and F. Lynen, *Eur. J. Biochem.* **60**, 259 (1975).
[7] L. Hallmann, "Bakteriologische Nährböden," p. 70. Thieme, Stuttgart, 1953.

increased in steps. The bacteria are harvested at the end of the logarithmic phase. From 100 liters of medium, between 350 and 500 g of a wet paste are obtained. The bacteria are stored frozen.

Purification Procedure[1-3,6]

All steps of the purification procedure are carried out between 0 and 4°. During the calcium phosphate gel fractionation step, the radioactivity due to the biotin carboxyl carrier protein of acetyl-CoA carboxylase is removed. 3-Methylcrotonyl-CoA carboxylase remains as the only radioactive protein.

Step 1. Preparation of Crude Extract and High-Speed Supernatant.[8] From the frozen mass of bacteria, 200 g are cut into pieces and thawed in the presence of 550 ml of 0.01 M Na_2HPO_4. n-Octanol (1 ml) is added. The bacteria are broken for 3 min in 150-ml batches with 300 g of glass beads (31/10, Dragonwerk, Wild/Bayreuth, Germany) in a homogenizer as described by Merkenschlager *et al.*[9] The glass beads are removed from the homogenate by filtration through a sintered-glass funnel. The filtered homogenate is first centrifuged at 2000 g for 20 min to sediment the foam. It is then centrifuged at 44,000 g for 110 min. The clear supernatant can be kept overnight at 4° or stored frozen without loss of activity.

Step 2. Calcium Phosphate Gel Fractionation. The supernatant is adjusted to a concentration of 10 mg of protein per milliliter (measured by the biuret method) and to 13.5 mM phosphate[10] with potassium phosphate buffer pH 6.0. A calcium phosphate gel suspension containing 40 mg dry weight per milliliter is added to yield a final proportion of 4 g dry weight of gel per gram of protein. The pH is adjusted to 6.0 with 2 M acetic acid. The enzyme is adsorbed to the gel under these conditions. After stirring for 10 min the gel is collected by centrifugation at 2000 g for 10 min. The supernatant, which contains biotin carboxyl carrier protein of acetyl-CoA carboxylase, is removed, and the gel is washed twice with 0.01 M potassium phosphate buffer, pH 6.0. The volume of buffer used for each washing is twice the applied volume of calcium phosphate gel suspension. Finally, the enzyme is eluted twice with 0.04 M potassium phosphate buffer, pH 6.4. The volumes of the elution buffer are two-thirds and one-third of the applied volume of calcium phosphate gel suspension. 3-Methylglutaconase remains bound to the calcium phosphate gel pellet (E. von Stetten, personal communication).

---

[8] The crude extract can be prepared in the presence of phenylmethanesulfonyl fluoride (PMSF).

[9] M. Merkenschlager, K. Schlossmann, and W. Kurz, *Biochem. Z.* **329,** 332 (1957).

[10] This refers only to added phosphate. The phosphate content of the bacteria is neglected.

Step 3. Precipitation with Ammonium Sulfate. The combined fractions of the eluted enzyme are stirred with a magnetic stirrer while solid ammonium sulfate is added slowly to a saturation of 60%. The pH is kept at 7.0 by addition of 2 M ammonia. Stirring is continued for 45 min. The precipitate is then collected by centrifugation at 10,000 g for 40 min.

Step 4. Fractionation with Ammonium Sulfate. The precipitate is dissolved in 0.03 M potassium phosphate buffer, pH 7.2, and adjusted to 8.5 mg of protein per milliliter. The solution is brought to 30% saturation with solid ammonium sulfate as described above. The precipitate is removed by centrifugation. The enzyme is precipitated from the supernatant by addition of further ammonium sulfate to 50% saturation and is collected by centrifugation.

Step 5. Freezing and Thawing. The precipitate is dissolved in a small volume of 0.03 M potassium phosphate buffer, pH 7.2, and frozen at −30°. In many cases a precipitate forms when the solution is thawed. The precipitate is removed by centrifugation at 10,000 g for 20 min. The enzyme remains in the supernatant.

Step 6. Gel Filtration with Sephadex G-200. The enzyme solution (not more than 1 g of protein) is layered on a Sephadex G-200 column (4.2 × 120 cm, bead diameter 40–120 μm) equilibrated with 0.03 M potassium phosphate buffer, pH 7.2, and 0.02% NaN_3. The enzyme is eluted with the same buffer system. The peak of the 3-methylcrotonyl-CoA carboxylase appears shortly after the void volume.

Step 7. Chromatography on DEAE-Cellulose. 3-Methylcrotonyl-CoA carboxylase-containing fractions are applied to a tightly packed DEAE-cellulose column which is equilibrated with 0.1 M NaCl in 0.03 M potassium phosphate buffer, pH 7.2. At least 1 ml of DEAE-cellulose is used for every 2 mg of protein. The enzyme is eluted with five column volumes of a linear gradient from 0.1 M to 0.5 M NaCl in 0.03 M potassium phosphate buffer, pH 7.2. The peak of 3-methylcrotonyl-CoA carboxylase appears between 0.2 and 0.3 M NaCl.

The fractions with 3-methylcrotonyl-CoA carboxylase activity are combined. The solution is brought to 70% saturation with ammonium sulfate as described above. After about 10 hr the precipitate is collected by centrifugation at 10,000 g for 40 min. The precipitate is dissolved in a small volume of 0.03 M potassium phosphate buffer, pH 7.2. The enzyme solution is kept frozen at −30°.

A typical purification is summarized in the table.

Isolation of Subunits[6]

Isolated subunits of the enzyme, unlike the intact complex, are very sensitive to oxygen. Moreover, when the isolated subunits are brought

PURIFICATION OF 3-METHYLCROTONYL-CoA CARBOXYLASE FROM *Achromobacter*[a]

	Volume (ml)	Total protein (mg)	Total activity (units)	Specific activity (units/mg protein)	Recovery (%)
Crude extract	750	14860	863	0.058	100
Calcium phosphate gel fractionation	1012	1240	475	0.42	55
$(NH_4)_2SO_4$ precipitation	107	946	435	0.46	50
$(NH_4)_2SO_4$ fractionation	12	515	421	0.83	49
Sephadex G-200 gel filtration	200	105	313	2.98	36
DEAE-cellulose chromatography	58	27	255	9.36	30

[a] This purification was done according to Apitz-Castro *et al.*[2] The section entitled Purification in this chapter contains some smaller modifications. The yields were between 15 and 49 mg. We obtained specific activities of up to 11.5.

back to an environment where the complex is stable (neutral pH, low concentration of urea), they show a strong tendency to bind to glass and other materials. Polyethylene or polypropylene containers have proved to be the best materials to store the subunits.

Isolation of the Biotin-Containing Subunit (B-Subunit). Isolation of the subunits is carried out between 0 and 4°C. 3-Methylcrotonyl-CoA carboxylase (2–6 mg) in not more than 0.5 ml is dialyzed under nitrogen against 400 ml of 30 mM cysteine in 30 mM Tris-HCl, pH 7.0, for 5 hr. The dialysis tube (Sartorius Membranfilter) is attached to the beaker that contains the dialysis fluid. The dialysis is performed in a desiccator, which is evacuated twice until most of the dissolved gas has been removed from the fluid and is subsequently filled with purified nitrogen. The dialysis fluid is stirred during this process. The desiccator is connected to a vacuum pump and a nitrogen bottle via a three-way tap. A small balloon connected via a T-shaped tube between the nitrogen bottle and the desiccator begins to blow up when the inside pressure reaches atmospheric pressure.

Urea, 8 M, is added to the dialyzed enzyme to yield a 5 M solution. The 8 M urea is previously chromatographed over Dowex-1 (X2) in the chloride form to remove cyanate. The enzyme is kept in the urea solution for 1 hr and is then adsorbed onto a DEAE-cellulose column that is equilibrated with 5 M urea and 30 mM cysteine in 30 mM Tris-HCl, pH 7.0. For every 1 mg of protein 6.5 ml of DEAE-cellulose is used. The adsorbed protein is eluted with 20 column volumes of a linear gradient from 30 mM to 300 mM Tris-HCl, pH 7.0, in 5 M urea and 30 mM cysteine. The mixing vessels and collecting tubes are covered with about 1 cm of liquid paraffin or heptane to minimize access of oxygen. The biotin-containing subunit (B) appears as a sharp peak, whereas the biotin-free subunit (A) is distributed over many fractions. Possibly the biotin-free subunit is partly associated under these conditions. The fractions with the biotin-containing subunit are concentrated with an Amicon PM-10 filter. Generally this solution is stirred into 10 volumes of 30 mM cysteine in 30 mM potassium phosphate buffer, pH 7.0. This mixture is kept for 12 hr at 4° before testing. When assayed directly after dilution, only low activities are found.

Isolation of the Biotin-Free Subunit (A-Subunit). 3-Methylcrotonyl-CoA carboxylase (1–4 mg) in up to 0.75 ml is dissociated by dialysis against 400 ml of 10 mM cysteine in 10 mM glycine–NaOH buffer, pH 9.8, for 3 days. The dialysis is performed in a desiccator that is filled with nitrogen as described above. The dialysis liquid is changed at least once; otherwise the complex does not dissociate. The dissociated subunits are separated by chromatography on avidin–Sepharose 4 B prepared according to

Bodanszky.[11] The biotin-containing subunit (B) is bound by the avidin, but the biotin-free subunit (A) is not retarded on the column. We used a column with a biotin-binding capacity 16-fold higher than the amount of protein-bound biotin applied. Because of steric hindrance only one of the four binding sites of the avidin molecule should be effective; so there remains a threefold excess of effective binding sites. The chromatography is performed with the buffer system used for the dissociation of the enzyme. Fractions are collected in polyethylene tubes that contain sufficient 30 mM cysteine in 0.5 M potassium phosphate buffer, pH 6.4, to produce a final pH of 7.5. The column and the fractions are covered with a layer of about 1 cm of liquid paraffin or heptane to minimize access of oxygen.

Properties

Specificity.[1,3] The enzyme requires Mg^{2+} or Mn^{2+} for activity. It produces the *trans*-isomer of 3-methylglutaconyl-CoA. Besides 3-methylcrotonyl-CoA it accepts crotonyl-CoA and acetoacetate as substrates. The reaction rates are 20% and 14%, respectively, in comparison to 3-methylcrotonyl-CoA. Carbon dioxide is accepted in the HCO_3^- form. The enzyme has the ability to carboxylate free biotin, showing absolute specificity for the naturally occurring D isomer. DL-Oxibiotin, D-norbiotin, D-homobiotin, and biocytin are also active, the latter three to a smaller degree. D-Biotin sulfoxide, D-biotin sulfone, D-thiobiotin,[12] DL-dethiobiotin, and the diamino acid produced by hydrolysis of the urea moiety of the biotin are not active.

Exchange Reactions. The enzyme catalyzes the following exchange reactions:[3]

$$ATP + {}^{32}P_i \rightleftarrows [{}^{32}P]ATP + P_i$$

only in the presence of ADP, HCO_3^-, and Mg^{2+},

$$ATP + [{}^{14}C]ADP \rightleftarrows [{}^{14}C]ATP + ADP$$

only in the presence of P_i, Mg^{2+}, and HCO_3^-, and

[3-^{14}C]Methylglutaconyl-CoA + 3-methylcrotonyl CoA $\rightleftarrows$
3-methylglutaconyl-CoA + [3-^{14}C]methylcrotonyl-CoA

also in the absence of Mg^{2+}.

All of these exchange reactions are inhibited by avidin. The exchange rates with 1 unit of enzyme are approximately 0.09, 0.06, and >0.06 μmol/min, respectively.

[11] A. Bodanszky and M. Bodanszky, *Experientia* **26**, 327 (1970).
[12] H. P. Blaschkowski, Doctoral thesis, University of Heidelberg, 1969.

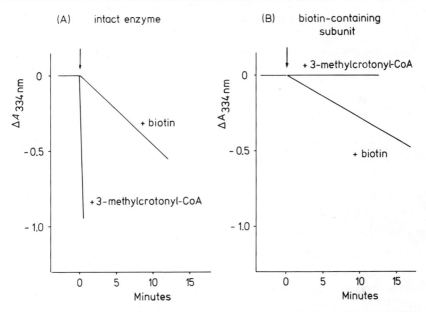

FIG. 1. Enzymatic activities of the intact enzyme and the biotin-containing subunit. In (A) the cuvette contained 3-methylcrotonyl-CoA carboxylase (66 μg, 2620 cpm), and in (B) an equivalent amount of biotin-containing subunit (2530 cpm). The reaction conditions were as described under Assay Method. 3-Methylcrotonyl-CoA or biotin were added as indicated by the arrow.

Kinetic Properties[1,3] The apparent K_m values are as follows: 3-methylcrotonyl-CoA, 12 μM; ATP, 83 μM; HCO_3^-, 2.17 mM. The forward reaction is about 10 times faster than the reverse reaction. The pH optimum of the enzyme is in the range of 7.9–8.3.

Physical Properties.[1,2] The $s_{20,w}$ value is 19.4 S[1] and 20.7 S.[2] The diffusion coefficient $D_{20,w}$ is 2.8 × 10^{-7} cm²/sec. The partial specific volume is 0.765 ml/g. From the amino acid composition, a volume of 0.746 ml/g is calculated. The ratio of the absorbance at 280 nm to the absorbance at 260 nm is 1.60–1.64.

Structure. As in all other biotin-containing enzymes investigated so far, the biotin is bound as an acid amide to the ε-amino group of a lysine side chain.[13,14] In bacteria grown without biotin, up to about 20% of the enzyme is in the apo form. When sufficient biotin is added to the growth medium, the synthesis of holoenzyme is practically complete.[12,15]

[13] J. Knappe, K. Biederbick, and W. Brümmer, *Angew. Chem.* **74**, 432 (1962).
[14] J. Knappe, B. Wenger, and U. Wiegand, *Biochem. Z* **337**, 232 (1963).
[15] T. Höpner and J. Knappe, *Biochem. Z.* **342**, 190 (1965).

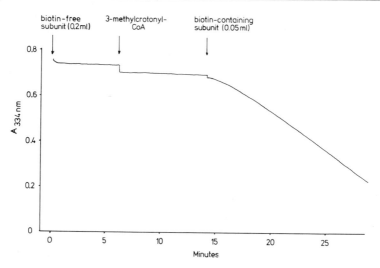

Fig. 2. Reconstitution of active 3-methylcrotonyl-CoA carboxylase from the separated subunits. The following additions were made at the times indicated by the arrows: 35 μg of biotin-free subunit, 0.2 μmol of 3-methylcrotonyl-CoA, and 4 μg of biotin-containing subunit. The optical test was performed as described under Assay Method. On the basis of the biotin-containing subunit 35% of the original enzymatic activity was regained in this experiment.

Sodium dodecyl sulfate (SDS)–gel electrophoresis of the 3-methylcrotonyl-CoA carboxylase reveals two distinct bands of equal density, referred to as A and B, corresponding to molecular weights of about 78,000 and 96,000, respectively. This leads to a structure A_4B_4 for the enzyme complex. The slower moving band contains all the biotin. The isolated subunits migrate as single bands during SDS–gel electrophoresis.[6]

Regulation of the Degree of Dissociation of the Enzyme Complex.[16] At neutral pH the 3-methylcrotonyl-CoA carboxylase complex is quite stable. There is no tailing during Sephadex chromatography or sedimentation. Differences in the degree of dissociation can nevertheless be detected by trapping dissociated subunits with iodoacetamide. The resulting subunits cannot reassociate. Thus the rate of inactivation of the enzyme in the presence of a surplus of iodoacetamide is proportional to the degree of dissociation.

3-Methylcrotonyl-CoA decreases the degree of dissociation of the enzyme considerably. Since acetyl-CoA does not influence the degree of dissociation very much, the effect of 3-methylcrotonyl-CoA must primarily be due to the 3-methylcrotonyl moiety of the compound. The efficiency of 3-methylcrotonyl-CoA as a protective agent against inactivation

[16] U. Schiele and M. Stürzer, *Eur. J. Biochem.* **60,** 267 (1975).

by iodoacetamide is higher than would be expected from its affinity for the enzyme, indicating a cooperative effect of the 3-methylcrotonyl-CoA-bearing subunits on those with free 3-methylcrotonyl-CoA-binding sites. ATP does not decrease the degree of dissociation of the enzyme. It seems probable that these properties serve in the regulation of the intracellular degradation of the enzyme if the dissociated subunits are the preferred targets of proteolytic attack as proposed by Dehlinger and Schimke.[17] From these experiments it can also be concluded that dissociation and reassociation occur very rapidly in the absence of 3-methylcrotonyl-CoA. Most of the subunits are set free within a matter of minutes.

Properties of the Subunits.[6] The isolated biotin-containing subunit (B) does not catalyze the overall reaction but can still carboxylate free biotin. This is shown in Fig. 1, where the enzymatic activities of the intact enzyme and an equivalent amount of biotin-containing subunit (B) are compared. The isolated biotin-free subunit (A) also does not catalyze the overall reaction. A transfer of CO_2 from 3-methylglutaconyl-CoA to free biotin in the presence of subunit (A) could not be demonstrated, but this has not been shown with the intact complex either. When biotin-free subunit (A) is added to biotin-containing subunit (B), however, overall activity is gradually regenerated (Fig. 2). Since the biotin-containing subunit (B) carries the biotin carboxylase activity, the transcarboxylase activity can be ascribed to the biotin-free (A) subunit. 3-Methylcrotonyl-CoA increases the rate of reconstitution considerably.

[17] P. J. Dehlinger and R. T. Schimke, *Biochem. Biophys. Res. Commun.* **40**, 1473 (1970).

[91] 3-Methylcrotonyl-CoA and Geranyl-CoA Carboxylases from *Pseudomonas citronellolis*

EC 6.4.1.4 3-Methylcrotonoyl-CoA : carbon-dioxide ligase
(ADP-forming)
EC 6.4.1.5 Geranoyl-CoA : carbon-dioxide ligase (ADP-forming)

By R. RAY FALL

Pseudomonas citronellolis contains the inducible acyl-CoA carboxylases, 3-methylcrotonyl-CoA carboxylase[1] and geranyl-CoA carboxylase,[2]

[1] M. L. Hector and R. R. Fall, *Biochemistry* **15**, 3465 (1976).
[2] W. Seubert, E. Fass, and U. Remberger, *Biochem. Z.* **338**, 265 (1963). As shown in this paper and confirmed by us (unpublished data), the substrate for geranyl-CoA carboxylase is the 2Z isomer (*cis*-geranyl-CoA), not the 2E isomer (*trans*-geranyl-CoA). The systematic name for the enzyme (EC 6.4.1.5) should probably be (2Z)-3,7-dimethyl-2,6-octadienyl-CoA : carbon-dioxide ligase (ADP-forming).

two enzymes that catalyze analogous biotin-dependent CO_2-fixation reactions at the β-methyl group of their respective acyl-CoA substrates as follows[2,3]:

$$\begin{array}{c} R-\underset{\underset{CH_3}{|}}{C}=CHCOSCoA + HCO_3^- + ATP \underset{\overline{\qquad}}{\overset{Mg^{2+}}{\rightleftharpoons}} R-\underset{\underset{CH_2CO_2^-}{|}}{C}=CHCOSCoA + ADP + P_i \end{array}$$

where R = CH$_3$ (3-methylcrotonyl-CoA carboxylase) and
R = (CH$_3$)$_2$C$=$CHCH$_2$CH$_2$ (geranyl-CoA carboxylase)

The two enzymes are structurally very similar and copurify in a variety of protein fractionation procedures.[4,5] The isolation of each enzyme free from the other is made possible by inducing them separately—3-methylcrotonyl-CoA carboxylase as part of the leucine–isovalerate degradative pathway[6] and geranyl-CoA carboxylase as part of the citronellol degradative pathway[7]—and then purifying each enzyme individually, but by an identical procedure.[5]

Assay Methods

Principle. The enzymes are routinely assayed by a [^{14}C]bicarbonate fixation procedure similar to that developed for the assay of other acyl-CoA carboxylases.[8,9] The assay measures the production of an acid-stable carboxylation product (i.e., [^{14}C]3-methylglutaconyl-CoA or [^{14}C]3-iso-hexenylglutaconyl-CoA) after acidification of the reaction mixture and liberation of excess [^{14}C]bicarbonate as [^{14}C]CO$_2$.[2,3] Alternatively, after step 4 of the purification a spectrophotometric assay can be used.[2,10] This assay measures ADP formation by coupling to the pyruvate kinase and lactate dehydrogenase reactions and monitoring NADH oxidation.

[^{14}C]Bicarbonate Fixation Assay

Reagents

N-2-Hydroxyethylpiperazine-N'-2-ethanesulfonic acid (HEPES) buffer, 1.0 M, pH 8.0; pH adjusted with KOH
MgCl$_2$, 0.1 M

[3] R. H. Himes, D. L. Young, E. Ringelmann, and F. Lynen, *Biochem. Z.* **337**, 48 (1963).
[4] M. L. Hector and R. R. Fall, *Biochem. Biophys. Res. Commun.* **71**, 746 (1976).
[5] R. R. Fall and M. L. Hector, *Biochemistry* **16**, 4000 (1977).
[6] L. K. Massey, J. R. Sokatch, and R. S. Conrad, *Bacteriol. Rev.* **40**, 42 (1976).
[7] W. Seubert and U. Remberger, *Biochem. Z.* **338**, 245 (1963).
[8] M. Flavin, H. Castro-Mendoza, and S. Ochoa, *J. Biol. Chem.* **229**, 981 (1957).
[9] See this series, Vol. 14 [2].
[10] See this series, Vol. 1 [66].

ATP, 0.1 M, pH 7.0

[^{14}C]NaHCO$_3$, 0.25 M, 2 mCi/mmol

Bovine serum albumin, 10 mg/ml

3-Methylcrotonyl-CoA or *cis*-geranyl-CoA, 10 mM. Each is prepared
by a mixed anhydride method[5,11] and quantitated by a hydroxamate
procedure.[12] Geranic acid is prepared by oxidation of citral[5] or by
synthesis from 6-methyl-5-hepten-2-ol.[13] The *cis* and *trans* isomers
are separated as the methyl esters, and *cis*-geranic acid is isolated
after saponification.[13]

HCl, 3 N

Scintillation fluid; a variety of commercially available detergent
based fluids can be used, or the following can be prepared: toluene,
2 liters; Triton X-100, 1 liter; 2a 70 (Research Products Int.), 18 g.

Procedure. The assay mixture contains, in a final volume of 100 μl: 0.1
M HEPES, 25 μg of bovine serum albumin, 1.5 mM MgCl$_2$, 500 μM
3-methylcrotonyl-CoA or geranyl-CoA, 12.5 mM [^{14}C]NaHCO$_3$ (2.5 μCi),
appropriate amounts of enzyme and 2 mM ATP. An identical reaction
mixture without the acyl-CoA substrate, serves as a control. The assay
mixture, without ATP, in 6 × 50 mm culture tubes, is equilibrated to 30°
for 2 min. The reaction is started by adding the ATP solution and is
allowed to run at 30° for 5 min. It is stopped by the addition of 10 μl of 3 N
HCl. Each sample tube is cut in half and the reaction mixture plus tube is
inverted into a scintillation vial, and then heated to dryness in an oven
90–100° for 15 min. The residue is dissolved in 0.5 ml of water and is
counted for radioactivity in 4.5 ml of scintillation fluid.

Spectrophotometric Assay

Reagents

Tris-HCl buffer, 1.0 M, pH 8.0
MgCl$_2$, 0.5 M
Bovine serum albumin, 10 mg/ml
ATP, 0.1 M, pH 7.0
3-Methylcrotonyl-CoA or *cis*-geranyl-CoA, 10 mM
KHCO$_3$, 1.0 M
Phosphoenolpyruvate, 0.02 M (prepared fresh daily)
NADH, 10 mg/ml (prepared fresh daily)
Pyruvate kinase, 10 mg/ml, ~465 units/mg protein
Lactate dehydrogenase, 10 mg/ml, ~550 units/mg protein

[11] See this series, Vol. 3 [137].
[12] B. Shapiro, *Biochem. J.* **53**, 663 (1953).
[13] S. G. Cantwell, E. P. Lau, D. S. Watt, and R. R. Fall, *J. Bacteriol.* **135**, 324 (1978).

Procedure. This assay is carried out essentially as described by Seubert *et al.*[2] with slight modifications. The reaction mixtures contain 100 mM Tris-HCl, 10 mM MgCl$_2$, 0.5 mM ATP, 10 mM KHCO$_3$, 0.2 mM phosphoenolpyruvate, 6.3 units/ml of pyruvate kinase, 13 units/ml of lactate dehydrogenase, 0.1 mg/ml of NADH, 0.5 mg/ml of bovine serum albumin, 500 μM 3-methylcrotonyl-CoA or geranyl-CoA, and the appropriate amount of enzyme (0–5 mU) in a total volume of 0.2 ml. The reaction is started by the addition of enzyme. Reaction mixtures containing no acyl-CoA serve as controls. Reactions are carried out at 30°, and the decrease in absorbance at 340 nm due to the oxidation of NADH is monitored.

Units. One unit of activity is defined as 1 μmol of product formed per minute (^{14}C-carboxylated product or ADP, depending on the assay used). Specific activity is defined as units per milligram of protein. Protein is determined after deoxycholate–trichloroacetic acid precipitation as described by Bensadoun and Weinstein.[14]

Purification Procedure[4,5]

The procedure described below is suitable for the isolation of 3-methylcrotonyl-CoA carboxylase from cells grown on isovalerate as carbon source, or geranyl-CoA carboxylase from cells grown on citronellic or geranic acids.[15] We have found it convenient at times to grow cells in the presence of [^{3}H]biotin to provide a covalent radiolabel on the prosthetic group of these biotin enzymes.[1,5] All purification procedures are carried out at 4°, and all buffers contain 0.1 mM dithiothreitol.

Step 1. Bacterial Growth and Preparation of the Crude Extract. *Pseudomonas citronellolis* 13674 is obtained from the American Type Culture Collection. Cultures are grown on the following basal medium (medium T): 50 mM Tris-HCl, pH 7.4; 2 mM potassium phosphate, pH 7.4; and the following (in grams per liter): (NH$_4$)$_2$SO$_4$, 2; KCl, 0.74; MgSO$_4$, 0.2; CaCl$_2 \cdot$ 2 H$_2$O, 0.015; FeSO$_4 \cdot$ 7 H$_2$O, 0.015; Na$_2$MoO$_4$, 0.0002; MnSO$_4 \cdot$ H$_2$O, 0.0002; yeast extract (Difco), 0.5. For induction of 3-methylcrotonyl-CoA or geranyl-CoA carboxylase the carbon source is 0.3% (w/v) isovaleric acid or 0.15% (w/v) citronellic acid,[15] respectively. The pH of the medium is adjusted to pH 7.4 with NaOH. Cells are grown aerobically with shaking at 33° to late-logarithmic state and are harvested

[14] A. Bensadoun and D. Weinstein, *Anal. Biochem.* **70**, 241 (1976).
[15] Citronellic acid or geranic acid can be easily prepared by oxidation of commercially available citronellal or citral, respectively.[5] Alternatively, cells can be grown on citronellol or geraniol as carbon source after growth adaptation to these isoprenols (see Cantwell *et al.*[13]), and then used as a source of geranyl-CoA carboxylase.

by centrifugation at 4°. Cells are washed twice with 0.02 M potassium phosphate, pH 7.0, and the cell pellets are stored frozen.

Frozen, washed cells are suspended in 2–3 volumes of buffer K (0.02 M potassium phosphate, pH 7.5) and ruptured by passage through a French pressure cell (20,000 psi). The homogenate is centrifuged at 48,000 g for 20 min, and the resulting supernatant is used as the source of crude extract.

Step 2. Ammonium Sulfate Fractionation. The protein concentration of the crude extract is quickly determined by a microbiuret procedure,[16] and the extract is diluted with buffer K to 10 mg of protein per milliliter; solid ammonium sulfate is slowly added with stirring to 40% saturation (226 g/liter). After sitting for 1 hr, the precipitate is collected by centrifugation at 27,000 g for 20 min and stored at 4°. The supernatant is discarded.

Step 3. DEAE-Cellulose Chromatography. The ammonium sulfate precipitate is dissolved in a minimal volume of KG buffer (buffer K containing 20% v/v glycerol) and dialyzed against 100 volumes of KG with three buffer changes for a minimum of 1 hr each. The dialyzed fraction is applied to a column of DEAE-cellulose (Whatman DE-52) equilibrated with KG; the bed volume of the column is 1 ml per 20 mg of protein. The column is washed with 1 column volume of KG and then eluted with a linear gradient of KG vs KG containing 0.3 M KCl, using a total gradient volume 10 times the column volume. (Typical elution profiles of 3-methylcrotonyl-CoA or geranyl-CoA carboxylase are shown by Hector and Fall[4]). Both enzymes elute from the columns at the same conductivity (~7.5 mmho). The peak enzyme fractions are pooled and precipitated with ammonium sulfate as described in step 2.

Step 4. Sepharose 4B Chromatography. The precipitate is dissolved in a minimal volume of KG and applied to a column of Sepharose 4B (2.5 × 80 cm) equilibrated with KG; a maximum of 250 mg of protein in a 5 ml volume is applied. The column is eluted with KG and 3–4 ml fractions are collected. (Typical elution profiles are shown by Hector and Fall[4].) This step separates a very active ATPase activity from the carboxylase peak and allows the use of the spectrophotometric assay.

Step 5. DEAE-Agarose Chromatography. The peak enzyme fractions from step 4 are applied to a column of DEAE-agarose (1 × 30 cm; Bio-Rad) equilibrated with KG. The enzyme is eluted with a linear gradient as described in step 3; the total volume of the gradient is 200 ml, and 3-ml fractions are collected. The active enzyme fractions are pooled and concentrated by ultrafiltration (PM-10 membrane, Amicon) to a protein concentration of 2 mg/ml.

---

[16] A. L. Koch and S. L. Putnam, *Anal. Biochem.* **44**, 239 (1971).

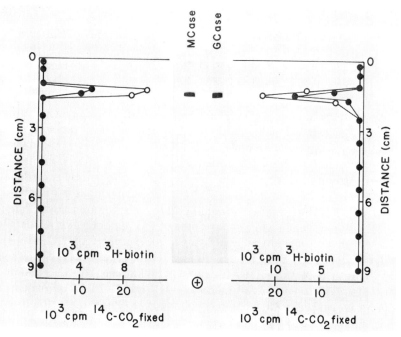

F IG. 1. Polyacrylamide gel electrophoresis of 3-methylcrotonyl-CoA carboxylase (MCase) and geranyl-CoA carboxylase (GCase) at pH 7.9. Acrylamide gels (4.5%) were run as described by Fall and Hector.[5] Duplicate gels containing 10 μg of [³H]biotin-labeled 3-methylcrotonyl-CoA or geranyl-CoA carboxylase were run, and one was stained for protein and photographed. The other gel was sliced and assayed for enzyme activity and [³H]biotin as described in the text. ○——○, [¹⁴C]CO₂ fixed; ●——●, [³H]biotin. Reproduced from Fall and Hector[5] with the permission of the American Chemical Society.

Step 6. Hydroxyapatite Chromatography. The concentrated enzyme fractions from step 5 are applied to a column of hydroxyapatite (0.8 × 24 cm; Bio-Rad) equilibrated with KG. Neither 3-methylcrotonyl-CoA carboxylase nor geranyl-CoA carboxylase is adsorbed on the column under these conditions. The nonadsorbed fractions are pooled, concentrated by ultrafiltration to a protein concentration of approximately 1 mg/ml, and stored at −80°.

The results of the purification procedure are summarized in the table.

Properties

Stability. The purified geranyl-CoA carboxylase is stable at −80° for several months. The purified 3-methylcrotonyl-CoA carboxylase is much less stable, losing approximately 25% activity per month at −80°. Corre-

PURIFICATION OF 3-METHYLCROTONYL-CoA AND GERANYL-CoA CARBOXYLASES FROM *Pseudomonas citronellolis*

Step	3-Methylcrotonyl-CoA carboxylase[a]				Geranyl-CoA carboxylase[a]			
	Total protein (mg)	Total units	Specific activity[b]	Yield (%)	Total protein (mg)	Total units	Specific activity[b]	Yield (%)
1. Cell extract	8900	267	0.03	(100)	8148	220	0.027	(100)
2. (NH₄)₂SO₄ precipitate	1369	295	0.18	110	1727	257	0.15	117
3. DEAE-cellulose	199	189	0.95	71	182	162	0.89	74
4. Sepharose 4B	54	105	1.93	39	59	120	2.02	55
5. DEAE-agarose	15.3	73	5.48	27	19.9	87	4.37	40
6. Hydroxyapatite	3.8	40	10.54	15	6.6	55	8.38	25

[a] Cells were grown in T medium containing either 0.3% isovalerate or 0.1% citronellate described in the text. Approximately 80 g (wet weight) of cells were used in each purification.

[b] Specific activity is defined as the carboxylation of 1 μmol of substrate per milligram of protein per minute at 30°. For steps 1–3, the [¹⁴C]bicarbonate fixation assay was used; for steps 4–6, the spectrophotometric assay was used.

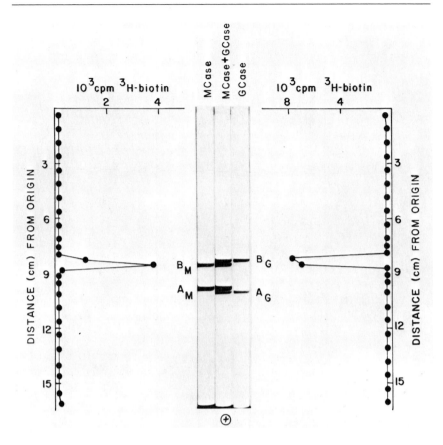

FIG. 2. Sodium dodecyl sulfate (SDS)–polyacrylamide gel electrophoresis at pH 8.9 of methylcrotonyl-CoA carboxylase (MCase) and geranyl-CoA carboxylase (GCase). Six percent acrylamide gels (5 × 170 mm) containing 0.1% SDS were run as described by Fall and Hector.[5] Gels containing 10 μg of [³H]3-methylcrotonyl-CoA or [³H]geranyl-CoA carboxylase, or 10 μg of both enzymes, were run, stained, and photographed. After photography the gels containing 3-methylcrotonyl-CoA or geranyl-CoA carboxylase were sliced, digested, and counted to locate the [³H]biotin-containing subunit. The designation of the subunits, A_M, A_G, B_M, and B_G uses the nomenclature of Schiele *et al.* (this volume [90]) where A = the biotin-free subunit and B = the biotin-containing subunit. Reproduced from Fall and Hector[5] with the permission of the American Chemical Society.

spondingly, 3-methylcrotonyl-CoA carboxylase is much more sensitive to heat denaturation than geranyl-CoA carboxylase; indeed this property was used to distinguish the 3-methylcrotonyl-CoA carboxylation activity of geranyl-CoA carboxylase from 3-methylcrotonyl-CoA carboxylase.[4]

Molecular Characteristics.[5] Each enzyme has an approximate molecular weight of 520,000–580,000 as measured by gel filtration, and from exclusion limits by polyacrylamide gel electrophoresis at pH 7.9. In the latter procedure each [³H]biotin-labeled enzyme reveals a single identically migrating protein band with coincident [³H]biotin and enzymatic activity as shown in Fig. 1. Polyacrylamide gel electrophoresis in the presence of sodium dodecyl sulfate (SDS) reveals that each enzyme contains two subunits, including a smaller biotin-free subunit (A subunit) and a larger biotin-containing subunit (B subunit). Molecular weights determined by electrophoresis in SDS at pH 7.2 are 63,000 and 73,000 for the 3-methylcrotonyl-CoA carboxylase subunits, and 63,000 and 75,000 for the geranyl-CoA carboxylase subunits. Electrophoresis of a mixture of the two purified enzymes in SDS at pH 8.9 resolves their respective A and B subunits as shown in Fig. 2, demonstrating that these two enzymes do not share common subunits. The biotin contents of the two enzymes are very similar, ranging from 148,000 to 157,000 g of protein per mole of biotin. It is concluded that these two enzyme complexes contain very similar, but not identical, substructures, probably with an A_4B_4 stoichiometry.

Other Properties. 3-Methylcrotonyl-CoA carboxylase is relatively specific for its 3-methylcrotonyl-CoA substrate ($K_m = 43 \ \mu M$)[17]; carboxylation of *cis*-geranyl-CoA by the purified enzyme is not detectable using standard assay techniques. The enzyme will carry out the carboxylation of *trans*-crotonyl-CoA ($K_m = 138 \ \mu M$; relative rate = 44%)[17] similar to results obtained with 3-methylcrotonyl-CoA carboxylase from *Achromobacter*.[3] Purified geranyl-CoA carboxylase will utilize 5- to 15-carbon acyl-CoA substrates, including 3-methylcrotonyl-CoA (K_m 24 μM; relative rate 4–8%)[2,17] *cis*-geranyl-CoA ($K_m = 10 \ \mu M$; relative rate 100%),[17] and farnesyl-CoA (relative rate 36%).[2] As shown by Seubert *et al.*[2] and confirmed by us, geranyl-CoA carboxylase utilizes only *cis*- but not *trans*-geranyl-CoA.[18] However, the enzyme will carboxylate both *cis*- and *trans*-ethylcrotonyl-CoA at low rates, and will carry out these latter reactions under physiological conditions during growth of *P. citronellolis* on 3-methylpentanoic acid.[17]

Geranyl-CoA carboxylase is present in other bacterial strains that can metabolize citronellic and geranic acids, including *P. aeruginosa* and *P. mendocina*, and *Acinetobacter* species.[13]

Acknowledgments

The studies described here were supported by research grants from the National Institutes of Health (HL-16628) and the National Science Foundation (PCM-16251).

[17] E. P. Lau, and D. S. Watt, and R. R. Fall, *Fed. Proc.* **38**, 828 (1979).
[18] M. L. Hector and R. R. Fall, unpublished observations.

[92] 3-Methylcrotonyl-CoA Carboxylase from Bovine Kidney

EC 6.4.1.4 3-Methylcrotonoyl-CoA : carbon-dioxide ligase
(ADP-forming)

By Edward P. Lau and R. Ray Fall

$$CH_3C{=}CHCOSCoA + HCO_3^- + ATP \xrightarrow{Mg^{2+}} CH_3C{=}CHCOSCoA + ADP + P_i$$
$$\quad\ \ |\qquad\qquad\qquad\qquad\qquad\qquad\qquad\qquad\qquad\qquad |$$
$$\quad\ \ CH_3\qquad\qquad\qquad\qquad\qquad\qquad\qquad\qquad\quad CH_2CO_2^-$$

Assay Methods

Principle. The enzyme is assayed by a [14C]bicarbonate fixation procedure or a spectrophotometric assay as described elsewhere.[1]

Reagents. Most of the reagents are listed in this volume [91]. In addition the following are needed.

Tris-HCl buffer, pH 8.0, 1.0 M
MgCl₂, 0.5 M
KCl, 1 M

Procedure. The [14C]bicarbonate fixation assay is carried out as described in this volume [91], with the following changes. (*a*) The assay mixture contains, in a final volume of 100 μl: 0.1 M Tris-HCl, 100 mM KCl, 50 μg of bovine serum albumin, 7.5 mM MgCl₂, 500 μM 3-methylcrotonyl-CoA, 12.5 mM [14C]NaHCO₃ (2.5 μCi), an appropriate amount of enzyme, and 2 mM ATP. (*b*) The assay is carried out at 37°.

The spectrophotometric assay is carried out essentially as described in this volume [91] with slight modifications. The reaction mixtures contain 100 mM Tris-HCl, 10 mM MgCl₂, 100 mM KCl, 0.5 mM ATP, 10 mM KHCO₃, 0.2 mM phosphoenolpyruvate, 6.3 units/ml of pyruvate kinase, 13 units/ml of lactate dehydrogenase, 0.1 mg/ml of NADH, 0.5 mg/ml of bovine serum albumin, 500 μM 3-methylcrotonyl-CoA, and the appropriate amount of 3-methylcrotonyl-CoA carboxylase (0–5 mU) in a total volume of 0.4 ml. Reactions are carried out at 37°.

Units. One unit of activity is defined as 1 nmol of product formed ([14C]3-methylglutaconyl-CoA or ADP, depending on the assay) per minute. Specific activity is defined as units per milligram of protein. Protein

---

[1] R. R. Fall, this volume [91].

is determined after deoxycholate–trichloroacetic acid precipitation as described by Bensadoun and Weinstein.[2]

Purification Procedure[3,4]

Bovine kidneys, obtained fresh after slaughter, were defatted, diced, washed with 0.2 M NaCl, and frozen. All subsequent steps were carried out at 0–4°.

Step 1. Preparation of Crude Extract. Frozen bovine kidney, 500 g, is passed twice through a meat grinder fitted with a fine grate followed by homogenization in a Waring blender at the high setting for 2 min in 800 ml of 20 mM potassium phosphate, pH 7.0, containing 0.1 mM EDTA and 0.1 mM dithiothreitol (buffer K). The suspension is centrifuged at 10,000 g for 15 min. The supernatant (930 ml) is centrifuged at 48,000 g for 20 min. This supernatant is the crude extract (860 ml).

Step 2. Polyethylene Glycol 6000 Fraction. Fifty percent (w/v) polyethylene glycol 6000 (Carbowax 6000) is added slowly with stirring to the crude extract to a final concentration of 3% (w/v). After stirring the suspension for an additional 10 min, the suspension is centrifuged at 14,000 g for 20 min. The pellets are resuspended in 150 ml of buffer K using a glass–Teflon homogenizer. The suspension is centrifuged at 17,000 g for 20 min. The supernatant (162 ml) is used for further purification.

Step 3. DEAE-Sephacel Chromatography. The supernatant from step 2 is applied to a column of DEAE-Sephacel (Pharmacia) (5 × 14.5 cm) equilibrated with buffer K. The column is washed with 1 liter of buffer K. 3-Methylcrotonyl-CoA carboxylase activity is eluted using a linear gradient of 0 to 0.2 M KCl in buffer K (1 liter each). Fractions containing the enzyme are pooled and concentrated to ~150 ml in an ultrafiltration cell using a PM-10 membrane (Amicon).

Step 4. Blue Dextran–Sepharose Chromatography. The concentrated enzyme solution is dialyzed against buffer KG (10% glycerol in buffer K) with three changes of buffer. The dialyzed sample is applied to a column of Blue dextran–Sepharose 4B (1.5 × 28 cm) equilibrated with buffer KG. Blue dextran–Sepharose is synthesized by the procedure of Ryan and Vestling.[5] After washing the column with 200 ml of the same buffer, a stepwise elution with 0.1 M KCl, 0.19 M KCl, and then 0.3 M KCl in buffer KG is used. 3-Methylcrotonyl-CoA carboxylase activity is present in the 0.3 M KCl wash.

[2] A. Bensadoun and D. Weinstein, *Anal. Biochem.* **70**, 241 (1976).
[3] E. P. Lau, B. C. Cochran, L. Munson, and R. R. Fall, *Proc. Natl. Acad. Sci. U.S.A.* **76**, 214 (1979).
[4] E. P. Lau, B. C. Cochran, and R. R. Fall, *Arch. Biochem. Biophys.*, in press.
[5] L. D. Ryan and C. S. Vestling, *Arch. Biochem. Biophys.* **160**, 279 (1974).

Step 5. Hydroxyapatite Chromatography. The fraction containing the enzyme is concentrated and applied to a column of hydroxyapatite (Bio-Rad HTP; 1.1 × 12 cm) equilibrated with buffer KG.

After washing the column with 250 ml buffer KG, a linear gradient of 0.1 to 0.4 M potassium phosphate, pH 7.0, both containing 10% glycerol, 0.1 mM EDTA, and 0.1 mM dithiothreitol (250 ml each), is used to elute the enzyme. The enzyme peak (at 0.18–0.19 M potassium phosphate) is analyzed by sodium dodecyl sulfate (SDS)-polyacrylamide gel electrophoresis (see below), and fractions containing only the two subunits of the enzyme[3] are pooled, concentrated to about 0.25 mg of protein per milliliter as in step 3, and stored frozen at −80°.

We have noted that some 3-methylcrotonyl-CoA carboxylase preparations show evidence for limited proteolytic degradation of the smaller subunit when subjected to SDS gel electrophoresis. This degradation can be prevented by inclusion of the protease inhibitor diisopropylfluorophosphate (0.1 mM) during the first three steps of the purification procedure.[4]

The results of the purification procedure are summarized in the table.

Comments about the Purification Procedure. An alternative procedure for the purification of the enzyme has been outlined in a previous publication.[3] This procedure involves more steps than the one described above.

PURIFICATION OF 3-METHYLCROTONYL-CoA CARBOXYLASE FROM BOVINE KIDNEY[a]

Step	Total protein (mg)	Total activity (units)	Specific activity (units/mg)	Yield (%)
1. 48,000 *g* crude extract	25860	5077	0.20	(100)
2. PEG-6000 fraction	5119	8004	1.56	158
3. DEAE-Sephacel	88	14218	162	280
4. Blue dextran–Sepharose	12	3680[b]	307[b]	73
5. Hydroxyapatite	1.36	5227	3843	103

[a] Kidney, 500 g, was used.

[b] The presence of a blue color in the 0.3 M KCl wash of the Blue dextran–Sepharose column may be responsible for the low enzyme activity in this step due to inhibition of the enzyme by the blue dye. The blue material sticks to the top of the hydroxyapatite column in the next step and is effectively removed from the enzyme preparation. Repeated washing of the column with high salt (0.3–1.0 M KCl) does not prevent reappearance of the blue dye when the enzyme preparation is applied, suggesting that release of the dye is an enzymatic process.

Isolated bovine kidney mitochondria[6] are disrupted either with a French pressure cell or by sonication to release the enzyme, which is purified sequentially by polyethylene glycol 6000 fractionation, DEAE-cellulose, Sepharose 4B, DEAE-agarose, and hydroxyapatite chromatography. Another variation, useful for small-scale work, also starts with isolated mitochondria from fresh bovine kidney.[6] Sonification of the mitochondria and subsequent centrifugation at 48,000 g for 20 min yields a supernatant similar to the one produced at the end of step 1 with considerably less contaminating protein (specific activity = 1.3 nmol/mg/min). 3-Methylcrotonyl-CoA carboxylase can then be isolated from this supernatant by the procedures described in steps 2–5. The enzyme purified by any of these procedures appears to be very similar.

Properties

Molecular Properties. Purified 3-methylcrotonyl-CoA carboxylase exhibits a single protein band upon electrophoresis in polyacrylamide gels of 4–6% acrylamide; from the mobility of the enzyme on these gels a molecular weight of 835,000 can be estimated by the procedure of Hedrick and Smith.[7] The SDS–polyacrylamide gel electrophoresis reveals nonidentical subunits with molecular weights of 73,500 and 61,000.[3,4] The larger subunit contains the biotin prosthetic group.[3] The biotin content found by bioassay with *Saccharomyces cerevisiae*[8] is 1 mol per 157,000 g of protein. Isoelectric focusing in granulated Sephadex G-75[9] reveals an isoelectric point of 5.4.

Kinetic Properties. Varying the pH while maintaining all other conditions constant reveals maximal activity at pH 8.0. When only the assay temperature is varied, at pH 8.0, the maximal activity occurs at 38°. The apparent K_m for 3-methylcrotonyl-CoA, ATP, and HCO_3^- are 78 μM, 82 μM, and 1.8 mM, respectively. The enzyme is activated four- to fivefold by K^+ or NH_4^+ ions (at a concentration of 50–100 mM).

Other Properties.[6] Bovine kidney 3-methylcrotonyl-CoA carboxylase is localized in mitochondria and appears to be bound to the inner mitochondrial membrane. Release of the enzyme is effected by treatment with polyethylene glycol, and after "solubilization" the enzyme is stabilized in glycerol-containing buffers.

[6] M. L. Hector, B. C. Cochran, E. A. Logue, and R. R. Fall, *Arch. Biochem. Biophys.* **199**, 28 (1980).

[7] J. L. Hedrick and A. J. Smith, *Arch. Biochem. Biophys.* **126**, 155 (1968).

[8] H. C. Lichstein, *J. Biol. Chem.* **212**, 217 (1955).

[9] P. G. Righetti and J. W. Drysdale, *in* "Laboratory Techniques in Biochemistry and Molecular Biology" (T. S. Work and E. Work, eds.), Vol. 5, p. 424. North-Holland Publ., Amsterdam, 1976.

Using the [^{14}C]bicarbonate fixation assay, 3-methylcrotonyl-CoA carboxylase has been detected in the following mammalian tissues[10]: bovine kidney, liver, adrenals, lung, skeletal muscle; rat kidney and liver; and human placenta. Antiserum produced in rabbits after injection of the purified bovine kidney enzyme cross-reacts with the enzyme from bovine liver, rat liver, and human placenta[3,10]; it has not been tested with other enzyme preparations.

Acknowledgments

The studies described here were supported by research grants from the National Institutes of Health (HL-16628) and the National Science Foundation (PCM-16251).

[10] B. C. Cochran, M. L. Hector, D. M. Tasset, and R. R. Fall, unpublished observations.

[93] Enzymes of Wax Ester Catabolism in Jojoba[1]

By Robert A. Moreau and Anthony H. C. Huang

Wax Ester Catabolism in Jojoba

The jojoba plant is native to arid regions of southwestern North America. It is the only known plant species whose seeds contain a large amount (50–60% of the fresh weight) of intracellular wax esters in the cotyledons.[2-5] During germination, this reserve wax ester is mobilized to support the growth of the embryonic axis.[6] The carbon skeleton of the wax ester is converted efficiently to carbohydrate. The wax ester in the storage wax bodies is hydrolyzed to a fatty acid and a fatty alcohol. The latter is oxidized to produce another fatty acid. The fatty acids are metabolized first to acetate by the β-oxidation reaction sequence and then to succinate by the glyoxylate cycle in the glyoxysomes. Succinate is metabolized to malate by the tricarboxylic acid cycle in the mitochondria. Finally, malate is converted to sucrose by gluconeogenic and other enzymes in the cytosol. Except for the conversion of 1 mol of wax ester to 2 mol of fatty

[1] Supported by the National Science Foundation.
[2] R. S. McKinney and G. S. Jamieson, *Oil Soap*, **13**, 289 (1936).
[3] T. K. Miwa, *J. Am. Oil Chem. Soc.* **48**, 259 (1971).
[4] L. L. Muller, T. P. Hensarling, and T. J. Jacks, *J. Am. Oil Chem. Soc.* **52**, 164 (1975).
[5] D. M. Yermonos, *J. Am. Oil Chem. Soc.* **52**, 115 (1975).
[6] R. A. Moreau and A. H. C. Huang, *Plant Physiol.* **60**, 329 (1977).

acids, the pathway in jojoba is identical to that in the triacylglycerol-storing fatty seedlings of castor bean and other species.[7]

The conversion of wax ester to fatty acids in jojoba involves three enzymes.[6,8,9] The first is a wax ester hydrolase that catalyzes the hydrolysis of the wax ester. The fatty alcohol is oxidized to a fatty aldehyde by a fatty alcohol oxidase (fatty alcohol: O_2 oxidoreductase) that requires molecular oxygen as the electron acceptor. The fatty aldehyde is oxidized to a fatty acid by fatty aldehyde dehydrogenase (fatty aldehyde: NAD^+ oxidoreductase) with NAD^+ as the electron acceptor. All three enzymes are associated with the membrane of the storage wax bodies, and they share many similar enzymatic properties. The activities of the three enzymes are absent in the dry seeds and increase at parallel rates during germination.

Other Organisms that Metabolize Wax Esters or Fatty Alcohols

Intracellular wax esters are very important in lipid metabolism of marine organisms.[10,11] They are the major type of lipid, often comprising more than 20% of the total dry weight, in at least 120 species of marine invertebrates and vertebrates distributed among 17 orders and 9 phyla.[12] It has been estimated that at least half of all organic substances synthesized initially by phytoplankton is converted into wax at some points in the marine food chain.[13] The activities of wax ester hydrolase in the digestive juices of surf clams and several species of teleost fishes have been partially characterized.[14-16] However, these hydrolases are probably involved in the digestion of dietary wax esters only, not in the hydrolysis of the intracellular wax reserves in these animals. Although wax esters are very important in the overall metabolism of marine organisms, the biochemical steps involved remain to be elucidated. The techniques described herein should be useful as a guide in the future study of wax ester hydrolysis and fatty alcohol oxidation in marine organisms.

Microorganisms are indispensable in the catabolism of large quantities

[7] H. Beevers, *Ann. N. Y. Acad. Sci.* **169**, 313 (1969).

[8] A. H. C. Huang, R. A. Moreau, and K. D. F. Liu, *Plant Physiol.* **61**, 339 (1978).

[9] R. A. Moreau and A. H. C. Huang, *Arch. Biochem Biophys.* **194**, 422 (1979).

[10] J. C. Nevenzel, *Lipids* **5**, 308 (1970).

[11] A. A. Benson, R. F. Lee, and J. C. Nevenzel, *Biochem. Soc. Symp.* **35**, 175 (1972).

[12] J. R. Sargent, R. F. Lee, and J. C. Nevenzel, *in* "Chemistry and Biochemistry of Natural Waxes" (P. E. Kolattukudy, ed.), pp. 49–91. Elsevier, Amsterdam, 1976.

[13] A. A. Benson and R. F. Lee, *Sci. Am.* **232**, 77 (1975).

[14] J. S. Patton and J. G. Quinn, *Mar. Biol.* **21**, 59 (1973).

[15] J. S. Patton and A. A. Benson, *Comp. Biochem. Physiol.* **52B**, 111 (1975).

[16] J. S. Patton, J. C. Nevenzel, and A. A. Benson, *Lipids* **10**, 575 (1975).

of wax esters and alkanes of dead plant materials. Again, the biochemical steps involved have been studied inadequately. Some microbes grown in alkanes produce fatty alcohols as intermediates during the ω-oxidation of alkanes. The fatty alcohols are then oxidized by a NAD specific fatty alcohol dehydrogense.[17-19] The resulting fatty aldehydes are oxidized to fatty acids by a NAD specific fatty aldehyde dehydrogenase.[19,20] Discovery of the fatty alcohol oxidase (fatty alcohol: O_2 oxidoreductase) raises the possibility that the reported microbial NAD-fatty alcohol dehydrogenase (fatty alcohol: NAD^+ oxidoreductase) might actually be an oxidase. In the jojoba enzyme preparation containing both fatty alcohol oxidase and fatty aldehyde dehydrogenase, substrate-dependent NAD^+ reduction can be demonstrated using either fatty alcohol or fatty aldehyde as a substrate. This apparent "NAD^+-linked dehydrogenase" activity with fatty alcohol is due to the activities of both the oxidase and the dehydrogenase and is inhibited by eliminating oxygen from the system. The microbial fatty alcohol dehydrogenase has not been purified to homogeneity or separated from the fatty aldehyde dehydrogenase, and the effect of anaerobiosis on its activity has not been tested. Although this microbial fatty alcohol oxidoreductase may indeed utilize NAD^+ as the electron acceptor, the possibility exists that it actually utilizes oxygen as the electron acceptor, since its activity was assayed together with the NAD^+-fatty aldehyde dehydrogenase.

Isolation of Wax Bodies and Their Membranes

Seeds of jojoba (*Simmondsia chinensis*) are dusted with Phaltan (Chevron Chem. Co., Richmond, California) to retard fungal growth and allowed to germinate in moist vermiculite in darkness at 28°. A small quantity of seeds can be obtained from the Office of Arid Lands Studies, University of Arizona, Tucson, Arizona, and large quantities can be purchased from American Jojoba Industries Inc., Bakersfield, California. All chemical reagents are purchased from Sigma Corp. St Louis, Missouri, except as otherwise noted.

Wax bodies are isolated from the cotyledons of seedlings 15–20 days old. All steps are performed at 0–4°. Ten grams of cotyledons are chopped with a razor blade in a petri dish containing 20 ml of grinding medium consisting of 0.6 M sucrose, 1 mM EDTA, 10 mM KCl, 1 mM $MgCl_2$, 2

[17] E. Azoulay and M. T. Heydman, *Biochim. Biophys. Acta* **73**, 1 (1963).

[18] B. Roche and E. Azoulay, *Eur. J. Biochem.* **8**, 426 (1969).

[19] J. M. LeBeault, B. Roche, Z. Duvnjak, and E. Azoulay, *Biochim. Biophys. Acta* **220**, 373 (1979).

[20] M. T. Heydman, and E. Azoulay, *Biochim. Biophys. Acta* **77**, 545 (1963).

mM dithiothreitol, 0.15 M Tricine buffer, adjusted with KOH to a pH of 7.5, until pieces 1–2 mm² in size are obtained. The pieces are further homogenized with a mortar and pestle. The homogenate is filtered through a piece of Nitex cloth with a pore size of 44 μm² (Tetko Inc., Elmsford, New York). The filtrate is centrifuged at 10,000 g for 30 min, and the resulting lipid pad is removed with a spatula. The lipid pad is resuspended in 10 ml of grinding medium and shaken with a Vortex mixer; the resuspended material is centrifuged at 10,000 g for 30 min. The lipid pad is then removed and resuspended with grinding medium to make a total volume of 5 ml. Examination under the electron microscope reveals that the resulting suspension contains essentially only wax bodies.[6]

Wax bodies are spherical organelles approximately 1 μm in diameter. Each organelle is packed with wax ester and is surrounded by a membrane. The membranes of isolated wax bodies are obtained by removing the wax with diethyl ether. The isolated wax bodies, in a suspension of 5 ml prepared as described in the preceding paragraph, is mixed with 10 ml of cold diethyl ether. After shaking with a Vortex mixer, the mixture is allowed to settle for 1 hr at 0–4°. The upper diethyl ether layer is removed by aspiration. The ether extraction procedure is repeated twice. After extraction, the trace amount of diethyl ether remaining is evaporated under a stream of nitrogen. The resulting aqueous resuspension of wax body membranes contains approximately 5 mg of proteins in 5 ml. The wax body membranes (or in some cases intact wax bodies) are used as a source of enzymes for the following enzymatic studies.

Wax Ester Hydrolase

$$R-\overset{\overset{\displaystyle O}{\|}}{C}-O-R_2 + H_2O \rightarrow R-\overset{\overset{\displaystyle O}{\|}}{C}-OH + R_2-OH$$

Wax ester   Fatty acid   Fatty alcohol

Assay Methods

Fluorometric Assay

Wax ester hydrolase activity is routinely assayed by a fluorometric lipase assay.[21,22] The activity is measured at room temperature in a reaction mixture of 4 ml, containing 0.1 M Tris-HCl buffer, pH 9.0, 2 mM

[21] G. G. Guilbault and J. Hieserman, *Anal. Chem.* **41**, 2006 (1969).
[22] S. Muto and H. Beevers, *Plant Physiol.* **54**, 23 (1974).

dithiothreitol, and 10–50 μl (10–50 μg of protein) of enzyme. The reaction is initiated by the addition of 0.83 mM N-methylindoxylmyristate (from I.C.N. Pharmaceuticals) dissolved in 0.1 ml of ethylene glycol monoethyl ether. Fluorescence measurements are made with a Turner Model 111 fluorometer with excitation filter No. 405 (405 nm maxima) and emission filter No. 2A-12 (>510 nm), attached to a X-Y recorder (Model 7034A, Hewlett-Packard Co.). The reaction rate is linear for the first 10 min.

Spectrophotometric Assay

A modified colorimetric assay[22] is used to measure hydrolase activity when various wax esters and glycerides are tested as potential substrates. In this method, the fatty acids produced are converted to copper soaps and measured using sodium dithiocarbamate. The reaction is performed at room temperature in a 15-ml tube. The reaction mixture contains in a final volume of 1 ml: 0.1 M Tris-HCl, pH 9.0, 5 mM dithiothreitol, 10 mM substrate, and 100 μl (100 μg of protein) of enzyme. Substrates (100 mM) are first emulsified in 2 ml of 5% gum acacia for 1 min at low speed with a Bronwill Biosonic IV ultrasonic generator fitted with a microprobe. For each substrate, two enzyme concentrations (50 and 100 μg of protein) are used, and the reaction is stopped at time intervals of 1–2 hr each to ensure that proper kinetics are observed. The reaction is stopped by the addition of 1 ml of copper reagent (0.9 M triethanolamine, 0.1 M acetic acid, 5% cupric nitrate). Copper soaps are extracted into 4 ml of chloroform by shaking the tubes closed with Teflon screw-caps horizontally for 30 min. Two milliliters of the chloroform layer are removed and added to 0.2 ml of 0.1% sodium dithiocarbamate in 1-butanol. The absorbance is measured immediately at 440 nm. Palmitic acid is used to produce a standard curve that is linear up to a concentration of 0.2 μmol per 2 ml of chloroform. Stearic acid and arachidic acid exhibit the same molar extinction coefficient.

Assessment of the Two Methods

The fluorometric assay is by far the more convenient one. The activity can be continuously monitored on a recorder. Each assay requires about 5 min. The major drawback of this assay is that the substrate, N-methylindoxylmyristate, is an artificial one, and caution must be used in interpreting the physiological significance of results obtained with this substrate.

The colorimetric assay is relatively time-consuming and less accurate, since the activity cannot be measured continuously. Furthermore, the amount of enzyme required for this assay is 10 times that required for the fluorometric assay. However, the colorimetric assay is essential in testing the activity of the hydrolase with all of the natural, nonfluorescent substrates.

Properties

Wax ester hydrolase activity is absent in the dry seeds, but increases at a nearly linear rate during the first 20 days of germination. Most of the cellular activity is associated with the membrane of the wax bodies. The enzyme has an optimal activity at pH 9. The apparent K_m value for N-methylindoxylmyristate is $9.3 \times 10^{-5}\, M$. It is stable at 40° for 30 min but is inactivated at higher temperatures. Various divalent and monovalent cations at a concentration of 1 mM, such as $CaCl_2$, $MgCl_2$, NaCl, and KCl, and 1 mM EDTA have little effect on the activity. p-Chloromercuribenzoate at 0.1 mM inhibits 81% of the activity, and its effect is reversed by subsequent addition of 5 mM dithiothreitol. The enzyme exhibits a broad substrate specificity with natural lipids, having high activities on monoglycerides, wax esters, and the native substrate (jojoba wax), and low activities on diacylglycerols and triacylglycerols. It hydrolyzes the artificial substrate, N-methylindoxymyristate, about 50 times faster than any of the natural substances.

Fatty Alcohol Oxidase (Fatty Alcohol : O_2 Oxidoreductase)

$$RCH_2OH \quad + \tfrac{1}{2}O_2 \rightarrow \quad RCHO \quad + H_2O$$
$$\text{Fatty alcohol} \qquad\qquad \text{Fatty aldehyde}$$

Assay Methods

Oxygen Uptake Assay

Fatty alcohol oxidase activity is routinely assayed with an oxygen electrode (Yellow Springs Instrument Co., Model 53 Biological Oxygen Monitor, attached to a Hewlett-Packard X-Y recorder, Model 7034 A). The reaction mixture contains in a volume of 4 ml: 13 mM fatty alcohol, 0.1 M Tris-HCl buffer, pH 9.0, 5 mM dithiothreitol, 1 mM KCN, and 0.5 ml (0.5 mg of protein) of enzyme. The fatty alcohol (130 mM) is first emulsified in 0.5% Tween 80 (2 ml) for 1 min at low speed with a Bronwill Biosonic IV ultrasonic generator fitted with a microprobe. The temperature of the reaction mixture is allowed to equilibrate for 5 min at 30° in the water bath before the reaction is initiated by addition of substrate. Oxygen uptake is recorded continuously for 10 min.

Radioisotope Assay

Fatty alcohol oxidase activity is assayed using [1-^{14}C]lauryl alcohol as a substrate, and the fatty aldehyde formed is identified by thin-layer chromatography. Substrate emulsion is prepared by sonicating 0.33 μmol of [1-^{14}C]lauryl alcohol (12 μCi, from Amersham/Searle or I.C.N. Chemical and Radioisotope Division) and 230 μmol of unlabeled lauryl alcohol in 1 ml of 0.5% Tween 80 for 1 min at low speed with a Bronwill Biosonic IV ultrasonic generator fitted with a microprobe. The reaction mixture contains in a final volume of 4 ml: 12 mM lauryl alcohol (200 μl), 0.1 M glycine–NaOH buffer, pH 9.0, 5 mM dithiothreitol, and 0.8 ml (0.8 mg of protein) of enzyme. The reaction mixture is shaken at room temperature. After 0, 3, 6, and 9 hr the reaction is stopped by extraction of the fatty components according to the method of Bligh and Dyer.[23] To each 0.8-ml aliquot of the reaction mixture in a tube are added 2 ml of methanol and 1 ml of chloroform. The tube is capped, shaken, and incubated at room temperature for 30 min; then 1 ml of chloroform and 1 ml of water are added. The tube is shaken and set aside until the liquid phases separate clearly. The chloroform layer is removed, dried, and resuspended in 500 μl of chloroform; 100 μl of this chloroform suspension is spotted onto a thin-layer chromatography (TLC) plate coated with silica gel G (200 μm). Then 0.2 mg each of dodecyl alcohol and dodecyl aldehyde dissolved in 50 μl of chloroform are added to each spot as carriers. The plate is developed in 80 : 20 : 1.5 (v/v/v) hexane/diethyl ether/acetic acid. The lipids are detected by spraying with 0.2% 2,7-dichlorofluorescein. The R_f values for fatty alcohol and fatty aldehyde are 0.50 and 0.95, respectively. The identified spots are scraped into a scintillation vial of 10 ml of 0.5% PPO and 0.01% POPOP in toluene and counted.

Assessment of the Two Methods

The oxygen uptake measurement is more convenient than the radioactive assay. A larger quantity of enzyme is required for the oxygen uptake assay because of the low sensitivity of the apparatus. The radioactive assay, aside from its higher sensitivity, is also useful for measuring the combined activities of fatty alcohol oxidase and fatty aldehyde dehydrogenase. The substrate (fatty alcohol), the first product (fatty aldehyde), and the second product (fatty acid), can all be separated by TLC. This

[23] E. G. Bligh and W. J. Dyer, *Can. J. Biochem.* **37**, 911 (1959).

system can be used to identify the electron acceptors involved in the two oxidative reactions catalyzed by the oxidase and the dehydrogenase (see next section). In the presence of oxygen alone and the absence of NAD^+, fatty alcohol is converted to fatty aldehyde only. When oxygen and NAD^+ are both present, fatty alcohol is converted to fatty acid with little buildup of fatty aldehyde. Under anaerobic condition, fatty alcohol is not oxidized, even when NAD^+ is supplied.

Properties

Fatty alcohol oxidase activity is absent in the dry seeds, but increases at a nearly linear rate during the first 20 days of germination. As for the hydrolase, most of the cellular activity is associated with the wax body membrane. The enzyme utilizes molecular oxygen as the electron acceptor, and has an optimal pH of 9.0 with dodecyl alcohol as substrate. It has an apparent K_m value of 4 mM for dodecyl alcohol. Dodecyl alcohol gives the highest activity. Other alcohols give the following activities relative to dodecyl alcohol (100%): myristyl (74%), decyl (51%), palmityl (26%), and stearyl (16%). Arachidyl and behenyl alcohols give rates of oxygen consumption that are barely detectable. The physiological substrates eicosenol (19%) and docosenol (13%) are oxidized at rates higher than those of their saturated analogs. Ethanol and palmitic acid are not substrates. The oxidase is more labile than the hydrolase; freezing overnight results in 50% reduction of its activity. Oxidase activity is cyanide insensitive and is not stimulated by NADH or NADPH, suggesting that the enzyme is not a mixed function oxidase. Measurement of the stoichiometry of the oxidase reaction with and without KCN or commercially purified catalase, suggest that the electrons and O_2 combine to form H_2O, but not H_2O_2.

Fatty Aldehyde Dehydrogenase (Fatty Aldehyde: NAD^+ Oxidoreductase)

$$\text{R-CHO} + NAD^+ + H_2O \rightarrow \text{RCOOH} + NADH + H^+$$

Fatty aldehyde   Fatty acid

Assay Methods

Fluorometric Assay

Fatty aldehyde dehydrogenase is assayed by measuring the reduction of NAD^+ fluorometrically. The reaction mixture contains, in a total volume of 4 ml: 3.25 mM fatty aldehyde, 0.1 M Tris-HCl buffer, pH 9.0, 5 mM dithiothreitol, 1 mM NAD$^+$, and 1–50 μl (1–50 μg of protein) of enzyme. The fatty aldehyde (32.5 mM decyl aldehyde) is first emulsified in 2 ml of 0.5% Tween 80 by sonicating for 1 min at low speed with a Bronwill Biosonic 1V ultrasonic generator fitted with a microprobe. The reaction is initiated by addition of substrate. Fluorescence measurements are made on a Turner Model 111 fluorometer with excitation filter No. 7-60 (360 nm maximum) and emission filter No. 2A-12 (> 510 nm) attached to a Hewlett-Packard X-Y recorder Model 7034A. The reaction rate is linear for at least 5 min.

Spectrophotometric Assay

The enzyme activity is assayed by measuring the production of fatty acid with the colorimetric method exactly as described for the wax ester hydrolase. The substrate used in routine assays is 3.25 mM decyl aldehyde.

Assessment of the Two Methods

The fluorometric assay is sensitive and convenient. It is at least 10 times more sensitive than the conventional spectrophotometric assay of NAD$^+$ reduction at 340 nm. More important, unlike absorbance, relative fluorescence measurements are not seriously affected by the high turbidity of the reaction mixture. The colorimetric assay can be used to establish the stoichiometry of the reaction; one fatty acid being formed for each NAD$^+$ reduced. It is less convenient than the fluorimetric assay, since it necessitates extracting the fatty acid soaps prior to photometric assay.

Properties

Like the hydrolase and oxidase previously described, fatty aldehyde dehydrogenase activity is absent from dry seeds and increases during germination to a maximum after about 20 days of germination. The enzyme is also associated with the membrane of the wax bodies. It is stable for several months when frozen. Eighty percent of the activity is destroyed by heating the enzyme at 55° for 30 min. The optimum pH for activity is 9.0. The enzyme has an apparent K_m of 4 × 10^{-6} M for decyl aldehyde, and an apparent K_m of 2.5 × 10^{-4} M for NAD$^+$. Of the various substrate analogs, dodecyl aldehyde exerts the highest activity. Other aldehydes give the following activities relative to dodecyl aldehyde (100%): decyl (68%), myristyl (59%), palmityl (32%), and stearyl (14%).

Acetaldehyde is not a substrate. NAD^+ is a much better electron acceptor than $NADP^+$, FAD, or flavin mononucleotide. Divalent and monovalent cations at a concentration of 1 mM, such as $CaCl_2$, $MgCl_2$, NaCl, and KCl, and 1 mM EDTA exert little effect on the activity. On the other hand, 1 mM $CuSO_4$ inhibits 60% activity, and 1 mM $MnCl_2$ causes 24% increase in activity. p-Chloromercuribenzoate at 0.1 mM causes a complete inhibition of the activity, but its effect is overcome by the subsequent addition of 5 mM dithiothreitol.

Author Index

Numbers in parentheses are reference numbers and indicate that an author's work is referred to although the name is not cited in the text.

Subject Index

A

Acetate
  incorporation into long-chain fatty
    acids by plants, 283–288
  substrate for acetate kinase, 311
Acetate:CoA ligase (AMP-forming), *see*
  Acetyl-CoA synthetase
Acetate kinase
  from *Veillnella alcalescens,* 311
    ATP formation assay, 312
    hydroxamate assay, 311
    properties, 315
    stability to heat, 313
    stimulation by succinate, 316
    substrate specificity, 316
Acetoacetate
  product of HMG-Coa cleavage, 502
  substrate for 3-methylcrotonyl-CoA
    carboxylase, 788
Acetone powder preparation
  from wound-healing potato tuber
    disks, 414
Acetylcholinesterase, 745
  assay, 733
  release by phospholipase C, 733, 740
Acetoacetyl-CoA
  substrate for 3-ketoacyl-CoA thiolase,
    402
Acetoacetyl-CoA reductase
  activity of fatty acid synthase, 96
Acetoacetyl-CoA thiolase, 398, 399
Acetyl-CoA:carbon-dioxide ligase (ADP-
    forming) *see* Acetyl-CoA carboxylase
Acetyl-CoA carboxylase
  activators and inhibitors, 12, 42, 58
  biotin carboxyl carrier protein, 784
  biotin content, 10, 42, 60
  cell-free translation, 15, 43
  effect of monovalent cations on
    activity, 58, 59
  estimation of active and inactive
    forms by immunotitration, 292

from *Candida lipolytica,* 37
  kinetic properties, 42
  pH optimum, 43
  radioactive assay, 37, 38
  regulation of cellular content and
    synthesis, 43
    of mRNA, 43
  repression, 43
  specificity, 43
  spectrophotometric assay, 37
  stability and storage, 42
from chloroplasts and cytosol of
    higher plants, 44
  assay, 47
  cytosolic form, 46
  instability, 46
  measurement of carboxybiotin, 48
  prokaryotic form in chloroplasts,
    46
  transcarboxylation assay, 49
from *Euglena gracilis,* 60
  assay, 61
  dissociation of complex, 70
  properties, 72
from lactating rabbit mammary
    gland, 26
from lactating rat mammary gland,
    16, 26
  assay, 27
  biotin and phosphate content, 26
  contamination with proteolytic
    activity, 25
  effect of S-4-bromo-2,3-dioxobutyl-
    CoA, 25
  heterogeneous population of
    filaments, 22
  hypersharp peak, 24
  immunochemical studies, 25
  inhibition by biotin-binding
    antibodies, 26
  polymeric form, 21
  properties, 24, 32
  protomers, 22

C

M